Geochemical Abundances of Selected Elements in Earth's Crust (parts per million)

Element	*Symbol*	*Atomic Number*	*Atomic Weight**	*Average Continental Crust*	*Seawater*
Aluminum	Al	13	26.98	82,300	0.01
Antimony	Sb	51	121.75	0.2	5×10^{-4}
Arsenic	As	33	74.92	1.8	0.003
Barium	Ba	56	137.34	425	0.03
Beryllium	Be	4	9.01	2.8	6×10^{-7}
Bismuth	Bi	83	208.98	0.17	2×10^{-5}
Boron	B	5	10.81	10	4.6
Bromine	Br	35	79.91	2.5	65
Cadmium	Cd	48	112.40	0.2	1×10^{-4}
Calcium	Ca	20	40.08	41,000	400
Carbon	C	6	12.01	200	28
Chlorine	Cl	17	35.45	130	19,000
Chromium	Cr	24	52.00	100	5×10^{-5}
Cobalt	Co	27	58.93	25	1×10^{-4}
Copper	Cu	29	63.54	55	0.003
Fluorine	F	9	19.00	625	1.3
Gallium	Ga	31	69.72	15	3×10^{-5}
Germanium	Ge	32	72.59	1.5	6×10^{-5}
Gold	Au	79	196.97	0.004	$<1 \times 10^{-5}$
Hydrogen	H	1	1.008	1400	108,000
Iodine	I	53	126.90	0.5	0.06
Iron	Fe	26	55.85	56,000	0.01
Lead	Pb	82	207.19	12.5	3×10^{-5}
Lithium	Li	3	6.94	20	0.17
Magnesium	Mg	12	24.31	23,000	1350
Manganese	Mn	25	54.94	950	0.002
Mercury	Hg	80	200.59	0.08	3×10^{-5}
Molybdenum	Mo	42	95.94	1.5	0.01
Nickel	Ni	28	58.71	75	0.002
Nitrogen	N	7	14.01	20	0.5
Oxygen	O	8	16.00	464,000	857,000
Palladium	Pd	46	106.4	0.01	$<1 \times 10^{-5}$
Phosphorus	P	15	30.97	1050	0.07
Platinum	Pt	78	195.09	0.005	$<1 \times 10^{-5}$
Potassium	K	19	39.10	21,000	380
Rubidium	Rb	37	85.47	90	0.12
Selenium	Se	34	78.96	0.05	4×10^{-4}
Silicon	Si	14	28.09	282,000	3.0
Silver	Ag	47	107.87	0.07	4×10^{-5}
Sodium	Na	11	22.99	24,000	10,500
Sulfur	S	16	32.06	260	885
Thorium	Th	90	232.04	9.6	5×10^{-5}
Tin	Sn	50	118.69	2.0	8×10^{-4}
Titanium	Ti	22	47.90	5700	0.001
Tungsten	W	74	183.85	1.5	1×10^{-4}
Uranium	U	92	238.03	2.7	0.003
Vanadium	V	23	50.94	135	0.002
Zinc	Zn	30	65.37	70	0.01

*Source of atomic weights: International Union of Pure and Applied Chemistry, *Compte. Rendu.*, XXIII Conf., pp. 177–178, 1965. Based on atomic mass of $C^{12} = 12$; rounded to two decimal places.

EARTH RESOURCES AND THE ENVIRONMENT

Fourth Edition

EARTH RESOURCES AND THE ENVIRONMENT

James R. Craig

Virginia Polytechnic Institute and State University

David J. Vaughan

The University of Manchester

Brian J. Skinner

Yale University

Prentice Hall

Boston Columbus Indianapolis New York San Francisco Upper Saddle River
Amsterdam Cape Town Dubai London Madrid Milan Munich Paris Montréal Toronto
Delhi Mexico City São Paulo Sydney Hong Kong Seoul Singapore Taipei Tokyo

Acquisitions Editor: Andrew Dunaway
Editor in Chief, Geosciences and Chemistry: Nicole Folchetti
Marketing Manager: Maureen McLaughlin
Assistant Editor: Sean Hale
Editorial Assistant: Kristen Sanchez
Marketing Assistant: Nicola Houston
Managing Editor, Geosciences and Chemistry: Gina Cheselka
Project Manager: Edward Thomas
Senior Manufacturing and Operations Manager: Nick Sklitsis
Operations Specialist: Amanda Smith
Art Director: Jayne Conte
Cover Designer: Bruce Kenselaar
Senior Technical Art Specialist: Connie Long
Production Service: Elm Street Publishing Services
Compositor: Integra Software Services Pvt. Ltd.
Art Studio: Laserwords
Cover Photo Credits: Main photo of Bingham Canyon copper mine, Utah, from Lee Prince/Shutterstock. Left inset photo of spray irrigation patterns courtesy of Valmont Industries. Center inset photo of banded iron formation by H. L. James, *Economic Geology.* Right inset photo showing burning oil fields during Gulf War 1991 by Jonas Jordan, U.S. Army Corps of Engineers.

Library of Congress Cataloging-in-Publication Data
Craig, James R.
Earth resources and the environment / James R. Craig, David J. Vaughan, Brian J. Skinner. — 4th ed.
p. cm.
Includes index.
ISBN 978-0-321-67648-1
1. Natural resources. 2. Environmental policy. I. Vaughan, David J. II. Skinner, Brian J. III. Title.
HC21.C72 2011
333.7—dc22

2009043570

Pearson Prentice Hall
Pearson Education, Inc.
Upper Saddle River, New Jersey 07458

Printed in the United States of America
10 9 8 7 6 5 4 3 2 1

ISBN-10: 0-321-67648-3
ISBN-13: 978-0-321-67648-1

Prentice Hall
is an imprint of

www.pearsonhighered.com

We dedicate this book to Lois, Nancie, and James Craig; Emlyn Vaughan; and Catherine, Adrienne, Stephanie, and Thalassa Skinner.

CONTENTS

PREFACE

There is a mine for silver
and a place where gold is refined.
Iron is taken from the earth,
and copper is smelted from ore.
Man puts an end to the darkness;
he searches the farthest recesses
for ore in the blackest darkness.
Far from where people dwell he cuts shafts
in places forgotten by the foot of man;
far from men he dangles and sways.
The earth, from which food comes,
is transformed below as by fire;
Sapphires come from its rocks,
and its dust contains nuggets of gold. . . .
Man's hand assaults the flinty rock
and lays bare the roots of the mountains.
He tunnels through the rock;
his eyes see all its treasures.
He searches the sources of the rivers
and brings hidden things to light.
But where can wisdom be found?

(Job 28:1–6, 9–12, Revised Standard Version)

The Earth may be likened to the largest, most complicated, most accurate, and longest-working majestic old Swiss clock. The analogy falls far short of reality because everything about Earth is on such a larger scale. Furthermore, Earth is a one-of-a-kind, precariously balanced, life-sustaining sphere that speeds through the vacuum of space and holds all of the known life of the Universe.

Like the grand Swiss clock, Earth contains myriad parts moving in a vast array of directions and cycles. Some move rapidly; others so incredibly slowly that they appear fixed. Some move only minute distances; others circle Earth. Some processes stimulate and feed the living organisms, including humankind; others can be lethal.

Prior to the appearance of mankind, Earth's processes existed, at least more or less, in equilibrium or in steady states in which most elements, materials, and organisms remained approximately constant through time. The vast variety of environments changed only slowly with new and different species appearing over millions of years. There was no "pollution," merely gradual changes to which organisms adapted or died out. This scenario was changed forever with the appearance of humans who had the desire and ability to alter Earth processes as they sought out the materials to provide food, build dwellings, and develop technology.

The Earth was large and contained a great bounty; humans were few and knew little of technology—hence, environmental impacts were few, local, and relatively minor for the humans' first million years or so. But the onset of the Industrial Revolution about 1700 A.D. followed by new crops, and the slow development of modern medicine, brought about rapid increases in population and technology and would change Earth forever.

Humans rapidly progressed from being largely passive wanderers and minor users of Earth's resources to become the primary force extracting the bounty and even altering the very shape of the planet. In the process, humans interrupted Earth's processes and cycles, rather like altering the fine-tuning of the grand clock. Furthermore, humans altered the environments, spread toxic materials, and threatened all living species, like corrosion building on the delicate mechanisms of the clock.

The step into the twenty-first century is accompanied by the largest population explosion and greatest demand for resources humankind has ever experienced. The continuing rapid growth in human population and the quickly changing nature of technology are simultaneously aiding in the discovery of resources, and in creating more and farther-ranging impacts on Earth's environment. Humans have become the principal agents of modification of Earth as we increase in numbers and develop new and novel ways to search for, extract, and use resources. Despite the optimism of large reserves of many resources and the ability to discover yet more, numerous economic, political, environmental, and societal problems limit the availability of those resources to many populations.

Earth Resources and the Environment offers an objective view of Earth's resources, where they occur, how they were concentrated, how they are extracted and used, and how all aspects of human activities impact the environment. The text has been written at a level for first-year college courses that deal with geology, resources, and human impacts on the environment. It is written in a manner to provide geological information on the origins and occurrences of the resources so that no prerequisite courses are required. It then places the resources in a context of human usage and history and considers the effects that resources have on the world around us. Where applicable, environmental impacts are intertwined with economic, political, and social factors.

The fourth edition of *Earth Resources and the Environment* offers an updated discussion of the mineral resources as humankind moves into the twenty-first century. The text is divided into parts in which the related chapters (i.e., those dealing with energy, or those dealing with metals) are introduced with a broader overview to place them in perspective. The information and discussions within the chapters have been updated to reflect the early years of a new century. This is especially important in terms of resources for energy and metals where concerns of adequacy have been raised in the recent past. Some of the numerous changes in the new edition include:

NEW TO THIS EDITION

- update of all information to the most recent data before publication
- increased emphasis on environmental impacts throughout
- much expanded discussion of rising carbon dioxide levels in the atmosphere including sources, levels, and consequences
- discussion of "carbon footprints," their meaning, impact, measurement, and reduction
- incorporation of several new diagrams designed to demonstrate process over specifics of data
- update of discussion of oil economics and politics
- simplification of several data—saturated tables
- use of trend curves in place of dry data tables
- increased emphasis on water shortages/excesses
- increased discussion of the "greening" of society and more environmentally friendly application of resources
- increased discussion of renewable energy sources—wind, solar, hydro, biofuels, nuclear, geothermal, ocean generated
- update of the Resources Calendar, which is presented as an appendix

The term *resources* is used to mean those chiefly inorganic materials that are extracted from Earth's crust—lithosphere, hydrosphere, and atmosphere—and whose use may impact these parts of the crust. Thus, metals, industrial rocks and minerals, chemical minerals, water, and soil are discussed at length. Major sections are devoted to the fossil fuels (peat, coal, petroleum, natural gas), to nuclear energy (fission and fusion), and to renewable energy sources (wind, water, solar, biomass, hydrogen, tidal and wave, and geothermal). Additional chapters deal with the history of resource usage, from earliest human efforts to the twenty-first century, and with the manner in which those uses have impacted the environment. The text concludes with a realistic assessment of the resources for the future.

This edition, like the earlier ones, addresses the objective so well summarized by our friend Dr. Paul B. Barton, Jr., of the United States Geological Survey, in his presidential address to the Society of Economic Geologists in 1979. He stated, "It is as important to the future voter to appreciate the realities of our resource-environment situation as it is to be able to read the ballot. I believe that our principal hope is in education." The great American jurist Oliver Wendell Holmes observed: "A man's mind, once stretched by a new idea, can never return to its original dimension." We have written this text to educate and to stretch the minds of the students using this book in their courses; they are the future.

WEB RESOURCES

The sources of specific information on Earth resources are varied and vast, ranging from textbooks to company reports, to professional journals, to governmental agencies. The development and expansion of the World Wide Web has been especially important such that it is now the most easily assessed initial source for information on most topics. Because specific web addresses are constantly being added and modified, students or researchers seeking additional information are advised to use their computers' search engines. Insertion of a mineral commodity's names, an environmental impact, a geological process, a specific event, a commodity price, or even a commodity reserve status will result in a host of suitable starting points.

Unfortunately, the web does not always present references in well-organized forms and cannot easily decipher scientifically sound material from that which might be incorrect, incomplete, or biased in content. Furthermore, there are large numbers of web sites that are not maintained and have rather short life spans.

ACKNOWLEDGMENTS

The authors gratefully acknowledge the assistance of many individuals whose ideas, comments, questions, and criticisms were helpful in the preparation of this text in the first three editions. Countless students and professional colleagues have stimulated our thinking and either provoked us with questions or educated us with answers. Numerous companies and governmental agencies have been kind in providing information and diagrams and have been generous in allowing us to include them in this text.

We would like to thank the reviewers of the Third Edition: Abdolali Babaei, *Cleveland State University;* J. Allen Cain, *University of Rhode Island;* John Callahan, *Appalachian State University;* Wolfgang E. Elston, *University of New Mexico;* Udo Fehn, *University of Rochester;* David T. Fitzgerald, *St. Mary's University;* A. Kern Fronabarger, *College of Charleston;* Andrew MacFarlane, *Florida International University;* Jill D. Pasteris, *Washington University;* Joseph L. Tinsley, *Clemson University;* and Stephen Van Horn, *Johnson State College.* We are also grateful to individuals who have critically reviewed the manuscript at various stages in the three editions: Diane Burns, *Eastern Illinois University;* John E. Callahan, *Appalachian State University;* Barbara Dexter, *SUNY at Purchase;* Udo Fehn, *University of Rochester;* George McCormick, *University of Iowa;* Lawrence D. Meinert, *Washington State University at Pullman;* David Ostergren, *Northern Arizona University;* J. D. Rimstidt, *Virginia Tech;* Brandon Schwab, *Humboldt State University;* Andrew Stack, *Georgia Institute of Technology;* and Half Zantop, *Dartmouth University.*

Assistance in the first two editions was provided by Peggy Keating, Christine Gee, Margie Sentelle, Mary McMurray, Llyn Sharp, and Mark Fortney. Our editors, especially Robert McConnin, Patrick Lynch, and Ed Thomas, have been encouraging, patient, and generous in their wise counsel. We also acknowledge the assistance of Kevin Geraghty with computer graphics and the help and guidance of Sean Hale and Kristin Jobe in this latest edition.

James R. Craig
David J. Vaughan
Brian J. Skinner

Earth Resources and the Environment

Introduction and the Origins of Resources

Our entire society rests upon—and is dependent upon—our water, our land, our forests, and our minerals. How we use these resources influences our health, security, economy and well-being.

JOHN F. KENNEDY, FEBRUARY 23, 1961

MODERN SOCIETY AND EARTH RESOURCES: THE COMPLEX NETWORK

President Kennedy's words are as true today as they were years ago. Furthermore, they are as relevant to the world as they are to the United States. All of the materials needed for the well-being of humans in our complex society come from the Earth. Food and water, clothes and dwellings, automobiles, aircraft, televisions, and computers all contain materials drawn from the Earth. Even the paper on which these words are printed contains cellulose from wood, plus clay and barium sulfate to give body and an absorbent, tough surface. Almost every material we use actually employs many other materials, as well as energy, in its extraction and any subsequent processing. Now, early in the twenty-first century, we have finally come to recognize that there is much truth in the old adage that "there is no such thing as a free lunch." Almost everything that humans do impacts the world around us. In recent decades, we have become aware of the many unintended and undesired impacts on the quality of our air, water, or living space that result from resource extraction and use. In this context, we have also seen the emergence of new terms such as *greening* of society, *carbon footprint*, *global warming*, and *sea level rise.* One can envision the whole of Planet Earth as a large, interconnected network. When something disturbs it in one place, there is a reverberation throughout the entire network and, if any part is broken, the network is weakened. A disruption in one part can have unforeseen consequences elsewhere and, thus, we must proceed with caution if we are not to damage irreparably this network on which our very survival depends.

THE CHANGING WORLD

Profound changes have occurred in the fifty years since President Kennedy spoke his insightful words. The world population has more than doubled to 6.7 billion, fertilizer use has more than tripled, energy production has tripled, the number of automobiles in the world has quadrupled, and nuclear power has grown from a tiny fraction of the energy supply to the world's second leading source of electricity. This growth has not been without consequences for the environment; since 1961, the amount of carbon dioxide in the atmosphere has risen by about 20 percent, and the world's rain forests have been reduced by one-third. We hear reports of the over-exploitation of the world's resources and, at the same time, we learn of new technologies that can enable us to discover new resources, or more efficiently exploit old ones. We hear of growing levels of atmospheric pollution, and of damage to the ozone layer that protects us from cancer-causing ultraviolet radiation, but we also hear of cleaner water in once polluted rivers. We hear reports of vanishing plant and animal species in the wild, and of new, genetically altered, and cloned species in human-made environments. In the United States, concerns about reliance on imported fossil fuels, especially petroleum from the Middle East, have provided a stimulus for the generation of domestically produced biofuels. But this created the largely unforeseen dilemma of driving up prices for food commodities as farmers diverted their efforts to grow corn for ethanol production rather than for human consumption. The end result was little change in environmental impact, insufficient ethanol production to offset much imported oil, reduced grain for exports, increased grain prices that especially hurt the world's poor and, hence, anger toward the United States, and ultimately calls to reduce the use of corn for ethanol production. It is clear that in nearly every aspect of life more changes are occurring; the rates of those changes accelerating. Such changes make it more difficult to predict future needs in terms of the nature and the amounts of resources. A general view has developed that more of everything will be needed in the future because of the increase in world population. This is true for many commodities such as petroleum, iron, aluminum, and crushed stone. However, there have been reductions in the need for commodities such as asbestos, lead, arsenic, and mercury, partly due to the development of new technologies, but mainly due to the recognition of the damage to human health and the environment caused by these materials. There is no doubt that world population will continue to grow in the twenty-first century, and may even double again to 12 billion people, which alone would drive the need for vast quantities of additional resources. The result of this growth will inevitably be to increase the extent of the human impact on our global environment.

INTERDEPENDENCE AND COMPLEXITY

Our ancestors learned long ago that the materials needed for food, clothing, dwellings, fuel, and later for commercial activities are not uniformly distributed over Earth. When demand locally exceeded supply, three options were open to our ancestors: substitutes could be developed, trading could be started with people who had ample supplies, or the people in need could move in order to find new supplies. Even where quantities were once sufficient, the demands of increasing populations would often eventually exceed the limits of supply. The result was the development of local, regional, national, and international trade routes. There is now a vast network of interdependence that becomes evermore complex. This increasing complexity, along with the increasing quantities of materials being transported around the world, has made more materials available to more people, but it has also increased the vulnerability of supply systems to disruption by both natural and human forces. It is important to remember that the availability of materials is not just a function of resources—there are also economic, social, political, and environmental considerations. The availability of a commodity may be dictated by politics more than by any other factor; examples include the prevention of sales of oil from Iran and Iraq to the United States, and the long-standing embargo of importation of a variety of goods from Cuba to America. In recent years, environmental concerns have also been ranked in importance above economics, as demonstrated by the removal of hydroelectric dams to allow salmon to migrate up the Snake River in the western United States.

THE EARTH—OUR ONLY HOME

During the latter part of the twentieth century, we humans reached out and touched the Moon and our two nearest planets, Mars and Venus; we also sent unmanned probes close to the other planets. These explorations have confirmed what we have long suspected: Earth is the only planet in our solar system that is capable of providing the resources needed to build and nurture human societies. Our ability to reach these other celestial bodies has also led to speculation that we might one day inhabit such planets and even extract valuable resources from them. But despite our extraterrestrial explorations and our advancing technology, there is no serious expectation that humans will extract resources from other planetary bodies within the next half century (and quite possibly well beyond that). We humans, for the next several generations, are Earth-bound; we must be mindful that all of the resources we shall need for the foreseeable future must be found on this planet, and that any negative environmental impact associated with the extraction and use of any resource will be in our own "backyard". We must remember that all known life in the Universe lies on or near Earth's surface. Furthermore, we must all understand that we are enmeshed in a complex, interdependent network where actions often have consequences far beyond the obvious, the local, and the immediate.

PLATE TECTONICS AND THE FORMATION OF EARTH RESOURCES

If the Earth were homogeneous (that is, if all of the chemical elements were evenly distributed), the only two solids would be ice and an "average" rock. There would be no fossil fuels, no ores, no fertilizers, no gems, nor any of the other mineral resources that we now use in such large quantities. Fortunately, Earth is a dynamic body in which there are great cyclical movements of materials that are ever rearranging and concentrating the chemical elements and their compounds. Some cycles are relatively rapid and easily visualized; an example is the hydrologic (water) cycle in which rainfall is followed by runoff and then evaporation returning the water back into the atmosphere from which it will fall again as rain to continue the cycle. Other cycles, such as the subsurface movement of fluids that deposit metal-rich veins, are of much longer duration and are hidden from view. We do not see these moving fluids because they are deep within the Earth, but we find the veins exposed in outcrops of rock, or in mines, long after they have formed.

The most important of Earth's cycles is the one referred to as "the plate tectonic cycle"; this involves the slow transport of large blocks, or plates, that make up Earth's outermost solid layer. The plates are carried along atop convecting cells within the Earth's underlying mantle. The development of plate tectonic theory in the 1960s provided the key to understanding the formation of the Earth's crust, the expansion of ocean basins, and the movement of continental masses throughout geologic time. Earth's internal heat provides the driving force for the large convecting cells within the mantle. The upwelling zones, such as the mid-Atlantic Ridge and the East Pacific Rise, are spreading centers where hot mantle rock rises upward to the surface, adding new material to Earth's crust along the plate boundary. As new material is added, older rock is transported laterally away from the boundary. At spreading centers, there is much seismic activity, volcanism, and the release of hot aqueous fluids that carry dissolved metals. As the fluids are expelled into the cold ocean water, there is immediate precipitation of metal-rich sulfide deposits at hot vents called "black smokers." Although mining of these deposits is presently impractical because of their deep-water locations, similar deposits on the continents (and which are the fossil equivalents of present-day black smokers) have been mined for hundreds of years. Elsewhere, plates collide, commonly with one plate sinking downward under another. These convergent or collision boundaries are also sites of extreme seismic activity and volcanism; there too, there is a release of hot metal-bearing fluids and the formation of even richer and more diverse types of metal ores, often intermixed with a variety of volcanic rocks.

Although many rich resources have been formed along plate boundaries, others that were formed far from the boundaries also owe their locations to plate tectonics. For example, movement of the plates has played a critical role throughout geologic history in the positioning of the continents and the formation of great sedimentary basins. This, in turn, has dramatically affected climatic conditions, and rates of erosion and sedimentation, which, in turn, have controlled the formation of many types of resources, especially oil and gas. Every type of resource forms by one or more geological processes, and every process in the Earth's crust and mantle can concentrate, alter, or disperse resources. Igneous intrusions drive fluids which concentrate and precipitate metal-rich veins. Erosion of those veins disperses the concentrations of copper, lead, gold, or silver, but can result in the subsequent concentration of gold nuggets in a river bed. Evaporation removes water but leaves behind rich salt deposits. The gradual burial of peat bogs on the land, or of marine plankton in the sea, results in the formation of coal and oil, respectively. Earth's resources are formed by myriad processes, which have been acting over vast periods of time. The result is the accumulation of large numbers of deposits of differing richness and size in various places and at various depths. The challenge of mineral exploration is to understand where resources are hidden so that we can exploit them for the sustenance of modern society.

CHAPTER 1

Minerals: The Foundations of Society

Earth's resources used by humans are constantly changing in response to the technologies of particular societies. This photograph contrasts the natural stone used in the construction of the great Pyramids in Egypt more than 4000 years ago with the wide variety of special metals used in the construction of a modern U.S. Air Force B-1B bomber and its escort fighter jets. (Photograph by Staff Sgt. Jim Varhegyi; used with permission of U.S. Air Force.)

Resources are like air, of no great importance until you are not getting any.

ANONYMOUS

FOCAL POINTS

- All materials needed for modern society are derived from Earth, directly or indirectly.
- World population grew slowly until about A.D. 1500; increasingly rapid growth from around 1800 raised population to 2 billion by 1930, to 4 billion by 1975, to 6 billion by 2000, and to an expected 8 billion by 2035.
- Human population, presently exceeding 6.7 billion, is projected to rise to at least 12 billion before stabilizing by about A.D. 2100.
- The rates of population growth are much higher in less developed countries than in developed countries.
- Renewable resources, such as plants, flowing water, and sunlight, are replenished on timescales that are short compared to human life spans; nonrenewable resources, such as mineral deposits, are formed over vast timespans and are being used much faster than they are replaced. (Some resources such as tropical rain forests, while clearly renewable, are also being consumed at rates far greater than natural processes are capable of replacing them.)
- Earth's crust is like a large "machine" with energy input from Earth's interior and from the Sun; the energy fluxes result in movement of material, or *geochemical cycles*, that are constantly forming, concentrating, redistributing, and altering the mineral resources.
- "Resources" are naturally occurring concentrations of mineral substances from which economic extraction may occur.
- "Reserves" or "ores" are those concentrations of mineral substances that are presently economical to extract.

THE WORLD'S RESOURCE NEEDS

All the materials needed for the health and prosperity of human societies have come from Earth. In primitive societies, it was simply the food, water, and shelter necessary for survival. Today, the list is much longer and includes automobiles, planes, televisions, and computers, but we are just as dependent on Earth for raw materials. The major differences are that today we

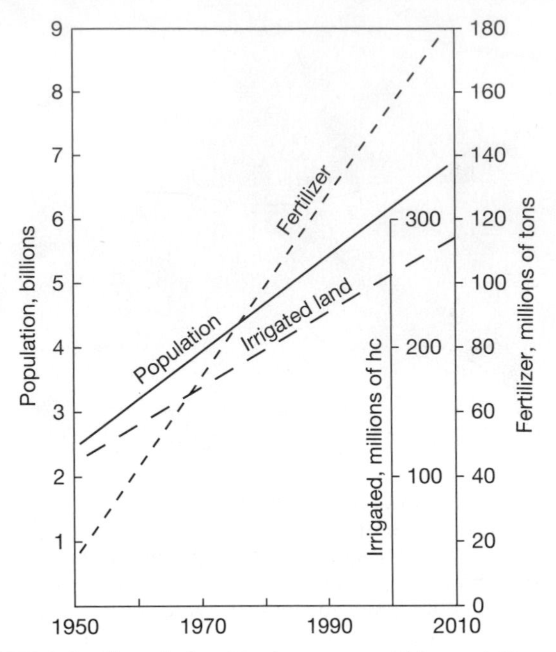

FIGURE 1.1 The relationships between world population, irrigated farmland, and the consumption of fertilizers needed in the second half of the twentieth century in order to produce sufficient food for the world. As world population has more than doubled, agricultural production has been able to keep pace, but this required more than a doubling of irrigated land area, and a tenfold increase in the use of fertilizers to promote crop growth. (Data from FAO of the United Nations.)

use much greater amounts than did individuals in the past; we use a greater variety of materials; we use materials from more sources; and we use many materials that we have synthesized from those raw materials. How straightforward it would be if we could consider each raw material or end-product irrespective of all the others. This is not possible because the increasing use of materials has been associated with increasing complexity in the systems needed to extract and deliver materials, and increased interdependence in the supply of materials. Our most basic need is for food, and much concern has been expressed about the world's ability to provide food for a growing population. Consider Figure 1.1, which addresses food production and some of its complexities. In the second half of the twentieth century, world population increased from 2.5 billion to 6 billion with food production just keeping pace. However, in order to feed the world's growing population, the need for water to irrigate crops increased at approximately the same pace, while the need for tractors manufactured using metals and many other raw materials and using fossil fuels to propel them, increased at a faster pace. Finally, it shows that need for fertilizer to produce that food increased more than four times faster than the population. Thus, the production of food depended not only on seed and soil, but also upon water, metals, fossil fuels, and fertilizers (and this is only a very simplistic view). Whether the growth in agricultural production can continue to keep pace with increasing world population over the next century (when population will grow toward 12 billion) is not clear. What is clear is that the terrible scenes of famine in Africa, caused by disruptions in food supplies, are all too common in the newspapers and on television. Such local famines, whatever their cause, are only minor examples of what would happen on a much larger scale were the global food network to be disrupted. Because adequate food production now requires adequate mineral and energy production, minerals and energy have also become part of the foundation of all societies. If important minerals and energy sources were to run out, or for some reason be denied us, social chaos would ensue. The result could be a drastic reduction in the world's population.

POPULATION GROWTH: THE FORCE THAT DRIVES RESOURCE CONSUMPTION

Many millennia ago our ancestors were hunters and gatherers for whom nature's random production of grains, fruits, and animals were sufficient. When local population needs began to exceed the natural yield of materials, farming was developed out of the necessity to control, and thereby to increase, the production of fruits, grains, and meats. Archeologists suggest that farming began in the Middle East about 10,000 years ago. From that time onward, the world's population has not only increased, but has grown ever more dependent on the controlled production of food and clothing and, thereby, on all the other materials we draw from the Earth. As far as historians and archeologists have been able to decipher, world population grew slowly but more or less steadily to the end of the sixteenth century A.D. (Figure 1.2), interrupted only by occasional outbreaks of plagues, pestilences, and famines. One of the worst epidemics started in Europe when Crusaders, returning from Asia Minor in 1348, inadvertently brought back rats bearing plague-carrying fleas. Starting in Italy, the bubonic plague swept through Europe over a period of two years. Between a third and a quarter of the entire population of Europe died; the populations of some cities were reduced by half. This particular outbreak was known as the "Black Death."

Near the end of the sixteenth century, the world population began to grow rapidly. The initial causes for the increase in rate were primarily due to advances in hygiene in the cities, but a marked improvement in the diets of Europeans also occurred when potatoes and maize were introduced from the Americas. World population reached 1 billion in about A.D. 1800. One hundred and thirty years later—by 1930—the population had reached 2 billion (Figure 1.2). A mere 45 years was all the time it took before the population doubled again to reach 4 billion in 1975; another 2 billion people had been added by 1999, bringing the total to 6 billion.

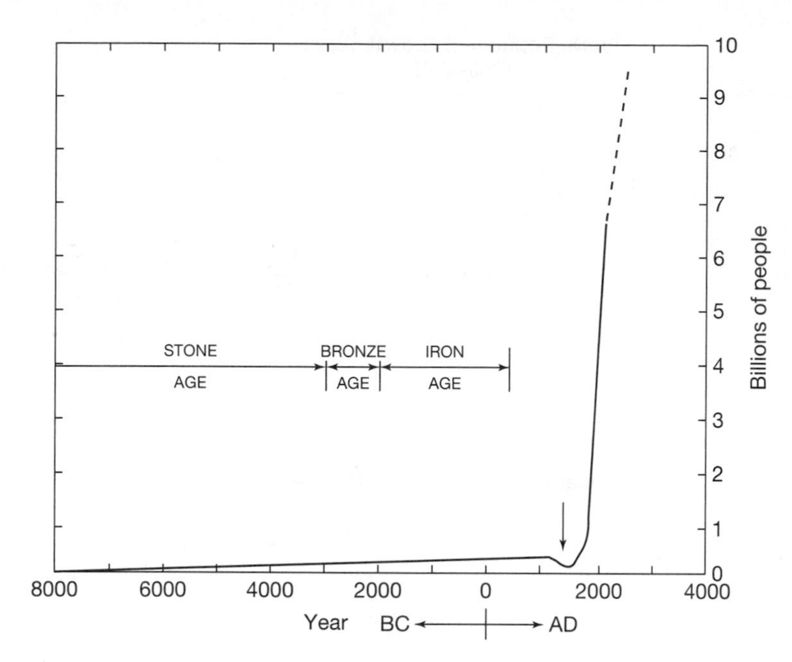

FIGURE 1.2 Growth of the world's population through history. Notice the sharp drop due to the Black Death that struck Europe in 1348 and the sharp rise that occurred in the nineteenth and twentieth centuries.

During the 1960s, the world population grew at a rate of up to 2.2 percent a year. However, by 1980, the overall growth rate had declined to about 1.7 percent annually, and by 2000, it was down to about 1.3 percent (Figure 1.3a). Nevertheless, in some regions, such as parts of Africa, populations are still growing at 2.5 to 3 percent per year. Despite a slowing of the overall growth rate, the world population will continue to grow by more than 60 million people per year until 2030 (Figure 1.3b) and is expected to number 9 billion to 9.5 billion by the year 2050 and 11 billion to 12 billion by the end of the twenty-first century. These stupendous numbers, which almost defy imagination, mean that humankind has a task of Herculean proportions ahead if all of the new members of the human race are to be well fed and given the chance to enjoy a decent life. Many of the problems will be social and political in nature, but underlying everything will be the needed scientific and engineering expertise to exploit Earth's resources. Not only must the supplies be fairly divided, but exploitation must be carried out in such a way that the environment is not so fouled and irretrievably spoiled that we ruin the planet upon which we live. This will be difficult because more people will use more space and more resources, leaving less space and resources for wildlife, and giving rise to even more environmental problems.

Neither the densities of populations nor growth rates of populations are uniform around the world. The most populous countries today are China, with about 1.3 billion people, and India with approximately 1 billion; these two countries, plus the United States, Russia, and Indonesia, account for more than half of the present world population. But the situation is changing rapidly. The rates of population growth in different countries seem to vary inversely with the extent of their industrial development and standard of living; as a result, the technologically advanced nations tend to have both low birth rates and populations that have reached or are now approaching zero growth. In such countries, it is not necessary for people to have large families to support them in their old age. Less developed countries, in contrast, tend to have higher birth rates and populations that are still expanding rapidly. Such countries have agriculturally based economies, and large families are a way to provide the needed manual labor to assure security in old age. The difference between the two kinds of society can be seen in the **age-sex pyramids** shown in Figure 1.4. The near stability of the more developed regions is reflected in the fact that the number of people below childbearing years is nearly identical to the number of adults producing children. Hence, there is an approximate one-for-one replacement. By contrast, adults in less developed regions of the world are bearing many more children than are needed to replace themselves; as long as this trend continues, the pyramid will get broader and the population of such a region will grow ever larger. Even if the adults in these populous regions started today to have just enough children to ensure a one-for-one replacement, the populations of these regions would continue to grow for at least two generations because of the present ratio of children to adults. Somewhat ironically, the small families now characteristic of developing countries, combined with medical advances leading to a significant increase in average life span, may lead to a different problem; namely, that the wealth generated by the younger and smaller working population may become insufficient to provide adequately for the greatly increased population of the elderly.

What will the ultimate size of the world's population be? Obviously, the population cannot continue to grow unchecked forever. This is true, if for no other

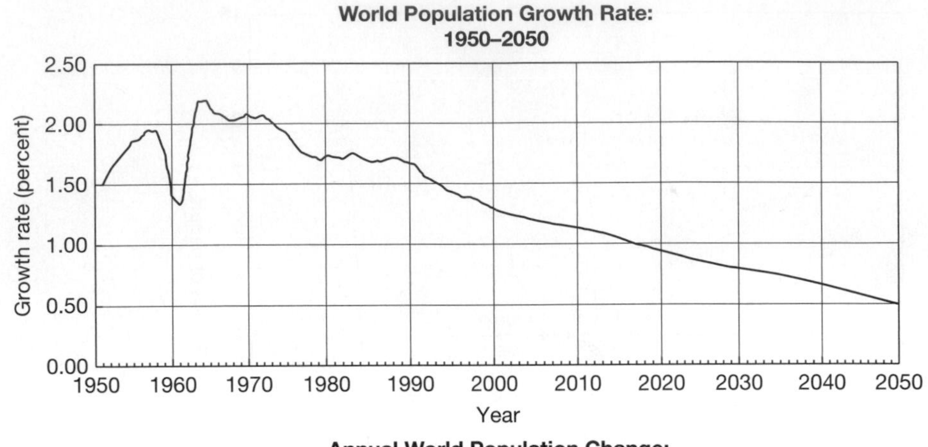

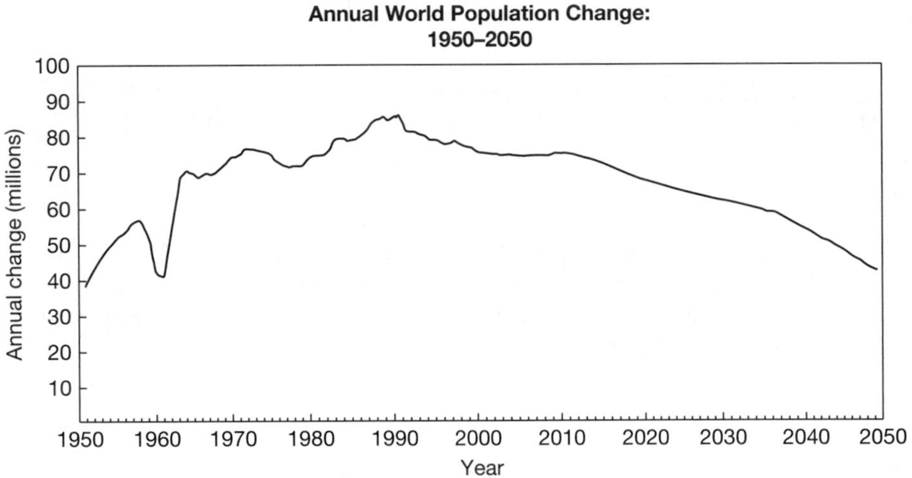

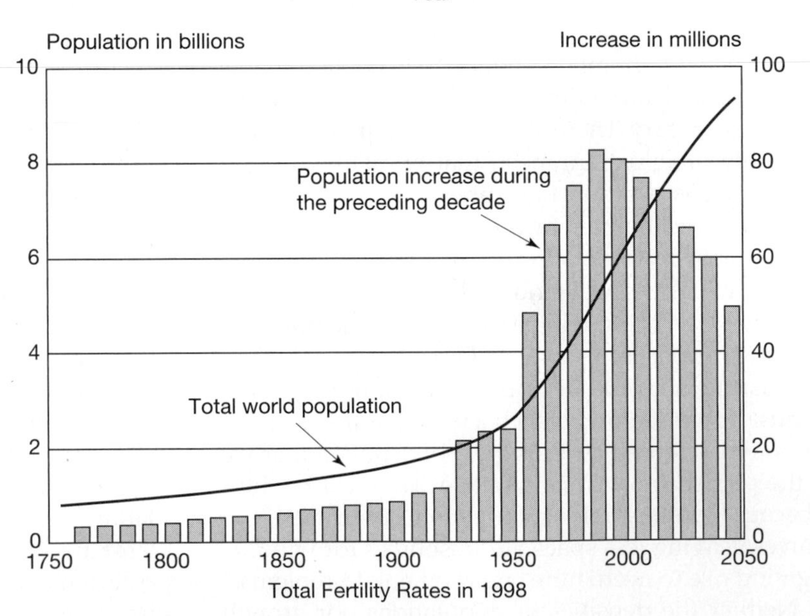

FIGURE 1.3 Changes in the world's population and growth rates. (a) The world's growth rate peaked above 2.0 percent in the early 1960s and has been slowly declining since then. (b) The actual annual change in population peaked in the late 1980s at about 85 million and has now dropped to about 70 million and is now slowly declining. (Data from the U.S. Census Bureau.)

reason than the fact that in a continually growing population, a point must be reached when there is no longer enough room for everyone to stand up. Studies by the Population Council, the World Bank, and the United Nations have all drawn similar, but less than comforting, conclusions about the future of the world's population. Demeny's most encouraging conclusions, published in 1984 and supported by others since then, are shown in Figure 1.5; they appear to be proving accurate in indicating a pattern of slowing growth rates. Some countries are further advanced in the pattern than are others. Demeny suggests that the populations of today's less developed countries will eventually level off despite the present high birth rates. The indication is that rates of population growth for all countries will decline and approach

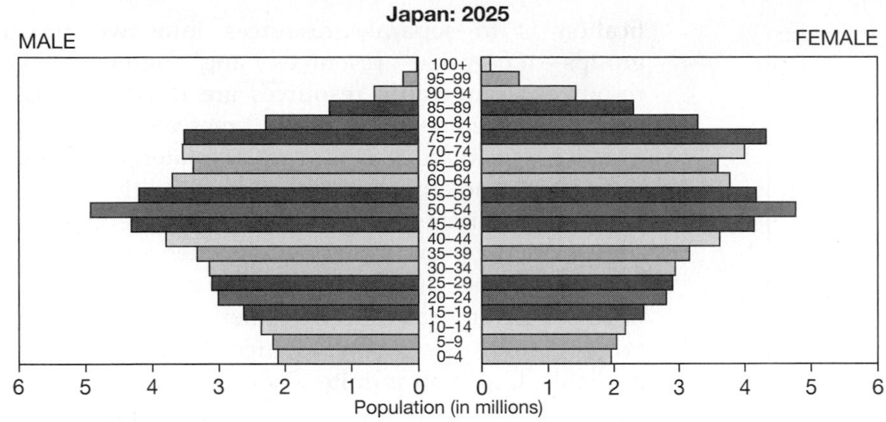

Source: U.S. Census Bureau, International Data Base.

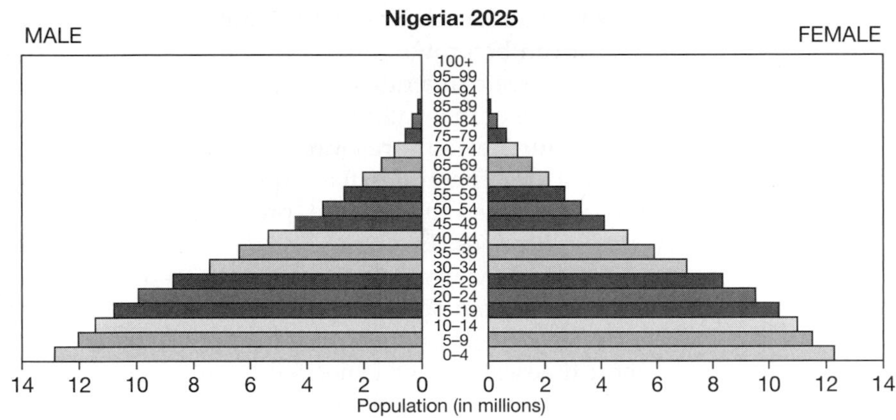

Source: U.S. Census Bureau, International Data Base.

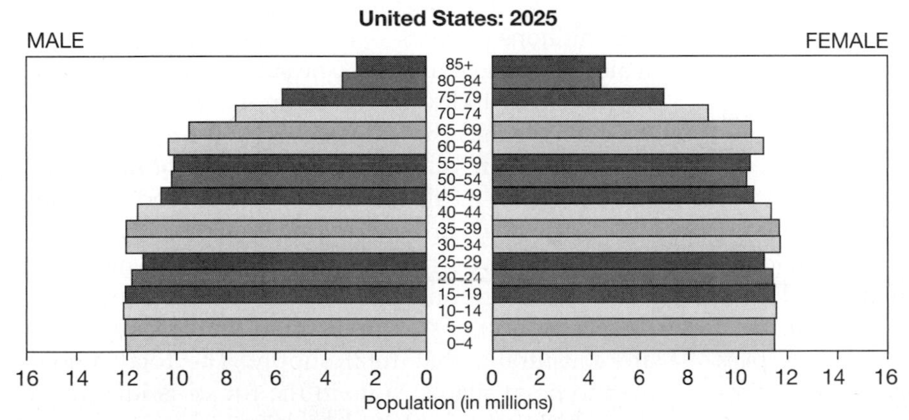

Source: U.S. Census Bureau, International Data Base.

FIGURE 1.4 The age and sex distributions in the populations of three different countries as projected to 2025. Japan's population distribution, with fewer young people below 25 years of age than adults between 25 and 50, is typical of a country that will decline in population during the next century. The United States, with the number of young people below 25 years of age similar to the number of adults between 25 and 50, is typical of a country that is approaching a one-for-one replacement situation. In contrast, the population distribution of Nigeria, with many more young people under 25 years of age than adults between 25 and 50, is typical of a country whose population will continue to increase during the next century. (Diagrams from the U.S. Census Bureau.)

zero sometime during the next 120 years. Projecting demographic trends far into the future is an uncertain process at best. Nevertheless, present trends suggest that by about the year 2100, the world's population will have leveled off to about 12 billion people. The six most populous countries in the year 2100 are predicted to be, in decreasing order of size, India, China, Nigeria, Bangladesh, Pakistan, and Indonesia. These six countries will account for approximately 50 percent of the world's population. The times in the future when the populations of different countries level out will vary from country to country, but all will have reached a stable figure by about 2100.

Can the Earth supply all the materials needed for 12 billion people to enjoy a reasonable life? Some experts believe that 12 billion is far too large for a continuous and healthy balance to be attained in food and other material supplies. Others are convinced that economic and technological growth will help us find ways to meet the challenge, and that the world's population can safely grow to even surpass 12 billion. Possibly both sets of experts are too extreme in their conclusions, and the final answer obviously lies in the future. Despite uncertainties, many of the problems that must be faced are already apparent—problems such as a disparity in living standards between regions that are industrially advanced but poor

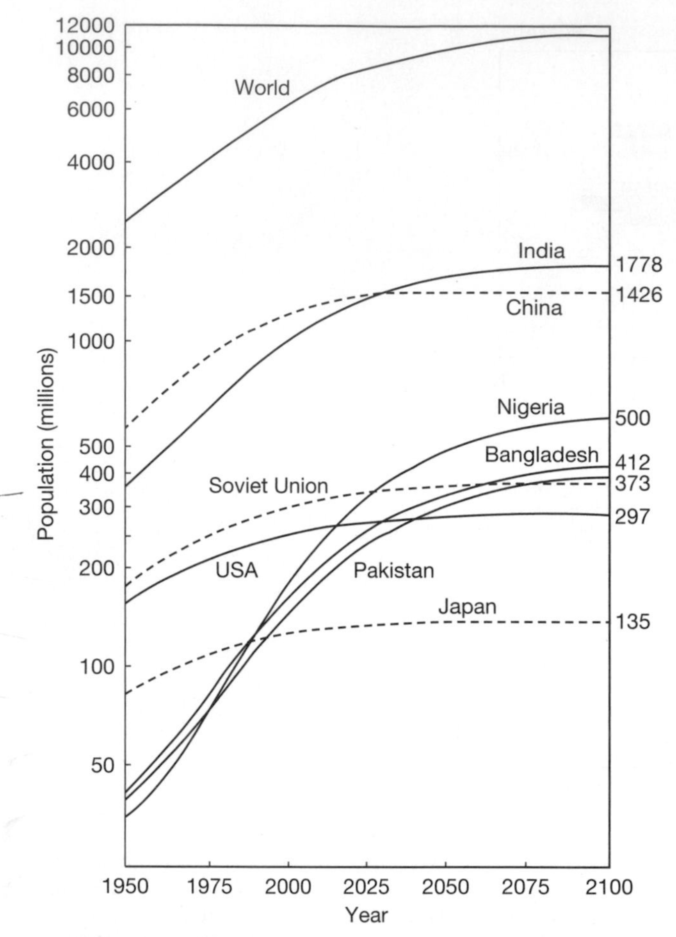

FIGURE 1.5 Projected growth of populations in several large countries and for the world as a whole to the year 2100. Demographers suggest that by 2100, the world will have attained a constant-sized population. (From Demeny, *Population and Development Review,* vol. 10, no. 1 [1984], p. 102.)

in mineral resources, as opposed to regions that are rich in resources but underdeveloped industrially. The forces that drive resource consumption are human needs and desires. But the controls underlying resource availability and use are geological and environmental. The interplay between availability and need is probably the most difficult problem the human race has to face; it is one of the basic issues that must be confronted in trying to end the continuing scourge of war.

This book addresses the environmental and geological questions involved with resource exploitation and use. However, we must remember that use and production of resources are inevitably intertwined with, and immensely complicated by, economic, social, political, and strategic issues.

MATERIALS WE USE

No classification of natural resources is completely satisfactory, but one convenient way to start the classification is to separate resources into two broad groups—renewable resources and nonrenewable resources. **Renewable resources** are those materials that are replenished on short timescales of a few months or years, such as the organic materials derived from plants and animals. However, in addition to food plants and animals, they also include the energy we draw from the wind, from flowing water, and from the Sun's heat. Use of renewable resources raises questions concerning rates of use rather than the ultimate total quantity of a given resource that shall ever be available. Given an infinite amount of time, it would be possible to grow infinitely large amounts of food and draw infinitely large amounts of water from a flowing stream. However, we cannot eat food faster than it can be grown, nor can we draw water from a flowing stream at a rate faster than a limit imposed by the volume of the water flowing in the stream.

Nonrenewable resources are those materials of which Earth contains a fixed quantity and which are not replenished by natural processes operating on short timescales. Examples are oil, natural gas, coal, copper, and the myriad other mineral products we dig and pump from the Earth. Notice that the definition includes a qualification concerning replenishment on short timescales. This is needed because new oil, gas, and certain other resources are continually being formed inside Earth, but the processes of formation are so slow that sizable accumulations only develop over millions of years—vastly slower than the rates at which we mine materials. The substances we dig from Earth's crust today have accumulated over the past 4 billion years. The rates of replenishment of fuels derived from fossil organic matter, and of metals distilled from the Earth's interior, are so exceedingly slow that the crop we are mining today is the only crop we will ever have—hence, the term "nonrenewable." Questions surrounding nonrenewable resources are therefore questions of total supply and of how fast we are consuming that total supply. The total amount ever to be available to us in the future is identical to that which is available today.

Most nonrenewable resources are also mineral resources, which means they are nonliving, naturally occurring substances that are useful to us, whether they are organic or inorganic in origin. We use this broad definition to include all natural solids, plus liquids such as petroleum and water, and gases such as natural gas and the gases of the atmosphere. A possible confusion in terminology becomes apparent when one considers a resource such as water. Water in a flowing stream is a renewable resource because it is replenished on a short timescale by rainfall; by contrast, water in a deep aquifer in a desert area, as in Israel, central Australia, or the High Plains of the western United States, is a nonrenewable resource because

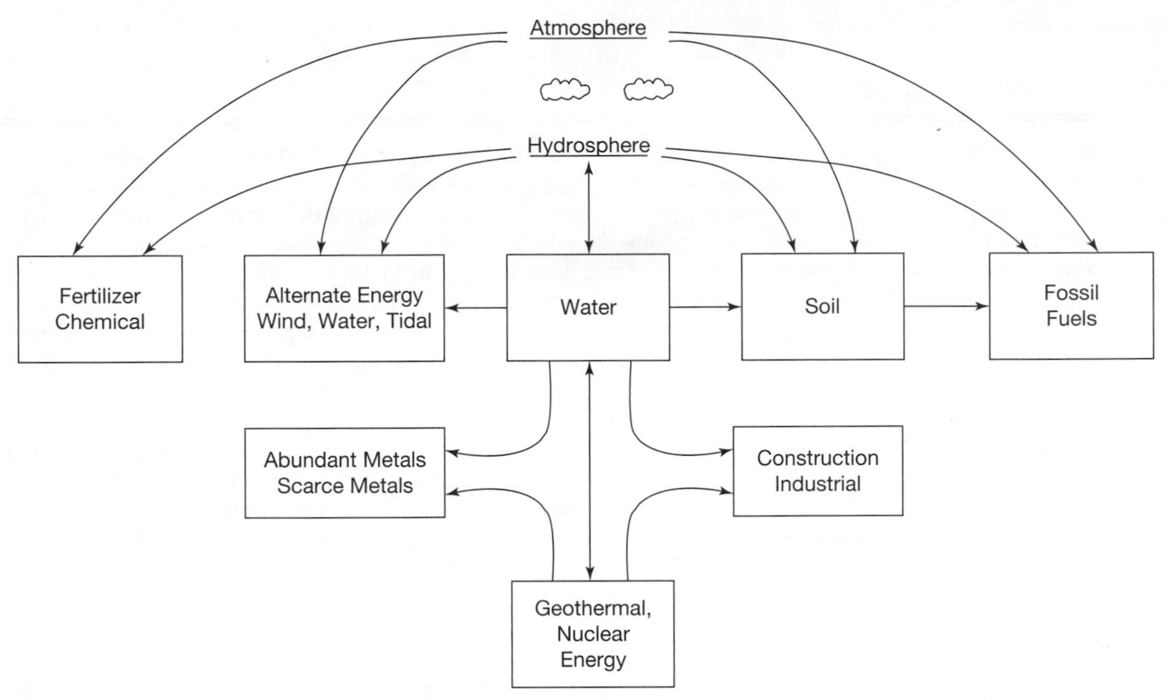

FIGURE 1.6 General categories of Earth's resources and their relationships. The arrows indicate major processes that are active in the development of the resources. For example, fossil fuels form as plants take up carbon dioxide from the atmosphere and water from the hydrosphere; metals are deposited by subsurface waters that have been heated by geothermal energy. The eight major categories of resources in the boxes are discussed in Chapters 5–11.

it is only replenished over timescales of thousands or tens of thousands of years.

The first major group of resources of those depicted in Figure 1.6 are the resources upon which we now, or in the future, draw energy. Some of the resources, such as the fossil fuels and uranium, are nonrenewable resources. Other energy resources, such as running water and solar heat, are renewable. The importance of the energy resources which are vital to the operation of modern society, was brought into focus by the energy crisis of 1973 when Middle Eastern oil was withheld from Europe and the United States by the producers. Since then, worldwide attention has been focused on the cost and sufficiency of energy resources for the future.

The second group are metals, the chemical elements that either singly or in combination as alloys have special properties such as malleability, ductility, fusibility, high thermal conductivity, and electrical conductivity that allow them to be used in a wide range of technical applications. Metals have been the key materials through which humans have developed the remarkably diversified societies we now enjoy and through which we have managed to proceed from the primitive societies of antiquity to the present. It is not surprising that the metal-winning and metal-working skills of ancient communities have been used as a measure of societal development; hence, terms such as *Bronze Age* and *Iron Age* have been adopted. Metals can be divided into two classes on the basis of their occurrence in Earth's crust. The geochemically **abundant metals** are those that individually constitute 0.1 percent or more of Earth's crust by weight; they are iron, aluminum, silicon, manganese, magnesium, and titanium. The term *abundant* is also appropriate for two reasons. First, these metals occur in so many diverse forms in the Earth that reserves of rich mineable ores are truly enormous. Even though rich deposits are not uniformly distributed around the world, the question of sufficiency for future generations is not in doubt. Second, the geochemically abundant metals form most of the common minerals and, as a consequence, they influence many of the geologic processes that shape the planet.

Geochemically **scarce metals**, by contrast, are those that individually constitute less than 0.1 percent by weight of Earth's crust; they are metals such as copper, lead, zinc, molybdenum, mercury, silver, and gold. The scarce metals are present in such tiny concentrations that they play very minor roles in geological processes; very special (even rare) circumstances are needed for local concentrations to form. Unlike the abundant metals, therefore, mineable deposits of scarce metals tend to be smaller and less common. As a consequence, the question of sufficiency for future generations is a more important one where scarce metals are concerned.

The third group of resources includes all of those mineral and related substances used in one way or

BOX 1.1

CO_2 and the Greenhouse Effect

Both the greenhouse effect and its potential long-term implications for the Earth's climate have been much in the news for the past several years. The term *greenhouse* is applied by analogy to the way that the air in glass greenhouses becomes warmer than outside air, and refers to the gases that absorb long wavelength (infrared) radiation from Earth and, hence, act like a blanket. Carbon dioxide (CO_2) is generally considered the greatest problem gas, but water vapor, methane (CH_4), nitrous oxide (N_2O), and chlorofluorocarbons also contribute. These gases are transparent to short wavelength solar radiation as it enters Earth's atmosphere from above, but they absorb and are heated by long wavelength radiation emitted by Earth below (Figure 1.A).

In 1896, the distinguished Swedish chemist Svante August Arrhenius recognized that Earth's atmospheric temperature was related to the amount of CO_2 it contained and predicted that a tripling of the CO_2 content would lead to an average global temperature increase of 9°C (16°F). Since that time, many warnings have been voiced, but there is still much disagreement as to the magnitude of the change that could result from an increase in atmospheric CO_2. It is clear that the CO_2 concentration in Earth's atmosphere is increasing (see Figure 1.8) as a consequence of the combustion of massive quantities of fossil fuels (coal, gas, oil). Studies of atmospheric bubbles that were trapped in glacial ice before the Industrial Revolution, when large-scale fossil fuel burning began, indicate that CO_2 levels were only approximately 260 ppm by volume (0.026%) or about two-thirds of today's values. It is important to note that the 260 ppm CO_2 level is critical in the maintenance of the Earth's climate. It has been estimated that this level raises the average ambient Earth temperature from about −19°C (−2°F), which would make the planet essentially uninhabitable, to about +7°C (+45°F), which permits the growth of the vast variety of life-forms on which we depend for food. Detailed modern atmospheric analyses began in 1957 at stations on Mauna Loa, Hawaii, and at the South Pole. These data have revealed a steady increase in CO_2 from 316 ppm in 1959 to 350 ppm in 1990, and to an estimated 390 ppm in 2010. This is an average annual increase of 1.14 ppm, but the rate is increasing and is now approaching 3 ppm per year.

Disagreements arise when trying to determine where the added CO_2 will go and what the effects will be. The atmosphere actually holds relatively little of the total CO_2 involved in the atmosphere–biosphere–hydrosphere circulation system. The deep oceans contain sixty times more CO_2 than the atmosphere, but the rate of absorption is slow. The biosphere constantly removes CO_2 by photosynthesis, and many plants grow faster if given more CO_2. However, large-scale deforestation, especially in the tropics, is reducing the most effective natural CO_2 removal process. If most of the extra CO_2 from fossil fuel combustion remains in the atmosphere, the level of CO_2 is projected to rise to about 600 ppm by the year 2050. There is debate as to the exact effect that this will have on temperature, but several workers have modeled projections of an increase in average global temperature of 4–5°C (7–9°F). The changes would be less (2°C; 4°F) at the equator and more (7°C; 13°F) in polar regions.

The ultimate effects of such a temperature rise are not clear, but many scientists agree that there could be significant changes in growing seasons and rainfall patterns, and

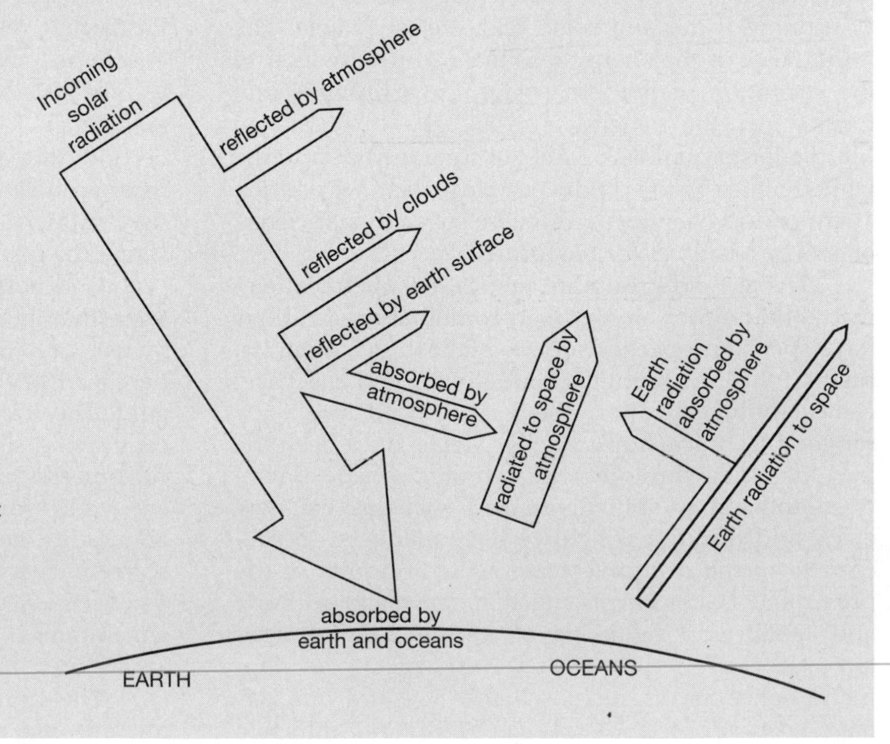

FIGURE 1.A The "greenhouse" effect occurs when incoming solar radiation and outgoing terrestrial radiation are absorbed by the atmosphere. Some incoming radiation is absorbed directly by the atmosphere but most passes through the atmosphere and is absorbed or reflected by Earth's surface. The surface then emits infrared radiation that is also absorbed by greenhouse gas molecules, especially carbon dioxide, in Earth's atmosphere. This raises the atmospheric temperature and alters the climate.

that there would be a rise in sea level. That rise is likely because of melting of the West Antarctic ice sheet and the expansion of the surface layers of the oceans as they are warmed. A rise in the ocean's water levels of up to 1 meter in the next 100 years and of 4 to 6 meters over a 500-year span could have devastating effects on coastal cities and low-lying regions around the world, including the Netherlands in Europe and the Gulf and Atlantic states in the United States. Presently, there appear to be no adequate alternatives to energy production by combustion of fossil fuels, and no efficient methods of limiting the increases in the levels of CO_2 in Earth's atmosphere.

another for reasons other than their metallic properties or their energy content. Such resources include the minerals used as sources of chemicals—such as halite (NaCl) and borax ($Na_2B_4O_7 \bullet 10H_2O$)—plus minerals used as the raw materials for fertilizers. Also falling into this category are the wide range of industrial minerals that are used in everything from smelting of metals to drilling for oil, and in such diverse products as paints, fillers, and abrasives. The construction and building industries employ large volumes of many different materials drawn from Earth—for example, crushed stone, sand, gravel, and the raw materials for cement and concrete.

The fourth group of resources includes water and soil. These ubiquitous resources are the most vital for sustaining life. Despite the description of Earth as the water planet, with 70 percent of the surface covered by oceans, only slightly more than 2 percent of all of the water on Earth is fresh enough for human use. Furthermore, only about one-quarter of the fresh water is actually accessible. Soils may cover nearly all areas of the land surface, but only about 10 percent of the total surface is actually suitable for growing crops.

CONSEQUENCES OF RESOURCE EXPLOITATION

Earth can be envisioned as a huge engine with two sources of energy. The first source is the Sun's heat. This is responsible for the turbulence in the atmosphere that we call wind, for temperature variations across the face of Earth that make equatorial regions warm and polar regions cold, for ocean currents, for evaporation of the water that forms clouds and leads to rain, and for many of the other phenomena, including growth of plants, that happen at Earth's surface. Flowing water, moving ice, blowing wind, and the downhill sliding of water-weakened rocks and mud are the main agents of erosion and transportation by which materials are moved as solids, liquids, and gases around the globe. If the Sun were the only source of energy, Earth would by now be a nearly smooth globe devoid of mountains. The reason that Earth is not a smooth globe is that the second major energy source, the Earth's internal heat, causes slow horizontal and vertical movements in the seeming solid rocks of the mantle and crust; these slow movements produce the crumplings and bucklings of the surface that thrust up mountains, split continents apart to form new ocean basins, and cause continents to collide and destroy old ocean basins.

The two systems of forces—those driven by the Sun's heat energy and those driven by Earth's internal heat energy—maintain a dynamic balance. The balance involves myriad natural transfers of material through streams, oceans, atmosphere, soils, sediments, and rocks. As new mountains like the Alps are thrust up, erosion slowly wears them down. If a balance were not maintained, not only would the face of Earth be smooth, but the compositions of the oceans and atmosphere would be very different. The movement of materials from rocks to soils to streams to oceans and back to rocks is called **geochemical cycling**, and the dynamic balance that results is called the **geochemical balance**. Formation of many of the Earth's resources (such as fossil fuels, metals, industrial minerals, and rocks) is a consequence of these geochemical cycling processes, which will be further described in Chapter 2.

One of the major consequences of our exploitation of natural resources is that we are interfering with some of the natural geochemical cycles. An example is shown in Figure 1.7, where the major reservoirs and flow paths for carbon at the Earth's surface are shown. There are five main reservoirs of carbon: (1) carbon dioxide (CO_2) in the atmosphere; (2) carbon tied up in the cells of living plants and animals (the biomass); (3) organic matter buried in sediments and sedimentary rocks, which includes oil, natural gas, and coal as well as the small percentage of carbon compounds found in all sediments and sedimentary rocks; (4) carbon dioxide dissolved in the oceans; and (5) carbon tied up in shells, limestones, and marbles as calcium carbonate ($CaCO_3$). The fluxes of carbon between the five major reservoirs are nicely controlled so that, on timescales of thousands of years, the system remains in balance.

Human involvement with the geochemical cycling of carbon involves the rapid removal of organic carbon (in the form of fossil fuels) from sedimentary rocks, and the conversion of that carbon to CO_2 through burning. Eventually, as burning is continued, the other reservoirs and fluxes must readjust, and the system moves toward a new dynamic balance. But the rate of readjustment is slow when considered

BOX 1.2

The Lessons of Busang and Bre-X

No metal has been the focus of so much attention as gold. It was one of the first metals recognized and used by our ancestors. Its rarity, combined with its noble nature that always kept it lustrous, enhanced its desirability and value and led to its choice as a monetary standard. Through the millennia, gold has adorned monarchs, both living and dead. It has been fashioned into jewelry, been used to finance wars, and stimulated great gold rushes. By the 1990s, it seemed to nearly everyone that the glory days of great gold discoveries had long passed. Imagine, then, suddenly learning that it was possible to invest in what promised to be the largest gold discovery in more than a century—and perhaps of all time. Would you invest?

Let us set the scene. The evolution of the theory of plate tectonics over the past forty or so years aided in the discovery of scores of large copper-bearing porphyry-type intrusions associated with the subducting zones around the Pacific Basin. Increasingly, these deposits were also recognized as major reserves of gold, with some deposits containing more gold value than copper. During the 1980s and 1990s, much exploration was focused on Indonesia in the southwestern Pacific. Thus, it seemed like "business as usual" when a small Canadian mining company, Bre-X purchased, for $180,000 in October 1993, a property called Busang, in Kalimantan on the island of Borneo (Figure 1.B). Bre-X began on-site exploration followed by some systematic drilling, all of which was reported in a routine manner in *The Northern Miner,* a well-known and highly reputable mining newspaper. The early results, announced in March 1995, indicated reserves of 1.1 million ounces of gold. This was more than many expected from the Busang property but on a par with numerous other deposits in the southwestern Pacific.

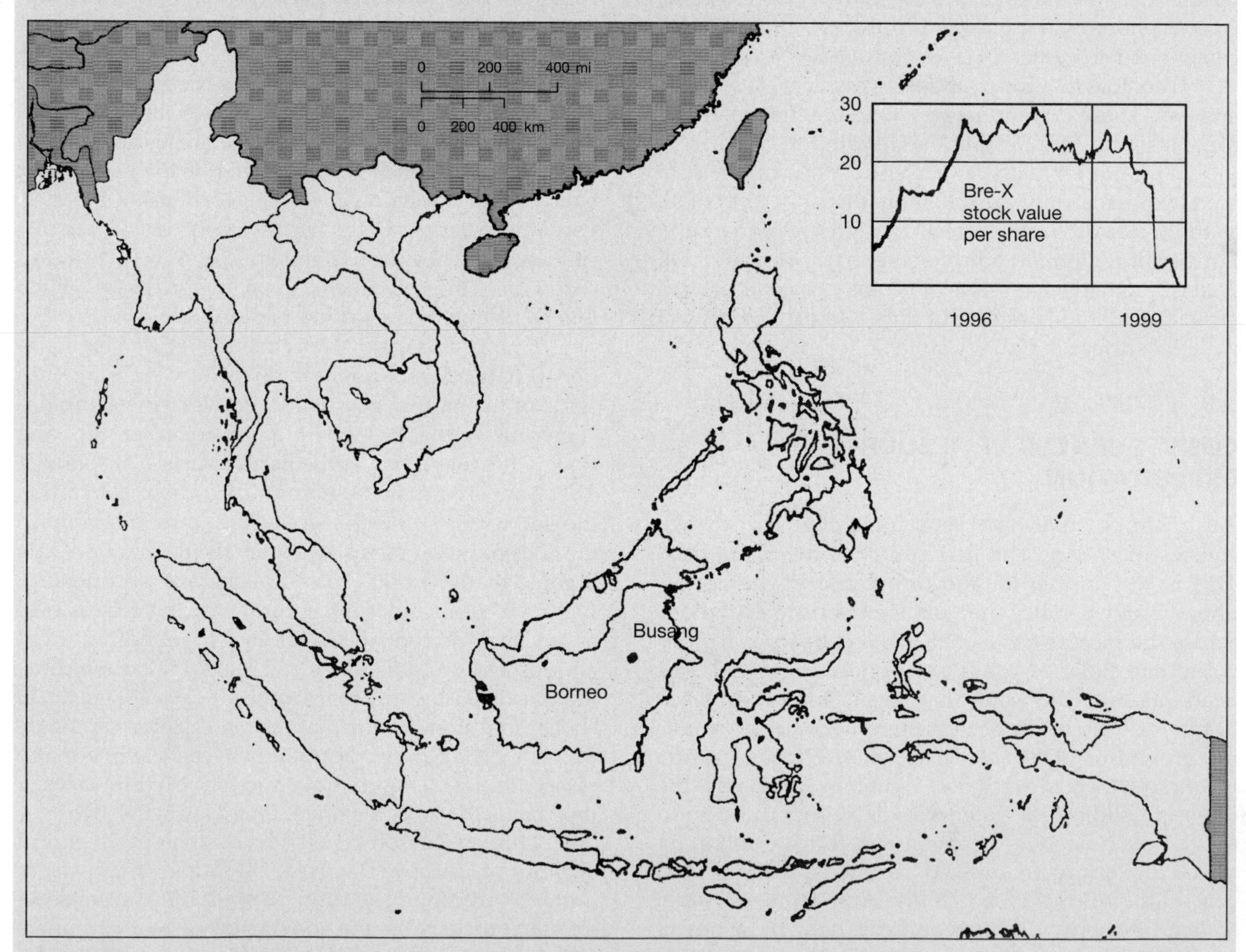

FIGURE 1.B The Busang deposit occurs in northeastern Borneo, as shown on the map, and was the supposed site of the world's largest gold deposit. A swindle ("scam") was uncovered and the deposit was shown to have little or no value; it had been salted to make it appear rich in gold. The inset shows how the stock value rose in 1996, but fell to near zero when the scam was revealed.

Even though this would amount to more than $300 million in recoverable gold at 1995 prices, the relatively low gold concentration ("grade") and the remoteness of the deposit did not promise great profits.

Subsequent reports from additional drilling began to attract more attention. In May 1995, the company reported that gold reserves had risen to 6 million to 8 million ounces; the value of the stock began to climb. The company reconfirmed the amount of gold as reserves in a September 1995 release, but then reported a remarkable increase to 30 million ounces in January 1996. This made headlines, and investors began rapidly to push up stock prices. Over the following year, the quantities of gold reserves reported continually increased: March, 40 million ounces; July, 50 million ounces; December, 57 million ounces; February 1997, 71 million ounces. With the last announcement, the president of the company stated there "could be 200 million" ounces—the largest announced gold discovery ever! By mid-1996, Bre-X had announced that it could not possibly mine such a giant deposit and wanted to become partners with, or be bought out by, a much larger mining company. Despite offers of up to $4.5 billion by two other large companies, McMoRan Inc. was granted the concession to take over the mine for $1.5 billion and development costs by the Indonesian government. Everything seemed to be wonderful for Bre-X and its investors; the stock price had risen from $0.08 to more than $210 per share. The billions invested in Bre-X stock made paper millionaires of many individuals, especially in one small town in Canada where nearly everyone had invested heavily.

In March 1997, prior to actually paying Bre-X for its share of the deposit, McMoRan began to conduct a routine *due-diligence* exercise, a testing of the validity of the claims made by Bre-X. Word traveled rapidly as reports were released: on March 19, the chief geologist for Bre-X was killed falling from a helicopter on his way to the due-diligence meeting; on March 24, McMoRan announced that there was far less gold than had been expected; on March 27, a sell-off of shares cut more than $2 billion in stock value in 30 minutes, and on May 4, an independent analytical report stated there was no possibility of an economically viable deposit and that there had been "salting" of ore samples on a massive scale. (Salting is the intentional addition of gold to worthless rock to make it seem as though it really contains gold.) The entire project was immediately terminated, and the Bre-X executives fled to foreign countries to avoid prosecution.

Many people had said that today the world was too alert and mining technology too sophisticated to permit a great mining scam. How wrong they were. Those who set up the Bre-X scam knew enough geology to make the Busang deposit sound plausible, and they knew that there would always be investors eager to get rich quickly. Unfortunately, they brought truth back to the old adage that "a gold mine is a hole in the ground with a liar at the top!" The lesson for all of us is to be wary of situations that promise so much and to remember the value of due-diligence.

in terms of a human life span. Seen from our perspective, the CO_2 content of the atmosphere is slowly but steadily increasing (Figure 1.8). Because CO_2 plays a major role in the thermal properties of the atmosphere, a change in its CO_2 content may cause changes in global temperatures (global warming) and in other geochemical cycles. For example, changes in climate could alter sea levels and rainfall patterns that, in turn, could alter the availability of water. Changes in the water supply could, in turn, affect the use of soils and

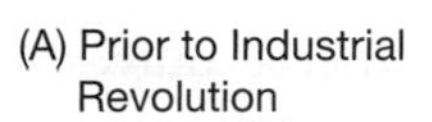

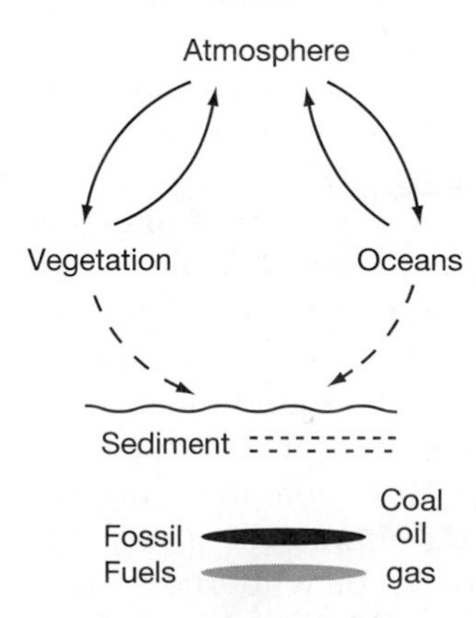

More or less a steady state with slow accumulation of organic matter in sediment forming fossil fuels

(B) Present Day

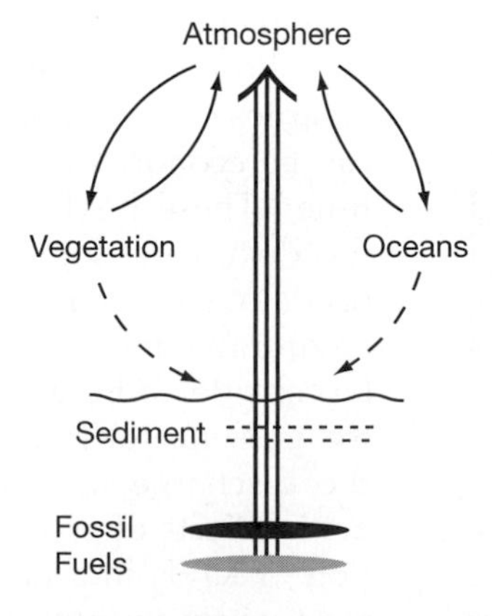

Vegetation and ocean exchange of carbon still more or less the same, but now a huge input of carbon into atmosphere by burning of fossil fuels

FIGURE 1.7 The carbon cycle, shown here in very simple form, provides an example of the way in which geochemical balances are maintained. There are several different reservoirs, and carbon is constantly being exchanged between them in processes such as photosynthesis and respiration. (a) Prior to human intervention and the large-scale use of fossil fuels, the carbon cycle was more or less balanced, with approximately equal amounts of carbon absorbed by vegetation and the oceans as was being released. There was also the slow accumulation of some carbon in the form of fossil fuels. (b) Since the onset of the Industrial Revolution, the burning of massive amounts of fossil has contributed vast quantities of carbon dioxide into the atmosphere at a rate much faster than it is being extracted by natural processes. This has led to a large increase in the amount of carbon dioxide in the atmosphere as shown in Figure 1.8.

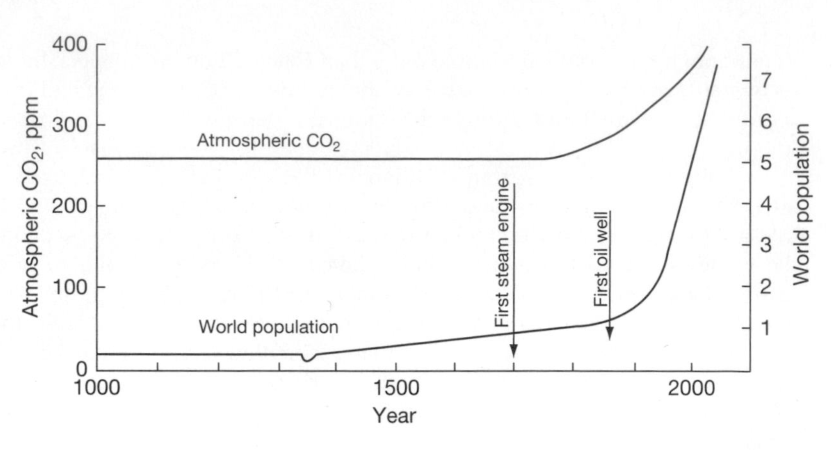

FIGURE 1.8 Carbon dioxide changes in the atmosphere. (a) The pre-industrial levels of carbon dioxide are estimated to have remained constant at approximately 280 parts per million (ppm) by volume, but have now risen to levels about 370 ppm. Note the rapid increase in the carbon dioxide level in the twentieth century as the rate of fossil fuel consumption has accelerated. There are also small scale (5 to 8 ppm) yearly oscillations that are due to seasonal growth of plants in the Northern Hemisphere. (From the National Oceanic and Atmospheric Administration.)

the growth of crops. Recently, there has also been concern that the rise in atmospheric CO_2 is slowly changing the chemistry (particularly the pH) of the oceans. This change can affect the ability of some marine organisms to produce calcite for their shells and, thus impact the health and productivity of the oceans.

The production and use of every natural resource, from the clearing of forests and the tilling of land, to the mining of copper and the burning of coal, trigger changes in the natural geochemical cycles. These changes can be large or small, local or global, pleasant or unpleasant, and they can be given names such as "pollution" and "environmental degradation," or can even be called "environmental disasters," but they are all consequences of the exploitation and utilization of natural resources. Among the topics addresssed in this book, therefore, are the important environmental consequences of our increasing exploitation of natural resources.

RESOURCES, RESERVES, AND ORES

Few words seem to cause more confusion than the terms resource, reserve, and ore as they are applied to mineral deposits. Indeed, it is not uncommon to find the words used interchangeably, as if they all mean the same thing. The meanings are actually very different. In part, the confusion arises because *resource* and *reserve* are common words, and each has a range of meanings depending on the materials being discussed. But, to a greater extent, the confusion arises for another reason; even though mineral commodities are used in every aspect of our daily lives, few among us have actually seen a mineral deposit and thereby developed an understanding of how big they are, how they vary in richness, and what difficulties are involved in producing mineral raw materials. Further confusion results from the misuse of the words by those seeking to make financial investments in mineral resources. The terminology given that follows has been adopted by the United States Geological Survey and most other responsible geological organizations.

The use of standard and exact terms as shown in Figure 1.9 is necessary for valid estimates and comparisons of resource worldwide and for long-term public and commercial planning. To serve these purposes, the classification scheme is based on both (1) geological characteristics, such as **grade**, tonnage, thickness, and depth of a deposit, and (2) profitability assessments dependent on extraction costs and market values.

A **mineral resource** is defined as "a concentration of naturally occurring solid, liquid, or gaseous material, in or on Earth's crust, in such form and amount that economic extraction of a commodity from the concentration is currently or potentially feasible." In the geological sense, the resources are subdivided into those that have been identified and those as yet undiscovered. Depending upon the degree of certainty, the identified resources fall into the categories of *measured* (where volumes and tonnages are well established), *indicated* (where volume and tonnage estimates are based on less precise data), and *inferred* (where deposits are assumed to extend beyond known resources). In terms of profitability, resources are classed according to their current economic status, as shown at the left side of Figure 1.9.

When referring to metal-bearing materials, **reserves** or **ores** are "that part of the resources that can be economically and legally extracted at a given time." These are the materials that are mined or otherwise extracted and processed to meet the everyday needs of society. It is important to note that legal and environmental as well as economic constraints must be considered because issues such as land ownership, the presence of endangered animal or plant species, the discharge of mining wastes, potential carcinogenic effects of products of mining, or the incorporation of lands into national parks or wilderness areas may exclude otherwise mineable resources from reserve status. In such cases, those materials would continue to be considered as resources; their potential for extraction would be high, but they would not become reserves unless laws or other restrictions were changed.

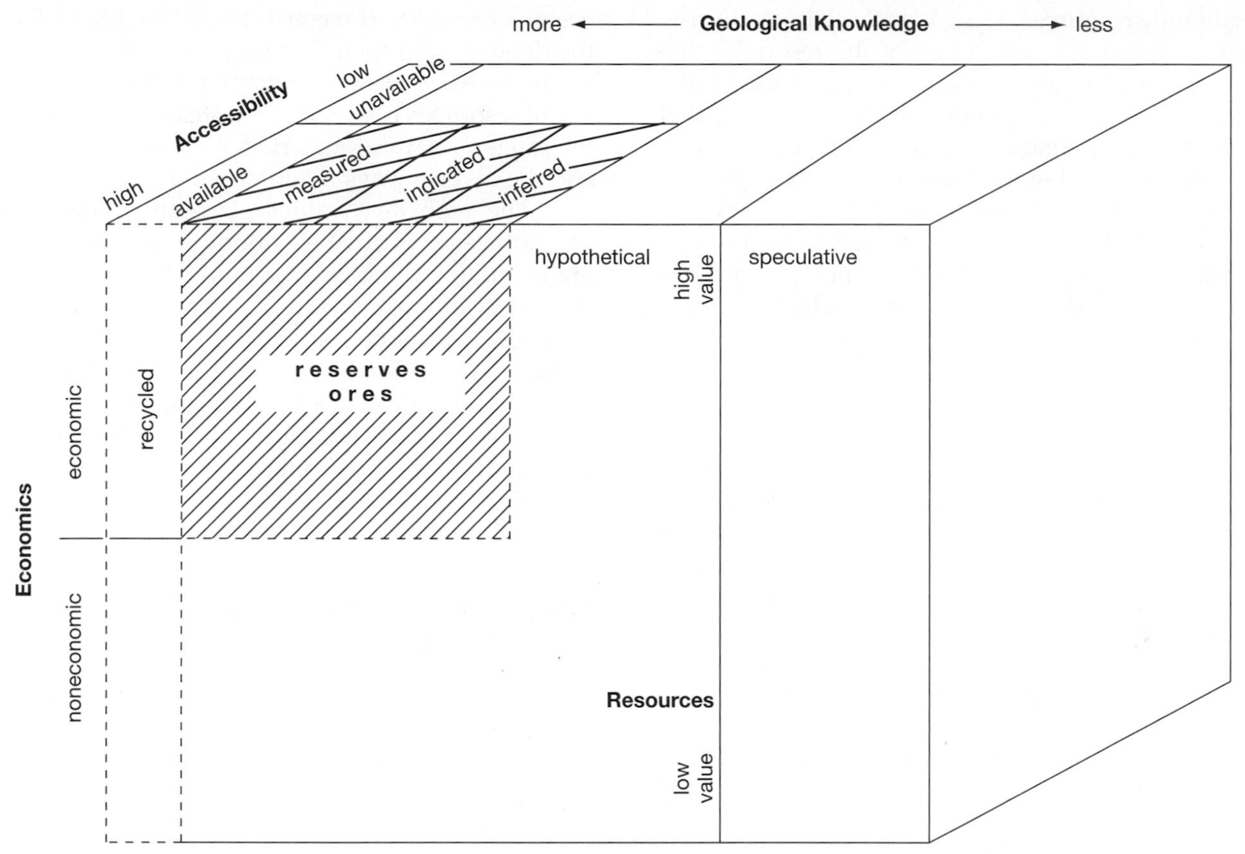

FIGURE 1.9 Resources may be classified according to geological knowledge, economic viability, and accessibility. The best-known and most profitable of the resources fall into the category of reserves (commonly called *ores* when referring to metal-bearing deposits) and constitute our present sources of mineral commodities. The term *reserve base* has sometimes been applied to the reserves and to some additional currently noneconomic resources that might become economic in the near future. Recycled materials are resources that have already been extracted from the soil and used at least once; the metals currently represent the most visible recycled materials. (Based on diagrams prepared by the U.S. Geological Survey.)

The quantities of reserves at any time are well-defined but change constantly; they decrease as ores are mined out but increase as new discoveries are made or as technological advances occur. Also, they increase as the market value of mineral products rise, and they decrease when the market value falls.

Many historical examples exist of resources becoming ores. A famous one occurred near the beginning of the twentieth century when two young mining engineers, D. C. Jackling and R. C. Gemmell, discovered that copper deposits previously ignored because of their very low grades could be worked at a profit by using new bulk-mining processes. This greatly increased the supplies of copper available for world-wide use. A second famous example occurred soon after the end of World War II, as the richest portion of the iron ores of the Great Lakes region were running out. New mining and processing technologies allowed the leaner and formerly unworkable low-grade deposits called "taconites" to be worked. Taconites are now highly desirable ores and supply most of the iron used in the United States. The third example occurred in the mid-1960s when the "invisible" and low-grade gold ores of the Carlin District in Nevada were brought into production. Today, many of the world's largest gold mines extract gold from rocks previously considered as waste.

There are many examples, too, of ores becoming too expensive to be mined and thereby slipping back again into resource status. The 1980s and 1990s brought the example of backward slipping of numerous gold mines around the world; when gold was selling for close to \$800 an ounce in 1980, it was possible to work very low-grade ore—some grades were so low, in fact, that it cost more than \$700 per ounce to recover the gold. The reserves of all gold mines were expanded as a result of the high price. When the price of gold dropped below \$400 an ounce in the mid-1980s, and then below \$300 an ounce in the late 1990s, reserves declined as the previously low-grade ores once again reverted back to the status of resources. But the increase in the price to over \$900 per ounce in 2007 once again moved the low-grade deposits into the reserve status.

In recent years, a broad term called the **reserve base** has been introduced to include the previous reserves, **marginal reserves**, and a portion of

subeconomic resources. Thus, it encompasses not only the reserves, but also the "parts of the resources that have a reasonable potential for becoming economically available within planning horizons beyond those that assume proven technology and current economics." Although this is a less well-defined quantity, it does take into account the resources that, although not now mineable, will likely become available for our use.

Traditionally, reserve status has been shown in a two-dimensional diagram with geological and economic axes. Today, however, it is appropriate to consider the third dimension of availability and to add recycled material on the left side of the diagram, as shown in Figure 1.9. The recycled material has already been extracted, refined, and used at least once. Some of it is still economically available and can find immediate reuse, whereas other material is accessible through community or commercial recycling efforts; still other material is buried in landfills and remains uneconomic.

It is all too easy to overlook the fact that quite large concentrations of materials can, for one reason or another, be too expensive to exploit. We must always keep in mind the fact that while mineral resources are the products of processes operating in the past, they become ores only if we are clever enough to discover the deposits and then find ways to exploit them profitably. As we shall see in Chapter 3, the history of our use of resources is a record of a steady increase in both the range of natural materials we have learned to use and the diverse ways in which we use them.

The monetary values of resources range from very little—as for crushed stone or sand—to extremely high— as for diamonds or gold. The general public hears little of most resources, but occasionally rich discoveries of resources such as oil, diamonds, or gold make news. Usually, the news is positive and investors make money. Unfortunately, there are also instances where dishonest individuals have swindled large sums of money from the public (see Box 1.2, *The Lessons of Busang and Bre-X*).

WHERE DO EARTH RESOURCES COME FROM?

Of the 90 or so elements known to occur naturally on Earth, at least 86 are used in various ways by modern societies; these are shown in Figure 1.10A. In some cases, the use is obvious: the iron used to make steel for

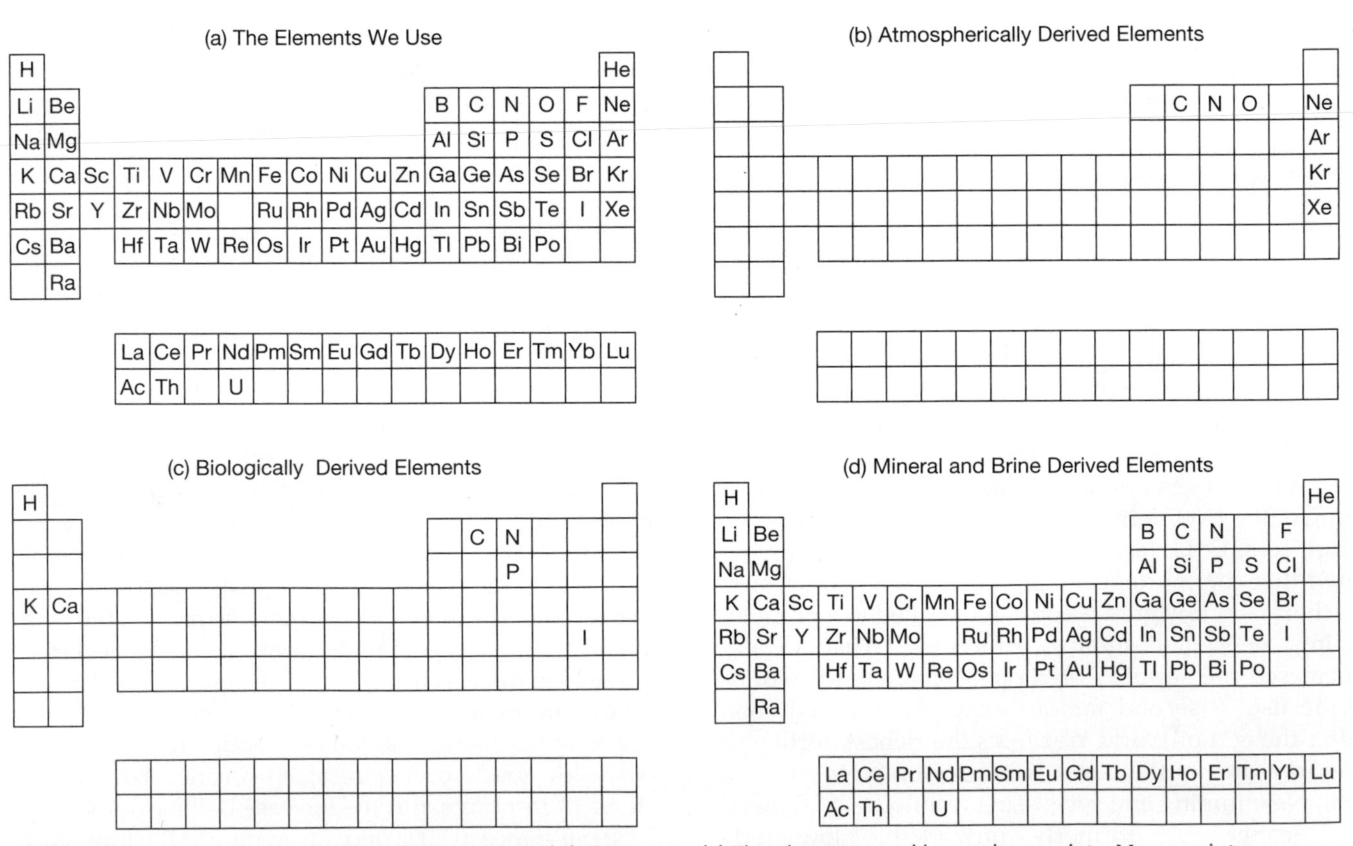

FIGURE 1.10 The chemical elements that we use and their sources. (a) The elements used by modern society. Many go into compounds in which their presence is not evident. (b) Elements that are extracted from the atmosphere. (c) Elements that are derived from biological materials. (d) Elements that are derived from the mineral resources and brines (ocean water and other salt-bearing waters). It is apparent that most of the chemical elements used in modern society are derived only from minerals and brines.

automobiles or refrigerators, and the sodium and chlorine that combine in nature to give us common salt. In other cases, the use is not so clear, as when vanadium is alloyed with iron to make special steels, or when rare earth elements (lanthanum, cerium, etc.) are used to make color phosphors for TV sets. As shown in Figure 1.10 B–D, only seven of these elements are ever derived from the atmosphere and only a further seven from biological systems. The rest of the elements (and even some of those already mentioned) are derived from minerals mined from Earth or brines extracted from the Earth (groundwaters) or the seas. The answers to the questions "Do we really need all of the mines, quarries, and wells that we have?" and "Could we obtain the materials we want from other sources?" are clearly "yes" in the first case, and "no" in the second case, given that we wish to maintain our present standards of life.

Volcanic activity as expressed in the great explosion at Mt. St. Helens in Washington on May 18, 1980, is evidence of the heat within Earth and the movement of fluids in the crust. The physical and chemical processes within and on Earth are constantly generating and modifying the materials of Earth and locally creating the resources that humans use every day. (From the U.S. Geological Survey.)

CHAPTER 2

Plate Tectonics and the Origins of Resources

People benefit from, and are at the mercy of, the forces and consequences of plate tectonics. With little or no warning, an earthquake or volcanic eruption can unleash bursts of energy far more powerful than anything we can generate. While we have no control over plate-tectonic processes, we now have the knowledge to learn from them. The more we know about plate tectonics, the better we can appreciate the grandeur and beauty of the land upon which we live, as well as the occasional violent displays of the Earth's awesome power.

THIS DYNAMIC EARTH *BY* W. J. KIOUS *AND* R. I. TILLING. U.S. GEOLOGICAL SURVEY ONLINE EDITION (2000)

FOCAL POINTS

- If Earth were homogeneous, most of the mineral resources used to develop and maintain modern society would not exist.
- Plate tectonics, driven by convection in Earth's mantle, has controlled the locations of the continents and of the processes responsible for the formation of mineral resources.
- Mineral resources are natural concentrations that result from the physical and chemical processes active in Earth's crust.
- Nearly all geologic processes form, modify, or destroy some Earth resources.
- Igneous activity, generated by radioactive heating, may concentrate resources by means of circulating hydrothermal fluids or by means of selective precipitation of layers of metal oxides or metal sulfides in a magma.
- Regional metamorphism modifies the properties of preexisting rocks and converts some of them into construction materials such as slate and marble.
- Contact metamorphism, resulting from the heat and fluids released by igneous intrusions, may generate metal ores and even result in the formation of gems.
- The shallow subsurface zone contains valuable groundwater resources and is the site of the initial generation of fossil fuels when buried organic matter is altered by increasing temperature and pressure.
- Weathering generates not only soils, but also specific resources such as clays and bauxite (the primary ore of aluminum).
- Processes of weathering and erosion break down rocks and transport the residual materials; this may

create large quantities of sand and gravel and may concentrate gold, tin, and titanium minerals as placer deposits.
- Evaporation in arid regions concentrates soluble salts that may form deposits of halite (rock salt), potassium salts, gypsum, and, in rare instances, nitrates.
- Burial, and subsequent compaction and heating, may convert terrestrial organic debris into coal.
- The ocean basins are the ultimate site of deposition of the sediments that erode from the continents; thus, they contain vast quantities of sand as well as some placer ores.
- Evaporation of shallow seas and ocean margin lagoons, over long periods of time, formed the largest evaporite deposits.
- Burial of marine plankton, later subjected to compression and heating, forms petroleum and most natural gas.
- Manganese and iron-bearing nodules form slowly on the deep ocean floor.
- Modern submarine hydrothermal vents are generating iron-, copper-, and zinc-rich sulfide deposits analogous to the large ore bodies being mined in ancient rocks.

INTRODUCTION

This chapter provides an overview of the processes involved in resource generation. The processes discussed are those that have generated the resources exploited today; however, the nature of the resources that we use may well change in the future (e.g., today we use bauxite as the source of aluminum, but in the future we may be able to use the feldspar found in many common rocks for the same purpose). Here we focus only upon the most important resource-forming processes; additional details are provided in later chapters, along with discussions of the specific types of resources.

Two very important points should be kept in mind throughout the reading of this book and in most discussions of the origins of Earth's resources:

1. All Earth resources have been generated by one or more geologic processes.
2. All geologic processes are forming, modifying, or destroying some Earth resources.

PLATE TECTONICS

The large scale of many geological features, and the very slow rates at which they change, gave our ancestors a sense that continents must remain permanently fixed on Earth's surface. Observations of rocks containing fossils high up on mountains made it clear that either sea level or vertical changes in elevation of the land surface had happened numerous times in the geological past, but such discoveries offered no compelling evidence for lateral motions and changing continental positions. The search for resources was undoubtedly quite random at first; but as experience grew, it began to focus on certain types of rocks tending to serve as hosts to certain types of resources. For example, tin-bearing veins were associated with granites, but nickel and chromium deposits were associated with gabbros and basalts. The gradual recognition of resource–host rock relationships aided greatly in the exploration for new deposits, but still the overall distributions of resources could not be explained by traditional geological patterns.

This all changed rather abruptly in the 1960s when plate tectonics brought a new paradigm of understanding to Earth and the forces within it. Plate tectonics recognizes that Earth's outermost layer consists of a number of irregularly shaped "plates" of lithosphere, and that these plates have slowly changed in dimensions, locations, and orientations throughout geologic time. The temperature of Earth's interior is maintained as a result of heat given off by the natural decay of radioactive elements such as uranium, thorium, and potassium. Earth dissipates this heat through the generation of large convection cells within the mantle, much in the way that water convects in a pan when it is heated from below (Figure 2.1). The cells carry the heat upward, releasing much of it in the form of vulcanism at upwelling, or spreading zones (called *divergent plate boundaries*), such as the mid-oceanic ridges (Figure 2.4). The mantle is composed of rocks that are related to basalts in composition; the volcanic activity at these boundaries is nearly all basaltic. As new magma is brought upward, earlier, now-cooled volcanic rocks of the plate are moved laterally away from the boundary. At other boundaries, called *subduction zones,* cooled rocks sink back into the mantle, as shown in Figures 2.1 and 2.4. The continents are composed primarily of lower density rocks, such as granites, that are carried along atop the convecting mantle rocks. Individual convection cells are not permanent, so neither are the plates that they transport. Instead, the cells and plates are constantly changing shape and size as old cells die and new ones develop. Plate tectonics has become the great unifying theory for the geological sciences because it provides an explanation for many geological processes; it also explains the types of rocks, minerals, and energy resources found at Earth's surface and their distributions.

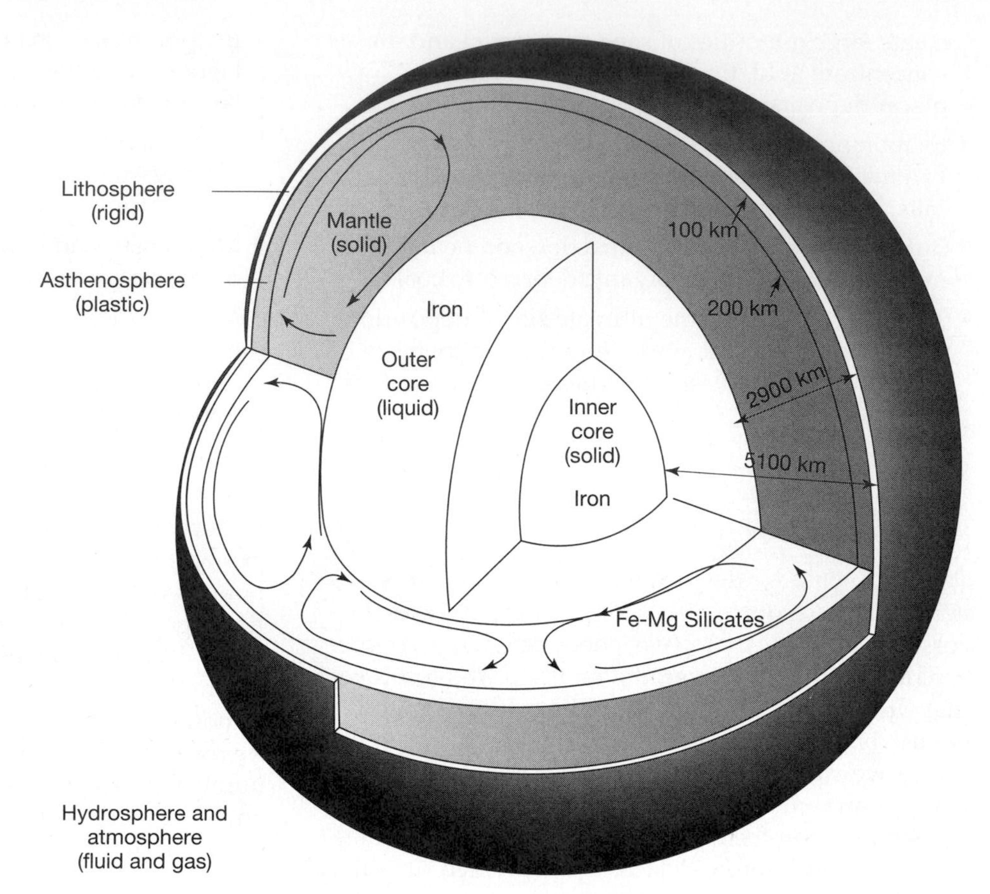

FIGURE 2.1 Earth is composed of a series of concentric shells with approximate thicknesses as shown. The interior, like the hydrosphere and atmosphere, is dynamic but its movements are generally slower and episodic. Large convection cells, moving in the manner shown by the arrows, shift the continents and are responsible for the generation and location of many of Earth's resources.

Prior to the development of the plate tectonic theory, the distribution of resources seemed to be random, even if there were clear relationships to rock types. It appeared that local factors had determined the origins of resources. Plate tectonics, however, has provided a much larger and more comprehensive picture of resource occurrences and an understanding of how resources were generated. Two examples will help to demonstrate this point. The world's largest copper deposits, porphyry-type ores, occur in large numbers around the margins of the Pacific Ocean Basin. The reason for their distribution became clear when it was realized that such deposits form at subducting plate boundaries (as discussed in greater detail in Chapter 8) due to partial melting of subducting plates as they slide beneath the edges of the continents. The presence of large salt deposits under the Michigan Basin and the western part of New York State was difficult to understand given the temperate and moist climate of that part of Earth today. It is now apparent that, as a consequence of plate tectonic movements, those deposits formed when the North American continent was much farther south, in a climatic regime where warmer and drier conditions prevailed. Only after the deposits formed did plate tectonics move the continent to its present location.

Distribution of Earth's resources has been controlled by two principal factors: (1) global plate-tectonic processes, which established the major geological settings (e.g., subduction zones) and rock types (e.g., granites, basalts), and (2) local processes that resulted in the formation of the resources (e.g., fluids emitted by igneous rocks; local swamps where peat accumulated). Both types of processes are difficult to observe and fully understand. Nevertheless, it is clear that the heat escaping from Earth's interior generates magmas and determines their locations in Earth's crust, thus controlling the locations of many of the major metal deposits. Furthermore, the movement of plates creates great stresses within Earth's crust and generates major fractures, which, in turn, serve as major conduits through which large quantities of metal-bearing fluids flow. Plate movements shift continents through different climatic zones; these zones may dramatically affect the formation of resources, with evaporates forming in arid zones, bauxites in warm, heavy rainfall zones, and coal swamps in moist temperate zones. Subsequent plate movements then may have moved the resources into regions where the climate is now such that they could not have originally formed.

We shall return again to specific aspects of plate tectonics during the discussions of individual types of resources throughout this book. The remainder of this chapter will discuss some of the specific types of relationships between resources and the local geologic processes responsible for their formation. There are many ways to classify these processes; a simple approach is:

1. Subsurface Igneous and Metamorphic Processes
   a. formation of granites and other relatively silica-rich rocks
   b. formation of basalt and other relatively silica-poor rocks
   c. regional metamorphism
   d. contact metamorphism
2. Surface Processes
3. Shallow Subsurface and Diagenetic Processes
4. Marine Processes

These are somewhat arbitrary divisions, and the boundaries between them are not sharply defined. Furthermore, more than one process has commonly been involved in the generation of a particular resource (e.g., coal originates as debris of land plants that became compacted to form peat, then lignite, then bituminous coal, and finally anthracite; metallic gold that has been concentrated and deposited by hydrothermal fluids can be released by weathering, then transported by streams, and finally deposited in beach sands at the ocean margin as placer deposits). Nevertheless, these subdivisions help us to recognize the relationships between processes and particular resources; that recognition helps us to understand where to search for additional resources as they become needed.

The various processes are all parts of the **rock cycle,** and because the rock cycle (Figure 2.2) is continuous, there is no obvious point at which to start a discussion of it. We shall begin with an examination of igneous and metamorphic processes, followed by a discussion of surficial processes, and then marine processes.

SUBSURFACE IGNEOUS AND METAMORPHIC PROCESSES

The continents are composed primarily of granite, whereas the floor of the ocean is composed primarily of basalt. These *bedrocks* are mostly covered with a veneer of sediments that are the products of the weathering and erosion of the granites and basalts caused by interactions with the atmosphere and hydrosphere. There are many other specific types of igneous rocks; a thorough discussion of their origins is far beyond the scope of this book. The discussion that follows focuses upon two end-members of the range of igneous rocks—granite (and other silica-rich rocks) and basalt

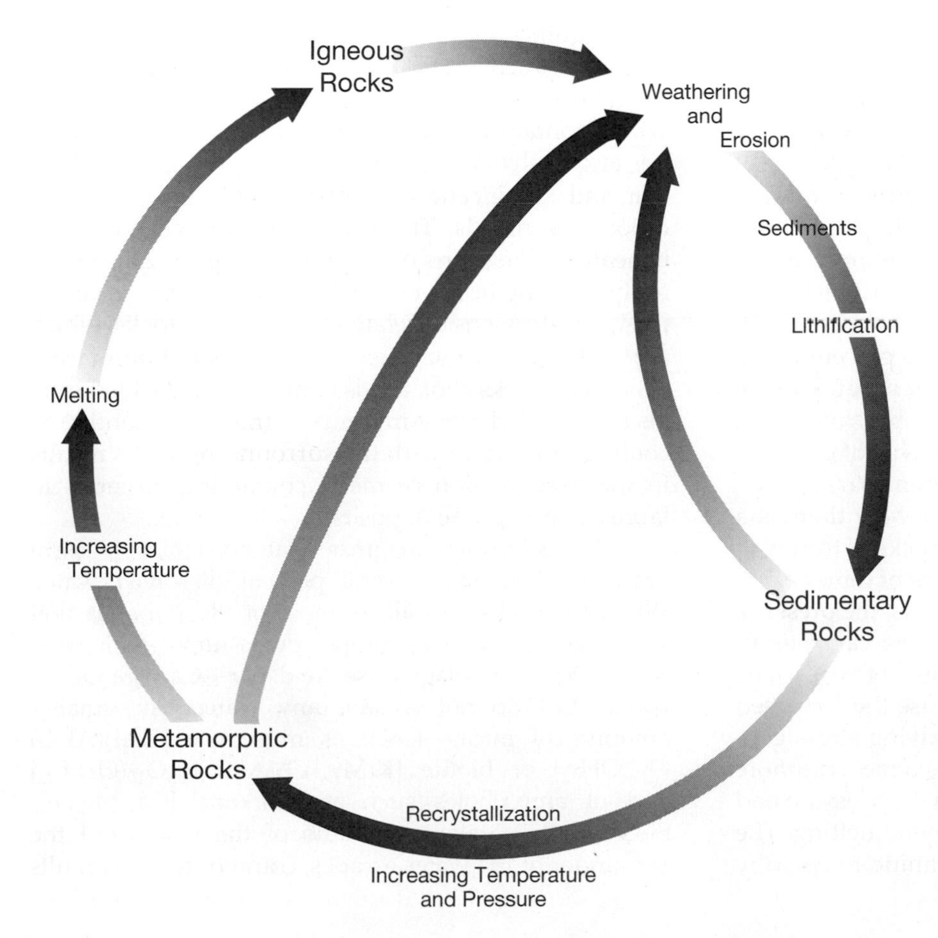

FIGURE 2.2 The outer circle of the rock cycle schematically illustrates the changes experienced by rock matter in its uninterrupted cycle. Crystalline igneous rocks may weather into sediments that, in turn, can be consolidated, buried, metamorphosed and, ultimately, remelted. Breaks in the cycle are indicated by the internal arrows.

(or gabbro, the coarse-grained equivalent of silica-poor rocks)—and the processes of their formation.

From prehistoric times, igneous rocks have been used as construction materials either as crushed stone (because of their strength and durability), or as **dimension stone** (because of their ease of shaping and polishing). Many other resources of Earth are directly generated during the formation of igneous rocks, or by the movement of the significant amounts of hot fluids associated with their formation that dissolve, transport, and precipitate metals as ores. Similar fluids, even when devoid of metals, may prove important as sources of geothermal energy.

Formation of Granites and Other Relatively Silica-Rich Rocks

Granites, like all igneous rocks, form by melting of preexisting rocks due to heat buildup within the Earth. The primary source of heat is radioactive decay of the isotopes of uranium, thorium, and potassium. Because rocks are poor conductors and the isotopes are irregularly distributed, the heat generated by radioactive decay gradually builds up locally and may cause temperatures high enough to melt rocks and form magmas. It was this buildup of heat and the subsequent melting that produced the internally layered structure (Figure 2.1) of Earth, with the partitioning of iron into the *core*, iron and magnesium silicates into the *mantle*, and the lighter components into the *crust*. Despite this partitioning, many inhomogeneities persist and radioactive decay continues. The overall rate of decay is now less than during earlier periods of Earth's history because the radioactive isotopes are now diminished in quantity. For example, the amount of ^{235}U now contributing to this heat is only about 1.5 percent of that present at the time of Earth's formation; ^{238}U has been reduced to about 50 percent, ^{232}Th to about 75 percent, and ^{40}K to about 10 percent of the original amounts. Despite reduced heat input, Earth is still a dynamic planet as evidenced by volcanism, earthquakes, and hot springs, but the level of dynamic activity is lower than it was a few billion years ago.

In simple terms, granites form where there is a sufficient buildup of heat to melt the rocks of the crust. Different minerals and different groups of minerals together have very different melting points; the presence of water and other volatile substances can significantly reduce the melting temperatures of rocks. The situation is further complicated because the increased pressure placed on the rocks by overlying strata generally raises the melting temperature. Hence, the more deeply a rock is buried, the greater is the pressure and the more the rock must be heated before melting. The minimum temperature to form a granitic magma by the melting of rocks is about 675°C. When magmas form, they are able to move upward in response to pressure and density differences, and they may emerge at the surface. The most obvious example of this is lava flowing from a volcano. Lavas, however, are nearly always basalts and are quite different chemically from granites. Basalts, because of low silica contents (that is, low amounts of silica as a part of their total chemical composition—not silica as quartz), are much less viscous (that is, they flow more easily) than are granite melts with high silica contents. Granite magmas move slowly, but can travel upward and laterally through Earth's crust for kilometers (Figure 2.3). They may be forced into regions of the crust where they physically push aside other rocks, or they may gradually melt their way upward incorporating other rocks (a process called *stoping*). When magmas reach their limit of movement because of heat loss, or a balance based on their density, or because they encounter some physical barrier, they slowly crystallize. Subsequent erosion can remove up to several kilometers of overlying rocks and expose the granites. The removed material becomes the sediment that is transported by rivers and deposited on the continental shelves.

The development of plate tectonic theory since the 1960s and the recognition of the mechanisms of magma generation along subducting plate boundaries have significantly increased our understanding of the ways in which many types of resources are formed. This is especially true for porphyry-type deposits, which contain the world's greatest quantities of copper and molybdenum, large amounts of gold and silver, and significant amounts of zinc, lead, tin, and other base metals. The subduction of oceanic plates beneath the margins of continents (Figure 2.4) results in the melting of lower crustal rocks and the development of intrusions of relatively silica-rich rocks, which are emplaced in the crust near the plate boundaries. As molten rocks cool, fluids containing dissolved metals are released along myriads of fractures. Continued cooling or reaction with the surrounding rocks results in the precipitation of metal containing minerals in large but low-grade deposits.

The silica-rich magmas that crystallize to form granites all contain several percent dissolved water. When the rocks crystallize, most of the minerals that form (e.g., quartz—SiO_2; potassium feldspar—$KAlSi_3O_8$; and plagioclase feldspar—$CaAl_2Si_2O_8$ to $NaAlSi_3O_8$) do not contain any water. The smaller amounts of micas—(such as muscovite $[KAl_2(AlSi_3)O_{10}(OH)_2]$ or biotite $[K(Mg,Fe)_3(Al,Fe)Si_3O_{10}(OH,F)_2]$ and of amphiboles such as actinolite $[Ca_2(Mg,Fe)_5Si_8O_{22}(OH)_2]$)—take up a little of the water, but the remainder of the water escapes, commonly along faults

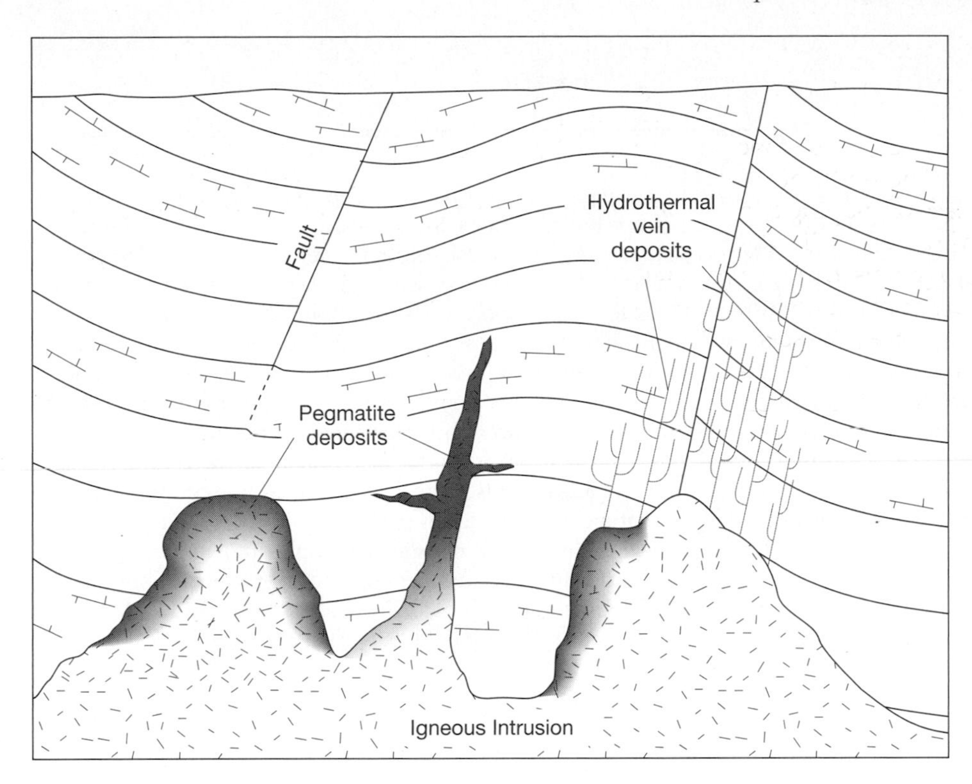

FIGURE 2.3 Hydrothermal veins commonly occur near igneous intrusions where fluids are given-off from the intrusions or where groundwater is heated by the intrusions. As the fluids move upward along fractures, the pressure decreases and the fluids are cooled; this results in the deposition of the metal-bearing minerals in veins. Pegmatites usually form at the margins of intrusions where fluids accumulate and where some minor and rare elements are enriched.

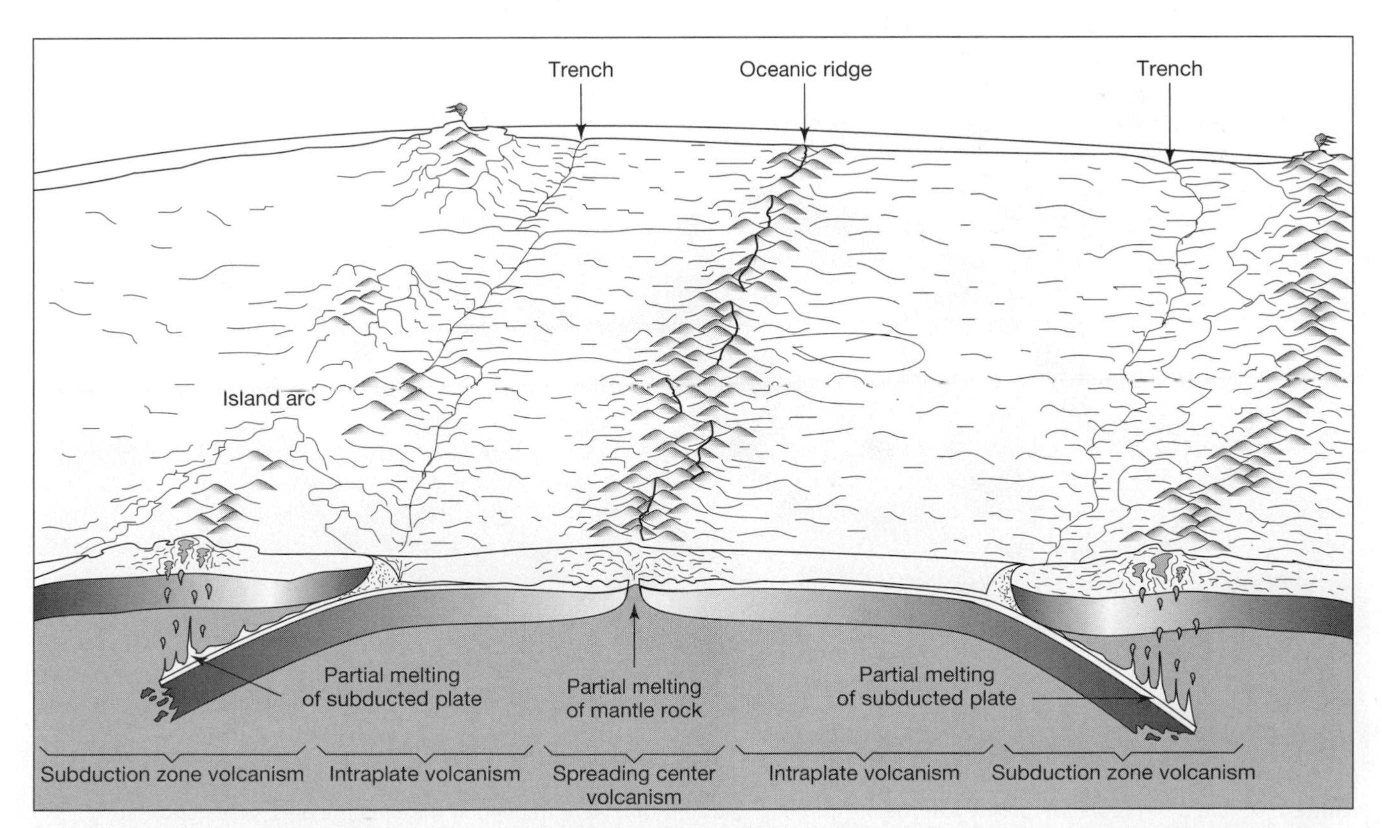

FIGURE 2.4 Basaltic oceanic crust is emplaced at spreading centers (mid-ocean ridges). At subduction zones, oceanic crust slides beneath the granitic masses of the continents; these are active seismic zones and the sites of magma formation by partial melting. The magmas appear at the surface as volcanoes, and beneath the surface as intrusions. These intrusions serve as major sources of copper, molybdenum, and gold.

BOX 2.1
Fluid Inclusions

The search for new ore deposits to replace those being mined away requires a detective-like approach to the examination and interpretation of clues in the rocks. On the large scale, geologists frequently use satellite images and high-level aerial photographs to look for structures or changes of rock-type that might indicate the presence of mineralization. At the other end of the scale are features so small that they can only be seen through a microscope. Among the most valuable of these tiny clues are *fluid inclusions,* small droplets of the fluids that deposited the ores and that we now find trapped in the ore and host rock minerals. They provide evidence of the processes and conditions under which an ore was formed, even though that event may have occurred millions of years ago at high temperatures and deep within Earth's interior.

Hot water solutions, especially those with dissolved chloride, sulfate, sodium, and potassium salts, are effective in dissolving and transporting the metals we find in ore deposits. In the exploration for new ore deposits, it is important to determine the conditions, the causes, and the locations where the valuable metals in these solutions precipitate to form the mineable concentrations. Because fluid inclusions hold some of the clues to answer those questions, both their examination and their analysis have become integral parts of mineral exploration. Most ore minerals form when the metal-bearing solutions cool, when they emit dissolved gases, or when they react with the rocks through which they are passing. Under ideal conditions, the ore and associated minerals would form as perfect, flawless crystals. However, under the real condition of ore formation, many small imperfections in the growing crystals trap tiny droplets of the metal-bearing fluids. Once trapped, the fluids often remain unchanged except for some shrinkage as the temperature drops; shrinkage results in the development of a small vapor bubble, as seen in Figure 2.A.

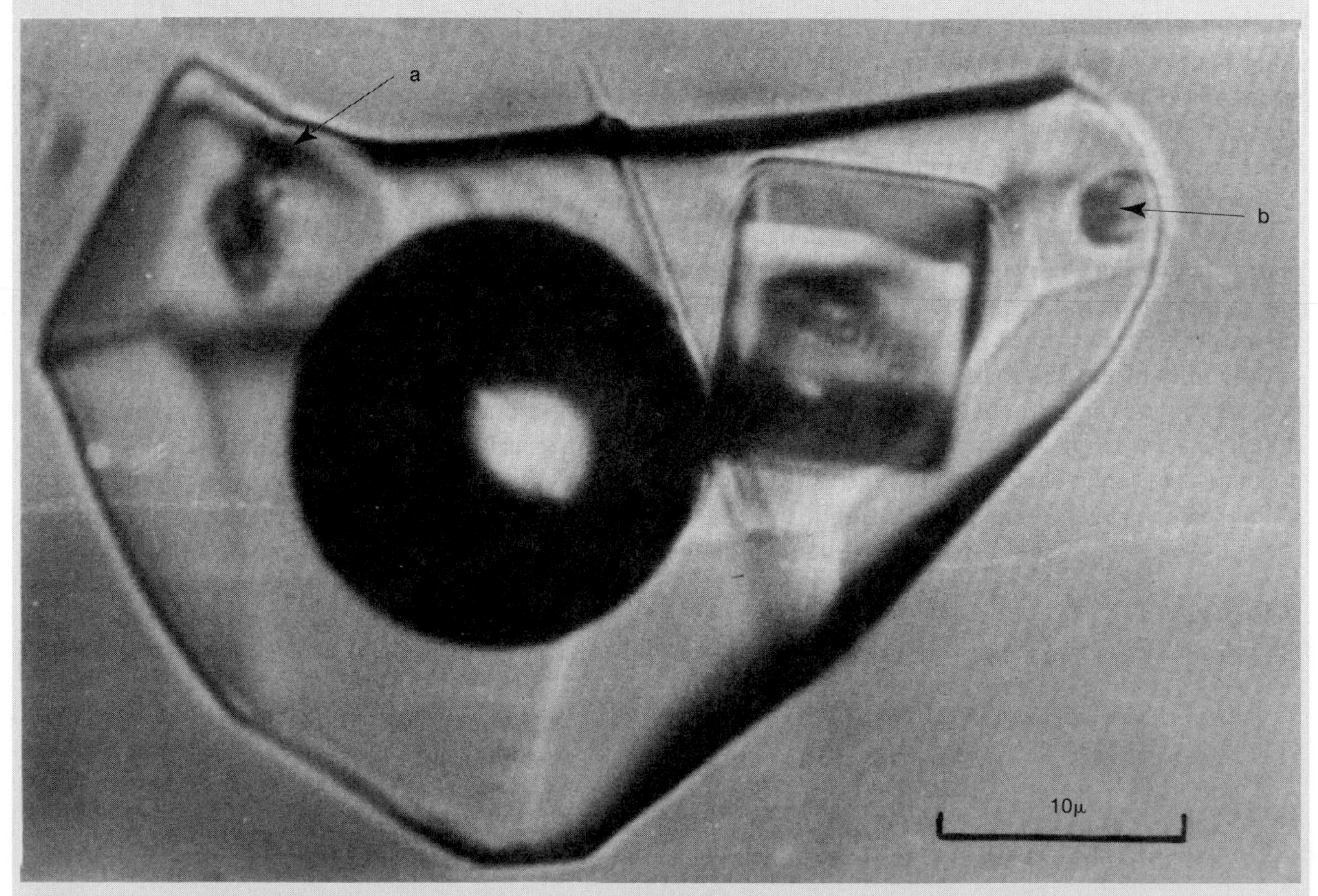

FIGURE 2.A Fluid inclusions are droplets of the ore-forming solutions that were trapped at the time the ores were deposited. This inclusion from the Laramesta tin and tungsten deposit in Bolivia contains a round vapor bubble that formed on cooling from its formation temperature of 430°C. The large white salt crystal and the small crystals, at (a) and (b), crystallized from the saline solution as it cooled. (Reproduced from W. C. Kelly and F. C. Turneaure, *Economic Geology,* vol. 65, [1970] p. 651. Used with permission.)

Fluid inclusions are usually studied by cutting or breaking small pieces of rock and examining them with a microscope. By carefully heating and cooling the specimens during observation, it is possible to determine the temperatures at which they were trapped and the salinities of the fluids. Further analysis using spectrometers sensitive to specific elements in solution permits the determination of the concentrations of metals and other compounds in the inclusions. (Note the small crystals in the inclusion shown in Figure 2.A.) Much can be learned about the types of fluids, the temperatures and pressures at the time of entrapment, and their sequences of introduction or evolution. These data, combined with field geologic observations, can provide powerful insights into the locations and mechanisms of ore formation and, hence, into the search for additional ores.

and fractures that form during and after emplacement of the intruding magma. This water often contains dissolved elements (Cu^{2+}, Pb^{2+}, Zn^{2+}, Ag^{+}, S^{2-} as listed in Table 2.1) and compounds (such as NaCl) that do not get incorporated into the principal granite minerals. As these water-rich solutions, generally termed *hydrothermal fluids* ("hydro" meaning water and "thermal" meaning hot), move outward from the crystallizing granite, they cool, undergo reduction in pressure, and may react with other rocks. Each of these changes can decrease the solubility of the dissolved species and cause precipitation of minerals in the faults or fractures to form "veins" (Figure 2.5). The veins range widely in complexity, richness, mineralogy, and size, but those that are most valuable as resources may contain several percent of one or more sulfide minerals of copper, lead, zinc, silver, and sometimes, traces of native gold. For lower-value metals such as lead and zinc, veins may need to be one-half meter or greater in thickness to be economic to mine. In contrast, the much higher value of gold allows mineable gold-bearing veins to be only millimeters in thickness, or to contain only tiny grains disseminated along the vein.

There is no clear boundary between the deeper and the shallower subsurface zones because veins and igneous intrusions that originally formed at considerable depths are commonly exposed at the present ground surface by erosion. Furthermore, igneous activity with veins, hot springs, and other associated phenomena may occur at Earth's surface in some areas, whereas other areas (e.g., the Mississippi Delta) contain thousands of meters of sediments with no evidence of igneous rocks except in the underlying *basement* at depths greater than 20,000 meters.

Sometimes the fluids generated by crystallizing magmas concentrate minor or unusual elements at the margins of the intrusions, forming very coarse-grained masses called **pegmatites**. Pegmatites consist primarily of three common minerals—feldspar, quartz, and mica—but they can also host minerals containing fluorine, beryllium, lithium, or rare earth elements. Consequently, pegmatites serve as major

TABLE 2.1 Analyses of hydrothermal solutions, weight percent

Chemical Element	(1)	(2)	(3)	(4)
Chlorine	15.50	15.70	15.82	4.65
Sodium	5.04	7.61	5.95	1.97
Calcium	2.80	1.97	3.64	0.750
Potassium	1.75	0.041	0.054	0.370
Strontium	0.40	0.064	0.111	—
Magnesium	0.054	0.308	0.173	0.057
Bromine	0.12	0.053	0.087	—
Sulfur*	0.005	0.031	0.031	0.160
Iron	0.229	0.0014	0.030	—
Zinc	0.054	0.0003	0.030	0.133
Lead	0.010	0.0009	0.008	—
Copper	0.0008	0.00014	—	0.014

*Sulfur analyzed as $(SO_4)^{-2}$

(1) Salton Sea Geothermal brine (Muffler and White, 1969).

(2) Cheleken geothermal brine (Lebedev and Nikitina, 1968).

(3) Oil field brine, Gaddis Farms D-1 well, Lower Rodessa reservoirs, central Mississippi, 11,000 ft (Carpenter et al., 1974).

(4) Fluid inclusion in sphalerites, OH vein, Creede, Colorado (Skinner and Barton, 1973).

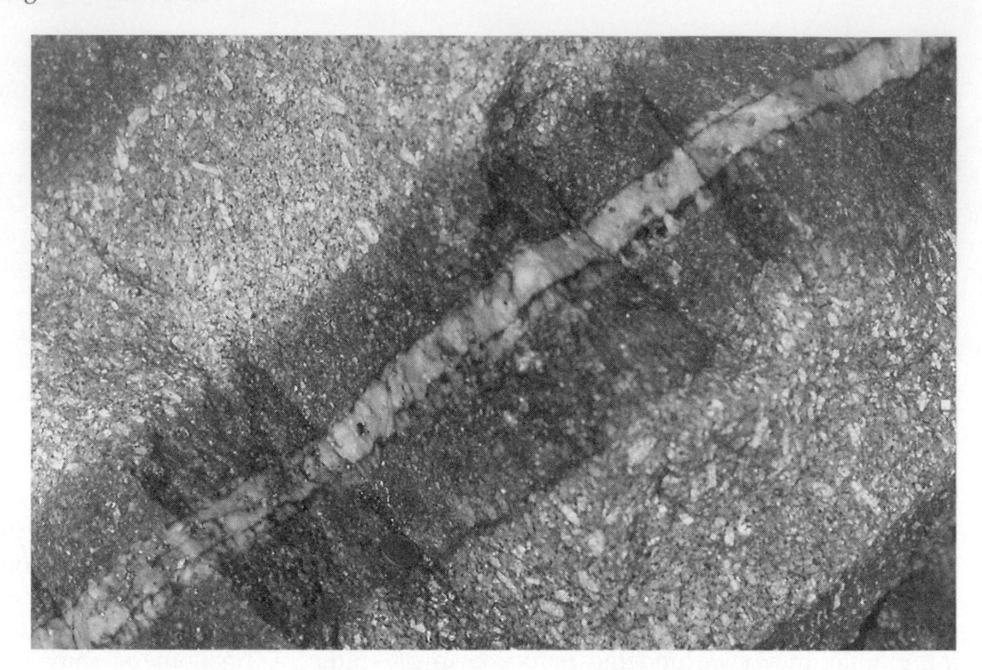

FIGURE 2.5 Hydrothermal vein cutting across granite at Cligga Head, on the southwest coast of England. The vein is filled with white quartz and small amounts of tin, copper, and tungsten minerals. The dark zones on each side of the vein are where the hot vein-forming fluids have altered the surrounding rocks. (Photograph by J. R. Craig.)

sources of minerals containing these elements as well as several types of gemstones, such as emerald and aquamarine.

Formation of Basalts and Other Relatively Silica-Poor Igneous Rocks

Silica-poor magmas, including basalt magmas, are generated by partial melting of Earth's mantle. Such magmas contain only tiny amounts of dissolved water. Localized melting appears to result from a differential buildup of radioactive heat and may be followed by a rapid upward flow of magma along faults or other zones of weakness, with surficial expression in the form of volcanoes or basalt lava flows. Although these events can be spectacular and life-threatening, their effects on resources are usually localized and negative, such as the loss of cropland to volcanic ash fall. The rapid cooling of the lavas prevents the concentration of ore minerals into economically recoverable resources, instead leaving them dispersed throughout the rocks. In many areas, including some beneath volcanoes, large masses of basaltic magmas are intruded and crystallize at depths of many kilometers. These magmas may "stope" their way upward into the crust by melting overlying rocks, or move upward or laterally along major faults. Regardless of their mode of emplacement, these large bodies (which can be hundreds of kilometers across and thousands of meters in thickness) ultimately reach positions within the crust where they slowly crystallize (Figures 2.6 and 2.7).

Such large masses of magma may be emplaced at 1100°C or more and surrounded by rocks that are such poor conductors of heat that millions of years may be required for the magma to crystallize to rock. The slow loss of heat, the incorporation of foreign matter as a result of the melting of surrounding rocks, and the sequential crystallization of minerals, can trigger major chemical changes in the magmas and may result in the formation of important resources. Crystallization proceeds, in general, following the principles of **fractional crystallization**. For example, the earliest minerals to crystallize are usually olivines, $(Fe,Mg)_2SiO_4$, which settle to the bottom of the magma chamber and form layers of dunite, a rock composed primarily of olivine. The next minerals to form are the pyroxenes and calcic plagioclase feldspars that then settle as layers of gabbro overlying the layers of olivines. As certain elements are extracted to form the early minerals, the composition of the remaining melt changes such that subsequent minerals are of a different composition. The final result is a coarse-grained layered igneous rock; the layers are made up of minerals that are iron- and magnesium-rich and silica-poor at the base, but grade into increasing iron- and magnesium-poor and silica-rich toward the top.

The layers of olivine-rich rocks are valuable resources because the olivine is widely used in refractory bricks and as casting sand. Usually much more valuable are the layers rich in metal sulfide and oxide minerals that may form at particular stages during crystallization of the magma. For example, if the limit of sulfur solubility in the magma is exceeded as it cools, the sulfur expelled from the magma is very efficient in combining with iron, nickel, and copper forming small droplets of a metal sulfide liquid. This liquid is more dense than the magma and, if there is sufficient time, the droplets settle in response to gravity

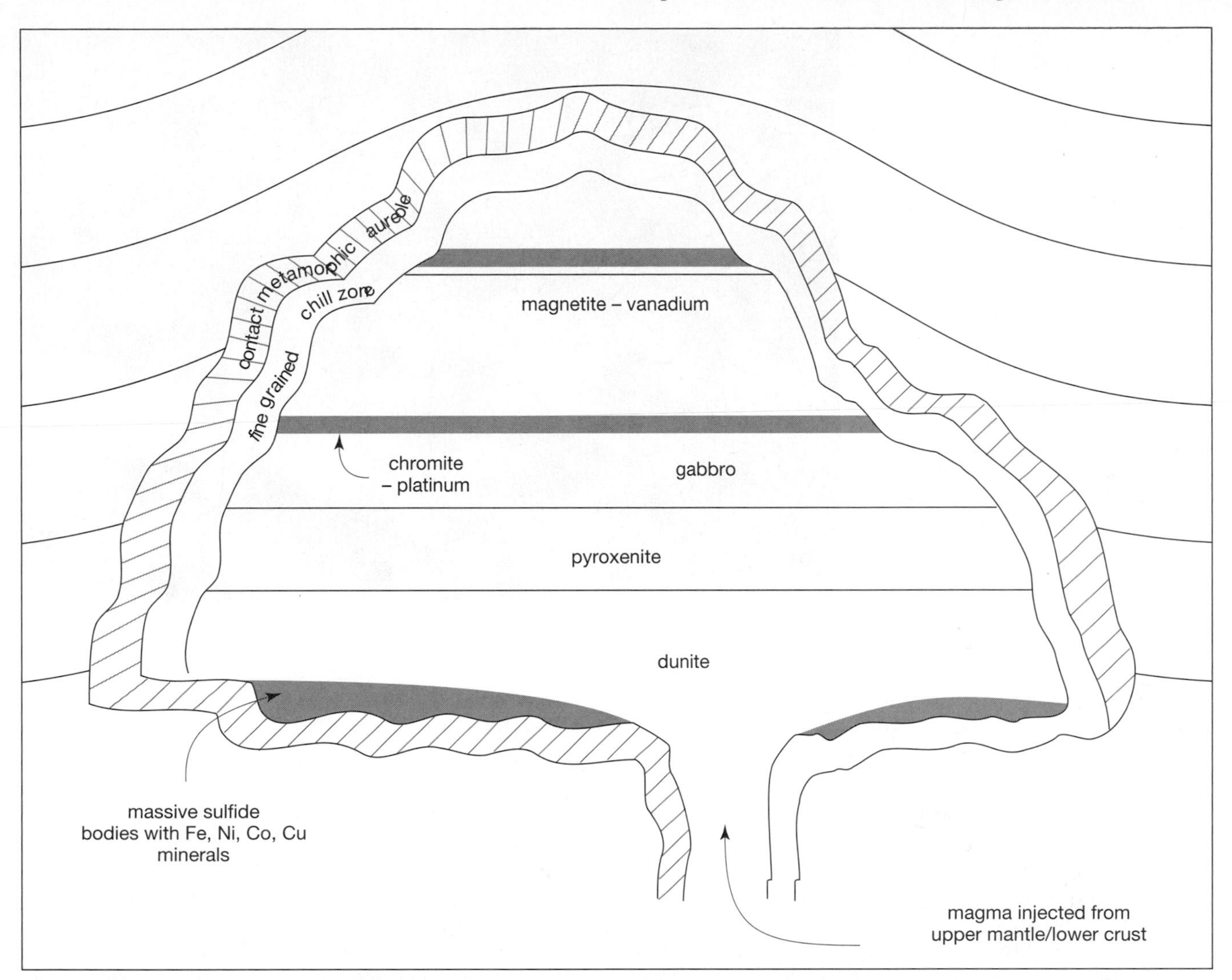

FIGURE 2.6 After large masses of silica-poor magmas are emplaced or melt their way into Earth's crust, they may take tens of thousands to millions of years to crystallize. During crystallization, these intrusions acquire a distinctly layered structure, with some zones being rich in minerals useful as mineral resources.

and can form masses of millions of tons of metal sulfide ores, as are found at the great nickel mining centers at Noril'sk in Russia and at Sudbury, Ontario, Canada.

Under other chemical conditions, the magma may become saturated in chromium so that the mineral chromite, $FeCr_2O_4$, begins to crystallize and settle, forming layers or *beds* (see Figure 2.7 and Figure 8.6) up to several meters thick that may persist for tens to hundreds of kilometers. Such layers constitute the world's richest chromium ores. Either simultaneously with chromite precipitation, or as a separate event, platinum metal minerals may begin to crystallize and settle out of the magma. The simultaneous precipitation of chromite and platinum minerals formed the UG-2 bed, a 1.3-meter-thick layer that is now mined for both metals in the Bushveld Igneous Complex in South Africa. Separate precipitation of platinum minerals formed the Merensky Reef, a 30–45-centimeter-thick layer that is traceable for at least 300 kilometers across the Bushveld Complex.

The chromite layers generally form early in the crystallization of these large igneous bodies. In contrast, similar magnetite (Fe_3O_4)-rich layers form somewhat later during crystallization. These layers (see Figure 8.7) resemble the chromite layers in thickness and extent, but may contain large quantities of vanadium that occurs bound up in *solid solution* in the magnetite.

Regional Metamorphism

Regional metamorphism is the large-scale modification of rocks in response to rising pressure and temperature when they have been buried to depths as great as

FIGURE 2.7 Chromite layers (black zones) crystallized in the large layered intrusion called the *Bushveld Complex* in South Africa. As the result of uplift and erosion, the chromite layers are now exposed along the Dwaars River. (Photograph by B. J. Skinner.)

BOX 2.2

Placer Deposits: Panning Gold and Mining Gravel

The processes of weathering and erosion are constantly at work decomposing rocks at Earth's surface and depositing the decomposed material in rivers, lakes, and the oceans. These processes destroy many deposits of mineral resources by dispersing them. At the same time, however, new deposits may be created—so-called *placer* deposits. The term "placer" is derived from the Spanish *plaza,* meaning "place."

During the transport of rock fragments by running water, there is a systematic sorting out of the particles according to size and density. This process can be extremely efficient and can result in the concentration of certain kinds of minerals into economically recoverable deposits. It was discoveries of placer gold in streams by prospectors, who traced sources upstream that led to the establishment of every major gold mine until the development of sophisticated geochemical exploration techniques in the 1970s. The discoveries of placer gold led to the first American mines in North Carolina in 1803, the California gold rush in 1849, the Australian gold rush of 1851, the Black Hills mines in 1876, the discovery of the world's largest gold deposits in South Africa in 1886, and the Klondike and Yukon finds in 1896. Placer mining has now been overshadowed by hard rock mining, but placer techniques are still widely used in exploration, and gold panning for placer grains is a widespread hobby.

Placer gold deposits form because the density of the gold (19.3 grams per cm^3 for pure gold) is so high relative to that of common minerals (about 2.6 grams per cm^3 for quartz and feldspar) that running water will constantly winnow out the lighter minerals and leave behind concentrations of gold and other heavier minerals. In essence, the gold panner tries to copy the action of the flowing stream by creating a water turbulence to separate minerals, and a water flow to carry the unwanted minerals away (Figure 2.B). Gold pans vary in size, shape, and complexity, but all are basically shallow

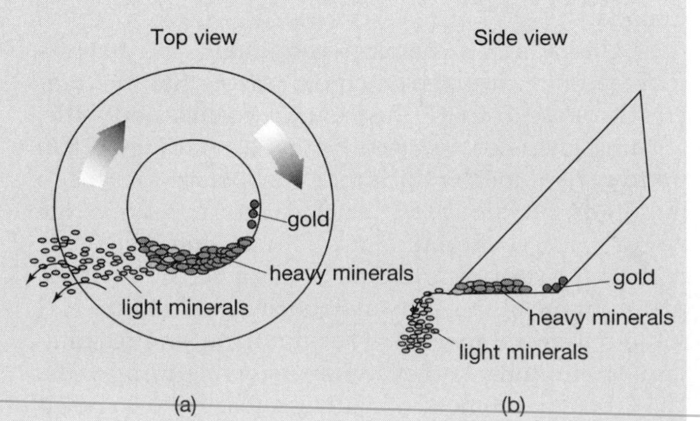

FIGURE 2.B The gold pan has long been used to separate nuggets of placer gold from other sediments. (a) Top view of the gold pan showing the circular motion of water in the pan and how it is used to push the light minerals ahead of the *heavy minerals* and the gold grains. (b) Side view of the pan showing how the light minerals are pushed outward and over the edge of the pan, leaving the heavy minerals and gold behind.

conical dishes. Once gold-bearing sediment has been placed in the pan with water, the panner gently tips the pan and attempts to create a circular motion that simulates stream flow. The movement of the water washes the lighter grains of quartz and feldspar ahead and up the sides of the pan, while the gold and other more dense minerals remain behind at the lowest point in the pan. By careful swirling, the lighter minerals are selectively washed out to and over the edge of the pan. The first to be separated are the least-dense grains (quartz, feldspar, mica, and clay) leaving the gold with the more dense minerals, or so-called black sand (minerals that are dark in color and which have specific gravities of greater than 3.0 grams per cm^3). The most common minerals in the black sands are usually ilmenite ($FeTiO_3$, black sand that is shiny and nonmagnetic), magnetite (Fe_3O_4, black, shiny, and magnetic), and zircon (tan to pink, glassy, nonmagnetic); these are usually discarded but may be of economic value if quantities are sufficient. (See especially the discussion of placer titanium ores in Chapter 7.) Continued careful swirling of the mixture of heavy minerals and gold in the water will then push the black sands ahead, leaving the gold grains behind where they can be seen and removed.

The same processes that formed placer gold deposits formed the much larger placer sand and gravel deposits. The volume of sand and gravel in a deposit is generally at least 1 million times greater than the gold present, and many sand and gravel deposits do not contain any gold at all. In the United States, at the beginning of the twenty-first century, the total value of the sand and gravel mined was greater than the value of all of the gold mined from all types of gold mines (virtually all of it from hard rock mining). Furthermore, the country mined 3 million times more sand and gravel by weight than gold each year. Thus, placer deposits, although generally thought of in terms of gold, are actually much more valuable for their sand and gravel contents.

10 kilometers or more. As depths of burial increase, the pressure from overlying rocks increases at a rate of about 230 kilograms on each square centimeter for each kilometer of depth, and temperatures increase at about 25°C for each kilometer of depth. Thus, at depths of 10 kilometers, the pressure is about 2300 kilograms per square centimeter (or about 16 tons per square inch) and the temperature is about 250°C (or 480°F). At burial depths of 20 kilometers, the temperature and pressure are about twice these values. Under the elevated pressure and temperature conditions of regional metamorphism, the common minerals of sedimentary rocks recrystallize, and the rocks are transformed mineralogically and texturally, as shown schematically in Figure 2.8.

Commonly observed effects of regional metamorphism include the conversion of shales into slates (and at higher grades into schists or gneisses), limestones into marbles, and sandstones into quartzites. All of these can be used as building materials, but the deeper a rock has been buried, the harder and more durable a metamorphic rock tends to be. Particular initial compositions of sedimentary rocks, or compositional changes brought about by fluids squeezed out of the rocks during metamorphism, can result in formation of some resources. Small amounts of garnet are common in high-grade metamorphic rocks, but where conditions have been ideal, the garnets may become very abundant or very large (e.g., as large as basketballs at Gore Mountain, New York; see Chapter 10).

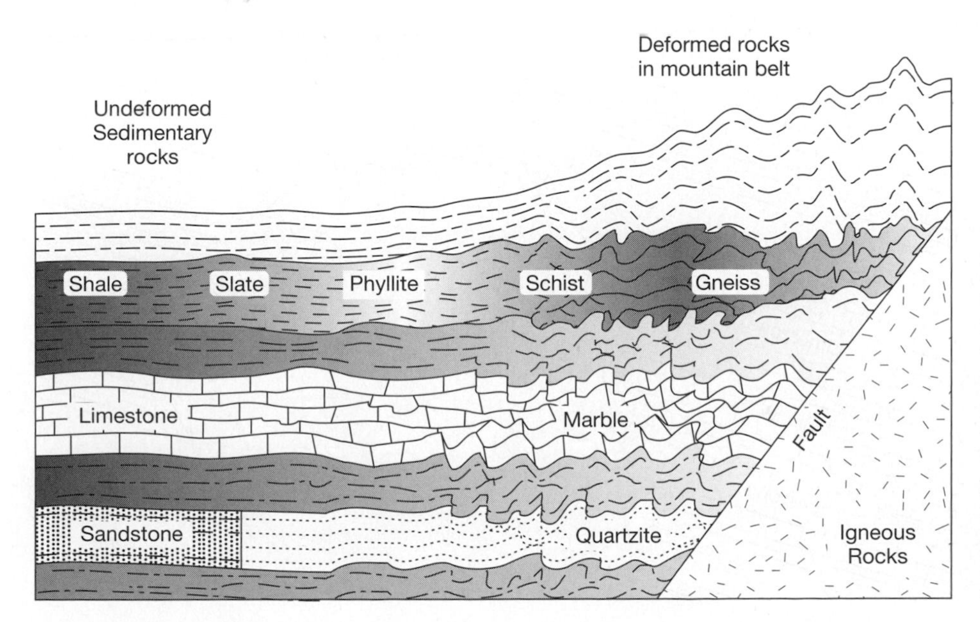

FIGURE 2.8 Idealized cross section through an area of regionally metamorphosed rocks. Progressing from left to right, the rocks have been subjected to higher temperatures and pressures. Consequently, shales are converted to slates, then to phyllites, then to schists, and ultimately to gneisses. Limestones are converted to marbles during the recrystallization that accompanies regional metamorphism, and sandstones are converted into quartzites.

Similarly, kyanite (Al_2SiO_5) is common in small amounts in many high-grade regional metamorphic rocks, but locally (as at Willis Mountain, Virginia) it may become the dominant mineral. These deposits are rare, but they are very valuable as sources of materials for the preparation of many kinds of ceramics. Some silica-poor but aluminum-rich zones of high-grade metamorphic rocks contain sufficient amounts of corundum (Al_2O_3) to be mineable for use as abrasives.

Contact Metamorphism

Contact metamorphism is the transformation of minerals in response to the heat and fluids released by igneous intrusions. It occurs in localized zones adjacent to the intrusions and can vary in thickness from a few meters to several kilometers, depending upon the size and temperature of the intrusion and the amount of fluid released. The baking effects on rocks adjacent to the intrusions are often similar to the thermal effects of regional metamorphism, except that they are more localized. Shales adjacent to intrusions often resemble baked ceramic materials; limestones are locally recrystallized into marbles, and sandstones are converted into quartzites.

The most important resources formed by contact metamorphism are metalliferous ores, called *skarns*, that may occur at the margin of the intrusion or extend into the surrounding rocks (Figure 2.9). Skarns are especially well-developed around the margins of many of the large igneous intrusions that have developed along the subducting margins of plates. These skarns commonly contain iron oxides as well as sulfides of copper, lead, zinc, and iron; gold and silver are usually present and may locally be rich. The ores are deposited by fluids given off by the cooling igneous intrusions or by fluids in the adjacent rocks. Although skarn deposits can occur in many types of rocks, they are generally most extensively developed adjacent to intrusions of silica-rich rocks because silica-rich magmas usually contain more fluids than do silica-poor ones. They are also usually better developed in limestones than in shales or sandstones because limestone is more reactive and is relatively easily replaced by the sulfide or oxide ore minerals. A good example of limestone replacement by iron minerals is the Cornwall iron deposit in Pennsylvania, which is discussed and illustrated in Chapter 7.

Contact metamorphic zones and the vein deposits that extend outward from them are, in some parts of the world, major sources of the most important gemstones. These are, however, usually very special environments where unusual elements concentrate (e.g., beryllium to form emerald, $Be_3Al_2Si_6O_{18}$) or where aluminum is more abundant than silica (e.g., to form the colored forms of corundum, Al_2O_3, that we call sapphire and ruby). These minerals are very resistant to weathering, and they are commonly extracted from weathered zones or from sediments into which they have been transported during erosion.

SURFACE PROCESSES

Earth's surface is a dynamic environment that is subjected to a wide variety of climatic and geologic processes. Some processes occur at very rapid rates, such as volcanic eruptions, floods, and tornadoes; others, such as glacial advances and continental drift, are so slow as to be imperceptible on the scale of a human lifetime. Some of these processes are discussed in general here and in more detail in relevant sections later in the book.

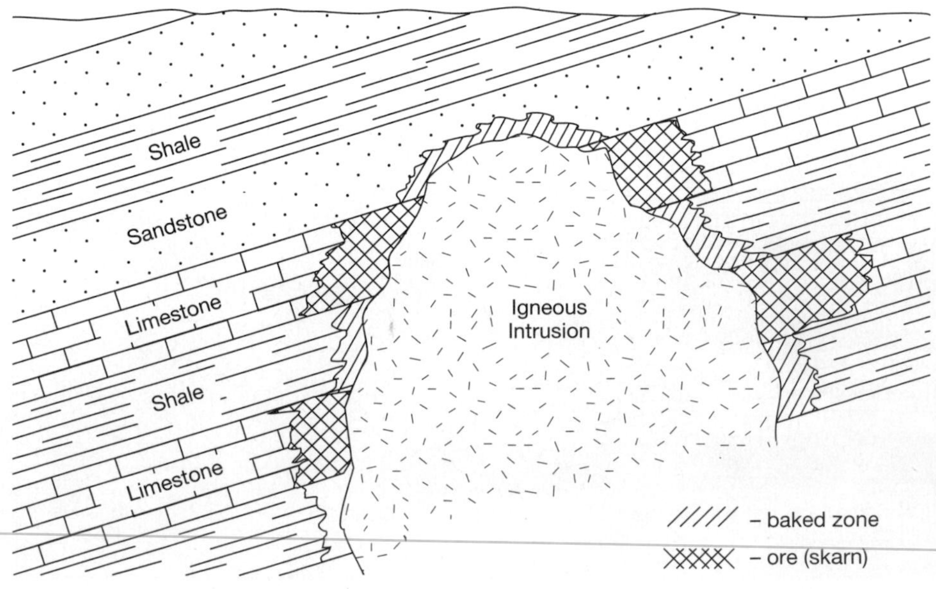

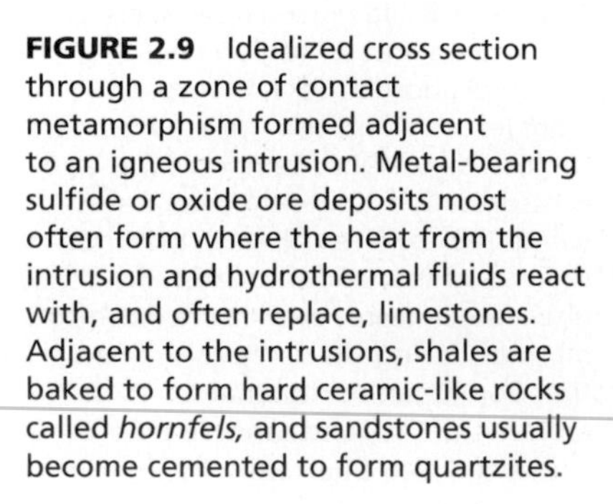

FIGURE 2.9 Idealized cross section through a zone of contact metamorphism formed adjacent to an igneous intrusion. Metal-bearing sulfide or oxide ore deposits most often form where the heat from the intrusion and hydrothermal fluids react with, and often replace, limestones. Adjacent to the intrusions, shales are baked to form hard ceramic-like rocks called *hornfels,* and sandstones usually become cemented to form quartzites.

There are no sharp boundaries between the surface processes and those of the shallow subsurface, or between surface processes and those of the marine environment. Processes that act directly on the continental surface transport materials into the marine environment and result in burial and replacement of materials into the shallow subsurface zone. Furthermore, the rain that falls on the continents becomes groundwater, and ultimately the water of the sea. Hence, almost every geological process links with other geological processes in the dynamic evolution of Earth and the mineral resources it contains. It is also important to remember that the present environment of a region does not necessarily represent the conditions of millions of years ago. One of the important discoveries of plate tectonics is the recognition that continents have moved long distances upon Earth's surface. Thus, resources formed in the tropics may now be found in the arctic and vice versa. Understanding the distribution of resources often requires understanding the movement of the continents through geologic time.

Weathering and Erosion

Earth's surface is characterized by constant changes and movement. The chemical and physical changes to rocks and minerals in response to agents such as rain, wind, frost, and biological activity are known as

BOX 2.3

Seabed Ownership

History is filled with examples of human disagreements, legal debates, and armed conflict over the ownership of mineral resources. Although greed has led to many disputes over resources, the ownership of mineral resources on continents is usually straightforward. The owner of the land owns the resources unless there has been some formal documentation to show otherwise, such as the resources being held by the government (the "crown" in some European countries) or being separated from the land surface (separation of mineral rights, especially of coal and oil).

Ownership of the oceans and their resources has always been less clear. For long periods of time, the resources of the oceans, primarily fish, were viewed as free for anyone who could take them. Inevitably, conflicts arose when fishing fleets from differing nations tried to work the same waters, or when fishermen from one country approached the shores of another country too closely. Nevertheless, there has arisen a consensus that countries bordering on bodies of water retain the rights to the resources within 12 miles (19.2 kilometers) of their shores. This still leaves the vast majority of the oceans, which cover 70 percent of Earth's surface, open to anyone who can extract the resources. Knowledge of the ocean basins and their resources remained very limited through the first half of the twentieth century, but grew rapidly beginning in the 1960s as serious efforts to explore the seafloor and understand plate tectonics came together. At the time, offshore drilling techniques, first applied in very shallow water platforms in the 1930s and 1940s, demonstrated the potential value of rich, offshore petroleum resources on the continental shelves. Concerns for resources were accentuated by the discovery that the deep seafloor contains vast quantities of manganese nodules, some of which contain significant amounts of metals such as nickel, cobalt, and copper (see discussion in Chapter 7) and the discovery of rich metal-bearing black smoker deposits on spreading ridge crests (see Chapter 8). Clearly potential riches lie in and on the seafloor.

Who owns or who can access these resources? Whose permission, if any, is required to extract them? Are royalties due to someone? What about the environmental impact of exploitation? Should we worry about submarine disturbances, or should they remain out of sight, out of mind? The International Law of the Sea Conference (which became effective in 1994) was convened to address these and many other questions, to confront conflicting views of ownership, and to establish an equitable set of rules to govern ownership and use of the oceans throughout the world. Discussion took place over a span of more than ten years and finally established the following guidelines:

1. Countries retain sovereign rights to ownership and even to the passage of ships out to 12 miles (19.2 km) from the shoreline.
2. Countries retain all rights to all mineral resources out to 200 miles (384 km).
3. Countries retain specific oil and gas rights out to 350 miles (740 km).

This clarified ownership of near-shore resources, but many areas of dispute remain because of the irregularities of coastlines, overlapping claims on uninhabited islands, and a few disputed coastal regions. Furthermore, establishing these limits did nothing to settle the concerns about the greater part of the ocean basins that lie more than 350 miles from shore. Both the remoteness of these unclaimed areas and the difficulty of extracting resources in such deep waters leave them unexploited at present. However, technological advances in the twenty-first century may well permit economic extraction of resources from the deep oceans; at that time, it will be necessary to agree upon ownership and environmental responsibilities of resource recovery from the deep oceans. Furthermore, melting of the polar ice pack in recent years (attributed in part to global warming) has also raised interest in Arctic Ocean floor ownership because it will become increasingly accessible for resource extraction.

weathering; the movements of materials down slope in response to gravity, and downstream by flowing water, is **erosion**. Weathering can form or destroy the rocks or minerals that constitute resources, and erosion can concentrate or disperse these materials. A few examples are discussed below to demonstrate how these processes influence resources.

Weathering of igneous and metamorphic rocks is a continuous process, but weathering rates vary widely and are dependent upon factors such as rainfall, temperature, and the presence of organic matter. Furthermore, the common minerals that make up igneous rocks respond quite differently to weathering. Quartz grains are resistant to chemical attack and withstand weathering to accumulate in soils, in streams, and on beaches. Feldspar minerals are transformed by removal of cations (K^+, Na^+, Ca^{2+}) and by hydration (addition of water as OH^- into their structures) into clay minerals. Micas are similar to clay minerals in structure, and are readily converted into clays by weathering. Ferromagnesian minerals, such as olivine and pyroxene, decompose relatively rapidly with the release of Fe^{2+}, Mg^{2+}, and silica, which may dissolve in water or be in a colloidal form. These collective transformations result in the formation of a vital resource, the soil. The minerals of igneous rocks, even if broken into very fine fragments, serve as a poor base in which to grow crops, but the minerals of soils, especially the clays with their ability to store and to release nutrients to plants, provide the base for the world's agricultural production.

Soils and related weathering products vary widely in composition and texture, and some can be directly exploited to provide valuable resources. The most widespread resource of this type is **clay**. Depending upon the type of clay and its color and purity, it may be extracted for making bricks, ceramics, paint extenders, paper coatings, or even fine china. Another resource resulting from the weathering process is **bauxite** (discussed in Chapter 7), the world's major source of aluminum. Under conditions of subtropical to tropical weathering, rocks and soils may be subject to extreme leaching that extracts all but the least soluble constituents, aluminum hydroxides, which can be mined and processed for aluminum. Under only slightly different conditions, it is not aluminum hydroxides, but iron hydroxides that accumulate as the other elements are removed; the result is a **laterite**, which may constitute a low-grade iron ore.

Weathering can also destroy potential resources. The deposits of metal sulfide minerals that form as a result of hydrothermal activity are rapidly attacked and decomposed by weathering processes when exposed by erosion. The iron sulfide pyrite, FeS_2, usually abundant in such deposits, rapidly decomposes and forms sulfuric acid, which then dissolves away the valuable metals, leaving only a spongy mass of iron hydroxides of little or no value.

Once weathering has reduced the original rocks to fragments or individual mineral grains, the processes of erosion transports the materials. Erosion can disperse previously concentrated minerals, can transport materials in roughly their same proportions, or it can selectively concentrate minerals on the basis of their size, durability, or density. The resources most commonly formed by weathering and erosion are **sand** and **gravel**. The processes that break down rocks allow them to be moved by flowing streams and rivers, and the abrasion during transport rounds the particles. As a result of the sorting of particles by size and weight, fragments ranging from cobbles and large pebbles down to fine sands and silt, may be accumulated. The finest and most resistant particles accumulate where rivers enter lakes or the sea, and where wave action forms beaches (Figure 2.10). The accumulation of sediments as sand and gravel bars in rivers and as beaches along the shores of lakes and oceans constitutes one of the largest resources, in terms of volume, used by modern society. Construction of roads and buildings is dependent upon the very large quantities of sand and gravel that have been deposited in ancient rivers and oceans as well as those still forming today.

The erosional processes that sort sands and gravels are also efficient in concentrating certain other valuable resources such as gold, tin, and titanium. The minerals that make up these concentrations, known as *placer* deposits, are all characterized by resistance to weathering and abrasion, and by having a high density. Consequently, they may be transported long distances in streams and rivers after weathering out of

FIGURE 2.10 Beaches, such as the one shown above, constitute huge potential masses of sand for construction purposes, and they may be the sites from which gold, tin, diamonds, and titanium minerals are extracted. Most beaches are, however, highly valued as recreational areas and will never be used as sources or building materials or mined for the other placer minerals they contain.

their original rocks. Their high densities result in their being segregated from quartz sand grains of the same size but lower density, and this allows for the use of relatively simple separation procedures to recover them (see discussion on pages 30–31 and Figure 2.B). Placer minerals weather out intact as the surrounding minerals of the original rocks dissolve or decompose. Once liberated, their downstream movement is usually very irregular, with long periods resting in sediments on the bottom of a river or stream, interspersed with brief movements at times of heavy rains. When valuable mineral grains reach the sea, they may be deposited on beaches where wave action and longshore currents will gradually concentrate them and form beach placers. The tin and titanium oxide minerals in placers are gradually worn and chipped by erosion, and thus get smaller as they proceed downstream. Gold grains are malleable and become rounded, flattened, and slowly reduced in size during transport. Although diamonds are formed within Earth's mantle, they are brought to the surface in volcanic rocks called kimberlites (see Chapter 10). Diamonds released by weathering of kimberlites, can be transported significant distances by rivers. Because of their extreme hardness, diamonds survive the erosional processes very well, and many are now recovered from placer deposits.

Evaporation

The *evaporation* of water at Earth's surface (or from soils) concentrates dissolved salts and may form crusts or beds of salts (Figure 2.11). Evaporation occurs everywhere, but the effects become most significant and noticeable in arid regions. The Dead Sea in Israel and the Great Salt Lake in Utah are two well-known examples where the evaporation of initially *fresh* water has produced lakes that are now so concentrated in salts that the water is saturated and salt is precipitating out around the shorelines. However, smaller and less well-known saline lakes and deposits occur in the arid regions of all continents. In most places, the principal dissolved constituent to precipitate is sodium chloride (NaCl, the mineral halite, or common salt) along with lesser amounts of gypsum ($CaSO_4 \cdot 2H_2O$). In some places, where evaporation is more intense, precipitation of potassium, magnesium, or calcium salts may occur.

Along the margins of some enclosed seas (e.g., the Persian Gulf), evaporation can form broad salt flats. Here, episodic flooding with seawater results in significant thicknesses of salt and gypsum. The geologic record contains numerous very thick sequences of salt that apparently formed, not as a result of repeated periods of total evaporation of water, but in slowly subsiding basins that episodically received an influx of seawater. Continuous evaporation resulted in the waters becoming saturated, with consequent precipitation of salts on the floor of the basin. Rates of salt precipitation corresponded roughly with rates of subsidence, and this resulted in the development of salt beds more than 1000 meters thick. In a somewhat similar way, the evaporation of water in lakes with unusual chemistries, commonly influenced by local volcanic exhalations, has episodically formed thick sequences of sodium sulfate. Such lakes, although not common in the geologic record, have left us with valuable resources.

Nitrates are rare in the geologic record because they are so very soluble and are easily dissolved away

FIGURE 2.11 The evaporation of water from lakes and marginal basins along oceans results in the concentration of dissolved substances and may result in the precipitation of salts, such as the halite (NaCl) shown here at Don Juan Pond in Antarctica. Such salt deposits occur in the arid regions of all continent and are commonly exploited. This same process has occurred throughout geologic time and has resulted in the formation of some salt deposits up to nearly 2 kilometers (5000 feet) thick. (Photograph by J. R. Craig.)

by even modest amounts of rain. However, the extreme dryness of the Atacama Desert of Chile has allowed for nitrate-rich sea spray to form outcrops of nitrates along the shoreline that once served as the world's most valuable fertilizer resources. These are discussed in Chapter 9.

SHALLOW SUBSURFACE AND DIAGENETIC PROCESSES

The *shallow subsurface region* is a zone in which rocks and minerals are influenced from above by downward percolating **meteoric** water, from adjacent areas by migrating fluids (Figure 2.12), and locally from below by rising thermal waters. It is a complex zone that may consist of freshly deposited sediments, or of previously deeply buried igneous, metamorphic, or sedimentary rocks that have been exposed by erosion. Where it consists of young sediments, it is a zone in which there is often significant **diagenesis**—the low-temperature modification, recrystallization, and cementation of sediments—because original trapped fluids are being squeezed out of the sediments as they become buried. In contrast, where the zone consists of metamorphic or igneous rocks, it is commonly a zone in which the preexisting minerals are breaking down to form new minerals, especially clays.

In many parts of the world, the most valuable resource in the shallow subsurface zone is fresh water. This zone contains most of the accessible fresh water for human use, much of it in water-saturated beds called *aquifers*. The water moves between particles (e.g., in sands and gravels), along fractures (e.g., in igneous and metamorphic rocks), or in open solution tunnels (e.g., in some limestones). The movement is in response to gravity; hence, the water flows from higher elevations (recharge areas) to lower elevations (discharge zones). The flow is usually slow because the openings through which the water passes are small. Thus, aquifers act as strainers to remove any large particulate matter, and bacterial actions decompose most suspended organic and other contaminant material. At the same time, there may be dissolution or precipitation of other materials, such that the waters have a changing chemistry and they, in turn, slowly alter the rocks.

Among the most important processes active in the shallow subsurface zone are those that begin the transformation of buried organic matter into fossil fuels as discussed in detail in Chapter 5. (See also Figures 2.13 and 5.6.) Most organic material is rapidly decomposed by the action of a variety of predatory creatures and bacteria, and it leaves no record of its former existence. Under some conditions, there is either a long-term buildup of organic debris (e.g., a peat swamp that is slowly subsiding with continuous overgrowth of new plants), or a rapid burial of organic matter in sediment. If organic matter is buried in a subsiding package of sediments, as in the delta of a large river (e.g., the Mississippi), it becomes subjected to a slow increase in pressure due to the overlying sediments, and a slow rise in temperature due to the heat escaping from Earth's interior. This burial process can continue for millions of years; the overall effect of the rising temperature and pressure is to break down the original organic structure and increase the ratio of carbon to hydrogen in the material. Land plant material, rich in cellulose, is transformed slowly through the stages we recognize as the **ranks** of coal (lignite, bituminous, anthracite). In contrast, marine debris, which

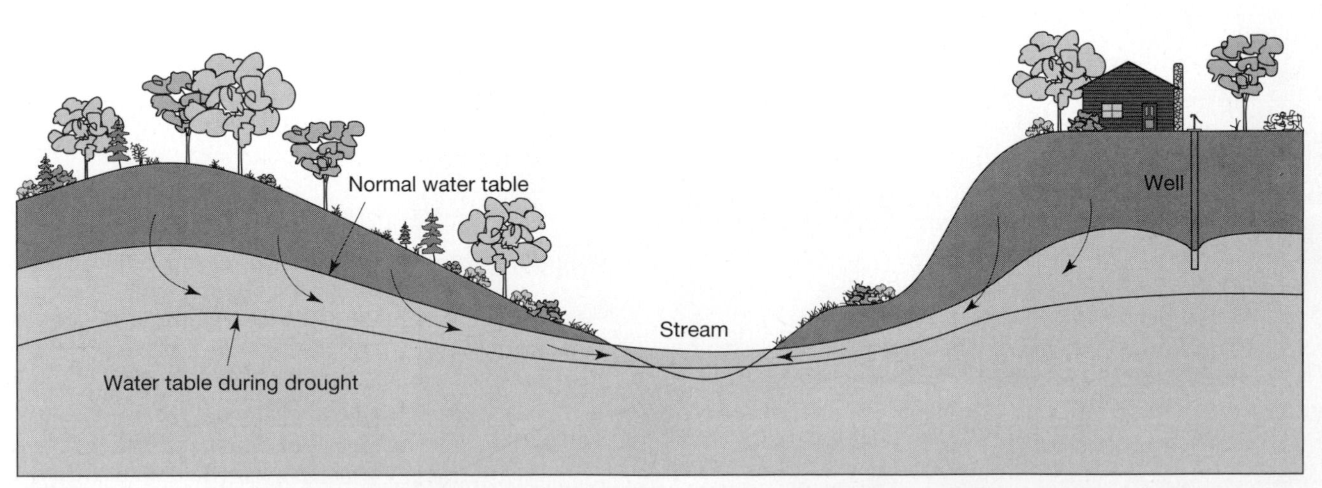

FIGURE 2.12 Groundwater is a valuable resource and provides the water supplies for much of the population of Earth. Depending upon the types of rock and soil, the vegetation cover, and the slope of the land surface, some portion of rainfall percolates into the subsurface. The boundary between the overlying unsaturated soil zones and the underlying saturated zone is the water table. The shape of the water table generally is similar to the shape of the ground surface, and the water table intersects the ground surface at permanently flowing streams and lakes.

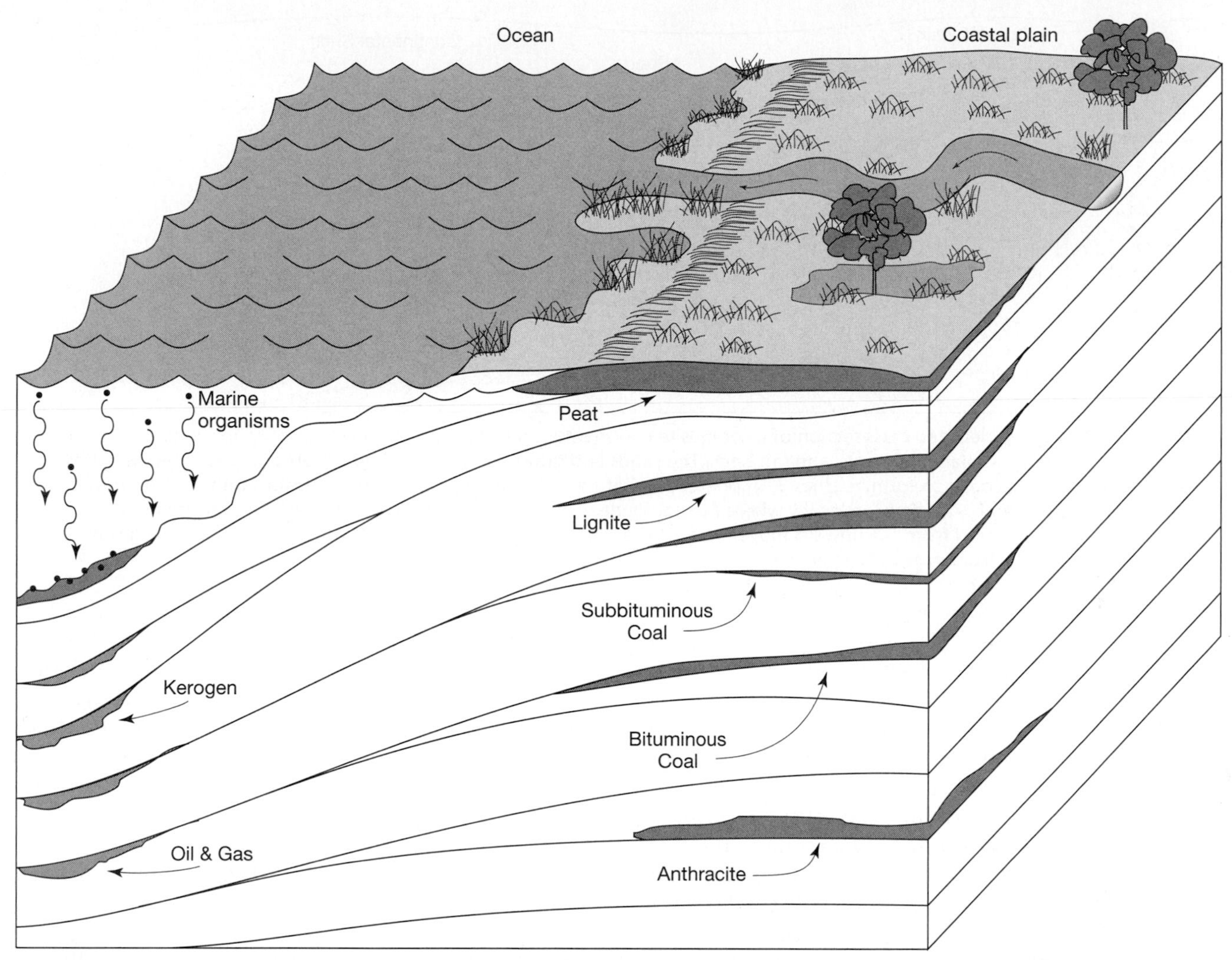

FIGURE 2.13 Schematic diagram illustrating how the burial of terrestrial organic matter can result in the formation of coal, and the burial of marine organic matter can result in the formation of oil and gas. The increasing temperature and pressure at greater depths of burial both compact and modify the terrestrial matter as it progresses through the various ranks of coal. Marine organic debris is converted into a waxy material called *kerogen;* upon additional heating, the kerogen is converted into petroleum.

is poor in cellulose, is transformed first into a waxy material called **kerogen,** and then, gradually, into the viscous liquid we call *petroleum.*

Natural gas is generated at two stages as a result of two very different processes. Immediately after burial, organic matter is subjected to attack by methanogenic bacteria that consume organic matter and release *biogenic* methane (CH). The gas may slowly escape, as in the bubbles seen coming to the surface in swamps and bogs, or it may remain trapped in sufficiently large quantities to warrant extraction. Bacteria, although ubiquitous in near-surface environments, do not survive the higher temperatures encountered during deeper burial. Rising burial temperatures, which conform with the average geothermal gradient of 25°C/kilometer, result in breakdown of large organic molecules into smaller ones, plus the release of methane. This *thermogenic* methane, so-called because it forms as a result of temperature increase, can be a major resource in petroleum-bearing strata, or in coal beds, where it may also be a resource or a danger to underground mining.

MARINE PROCESSES

It is not surprising that geologic processes active in the marine environment play a major role in the development of natural resources (Figure 2.14). After all, oceans presently cover about 70 percent of Earth's surface, and more than half of the exposed continental land area is covered by sediments originally deposited when shallow seas invaded the land. It is not possible here to review all known marine processes, but some of the most important are outlined next. Furthermore, many marine processes are

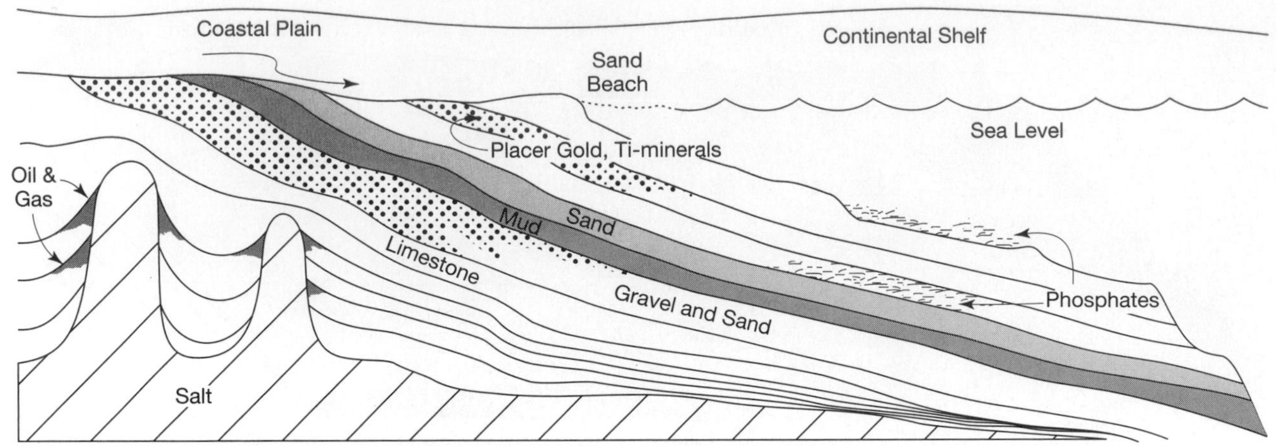

FIGURE 2.14 Highly idealized cross section of a composite continental shelf showing the presence of sands, gravel, limestones, phosphates, placer minerals, and salt beds. The sands and gravel are washed down rivers and deposited as deltas and spread along the coasts to form beaches. Phosphates precipitate by inorganic and organic processes on the outer portions of the shelves. Salt domes can develop locally where previously deposited salt beds are deeply buried. The upturning of sedimentary beds resulting from the upward movements of the salt provides good sites for the migration of oil and gas that form in the sediments from organic matter.

similar in principle to those active on land, or they represent the final stages of processes that were initiated on land.

The largest evaporite deposits are marine and clearly formed in marginal marine basins that episodically received an inflow of large quantities of seawater. A well-known and important example is the salt from the Jurassic period that underlies the Gulf Coast region of the United States. Here a 1000-kilometer region from Florida to Texas is underlain by beds that are nearly 2000 meters in thickness and that contain billions of tons of salt. The warm, shallow seas along the margins of subtropical and tropic continental areas are today the sites of great coral reefs and limestone beaches (Figure 2.15). The organic and inorganic processes active today in forming these limestones are essentially the same as those that have precipitated the great thickness of carbonate rocks over the past 600 million years. These limestone beds, which are abundant on nearly all continents, constitute a great resource for construction materials (crushed stone and dimension stone) and for making cement. Many of the largest limestone regions are far removed from the tropical climates in which they formed, providing another example of how plate tectonics has shifted continental masses from near tropical to temperate or even arctic locations over millions of years.

Rich phosphate deposits are more localized than the limestones but currently serve as the world's major source of fertilizers. The processes involved in the formation of phosphate deposits are not thoroughly understood, but it is clear that upwelling phosphate-rich ocean waters washing across shallow continental shelves for significant spans of geologic time must be involved. During these times, phosphatic debris, such as fish teeth and bones, accumulates and is mixed with nodules, grains, and crusts of phosphate minerals in beds up to tens of meters in thickness. It is likely that organic activity influences much of the precipitation of the fine-grained crusts and small rounded grains.

The deep ocean floor is the site of **manganese nodule** formation. These nodules, which usually contain as much iron as manganese, occur in vast quantities over tens of thousands of square kilometers of the deep ocean floor. The nodules range from about the size of peas up to the size of grapefruit, are roughly spherical, and internally consist of fine but irregular concentric layers of iron and manganese hydroxides. The rates of growth appear to be very slow (~1 millimeter/1000 years), and the precise mechanisms of growth remain unknown. However, it is believed that microbial activity causes much of the precipitation of the metal-bearing minerals.

The mid-ocean ridges represent divergent plate boundaries and have long been known as the sites of volcanic activity, such as the extrusion of basaltic lavas, and of associated geothermal activity (as is seen in Iceland). In recent years, these ridges have been found to be the sites of active metalliferous mineral deposit formation (see Figures 2.16 and 8.22). Examination of the ridges using submersible exploration vessels led to the discovery of *black smokers*, which are vents on the seafloor from which hot (up to 350°C) hydrothermal fluids are issuing.

FIGURE 2.15 Shallow, warm water marine environments such as that at Lee Stocking Island in the Bahamas are the sites for the deposition of limestone reefs and muds. The thick beds of limestone found throughout the world are evidence of former marine conditions similar to those shown here. (Photograph courtesy of the Caribbean Marine Sciences Institute.)

Upon entering into the cold ocean waters, these fluids rapidly mix, cool, and precipitate very fine iron and other metal sulfides that appear as *black smoke*. The hydrothermal fluids issuing from many of these vents have formed large mounds and "chimneys" that consist of ore minerals of zinc, copper, and iron, in textures very similar to those found in large ore deposits of much greater age. It is apparent that the processes active here are hydrothermal marine processes and that they are similar to the processes that formed many of the world's largest metalliferous deposits through geologic time.

CONCLUSIONS

Earth resources, and the mineable portions we term *reserves*, are formed by a very wide range of geologic processes. Such processes extract or concentrate chemical elements and compounds to create the rich

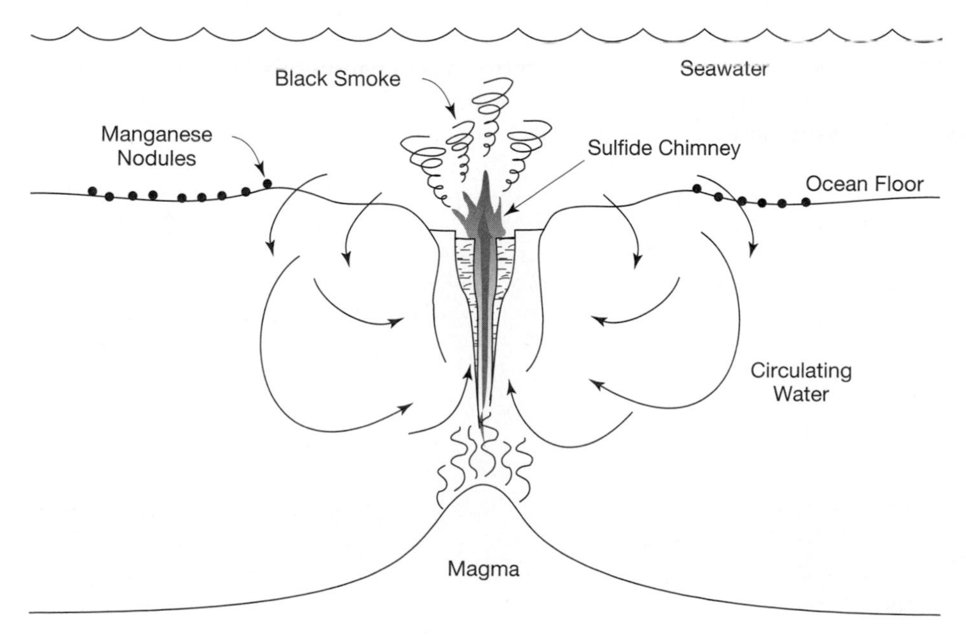

FIGURE 2.16 Ocean spreading zones where crustal plates diverge are commonly the sites of volcanic activity and may be the sites of sulfide ore formation at black smokers. Hydrothermal fluids, generated as seawater circulates through the cracks in the oceanic crustal rocks, dissolve and transport metals and sulfur. When the hot fluids reemerge along faults at the spreading centers on the ocean floor, the clear fluids are cooled rapidly, and very fine grained sulfide minerals are formed and appear as a black smoke (see Figure 8.22). The deep ocean floor is also the site of deposition of manganese nodules as the result of organic and inorganic processes; these are potential resources of several metals.

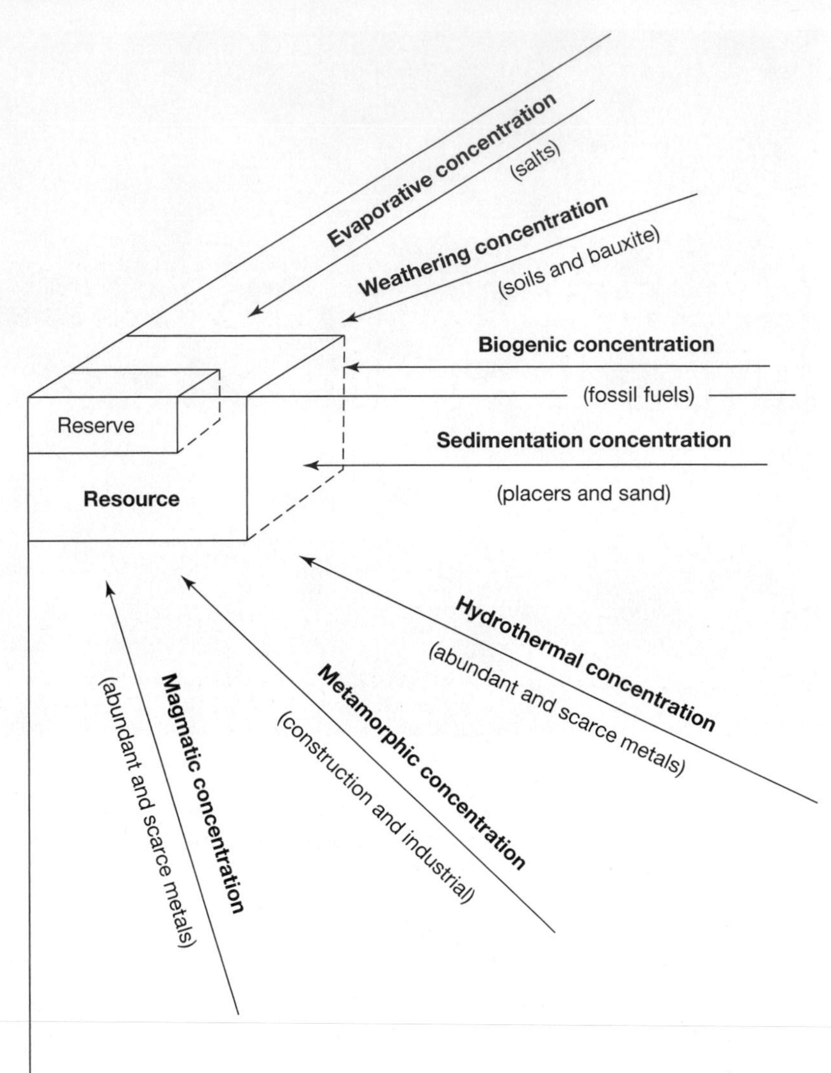

FIGURE 2.17 The concentrations of mineral resources that constitute reserves and resources have been formed by a wide variety of geological and geochemical processes, as schematically shown here. For example, the metals are primarily concentrated by magmatic and hydrothermal concentration processes, whereas the fossil fuels are formed through biogenic concentrations.

deposits that meet our resource needs (Figure 2.17). Particular resources may be formed by one or more processes, and virtually all geologic processes are active in the generation, modification, or destruction of resources. The most important processes are those that concentrate valuable minerals to levels far greater than their normal levels of occurrence in average crustal rocks.

History and Environmental Impact of Resource Usage

HISTORICAL CHANGES

Human history is inextricably linked with the use of resources and consequent changes to the environment. Both the migration and the settlement patterns of our ancestors were dictated by the availability of the two most essential resources—water and arable land. As the world's population grew and agriculture expanded, Earth's surface was modified to provide adequate resources for sustenance. When demands exceeded the locally available resources owing to population growth, climatic change, or the exhaustion of supplies, people had either to move or to develop ways to import the needed resources. The pressure of meeting resource needs was one of the factors that led to the growth of technology.

The development of technology was accompanied by new, and often more complex, resource needs. There was a progression from the *Stone Age* to the *Copper Age*, and then to the *Bronze Age*, to the *Iron Age,* and finally to the *Nuclear Age.* These timespans represented advances from the use of materials in their natural state, such as stone in arrowheads and spear points, to the use of technologies to smelt and shape metals, and now to the use of metals for their electronic properties. The usefulness of metals stimulated the desire for greater quantities of them, which, in turn, led to greater efforts to explore for ore deposits and to carry out mining operations. The mining and smelting of ores required the construction of processing facilities and drove the need to obtain wood, then later charcoal, coal, and petroleum, to fuel the new facilities. Each step represented an increase in complexity, with greater interdependency between those providing and those using the goods and services, and a greater and more widespread environmental impact.

The two principal driving forces in the use of Earth's resources have been a steadily rising population and an expanding technology. World population has progressively increased through time—slowly until about 1750 and then accelerating rapidly, especially throughout the twentieth century. The increasing use of many basic Earth resources—water, land, metals, fuels—has paralleled or exceeded the growth of population.

ENVIRONMENTAL IMPACT

The history of the exploitation and use of natural resources is also the history of the human impact on the environment. Although all organisms modify their environments, humans have had a vastly greater effect on both the local and global environments than any other species. Some effects, such as the clearing of a wooded area to build a house, are small; others, such as the construction of a dam to create a reservoir hundreds of miles in length, are vast. Some effects, such as an oil spill, have an immediate impact; others, such as the collapse of a house or a road into an old mine, may be long delayed and take place over a protracted time period. Some effects are local, such as the digging of an open pit or quarry; others are regional, such as downstream flooding resulting from an increase in rainwater runoff, or global, as in the increase in atmospheric carbon dioxide. And some effects are temporary, such as an oil spill or a flood, whereas others are effectively permanent, as in a large open pit mine or a long-term nuclear waste storage facility.

It is self-evident that a population that continues to grow will require increasing amounts of resources; it will use more energy, occupy more space, and create more waste. Earth, however, is fixed in size, in land area, and in the total amounts of the resources it contains, even though demands for those resources continue to rise.

Stanford professor Peter M. Vitousek and his co-workers have made the following blunt observations about Earth's population growth in their article, "Human Domination of Earth's Ecosystems":

> *Human alteration of Earth is substantial and growing. Between one-third and one-half of the land surface has been transformed by human action; the carbon dioxide concentration in the atmosphere has increased by nearly 30 percent since the beginning of the Industrial Revolution; more atmospheric nitrogen is fixed by humanity than by all natural terrestrial sources combined; more than half of all accessible surface freshwater is put to use by humanity; and about one-quarter of the bird species on Earth have been driven to extinction. By these and other standards, it is clear that we live on a human-dominated planet. (Science 1997, v. 277, p. 494)*

However, as these authors note, not all human impacts result in environmental disasters; much of the damage from past negligence has been remediated. Nevertheless, the vast needs of a world population that in the year 2009 is 6.7 billion, and that is expected to grow to as many as 12 billion during the twenty-first century, give much cause for concern regarding the effects of resource extraction, use, and disposal.

The continued growth of human population with its demands for greater living space, increased water usage, increased fossil fuel extracion, and increased waste production, creates ever more conflicts with the natural environment. This is like a "zero sum game" in which there is a set amount of space, and each new human means less space left for a fish, a bird, a gorilla, or an elephant. Humans with their technology overwhelm and displace the other creatures of the world. Some environmental impacts are intentional; others, sheer negligence—but the final effects may be the same. One example makes the point. In the early 1800s, the passenger pigeon, with a total population of perhaps 5 billion, was the most abundant bird on Earth. Within one hundred years, as a result of intentional slaughter and habitat destruction (mostly through new farming technology), the species had become extinct. Its extinction provides a powerful lesson for all of us.

The environmental consequences of energy production are of special concern and are well summarized by the following quote from the *Report to the President on Federal Energy Research and Development for the Challenges of the Twenty-First Century,* which was prepared for President Clinton in 1999.

> *Energy is no less crucial to the environmental dimensions of human well-being than to the economic ones. It accounts for a striking share of the most trouble-some environmental problems at every geographic scale—from wood smoke in Third World village huts, to regional smogs and acid precipitation, to the risk of widespread radioactive contamination from accidents at nuclear-energy facilities, to the buildup of carbon dioxide and other greenhouse gases in the global atmosphere.*

Throughout most of human history, resources were considered to be inexhaustible, to be "fruits of the Earth" that could be plucked and consumed as they were discovered. In the second half of the twentieth century, the rise in environmental awareness and the recognition that there were no longer new areas on Earth to settle and exploit, have forced us to consider how we have caused so much damage to our planet, and what we might do to prevent further problems in the future. Environmental protection agencies have been created in many countries to prevent or limit further degradation of nature's systems; restoration programs, such as the Super Fund in the United States, have been charged with cleaning up problems arising from past resource extraction and use.

We have also come to recognize that many environmental problems do not respect national boundaries; many are international or even global in their extent. We have seen the development of the Kyoto Protocol in 1992, which brought together the nations of the world to address the problem of global environmental degradation, and to implement the means of restoring environments and mitigating future damage. The best-known of the Kyoto accords is the recommendation to contain carbon dioxide emissions, and as rapidly as possible to reduce them for fear that they will adversely impact the world climate in the future. There can be little doubt that from now on there will be increasing calls for globally coordinated action to solve environmental problems.

Hadrian's Wall, built between A.D. 122 and 136, was a Roman defensive barrier guarding the northern frontier of the Province of Britain until the end of the fourth century. It extended 118 kilometers (73 miles) across the narrowest portion of Britain and was for most of its length 6 meters (20 feet) high and 3.3 meters thick. (Photograph courtesy of the British Tourist Authority).

CHAPTER 3

Earth Resources Through History

While it won't be [easy] to reduce our impact, it won't be impossible either. Remember that impact is the product of two factors: population, multiplied times impact per person.

JARAD DIAMOND, COLLAPSE

FOCAL POINTS

- Earth's natural resources have been used by all cultures throughout history.
- The earliest uses of Earth's resources involved water, salt, and simple but effective tools made from rocks.
- The first metals used by humans, before 15,000 B.C., were gold and copper that were found in their native (metallic) states.
- A steady increase occurred in the use of resources and reached a peak about the time of the Greek and Roman empires; this was followed by a prolonged period in Europe, from about A.D. 400 until the late 1400s, when there were few new developments.
- The voyages of Columbus and others opened a nearly 400-year period of global exploration and colonialism, involving several major European countries. This brought great wealth to those European nations (e.g., gold and silver to Spain) and imposed their cultural influences on the rest of the world.
- The Industrial Revolution of the 1700s and 1800s transformed the economies of the more advanced nations from agrarian and rural to industrial and urban, and greatly expanded the use of mineral resources, especially iron and coal.
- The development of the modern science of chemistry led to the discovery of many new metals and their subsequent utilization, beginning in the late eighteenth century.
- Today, no country is self-sufficient in all its resource needs; every country is dependent on others for supplies of mineral resources, and much of the world's production is controlled by large multinational groups or companies.
- The most famous organization involved in the production and control of a vital resource is the Organization of Petroleum Exporting Countries (OPEC).
- The demand for petroleum-based fuels has driven up the price and the political importance of oil and has raised the cost of almost every commodity and service.
- The control of strategic world resources continues to play a major role in world politics.

INTRODUCTION

Natural resources are the raw materials from which, directly or indirectly, all products used by our societies are made. The utilization of Earth's resources, either in natural or processed form, dates from our early ancestors' dependence on drinking water and salt for their diets, the shaping of stone tools, and the use of natural pigments for decoration. From such humble beginnings, mineral resources have grown in national and international importance and now are essential materials of trade, and the basis for profit and power. This is best demonstrated by petroleum, which is the most valuable and most vital of the mineral commodities exchanged between nations.

The quantities of various mineral resources used by particular societies vary widely, but they generally correspond, on a *per capita* basis, to the standard of living in each society. Figure 3.1 illustrates the annual per capita consumption of a variety of mineral resources in the United States; the quantities would be similar for other highly industrialized countries such as Canada, Britain, Germany, France, Sweden, or Australia. Of course, individually very few of us use 4700 kilograms (10,300 pounds) of stone or 165 kilograms (360 pounds) of salt annually, but for our society to provide the vast array of products and services we enjoy, these quantities are used by various industries on behalf of each of us.

The international importance of mineral resources is evidenced by the value of world crude mineral (not including any fuels) production, which exceeds $500 billion (£300 billion) annually; processing raises the value of the commodities to more than $1000 billion (£600 billion) annually, which accounts for about 30 percent of the total of all traded materials. The virtual explosion in the prices of fuels to unimagined levels in the last few years has raised the price of nearly every commodity and altered the lifestyles of many people. By 2008, the world was extracting and consuming about $12 billion worth of petroleum per day (when the price hit $150 per barrel and 80 million barrels per day or more than $4 trillion per year). This chapter summarizes the changing and growing uses of mineral resources through history as well as some of the influences they now have on the political and economic aspects of modern society.

RESOURCES OF ANTIQUITY

The beginnings of our use of Earth's resources are lost in antiquity, but it seems likely that our ancestors' earliest concerns were for water, salt, and suitably shaped rocks to aid in hunting. The constant need for water was the dominant factor in the choice of dwelling sites and determined early migration routes. Some things never change, for despite our technological advances, water remains a key factor in the location of major population centers. Salt was originally provided by meat-rich diets, but the development of societies with cereal-based diets required the acquisition of salt to use as a food additive. Beyond being a necessary dietary component, the addition of salt was also the cheapest and easiest way to preserve food and to enhance its taste. Consequently, salt became a commodity

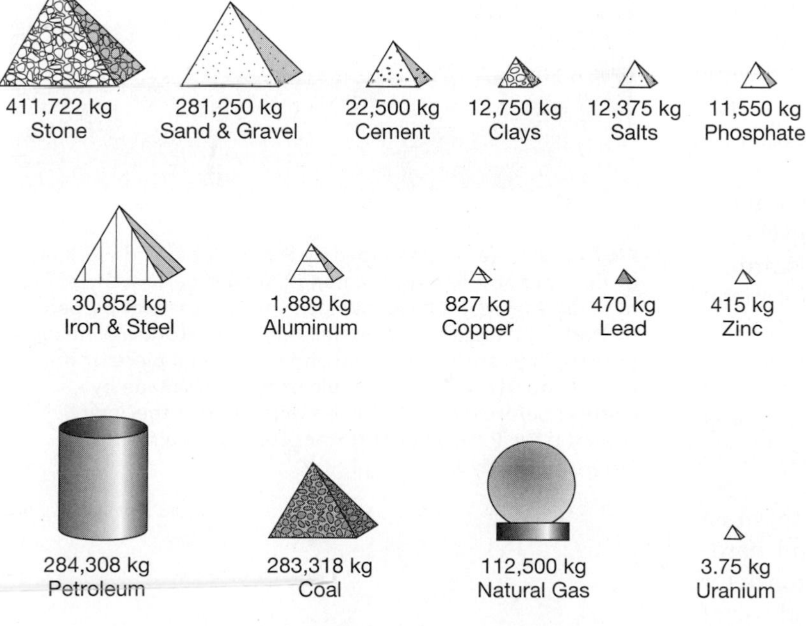

FIGURE 3.1 The lifetime per capita consumption of mineral resources is both varied and large. On average, an American living to 75 years of age will consume about 800 metric tons of nonfuel resources and use about 400 metric tons of fossil fuels. Thus, the present U.S. population of about 308 million in 2010 consumes more than 330 billion metric tons of resources in their lifetimes. (Data from the United States Geological Survey.)

of exchange long before recorded history, and salt routes crisscrossed the globe many thousands of years ago. Salt (in Latin, *sal*) was a good antiseptic; hence, Salus was the Roman goddess of health. A Roman soldier's pay, consisting in part of salt, was known as *salarium*, from which we derive the word *salary*. From this, and the payment of salt for slaves, came the expression of a worthless individual being "not worth his salt."

The use of rocks as tools extends back at least 2 million years. At first, they were crudely chipped into useful shapes; subsequently, numerous prehistoric peoples developed techniques to shape **flint**, **obsidian**, and other tough rocks with uniform properties into delicate implements and tools (Figure 3.2). A major societal advance occurred just prior to 9000 B.C. when our ancestors began to fire clay to make pottery; the pottery, which represented the first fabrication of new materials from minerals, provided a better means for storage and transport of food and water, thus helping in the struggle for survival. This led to the development of the ceramic arts of brick-making, glazing, the making of mineral pigmented paints, and even glass-making by about 3500 B.C.

The first metals were utilized by humans before 15,000 B.C.—they were gold and copper because these are the two metals that most commonly occur in the metallic, or *native*, state. No doubt the first finds were treated as curiosities because the metals felt, looked, and behaved differently from brittle rocks; however, the ability to shape the metals into useful and desirable forms developed rapidly (Figure 3.3). By 4000 B.C., our ancestors had learned that copper could also be extracted from certain kinds of rocks using primitive smelting techniques in which charcoal supplied the heat and also the means of chemically reducing copper ores to free the copper metal. By 3000 B.C., other metals, including silver, tin, lead, and zinc, were also being extracted and combined to form alloys such as brass (copper and zinc), bronze (copper and tin), and pewter (tin and other metals such as lead, copper, or antimony). It should be noted that the use of these early metals often involved the first "recycling". Although stone tools, once broken, were discarded, broken metal tools could be remelted and the metal used over and over again.

Iron, though much more abundant in Earth's crust than most other metals, is more difficult to extract from its ores, and hence its use came somewhat later. It is thought that the first iron used came from meteorites; it is easy to imagine that the iron, especially if a meteorite were seen to fall to Earth, must have evoked much wonder. The strength and hardness of iron made it superior to copper and bronze for weapons; this eventually led to the widespread use of iron, even generating numerous myths concerning its

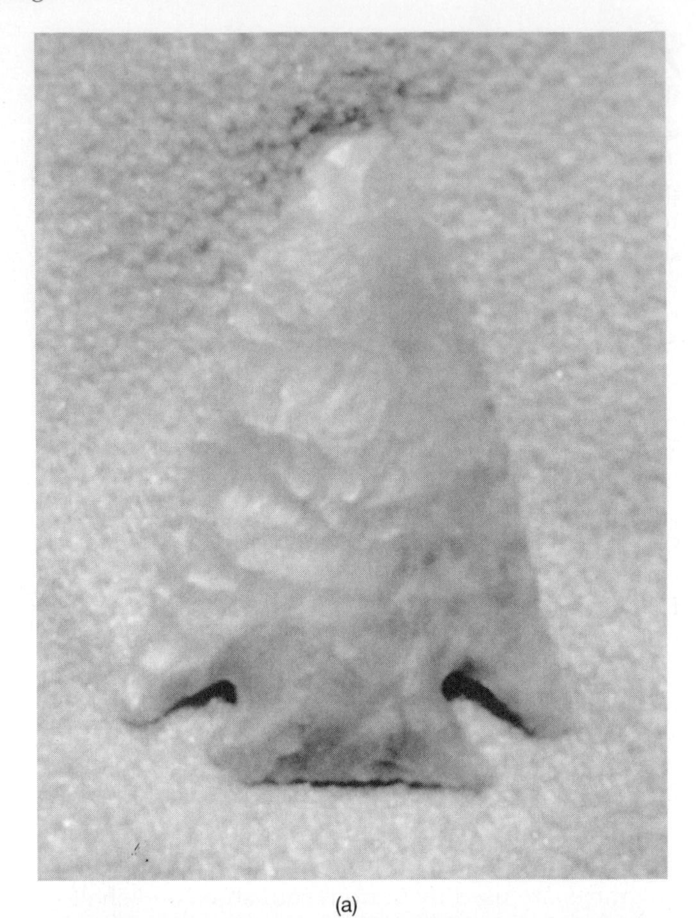

(a)

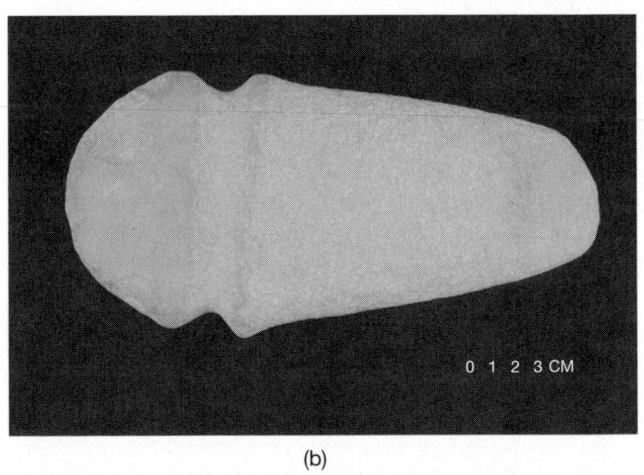

(b)

FIGURE 3.2 (a) Tools shaped by the chipping and working of flint and obsidian were among humans' earliest uses of mineral resources. The arrowhead shown here was prepared by Native Americans in West Texas. (b) This stone axe head, prepared by carefully shaping and polishing a piece of fine-grained quartzite, is an example of the tools made by cultures before metals became available. This specimen is from the Tye River area of Nelson County, Virginia. (Photographs by J. R. Craig.)

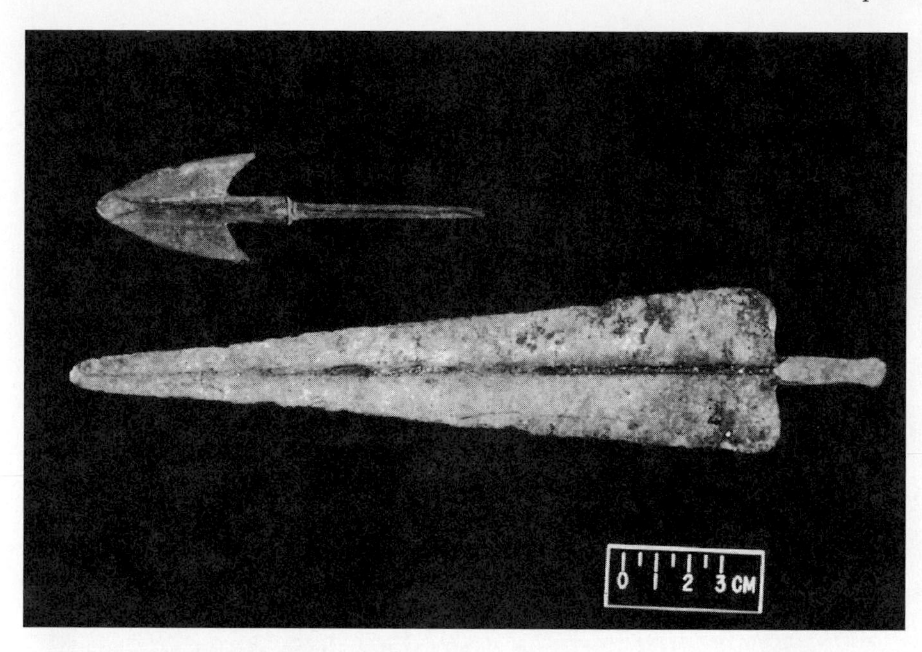

FIGURE 3.3 Early metal tools were found to be much less brittle and superior to stone tools. Hence, the use of metal tools spread rapidly, first involving the use of single metals and then of alloys. The arrow and spear points would have been attached to wooden shafts that have decomposed. (Photograph by J. R. Craig.)

magical powers when shaped into weapons. The Roman writer, Pliny, described iron as the "most useful and most fatal instrument in the hand of man," but perhaps its usefulness is best summarized by the lines of Rudyard Kipling's poem "Cold Iron":

> Gold is for the mistress—silver for the maid—
> Copper for the craftsman cunning at his trade.
> Good! said the Baron, sitting in his hall,
> But iron—Cold Iron—is master of them all.

Although we commonly center our attention upon particular metals by using terms such as **Bronze Age** or **Iron Age**, our ancestors learned to use an increasingly broad range of mineral resources as the ages passed. Simple, crudely shaped rock fragments were replaced by carefully shaped knives, arrowheads, and spear points. The use of animal hides for storage gradually gave way to the use of pottery and ceramics; shelters fashioned from plant materials and animal skins were replaced by dwellings of more permanent and protective materials, namely stones, bricks, and mortar.

The rise and expansion of the Egyptian, Greek, and then the Roman Empires saw the extensive development of mining and stoneworking industries around the Mediterranean Sea to provide the building materials for their great palaces, stadiums, theaters, temples, roads, and aqueducts. These cultures not only used great volumes of mineral resources, but they also vastly expanded the range of materials used. They also began to use large quantities of processed resources such as cement and plaster to bond bricks or cut stone (Figure 3.4). The Greeks developed domestic metal mines and used silver mined near Athens to finance the fleet that defeated the Persians at Salamis in 480 B.C.; they also used gold from northern Greece to support the army of Alexander the Great. As the Romans expanded their empire throughout the Mediterranean and beyond, they extracted metals first by plunder, then by tribute, and finally by mining. Examples include mercury from Spain, copper from Cyprus, and tin from the British Isles.

FROM ROME TO THE RENAISSANCE

The collapse of the Roman Empire resulted in a breakdown of organized society in southern Europe, including in the production, transportation, and marketing of mineral resources. The onset of the Dark Ages in Europe saw trade decline and mines close; people turned to subsistence agriculture. Mineral resource needs were met by reusing the materials at hand, and mining was confined to exploitation of salt needed for food, some alluvial gold recovery, and mining for other metals at a few centers such as Cornwall, Devonshire, and Derbyshire in England, and Saxony in the Erzgebirge in Germany.

The emergence of Europe from the Dark Ages began in about A.D. 800 and coincided with the discovery and development of mineral deposits in southern Germany. New discoveries and the reopening of old mines provided metals, especially silver and gold, to trade for spices, gems, and silks from China and India. The overland trade routes through Assyria and Persia, dating from ancient times, were replaced by new routes through Egypt, down the Red Sea, and across the Indian Ocean. Metals mined in northern and central Europe were carried south and traded through Mediterranean ports such as Venice, which grew from a small fishing village to a major trade center. Europe experienced a relatively rapid

FIGURE 3.4 The Romans were masters of construction, as evidenced by the carefully cut and fitted limestone blocks of the Colosseum in Rome. (Photograph courtesy of Istituto Italiano di Cultura, New York.)

expansion or "rebirth"—now called the *Renaissance*—of the arts, culture, and engineering beginning in the late fourteenth and early fifteenth centuries as people emerged from the bleak economy of the Dark Ages.

Spain and Portugal rose to new importance when Christopher Columbus opened the seas westward to the New World in 1492, and when Vasco da Gama found the eastward sea route around the Cape of Good Hope at the southern tip of Africa in 1498, opening the way to India. This shifted the trade center for metals as well as other commodities from Venice to the Iberian Peninsula. Spain's fortune grew rapidly as significant quantities of gold and silver from the New World flowed into her coffers. This wealth helped finance the Renaissance in Europe as well as Spain's participation in several wars (Figure 3.5).

GLOBAL EXPLORATION AND COLONIALISM

Humankind's curiosity and sense of adventure, combined with a desire for riches and a need for resources, made our ancestors explorers before history was recorded. The Phoenicians, who sailed throughout the then known world of the Mediterranean, and the Romans, whose empire extended from Britain to the Orient, were among the first great explorers and colonizers to exploit resources from very wide areas. It was, however, the explorations of the Europeans from the fifteenth until the nineteenth centuries (Figure 3.6) that left their mark upon the ownership and exploitation of mineral resources in the twentieth century. Portugal and Spain became the first of the modern European nations to send out explorers in search of sea routes to India and the Far East in the late 1400s. Exploration culminated in the discovery of America by Christopher Columbus. The new lands and new trade routes were encouraging, but Vasco da Gama's discovery of gold in the hands of Africans, and Columbus' discovery of gold owned by natives in the West Indies provided a strong incentive to explore further. On his return, Columbus reportedly said, "The gate to the gold and pearls is now open, and precious stones, spices, and a thousand other things may surely be expected."

Conflict between the two major Catholic sea powers, Spain and Portugal, over rights to explore and claim the New World seemed inevitable, so the Pope intervened in 1494 through a proclamation called the Treaty of Tordesilla. This edict drew a north-south boundary (often called the *Line of Demarcation*) 100 leagues (later moved to 360 leagues) west of the Cape Verde Islands (a league is an old unit of measurement equal to about 3 miles). Portugal was granted the rights to lands east of this line and Spain the lands to the west. The long-term consequence has been that the bulge of South America projecting east of this line is now Brazil, where Portuguese is spoken; in nearly all other countries of South and Central America, Spanish is spoken.

The discovery of gold in the Americas was a powerful incentive to the Spanish, whose conquistadors under Pizarro and Cortez rapidly subdued the large indigenous empires centered in Peru and Mexico, plundering their gold and silver. In a letter to Pizarro, King Ferdinand wrote, "Get gold, humanely if you can, but get gold" and Cortez stated, "I came to get gold, not to till the soil like a peasant." Both the King's and

(a)

(b)

FIGURE 3.5 (a) Spain's recovery of large amounts of gold and silver from the lands conquered in the Americas resulted in growing animosity between Spain and England, because the English desired a share of the wealth. In the hope of ending English raids on Spanish ships and ports, Philip II of Spain assembled the Spanish Armada, a fleet of 130 ships that sailed for England on May 20, 1588. The defeat by the English on July 29 was a great blow to the prestige of the Spanish and reduced the influence of Spain on the high seas. (Courtesy of the Beverly R. Robinson Collection of the United States Naval Academy.) (b) The silver eight-reale coin, commonly called *a piece of eight,* was used throughout the Spanish-speaking world. These fragments show how the coins were commonly cut into smaller denominations, usually an eighth, called a *bit.* The most popular item was a quarter of a coin and led to the slang term "two bits" for the American quarter. (Courtesy of the Colonial Williamsburg Foundation.)

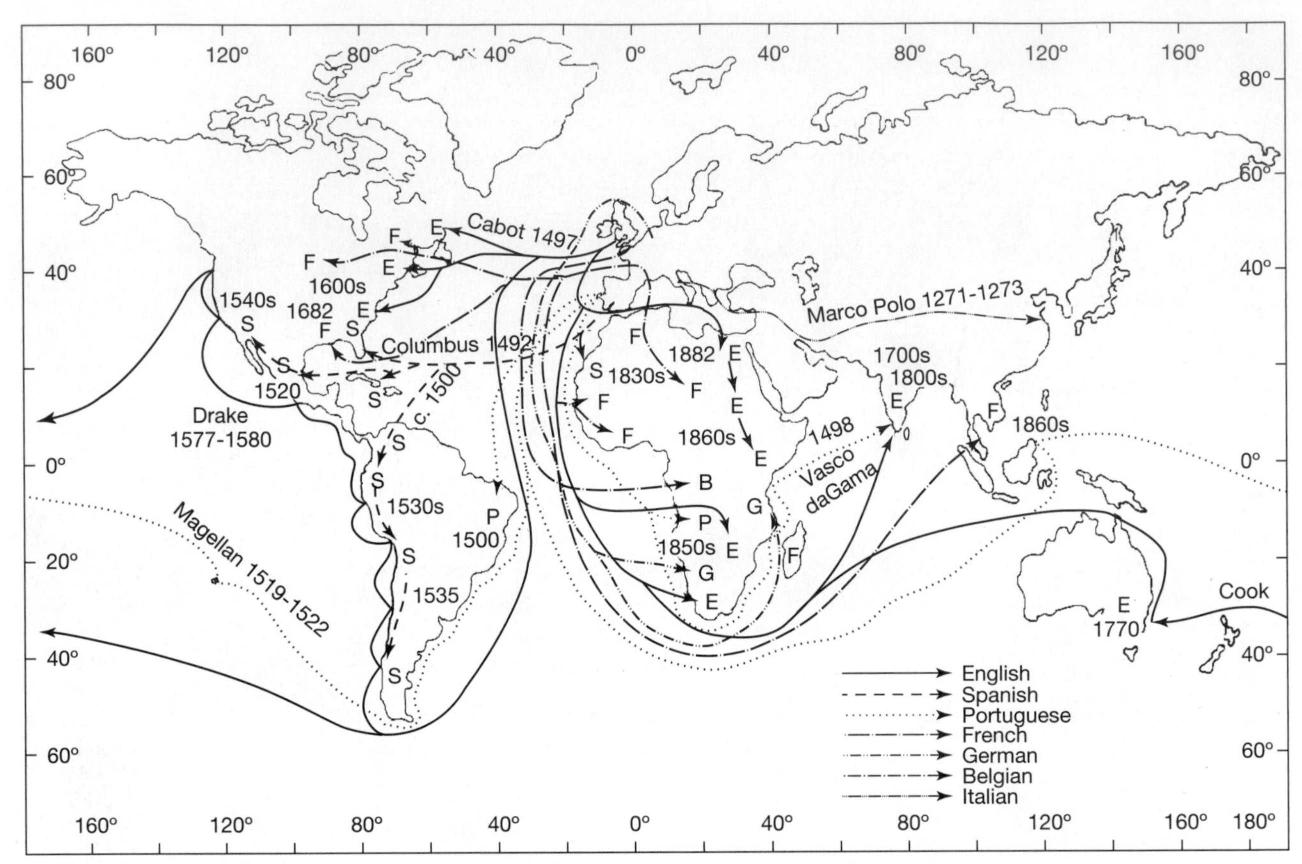

FIGURE 3.6 Major colonial routes and the extension of influence of the major European powers from the late 1400s through the late 1800s. There has been no attempt to include all European excursions nor to represent multiple, often successive, colonial influxes.

Cortez's desires were richly met as the Spanish galleons carried an estimated 181 metric tons of gold and 16,887 metric tons of silver back to Europe between 1500 and 1660. Although these quantities are only the equivalents of about one-tenth of a year's modern gold production and about one year's silver production today, they were huge amounts and of enormous value in the sixteenth and seventeenth centuries.

The British, Dutch, and French carried out the exploration of eastern North America, hoping to find gold and silver just as the Spanish had in South America. They, of course, encountered only forest-dwelling Indians who used little or no metal and who knew nothing of gold. As a result, the exploration and colonization of North America proceeded much more slowly and rather differently from that of Central America and South America. As described in more detail in the box on the California Gold Rush, it was also the search for gold that led to the great migration to California in 1849 and the consequent opening up of the American West.

While the Spanish made great inroads in South America, other European peoples explored and established colonies in the coastal regions of Africa. However, it was not until the 1800s, with the Industrial Revolution in full swing, that the interior of Africa was opened to colonialism. Driven by the desire to take possession of all available lands that could provide raw materials and potential markets, the European countries divided the rights of exploitation of all of Africa and parts of Southeast Asia among themselves. Since the middle of the twentieth century and out of concerns for human rights, the power of the European countries declined and the colonies gained independence, one by one, due to the rise of nationalism.

HUMANS AND METALS

The developing complexity of society linked to an increasing dependence upon a variety of resources is nowhere better illustrated than in the use of expanding numbers and quantities of metals through time. Archeological research indicates that "**Stone Age**" people relied only upon implements that were shaped from stone. The earliest evidence of metal usage dates from about 15,000 B.C. Native copper and gold were the first metals used in virtually all cultures. These metals were used for ornaments, amulets, tools, and weapons, for they can be hammered, carved or melted, and cast into many shapes. Furthermore, the metals could be recycled by remelting over and over again with no loss in strength or hardness.

BOX 3.1

The California Gold Rush

The California Gold Rush, like many other important events in world history, resulted from an unexpected discovery in an unlikely place. In the late 1840s, the U.S. population was concentrated along the eastern seaboard, with only modest numbers of farmers, explorers, and hunters venturing to the far west. The area known as California, which Spanish settlers first populated in 1542, was a neglected province of Mexico. The Russians had built a fort near San Francisco in 1812, and American settlers began to arrive in 1841. An American revolt in 1846 led to the Mexican War and, ultimately, to the purchase of the provinces of California, Nevada, Utah, New Mexico, and Arizona by the United States for $15 million.

On January 24, 1848, nine days before the treaty was signed, and hence before California actually belonged to the United States, gold was discovered near California's American River. James Marshall, foreman at John Sutter's sawmill, picked up two gold nuggets while examining a mill race under construction along the American River (Figure 3.A). Finding the nuggets to be malleable and not brittle like *fool's gold*, he realized what he had found and showed the nuggets to Sutter (Figure 3.B). Despite Sutter's desire to keep the discovery secret, the workers talked of their findings, and local merchants spread the word. By July, more than 4000 men, a quarter of California's total non-native population, were digging for gold along the tributaries of the American River. Word of the discovery reached the East Coast by the fall of 1848, and in the following spring, the most extensive immigration the world had ever seen occurred as the "Forty-Niners" trekked to California, all hoping to become rich. By January 1850, 90,000 adventurers had arrived from all parts of the world. Tragically, disease killed one-fifth along the way or shortly after their arrival. Wagon trains crossed North America, ships off-loaded passengers who walked across Panama to board other ships, and some vessels braved the long and stormy route around the tip of South America. The population of California soared to 269,000 by November 1852 and more than 500,000 by 1856. Many others did not complete the trip but stopped along the way. Consequently, the gold rush did more to populate the American West than any other event.

Most new settlers were unlucky in their quest for gold. Nevertheless, a few found enough to raise the total production of gold in the United States from less than 30,000 troy ounces per year in the late 1840s to more than 4,000,000 troy ounces in 1852. The value of the gold extracted in any one of the first ten years of the American ownership of California was more than twice the amount the government paid to Mexico in 1848; a good deal! The *Mother Lode* country of California, as this area was known, and adjacent areas were scoured over by prospectors trying to strike it rich. Mining towns sprang up with incredible speed, and the prices of food, lodging, and nearly everything else fluctuated wildly

FIGURE 3.A Map of California showing the location of Sutter's Mill on the American River, where James Marshall found the gold that led to the California Gold Rush of 1849.

FIGURE 3.B The original "discovery nugget" from Sutter's mill. (Courtesy of National Museum of American History, Smithsonian Institution.)

(continued)

BOX 3.1

The California Gold Rush *(Continued)*

depending on the gold available and the whims of the miners. Within a few years, the shallow diggings containing the placer gold began to be exhausted, the miners moved on, and many once busy towns became ghost towns.

By 1852, large-scale hydraulic mining had commenced, and it remained the major method of gold mining for about thirty years. These operations polluted the rivers, caused much silting of navigable portions downstream, and even resulted in the infilling of parts of San Francisco Bay. Court actions in 1884 stopped the devastation, but by then most of the damage had already been done.

Few names are remembered from the gold rush; Sutter never benefited from the gold found on his land and finally left; he died, heartbroken, in Pennsylvania. James Marshall lived on odd jobs and handouts for a few years until his death. The name that is best known is that of a man who probably never used a gold pan. His name is Levi Strauss, an Austrian tailor (Figure 3.C), who capitalized on the miners' needs for rugged work clothes. After all, the miners needed good pants whether or not they found any gold. Gold fever gradually waned, with placer operations giving way to hard rock miners; fewer people mined, and gold production declined, but the impact of the great immigration has continued to the present day. Not only that, many reading this book probably own a pair of Levi's jeans.

FIGURE 3.C Levi Strauss, the most famous name from the California Gold Rush, did not pan for gold, but made his fortune selling clothes to the miners. (Photograph courtesy of Levi Strauss & Co.)

By 4000 B.C., copper was being smelted from sulfide ores in Egypt and Mesopotamia. The discoveries leading to the intentional extraction of a metal from its ores are unknown, but they probably began before 4000 B.C. with the accidental smelting of metallic copper from copper-bearing minerals in a hearth, campfire, or pottery kiln. Once the association of the ore (many copper-bearing minerals such as malachite and azurite are brightly colored and easily recognized) with the metal was realized, together with the fact that charcoal in the fire both heated and reduced the ore to metal, techniques of smelting spread rapidly. Either through communication of the techniques or many separate discoveries, copper smelting was practiced throughout southwest Europe, the Middle East, and as far as India by 3000 B.C. The earliest copper samples were often impure because of the presence of small amounts of arsenic and antimony containing minerals that occur within some copper minerals. Thus, the smelting created unintentional, but nevertheless useful, alloys that were superior to pure copper in terms of hardness. Beneficial as the arsenic and antimony were, it was the impurity of tin, either from tin sulfide (**stannite**), or tin oxide (**cassiterite**), that effectively ended the Copper Age and ushered in the Bronze Age.

The addition of tin confers considerable strength upon cast copper objects without the need for **cold-working**. This discovery, probably made in Iran between 3900 and 2900 B.C., spread rapidly throughout southeast Europe, the Mediterranean area, and India, and resulted in the development of a tin trade. The first significant sources of tin were probably in Italy, Bohemia, Saxony, and possibly even Nigeria; this marked the first instance of foreign dependency on natural resources for many nations. The usefulness of bronze led to a large increase in the scale of metallurgical operations, so that ingots of bronze weighing more than 30 kilograms were being produced in the Mediterranean area by about 1600 B.C. The zenith of the Bronze Age was reached in the period between 900 and 750 B.C.

Throughout the Copper and Bronze Ages, gold was gathered from placer deposits and extracted from vein-type deposits in the Middle East. Because most naturally occurring gold is relatively pure (more than 80 percent gold with less than 20 percent impurity contents of silver and copper) and occurs as the native metal, the production of gold was primarily a question of manpower rather than smelting techniques. Silver also occurs in the native state, but it is probable that silver bars found with lead bars at Troy (2500 B.C.) were extracted from natural gold–silver alloys by a refining process known as **cupellation.** This process, still used today, employs lead, which is relatively easily smelted from the lead sulfide galena, to separate the silver from the gold. This silver was formed into a variety of ornaments, but lead was not widely used until the Romans found it useful for making pipes to carry water.

The earliest archeological iron implements have been found in the Middle East, the Americas, and in Greenland. In each case, the nickel contents indicate that the original sources were iron-rich meteorites that were found and worked by the native peoples. The rarity and uniqueness of iron led to its being highly prized; indeed, the knife that lay upon Tutankhamen's mummified corpse within its sarcophagus was made of iron.

The first working of terrestrial iron began about 1300 B.C. in Asia Minor and may well have resulted from the accidental building of a fire on iron-oxide rich rocks or even the unanticipated extraction of iron from rocks while trying to refine copper. The early production of iron was made by heating the iron ore in a hot charcoal fire. The iron was slowly reduced by the carbon reacting with it and removing the oxygen as carbon dioxide; the fires were not hot enough to melt the iron, but they did soften it so that it could be pounded or forged into wrought iron. The scale of iron production gradually increased from small items of jewelry to large-scale production of weapons by about 1200 to 1000 B.C. Knowledge of iron-working spread from Turkey and Iran to areas around the Mediterranean by about 900 B.C., to coastal Africa and Great Britain by about 500 B.C., and to India and possibly China by 400 B.C. By the early days of the Roman Empire, iron was being used for nails, hinges, bolts, keys, chains, and weapons. The small foundries that dotted the empire persisted until the Romans withdrew, after which time iron-making, like other forms of industry, was much reduced. Despite iron's usefulness, the difficulty of producing large quantities kept the supply limited until about A.D. 1340, when the invention of the blast furnace permitted iron workers to obtain temperatures high enough to make molten iron. This technological breakthrough has had a profound and lasting effect on civilization because it has made iron, and subsequently steel, cheap and widely available; inexpensive iron led to the Industrial Revolution.

Molten iron can be fabricated into useful cast iron objects (Figure 3.7) using preformed molds; this practice soon became widespread, but cast iron is brittle and relatively soft owing to the impurities of carbon and other elements. Nevertheless, there was a rapidly growing demand for the iron and, consequently, a seemingly insatiable appetite for wood to provide charcoal fuel. The British Admiralty became alarmed about the supplies of timber for ships, and as a result, royal edicts were issued in the mid-sixteenth century forbidding the use of certain forests for manufacturing charcoal. The demand for hardwood to make charcoal devastated forests in Europe, especially in England, and led to an "energy crisis." The shortage of charcoal put some iron-makers out of work and led others to seek alternative energy sources such as coal. It took many failures and more than 100 years, but finally in the early 1700s, iron was successfully smelted in England using coke produced from coal. The use of the coke opened up England's ample coal resources, led to the expansion of the mining industry, and placed England at the forefront of the Industrial Revolution.

However, there was one more major breakthrough to come—steel. Iron was strong and useful, but it had little flexibility, and iron castings were brittle. The desire to improve the properties of iron led to the discovery of steel; the first kind of steel produced, and still the most widely used variety today, is carbon

FIGURE 3.7 Cast iron became the primary metal for a vast variety of uses including cannon, such as the one shown here, which was recovered from a wreck site off the coast of North Carolina. The ship is believed to be the Queen Anne's Revenge, the flagship of the pirate Blackbeard. The ship sank in 1718. (Photograph by J. R. Craig.)

steel. It is formed by blowing enough air through the molten iron to lower the carbon content to less than 1 percent. The result is a harder, stronger, more workable and flexible metal. The date of the first steel-making is not known because, no doubt, some was accidentally produced from time to time in the normal melting and forging processes. The birth of the modern steel industry is usually dated to 1740, when a process was devised to produce a uniform quality carbon steel.

Although there were many early developments in techniques of mining and smelting, it was not until the Industrial Revolution of the eighteenth and nineteenth centuries that scientists and metallurgists discovered large numbers of new metals (Figure 3.8). Many were, at first, novelties with little practical use. For example, nickel was discovered as an element in 1751, and several metallic grains were isolated in 1804. Nickel remained a scientific curiosity and found no major usage until nickel-steels were developed in 1889. Similarly, aluminum was first discovered in 1827, but because of the difficulty of extraction, it was one of the most expensive of metals. Consequently, Napoleon III had aluminum forks and spoons for himself and honored guests while lesser guests ate with gold utensils. After the development of efficient electrical extraction techniques, the price of aluminum dropped from more than about $11,000 per kilogram to less than about $92 per kilogram and the metal began to find widespread usage.

The Industrial Revolution brought iron and carbon steel into a new prominence. They were used both in the machines of industry and many of their products. As previously noted, the first steels were carbon steels. In the latter part of the nineteenth century, the superior properties of steel prompted the search for other useful alloying agents among the newly discovered metals. New varieties of steels incorporating nickel, cobalt, titanium, niobium, and molybdenum were developed. Such innovations were accelerated by the development of the internal combustion engine, aircraft, and weaponry during the twentieth century.

The advent of aircraft spurred the development of new lightweight metals, especially aluminum and titanium. When the jet engine replaced the piston engine, new high-temperature alloys were needed; more emphasis was placed upon metals such as cobalt, vanadium, and titanium. With the dawn of the nuclear age in 1945, much of the world's attention turned toward two long-known but little used metals, uranium and thorium. In recent years, there have been some remarkable studies of the utilization of metals in the field of medicine (e.g., special barium dyes for X-ray diagnostic work and synthetic radioactive isotopes for cancer treatment), electronics (e.g., the use of gallium and germanium in transducers and the rare Earth elements in color TV phosphores), and energy production (e.g., platinum group metals as catalysts in gasoline production and as catalytic converters in automobile exhaust systems). The progression from the use of simple native metals to accidentally discovered alloys, to engineered compounds, to exotic rare metals, and even to artificial elements, is a measure of technological advancement. At the same

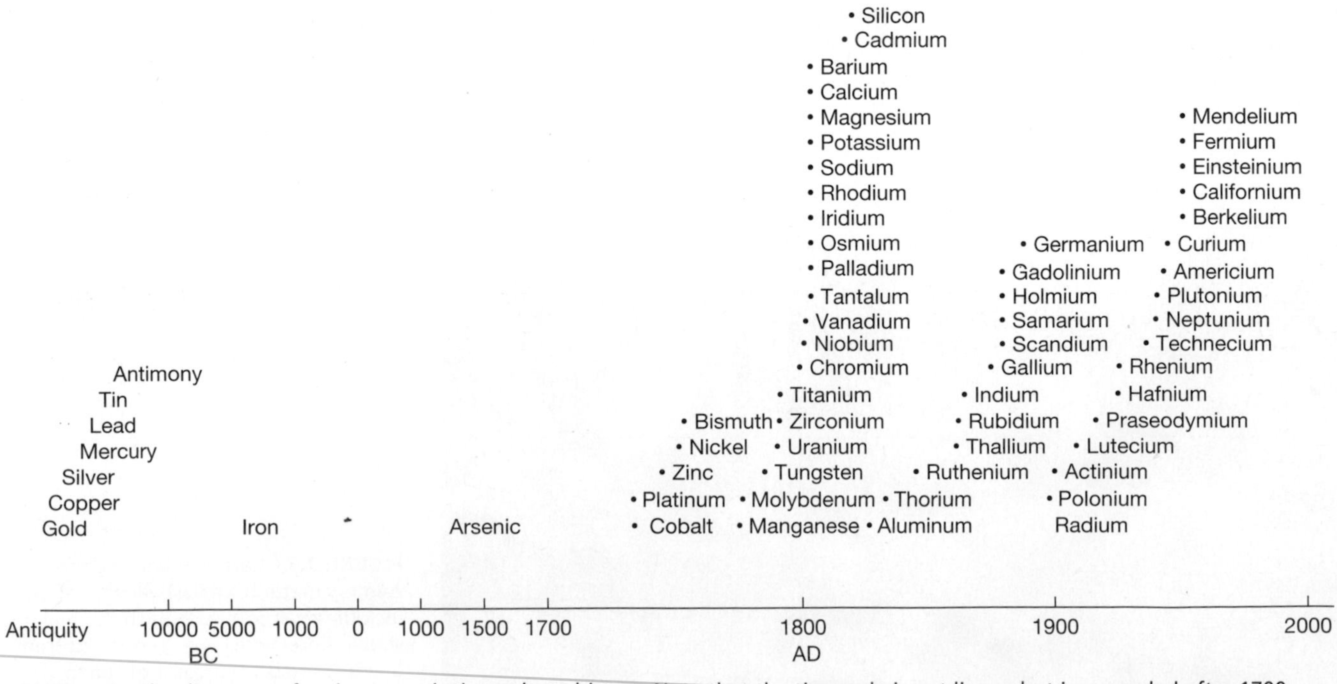

FIGURE 3.8 The discovery of various metals throughout history. Note that the timescale is not linear, but is expanded after 1700 when the growth of modern chemistry and the onset of the Industrial Revoultion led to the discovery of a large number of metals.

time, it is necessary to recognize that dependence on virtually all naturally occurring metals (and, indeed, all elements) results in a vast and complicated worldwide network of mining, processing, and supply. We now know that there are no more new naturally occurring metals to be discovered within Earth; thus, we must learn how to make the best use of those now available.

MODERN TRENDS IN RESOURCE USAGE

The modern era of resource extraction and usage began with the Industrial Revolution. The amount of any resource used before that time is negligible by comparison to today's consumption. The onset of the Industrial Revolution triggered the need for more coal, iron, and other metals to build and fuel the new machines, to supply the factories, and to develop the cities; the demands required larger and more efficient mining methods and more efficient transportation systems to move the products. The continued growth of industry, fed both by a growing world population and rising standards of living, has resulted in ever greater demands for Earth's resources to feed, house, and accommodate humankind. W. C. J. van Rensberg, a noted resource analyst, pointed out that in the period from 1770 to 1900, when world population approximately doubled, mineral production grew tenfold. From 1900 until 1970, when world population increased about 2.3 times, mineral production increased twelvefold. From 1970 to 2000, the population doubled again and mineral production tripled.

This worldwide trend is especially pronounced in the more developed countries such as the United States, as shown in Table 3.1, which compares increases in population and increases in the production and usage of some important mineral commodities. Table 3.1 reveals two very important points about the production and use of mineral resources by industrialized nations in general, and by the United States in particular:

1. The rate of mineral resource usage has generally risen more rapidly than has the rate of population growth.
2. The percentage of mineral resources being supplied domestically has decreased; conversely, the percentage of imported resources has increased.

The first point results from the rise in the standard of living, and the expansion of industry that relies upon the mineral raw materials; to a lesser degree, it also reflects the increased size of the population. It is important to note that the United States, with approximately 5 percent of the world population, uses approximately 30 percent of the mineral resources and that the *per capita* use of nearly every commodity in the United States dwarfs that of people in the developing countries. To bring all peoples up to the American level of mineral resource consumption would require a 700 percent annual increase in the production of each commodity. To do this by the year 2050 when world population is projected to have doubled, it would be necessary for annual production to increase by 1400 percent!

The second point illustrated by the numbers in Table 3.1 (updated to use 1875–2005 data) is that the United States (and many other highly developed countries), which was essentially self-sufficient in its production of most metals in 1875, now relies heavily upon imports and accumulated stocks. This lack of self-reliance is, in some cases, due to economics (i.e., U.S. mines cannot produce the material as cheaply as foreign sources because foreign labor is cheaper or because foreign governments subsidize mining), but in many cases, it is also the result of the depletion of America's richest ores. Furthermore, environmental restrictions now limit or prevent extraction of resources in some potentially rich areas. Increasing foreign dependence creates a drain on capital, a loss of jobs, and a lessening of security over the supply of strategic materials. The degree of American dependence on foreign sources for nonfuel commodities is illustrated in Figure 3.9a, which shows

TABLE 3.1 Comparison of the U.S. production and usage of some important metals and of population in 1875 and 2005 (U.S. production in metric tons)

Commodity	1875 (x 1000 mt) produced and used	2005 (x 1000 mt) Produced	2005 (x 1000 mt) Used	Increase in production	Increase in use
Aluminum	Not used	2500	6856	—	—
Copper	18.3	1150	2270	62.8x	124x
Lead	53.2	440	1500	8.3x	28.9x
Pig iron	2,057	92,400	120,000	45x	58x
Zinc	15.2	760	1120	50x	74x
Silver (million troy oz)	24.5	42	248	1.7x	10.1x
Gold (million troy oz)	1.6	8.0	6.3	5.0x	3.9x
Population (millions)	45.1	294	—	6.5x	—

2008 U.S. NET IMPORT RELIANCE FOR SELECTED NONFUEL MINERAL MATERIALS

Commodity	Percent	Major Import Sources (2004–07)[1]
ARSENIC (trioxide)	100	China, Morocco, Hong Kong, Mexico
ASBESTOS	100	Canada
BAUXITE and ALUMINA	100	Jamaica, Guinea, Brazil, Australia
CESIUM	100	Canada
FLUORSPAR	100	China, Mexico, South Africa, Mongolia
GRAPHITE (natural)	100	China, Mexico, Canada, Brazil
INDIUM	100	China, Japan, Canada, Belgium
MANGANESE	100	South Africa, Gabon, China, Australia
NIOBIUM (columbium)	100	Brazil, Canada, Estonia
QUARTZ CRYSTAL (industrial)	100	China, Japan, Russia
RARE EARTH	100	China, France, Japan, Russia
RUBIDIUM	100	Canada
STRONTIUM	100	Mexico, Germany
TANTALUM	100	Australia, China, Brazil, Japan
THALLIUM	100	Russia, Netherlands, Belgium
THORIUM	100	United Kingdom, France
VANADIUM	100	Czech Republic, Swaziland, Canada, Rep. of Korea
YTTRIUM	100	China, Japan, France
GALLIUM	99	China, Ukraine, Germany, Canada
GEMSTONES	99	Israel, India, Belgium, South Africa
BISMUTH	97	Belgium, Mexico, United Kingdom, China
DIAMOND (natural industrial stone)	92	Botswana, South Africa, Namibia, Ireland
PLATINUM	91	South Africa, Germany, United Kingdom, Canada
STONE (dimension)	89	Italy, Brazil, Turkey, China
RHENIUM	87	Chile, Germany, Netherlands
ANTIMONY	86	China, Mexico, Belgium
MICA, sheet (natural)	86	China, India, Belgium, Brazil
GERMANIUM	85	Belgium, Canada, Germany, China
COBALT	81	Norway, Russia, China, Canada
POTASH	81	Canada, Belarus, Russia, Germany
TIN	80	Peru, Bolivia, China, Indonesia
BARITE	79	China, India
TITANIUM MINERAL CONCENTRATES	77	South Africa, Australia, Canada, Ukraine
ZINC	73	Canada, Peru, Mexico, Ireland
PALLADIUM	72	Russia, South Africa, United Kingdom, Belgium
TUNGSTEN	61	China, Germany, Canada, Bolivia
SILVER	60	Mexico, Canada, Peru, Chile
PEAT	58	Canada
DIAMOND (dust, grit and powder)	56	China, Ireland, Russia, Rep. of Korea
SILICON (ferrosilicon)	56	China, Russia, Venezuela, Canada
CHROMIUM	54	South Africa, Kazakhstan, Russia, Zimbabwe
TITANIUM (sponge)	54	Kazakhstan, Japan, Russia, China
MAGNESIUM COMPOUNDS	52	China, Canada, Austria, Australia
MAGNESIUM METAL	50	Canada, Russia, Israel, China
NITROGEN (fixed), AMMONIA	48	Trinidad and Tobago, Canada, Russia, Ukraine
GARNET (industrial)	40	Australia, India, China, Canada
VERMICULITE	35	China, South Africa
NICKEL	33	Canada, Russia, Norway, Australia
COPPER	32	Chile, Canada, Peru, Mexico
SULFUR	28	Canada, Mexico, Venezuela
GYPSUM	27	Canada, Mexico, Spain, Dominican Republic
PERLITE	19	Greece
SALT	17	Canada, Chile, The Bahamas, Mexico
MICA, scrap and flake (natural)	16	Canada, China, India, Finland
CEMENT	12	Canada, China, Thailand, Rep. of Korea
PHOSPHATE ROCK	9	Morocco
IRON and STEEL	8	Canada, European Union, Mexico, China
IRON and STEEL SLAG	8	Canada, Japan, Italy, France
PUMICE	6	Greece, Italy, Turkey, Mexico
LIME	1	Canada, Mexico
STONE (crushed)	1	Canada, Mexico, The Bahamas

[1] In descending order of import share.

FIGURE 3.9 (a) The United States net import reliance for selected nonfuel mineral resources ranges from 100 percent down to zero. The reasons for high import reliance include availability of deposits, mining costs, and environmental regulations. (From the U.S. Geological Survey.)

some of the broad range of imported materials and the highly variable degree of import dependence. Although the United States is the largest importer of mineral resources, nearly all highly industrialized nations have import dependencies similar to those shown in Figure 3.9. Figure 3.9b provides some sense of the numbers of different resources being imported from some principal sources, but a complete listing of all of the countries from which the United States imports mineral resources would be a listing of most nations in the world. This demonstrates the vastness and complexity of the web of interdependency. The various strands in the web are, of course, "two-way streets," with resources moving one way and money and technology the other way.

The general trends in the changing number of working mines, in the amounts of domestically produced metals, and in the amounts of imported metals were outlined for industrialized nations as early as 1929 by Foster Hewett, as shown in Figure 3.10. The B curve defining the amount of metal produced annually starts at zero, when mining first commences in a country, and ends again at zero when all ore deposits have been depleted. The *area under the curve* is a measure of the total amount of metal produced in the lifetime of the mines. The A curve defining the number of mines is a measure of the rate of extraction of the metal; many small and easily extracted ore bodies are mined early in a country's development, but the bulk of the metal comes from larger, longer-lived mines. Ultimately, the mines become exhausted and domestic metal production drops; as this occurs, the country becomes increasingly dependent upon imports from foreign sources. The relative positions of the United States, Britain, and China are shown in terms of the three curves. It is fair to note that these curves are generalizations and do not fit all countries. Indeed, some industrialized countries such as Japan have never had a strong mineral base; many less developed countries such as Bolivia have not developed major industrialization.

The aging and depletion of mines in the major developed countries, coupled with high labor costs in developed nations and the discovery and development of mines in other parts of the world, has resulted in a dramatic decrease in the developed countries' share of world metal production, as shown for several metals in the United States in Figure 3.11. The one exception to the downward trend in metal production is gold. Most of the gold production comes from Nevada, where large deposits have been found and where very low cost extraction techniques have been very effective. The slight upturn in the proportion of many metals produced by the United States between 1985 and 1995 resulted mainly from increased efficiency of production at existing mines. This increase is unlikely to be sustained; in fact, mines in the United States are gradually becoming exhausted and the development of new mines to replace existing ones is unlikely because of high development costs and environmental restrictions. As a result of this, most jobs in the mining industry and most of the profits from the export of mineral commodities have shifted from the United States and the other major developed countries to the developing nations. Another consequence, discussed in a later section, is the increasing dependence of the major developed countries upon foreign sources for strategic mineral commodities.

In terms of impact on lifestyles and revenues generated, the most important modern trend in resource usage has been the rapid rise in the use of petroleum throughout the twentieth century. In the early 1900s, oil yielded the gasoline to power the growing number of automobiles, but coal continued to be the major energy source for industry. The Depression years of the 1930s saw only a slow growth in oil demand because of the difficult economic times, and the World War II years saw only controlled growth because of wartime constraints. However, after World War II, the rapid expansion of the world economy, the shift of industry from coal to oil as an energy source, the growth of the automotive industry, and the ready availability of cheap oil from the then recently opened and very rich Middle East oil fields led to a rapid rise in the demand for oil. This has led to an unprecedented dependence of much of the world upon one small geographic area for its major energy supplies and a high rate of flow of the world's money to the Middle East in order to buy the oil.

The increasing rate of resource usage, especially through the twentieth century, has raised concerns about the exhaustion of some types of mineral resources. Future availability must be considered separately for each resource; Figure 3.12 illustrates the situation for three different resources. Liquid petroleum reserves are declining over time as we extract and burn them. The rate of new petroleum formation, relative to the rate of consumption is so slow as to be negligible; hence, petroleum is not sustainable in the long term. Freshwater is fixed in quantity on Earth's surface, but is rapidly recycled; it is sustainable and we might be able to supplement current supplies with small quantities extracted through sea water desalinization. In contrast to petroleum or water, copper availability is increasing with time because the copper we mine is largely recycled and continuously adds to the total of the metal available. The Earth has a fixed amount of copper, but the deposits are widespread geologically and highly variable in grade. Because the copper is not consumed and technology has permitted the continuing discovery of new deposits and the extraction of ever lower ore grades, there is little concern for shortages.

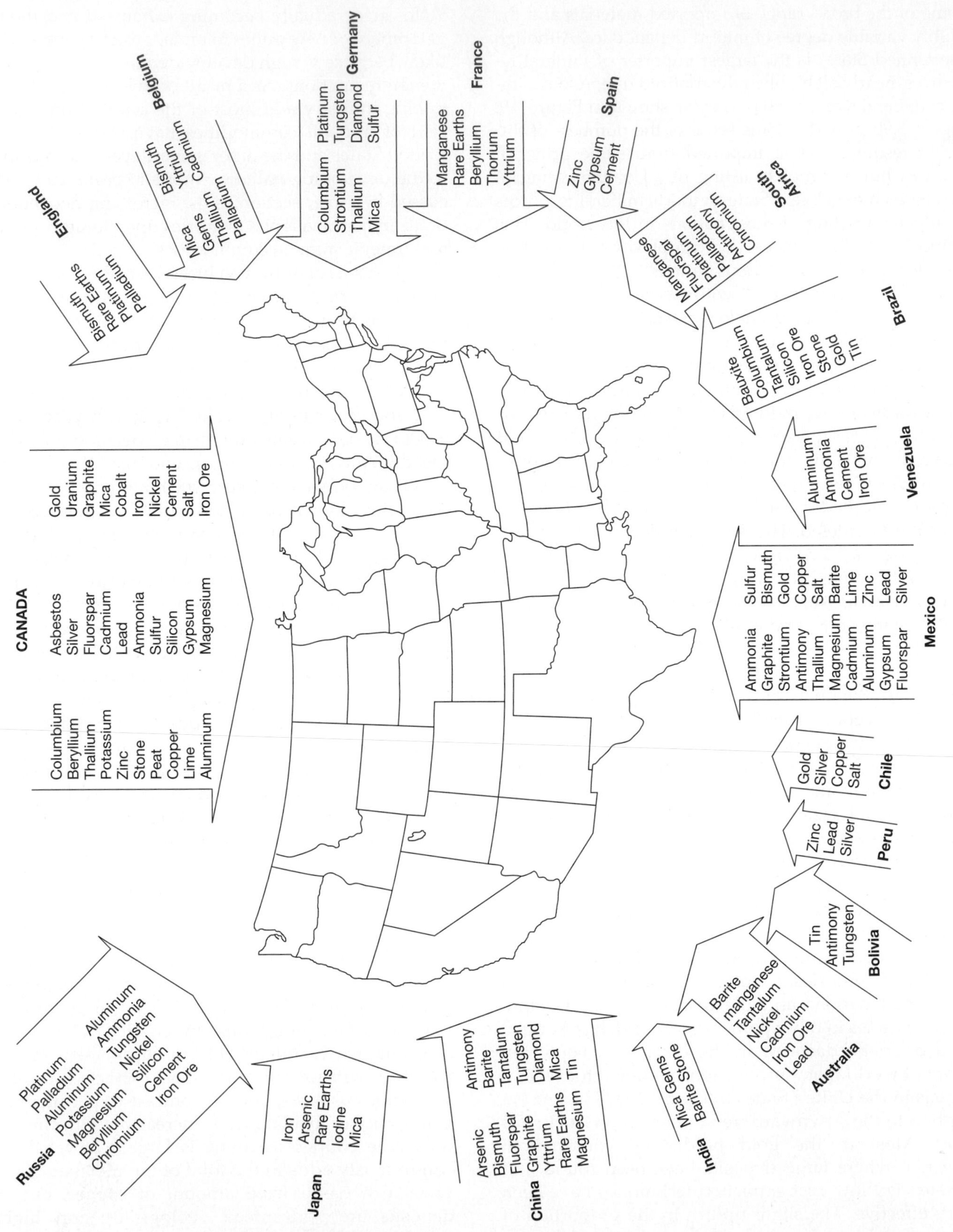

FIGURE 3.9 (b) The mineral resources listed within each arrow indicate those imported from the country named; the data are derived from the major import sources noted on the right side of (a). Canada and Mexico remain the principal sources of nonfuel mineral resources for the United States.

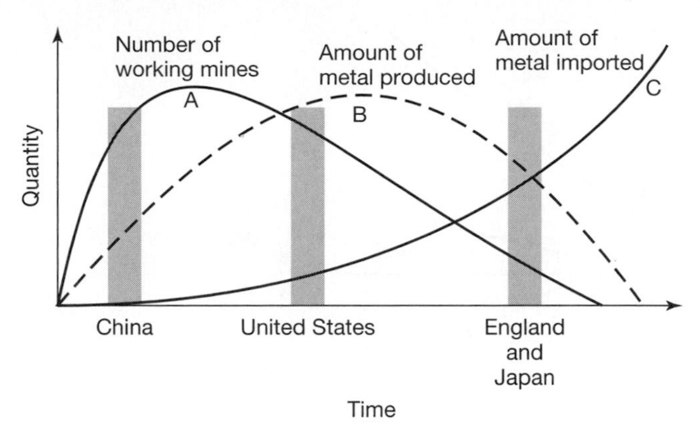

FIGURE 3.10 Traditional stages in mine development, metal production, and imports in industrial countries. Curve A, representing the number of working mines, rises rapidly as a new country is prospected, but it declines when the rate of mine exhaustion exceeds the discovery rate. Curve B, representing metals produced, also rises and falls as mines are worked and eventually exhausted. Curve C, representing metals imported, rises exponentially and expresses the increasing inability of a country to meet its own needs. The approximate present positions for four countries are indicated. With traditional development, each country moves along the time axis from left to right. For example, England was in about the position of the United States in the late nineteenth century, at which time the United States was at about the same stage of development as China is today. Consequently, China is self-sufficient in most metals; the United States is self-sufficient in a declining number of metals; and England and Japan are self-sufficient in very few metals or not any.

GLOBAL DISTRIBUTION AND THE INTERNATIONAL FLOW OF RESOURCES

The Irregular Distribution of Resources

The processes that concentrate resources have not acted uniformly throughout geologic time nor have they been even in their distribution around the globe. Consequently, mineral resources are irregularly distributed around Earth. Furthermore, only 0.0001 to 0.01 percent of any element in the crust is likely to be sufficiently concentrated to become extractable as an ore. Although the general geographical distribution of resources often seems, at first glance, to be random, geologically it is not. In fact, the distribution of resources is quite precisely controlled by the geologic processes that formed them. The processes are, however, often not visible because they occur beneath Earth's surface or are no longer active. Human settlement and political boundaries have generally been designed with regard to land area, water availability, and the locations of rivers and mountain ridges. Consequently, it is of little surprise that some countries have discovered that their lands are well-endowed with some types of mineral resources while others have very little (Table 3.2)

Although there is a randomness in the political distribution of mineral resources, there is a regularity in their availability. Hence, a good correlation exists between the crustal abundance of a metal and the quantity of its reserves. This is logical, because abundant metals are more available for geochemical processes to concentrate them. Thus, the reserves of abundant metals such as iron and aluminum are much larger than scarce metals such as nickel and cobalt and precious metals such as gold and silver. Fortunately, human usage patterns more or less parallel the abundance and reserves of metals. At current and projected rates of metal extraction, our reserves of these metals will last well into the twenty-first century.

Our estimation of the mineral resources available in any area is dependent upon our understanding of the geology of that area, and the degree to which exploration has been carried out. Bigger countries have a general advantage over smaller ones because they are likely to have rocks representing a greater variety of

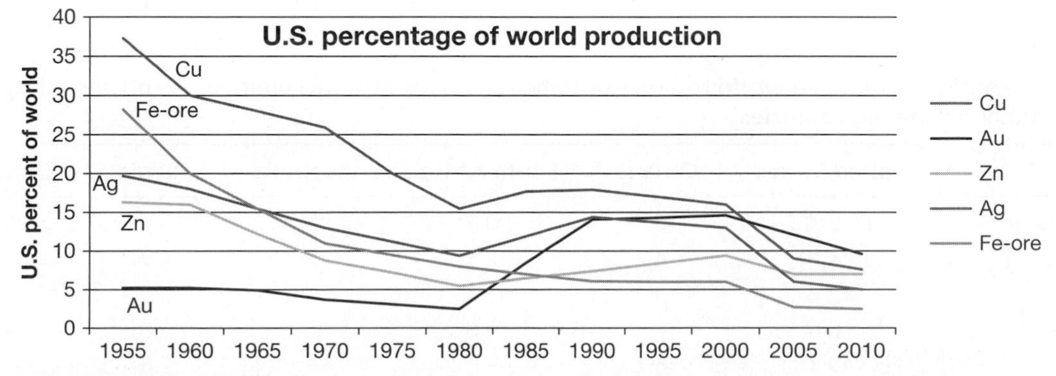

FIGURE 3.11 The United States' share of world production for most metals has declined markedly since 1950 as a result of the exhaustion of deposits, new environmental regulations, and the lower production costs in other countries. The only notable exception is gold because of the large deposits in Nevada and the new recovery technology that allows the extraction of gold for very low grade ores. (Curves based on data from the USGS.)

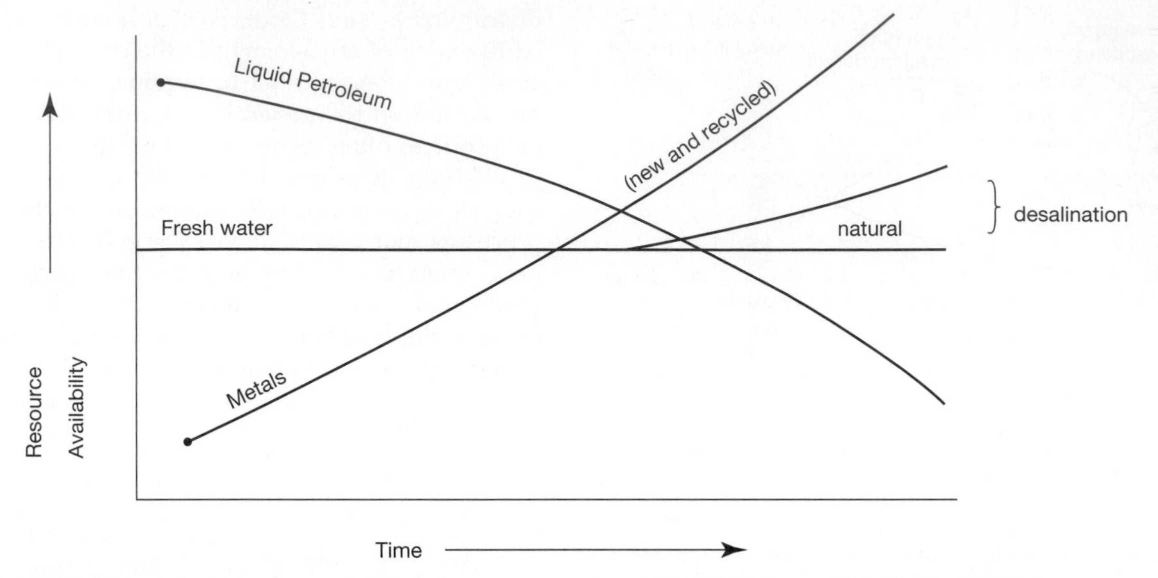

FIGURE 3.12 The changing availability of resources. As time passes, the amount of available freshwater remains essentially constant (perhaps with some small addition made by desalinizing seawater). The amount of remaining liquid petroleum decreases because it is consumed by being burned. In contrast, the amount of copper available has actually been rising because of technological advancements that allow lower-grade ores to be processed and because of the increasing amounts of copper available for recycling.

geological processes and the different types of resources formed by those processes. Thus, it is not too surprising to discover that a few of the larger industrial countries, where there has been more detailed geologic exploration, are the sites of many mineral resources and reserves; areas such as Antarctica—which has been little explored—have fewer known resources. It is reasonable to expect that intensive exploration of poorly known remote areas in the future will add to the quantity of mineral resources and reserves.

Even taking into account the disparities in the degree of our geological knowledge of different areas of Earth's crust, it is apparent that the various mineral resources display great geographical irregularities in distribution and abundance. These irregularities result from the nonuniform distribution of the geological processes that formed them and give particular economic and political significance to many types of resources. Nowhere is this more vividly seen than in the Middle East, where large reserves of petroleum have led to an enormous influx of the world's wealth and a constant vying among the world's powers for political favor.

The existence of mineral reserves within a country's borders has historically been a requirement for a minerals industry, although Japan has demonstrated an ability to develop such an industry on the basis of imported raw materials. However, the existence of mineral deposits alone is not sufficient to ensure a viable mineral industry, because other factors such as high labor costs or low productivity (tin in England), environmental restrictions (coal in the United States), absence of transport systems (Brazil), availability of

TABLE 3.2 Percentages of known world reserves of some important mineral commodities possessed by six major industrial mining countries

	United States	Canada	South Africa	Australia	Russia	China	Total
Percent of Land Area	6.4	6.7	0.8	5.2	15	6.4	40.5
Platinum	1.0	0.4	88.7	—	8.7	—	98.8
Gold	12.4	3.3	41.1	8.9	6.7	NA	72.4
Vanadium	0.4	—	29.8	—	49.8	19.9	99.9
Molybdenum	49.1	8.2	—	—	7.8	9.0	74.1
Potash	1.2	52.4	—	—	31.2	3.8	88.6
Chromium	—	—	81.1	—	0.1	—	81.2
Manganese	—	—	54.4	3.1	19.8	5.9	83.2
Zinc	13.1	7.4	2.1	18.9	6.9	17.4	66.8x

cheap imports (oil in the United States in the 1960s), high transportation costs (fluorspar in the United States), and political instability (many developing countries) may deter development.

Even if resources were uniformly distributed and exploited across the globe, the differences in population and, more importantly, the differences in demand would still result in their having great political and economic significance; political and economic aspects have dominated resource exploitation in the past—and will surely continue to do so in the future.

The International Trade of Resources

Both the irregular distribution of mineral resources, and the tendency of industrialized nations to use much larger quantities of resources than they produce, has resulted in massive movements of mineral commodities along the world's trade routes. In 2008, the annual value of mineral resources in world trade exceeded $5 trillion. By far, the most valuable traded commodity was petroleum; its value alone exceeded $3 trillion. The importance of mineral products to the economies of some developing countries is evidenced by noting that they account for about 70 percent of Bolivia's foreign exchange earnings, about 70 percent of Chile's total exports, and about 60 percent of the Democratic Republic of the Congo's foreign exchange. Even industrialized countries such as Australia and South Africa rely upon the export of mineral resources for about 40 and 50 percent of their total export values, respectively. Not surprisingly, several of the Middle Eastern oil-rich countries derive nearly all of their export earnings from petroleum. The United States, like many other industrialized nations, imports large quantities of many raw and processed mineral commodities, although much of this is subsequently exported as finished products.

Even though most of us are aware that countries such as the United States import many commodities, we tend to overlook the fact that the same commodities can also be exported. Thus, the United States imports and exports coal, oil, and numerous other mineral goods. This at first seems peculiar, but it commonly results from the differences of ship versus overland transportation costs. Thus, the city of Boston has, at times, found ship-transported coal from Europe cheaper than rail-transported coal from the nearer domestic Appalachian coal fields of Kentucky and West Virginia. Another cause for the importing and exporting of the same commodities is the **spot market**. Most large corporations require a stable long-term supply of raw materials and, therefore, often enter into multiyear contracts with suppliers at preagreed prices. When these companies or smaller noncontract companies need extra amounts of raw materials, they bid for them on an open or spot market in which prices may be higher or lower than long-term contract prices. These prices can also fluctuate rapidly, whereas contract prices tend to be more stable. The materials available on the spot market vary in quantity, in quality, and in place of origin from one day to the next.

The increasing dependence of industrialized nations such as the United States upon other nations for supplies of natural resources clearly emphasizes the need for international cooperation, and it highlights the impossibility of becoming isolationist.

The Control of Resources: Corporations, Governments, and Cartels

CORPORATIONS. Private companies and corporations have long been the traditional means of ownership of the mineral industries in capitalist societies. Most began either through single individuals or through small groups who put together *venture capital* to finance the extraction and processing of minerals; the more successful prospered, often expanding into large corporations, whereas the less successful went out of business or were bought by the larger companies. Today, there remain many small mining, drilling, and processing operations, but the great bulk of mineral commodities are produced by a relatively small number of large corporations. Indeed, it is not uncommon to read about large corporate mergers or about huge corporations purchasing mines and even mining companies for billions of dollars. Most mineral companies began with a single product but, in recent years, there has been a tendency to expand into multiple mineral commodities to have greater flexibility in changing markets. Typical examples are the large oil companies, many of which have expanded into coal and metal mining. However, as a result of the downturn in metal mining in the 1980s, many of these companies have closed or sold off their metal mining divisions.

Throughout the first half of the twentieth century, many American, European, Canadian, Australian, and South African corporations expanded into developing still colonial countries in Africa, South America, and Southeast Asia. Subsequently, especially in the 1960s and 1970s, the desire for independence and control of their own resources led many former colonies to alter the original terms of mineral exploitation agreements. Hence, either through nationalization (acquiring more than 50 percent control) or through expropriation, ownership of the mines and oil fields of many of the developing nations has been assumed by the local governments or companies. This has not only weakened the dominance of some of the large mineral companies, but it has also given greater political significance to the mineral commodities.

The first major act of nationalization to affect American companies and supplies occurred in 1938 when the Mexican government nationalized its oil industry and formed PEMEX, the state-run petroleum company. Nationalization of American oil interests also occurred in Peru, beginning in the early 1960s, but this was a more gradual process. More recently, a wave of nationalization has affected the world's copper industries. This began in 1967 when Chile announced plans to gradually nationalize the copper mines developed by major American companies; expropriation was finally announced in 1973. Between 1970 and 1975, the Zambian government assumed complete ownership of its major mines; during 1973 and 1974, Peru nationalized its major copper producers.

Although nationalization and expropriation have commonly been justified by the host countries on the grounds that foreign firms have been improperly exploiting the local resources, the actions have often backfired because the threat of future repetition has limited the willingness of foreign companies to participate further in the development of a country's resources. Without the expertise and venture capital of the major foreign companies, developing nations commonly do not have the capability to discover and exploit the resources for themselves.

The political situations in several nations, especially in central Africa, have been especially uncertain since the mid-1990s because of rebel insurrections and civil wars. These events have created difficulties for foreign-based mineral and petroleum companies because they do not know which factions will ultimately win control. Companies are faced with questions of how to acquire permits, how and to whom to pay taxes and royalties, and even how to safely transport personnel, machinery, and products. Often each one of the competing local political groups attempts to extort funds to support its own causes.

GOVERNMENTS. The degree of control over mineral resources exercised by governments varies widely from

BOX 3.2
The Industrial Revolution

The Industrial Revolution, which spread across Europe in the 1700s and 1800s, was made possible by the development of the coal and iron industries. The revolution stimulated a vast increase in the consumption of iron, coal, and other mineral resources. The Western world was converted, from rural and agricultural societies in which people produced most of their own food and made their own material goods, into largely urban and industrial societies. Two events occurring near the beginning of the eighteenth century in Great Britain played major roles at the onset of the Industrial Revolution. The first event was the manufacture of the first commercial steam engine in 1698 by Thomas Savery. In 1712, Thomas Newcomen improved on Savery's engine and built the steam engine that provided a previously unimagined power to remove water from coal, copper, and tin mines. The Newcomen engines, although widely used for more than fifty years, were inefficient, especially in the loss of steam, because there existed no way to bore the 40- to 100-inch diameter cylinders truly round. This problem was solved by the second event, when James Watt and John Wilkinson developed and first sold a new and more efficient steam engine in 1776. The importance of the Newcomen and Watt engines to early British mining is demonstrated by the presence of the numerous abandoned engine houses (Figure 3.D) that still dot the Cornish landscape.

The building of massive steam engines for use in mines and factories required a second major development—the use of coal to smelt iron and to fuel the steam engines. From earliest times through the 1600s, Great Britain's hardwood forests had provided fuel, as charcoal, necessary for iron-making as well as for manufacturing processes, construction, and home heating. By 1700, the British faced a fuel crisis because so

FIGURE 3.D Wheal (mine) houses that contained the steam engines used to drive machinery and pump water from the tin mines in southwest England in the 1700s and 1800s still dot the Cornish countryside. This restored structure and beam engine are near Camborne. (Photograph by J. R. Craig.)

much of the forest had been harvested. Although coal had been used locally as a fuel, its use had not become widespread. Not only was wood in short supply, but it also lacked sufficient heating capacity to drive some of the new steam engines. Coal proved to be an abundant substitute for wood, generating far more heat than an equal volume of wood. Furthermore, iron-makers discovered that coal could be converted to coke, which proved to be better than the charcoal for the smelting of iron. The use of coal and coke, combined with new smelting and iron-forging techniques, vastly expanded Britain's capability to produce more iron to make more machines. This, in turn, required more coal as a fuel.

The onset of the Industrial Revolution necessitated the development of transportation systems to move the coal, iron ore, and other freight. Until the early 1800s, waterways were the only inexpensive and efficient way to transport large quantities of materials. The British widened rivers and streams and built an impressive system of canals (Figure 3.E) that linked large cities with the coal fields. In the early 1800s, the steam engine was modified to drive locomotives, ushering in the great era of railroad transportation.

Although it took some time for the products and the ideas of the Industrial Revolution to reach the Americas and other parts of the world, the ideas arrived with a powerful impact, producing major changes in lifestyles, such as migration of populations to the cities, vast expansion in the mining of coal and iron ore, and the development of much more effective transportation systems.

FIGURE 3.E The English canal system, here shown at Paddington Junction where the Grand Junction Canal joins the Regents Canal, was developed to transport coal, iron ore, and finished products in the early 1800s. (This reproduction was drawn by Thomas Shepherd between 1820 and 1830; courtesy of the British Waterways Museum.)

one nation to another and within individual nations, depending upon the philosophy of the rulers or controlling party. Traditionally, the governments of capitalist countries have regulated mining methods and imposed taxes on earnings or profits but have left the extraction of minerals and fuels to private interests. In contrast, socialist and communist societies have tended to have state-operated or quasi-governmental companies. In a capitalist society, a company must mine at a profit or go out of business. Reductions in demand for a domestically produced mineral commodity resulting from economic recession or the importing of lower-priced foreign materials or other circumstances usually cause cutbacks in production and manpower and, if too severe, force a closure of the operation. In socialist or communist societies, mining at a profit is desirable but not essential to survival. Thus, state-run mines often continue to produce large amounts of mineral commodities even at a loss because the government always provides employment for all workers, and the government needs the mineral commodities for foreign trade.

In some nations, unprofitable mines have been subsidized by the government because such expenditure is cheaper than the welfare payments that would be required to support unemployed miners if the mines were closed. Other governments have provided cash subsidies, tax relief, or low-interest loans to companies in order to continue operation of unprofitable mines and to maintain jobs.

In today's world, many governments of developing countries now realize the importance of their mineral resources to their own national development. They need foreign capital to develop their natural resources, but are no longer willing to give up control to foreign companies. Accordingly, as the United States and the nations of western Europe become more dependent upon these countries for resources, the negotiation of mining rights, production quotas, taxes and royalties, and the prices of the minerals will become ever more delicate issues.

The breakup of the former Soviet Union and changes in policy by communist countries such as China and Cuba have led them to ask major international oil and mining corporations to help in developing their resources. Hence, firms such as Texaco, Exxon, and Chevron have been invited to develop partnerships with governments or newly created private companies to exploit the latest exploration and production technologies in producing resources to sell on world markets.

CARTELS AND SYNDICATES. Cartels and syndicates are groups of companies or individuals who join together to control or finance the production of some commodity. Their primary aims are usually to control the availability of their commodity and to maximize the income from its sale. Numerous cartels, syndicates, and trade organizations (less formal groups) exist in the mineral industries, but most remain relatively inconspicuous and little known to the general public. The one obvious exception in recent years is OPEC (Organization of Petroleum Exporting Countries), which, after achieving the dominant position in oil production, shocked the world in 1973 by announcing an embargo on shipments to the United States and several European countries. Ever since, most other cartels have wanted to control the prices of their commodities as effectively as OPEC controlled world oil prices through the 1970s. A less conspicuous, much longer-lived, and even more successful organization was the DeBeers syndicate, which for nearly a century controlled the distribution and pricing of the world's gem diamond supply.

Some of the major mineral commodity organizations are listed in Table 3.3. No others have had the success of OPEC or DeBeers because they have not controlled so important a commodity nor have they controlled it so dominantly. The increasing number of developing countries participating in cartels and other trade organizations suggests that such groups may play a more important role in the future availability of mineral resources. To understand cartels and syndicates better, we shall briefly examine the development of the two most successful ones—OPEC and DeBeers.

OPEC AND MIDDLE EASTERN OIL. OPEC has become a familiar name worldwide as this oil producers cartel has achieved international importance. Both the birth and development of OPEC were rooted in the early discoveries and subsequent partitioning of oil rights in the Middle East. The control of the oil resources of the Middle East did not become an important concern of Western nations until World War I because the energy needs of the industrial economies of Europe had been met by coal, and the United States had sufficient domestic oil. However, in 1901, a far-sighted British engineer, William D'Arcy, was granted the exclusive

TABLE 3.3 The major mineral cartels, syndicates, and trade groups

Name	Commodity	Membership
Organization of Petroleum Exporting Countries (OPEC)	Petroleum	Algeria, Angola, Ecuador, Indonesia, Iran, Iraq, Kuwait, Libya, Nigeria, Qatar, Saudi Arabia, United Arab Emirates, Venezuela
DeBeers	Diamonds	Operates in several countries but does not have members
Intergovernmental Council of Copper Exporting Countries (CIPEC)	Copper	Chile, Peru, Zambia, Zaire
International Bauxite Association	Bauxite	Australia, Guinea, Guyana, Jamaica, Sierra Leone, Surinam, the former Yugoslavia
Tungsten Producing Nations	Tungsten	Australia, Brazil, Bolivia, Canada, China, France, Peru, Portugal, South Korea, Thailand, Zaire
Association of Tin Producing Countries (ATPC)	Tin	Australia, Bolivia, Indonesia, Malaysia, Nigeria, Thailand, Zaire
International Tin Committee	Tin	All major producers and consumers

privilege to "search for, obtain, exploit, develop, render suitable for trade, carry away and sell natural gas, petroleum, [and] asphalt . . . throughout the whole extent of the Persian Empire" (modern Iran) for £20,000 ($2 million today) cash, £20,000 stock, 16 percent of annual net profits, and a rent of £1800 ($175,000 today) per year. After near bankruptcy, oil was finally discovered in 1908, and in 1911, the British Admiralty, under Winston Churchill, signed a 20-year supply contract. The seizure of control in a coup by the Reza Shah in 1921 required new agreements that provided new income, but at the same time extended exclusive rights to D'Arcy's company until 1993.

United States' interests entered the scene in the 1920s when British and American companies merged; the United States believed that it had an "energy crisis" and sought more foreign oil to supply its growing needs. The American involvement came through the purchase of oil rights throughout the Middle East, especially the Arabian peninsula, by the famed "seven sisters"—Standard Oil of New Jersey (Exxon), Texaco, Gulf, Mobil, Standard Oil of California (SOCAL or Chevron), Anglo-Persian, and Royal Dutch/Shell. Gulf bought up Saudi Arabian leases, which had originally been granted to a Major Holmes for £2000 a year; but Holmes found no oil. SOCAL obtained concessions in Saudi Arabia in 1933 for 60 years for £5000 per year, a £150,000 loan, and a royalty of 4 shillings per ton *for all time* and a promise of *no taxes*. Finally, in 1938, after much searching and drilling, the first of the large oil fields was discovered (Figure 3.13). The German threat to overrun North Africa and the Middle East in the early years of World War II ended with their defeat at El-Alamein in 1942. Despite consolidation of a patchwork of regional governments and some new negotiation of concessions, the "seven sisters" increased their control during and after World War II and began major oil exports to the nations that were rebuilding in Europe and also to Japan. By 1949, they controlled 65 percent of the world's oil reserves and 92 percent of reserves outside the United States, Mexico, and the former Soviet Union.

The 1950s proved a bonanza period for the international oil companies in the Middle East as production and profits rose to a total of nearly $15 billion. Production of this low-cost oil led to a surplus of crude oil and increased imports into American markets. Import quotas, which protected those markets for higher-priced U.S. domestic oil, increased the supply of crude oil, which was now forced to seek European markets and which led the international companies, without consultation with the producer governments, to reduce posted oil prices by 7-1/2 percent to about $1.80 per barrel; actual oil prices dropped to as low as $1.30 per barrel by the mid-1960s. This brought about a significant and unanticipated drop in the revenues to the Arab countries, which were enraged by such unilateral action. In September 1959, the oil ministers of Saudi Arabia, Kuwait, Iran, and Iraq were joined in Baghdad by the minister from Venezuela and formed the Organization of Petroleum Exporting Countries, which set as its objective the maintenance of stable oil prices at a restored pre-1959 level. Although OPEC failed to restore prices, it did succeed in preventing further cuts and gradually increased its share of the profits that went to the member countries from 50 percent to more than 85 percent. As shown in Table 3.4, three more major producers joined OPEC within three years and, ultimately, the membership grew to 13 nations. The only

FIGURE 3.13 The discovery well at Masjid-i-Sulaiman in Persia (present-day Iran) was a gusher and ushered in the major oil discoveries of the Middle East. (Photograph courtesy of B P America, Inc.)

TABLE 3.4 Membership of the Organization of Petroleum Exporting Countries (OPEC) [year joined]

Algeria [1969]	United Arab Emirates [1967]
Angola [2007]	Abu Dhabi
Ecuador [1973]	Fujairah
Indonesia [1962]	Sharjah
Iran* [1960]	Dubai
Iraq* [1960]	Ras al Khaimah
Kuwait* [1960]	Ajman
Libya [1962]	Ummal al Qaiwain
Nigeria [1971]	
Qatar [1961]	
Saudi Arabia* [1960]	Venezuela* [1960]

*Organizing members in 1960. Ecuador originally joined in 1973 but suspended membership from December 1992 until October 2007; Gabon resigned in 1998.

countries to leave OPEC have been Ecuador, which left in 1993 and returned in 2007, and Gabon, which left in 1998.

During the 1960s, political and economic divisions within the cartel and the availability of excess production capacity worldwide prevented OPEC from increasing oil prices. The OPEC countries relied upon the increase in oil produced (from 8.7 million to 23.2 million barrels per day in 1960 and 1970, respectively) to provide more oil revenues ($2.5 billion in 1960 to $7.8 billion in 1970). By about 1970, the oil scene was changing as the world's cushion of spare crude oil output capacity was dropping, especially in America; the Suez Canal remained closed as an aftermath of the 1967 Six-Day ArabIsraeli War; the 500,000-barrel per day Trans-Arabian pipeline (Tapline), which transported Saudi Arabian crude oil to Syrian ports, was ruptured; and the Middle Eastern Arab world grew hostile toward the West in general, and toward oil companies in particular. In 1971, OPEC began to form a united front and even threatened an embargo; consequently, it won price concessions that raised the price to about $3.00 per barrel. Arab frustration over the stalemated Arab–Israeli conflict and over the reluctance of oil companies to raise prices grew until late in 1973.

On October 6, 1973, Egypt and Syria moved militarily to dislodge Israel from land it had held since 1967. As a result, the atmosphere at the OPEC meeting that began in Vienna two days later was electric. Members of OPEC moved swiftly in rejecting oil company proposals to raise oil prices by 8 percent to 15 percent and countered with a staggering 100 percent increase proposal. On October 16, the price was finally pegged at a 70 percent increase ($5.12/barrel); on October 17, OPEC pronounced that oil-consuming countries were divided into four categories with the United States among the "embargoed." The communique said that the Arab oil cutback would let the United States know "the heavy price which the big industrial countries are having to pay as a result of America's blind and unlimited support for Israel."

Panic struck the oil industry and the Western world in general as suddenly, the principal energy source of the industrialized nations, an energy source previously assumed to be cheap and available, was suddenly scarce and expensive. By January 1979, OPEC raised oil prices to $11.65 a barrel, and the energy crisis led to lines of customers at gasoline stations and mandated rationing. The name "OPEC" had become a household word.

Although there have been no more embargoes, OPEC has been very effective in raising the price of world oil (see Figure 3.14) and has occasionally threatened to withhold oil. The rapid rise in the price of petroleum, especially between 1979 and 1981, stimulated the exploration for new oil fields, substitution of other fuels, and conservation measures.

The success of the new exploration was coupled with a price that was high enough to make previously known but uneconomic oil profitable, and this resulted in a significant increase in the world's oil supply. England, Norway, Mexico and many other nations became major exporters of oil, thus providing competition for OPEC. This resulted in a gradual slide in the price of petroleum after the peak of about $35 a barrel in 1980 and 1981. The culmination came at the end of 1985 when both OPEC and non-OPEC suppliers began undercutting prices to maintain or secure larger portions of the oil market. Markets saw a flood of excess oil, and prices tumbled, with Arab Light (a premium grade from Saudi Arabia) selling for $6.08 a barrel in late July 1986. The OPEC oil ministers had agreed since December 1985 that they should cut production in order to dry up the excess petroleum on the market and drive prices higher, but they had been unable to agree on how much each member country's production quota should be reduced. Economic considerations were constantly influenced by a long Iran–Iraq border war and the reluctance of either of these OPEC members to see the benefit from added oil revenues. Furthermore, Saudi Arabia, the dominant OPEC producer, was upset by the failure of the other members to follow its lead or recommendations on quotas.

The situation in the Middle East rose to new heights of tension when Iraq invaded Kuwait in 1990; at this time, the price of petroleum on world markets rose sharply with the fear of widespread shortages. The military response, by a coalition of American, European, and other Arab countries, quickly drove the Iraqi forces out of Kuwait, reassuring the world of continuing oil supplies from the Middle East and resulting in a return of oil prices to pre-"Gulf War" levels.

It is difficult to predict the future of oil production and the price of petroleum, but, as discussed in Chapter 5, the OPEC members control the bulk of the

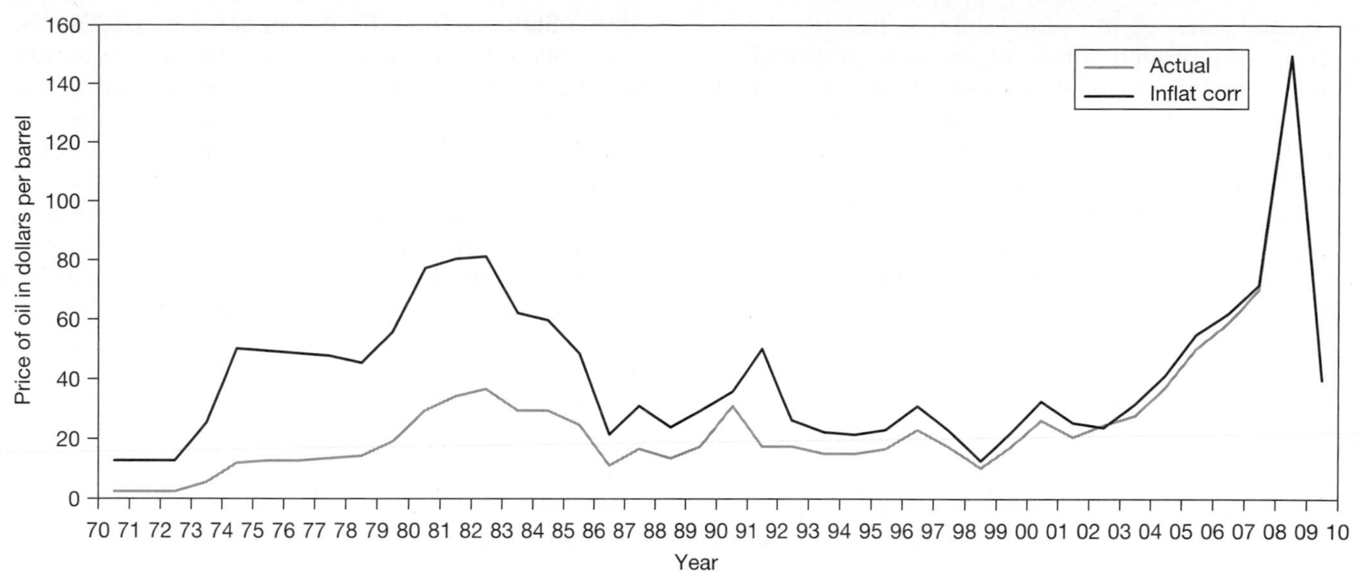

FIGURE 3.14 The cost of crude oil since the early 1970s. Arabian Light, used here as an example, is one of the premier petroleum crudes on the world market; there are many types of petroleum, but their prices all move more or less in the same pattern in response to world events. The sharp rise in 1973 was the time of the OPEC embargo; the rise from 1979 through 1980 resulted from OPEC's limiting production; the sharp fall in 1985 resulted from overproduction and price cutting; the effects of the 1990 Middle East war are evidenced by the brief increase in prices in 1990 and 1991. The price rise beginning in 1999 resulted from increasing world demand and OPEC's cuts in production.

world's known oil reserves; if they can again agree on policies and quotas, they will play a very significant role in world energy policy and politics in the decades to come. There is also a broad consensus that OPEC's control of oil prices will be limited because rapid price increases would result in increased competing production that would keep prices down. However, following the 2001 terrorist attacks on the World Trade Center "twin towers" in New York, there developed a "perfect storm scenario" to undo the most well reasoned predictions.

The United States attack of Iraq in Spring 2003 reduced Iraqi oil production, thus reducing world supplies. The investment of hundreds of billions to trillions of dollars (often quoted at $5,000 per second) into the Iraq war resulted in massive American budget deficits and vastly increased the U.S. national debt. Both of these caused the value of the U.S. dollar to drop against nearly all other world currencies. Because international oil prices have always been quoted in U.S. dollars, the drop in the value of the dollar encouraged other countries (especially OPEC members) to raise the price of oil to compensate. At the same time, between 2003 and 2008, Asian oil demand surged, especially in China, and speculators capitalized on the money to be made. As a consequence, oil prices began to rise and reached absolutely undreamed of values of nearly $150 per barrel in July 2008. For weeks, the commodities markets reported new oil prices every single day. Experts predicted that gasoline prices reaching $4.00 in the summer of 2008 would result in large-scale U.S. conservation—and they finally did. The high price of gasoline also rapidly affected the U.S. automobile market. In short order, sales for large gas-consuming vehicles dropped dramatically; U.S. automakers rapidly closed older manufacturing plants and turned to more fuel efficient and hybrid vehicles. The rapid reduction in the demand for gasoline in the fall of 2008, especially in the United States and the countries of western Europe, resulted in a rapid decline in gasoline and crude oil prices. By the end of 2008, the price of crude oil was back down to below $50 per barrel. Since that time, oil prices have risen and fallen many times depending on supply and demand as well as international tensions. This pattern is likely to continue well into the future.

DIAMONDS AND THE DEBEERS SYNDICATE. The earliest known accounts of **diamonds** are of Indian stones being transported to Greece in about 480 B.C. Throughout history, diamonds have generally been among the most valuable of mineral commodities per unit weight or size. Until late in the nineteenth century, diamonds were mostly found in alluvial deposits in riverbeds in India and Brazil and, rarely, in other parts of the world; they were scarce and valuable. This situation changed dramatically after a South African farmer's children found a "pretty pebble" in the sandy bed of the Vaal River in 1866. By the 1870s, prospectors had located rich alluvial deposits and primary deposits in diamond pipes in South Africa; the deep weathering allowed the rock to be easily removed by steam shovels, just like loose gravel. The sudden influx of millions of **carats** of diamonds onto a relatively small world market resulted in a price collapse that saw the value drop to less than a dollar a carat and the abandonment of many mines that were no longer profitable.

Cecil John Rhodes, a famous British colonial statesman and founder of the country originally called Rhodesia (now both Zimbabwe and Zambia) and sponsor of the Rhodes Scholarships at Oxford University, moved to South Africa in 1870 at the age of seventeen. He became a supervisor at his brother's diamond mine the following year. Over the next fifteen years, he gradually gained control over several additional mines, and in 1888, formed the DeBeers Consolidated Mines, Ltd. This consortium was established to control the production and sale of the world's diamonds, nearly all of which now came from Rhodes' mines (Figure 3.15). The basic operation was simple and exceedingly successful—the mines would release only the number of gem diamonds judged necessary to meet the demand, largely stones for the rings of European and American brides.

DeBeers continued to control the world market even as new mines outside of South Africa were opened by contracting to buy all gem-quality stones from the new operations. When the demand for diamonds dropped during the Great Depression years of the 1930s, DeBeers merely cut back on production and stockpiled stones. Subsequently, the American and European markets grew and stabilized through the 1960s. DeBeers then turned its attention to the nearly untapped market in Japan, where diamonds were not traditionally prized. As a result of the trend toward

FIGURE 3.15 The "Big Hole" at Kimberley, South Africa, was one of the early rich diamond mines brought into the DeBeers Syndicate by Cecil John Rhodes. The hole is 495 meters deep and is now partially filled with water. In its short life, it produced more than 14 million carats of diamonds. (Photograph courtesy by DeBeers Consolidated Mines Ltd.)

Westernization and aggressive advertising, the Japanese market rapidly expanded; in the decade from 1967 to 1977, the percentage of Japanese brides given diamonds rose from 5 percent to more than 60 percent.

After saturating America's bridal market, DeBeers has, since the mid-1980s, aggressively promoted the importance of "anniversary rings" and other diamond jewelry. This has been quite successful, and it has resulted in the sale of more gem diamonds to repeat customers.

Over the years, numerous observers have pointed out that the perceived value of diamonds, like that of gold, has been purely arbitrary and bears little relationship to their intrinsic value. It has generally been believed that DeBeers created and maintained an illusion through advertising that "diamonds are forever," that they are the best symbol of love, and that they have a market value far greater than nearly any other substance. The major diamond producers have cooperated with DeBeers in maintaining a limited availability of diamonds, because this is the only way that prices can be held high. Some have argued that if an open market for diamonds were to develop, the price would probably drop drastically and fluctuate widely. Since the late 1970s, the world's diamond market has lacked some of the traditional stability that the DeBeers syndicate had provided. Nevertheless, DeBeers has remained financially healthy, and the world demand for diamonds continued to grow.

The decline in value of the American dollar in the late 1970s led to much speculation in diamonds as investments and a rapid rise in their value. This trend was reversed in the early 1980s when high interest rates drew investment money out of diamonds and resulted in the "dumping" of many diamonds onto the world market. The availability of these stones and the flow of new diamonds from mines that were opened in Russia, Africa, Australia, and Canada reduced DeBeers' influence over the world market and stretched the company's financial resources as it tried to continue buying, up gem-quality stones. Over the past two decades, there have been several reports of DeBeers becoming overextended or losing control of the gem diamond market, but these reports have always proved incorrect. In fact, DeBeers has maintained its prominent position and has been able to develop long-term working relationships with the Russian producers and with the new Canadian mines. They all recognize the importance of high diamond prices to their economic well-being.

Since about 2000, DeBeers has undergone a significant transformation. It has been burdened with numerous lawsuits about unfair trade practices, accusations of selling "blood diamonds" (those whose sale profits rebel insurgencies), and competition from many new sources. DeBeers restructured itself into more than 15 complex companies that range from mining in Canada to retail sales in the U.S., Asia, and Europe. A significant development in the diamond trade has been the "Kimberly Process." This is an international governmental verification program that requires that governments and companies certify that shipments of rough diamonds are free from blood diamonds; there are now more than 70 signatory members. DeBeers today specializes in high-quality verified diamonds and still controls about 40 percent of the total world diamond market. The future of the diamond market is further clouded by the potential production of synthetic diamonds of gem quality; previously only industrial stones were produced synthetically, but even DeBeers has been an active participant in the attempts to synthesize stones of sufficient size and quality to be used in the gem market.

Resources in World Politics

STRATEGIC RESOURCES. The term *strategic* has been applied to a variety of mineral resources, especially metals, which have become important to certain key industries. There is no absolute definition of a strategic mineral, but the term is most often employed in referring to metals used in military and energy programs—metals such as chromium, cobalt, niobium, nickel, platinum, and tantalum. To this group can be added titanium, manganese, aluminum, and up to ten or more other mineral commodities, depending upon who compiles the list. Furthermore, we have come to recognize the vital importance of chemical minerals and fuels to feed and drive our industries and fertilizers to grow our foods.

What constitutes a strategic resource has changed throughout time. For earliest humans, the only strategic mineral materials were water and salt. In the Bronze Age, copper and tin assumed vital roles as the principal alloy metals for tools and weapons. The Romans are said to have required three basic metals for their society—iron for weapons, gold to pay the soldiers, and lead to make the pipes to transport water. The vast complexity of modern technology has greatly lengthened the list of important, if not always vital, minerals upon which we depend.

A primary concern of many governments now, as in the past, is that there should exist a reliable and adequate supply of strategic raw materials. The United States, Great Britain, Japan, and most of the other major industrialized nations do not have adequate domestic supplies of many of the mineral commodities they consider strategic. This is borne out in Figure 3.9, which shows the high degree of American reliance on foreign supplies for many important mineral commodities. The need to ensure the availability of strategic materials, in times when political, economic, social, and military factors can disrupt the flow of foreign supplies, affects governmental and industrial policies. Major mining and manufacturing companies in industrialized nations will often participate in joint ventures in foreign countries, especially in the developing areas of Africa, Latin America, and Southeast Asia, but only if they believe that their investments are safe. Consequently, U.S. government policies in trade, assistance, and even military presence are directly affected by the nation's needs for mineral resources and, to some extent, by the foreign investment of American companies. Over many years, official government policies have also prevented American firms from involvement with some foreign companies and in some foreign countries. Thus, through the 1990s, American oil companies were not permitted by the U.S. government to operate in Iran, Iraq, Libya, or Cuba. However, such government restrictions usually had little effect on those countries because the absence of the American firms merely allowed easier access for European or Canadian companies. American corporations often argued, to no avail, that the government's actions only harmed the American companies.

To prevent disruptions in supplies, especially from foreign sources, and from halting necessary industries, many governments have developed *stockpiles*. The concept of the stockpile as applied to foodstuffs as well as to mineral commodities is as old as recorded history; the biblical story in Genesis 41 relates how the stockpiling of grain during seven years of plenty permitted Egypt to survive the ensuing seven years of famine. Today, nearly all major industrialized nations have reserves of important materials.

The modern American stockpile of strategic mineral commodities was conceived during World War I when the United States found itself cut off from supplies of several minerals that had been imported from Germany. The first substantive action, however, was taken in 1938 when Congress, fearing the likelihood of another war, appropriated funds to initiate the procurement of materials. This was followed by the passage of the Strategic Materials Act of 1939 and the purchases of resources such as tin, quartz crystals, and chromite. At the end of World War II, the surplus government stocks of minerals were transferred to the Strategic Stockpile, and Congress enacted the Strategic and Critical Materials Stock Piling Act of 1946. This act, which is the basis of the present stockpile, reads in part "the purpose of this act . . . is to . . . decrease and prevent wherever possible a dangerous and costly dependence of the United States upon foreign nations for supplies of . . . materials in times of national emergency." The paramount importance of strategic minerals for economic prosperity and for waging war was recognized by both Germany and Japan during the 1930s; both countries acquired stocks of the mineral commodities they deemed necessary for their aggressive plans. The American stockpile was maintained into the 1990s. However, the changing needs of technology, the high cost of maintaining the stockpile, and the disappearance of the Cold War threat from the Soviet Union, brought into question the need for the stockpile. Consequently, the various commodities were gradually, and relatively quietly, auctioned off.

Although the American military establishment has maintained specific reserves of petroleum (such as the Naval Arctic Petroleum Reserve, which covers vast areas southwest of the North Slope oil fields in Alaska), the oil embargo of 1973 raised great concern regarding the nation's petroleum reserve status (see Box 5.2). Accordingly, the government authorized the development of a strategic petroleum reserve of 1 billion barrels of oil. It began to acquire oil and store it in large caverns carved into salt domes in Louisiana and Texas. Despite episodic development and the opposition of OPEC, which preferred not to have such stores that could be used to lessen the control of OPEC policies on world oil supply, the oil reserve rose to about 750 million barrels by 1994 and has been maintained at such levels ever since. Although the goal of 1 billion barrels is quite large, that alone could sustain America's oil needs for only about 60 days!

There can be no doubt that the strategic importance of mineral resources will grow in the years ahead as the industrialized nations increasingly turn to the developing countries for more resources. At the same time, the burgeoning populations of the developing countries and the increase in their own needs will stretch their abilities to provide such resources. This will be tempered by the needs of the developing countries for foreign revenues, generated largely through the export of mineral resources. It is clear, however, that the international flow of strategic minerals and fuels will continue to play a major role in the world's political and economic activities in the future.

RESOURCES AND INTERNATIONAL CONFLICT. Mineral resources are the raw materials and fuels with which modern industrialized societies function. To deprive a nation of them would rapidly lead to collapse of the nation's economy and industry. So vital are these resources that nations have before, and may again, go to war for them.

The primary mineral resource in the world today is oil, and the principal reserves lie in the Middle East. Accordingly, there has been considerable speculation that the next worldwide conflict could begin over control of the Middle East, because whoever controls that oil has a powerful edge in any conflict. Although the breakup of the former Soviet Union and the ending of the *cold war* has eased many fears of a conflict between *superpowers* over this oil, the rise of nationalism and ethnic and religious tensions have generated new concerns. These concerns have been justified by events in recent years, notably the Gulf War in 1991 and Operation Iraqi Freedom starting in 2003. Many argued that the Iraq invasion was strictly political, but many others held that it was primarily over the control of oil. Time will tell.

The importance of oil and other minerals has also been apparent in earlier times of conflict. Thus, prior to World War II, both Japan and Germany considered their needs for oil and mineral raw materials before beginning aggressive actions. In the 1920s and 1930s, Germany was rebuilding from defeat in World War I and clearly recognized the need for resources to run any future war machine. In 1936, Adolf Hitler ordered that Germany should be 100 percent self-sufficient in oil, steel, iron ore, synthetic rubber, and aluminum in the event of war. Because of the shortage of crude oil, Germany constructed synthetic oil plants that produced petroleum from coal. Once war had begun, Germany found that her iron ores were insufficient, so ores were imported from Sweden and the occupied areas of Austria, Czechoslovakia, and the Alsace-Lorraine region of France. Alloying metals were also in short supply, but the occupied countries often provided the sources: Norway, nickel; Ukraine, manganese; Balkans, chrome. Germany's copper, lead, and tin reserves were limited, but the seizure of stocks of these metals in occupied countries and energetic salvage drives provided what was needed.

Japan, which had been bogged down in a semicolonial war in China, waged in large part for control of the coal and oil there, for several years used the distraction of the German defeat of France and Holland to move into the rice fields of Indo-China, the rubber plantations of Malaysia, and the oil fields of the Netherlands East Indies to obtain the resources needed for the war being planned. In July 1940, President Franklin Delano Roosevelt reacted to this action by placing an embargo on the sale of top grades of scrap iron and of oil to Japan from the United States. Relations between the two countries deteriorated for a year until the United States broke off negotiations in July 1941, and Japan found itself in a total embargo of all strategic materials. The greatest concern was for oil; the Japanese navy had only 18 months' supply and the army only a 12-month supply. The military leaders argued vehemently for war as they saw their fuel supplies growing ever smaller. They believed that a swift attack to incapacitate the American Pacific Fleet would leave Japan free to exploit and import the oil she needed from Southeast Asia. Hence, Japan attacked Pearl Harbor on December 7, 1941, and the United States entered World War II. During the war, major campaigns, especially bombing raids, were directed at destroying oil-producing and oil-refining facilities, and iron and steel plants. Destruction of these and other vital mineral resources severely reduced the enemy's capacity to build and deliver weapons.

The wartime shortages emphasized the critical importance of many mineral resources, especially petroleum. Consequently, all of the major industrial nations have become more concerned about having stockpiles of strategic materials and maintaining access to the major supplies. Future technologies may redefine which mineral resources are vital, but they will not lessen our future dependence upon resources in general.

An accident during off-loading of oil from the tanker *Megaborg* south of Galveston, Texas, on June 8, 1990, released 5.1 million gallons of oil and resulted in a subsequent fire. This caused water pollution from the release of oil and air pollution from the smoke and fumes of the fire. Despite the fact that more than 99.999 percent of all oil is delivered safely, the rare accidents such as the *Megaborg* create local and severe environmental impacts. (From the Office of Response and Restoration, NOAA.)

CHAPTER 4

Environmental Impacts of Resource Exploitation and Use

The beginning of a new millennium finds the planet Earth poised between two conflicting trends. A wasteful and invasive consumer society, coupled with continued population growth, is threatening to destroy the resources on which human life is based. At the same time, society is linked in a struggle against time to reverse these trends and introduce sustainable practices that ensure the welfare of future generations.

GLOBAL ENVIRONMENT OUTLOOK 2000 *(GEO-2000)*, *UNITED NATIONS ENVIRONMENTAL PROGRAM, 1999*

FOCAL POINTS

- Environmental impacts result from the extraction of resources, the use of resources, and the disposal of resource products.
- Solid mineral resources are extracted either by underground mining or surface mining. Wells are used to extract fluids and gases such as oil, water, and natural gas.
- Underground mining is dangerous and more expensive than surface mining because of the potential for rock falls, water inflow, and gas buildup in the workings.
- Surface mining is less dangerous and less costly, but it commonly creates a greater environmental impact because a larger volume of rock is moved, and a large open pit is created, along with a large waste rock pile.
- Underground mines have little impact at the surface unless there is collapse into the mined-out areas, or unless the mining requires lowering of the groundwater table to prevent mine flooding.
- The processing of metalliferous ores to extract the relatively small concentrations of metals within them creates large quantities of rock waste, and the smelting and refining of ores can release atmospheric pollutants.
- The use of some resources, especially the burning of fossil fuels, releases large amounts of gases (CO_2, NO_x, SO_x) into the atmosphere.
- The increase in the atmospheric concentration of CO_2, an important greenhouse gas, is generally considered to be an important factor contributing to the overall increase in Earth's atmospheric temperature, commonly referred to as "global warming."
- Global warming has been blamed for changing rainfall patterns, increased storm intensity, melting of arctic and Antarctic ice and glaciers, and rising sea level.
- The release of sulfur and nitrogen oxide gases into the atmosphere as a result of burning fossil fuels,

especially coal, is recognized as a major source of acid rain.

- Acid rain may damage plants and trees directly, may lower the pH of surface waters so that fish are adversely affected, may leach nutrients from soils, and may damage some buildings directly.
- Nuclear power plants generate radioactive waste requiring special disposal sites that must safely encapsulate the waste for many thousands of years.
- The resource cycle for most materials ends in one of three ways: disposal, reuse, or recycle. For example, the United States generates about 200 million tons of municipal solid waste annually, with paper and cardboard constituting the largest proportion. About 28 percent of the total is recycled today.
- "Green" energy systems, such as solar, wind, and hydroelectric, may have a lower environmental impact when operational, but these systems use much energy and cause significant environmental disturbance during manufacturing and installation.

INTRODUCTION

The second half of the twentieth century saw a rapid growth in awareness by scientists, political leaders, and the public, of the importance and complexity of changes to the environment. It was realized that activities long viewed as beneficial—such as redirection of waterways, draining wetlands, clearing forests, or burning fossil fuels—are actually environmentally damaging. A well-publicized photograph from the 1940s shows a row of smokestacks at an American steel plant billowing out smoke—it was titled "Progress." Today, the same scene would bring cries of "pollution" and incur the wrath of state and federal environmental agencies. Early in the twentieth century, we hailed the development of the nearly indestructible material we call plastic and viewed it as evidence of a better life through advances in chemistry (see Box 5.4). Today, we use plastics in numerous products from automobiles to artificial heart valves, but we complain about plastic being a nearly indestructible pollutant. The chemical companies have made plastics so durable that they do not degrade; the public is now demanding new kinds of plastics that are biodegradable.

Nowhere have plactics been more visible than as the nearly universally used grocery store bag. Being cheap to manufacture, useful in advertising, very strong and durable, they have been produced by the billions and distributed nearly worldwide. As "throwaway" items, they have come to symbolize worldwide litter and are seen from the slopes of Mt. Everest to ocean flotsam where they represent a major threat to seabirds and sea turtles. Their unsightliness has led numerous cities to try to contral their use; China passed a law in 2008 outlawing their use altogether. Almost as common as the plastic grocery bag is the clear plastic water bottle. The worldwide explosion in use of bottled water has, no doubt, reduced the prevalence of some waterborne diseases, but it has left the landscape littered with billions of plastic bottles, a very large percentage of which go unrecycled.

Pollution problems are not new, but the scale of pollution in the modern world is vastly greater than in earlier times. The principal causes are advances in technology and a much larger human population; in 1900, the world population was only about one-fifth of what it is today, and even in 1960, it was only one-third that of today. As our ancestors learned to harness energy—from fire, to explosives, to the steam engine, to the internal combustion engine, to nuclear power—their ability to modify the environment vastly increased in scope. Until the twentieth century, the principal construction materials were rock and wood, neither of which created significant pollution problems. The rock structures persisted, often long after being abandoned, but they do not pollute the air, soil, or water; wood structures, in contrast, once unoccupied, rapidly decompose and disappear as vegetation restores areas to their previous condition. Today, the widespread use of concrete and asphalt to pave large areas, and the use of plastics and a host of other resistant synthetic materials, ensures that natural vegetation will not recover abandoned areas. Growing concerns about modern environmental problems has increased our recognition of problems created by our ancestors. For example, recent studies have found that the Romans, in their exploitation of lead-rich ores for their silver content, actually spread lead pollution over wide areas of Europe. In the seventeenth and eighteenth centuries, Great Britain suffered the environmental impact of two major developments—expanding population and the emerging Industrial Revolution. This called for supplies of energy that were largely met by cutting down the forests of England, Scotland, Ireland, and Wales. The energy crisis that arose following the destruction of these forests was alleviated by turning to coal as the energy source, but the burning of the sulfur-rich coals led to massive air pollution throughout the cities. To this day, most of the hillsides of Great Britain remain bare; only now are the reforestation efforts of the twentieth century beginning to be significant.

In many parts of the world, the use of mercury in the recovery of placer gold has left a legacy of mercury pollution in thousands of streams. Following the great California Gold Rush of 1849, extensive hydraulic mining (Figure 4.1) destroyed the natural channels of many rivers, flushed out fish and other aquatic life, and choked other channels and bays. Today, hydraulic mining is banned in most areas, and much effort has gone into the restoration of original river environments.

The recognition of detrimental human activities and the need to maintain a healthy environment for our survival has led to much research on environmental problems, to the teaching of courses on *environmental science* at colleges and universities, to the emergence of various groups or even political parties, united by *environmentalist* causes, and ultimately, to legislation to help prevent deterioration of the natural environment through human activities. This rising tide of activity has resulted from public awareness of our abused environment and the rapidly increasing levels of resource exploitation. The underlying causes are population growth and the growing rates of material consumption.

In this book, the nature and utilization of Earth's resources (fuels and other energy sources, metals, industrial rocks and minerals, fertilizers, water, and soils) are discussed along with appropriate evaluation of the environmental effects resulting from extraction and use. These are matters that rightly concern every one of us, and they carry serious implications for future generations. In this chapter, we first examine the ways in which the extraction of resources directly affects the environment—the effects of mining, quarrying, dredging, well-drilling, and also the effects of processing and smelting of ores. Then, we examine ways in which the use of resources affects the environment—the burning of fossil fuels, the utilization of nuclear fuels, the disposal of hazardous wastes, and the problems of **pollution** caused by other industrial processes. Finally, we examine the problems involved in disposal (or, where possible, of recycling) of the wastes that are produced in vast amounts, not only by manufacturing industries but by each one of us in our everyday lives.

In later chapters, each type of resource is discussed together with a consideration of how the resource is extracted and used, and what environmental consequences arise from its use. As we examine the individual resources, it is important to remember that all resource extraction and use is accompanied by some environmental impact; not all impacts are immediately apparent—some are subtle, some may be removed from the site of usage, and others may be long delayed in appearance. Regardless of the nature of the impact, it is important that it is carefully identified and assessed, that deleterious effects are minimized, and that we weigh these effects against the benefits derived from the resources.

FIGURE 4.1 Hydraulic placer mining in California in the 1860s yielded large amounts of gold, but dramatically altered the nature of the rivers and bays into which they flowed. Ultimately, hydraulic mining was outlawed because of the environmental damage and choking of the waterways. (Photograph courtesy of Levi Strauss and Company.)

ENVIRONMENTAL IMPACT OF RESOURCE EXTRACTION

The most obvious and severe disruptions of the natural environment have come from the exploitation of resources. Mining, quarrying, dredging, and the drilling and extraction of oil and gas from wells are all activities that have made irreversible impacts on the landscape and therefore on the environment. Directly linked to the extraction activities are issues concerned with the disposal of the waste products that accompany extraction. Further environmental problems may be caused at the site of exploitation when various extraction or concentration processes are employed. For example, most metal mines remove an ore that contains only a very small proportion (commonly less than 1 percent) of the metal being extracted. Various physical and chemical processes, often culminating in smelting, are then needed to extract the metal from the mined ore. For metal extraction, at least three aspects of the process create potential environmental problems—the mining operation, the disposal of very large quantities of waste rock, and the processing, smelting, and refining of the ore.

Mining and Quarrying: The Methods

The method used to extract minerals or rocks of any kind depends on the nature and location of the deposit. Depending on the size, shape, and depth of a deposit beneath the surface, and on the **grade** (percentage of valuable material or quality and purity), either a surface mining or underground mining technique is employed.

Surface mining, which accounts for about two-thirds of the world's solid mineral production, especially the production of sand and gravel, crushed stone, phosphates, coal, copper, iron, and aluminum, generally involves **open-pit mining** or a form of **strip mining.** The term **quarry** generally refers to an open pit mine from which building stone or gravel is extracted. Open-pit mining is an economical method of extraction where large tonnages of reserves are involved, high rates of production are desirable, and where the waste material overlying the deposit (**overburden**) is thin enough to be removed without making the operation costly. Surface mining is preferred over underground mining by mining companies, when at all practical, because it is less expensive, it is safer, and there are fewer complications with air, electricity, water, and rock handling. However, surface mining results in a greater environmental impact than underground mining. Many of the largest open-pit mines, such as those in the southwest United States and Chile, from which copper and gold are extracted (see Figure 4.2 and Plate 37), are developed as conical chasms with terraced benches that spiral downward to the bottom of the pit. These benches serve as haulage roads and working platforms on the steep, often 45°, sloping sides of the pit. Extraction proceeds by drilling, blasting, and loading material into large trucks that haul rock and ore out of the pit. The depth and diameter of the pit both increase as mining takes place, in some cases, reaching diameters of more than 2 kilometers and a depths of a kilometer.

The world's largest open-pit mine, at Bingham Canyon, Utah, has involved cutting away an entire

FIGURE 4.2 The giant smelter at Morenci, Arizona, is one of the world's largest copper-producing complexes and demonstrates the impact of mining, smelting, and waste disposal on the environment. Ore from an open-pit mine is brought by truck to the mill and smelter, which make up the building complex surrounding the two smokestacks. There the valuable copper minerals are separated from the rock, and the copper-rich concentrate is smelted down to copper metal. (Photograph by B. J. Skinner.)

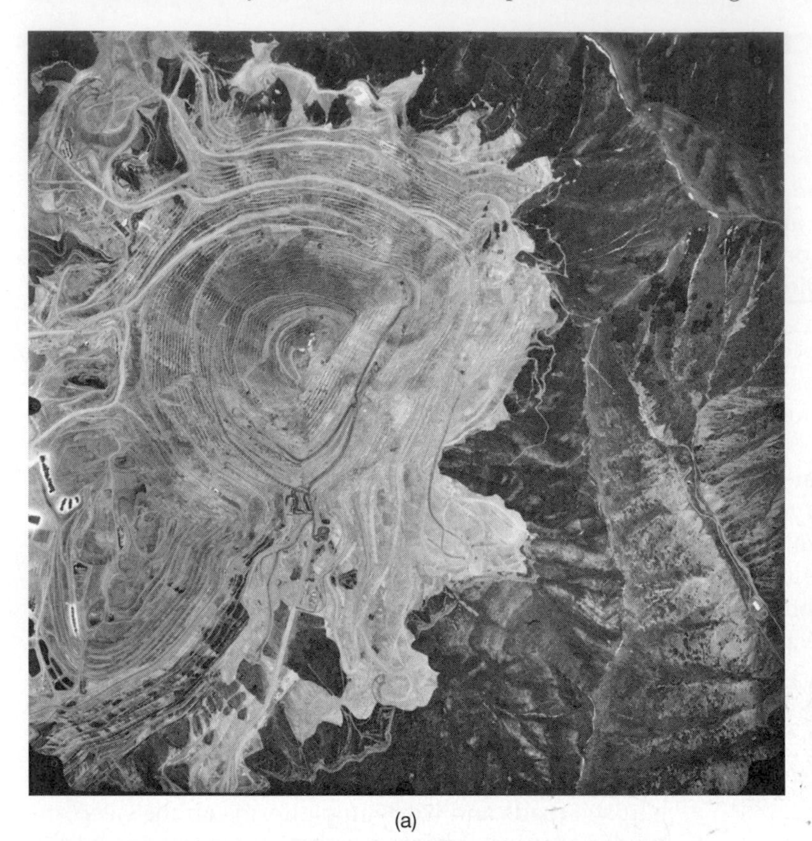

(a)

(b)

FIGURE 4.3 (a) The Bingham Canyon Mine near Salt Lake City, Utah, shown here in a near-vertical photograph, is about 4 kilometers across (2 miles) at the top and represents one of the largest mines in the world. Daily production of ore and waste rock at the beginning of the twenty-first century averages more than 400,000 metric tons; reserves will permit continued operation well into the twenty-first century and probably beyond. (b) The removal of more than 400,000 metric tons of ore and waste from the Bingham pit every day requires the use of huge power shovels and trucks, as shown in this photograph. Note the normal-sized pickup truck for scale. (Photographs courtesy of Utah Copper Corporation.)

mountain to form a pit roughly 4 kilometers across and 1 kilometer deep (Figure 4.3). At the peak of production, more than 100,000 metric tons of ore grading, about 0.5 percent copper, can be produced daily. With a waste rock-to-ore stripping ratio of 3 to 1, this involved drilling, blasting, and removing an average of 400,000 tons of material per day. Power shovels of 5 to 20 cubic-meter (m^3) capacity, railcars of up to 80 metric tons capacity, and diesel trucks capable of handling up to 140 metric tons were needed for this feat. Since the early 1900s, the extraction of nearly 6 billion metric tons of rock and ore at Bingham Canyon has yielded 15 million tons of copper, 19 million troy ounces of gold, 160 million ounces of silver, 600 million kilograms of molybdenum, and significant amounts of platinum and palladium. The record for open-pit mining was established in the mid-1990s when the Morenci Mine in Arizona moved the staggering quantity of 1.3 million metric tons of ore and waste in one 24-hour period.

Strip mining is employed when the material to be extracted is present as a flat-lying layer just beneath the surface. Many coal seams are exploited in this manner, but the method is also used in mining tar sands, phosphates, clays, and certain types of iron and uranium ores. These are all materials that can occur as thin, near-horizontal layers, often underlying enormous areas of country. The mining method involves removal of the overburden to expose the coal or other resource, which is then scooped up and loaded into trucks or trains. The waste rock is dumped to the rear (see Figure 4.4), and the mining continues along a strip that extends as far as practical. A new strip is then started parallel to the first, and waste from this strip is dumped back into the preceding strip. Large mechanical shovels and drag lines are used in removing overburden and the material being mined. Clearly, the ratio of the thickness of the overburden to the resource is a crucial factor in this type of operation.

Underground mining, involving a system of subsurface workings, is used to extract any solid mineral resource that is not near enough to the surface to allow for economic surface extraction. Most mines consist of one or more means of access via vertical **shafts,** horizontal **adits,** or inclined roadways

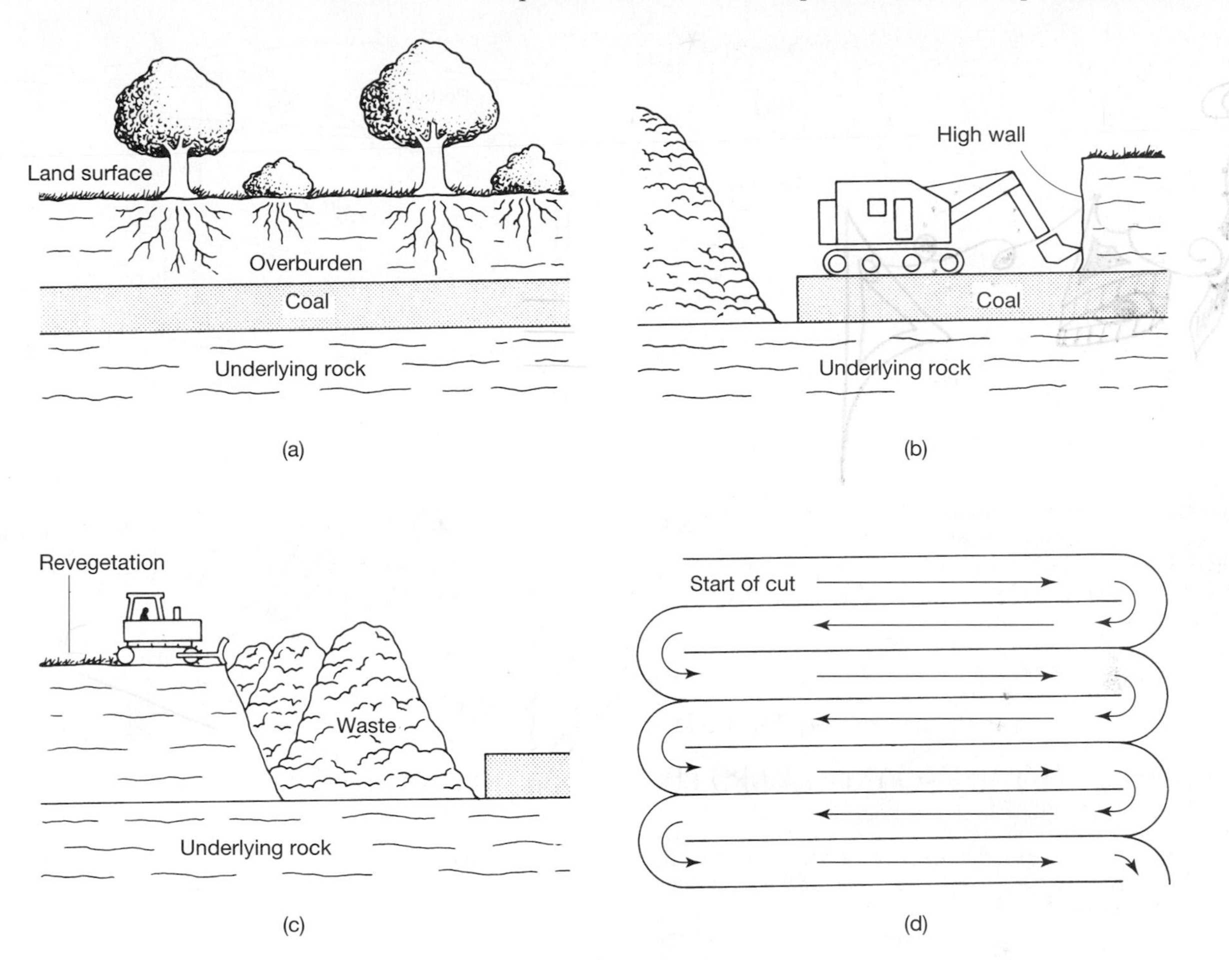

FIGURE 4.4 The sequence of steps involved in strip mining: (a) profile before mining; (b) profile during mining; (c) profile as land surface restoration begins; (d) plan view of the mining sequence.

(**inclines**) as shown in Figure 4.5. These allow for the movement of workers, machinery, materials, and the extracted ore and wastes. They also form access paths for ventilation and the control of underground water that is essential to mining operations. Indeed, in many underground mines, substantial amounts of energy are expended in pumping air in and water out to keep the working areas dry and well-ventilated. This is particularly the case where mining takes place in an extremely hostile environment.

For example, the zinc mine that operated for many years at Friedensville, Pennsylvania, was in a cavernous limestone terrain where water had to be pumped out routinely at rates of about 25,000 gallons (95,000 liters) per minute to prevent flooding of the workings (Figure 4.6). Immediately prior to the closing of the mine in the 1980s, the mine's monthly electric bill was more than $400,000, mostly for pumping water. When heavy rains fell on the mine area in February 1976, the pumping rate was raised to a peak of 60,000 gallons (227,000 liters) per minute to keep the mine from flooding.

At the Konkola Mine in Zambia, Africa, possibly the wettest mine operating in the world today, the pumping rate has reached 73,000 gallons (280,000 liters) per minute. In the deepest gold mines of the Witwatersrand, South Africa, mining levels reached 3500 meters (11,500 feet) below surface, where the rock temperatures are over 43°C (110°F). At the Magma Mine in Superior, Arizona, and in the Toyoha Mine in Japan, mining is routinely conducted where rock temperatures reached 50°C (122°F) with 100 percent humidity. Humans cannot survive in such conditions, so it is necessary to refrigerate the air; this is extremely expensive and requires special technology to cool and handle the air. One might think that it is only necessary to blow in large quantities of air, but mine ventilation systems are very complicated, and when air descends down shafts to depths of 3500 meters, it is compressed by gravity and heated nearly 6°C (10°F).

Underground workings usually consist of intersecting horizontal tunnels (**drifts** and **crosscuts**), often on several **levels** and joined by further vertical openings (**raises** or **winzes**). The region where ore is extracted in a metalliferous mine is referred to as a **stope** and the area that is actually drilled and removed is called a **face**. Examples of some of the many geometries

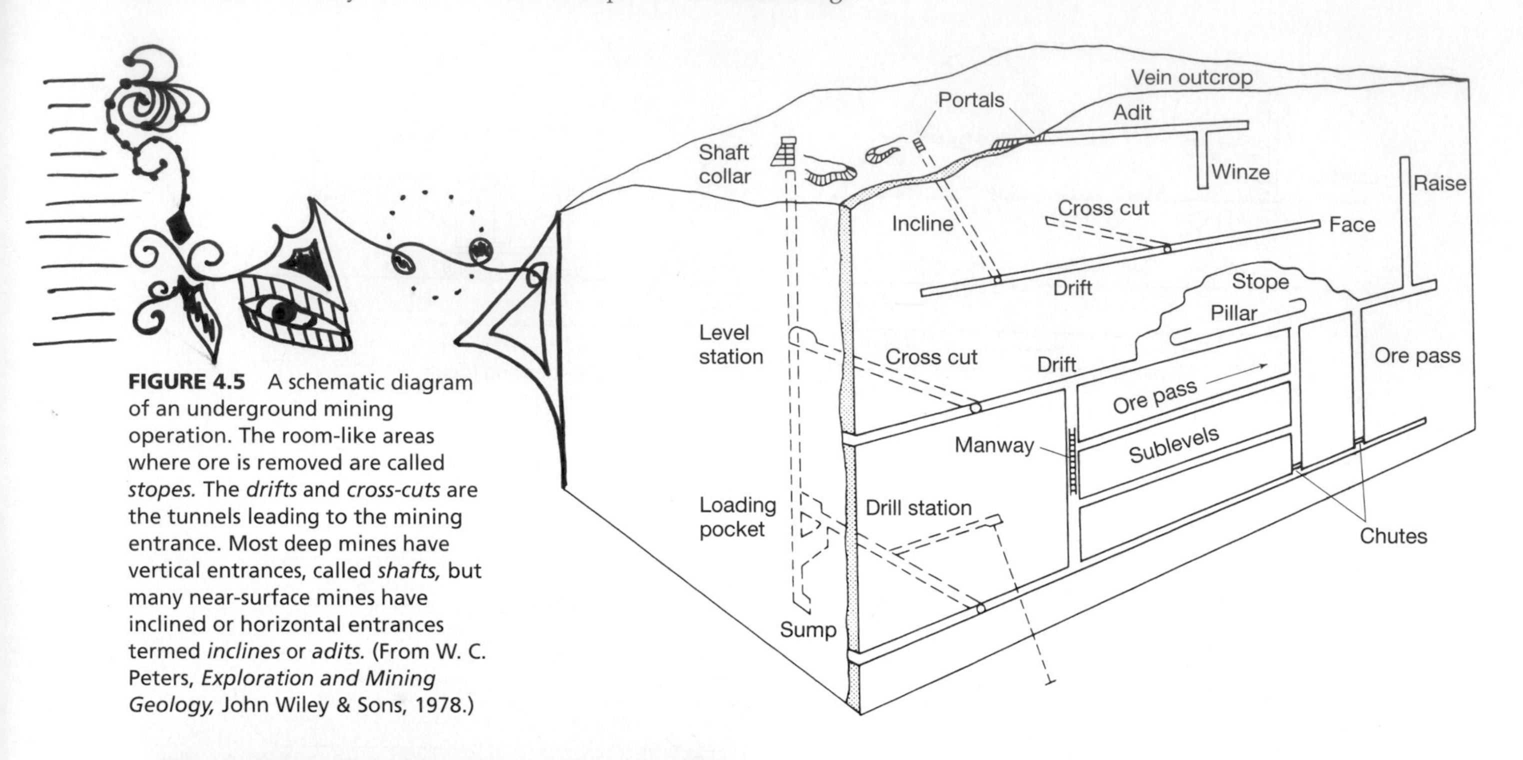

FIGURE 4.5 A schematic diagram of an underground mining operation. The room-like areas where ore is removed are called *stopes.* The *drifts* and *cross-cuts* are the tunnels leading to the mining entrance. Most deep mines have vertical entrances, called *shafts,* but many near-surface mines have inclined or horizontal entrances termed *inclines* or *adits.* (From W. C. Peters, *Exploration and Mining Geology,* John Wiley & Sons, 1978.)

FIGURE 4.6 The high rate of water flow into mines can make mining difficult, dangerous, and expensive. The flow of 95,000 liters (25,000 gallons) per minute into the Friedensville Mine in Pennsylvania lowered the water table throughout the area and increased the cost of electricity to operate pumps. (Photograph by R. Metzger.)

of underground mine workings are shown in Figure 4.7. The method employed in any mine will depend on the shape, size, and grade of the ore body or seam that is being worked. This, in turn, will determine the kinds of machinery used to break up and carry away the ores and waste rocks. In most mines, ore extraction and mine development involve drilling and blasting and removal with mechanical diggers onto underground railway cars or dump trucks that reach the surface via a shaft, incline, or adit. Some coal mines today are only about 1 meter high, just high enough to allow the use of continuous mining machines (see discussion of coal mining in Chapter 5) that cut the coal with rotating teeth and then feed it back to transport cars or conveyor belts. In contrast, some large stopes, such as that shown in Figure 4.8, which have been excavated in the zinc mines of the Appalachians, stand as open galleries more than 70 meters high and 100 meters long. Other factors that will affect the mining methods are related to ground conditions, such as the strength of rocks encountered underground, fractures in the rocks, and the amount of groundwater. For example, drifts, crosscuts, and stopes in a tin mine in granite may require little or no roof support, whereas in the much weaker sedimentary rocks of a coal mine, extensive roof support using timbers or other props might be needed.

Underground mining is inherently dangerous because of the potential for rock falls and cave-ins, but probably the most feared danger is the buildup of poisonous or explosive gases. Mine systems are carefully designed so that large volumes of air are moved from the surface to the working areas to flush out natural gases

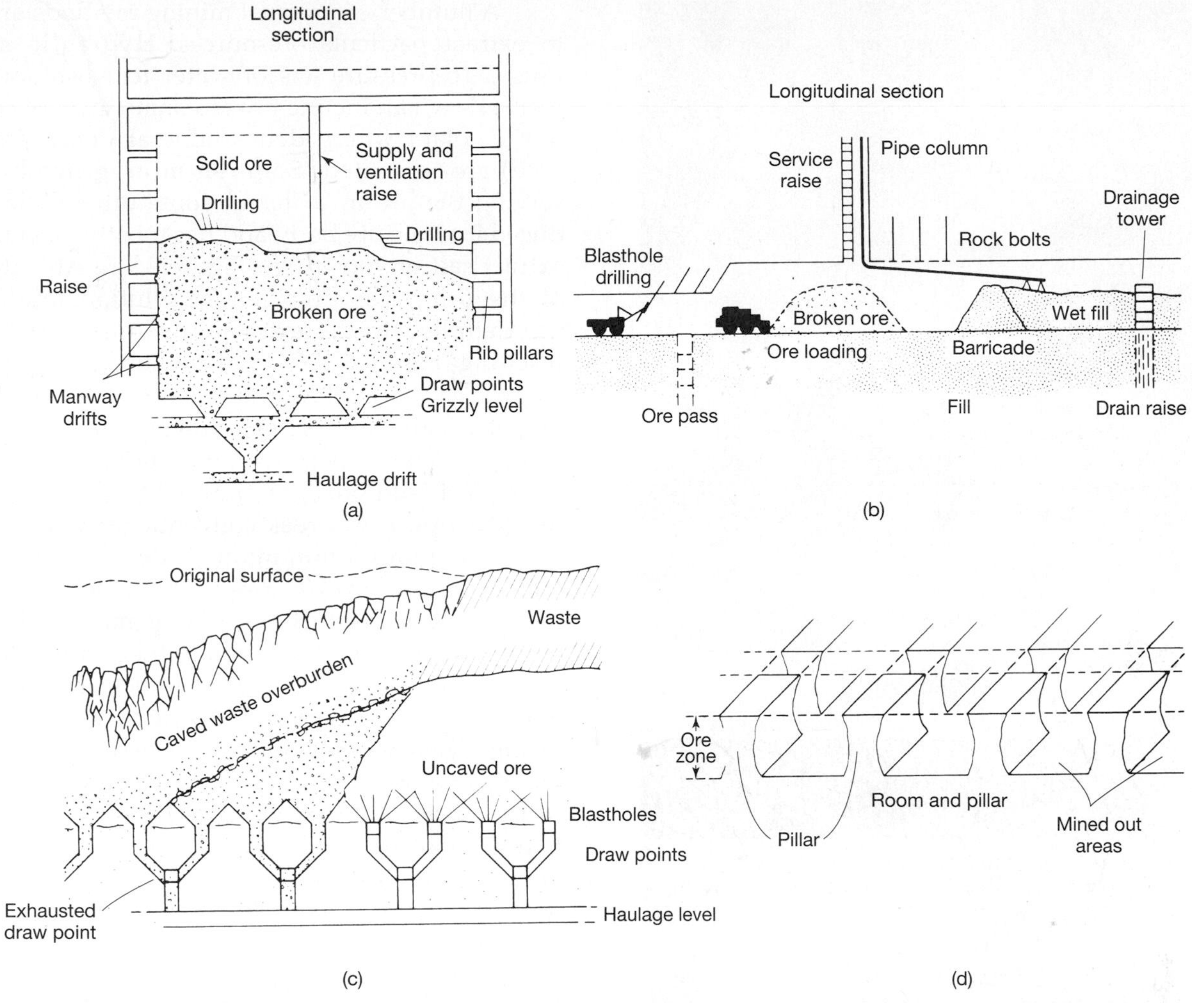

FIGURE 4.7 Schematic representation of the major mining methods. (a) Shrinkage stoping—ore is drilled and blasted from the ceiling (called the *back*) and allowed to fall; subsequent drilling and blasting are carried out by working on top of the broken ore. (b) Cut and fill stoping—as ore is removed, the open space left is refilled with waste materials. This method has the advantage of filling the open mined-out stopes and disposing of the wastes below ground. However, it is very expensive. (c) Block caving—an entire mass of ore is blasted from below, allowing it to flow slowly through the draw points. (d) Room and pillar—ore is mined in a series of rooms, leaving pillars to support the overlying rocks. (a, b, c from W. C. Peters, *Exploration and Mining Geology,* John Wiley & Sons, 1978.)

and any fumes arising from blasting and from operating equipment. All underground vehicles and equipment operate on electricity, compressed air, or diesel fuel; gasoline is forbidden in most mines because of its higher volatility and the potential for causing fires or explosions. Despite the best efforts to ventilate underground mines, methane gas is sometimes rapidly evolved from coal seams and can reach explosive levels (see Plate 24). Miners once carried canaries into the mine workings to test for bad air, and miners even sent workers wrapped in wet clothes into areas to burn out small methane accumulations (see Figure 4.9). Gas-generated explosions have been the major cause of death in coal mines for hundreds of years; tragically, despite modern testing devices, explosions still kill many miners every year.

One of the highest mining death rates in recent years has been in the Ukraine where inadequate safety measures have resulted in 300 to 400 deaths annually. Most of these are from methane gas explosions; as recently as March 2000, some 80 miners were killed in a single explosion. The situation in China is even worse, with more than 3000 miners dying in coal mines every year. Metal mines rarely contain methane gas because there is no concentrated organic matter from which it can be derived, but there can be problems of carbon dioxide or carbon monoxide buildup. Underground uranium mines actually have very low levels of emitted radiation, but they are always monitored for radon gas levels because radon is released during the natural radioactive decay of uranium. Usually ventilation air flow eliminates any problems, but the radiation and radon levels are so high at the very rich McArthur River and Cigar Lake uranium mines in Canada, that mining involves using remotely controlled equipment and robots.

FIGURE 4.8 A large hydraulic drill cuts blast holes in a zinc ore between the pillars that support the overlying rock more than 15 meters (45 feet) above. (Courtesy of ASARCO, Inc.)

A number of unusual mining methods are used to extract particular resources. **Hydraulic mining** uses high-pressure jets of water to wash soft sediments down an incline toward some form of concentration plant where dense mineral grains (such as gold) are separated. **Solution mining** involves dissolving the ore in water or some other fluid introduced into the ore body and has mostly been used to extract salt or sulfur (see Figure 4.10). An extension of the principles involved in solution mining has been applied to the extraction of metals by *in situ* **leaching** (Figure 4.10C). The method is being used to recover copper, gold, and uranium from low-grade ores. The value of *in situ* leaching is evidenced by the fact that today it permits the economic open-pit extraction of gold ores grading 0.02 troy ounce/metric ton (0.6 ppm), whereas conventional underground mining requires a minimum grade of about 0.2 troy ounce/metric ton (6 ppm). *In situ* leaching is thought by some to be a mining method likely to have much wider application in the future. One proposal includes the detonation of nuclear blasts at the bottom of a large ore body to cause extensive fracturing. Then wells would be drilled into this mass for injection and recovery of leach liquors such as sulfuric acid that will dissolve the valuable metals. Many problems still remain with such futuristic mining methods.

BOX 4.1

Acid Rain

Rainfall is generally thought of as beneficial, washing and watering Earth's surface. While this is still true in many places, the image of rainfall has become tarnished because we now know that rainfall is not always pure water and its effects are not always beneficial. As early as 1872, the British chemist Robert Smith coined the term *acid rain* to describe precipitation containing significant amounts of sulfuric acid. He attributed the acid to the burning of the coal that fueled the furnaces of the Industrial Revolution. Little more was said and nothing done until the 1950s when European scientists reported abnormally low pH values in Scandinavian lakes where there was no source of acids except from rainfall. Subsequently, in many parts of the world, acidic precipitation has been recognized, its origins debated, and remedies sometimes sought.

So, what is acid rain? This term describes precipitation—including sleet, hail, snow—that is more acidic than rainfall was in pre-Industrial Revolution times. In general, this means precipitation with a pH lower than 5.0 (Figure 4.A). The pH of a substance, the conventional measure of the hydrogen ion content (acidity) of aqueous solutions, ranges from 0 to 14, with neutral being 7. This actually means that the concentration of hydrogen ions is 10^{-7}mol/liter. Rainwater naturally contains dissolved carbon dioxide from the atmosphere that forms weak carbonic acid, H_2CO_3. This acid ionizes, forming H^+ and HCO_3^- ions, resulting in rainfall that is normally slightly acid with a pH of about 5.6. The increase of carbon dioxide due to the

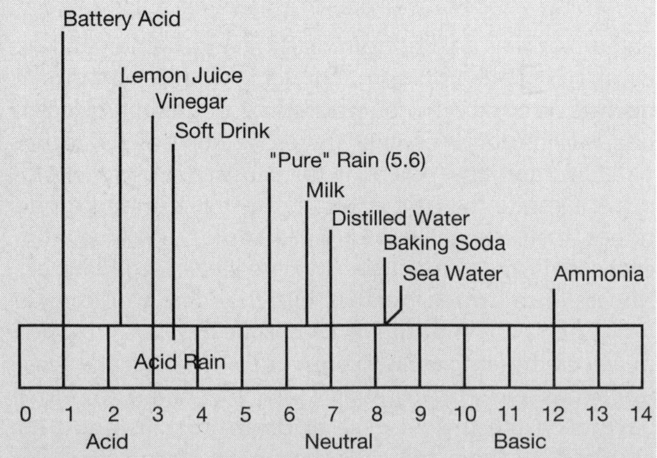

FIGURE 4.A Acid Rain. The pH scale shows the pH of several common compounds. Rain is generally considered to be acidic when it falls below a pH of 5.0.

burning of fossil fuels has some significant effects on atmospheric warming, but lowers the pH only slightly.

However, the introduction of large amounts of sulfur oxides and nitrogen oxides into the atmosphere by the combustion of fossil fuels has a significant impact on the acidity of rainwater. Sulfur oxides, which combine with water vapor to form sulfuric acid (H_2SO_4), are created when sulfur-bearing minerals in coal are burned. Nitrogen oxides, which react with the water vapor to form nitric acid (HNO_3), are created as an unintentional by-product of high temperature combustion in an atmosphere containing 78 percent nitrogen. Small amounts of these two strong acids in rainfall can sharply lower the pH of rainwater. In some places in the United States, rainfall has been reported to have a pH as low as 4.0.

On a worldwide basis, it appears that natural phenomena such as volcanoes, forest fires, and the decay of vegetation probably put more sulfur and nitrogen oxides into the atmosphere than do the emissions resulting from human activities. These natural emissions are very widespread and appear to drop the pH of much rainwater into the range of 5.0 to 5.6. Accordingly, most researchers now consider precipitation to be acidic if the pH is below 5.0. Using that threshold, it is apparent that considerable portions of North America and Europe are subject to acid rain. Local effects of smokestack emissions have long been evident. However, the first recognition of a large-scale problem came from European scientists in the 1950s. As a result, large numbers of monitoring stations have been established in North America and Europe to define the scope of the problem. The map shown in Figure 4.B, showing the distribution of acid rain in recent years, indicates that the pH of precipitation for most of the eastern United States is below 5.0, and for large areas of the upper Mississippi Valley and northeast is below 4.5.

Most of the acid rain effect is attributed to emissions from power plants that burn coal containing 1 to 2.5 percent sulfur. Nitrogen oxides come from both power plants and motor vehicles. Acid rain damages vegetation, limits fish growth by acidifying streams and lakes, causes deterioration of buildings and similar structures, especially those made of metals or rocks such as limestones, or coated in paints (see Figure 4.28), and increases the leaching of nutrients from soils. Acid rain in the United States resulted in the passage of the 1990 Clean Air Act. Implementation of this legislation involves capture of sulfur and nitrogen oxide emissions, and is estimated to cost \$25 billion to \$35 billion per year, but it had cut American power plant sulfur oxide emissions in half (to 10 million tons annually), and nitrogen oxides emission by one-third (to 4 million tons annually) by 2000. In addition, passenger car emissions of nitrogen oxides were cut by 60 percent by 2003. There are those who suspect that the detrimental effects of acid rain may have been exaggerated and that the costs of implementing the controls may have been too high. On the other hand, increasing public concern over environmental pollution will likely continue and lead to a push for even tighter controls on atmospheric emissions.

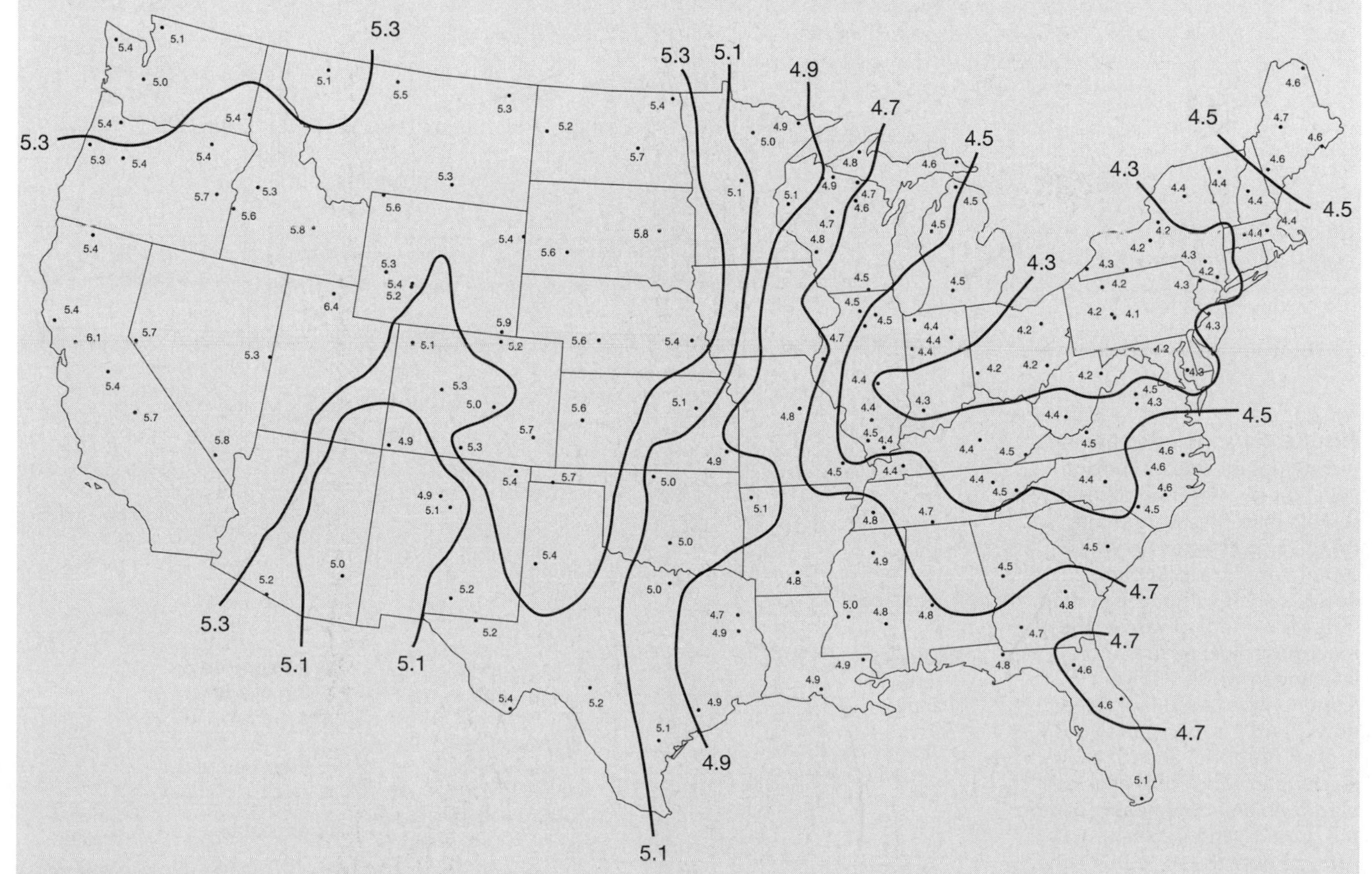

FIGURE 4.B The average pH of rainfall in the United States in recent years measured shows that the lowest values are concentrated in the Ohio River Valley. (Map from U.S. Environmental Protection Agency.)

FIGURE 4.9 In nineteenth century coal mines, a miner called a "penitent" was wrapped in water-soaked rags and carried along a candle-tipped stick to try to burn off pockets of methane gas so that the other miners could safely enter the work area. (Original wood engraving from Simonin 1869; from U.S. Geological Survey Circular 1115.)

Environmental Impact of Mining and Quarrying

Most of the impacts arising from mining and quarrying are obvious. They include the disruption of land otherwise suitable for agricultural, urban, or recreational use; the deterioration of the immediate environment through noise and airborne dust; and the creation of an environment that is among the most dangerous for its workers and is potentially also hazardous for the public. However, mining is a relatively short-term activity, and much can be done to both limit environmental damage during mining and to

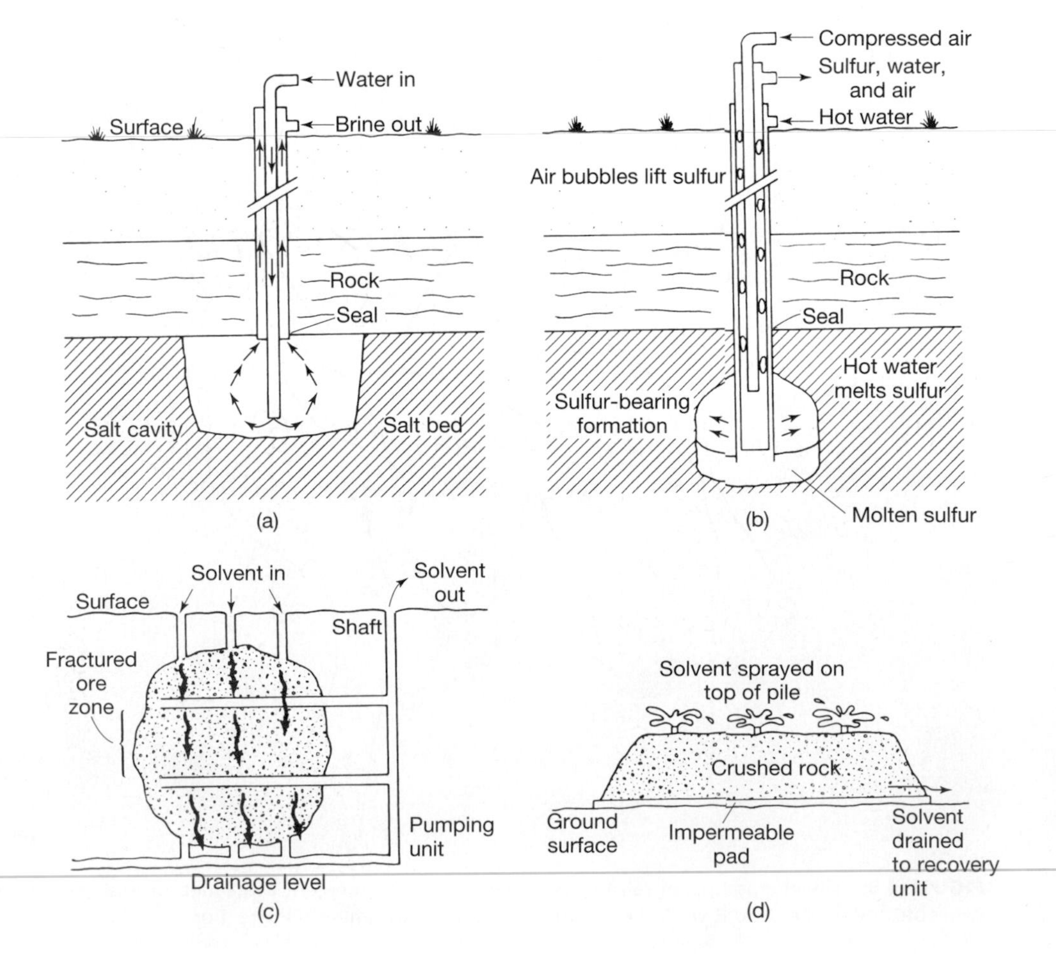

FIGURE 4.10 Solution mining techniques include: (a) bottom injection systems used to dissolve soluble materials such as salt; (b) the Frasch system in which superheated water is pumped down to melt sulfur; the molten sulfur is carried upward in the innermost pipe; (c) *in situ* ore leaching in which a solvent (commonly an acid) is drained downward through a previously broken ore zone; and (d) heap leaching in which broken rock is placed on an impermeable pad on the ground, and a solvent is sprayed over the rock; this is the method used today for most gold recovery.

restore the land when mining operations are complete. Today, in many countries, legislation has been enacted to ensure that these steps are taken. Of course, mining companies claim that laws that are too restrictive will make mining completely uneconomic and limit the availability of resources and jobs. Unfortunately, the absence of adequate controls over some mining activities in the past has left numerous scars on Earth's surface, and has led to a resistance among many members of the public toward new mining activities of any kind in their localities.

Fortunately, many underground mines leave little evidence of their presence even after the mining operations have ceased. The openings usually slowly fill up with percolating groundwaters, and the rocks are commonly strong enough to stand solidly despite the abandoned mine openings and passageways below ground. Sometimes, the old mines can even be put to very good use. Examples include the world's largest archive of grain seeds that have been placed in an old coal mine in Spitsbergen, Norway; the burial of nuclear wastes in abandoned salt mines in Germany; and the development of an extensive office and shipping complex in an old limestone mine in Kansas City, Missouri (Figure 4.11). In each of these examples, the old mines are watertight, and the air temperature and humidity are nearly constant.

When an open-pit mine closes, a large hole remains without any readily available waste rocks to fill it, except at very great cost. The pit slopes are often very steep and cannot be reclaimed by soil coverage and planting. If the groundwater table is high enough, the bottom of the pit may be flooded, creating an artificial lake. Furthermore, pits in sulfide-rich rocks may contain water that is extremely acidic as has developed at Butte, Montana (see Plate 18). The very large open-pit mines are therefore difficult, if not impossible, to reclaim. Smaller open-pit mines and quarries, in contrast, can often be filled with waste rock or, if the geologic and groundwater conditions permit, can be used for the disposal of refuse or other waste materials. Gravel pits or rock quarries can sometimes be filled with water and become recreational lakes.

Strip mining (see Figure 4.12A) can leave mounds of waste material too steep to build on or farm, and also without vegetation, and therefore easily eroded. Consequently, many thousands of acres of land in coal mining areas of Kentucky, West Virginia, Illinois, and other parts of the United States have been devastated in this way. Nevertheless, reclamation as a part of strip mining can be straightforward; nearly all modern mining operations are required to include this as a normal final stage of extraction. As shown in Figure 4.12B, waste mounds can be smoothed out, previously removed topsoil that has been stored can be returned, and ground covering plants, such as clover, planted until the soil is sufficiently restored to allow other crops to flourish. In some of the most mountainous areas of West Virginia, the horizontal benches left after reclaiming strip mines prove useful because they provide the only flat land available and serve as excellent mead-

FIGURE 4.11 The underground openings left after a mining operation has ceased can sometimes be put to good use. This former limestone operation in Kansas City, Missouri, used a room and pillar method and has been converted into a large underground storage area. Other parts of the same mine contain a trucking operation, an office complex, and a post office. The addresses are appropriately known as Underground Drive. (Courtesy of Hunt Midwest Real Estate Development, Inc.)

(a)

(b)

FIGURE 4.12 (a) An operating coal strip mine in eastern West Virginia. In the foreground, the overburden has been removed, exposing the coal bed (on which the front-end loader is sitting). Once the coal has been removed (as in the background where the pickup trucks are parked), reclamation will begin. The cliff at the right, known as the "high wall," reveals how much overburden had to be removed to excavate the coal bed. (Photograph by J. R. Craig.) (b) Proper reclamation procedures can restore previously mined areas to productive and attractive landscapes. (Photograph courtesy of Lee Daniels.)

ows for cattle and wild deer. Surface mines for sand and gravel, phosphates, and titanium have been reclaimed to form small lakes and wetlands that support fish, birds, and other wildlife (Figure 4.13). A relatively newer type of strip mining, called **mountaintop mining,** has recently become the focus of considerable controversy. This method allows mining companies to remove steep mountaintops to expose and mine coal seams and allows them to dump the waste rock into narrow natural stream valleys. It leaves flat tops on the mountains and raises the levels of the stream beds in the valleys. The deforestation of the areas leads to increased runoff and erosion and the choked valleys are prone to flooding during intense rain storms. This has led to numerous lawsuits and angry protests by those affected. Mining companies argue that this is the only economical manner to extract the coal, while local and downstream residents complain that the environmental degradation is too great a price to pay.

FIGURE 4.13 Reclaimed wetlands in the area of a former phosphate mine in Florida are now a productive wildlife habitat. (Photograph courtesy of Florida Institute of Phosphate Research.)

Underground mining does not usually lead to such drastic disruptions of the surface as do open-pit, strip, and mountaintop mining, but a new hazard encountered is that of **subsidence.** This problem occurs most often where underground mining has been shallow, and where the rocks are naturally weak or highly fractured. It is commonly seen where mining has been undertaken in soft sedimentary rocks, as in many coal mines, and where a method such as the "room and pillar" (see Figure 4.7d) was used. More than 8000 square kilometers (km^2) of land in the United States, for example, has subsided owing to underground coal mining, and many more areas are threatened. Subsidence of farmland or rangeland into old coal mines (Figure 4.14a) is a problem but, generally, has little impact on people's lives. Conversely, subsidence under towns can leave homes uninhabitable and under highways can severely disrupt transportation (see Plate 10, and Figure 4.14c). In March 1995, a coal mine that was last mined in the 1930s collapsed under a 600-meter (1800-foot) stretch of Interstate Highway 70 in eastern Ohio, closing the road for more than three months. Thus, a severe impact of the coal mining appeared sixty years after the mining activity ceased. Solution mining and the pumping of brine or water can also cause local subsidence. Subsidence is usually gradual and causes cracks, surface troughs, depressions, or bulges, but sometimes it is sudden and results in the destruction of houses and other buildings, roads, and farm areas.

An extreme example of this occurred at a zinc mine at Friedensville, Pennsylvania, in 1968. Gradual subsidence, with occasional minor episodic movement, had begun to be noticed in 1964. Careful monitoring, including the installation of acoustic **seismographs** (sounds of rock movements were often discernible to the unaided ear), allowed the mine geologist to define and isolate the problem area. At 10:41 a.m. on March 27, 1968, a block of rock 225 meters long by 115 meters wide and over 180 meters thick dropped catastrophically. The energy released by this 11 million metric ton block was equivalent to an earthquake of magnitude 3 on the **Richter Scale** and was recorded on several seismographs in the area. A vacant house and a portion of the state highway were destroyed, but there were no injuries or lost work time (Figure 4.14b).

A similar event occurred in March 1994 at the largest U.S. salt mine located in Retsov in western New York, when a collapse occurred 300 meters (1000 feet) below the surface and was recorded as a 3.5 magnitude earthquake. No immediate damage was done to buildings and no one was injured, but the fractures extended to the surface and the mine began to fill with groundwater at a rate of 76,000 liters (20,000 gallons) per minute. There was immediate concern over local draining of wells as the water table dropped and long-term concern that saltwater resulting from solution of the salt beds could contaminate groundwater supplies.

In addition to the impact that mining activities may have on the landscape, the environment can be disrupted over a wider area by changes in the distribution and chemistry of surface or groundwaters. An example of this is the **acid mine drainage** that is produced when the iron sulfide minerals, pyrite, and marcasite (both forms of FeS_2), or pyrrhotite ($Fe_{1-x}S$), are exposed to oxidation by moist air to form sulfuric acid, plus various other sulfate compounds and iron oxides. Pyrite and marcasite occur as minor minerals in many coals and are present with pyrrhotite in many

(a)

(b)

(c)

FIGURE 4.14 (a) The effects of subsidence in an old coal mining area are evident near Sheridan, Wyoming. Mining in the 1920s used a room and pillar method. Subsequently, there has been collapse of the overlying rocks into many of the rooms. (From C.R. Dunrud, U.S. Geological Survey Professional Paper 1164, 1980.) (b) A sudden collapse in the Friedensville Mine in Pennsylvania resulted in a small earthquake and severely damaged the overlying state highway. (Photograph courtesy of R.W. Metzger.) (c) Subsidence of the land surface into an abandoned anthracite mine in Scranton, Pennsylvania, has resulted in damage to this house, which has tilted on its foundation. (Photograph courtesy of the U.S. Bureau of Mines.)

metallic mineral deposits. The generation of sulfuric acid can occur when these minerals are exposed to air in underground mines or in open pits, or in the dumps of waste materials left by the mining operations. Water passing through the mines or dumps becomes acidified and then can find its way into rivers and streams or into the groundwater system. The U.S. Bureau of Mines estimated that by 1993, there were more than 60 billion tons of mine waste in the United States, a figure that increases by 3 billion to 4 billion tons annually. It has also been estimated that up to 10,000 miles of streams have been affected in this way in the United States alone, largely because of the impact of abandoned mine workings. The result is a major pollution problem giving rise to barren soils and rivers and streams devoid of living things.

Butte, Montana, in the United States has been called *the richest hill on earth* because of deposits of copper, lead, zinc, silver, gold, molybdenum, and several other metals; Butte has been the site of intense mining since the mid-1800s. However, mining has left large scars of abandoned open pits, large waste piles, and contaminated streams. The Berkeley Pit, shown in Plate 18, is now abandoned but is filling with groundwater. It will ultimately become one of Montana's largest lakes; unfortunately, the water contains 8000 parts per million (ppm) dissolved solids, is very acidic with a pH of 2.7, and is toxic to waterbirds unwary enough to land there.

Although not so widespread as the problem of acid drainage, other pollutants can also enter streams or groundwaters from mines and mineral processing plants. These include arsenic and various compounds of heavy metals such as lead, cadmium, and mercury. The mining of uranium leads to particular problems that are discussed later in this chapter. Problems resulting directly from mining and subsequent pollution have led some communities, states, or even countries to restrict or prohibit some mining endeavors. Because developing countries tend to have fewer laws and restrictions, mining companies often concentrate exploration and development efforts in those countries. Despite the jobs and tax benefits for these countries, environmental conflicts often develop sooner or later.

Disposal of Mining Wastes

All mining operations generate waste rock, often in very large amounts. If the method being used is strip mining, the waste can be used in reclamation; but for underground mining operations, and for most kinds of open-pit mining, an alternative method of disposal has to be found. Usually, this simply involves dumping the wastes in piles at the surface next to the mine workings. More rarely, the waste rock can be put back into the openings created by mining (**backfill**). However, the cost involved in doing this is commonly prohibitive, and backfilling may seriously restrict the development of the mine by making large areas no longer accessible.

Piles of waste rock are often unsightly, do not readily support plant growth, and may be dangerous. Commonly, these wastes have been through a process of crushing associated with the separation out of the coal, metalliferous, or industrial minerals being exploited. Crushing increases the volume of the rock by as much as 40 percent and produces a material that may be unstable when placed in steep piles. A tragic illustration of this occurred in 1966 in the Welsh mining village of Aberfan. There, a 400-foot-high pile of rock, which had been deposited on a mountainside during nearly a century of coal mining, slid down and engulfed many houses and a school, killing 144 people. This avalanche was estimated to have been made up of nearly 2 million tons of coal-mining waste.

Alternatives to the simple dumping of mining wastes, such as using the wastes for landfill, are likely to be expensive and impractical in most cases. However, waste dumps can certainly be made safe and often can be reclaimed as recreational or agricultural land. Such reclamation may involve lowering the slopes on dumps and encouraging vegetation growth through **hydro-mulching** or **hydro-seeding.** These processes consist of spraying the dumps with a pulp or mulch of organic material such as bark or hay mixed with a binding substance and with the seeds. The organic substance provides some bed material for germination of the seeds. Once this happens, the roots will hold the mulch in place and form a protective layer of vegetation to minimize water erosion. The more gentle the slope of the dump, the more likely that revegetation will be successful.

Numerous examples exist of successful reclamation of mining wastes, ranging from the planing down of the large piles of tailings outside Johannesburg, South Africa, to the seeding of the white quartz sand waste tips surrounding the china clay pits of Cornwall, England (see Figure 10.19).

Dredging and Ocean Mining: Methods and Environmental Impact

Dredging involves removal of unconsolidated material from the bottoms of rivers, streams, lakes, and shallow seas using machines such as the bucket-ladder dredge, dragline dredge, or suction dredge (see Figure 4.15). These methods are used extensively for the recovery of sand and gravel, and for minerals such as the oxide ore of tin (cassiterite), gold, and diamonds. The largest and most advanced dredgers are those employed in offshore diamond mining in Namibia and the northwest coast of South Africa, and in the tin fields of Southeast Asia, some of which can

FIGURE 4.15 A small dredge operating in Alaska recovers placer gold from river gravels. The bucket line at the left side cuts into the sediments and carries them into the dredge. The coarse rocks are dumped off a conveyer belt projecting out on the right side. Gold particles are recovered by sluices and jigs within the dredge, and the fine sediments are pumped back into the river. The dredge moves slowly by digging on one end of its own small lake and dumping waste rock on the other end. (Photograph by Ernie Wolford and Joe Fisher; courtesy of the Alaska Division of Mining and Geological Survey.)

handle up to 5 million cubic meters (m^3) per annum and recover material in water as deep as 45 meters (150 feet). In Sierra Leone and Ghana, large-scale diamond mining operations are undertaken using draglines, well suited to the swampy ground. There is no chemical pollution from dredging, but the process completely alters the bed of the river or seafloor where the dredge operates, and it disperses large quantities of fine sand and silt that can have damaging effects on fish and other aquatic wildlife.

Salts dissolved in seawater have long been a source of sodium, magnesium, and bromine. However, with the depletion of more conventional sources of other minerals, increasing attention is being paid to the mineral potential of the sea floor. Metal-rich sediments are known to occur in the Red Sea and along certain of the mid-ocean ridges. Interest in the past centered on **manganese nodules** (see Figure 7.16), pea- to cobble-sized, roughly spherical, nodules that cover large areas of the deep ocean floor. As well as large amounts of manganese and iron, they contain substantial nickel, copper, cobalt, and, in lesser amounts, a wide range of other elements. In fact, the total quantities of such metals as copper, nickel, and cobalt in manganese nodules probably exceed the totals in all known land deposits of these metals. They are very attractive as a resource, but mining the nodules poses technical, legal, and potential environmental problems. Most nodules lie at water depths of 4000 to 5000 meters, and various dredging or siphoning methods have been proposed for their recovery. Although the technical problems can probably be overcome (indeed, some systems have already been tested and proved feasible), rights to mine nodules and the distribution of profits gained in their mining are both matters of continuing international disagreement. This is so because the nodules largely lie in international waters, and all nations can claim some right to share their ownership.

Both dredging and any form of ocean mining involve significant disruption of the natural systems of rivers, lakes, shores, seas, or oceans, and this may result in the destruction of biological systems. There may also be long-term effects on river and ocean currents, sedimentation patterns, and patterns of erosion. The severity of the environmental impact of nodule mining is not known, but there are concerns that the disruption and dispersal of the sediment could harm delicate and little known deep-sea fauna.

There has been increasing interest in the past few years in the seafloor mining of massive sulfide deposits that occur along some of the major seafloor plate boundaries. This is especially true of the Woodlark Basin in the Western Pacific ocean where the ore grades are rich and the deposits large. Presently, the mining technology does not exist for large-scale excavation, but it will certainly be developed in the next few years. As with the mining of manganese nodules, there will likely be problems of ownership and concerns about the environmental effects on the deep ocean sea life.

Well Drilling and Production: Environmental Impact

The drilling of wells is a method of exploiting liquid and gaseous resources that dates back many centuries, even being mentioned in 1500-year-old Chinese manuscripts as a means of tapping underground strata for brine. Modern drilling methods are used in the exploitation of water and geothermal energy, and in the exploration and production of petroleum and natural gas. Apart from the relatively minor disruption of the environment caused in the actual drilling of water or geothermal wells, there is little danger of damage arising from these activities. The greatest impact is usually not the actual drill site, but the construction of the access roads built to bring in the equipment. Drilling for oil and gas, in contrast, does involve certain risks, although the disruption to the environment is generally much less than in major mining operations.

The technology involved in modern (rotary) drilling to locate and exploit petroleum and gas is discussed in detail in Chapter 5. The greatest hazard that might be encountered during drilling for petroleum is a **blowout**, which occurs when a high-pressure oil or gas accumulation is unexpectedly encountered, and the column of dense drilling fluid in the hole fails to contain the oil or gas that erupts from the wellhead (Figure 4.16). The fire hazard is great, and severe pollution of the surrounding area can occur very rapidly. Today, blowouts are quite rare because of improved equipment and monitoring of the drilling process. Thus, symptoms such as an increase in drilling rate accompanied by an increase in return flow of the drilling mud indicate that fluid from the hydrocarbon reservoir is entering the well bore. Valves at the surface, known as "blowout preventers," can then be closed, and corrective measures, such as increasing the density of the drilling mud, can be taken to enable drilling to continue. Despite these safety measures, major problems can still occur, as they did at the Ixtoc Number 1 well that blew out and caught fire in the Gulf of Mexico near the Yucatan Peninsula in 1979 (Figure 4.17). It spewed oil for more than nine months with a loss of oil estimated at more than 175 million gallons (4.2 million barrels).

FIGURE 4.16 A blowout such as that shown here is one of the greatest hazards that can be encountered while drilling for oil. Extremely high pressures have forced the drill rods out of the hole and destroyed the drill tower. (Courtesy of Shell International Petroleum Company Ltd.)

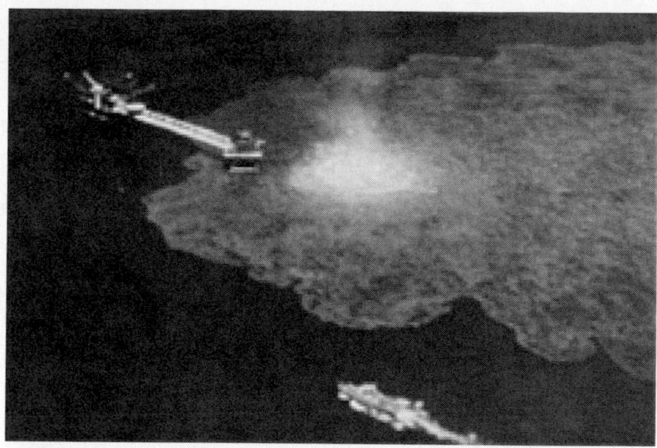

FIGURE 4.17 The blowout and subsequent fire that occurred at the Ixtoc no. 1 well in the Gulf of Mexico released more than 140 million gallons of oil over a six-month timespan until it could be capped. Most of the oil that escaped either evaporated or was burned, but many kilometers of beaches in northern Mexico and southern Texas were fouled by oil. This is one of the worst accidental oil spills in history. (From the Office of Response and Restoration, NOAA.)

Clearly, the dangers to human life and to the environment are greater in the offshore oil fields that have been so extensively exploited in recent decades. Nevertheless, major accidents have been relatively rare, although risks must increase as fields are developed in deeper offshore waters and even less hospitable seas. Some spillage of oil into the seas appears inevitable at such sites, and, as at all operations, care must be taken to prevent oil and brine seepage at the wellhead.

Although there have been many significant oil spills from tankers (see Table 4.1), one of the largest and most publicized was that of the *Exxon Valdez,* which ran aground in Prince William Sound, Alaska, on March 24, 1989 (Figure 4.18a). For about two months, more than 11 million gallons (260,000 barrels) of oil spread along nearly 800 kilometers (500 miles) of coastline. Besides the wastage of oil, an estimated 400,000 seabirds (including 900 bald eagles) and 4000 sea otters died as a result of oil on feathers and fur. Large numbers of workers cleaned beaches (Figure 4.18b) and skimmed up oil for months; some $100 million was spent on research, and more than $1 billion was paid in fines, but we shall not know the full ef-

TABLE 4.1 Largest Oil Spills

Name—Place	Date	Cause	Barrels Spilled
Ixtoc 1 well, Gulf of Mexico	June 3, 1979	Blowout	4,200,000
Nowruz field, Persian Gulf	Feb. 1983	Blowout	4,200,000
Persian Gulf War	Jan.–Aug. 1991	War act by Iraq	3,100,000
Atlantic Empress + Agean Captain, off Trinidad	July 19, 1979	Collision	2,100,000
Castillo de Bellver, South Africa	Aug. 6, 1983	Fire	1,750,000
Amoco Cadiz, France	Mar. 16, 1978	Grounding	1,500,000
Torry Canyon, England	Mar. 18, 1967	Grounding	833,000
Sea Star, Gulf of Oman	Dec. 19, 1972	Collision	805,000
Urquiola, Spain	May 12, 1976	Grounding	700,000
Exxon Valdez, Alaska	Mar. 24, 1989	Grounding	250,000

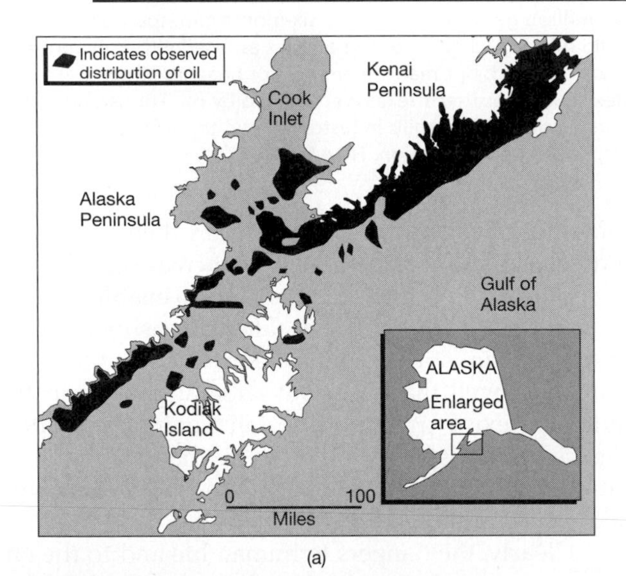

(a)

(b)

FIGURE 4.18 (a) The oil spilled when the *Exxon Valdez* ran aground in Prince William Sound in Alaska exceeded 11 million gallons (260,000 barrels) and spread along nearly 800 kilometers (500 miles) of coastline. Litigation continued for years, with an ultimate cost in the billions of dollars. (b) High-pressure hot water was used to clean the oil from many kilometers of rocky beaches in the belief that removal of the oil would hasten recovery of the natural flora and fauna. Subsequently, it has been realized that some of this cleaning sterilized the beach and probably slowed the rate of recovery. (From the Office of Response and Restoration, NOAA.)

fects for years to come. The National Oceanographic and Atmospheric Administration (NOAA) estimated that 50 percent of the oil spilled was biodegraded, 20 percent evaporated, 14 percent was recovered, 12 percent is at the bottom of the Gulf of Alaska, 3 percent unknown, and 1 percent still drifts in the seawater. Reevaluation of the spill area after ten and twenty years revealed that much of the marine life had been restored to pre-spill levels but that large quantities of heavy oil still lie in the beach sands and cobbles. It is apparent that it will take decades for the oil to completely decompose before the effects of the spill are completely gone. Much of the blame for the *Exxon Valdez* spill, and numerous other spills, has been attributed to single-hull ship design. The accident accelerated the changes in oil tanker design from the single to a double hull, in which there is a ten-foot open space between the two hulls (see Figure 5.42). Hopefully, in this design, any accidental puncture is confined to the outer hull without permitting leakage from the inner hull holding the petroleum. By 2009, double hulled ships accounted for more than 80 percent of oil transport, but between 2001 and 2008, two-thirds of the significant spills resulting from tanker hull ruptures came from single-hulled ships.

All wars cause environmental devastation, but the Gulf War of 1990–1991 (see Box 5.1) caused particularly severe mineral resource related environmental damage. The retreating Iraqi army blew up more than 700 oil wells, with the result that as much as 1 billion barrels of crude oil spread out over the Kuwait desert and into the Persian Gulf, and another 1 billion barrels burned from pools and uncontrolled gushers for several months. The effects on the desert sands were relatively slight, but the resulting groundwater contamination and the long-term effects on the aquatic life of the Persian Gulf will not be fully known for many years. In addition, the plumes of smoke from the burning oil wells blackened the skies and extended for hundreds of kilometers downwind, creating health and agricultural problems.

Processing and Smelting of Ores

The ores of all metals (see Chapters 7 and 8) and most of the industrial minerals (see Chapter 10) require some degree of processing. Such processing is usually done at or near the sites of mining in order to reduce the volume and weight of material that is shipped for metal extraction.

The percentage of metal in assayed mined ore ranges from about 30 percent in aluminum ores, to 55 percent in iron ores, and down to as little as 0.0001 percent in the case of gold (see Figure 4.19 and Table 4.2. Figure 4.19 has been updated with 2009 values). Even in copper mines, the percentage of metal in the raw ore as mined in many currently

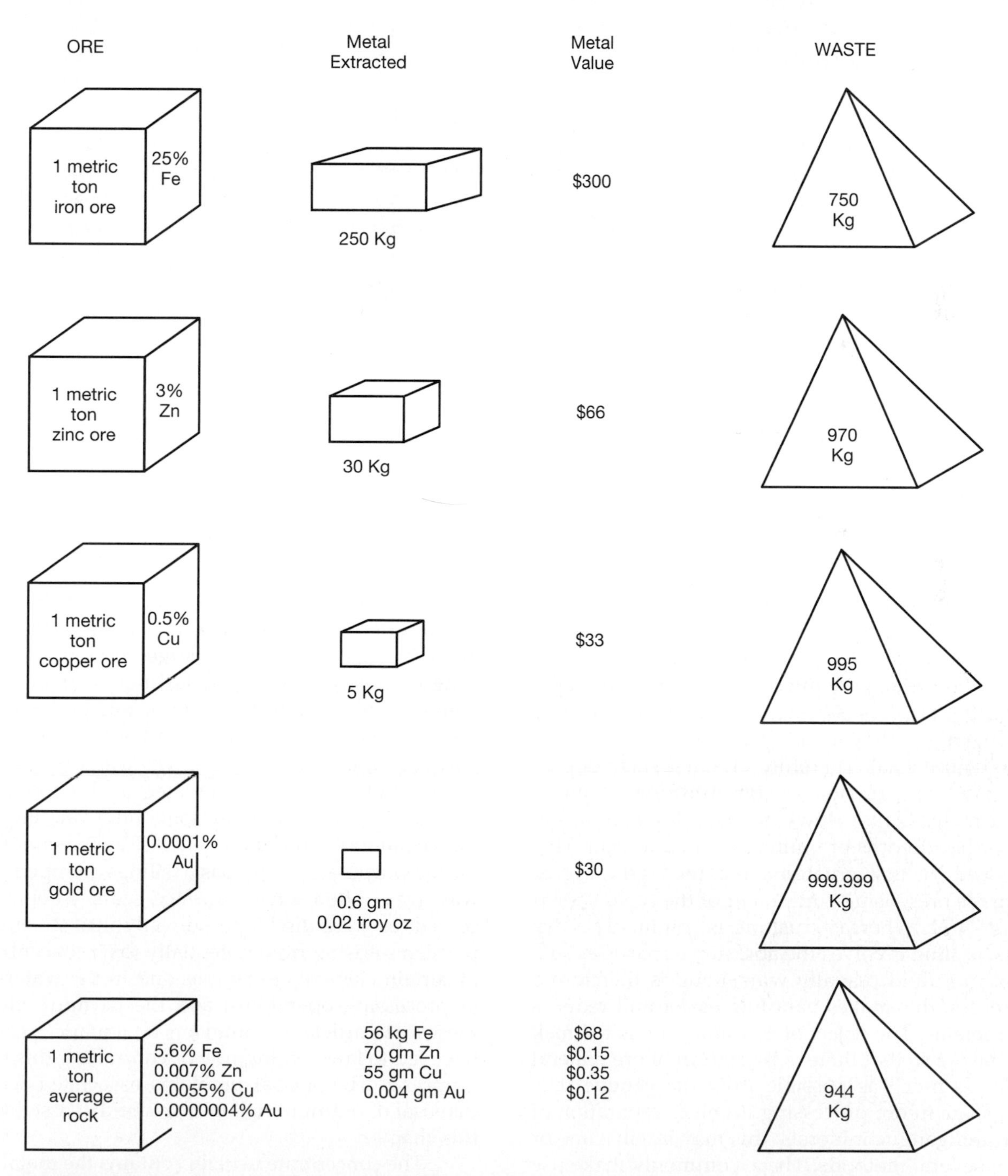

FIGURE 4.19 The amounts of metal extracted from a metric ton of typical ore ranges from as much as 250 kilograms for iron ores to less than1 gram for gold ores. The amounts of waste left for disposal are greater than the quantities of metals extracted; only occasionally is there a local market to make use of the waste rock. The approximate amounts of the same metals that could be extracted from average crustal rocks are shown for comparison.

TABLE 4.2 Concentrations of metals in Earth's crust, their minimum grades to be mined, and the degree of natural concentration required for exploitation

Metal	Crustal Abundance (%)	Approximate Minimum Grade to Be Mined (%)	Approximate Degree of Concentration Needed to Be Mined
Aluminum	8.2	30	4×
Iron	5.6	55	11
Titanium	0.57	1.5	25
Manganese	0.095	25	260
Vanadium	0.0135	0.5	35
Chromium	0.010	40	4000
Nickel	0.0075	1.0	130
Zinc	0.0070	2.5	350
Copper	0.0055	0.5	90
Cobalt*	0.0025	0.2	80
Lead	0.00125	3	2400
Uranium	0.00027	0.1	400
Tin	0.00020	0.5	2500
Molybdenum*	0.00015	0.1	660
Tungsten	0.00015	0.3	2000
Mercury	0.000008	0.1	12,500
Silver*	0.000008	0.005	625
Platinum	0.0000005	0.0002	400
Gold*	0.0000004	0.0001	250

*Much of these metals is recovered as by-products of the mining for other metals.

worked deposits averages only 0.5 percent or even less. In most ores, the metal-bearing minerals (commonly oxides, sulfides, arsenides, or more rarely alloys of the metals) are intergrown on a microscopic scale with valueless silicate or carbonate minerals (termed **gangue**) that act as "hosts" to the ore minerals (see Figure 4.20). Therefore, the first stage of processing is a size reduction (or **comminution**) of blocks up to a meter across down to particles only a few tenths or hundredths of millimeters in diameter. This is achieved by first **crushing** and then grinding or **milling** the ores, using equipment of the types shown in Figure 4.21. Whereas crushing is commonly a dry process, milling involves the abrasion of particles suspended in a fluid (usually water) and is therefore a wet process; this makes handling easier and reduces dust problems. The object of comminution is to break down the ore so that there is **liberation** of ore mineral particles as much as possible from the gangue. The second stage in ore processing involves separation of the ore and gangue minerals; this may involve one or more of several methods. These commonly make use of differences in density, magnetic, electrical, or surface properties between the ore and gangue minerals (see Table 4.3).

The end-product of such mineral processing operations (termed **beneficiation**) is a **concentrate** of the ore minerals, and a much larger quantity of waste gangue minerals known as **tailings**. Commonly, the tailings, which are in the form of fine-grained slurry, are dumped into an artificial pond or lake, and the fine particles allowed to settle out. The water may be recirculated as the quantities of water used in mining and beneficiation operations are sometimes very large, and laws commonly forbid the reuse of mine water for domestic purposes. In the past, tailings dumped in this way were often left as surface scars when mining ceased, or were discharged directly into streams. The problems arising from potentially toxic concentrations of certain elements in tailings and in the waters used in processing operations, and the harmful effects of these fine particle contaminants on aquatic organisms, have led to laws prohibiting or controlling this type of dumping. The special problems associated with the disposal of uranium mill tailings are discussed later in this chapter.

The concentrate usually contains the metal in the form of oxides, sulfides, or related compounds, and the traditional method of recovery of the pure metal is by **smelting**. The most familiar smelting process is the

(a)

(b)

FIGURE 4.20 These photomicrographs of polished surfaces of ore samples from Japanese volcanogenic ores are only 0.6 mm across and illustrate the fine-grained nature and intimately intergrown textures of ore minerals. To separate the various types of minerals, the ores must be crushed fine enough to free, or liberate, the individual grains. (a) Crystals of pyrite (FeS_2) in a matrix of chalcopyrite ($CuFeS_2$). (b) An intimate mixture of galena (PbS) and sphalerite (ZnS) with pyrite. (Photographs by J. R. Craig.)

(a)

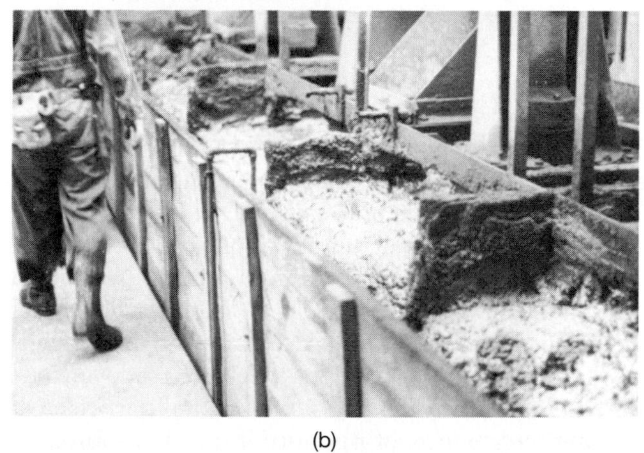

(b)

FIGURE 4.21 (a) The fine grinding required to liberate the ore minerals is often accomplished by large ball mills, such as these, which are approximately 4 meters in diameter. Coarse fragments of ore and steel balls 3 to 10 cm in diameter are fed into the mill, which revolves rapidly. The tumbling action of the balls and the ore grinds the ore to a fine powder. Water is added to prevent generation of dust and to allow the powdered ore to flow out as a slurry. (Photograph courtesy of Cleveland Cliff, Inc.) (b) After grinding, a process of selective flotation is commonly used to separate the various types of ore minerals. The ore minerals attach themselves to small air bubbles pumped through the pulp. Addition of an organic substance to produce a stable froth enables the ore minerals to float off the tops of the cells at the right side of the photograph. Once separated, the ore minerals are taken to a smelter to extract the metals. (Photograph by J. R. Craig).

recovery of iron from its oxide ores, which is discussed in detail in Chapter 7. In summary, carbon monoxide resulting from the incomplete combustion of coke reacts with iron oxide at high temperatures to form metallic iron and carbon dioxide.

Many other metals are produced by reducing oxides present in the ores or formed by **roasting** the ore prior to smelting. Roasting is a process involving heating in air without melting, so as to transform sulfide minerals (also arsenides, antimonides, etc.) into oxides by driving off the sulfur as gaseous sulfur oxides. Lead, zinc, copper, and nickel are examples of metals usually found as sulfides that are roasted before reduction. Copper (and nickel) sulfide ores, instead of being roasted and smelted directly, are often smelted to a **matte** that is a mixture of copper (or nickel) and iron sulfides and then *converted*. This is a process in which air is blown into the molten matte, oxidizing the sulfur to sulfur dioxide and changing the iron to an oxide that combines with a silica flux to form a slag, leaving the copper as impure metallic copper. For metals of a low boiling point, such as zinc, cadmium, and mercury, distillation as a vapor may be employed in the later stages of extraction.

TABLE 4.3 Mineral processing: Methods of mineral separation

Mineral Property Exploited	Method	Applicable to
High density (S.G.)	Mineral jig	Coarser-grained ores of lead (galena), barytes, etc.
	Shaking tables	Finer-grained ores of tin (cassiterite), gold, etc.
	Heavy media (liquid) separation	Preliminary separation of many denser ore minerals
Magnetism (chiefly ferro- or ferrimagnetism)	Magnetic separator	A small number of ferro- or ferrimagnetic minerals (magnetite, pyrrhotite, etc.)
		Some more weakly magnetic minerals (wolfram, ilmenite, etc.)
Electrical properties	High-tension separation	Dry particulate ores containing metallic conductors (e.g., ilmenite, cassiterite) and insulators (e.g., monazite)
Surface chemical properties	Froth flotation	A very wide range of metal sulfides and oxides as well as nonmetallic minerals

Smelting and other kinds of **pyrometallurgy** have been very significant sources of air pollution in the past because smelters emitted substantial amounts of gases such as sulfur dioxide and carbon dioxide, as well as particulate matter. Small quantities of toxic metals such as arsenic, lead, mercury, cadmium, nickel, beryllium, and vanadium were also released. Monitoring of the trace amounts of such metals in air, and in rainfall, shows that they travel over long distances and in considerable quantities, generally within fine particulates. The long-term effects of pollution from this source on human, animal, and plant life remain poorly understood. Many countries now have strict laws limiting releases of gases and metals. Unfortunately, inefficiency and lax laws have led to some severe cases of widespread contamination. In the late 1990s, studies of the Kola Peninsula region of northern Russia revealed nickel and copper contamination of thousands of square kilometers of land downwind from the major smelters that had processed ore for more than sixty years. A lack of funding available to either upgrade the smelters or to clean up the contamination means that it will remain unabated for years to come.

Gold recovery today is most commonly accomplished by dissolution of the gold using cyanide-bearing solutions in a method called **heap leaching**; this is an example of **hydrometallurgy** in which metals are dissolved from rocks (see Figure 4.10). The ore is extracted by conventional mining techniques (usually in open pits), crushed so that fragments are 1 to 3 centimeters in diameter, and piled onto large sheets of impermeable plastic. Then, dilute solutions of sodium cyanide are allowed to slowly percolate down through the ore to dissolve out the gold. The process might take several months, but it allows for the commercial treatment of ores containing as little as 0.02 troy ounce (0.6 ppm) of gold. Operated properly, gold dissolution using cyanide is very efficient and very safe. However, cyanide solutions are highly toxic and any accidental release into the environment can kill fish and birds. Responsible mining companies are very careful about cyanide use, but small releases have occasionally occurred. Probably the worst example, and one that has made the mining industry even more careful, happened at Summitville, Colorado. In 1984, Summitville became the first mine in Colorado permitted to use cyanide heap leach techniques to recover gold. Unfortunately, the leach pads were poorly designed and were constructed in winter against the advice of professionals; leaks were detected within six days and corrective actions were ordered. Nevertheless, mining and cyanide leaching continued until 1991, at which time the company went bankrupt, abandoning the mine and allowing large-scale cyanide release into the Alamosa River. The U.S. Environmental Protection Agency (EPA) has intervened to take over the water recovery system and to stop pollution, but has had to spend millions of dollars on remediation; it will take years to stop totally the cyanide leakage. The problems at Summitville are being analyzed carefully by the mining industry to avoid any recurrences.

The impact of pollution can be proportionately much larger for small countries, as seen in the 1995 accidental release of cyanide from the largest gold mine in Guyana, South America. Heavy rains resulted in the failure of a tailings dam and the release of relatively minor amounts of cyanide into the major river, which ran northward into the Caribbean Sea. The cyanide was sufficiently diluted that there were no measurable effects, but the news resulted in all other nations in the region refusing to buy what they feared would be contaminated seafood. The spill also led to the closure of the mining operation in order to evaluate the situation and to repair the dam; consequently, the mine stopped producing gold and paying taxes. As a result, Guyana lost its two largest sources of income and thousands were without jobs until the mine resumed operations and the fears of poisoned food subsided.

Cyanide contaminated runoff and cyanide-rich collection ponds at gold mines from the arid southwestern United States to South Carolina have sometimes served as unintentional traps for wildlife. The ponds, especially in areas where freshwater is scarce, attract wildlife that come to drink. The cyanide rarely kills immediately, but the effects build up until the birds and animals are sickened and ultimately die. Fencing can keep larger animals (e.g., deer) away, but birds and bats are harder to restrict, especially when ponds may be several acres in size. The Ridgeway mine in South Carolina employed cannon explosions to frighten away birds and even floated some large inflatable alligator toys. These helped with the birds but did little to discourage bats that would drink from the water at night.

ENVIRONMENTAL IMPACTS OF RESOURCE USAGE

Once resources have been extracted from Earth and processed, further disruption of the environment may be caused by their actual use. The principal example is the burning of fossil fuels in power stations, homes, and engines (particularly automobile engines) that result in emission of gases, particles, and, in some cases, excess heat into the environment. The use of nuclear fuels in power stations generates extremely toxic radioactive waste products requiring special means of disposal. The utilization of oil to manufacture a wide range of petrochemicals, and of metals and minerals to produce a wide range of industrial chemicals and products, also generates wastes and pollutants. Some of the effects of such resource utilization on the environment will now be discussed.

Burning of Fossil Fuels

The greatest problem caused by the burning of fossil fuels is air pollution. A good example of this is in urban areas in the United States and Europe (Figure 4.22 [a] a drawing showing the products of fossil fuel combustion; [b] diagram of the sources of the greenhouse

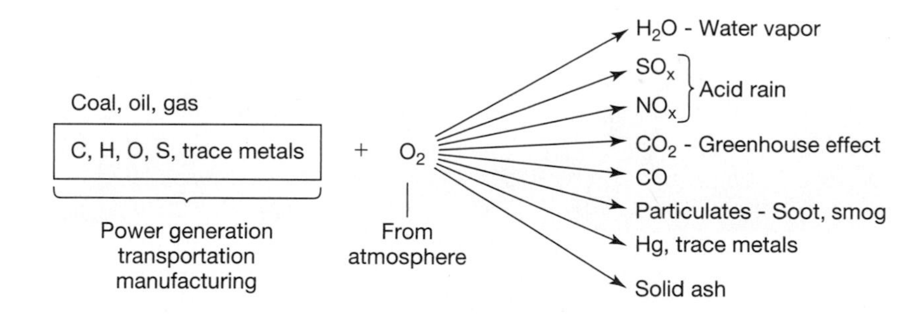

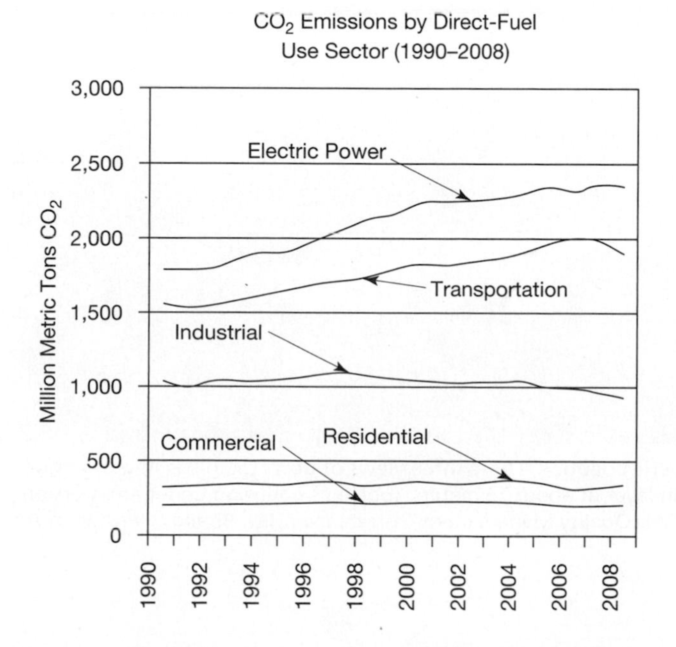

FIGURE 4.22 (a) Burning of fossil fuels releases carbon dioxide as the principal combustion byproduct. The other byproducts include water vapor, sulfur and nitrogen oxides, carbon monoxide, and particulate and solid ash. Coal is also the major source of mercury released into the atmosphere. (b) The major sources of greenhouse gases are electric power generation, transportation, and industry. Lesser amounts are released by residential, commercial, and agricultural activities. (Data from the U.S. EPA.)

FIGURE 4.23 The effects of atmospheric inversions on atmospheric pollution. These three views of downtown Los Angeles show: (top) a clear day; (middle) pollution trapped beneath an inversion layer at about 75 meters; (bottom) pollution under an inversion layer at about 450 meters. (Photographs courtesy of South Coast Air Quality Management District; from J.N. Birakos, "Profile of Air Pollution Control," County of Los Angeles, 1974.)

gases), where the pollutants come mainly from motor vehicles and from power plants, with substantial contributions from domestic and industrial heating systems. Industrial processes such as smelting can make significant contributions locally; the burning of solid wastes may add further to air pollution.

The main pollutants from these sources are carbon dioxide, carbon monoxide, hydrocarbon compounds, nitrogen oxides, sulfur oxides, and particulate matter. Complete combustion of fossil fuels yields mainly carbon dioxide (CO_2) and water vapor. These products are not really pollutants because they are already present in the atmosphere in significant amounts. Nevertheless, the burning of fossil fuels is leading to a steady increase in the CO_2 content of the atmosphere (see Figures 1.7 and 1.8) and consequent changes in other geochemical cycles. Furthermore, it is increasingly apparent that the rise in carbon dioxide is a principal cause of atmospheric warming (see Box 1.1).

Incomplete combustion, as is often seen from diesel trucks or buses, can release significant amounts of black particles or soot. Catalytic converters built into automobile exhaust systems are designed to convert the most harmful of gaseous pollutants into less noxious ones. Nevertheless, local and brief buildups of fumes, gases, and particulates in many metropolitan areas can create significant health problems. Undoubtedly, the worst example developed in London in 1952 when a severe smog lasting several days led to some 4000 more deaths than would have been expected at that time of year; the dangers were greatest for those already suffering from respiratory and cardiac diseases. That episode spurred enactment of clean air laws that have restricted the burning of certain types of fuels and have greatly reduced the problem in many cities. Figure 4.23 illustrates clearly how pollution can be trapped in the lower parts of the atmosphere by some air conditions.

The question of how the air pollution we create affects climate is more controversial. Dust particles reflect the Sun's rays and cause lower temperatures, but major volcanic eruptions introduce far more dust into the upper atmosphere than any human activity; the effects of even the largest such eruptions, which have certainly reduced temperatures, have generally been local and short-lived. Air pollution promotes fogginess, cloudiness, and even may affect rainfall on a local scale by providing nuclei for the condensation of drops of moisture; worldwide effects on precipitation are much more difficult to assess.

The burning of fossil fuels is certainly causing changes in the chemistry of the atmosphere. One such effect already mentioned is the worldwide increase in carbon dioxide (see Figure 1.8). Although carbon dioxide constitutes only about 0.03 percent of Earth's atmosphere (Figure 4.26), an increase in CO_2 is important because it, along with water and ozone, absorbs and therefore conserves heat given off by Earth. The trapping of heat in this way is called the **greenhouse effect** (see Box 1.1).

Just as the passage of an animal may be evidenced by the footprints it leaves on the ground, so the combustion of massive amounts of fossil fuels since the beginning of the Industrial Revolution is evidenced by the rising "footprint" of carbon dioxide in the Earth's atmosphere. The term "**carbon footprint**" found its way into environmental discussions by the time of the Kyoto Environmental Conference in the 1990s. It is now more-or-less defined as "the total amount of greenhouse gases produced to directly and indirectly support human activities, expressed in tons of carbon dioxide." Our direct footprint comes from the gasoline we burn in our cars, our home heating, and our travel in airplanes, buses, and trains (see Figure 4.24). Our indirect, or secondary, footprint comes from the CO_2

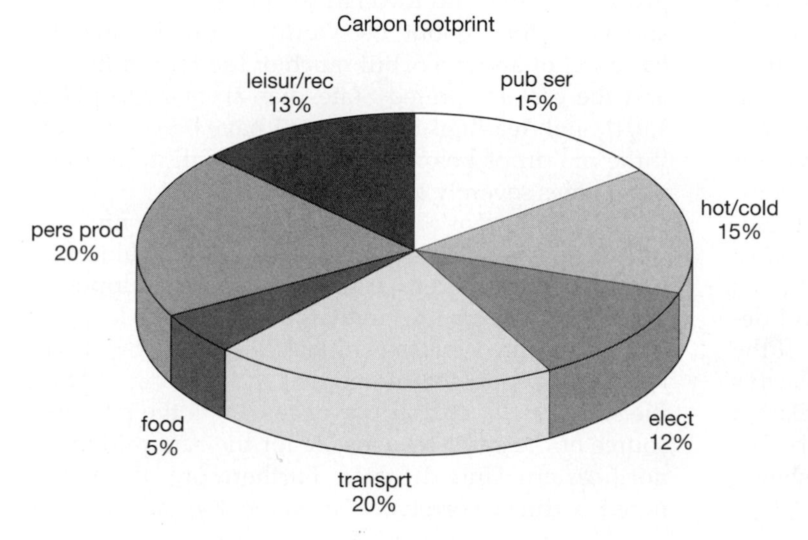

FIGURE 4.24 Contributions to the typical "carbon footprint" in terms of public service, heating/cooling, electricity, transport, food, personal products, leisure/recreation.

FIGURE 4.25 Very tall stacks such as this one, 370 meters (1216 feet) high, at the Homer City Electric Generating Station in Pennsylvania, are constructed so that waste gases and particles that escape entrapment in filters are released high enough in the atmosphere to allow for dilution and dispersal to minimize pollution problems. (Photograph courtesy of the Pennsylvania Electric Company, a member of the General Public Utilities System, and New York State Electric and Gas.)

emissions from the whole life cycle of the products we use—food, plastics, clothes, packaging. It is evident that citizens of more advanced and technological societies, especially those with vehicles that burn gasoline and use electricity from coal burning power plants, have much larger carbon footprints than do citizens of less developed countries. Because we now recognize the adverse effect of increasing CO_2 concentrations in the atmosphere, the challenge is to reduce the size of our carbon footprint by conservation or by a shift to non-CO_2 emitting sources of energy (e.g., solar, wind, nuclear).

There are other problems involving pollution of the upper atmosphere and particularly of the layer of ozone (O_3) that protects Earth from much of the Sun's ultraviolet radiation. The exhaust from high-flying jets damages the ozone layer and introduces particles, water, CO_2, and nitrogen oxides. Other sources of possible damage to the ozone layer are the compounds called chlorofluorocarbons and chlorofluoromethanes (or freons) such as $CFCl_3$ that are used as propellants in aerosol cans. When these compounds reach the upper atmosphere, they react with and destroy the ozone. Such changes in the chemistry of the upper atmosphere may cause an increase in Earth's surface temperature; increases in ultraviolet radiation can damage life forms and affect human health. The formation and the destruction of ozone is shown schematically in Figure 4.27.

During the combustion process of coal, trace quantities of mercury are released. It is now recognized that coal combustion in power plants has become Earth's greatest source of mercury pollution, annually distributing as much as 50 metric tons of mercury on a worldwide basis. In addition, the sulfur dioxide (SO_2) emitted when coal is burned (see Chapter 5), and also emitted by many smelters, is thought to be the cause of one of the most serious forms of pollution, **acid rain** (see Box 4.1). Sulfuric acid and nitric acid, produced by combinations of SO_2 and NO_x, respectively, with rainwater, has damaged plants and soils and lowered the **pH** in many streams and lakes throughout the world. Normal rainwater has a pH of about 5.6, but much of the rain in Europe and the eastern United States has an average pH of 4.0, though readings as low as 2.1 have been reported. If the pH drops below 5.0, fish usually disappear and plant life is severely affected.

The effect of the oxides of sulfur and nitrogen released by coal burning has been the subject of much debate, often from viewpoints supported primarily by opinions rather than proven facts. In 1983, the United States National Academy of Sciences released a report that noted that the burning of fossil fuels, especially coal in power plants, is the principal source of the gases responsible for the acid rain in the northeastern United States. Furthermore, the report noted a direct correlation between the sulfur oxide

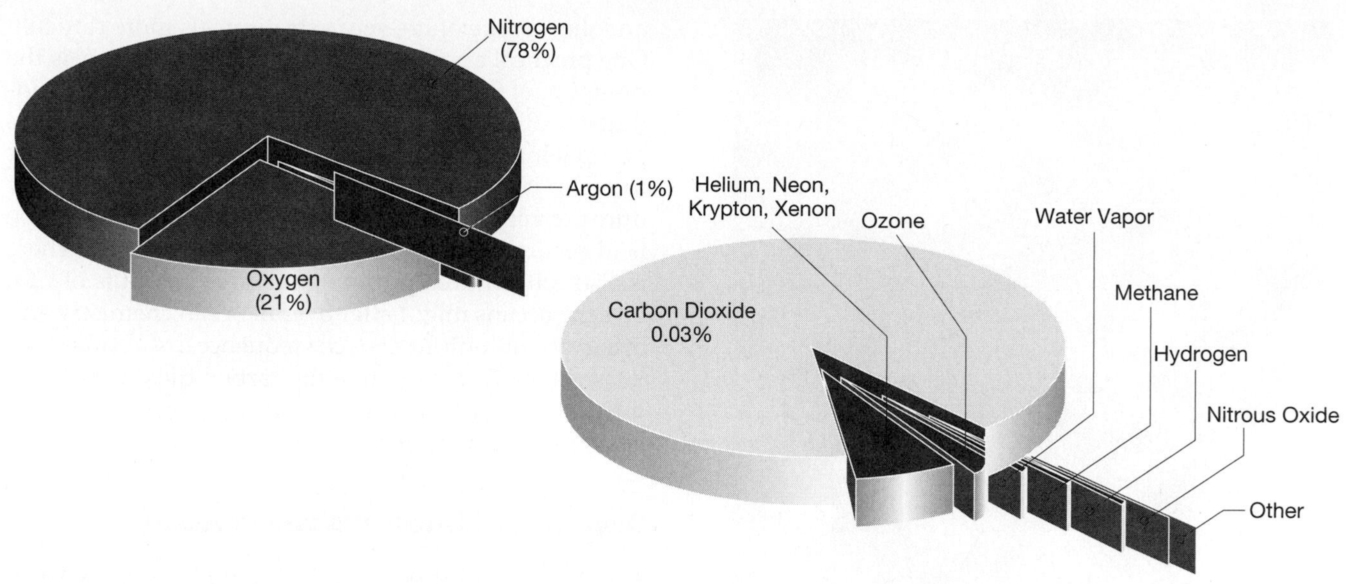

FIGURE 4.26 Composition of Earth's atmosphere.

content in acid rain and the sulfur content in the fuels. This report has been supported by numerous additional studies since that time. Reduction of the sulfur oxide emissions by half, the recommended target, can be achieved by only three methods: (1) a change to another type of fuel, such as from coal to oil, natural gas, or nuclear; (2) installation of expensive exhaust gas cleaner systems; and (3) a change to lower-sulfur coals. Critics of the National Academy's report point out, however, that there was as much coal burned in the United States and Europe in the 1950s as in the 1980s, but that there was much less acid rain in the 1950s. They suggest that either coal burning is not the culprit or that alkaline emissions that previously neutralized the acid are now being cleaned up too thoroughly. Clearly, the problem of acid rain is far from being completely understood or solved. Regardless of the debates, more than 15,000 lakes in Scandinavia and Canada have been damaged by acid rain, much of which is attributable to emissions in other countries. Acid rain also speeds the decay of buildings, sculptures, and other structures, particularly those made of **limestone** or **marble** (Figure 4.28).

The control of air pollution requires both legislation to limit emissions and the use of various devices to remove as many of the noxious substances as possible from exhaust gases and power plant or smelter emissions. Because polluted air may travel hundreds of miles, affecting areas far away from the pollution source, the difficulties of introducing adequate legal controls are compounded by the need for international agreements.

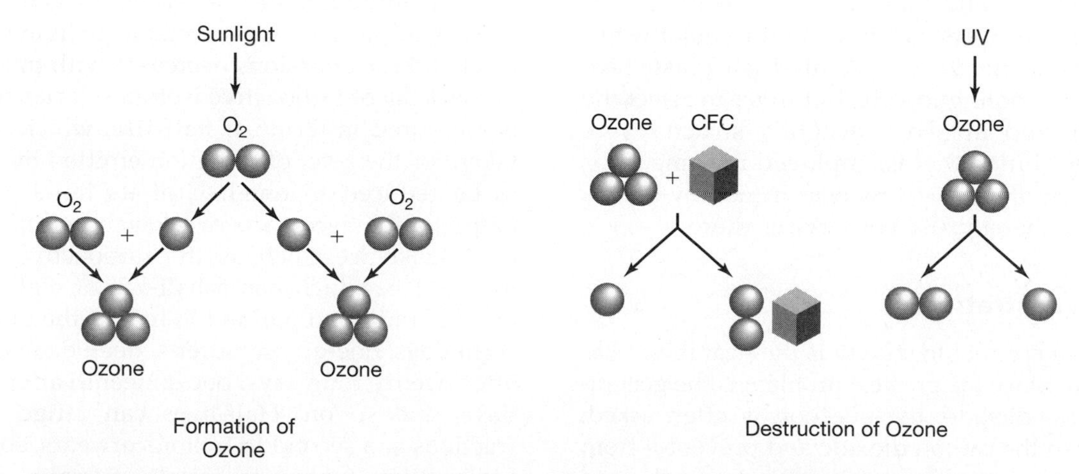

FIGURE 4.27 Ozone (O_3) is formed in Earth's atmosphere when sunlight breaks diatomic oxygen into single oxygen atoms. These then combine with other diatomic oxygen atoms. Ozone may be destroyed by ultraviolet (UV) radiation or by reacting with CFCs produced by humans.

FIGURE 4.28 The corrosive effects of acid rain and other atmospheric pollutants have eaten away part of this limestone ornamental feature on the Organization of American States Building in Washington, DC. (Photograph by E. McGee.)

A final and rather ironic form of pollution that results from power generation is **thermal pollution**. Commonly, power plants take water from rivers, lakes, and seacoasts, use it for cooling, and then return it at a higher temperature. This warmed water affects ecosystems and may cause the death of fish and other aquatic life accustomed to the cooler water conditions. One alternative is to dissipate waste heat into the air via cooling towers, but this can affect the local climate and involves unsightly structures. A much better solution that is employed in some areas is to make use of such excess heat in nearby homes and factories, thereby also conserving energy.

Carbon Sequestration

The great value of coal to society is the heat it releases when burned. Since the greatest problem is the generation of carbon dioxide, the question is often asked, "Can't we trap the carbon dioxide and prevent it from entering the atmosphere so that it will not contribute to global warming?" The answer is "yes", but at a very great cost. The capture of the carbon dioxide is difficult and the storage of it—**sequestration**—is more difficult. One practical application, already in limited use, is the pumping of CO_2 into old oil fields to help drive out additional oil. The two other principal proposals are to pump it into geologic formations deep underground or to pump it into the deep oceans. Costs for either procedure are very high and whether or not the gas could be held securely for the long term are problematic. There is also concern that pumping massive amounts of CO_2 into the oceans might alter overall ocean chemistry and bring about unforeseen consequences. Chemical sequestration, the reaction of the carbon dioxide with industrial or mineral waste to create inert solid waste, has not yet been perfected on a large scale.

Disposing of Nuclear Waste Products

The mining and processing of uranium ores, the fabrication of nuclear fuels from these ores, and the burning of the fuels in nuclear power stations all generate waste products requiring disposal. The safe disposal of these products of the nuclear fuel cycle (which is discussed in detail in Chapter 6) is understandably a matter of great public concern, particularly as small amounts of radioactive poisons that can be lethal cannot normally be detected by our senses. The manufacture and development of nuclear weapons also generate significant amounts of dangerous radioactive waste. The long-term disposal of the most dangerous of the radioactive wastes is still an unresolved problem. Before considering the possible answers to this dilemma, we need to discuss the various kinds of radioactive waste and the quantities in which they are produced.

CATEGORIES OF RADIOACTIVE WASTE. The nature of a radioactive waste material, in addition to whether it is in solid, liquid, or gaseous form, depends on the concentrations of radioactive **isotopes** that are present. Although such wastes emit gamma rays or subatomic particles that can damage living tissue, the level of these emissions decreases with time. The rate of the *decay* of radioactive isotopes varies widely and is measured in terms of **half-life**, which is the time taken for the level of radiation emitted by an isotope to be reduced to one-half of its initial value. The nature of the decay process is such that given 1 gram of a substance, such as the radioactive isotope of iodine, ^{181}I, which has a half-life of eight days, the emitted radiation will fall to half of the original after eight days, one-quarter after sixteen days, one-eighth after twenty-four days, one-sixteenth after thirty-two days, and so on. Half-lives can range from only fractions of a second to billions of years, so that some radioactive waste materials remain lethal for hundreds or thousands of years, whereas others become virtually harmless in a matter of days or months. Most radioactive wastes are mixtures of short- and long-lived isotopes. Radioactive wastes include isotopes that are produced directly by the burning of the fuel, together with others produced when the materials in close proximity to the fuel are irradiated. The principal radioisotopes involved, along with information on emitted radiation, half-life, and the units used to measure radioactivity, are given in Table 4.4.

The main subdivision of radioactive wastes as far as disposal is concerned is into **low-level**, **intermediate-level**, and **high-level** wastes. Low-level wastes are generally those in which the maximum level of radioactivity is up to 1000 times that considered acceptable in the environment; intermediate-level wastes have 1000 to 1 million times that considered acceptable; and high-level wastes have still higher levels of activity. The volume of high-level waste produced is a very small proportion of the total; however, it does account for 95 percent of the total radioactivity of the wastes with which we must deal.

LOW-LEVEL WASTES AND THEIR DISPOSAL. Large quantities of waste are produced at uranium mines as at other mining operations. Although the mined ores are processed by crushing, grinding, and separation of a concentrate rich in the uranium minerals, the residues (**tailings**) commonly still

TABLE 4.4 Radioactive isotopes occurring in nuclear wastes

Principal Fission Products		Products of Irradiation of Nonfuel Materials	
Isotope	**Half-life**	**Isotope (and Source)**	**Half-life**
Krypton-85	>9.4 yr	From air and water:	
Strontium-89	54 d	Tritium, H-3, (^{2}H)	12.3 yr
Strontium-90	25 yr	Carbon-14 (^{14}N)	5700 yr
Zirconium-95	65 d	Nitrogen-16 (^{16}O)	7.3 sec
Niobium-95	>35 d	Nitrogen-17 (^{17}O)	4.1 sec
Technetium-99	$>5 \times 10^{5}$ yr	Oxygen-19 (^{18}O)	30 sec
Ruthenium-103	39.8 yr	Argon-41 (^{40}A)	1.8 hr
Rhodium-103	57 min	From sodium (coolant):	
Ruthenium-106	1 yr	Sodium-24 (^{23}Na)	15 yr
Rhodium-106	30 sec	Sodium-22 (^{23}Na)	2.6 yr
Tellurium-129	>72 min	Rubidium-86 (^{85}Rb)	19.5 hr
Iodine-129	1.7×10^{7} yr	From metals and alloys:	
Iodine-131	8 d	Aluminum-28 (^{27}Al)	2.3 min
Xenon-133	>5.3 d	Chromium-51 (^{50}Cr)	27 d
Cesium-137	33 yr	Manganese-56 (^{56}Fe)	2.6 hr
Barium-140	12.8 yr	Iron-55 (^{54}Fe)	2.9 yr
Lanthanum-140	40 hr	Iron-59 (^{59}Co)	45 d
Cerium-141	32.5 d	Copper-64 (^{63}Cu)	12.8 hr
Cerium-144	590 d	Zinc-65 (^{64}Zn)	250 d
Praseodymium-143	13.8 d	Tantalum-182 (^{181}Ta)	115 d
Praseodymium-144	17 min	Tungsten-187 (^{186}W)	24 hr
Promethium-147	2.26 yr	Cobalt-58 (58Ni)	71 d
		Cobalt-60(^{59}Co)	5.3 yr

Units of Radioactivity

Unit of activity is the Curie (Ci); 1 Ci = 3.7×10^{10} disintegrations/sec

Unit of exposure dose for X-rays and γ-rays = Roentgen (R)

1 R = 87.8 erg/sec (5.49×10^{7} MeV/g) in air

Unit of absorbed dose = Rad; 1 rad = 100 erg/g (6.25×10^{7} MeV/g) in any material

Unit of dose equivalent (for protection) = Rem (Si unit equivalent = Sievert; 1 rem = 10^{-2} sieverts)

Rems (roentgen equivalents for people) = rads × QF where QF (quality factor) depends, for example, on type of radiation

(γ-rays QF ≅ 1, thermal neutrons QF ≅ 3, α particles QF ≤ 20)

BOX 4.2
Radon

Radon is a naturally occurring, odorless, colorless, and radioactive gas that forms during the normal radioactive decay of uranium and thorium. Many people believe that the higher rate of radon exposure experienced by uranium miners has contributed to their higher than normal cancer rates. The potential for deleterious radon effects on people was not even considered until December 1984 when an engineer from the Limerick nuclear plant in eastern Pennsylvania, was found to be radioactive—not on his way home, but on his way into the plant! Investigators finally determined that he was radioactive from exposure to high levels of radon gas in his home.

Subsequent studies have found that one home in fifteen in the United States has radon gas concentrations in excess of the level considered to be safe by the U.S. Environmental Protection Agency. The EPA has suggested that the radon gas exposure causes between 7000 and 30,000 lung cancer deaths annually in the United States. The effects, however, are slow, cumulative, and generally linked with the effects of smoking. Hence, it is difficult to isolate and document the impacts of radon on human health. However, we can still establish where the radon comes from and what can be done about it.

Radon, atomic element 86, forms as a result of the decay of radium, which, in turn, forms during the complex stepwise decay schemes of uranium 235, uranium 238, and thorium 232. All other isotopes in the decay schemes are solids, but radon is a gas that can move easily in the vapor form or can be dissolved in groundwater. When the radon decays, it becomes radioactive bismuth or polonium, which readily attaches to fine dust particles. The radioactive breakdown of the radon and its daughter products gives off high-energy gamma rays that can harm body cells.

Radon is generated in greatest amounts where underlying rocks such as granite contain the highest levels of uranium and thorium. As the uranium and thorium undergo natural radioactive decay, radon is released and can move along fractures in the rocks. If homes are built in such areas, some of the radon may enter through cracks in the basement, through drains, and even through water pumped from wells for household use (Figure 4.C). Radon has always entered homes in various ways, but if levels are low and there is adequate ventilation, there is usually no problem. In recent years, however, many people have become more conscious of energy loss from homes that are poorly insulated. In the process of sealing the cracks around doors and windows to conserve heat, the homeowners have effectively sealed in any radon present. Furthermore, before the radon incident in 1984, virtually no one tested houses for radon, so we really do not know if there were any earlier problems.

Since 1984, awareness of the radon problem has grown; many states in the U.S. now require radon testing of all homes when they are sold. The solution to the problem is to either prevent the radon from entering by sealing cracks or other points of entry or to install a ventilation system that removes radon before it builds up to harmful levels. Either solution can cost from a few hundred to a few thousand dollars. However, the cost is relatively small compared with that of a house. It is not possible to stop radon from being formed, but it is possible to prevent it from accumulating within a home.

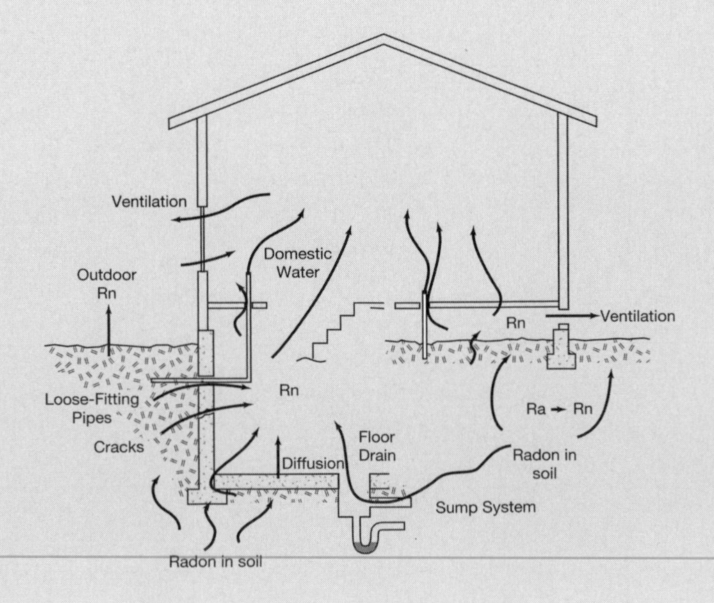

FIGURE 4.C Radon gas is generated in rocks and soils by the radioactive decay of uranium and thorium. Once generated, radon can move either as a free gas or dissolved in groundwater and enter a house. If the house is well ventilated, the radon disperses; but if a house is tightly sealed, the concentrations of radon can build up to hazardous levels.

contain significant radioactivity originally associated with the ore. This is so because beneficiation is never perfect and because the tailings are a far greater volume of material than the concentrate. The disposal of uranium mill tailings can give rise to human exposure at unacceptable levels of radiation because the material has been finely ground and can be transported by wind and water and because the radioactive gas radon, which is produced by the decay of uranium, can escape into the atmosphere more easily from this pulverized material.

A further point to note is that several of the radioisotopes present in the tailings are very long-lived (e.g., ^{230}Th with a half-life of 77,000 years). This means that the containment of these wastes must be designed to take account of geological processes of erosion and redistribution occurring over very long time periods. In practice, the wastes are usually dumped at or near the mine and subsequently stabilized by soil cover and vegetation. It is clearly important that water seeping through the wastes should not enter groundwater systems that provide water supplies for human or animal consumption. Concerns have been raised in recent years because some radioactive waste piles are located where some of the waste could be washed into major rivers. It is also important that the wastes are kept well away from human activities other than the mining operation. Unfortunately, this has not always been so in the past. For example, at Grand Junction, Colorado, more than 300,000 tons of radioactive tailings were used as fill material for land on which many buildings were later constructed; some of the waste was even used to make concrete blocks that went into the foundations of houses. The recognition of elevated levels of radioactivity in these buildings led to a cleanup program costing many millions of dollars.

Large volumes of low-level waste are also produced in nuclear power stations, in research laboratories, in hospitals, or in various nuclear industries. Typical items are contaminated laboratory equipment or protective clothing and even contaminated animal wastes. The lower-activity wastes are usually sealed in metal drums, commonly after burning in special incinerators to reduce their volume, and the drums are buried in shallow trenches beneath a meter or so of soil (Figure 4.29). The burial site needs to be carefully chosen based on geology because the radioactivity in the area must be regularly monitored to ensure that any contamination of plants, soils, and groundwater is within accepted limits. The upper limits of radiation exposure or dosage for workers in the nuclear industries, and for the general public, are governed by legislation based on the recommendations of the International Commission on Radiological Protection (ICRP).

Low-level wastes of somewhat higher activity have often been encased in concrete, enclosed in sealed drums, and dumped on the deep ocean floor. At an internationally agreed site in the Atlantic Ocean some 800 kilometers from southwest England, tens of thousands of tons of such wastes have been dumped in this way. The sealing in concrete enables handling and shipment of the material and ensures that it reaches the

FIGURE 4.29 Low-level nuclear waste is packed in drums and buried in shallow trenches where it is isolated from surface runoff and groundwater. (Courtesy of the U.S. Geological Survey.)

bottom of the sea intact, where it should remain for many years. The safety of this disposal method is based on the vast dilution of the activity as it slowly disperses in the ocean 5 kilometers below the surface.

Low-level liquid wastes arise at nuclear power stations and at plants where nuclear fuel is reprocessed. Such wastes are often treated and then discharged into rivers and into the sea. When this is done, the levels of radiation in waters and marine life are closely monitored and kept to acceptable levels. Nevertheless, this is a procedure that has understandably given rise to public concern and pressure from environmentalist groups. Radioactive gases are also emitted from power plants in very small amounts.

INTERMEDIATE-LEVEL WASTES AND THEIR DISPOSAL. Materials such as certain solid components from nuclear power plants, various liquids used in nuclear plants, and the flasks used to transport nuclear fuel have intermediate levels of activity. At present, these materials are stored in tanks or other containers at nuclear plants with a view to eventual disposal by methods similar to those used for the low-level wastes. Another possibility involves chemical treatment to remove long-lived active constituents so that the bulk of the material can be discharged or buried as low-level waste.

HIGH-LEVEL WASTES AND THEIR DISPOSAL—THE GREAT DEBATE. The high-level wastes from the nuclear power industry account for roughly 95 percent of the radioactivity but only about 0.1 percent of the volume of waste generated. These consist of large quantities of liquid wastes and used fuel rods that give off very high levels of radioactivity and substantial heat. The high levels of radioactivity, which are produced by a variety of isotopes, require that high-level wastes be isolated from human exposure for approximately 10,000 years. After that time, the level of radioactivity will have decreased to about one ten-thousandth of the original level (Figure 4.30). At present, used fuel rods are packed in stainless steel tanks, similar in design to the one shown in Figure 4.31. The drums are stored at sites of the fuel reprocessing plants or at the power plants where they were generated. The problem of their eventual disposal has been the subject of much research and even more debate; no acceptable solution has yet been agreed upon in the United States, in the United Kingdom, and in many other countries. Meanwhile, these lethal substances continue to accumulate. One estimate of their toxicity suggests that less than 4 liters of the hundreds of millions of liters held in storage would be enough to

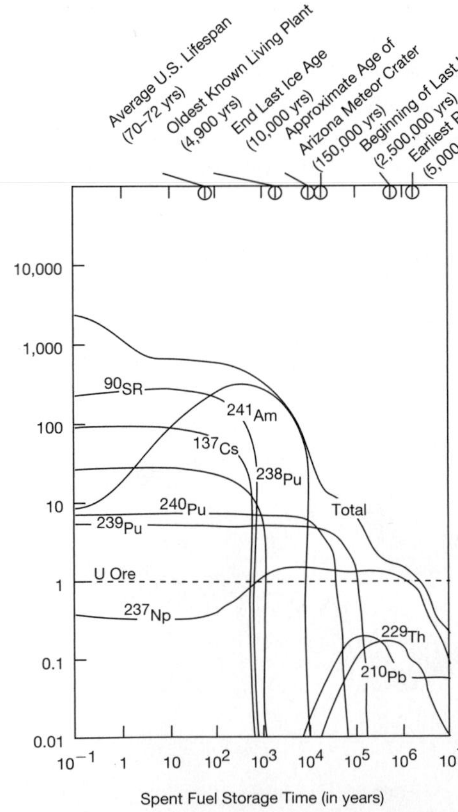

Average U.S. Lifespan (70–72 yrs)
Oldest Known Living Plant (4,900 yrs)
End Last Ice Age (10,000 yrs)
Approximate Age of Arizona Meteor Crater (150,000 yrs)
Beginning of Last Ice Age (2,500,000 yrs)
Earliest Primate Hominid (5,000,000 yrs)
10,000
1,000
100
10
1
0.1
0.01
^{90}SR
^{241}Am
^{137}Cs
^{238}Pu
^{240}Pu
^{239}Pu
Total
U Ore
^{237}Np
^{229}Th
^{210}Pb
10^{-1} 1 10 10^2 10^3 10^4 10^5 10^6 10^7
Spent Fuel Storage Time (in years)

FIGURE 4.30 The relative amounts of total radiation released from high-level nuclear wastes (upper curve) and the amounts from several major radioactive isotopes decrease with time. The storage time considered necessary for this high-level waste to drop to levels believed to be safe is about 10,000 years. (From the U.S. Department of Energy.)

FIGURE 4.31 Temporary above-ground storage of high-level nuclear waste in a mobile canister at the Surry Nuclear Power Plant in Virginia. (Courtesy of Virginia Power.)

bring every person in the world to the danger level for radiation exposure if it were evenly distributed. Although the safety record in the storage of these highly active wastes has been good over the nearly forty years in which they have been stored, leakage from the storage tanks has occurred, resulting in contamination of the ground around them.

Most long-term nuclear waste-disposal strategies require that the waste first be solidified. One method of doing this is to incorporate the waste into a glass, a process known as **vitrification**. The French have developed a method for doing this on an industrial scale using a glass containing boron and silicon. This process can be designed so that the glasses survive the effects of heating and radiation from the wastes, but there is concern about the ability of such glasses to survive attack by water or brine solutions at elevated temperatures. These are important considerations because the solid wastes are likely to be buried deep inside Earth where they may come into contact with groundwater. The alternative to vitrification is to incorporate the waste into ceramic or synthetic minerals; these are crystalline materials in which the radioactive elements are tightly and stably bound much as they were before mining. An example of such a material is the mixture of titanium oxide minerals (titanates) developed by an Australian group and known as **SYNROC**. The composition of this material is shown in Table 4.5 and, as the name suggests, it is actually a synthetic rock. Proponents of this method of waste processing point out that the radioactive elements are being returned to the sort of chemical environment in which they are most stable in Earth's crust. Certainly, experiments show that SYNROC remains stable in contact with brine solutions to much higher temperatures than glasses do, and to temperatures in excess of 700°C.

Solidification is only the first step in the ultimate disposal of nuclear wastes. Solidified waste can be stored in vaults or in reinforced and shielded buildings above ground, having first been sealed into concrete and stainless steel canisters. Such methods are used in France where the canisters containing vitrified waste are placed in a vault and cooled in a stream of air. Perhaps wastes in such stores could be kept safely for the indefinite future with minimal surveillance and maintenance. However, there are inevitable concerns over the vulnerability of such stores in wartime or through acts of terrorism or major disasters.

Various options have been considered for the ultimate disposal of the solidified wastes. The more extreme proposals, such as removal from Earth by rockets, or burial in the polar ice caps, have been rejected on technical or safety grounds. The remaining options are disposal on the bed of the ocean, burial in deep-ocean sediments, or deep burial on land. There are areas of the ocean floor where sediments have remained undisturbed for millions of years and are likely to remain very stable. Although these appear technically to be a good possibility for disposal sites, there are sure to be major legal problems in obtaining the necessary international agreements. Also, despite several decades of research, more needs to be known about dispersal mechanisms of the wastes before further ocean-bed dumping could be considered acceptable. Considerable research has also been done on suitable deep burial sites on land. Clearly, a very stable environment and one relatively impervious to groundwater is needed. Some have suggested burial in salt deposits because salt layers deform plastically rather than by fracture, and because salt is water soluble, it is evidence of a dry environment. However, all salt beds contain several percent water as small fluid inclusions, and some salt beds are known to contain pods of brine that could prove highly corrosive if brought in contact with solid wastes. Furthermore, some experiments have suggested that the 1 to 2 percent of water dispersed in salt beds may actually migrate slowly toward the hot area around the radioactive wastes. Other environments being considered include shales or crystalline rocks such as granites, the sort of rocks within which the uranium minerals would have originally occurred in nature.

TABLE 4.5 Mineralogical and chemical composition of SYNROC,* the titanate ceramic wasteform designed to incorporate radioactive wastes in a stable form for disposal by burial in Earth

Component (wt %)	Formula	Acts as Primary Host for
Hollandite (33)	$Ba(Al,Ti)Ti_6O_{16}$	Cs, Ba, Rb, K, Cr
Zirconolite (28)	$CaZrTi_2O_7$	Th, U, Pu and tetravalent actinides, Zr
Perovskite (19)	$CaTiO_4$	Sr, Na, trivalent actinides rare earths
Rutile (15)	TiO_2	
Alloy (5)		Tc, Mo, Ru, Pd, S, Te

(Data are from Ringwood, A.E., *Mineralogical Magazine,* vol. 49, pp. 159–176, 1985.)

*Actually the most common of several forms of SYNROC known as SYNROC-C.

Ideally, the geological environment should form a natural barrier to fluids that could disperse the waste or to any other mechanism of dispersal. It should be an environment as free as possible from any risk of earthquake or volcanic activity. As well as natural barriers to the escape of the wastes, there would certainly be artificial or engineered barriers. These might include sealing the solid wastes in corrosion-resistant canisters, perhaps of stainless steel, and surrounding the canisters with absorbent materials such as clays or zeolites. The wastes might be deposited in the site simply by drilling large-diameter deep holes as has been suggested for wastes locked up in SYNROC. In this case, it is proposed that 1-meter-diameter holes are drilled down to 4 kilometers of depth and the bottom 3 kilometers be filled with SYNROC-contained waste. Other sites are envisaged as more like mines, with shafts allowing access for workers and materials and the wastes stored in galleries (see Figure 4.32). Old mine workings would be evaluated for this purpose, for obvious reasons of savings in costs.

Several countries have made decisions on the geologic environment for long-term, high-level waste disposal (e.g., Belgium, clay; Germany, salt; Sweden, granite; India, granite), but both the United States and the United Kingdom have no functioning permanent repository. In the United States, experiments have been conducted in salt beds at the Waste Isolation Pilot Project (WIPP) site in New Mexico; WIPP has already received some low-level waste. The site selected as the best candidate in the United States for a permanent high-level waste repository is in volcanic tuffs at Yucca Mountain, Nevada (Figure 4.33 and 4.34; Plates 26 and 27). Unfortunately, the site originally chosen has been ruled out by the Obama administration leaving the U. S. with no repository to accept high level waste. Consequently, the highly radioactive spent fuel rods continue to accumulate at more than one hundred nuclear reactor facilities; when the site is finally opened, it is likely that the accumulated waste will exceed the available capacity. In the United Kingdom, the situation is even more uncertain following the failure of the organization charged

FIGURE 4.32 The Waste Isolation Pilot Project (WIPP site) in New Mexico has been cut into beds of salt and receives low-level nuclear waste for long-term storage. Barrels of waste have been packaged and placed within the storage chambers. The first remotely handled high-level waste was brought to the WIPP site from Oak Ridge, Tennessee, in March 2009. (Photograph by J. Rostro, courtesy of Department of Energy, WIPP Site.)

FIGURE 4.33 Entrance into the Yucca Mountain Nuclear Storage Facility. Despite an original plan that called for opening in 1998, revisions and delays pushed the earliest operational date to 2015 or later. However, new political controversies beginning in 2009 have made any use of Yucca Mountain problematic. (Courtesy of the U.S. Department of Energy.)

with developing a site (NIREX) to gain planning consent to construct an experimental underground facility near the Sellafield nuclear plant in Cumbria.

Ever since the construction of the first nuclear power plants, much effort has been put into the problem of disposing the very dangerous wastes they produce. There appear to be technical solutions to this problem that involve only small risks to humans and the rest of the living world, and certainly less than the risks involved in current methods of storage. There do remain many concerns about the secure long-term disposal of high-level nuclear wastes, however; many people feel that much of this material will ultimately be retrieved for recycling into fuels. Today, there is great uncertainty about the future of nuclear power in many parts of the world, as discussed in Chapter 6. This uncertainty adds to the confusion over establishing permanent high-level waste repositories.

Other Industrial Processes: Waste Products and Pollution

Many industrial processes based on mineral resources create and discharge substances that can have adverse environmental effects. The releases may be accidental as in the case of oil spills, or they may result from direct application, as in the case of fertilizers, pesticides, and herbicides used in modern agriculture. Such agricultural pollutants pose particular problems because they are widespread **dispersed sources** of pollution, unlike the waste dump, power plant, or factory that emits pollutants from a **point source** (i.e., a small and exactly defined area).

Although many of the substances that are deliberately released into the environment appear to enter into various chemical and biological cycles without harmful effects, some cause widespread concern. Certainly, our knowledge of the paths followed by

FIGURE 4.34 Spent fuel test chamber at the Climax test site about 40 miles from Yucca Mountain. Spent fuel rods are lowered into the receptacles cut into the floor of the passageway. (Courtesy of the U.S. Department of Energy.)

many of these substances after their release is inadequate. We know very little of the long-term effects that may result from prolonged buildup of many substances in particular environments. A well-known example of the problem comes from the chemical industry.

A chemical plant that started production in 1932 was emitting mercury-containing waste into Minamata Bay in Japan. This waste included a highly poisonous mercury compound, methyl mercuric chloride, which was not diluted to a harmless level as expected, but which preferentially accumulated in fish that constituted a major portion of the diet of the local inhabitants. As a result, more than 1500 people suffered mercury poisoning, referred to ever since as Minamata disease; many others were seriously disabled. Unfortunately, the disease was not identified until 1956, and although the toxic discharges were stopped in 1960, their effects still linger.

Sometimes naturally occurring elements can create problems as great as those from human-released materials—for example, selenium in groundwater in California or arsenic in groundwater in Bangladesh. These will be discussed in Chapter 11.

ENVIRONMENTAL IMPACTS OF WASTE DISPOSAL OR RECYCLING

Archeologists have learned much about past civilizations from studies of the waste products left behind. The archeologists of the future will surely have a rich supply of material from our age because advanced industrial societies produce vast quantities of waste each year, in addition to the wastes arising directly from mineral exploitation and the utilization of resources. These include agricultural wastes, domestic or municipal refuse, and the waste products of the manufacturing industries. Most of these materials are in the form of solid waste. There are lesser amounts of liquid waste material generated by industry and liquids in the form of sewage.

The disposal and treatment of these wastes are important matters in regard to Earth resources because many of them are potential sources of energy, metals, or other materials. Their disposal involves another disruption of the environment, whereas their recycling reduces somewhat the need for mining with its environmental disruption. After considering the nature and problems of disposal of solid wastes, the question of recycling will be addressed.

Disposing of Solid Wastes

In the United States, for example, over 5000 million tons of solid waste is generated every year, and this quantity continues to grow. Principal among these wastes are agricultural products and waste rock and tailings from mines. Domestic or municipal refuse accounts for less of the bulk, but contains within it many valuable raw materials and many potential pollutants. Manufacturing wastes may also contain valuable metals, fibers, and chemicals, as well as toxic by-products. The composition of typical domestic trash in the United States is roughly 33 percent paper and paperboard, 13 percent yard wastes, 12 percent plastics, 8 percent metals, 5 percent glass, and the remainder is a combination of materials including garden wastes, wood, food wastes, textiles, rubber, etc. (Figure 4.35a). The amount of municipal solid waste generated in the United States continues to increase, as shown in Figure 4.35b, and has surpassed 250 million metric tons annually.

Prior to the 1960s, the majority of solid wastes were disposed of by open dumping, with little material being incinerated or disposed of in sanitary landfills. The open dump was the most primitive and most widely used means of solid-waste disposal, accounting for disposal of more than half of the waste generated in the United States and the rest of the world. Municipal waste is often compacted after collection, hauled to the dump, and spread on the ground by bulldozers (Figure 4.36). In the past, this waste was commonly set on fire to help reduce the total volume, but such open burning is now prohibited because of fire hazard and air pollution. The open dumps themselves were unsightly; they served as breeding grounds for diseases carried by flies and rats, and they commonly contaminated rivers and groundwaters with hazardous chemicals or organisms. Many coastal cities, including New York City, formerly dumped their municipal wastes in the ocean; barges carried the waste out to sea and discharged it into a natural trench or canyon on the ocean floor. This caused disruption of the marine environment with the destruction of communities of the bottom-dwelling organisms and has now been banned.

The sanitary landfill is now the principal means for disposal of solid domestic trash in industrialized nations. The refuse is deposited at a carefully chosen site where contamination of surface water or groundwaters will not be a problem or where drainage is controlled. After the waste is brought to a site, it is compacted with bulldozers (Figure 4.36) or other heavy machinery, and each day the waste is covered with a layer of earth 15 to 30 centimeters thick, so as to exclude air and vermin. Three methods of sanitary landfill are commonly used. In the *area method*, which is well suited to flat ground or broad depressions, wastes are spread on the surface and covered with soil to form a cell that usually represents one day's waste (Figure 4.37). Additional waste cells can be deposited on top of finished cells and then all is covered with a thick layer of soil to form a cap. In the *trench*

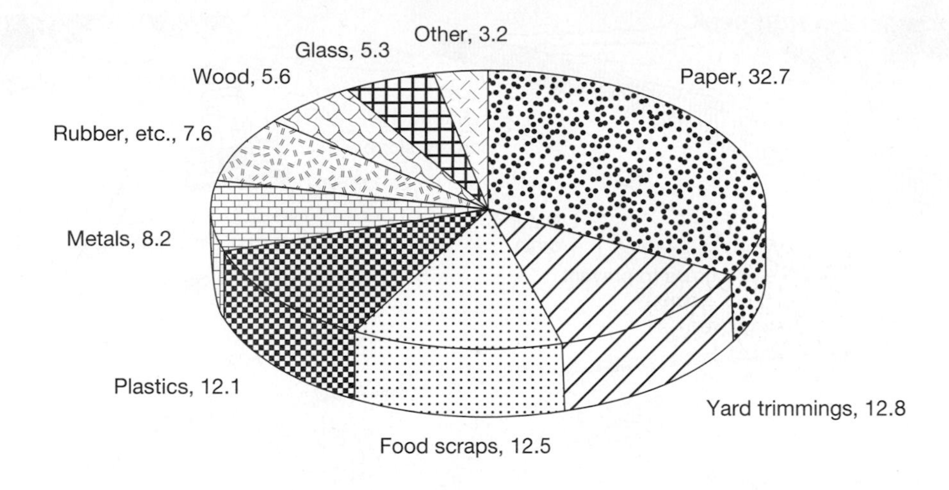

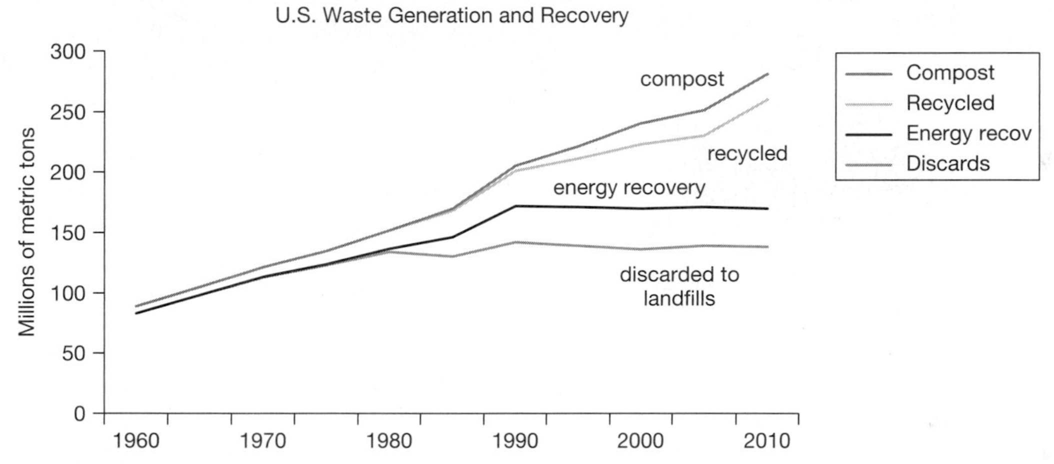

FIGURE 4.35 (a) Percentages of the various materials that composed municipal solid waste in the United States in 2007. (b) The generation and disposal of municipal solid waste in the United States since 1960. The amount of landfill waste increased until about 1989; since that time, the amount of waste recycled, composted, and burned for energy has been rising steadily. (From Franklin Associates Ltd., Prairie Village, KS; Characterization of Municipal Solid Waste in the United States: 1992 Update. Prepared for the U.S. Environmental Protection Agency.)

FIGURE 4.36 Municipal solid waste being compacted after being dumped into a landfill before the end of the day burial. Compaction reduces the volume of the waste and lengthens the life of the landfill. Some of the synthetic liner used to prevent groundwater contamination by leachate is visible in the foreground. (Photograph courtesy of Draper Aden Associates.)

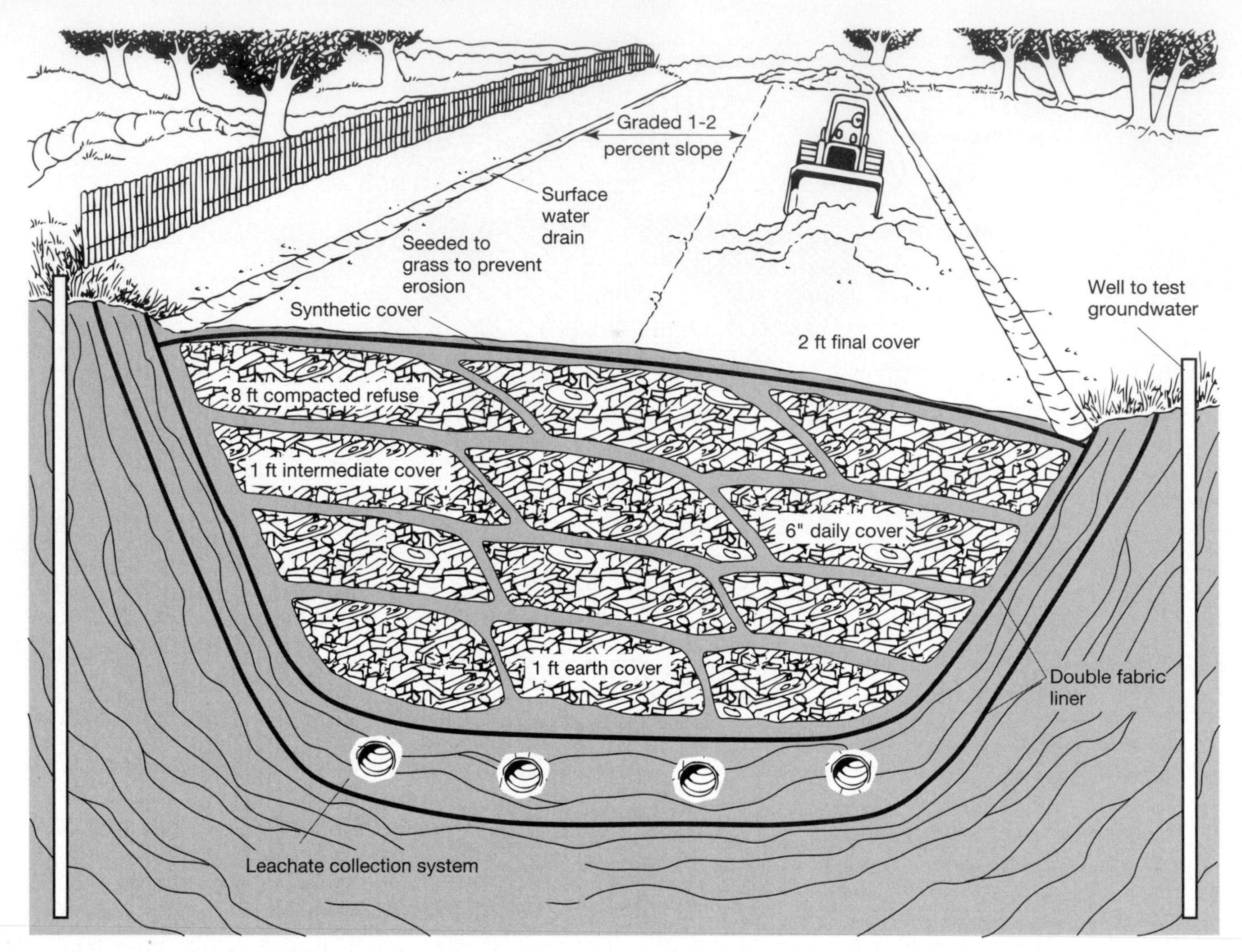

FIGURE 4.37 Modern sanitary landfills usually employ a cellular structure such as the one shown here. After cutting a large trench or pit, one or two impermeable liners (sometimes compacted clay, but usually plastic sheets) are placed in the bottom and covered with a layer of sand. Refuse is dumped into cells, compacted as shown in Figure 4.36 and covered at the end of each day. Drainage pipes are placed between the liners and at the base of the refuse fill zone so that leachate can be removed. In addition, test wells are drilled around the landfill to detect any contamination of groundwater. When a landfill is closed, a liner and a thicker layer of soil are placed over the refuse cells; vegetation is planted to prevent erosion. Monitoring of the closed landfill is usually mandated for at least thirty years. (After a diagram from the New York State Department of Environmental Conservation.)

method, a broad trench is excavated and filled with compacted refuse. In this method, the excavated material provides a soil cover that is well suited to level terrain. The *ramp method* befits sloping areas; refuse is spread over the hillside, compacted, and covered with soil that is often excavated from a cut at the base of the slope.

Present regulations in the United States require that the landfills have two impermeable layers of natural or synthetic materials (e.g., heavy plastic) beneath the landfill to prevent any leachate from leaking into the groundwater. There must also be a leachate removal system, some means of testing between the two underlying barriers, and wells adjacent to the landfill that can be tested for groundwater pollution. Upon closure, landfill sites must be covered by an impermeable layer and with 0.5 to 1.0 meter of soil. Surface slopes must be low enough that erosion will not be a problem. Furthermore, modern regulations require postclosure monitoring of sites for thirty years and the posting of some sort of bond or insurance to pay for any damages that might occur. Completed landfill sites are usually covered with a layer of soil one-half to 1 meter thick that may then be seeded and used for parkland or recreation areas. In some cases, hills have been constructed for use as ski slopes or other uses.

A good example is Mount Trashmore (Figure 4.38) at Virginia Beach, Virginia, where the shallow depth to the water table would not permit excavation of a pit to use as a waste dump site. There, 1000 tons of garbage per day were dumped, spread, compacted, and covered with soil to build a hill 100 by 260 meters and 22 meters

FIGURE 4.38 Subsurface burial of municipal waste may be impractical in coastal or other areas where the water table is very shallow. Virginia Beach, Virginia, solved the waste disposal problem by building 20-meter-high Mount Trashmore incorporating 575,000 metric tons of solid waste. After final sealing and landscaping, it is now a city recreation area. Because of its success, the Virginia Beach area is constructing even larger mounts in a similar manner. (Photograph courtesy of City of Virginia Beach.)

high and containing 580,000 metric tons of solid domestic waste. By the daily compaction of alternating layers of 45 centimeters of waste and 15 centimeters of soil, a structure was created free from problems of unpleasant odors, fires, vermin, and groundwater pollution. This site has been used to house a recreational area with a playground, amphitheatre, soapbox derby track, picnic areas, and, adjacent to the hill, a freshwater lake. The success of Mount Trashmore has now led to the development of a second similar site.

Landfills are generally not suited as sites for buildings because of possible subsidence and escape of gases or contaminated waters. Although the advantages of sanitary landfill over open dumping are obvious, they still present problems in terms of available space. On average, a well-managed landfill operation requires one hectare of land (about 2.5 acres) each year for 25,000 people. For an area such as greater New York with a population in excess of 12 million people, this could demand 480 hectares of land every year. Also, as the more obvious locations for landfill sites are used up, more distant sites have to be employed, increasing transport costs and the disruption of the surrounding countryside. The slow decay of the organic matter in landfill sites generates methane gas. This used to be vented to prevent potentially explosive buildup, but today it is usually collected and piped to commercial plants or utilities for use as a fuel.

Automobile tires have become a major problem for domestic solid-waste disposal; in the United States alone, there are about 250 million tires being discarded annually (Figure 4.39 and Plate 11). Landfill operators discovered that tires do not decompose in any reasonable length of time and that they gradually rise to the top of any landfill (probably as the result of air pockets and the episodic changes in temperature), unless the tires have been cut into pieces. The rubber and petroleum compounds within the tires burn very well and can serve as fuels, but they are difficult to handle and frequently create pollution problems when incinerated. Many landfills will no longer accept old tires, and there is no comprehensive recycling program in most countries; consequently, many illegal tire dumps have been created.

FIGURE 4.39 Approximately 250 million tires are discarded in the United States every year—probably 500 million worldwide. Many billions lie in refuse piles (see Plate 11). Like many materials made from Earth's resources, the tires are extremely durable. Most proposals to convert tires into fuel sources have failed because of the cost of processing the tires and the problems of air pollution when they are burned. Note the size of the tree that has grown through the tire while it lay in an unauthorized dump. (Photograph courtesy of the Virginia Department of Waste Management.)

Incineration has been used to simply reduce the volume of solid waste (up to a 90 percent reduction) and as a way to generate power while reducing the volume of waste. Although there has been increasing interest in incineration as a way to reduce landfill needs, this generally concentrates most of the trace metals from the burned waste. Consequently, this ash has been ruled a hazardous waste that requires special landfills for disposal. The burning of materials as complex as municipal trash is not simple. Substances that release poisonous gases or particles when burned are often present, so that costly pollution controls are needed to meet modern standards of air cleanliness. The capital cost of building a modern incinerator and the costs of operation are often much greater than those for a landfill operation.

In addition to air pollution, certain materials commonly found in refuse produce gases that corrode furnace interiors. For example, polyvinyl chloride (PVC), a plastic that is widely used in goods such as toys, containers, and CD's, produces highly corrosive hydrogen chloride gas when burned. However, one important factor that helps to compensate for these difficulties is the energy generated by the incineration of wastes. The heat given off by the process of incineration is increasingly being used for space heating of buildings or for electrical power generation (Figure 4.40). The increasing problems of refuse disposal combined with increasing energy costs have led to the construction of many energy-generating plants using solid wastes; this has occurred throughout the world since the mid-1990's.

The generation of energy in the form of heat by the burning of solid wastes leads to the general question of how energy can be extracted from solid wastes. A number of these alternative energy sources, in particular, the ways in which organic wastes and plant materials (including manure, sewage, and wastes from crop cultivation) can be used to produce a flammable gas such as methane or a liquid fuel are discussed in Chapter 6. A more direct method of extracting potential fuel materials from wastes is *pyrolysis,* a process that involves heating the wastes in the absence of air so as to drive off and collect volatile substances; this is further discussed on p. 118. Any type of recycling, in fact, leads to an overall saving in energy and can be regarded as an aspect of energy, as well as material, conservation. We discuss this next.

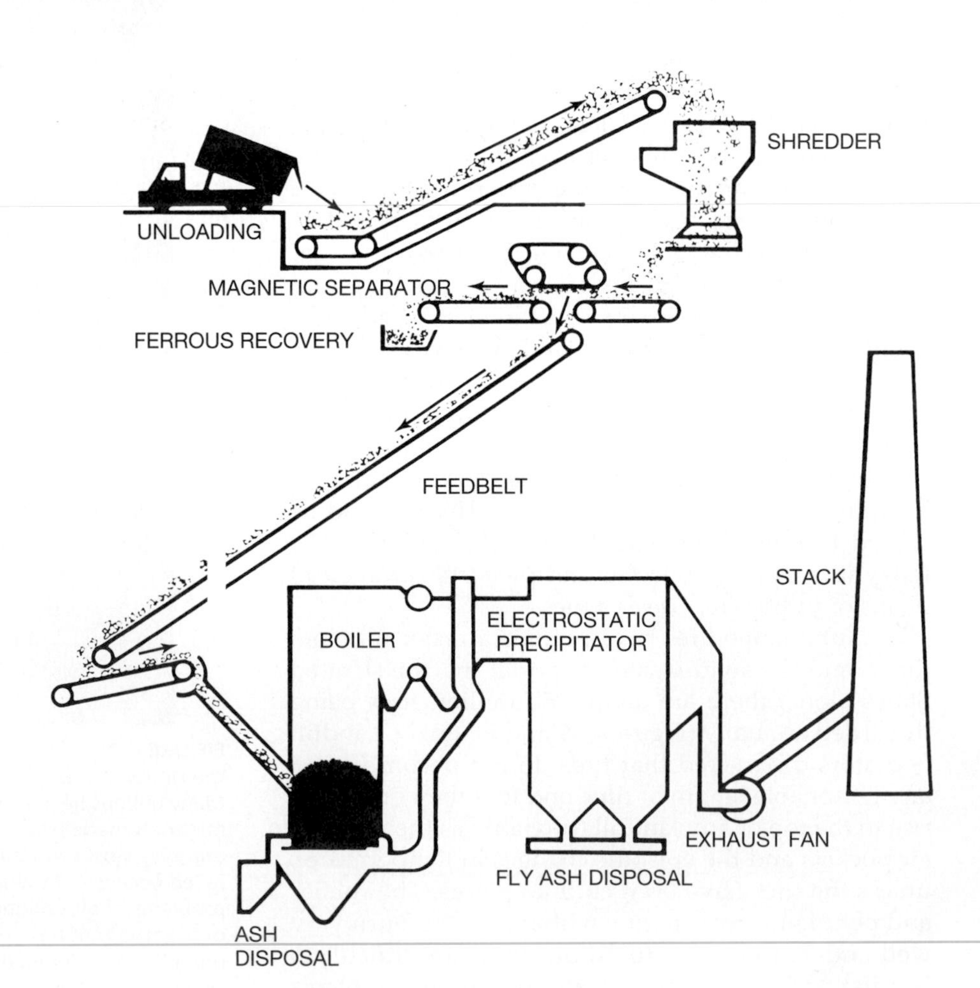

FIGURE 4.40 Schematic diagram showing a modern waste treatment system in which ferrous scrap is recovered and usable heat is generated by burning most of the other waste. The amount of ash for disposal is only a small fraction of the initial mass of the waste. (From R. Davies and P. Ketchum, "Energy From Waste," GEOS, vol. 15, no. 2, p. 18, Energy, Mines and Resources Canada, 1986. Used with permission.)

Recycling

Recycling of materials is at least as old as the use of metals. When our ancestors discovered how to smelt metals, they also found that metals such as copper and gold could be remelted and used over and over again. This recycling gradually extended to other metals such as lead, tin, and silver, and to alloys such as bronze, brass, and steel. In time, recycling expanded to paper goods, plastics, and many other products. In truth, most of the refuse discarded by people in modern industrial societies contains large amounts of material that could be reused or reprocessed to reclaim valuable raw materials. Such recycling (see Box 4.3) often conserves not just material resources but also fuel. For example, nearly twenty times the energy is required to produce aluminum from bauxite than is required to remelt aluminum scrap, and over twice as much energy is needed to manufacture steel from primary raw materials as from scrap. Recycling operations also generally emit fewer pollutants than the original process, as, for example, in the recycling of metals such as copper because the recycling does not require roasting or smelting.

The recycling of materials ranges from simple reuse or reclamation of an object (e.g., the reuse of bottles for beverages or milk), to complex processes of recovery by physical or chemical means (e.g., the revulcanization of rubber, which is discussed next). Certain finished products, parts of products, or raw materials are clearly more easily recycled than others. For example, scrap paper and related materials made up of cellulose fiber can readily be pulped and used again to make paper and cardboard. A junked automobile, however, presents a formidable recycling problem, because it contains many components made of materials ranging from rubber and plastic to glass, plus a whole range of metals and alloys.

It is possible to envisage a society in which most durable goods would be used for much longer than at present and then broken into their component parts to reclaim raw materials. Throughout the United States, the percentages of solid waste being recycled have been rising since the mid-1980s, and the United States overall recovery and recycling had reached more than 33 percent by 2007. Before considering the largely social and economic aspects of recycling, we shall consider the technology of recycling. Except for the larger items of machinery such as automobiles, most of the rubbish of modern society finds its way into municipal trash. There are also the more specialized forms of rubbish generated by agriculture and other industries. Most of this material cannot be reused and must be broken down into raw materials. Examples of the processes by which this is done include the following:

Melting of metals, glass, and some plastics which can then be purified and recast or remolded;

Revulcanizing of rubber, which is a material that cannot be simply heated and remolded. It must be shredded, broken down chemically, and then reacted with sulfur compounds;

Pulping to reclaim fiber from waste paper or other natural material containing cellulose fiber (wood, reeds, sugar cane stalks). The material is stirred and beaten to form a slurry, inks are dissolved, and the pulp put through the usual paper-making processes;

Pyrolysis, which is heating materials in the absence of air to about 1650°C so that they decompose into chemical compounds. There are also the processes by which organic wastes can be broken down; examples are the *composting* of vegetable matter to make fertilizers, *rendering* of animal wastes to make such products as soaps and glues, and *fermentation* to make alcohols, gases, and a variety of other products.

Although agricultural and industrial wastes may be fairly homogeneous and simply recycled through one of the above processes, domestic trash is a mixture of many kinds of material; thus, recycling is more complex. Although programs for the collection of bottles and cans were widely successful during World War II, such programs generally ended in the 1950s and 1960s. Since the 1980s, many regions in the United States and Western Europe have reinstituted curbside recycling and have required the separation of metals, glass, plastics, and papers. Notable success has been achieved in certain countries, particularly Japan, where a very high percentage of municipal waste is now recycled. Japan is a highly organized industrial country with very limited natural resources and landfill space; Japan has found it necessary to develop incentives as alternatives to the "throwaway society." In Hiroshima, for example, disposal of raw refuse has been reduced by 40 percent since 1976 by paying for separated wastes and saving the monies formerly used to operate landfill sites. Separation at the source is often undertaken by nonprofit groups such as student clubs and parent–teacher associations. A program involving source separation and computerized processing now enables the Japanese city of Machida to recycle 90 percent of its garbage.

Commonly, however, the problem has to be tackled by separation at the disposal facility. Various devices that have been employed include screens, magnets to remove iron, or the use of compressed air to separate light items from heavy ones. A series of devices of the type shown in the flow diagram of Figure 4.40 are needed to bring about complete separation. In this system, large items are removed manually and the remaining refuse fed with water into a pulper. In this, the fiber component is pulped, brittle material such as glass pulverized, but stronger solid objects are not affected. The fiber slurry is drained through a screen (2-centimeter mesh size) and large objects (over 2 centimeters) are removed and washed. From this, iron is separated magnetically, and other metals are recovered manually for sale as scrap. The fiber slurry is pumped to a cyclone that removes solids such as pulverized glass, metals, bone, and sand particles. The remaining slurry consists of paper, food wastes, cloth, and plastics; after further pulping and screening operations, paper fiber is extracted. Organic materials are then used as a fuel.

There is an irony in the fact that energy-producing incinerators compete with recycling systems. Increased recovery of paper goods and plastics by recycling reduces the combustible materials that provide most of the fuel for energy production. As a result, some waste-to-energy plants have been shut down for lack of fuel in favor of more efficient recycling.

Programs such as those just discussed for the recycling of domestic waste are unfortunately rare, whereas recycling of metals is much more widespread. Of course, in industries that use metals, such as in a metal fabricating shop, large quantities of uncontaminated scrap are generated, and this has a ready market. Consequently, in the United States, recycled metals include roughly 30 percent of copper and lead, 48 percent of steel, 43 percent of aluminum, and 25 percent of zinc, along with high percentages of the precious metals. Nevertheless, the quantities of such metals annually discarded in the United States are vast—over 11 million tons of iron and steel, 800,000 tons of aluminum, and 400,000 tons of other metals.

As already noted, the savings involved in recycling are not only of the raw materials themselves but also of energy. It has been suggested that recycling of the more than 60 billion aluminum cans alone that are used annually in the United States would save energy equal to the output of eight 500-megawatt power plants. Furthermore, discarding an aluminum beverage container wastes as much energy as pouring away such a can half-filled with gasoline. It is particularly the two metals, iron and aluminum, for which recycling offers the greatest benefits through savings; not only in raw materials but also in energy, and in much reduced environmental damage (recycling aluminum reduces air emissions associated with its production by 95 percent, for example). This is because of the vast amounts of these two metals consumed and large amounts of energy used in producing them. Despite the great advantages, however, only about one-quarter of the world's steel, aluminum, and paper is recovered for reuse.

Nevertheless, the three "R's" of **reduce, reuse, and recycle** are being increasingly applied by industries in a "waste hierarchy" to minimize impacts from the resources and energy used in manufacture. The ranking of environmental actions in increasing order of environmental benefit are:

1. To burn material for energy is better than sending it to a landfill.
2. To recycle it is better than burning it.
3. To reuse material is better than recycling it.
4. To reduce the amount needed is better than reusing it.
5. To eliminate the need for material is better than reducing it.

The electronic age has created a new category of waste generally referred to as "e-waste". The United States, for example, has at least 300 million discarded television sets, with tens of millions more being discarded every year. Added to that annual waste are millions of discarded computers, cell phones, and batteries. A conservative estimate places the e-waste at 1.5 to 2.0 million tons annually, most going directly to landfills. Most of the TVs and older computer monitors contain ten or more pounds of lead glass, installed to prevent radiation leakage during operation. The shattering of the glass into small pieces greatly increases the surface area of the particles and permits leaching of the lead into the waters at the landfills. In addition, countless printed circuit boards and electronic chips can release antimony, silver, chromium, zinc, tin, copper, and other metals. The ubiquitous CDs used for music, videos, and for countless computer programs have surface coatings that contain silver, indium, antimony, tellurium, cobalt, gadolinium, and other metals. Furthermore, vast numbers of discarded NiCad batteries can release nickel and cadmium, both of which are carcinogenic. Recycling of e-waste is practiced in some localities; India, China, and Nigeria actively seek to import large quantities of e-waste to process for the extractable metal components. Nevertheless, most of the e-waste is not yet being recycled and is becoming a great concern for the operators of landfills.

The one other recyclable material that occupies a special place is wood; it is a fuel, a building material, and a raw material for chemical industries and paper manufacture. Paper products use about 35 percent of the world's commercial wood harvest. Recycling just half of the paper used in the world today would meet nearly 75 percent of the demand for new paper and free 8 million hectares (20 million acres) of forest from paper production.

Why, then, is recycling not undertaken on a much larger scale? The factors that dictate this are not technical, but are economic and social. Thus, in the field of metals, the mining and processing industries are established groups operating on a large and highly organized scale. Scrap metal operations are usually small, labor-intensive, and incur heavier transportation costs because of their small scale. A domestic refuse reclamation plant is currently more costly than landfill disposal, a cost that will have to be borne by local citizens in their rates and taxes. Some citizens may be prepared to pay the few extra dollars a year to remove the need for unsightly landfill operations and save precious raw materials, but many people, not being directly affected, will not make this sacrifice to recycle materials.

Three factors could bring about more widespread recycling in the future. One is the need to use land area for more valuable purposes than that of a landfill. The second could be an increase in the price of primary raw materials. The third could be government-mandated use of more products containing recycled materials. To date, only the first of these factors has come into play. The benefits of recycling, in saving not only precious raw materials but also conservation of energy and reducing problems of waste disposal and pollution, are very clear. Unfortunately, in most countries, the economic and social incentives needed to set up large-scale recycling operations are still lacking.

BOX 4.3

The Move to Recycle

Recycling is not new to Americans, but over the past twenty-five years, it has received a new impetus. The long-established reason for recycling has been that a recycled resource such as iron, aluminum, glass, or paper could be provided to industry at a lower cost by recycling than through the use of virgin materials. Consequently, there has long been a viable market for scrap iron, lead, copper, and aluminum. Other materials were recycled locally and episodically, especially during national emergencies, such as World War II.

In the post–World War II era, the United States began to diverge from its European allies in terms of recycling. The rebuilding of Europe required enormous resources, and the resulting postwar shortages made recycling attractive. At the same time, the United States, which was buoyed by prosperity and the development of a suburban society, largely abandoned recycling efforts and rapidly became a *no deposit, no return* society. Despite the concerns of numerous individuals and environmental groups, recycling interest waned and the volume of household and commercial waste generated in the United States grew to massive proportions (see Figure 4.35). The U.S. Environmental Protection Agency estimated that in 2008, Americans were generating 4.6 pounds of solid waste per person every day; the total estimated for the country was more than 250 million metric tons.

During the 1980s, the dual impacts of a rising environmental consciousness and a shortage of landfill capacity suddenly brought recycling back into focus. Landfill regulations had become more stringent, requiring more protection of groundwater from leachate, requiring that solid waste be covered each day, and banning open burning of waste. However, landfills were still the least expensive ways to dispose of unwanted materials. The threat of federally imposed mandates issued by the U.S. Environmental Protection Agency in 1991 forced many states to develop their own regulations for better operation and environmental monitoring of landfills; this resulted in large and rapid increases in landfill "tipping fees," or the charges paid to dispose of wastes. This forced states and individual communities to reexamine the merits and costs of simply landfilling waste materials. Suddenly, it became clear that the recycling of waste not only saved resources, it also saved energy. In general, less energy was required to recycle than to manufacture a material from the raw mineral resources. It also reduced the potential for groundwater pollution, and it certainly reduced the volume of material going to landfills. Consequently, several state legislatures passed mandates for the recycling of 10, 25, or even 50 percent of their waste throughout the 1990s.

Two major problems arose. First, it became apparent that the infrastructure to recycle on a broad basis in suburban areas was very expensive; second, no adequate markets existed to absorb all of the recycled material. Fees for dumping into landfills varied widely and were generally rising, but in the mid-1990s, they typically were $20 to $50 per ton. In contrast, the cost of recycling frequently was $100 or more per ton. Aluminum cans were of sufficient value to pay for the effort of collection, but glass and plastic markets were inconsistent and rarely did better than break even. Newsprint, the largest volume solid-waste stream in America, commonly cost more to collect than could be obtained from selling it. In the late 1980s, there were relatively few companies reprocessing paper, and there was not enough market for the finished products. Some communities cut back or abandoned recycling efforts. By the mid-1990s, a significant increase arose both in demands for recycled goods and in the industrial capacity to process them. Prices for recycled commodities rose, helping to offset more of the high handling costs. The recession that developed in 2008 and 2009 unfortunately reduced prices and resulted in many communities cutting back on recycling in order to save money.

The amount and percentage of materials recycled remained constant through the 1960s and increased only slightly through the 1970s and early 1980s (see Figure 4.35). Since 1985, the recycling rate has increased so that in 2008, it was about 33 percent of the total waste produced in the United States. Many communities have taken a long-term view and recognize that, despite the high costs, there is great value in recycling. Future landfill costs could typically run from $50,000 to $250,000 per acre. Present tipping fees to operate on existing landfills may only be $50 per ton, but the future costs may raise the tipping fees in many places well in excess of $100 per ton. Thus, looking ahead, the extra costs of recycling now will save much cost and aggravation later.

PART 3

Energy

Earth can be thought of as a large, complex engine in which vast amounts of heat energy flow constantly from one part to another and matter is moved, often slowly, both within the planet and at its surface. Throughout Earth's history, the two major sources of energy that have driven these movements have been radiant energy from the Sun and geothermal energy from within Earth. Solar energy is responsible for driving the changes within the atmosphere, the hydrosphere, the biosphere, and the top few meters of the lithosphere. Geothermal energy comes in part from heat that still remains deep in the interior from Earth's fiery beginnings, and in part from the release of heat by natural decay of radioactive isotopes of elements such as uranium, thorium, potassium, and strontium that are distributed in small amounts throughout the crust and mantle (Figure Pt. 3A).

Solar energy is manifested in many ways, from the winds in the atmosphere and ocean currents in the hydrosphere, to photosynthesis by plants and, hence, maintenance of the modern biomass. The transformation of solar energy into moving air in the form of wind and moving water in rivers presents us with forms of kinetic energy that can be converted directly into useful operations such as turning windmills or spinning hydroelectric turbines. Photosynthesis stores some of the solar energy in

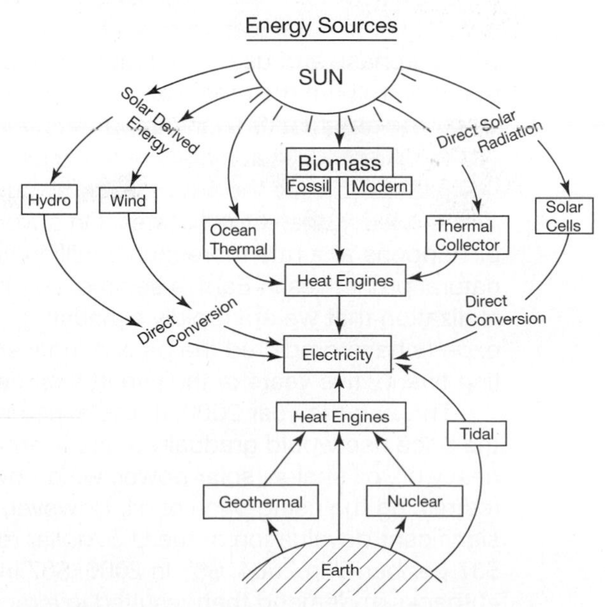

FIGURE Pt. 3A The ways in which electricity may be obtained from the energy of the Sun, either directly or indirectly, and from Earth.

the form of the hydrocarbon molecules of which plants are composed. We can release that energy for useful applications by burning the biomass and then using the resulting gases to propel engines, or using the heat generated to convert water into steam, which can then be used to drive a turbine.

Earth's natural nuclear engine generates heat by the spontaneous decomposition of the radionuclides. The rocks comprising most of Earth's upper mantle and crust are quite good insulators that allow the heat to build up more quickly than it can be transmitted away. Irregular distributions of the radionuclide elements within Earth, and variations in overall compositions and rock strengths, result in temperature gradients and convective movements within the mantle. On a large scale, this dynamism causes large convecting cells that drive plate tectonics; on a smaller scale, it produces igneous intrusions, volcanoes, geothermal fields, and hot springs.

FOSSIL FUELS

Fossil fuels, so called because they are the remains of biological materials, are the dominant source of energy being used by humans in the world today—some 85 to 90 percent of the total! It was not always so, nor will it be so for long in terms of Earth history or even human history. The massive use of the fossil fuels came about in stages. It began with coal, used in small amounts at first, but coming into prominence with the beginning of the Industrial Revolution in the eighteenth and nineteenth centuries. The modern petroleum era began with a single well drilled for oil in northwestern Pennsylvania in 1859. But oil did not make any significant inroads as a source of the world's energy until automobiles appeared in large numbers in the early twentieth century. Natural gas, the second major component in petroleum reservoirs, though long known, was very limited in usage until distribution systems became available in the era following World War II. Although oil and gas fields exist in most of the major sedimentary basins of the world, the largest petroleum occurrences exist around the Persian Gulf.

The rapid rise in the use of petroleum around the world and the emergence of the cartel known as the Organization of Petroleum Exporting Countries (OPEC) has made petroleum both a political and economic weapon. Now, early in the twenty-first century, world oil reserves are estimated to be about 1000 billion barrels; oil consumption has risen to 85 million barrels per day (or about 31 billion barrels annually) and continues to rise. This extraordinary rate of consumption costs those who purchase and use the oil more than six billion dollars per day worldwide; it has made petroleum the most valuable natural resource in the world. The United States is by far the greatest user and importer of oil. Although the United States was energy self-sufficient for many years, it no longer is, and its dependence on imported oil has risen steadily since the 1960s. Imports now account for about 75 percent of the oil that is used in the United States. On a worldwide basis, the enormous production of oil happens at a rate more than a million times faster than oil is formed through natural processes in Earth's sedimentary basins. This disparity has led to the realization that we are rapidly exhausting the world's supply of petroleum. Many experts have predicted the peak of conventional oil production to be some time in the first twenty five years of the twenty-first century.

Through the year 2000, it was expected that oil prices would rise slowly and that the price rise would gradually make alternative energy sources such as tar sands, heavy oil, oil shales, solar power, wind power, and biofuels competitive, thus restraining the rising price of oil. However, the rapid growth in demand for oil and the significant devaluation of the U.S. dollar resulted in a steady rise in world oil prices to $37 per barrel in 2004, $67 in 2006, $87 in 2007, and $148 in 2008. The resulting cutbacks in demand then resulted in a rapid decline back to around $50 per barrel in Fall 2008 and to below $40 per barrel in the early part of 2009. The rapid increase in the price of oil promoted the search for petroleum in remote localities such as in deeper, offshore waters and in arctic regions. It has also raised the oil in tar sands, such as those in western Canada, from resources into reserves and stimulated their exploitation. Renewable energy sources, such as wind and solar power, which had advanced slowly because of their cost and long payback times, have become much more competitive and are now receiving much more attention. Even nuclear power is being reconsidered. Unfortunately, the rapid decline in oil value in the early months of 2009 has removed some of the impetus toward developing renewable energies; further rapid price swings will make planning for the most economical and environmentally acceptable energy sources difficult.

Natural gas, which occurs with petroleum deposits, has risen in importance as a fuel since the middle of the twentieth century as the infrastructure needed for gas distribution has been developed. Natural gas is a "greener" fuel than either oil or coal because it is relatively easily cleaned of impurities and because it creates less carbon dioxide per unit of heat than the other fossil fuels when burned. Increasingly, natural gas, which was once just vented or burned-off as a waste, is now captured and even liquefied for worldwide distribution. Other types of natural gas are now extracted from coal seams to promote safety in coal mining and for its economic potential as a fuel or captured from landfills where it forms as a natural product of waste decomposition.

Coal, which powered the beginnings of the Industrial Revolution, remains an essential energy source in many countries, including the United States and in much of Europe, because it is the chief fuel of the electric power industry. The mineable reserves of coal are very large and are considered to be sufficient to power the world for several hundred years, but there are many environmental concerns attached to its use. These include the difficulties of underground coal mining, the release of carbon dioxide into the atmosphere, and the contribution of gaseous effluents to acid rain. Much effort has been expended in attempts to make "clean coal" by removing sulfur-bearing compounds and making liquid fuels from coal as petroleum substitutes. In addition, the mining of coal leaves much larger and more visible environmental scars than does drilling the wells through which oil and natural gas are removed.

NUCLEAR ENERGY

Radioactivity was discovered at the end of the nineteenth century but was not harnessed as a means of producing commercial electrical energy until the 1950s. The vast potential of radioactivity to generate electricity by using heat from controlled fission reactions got off to a promising start with twenty years of nearly accident-free success; this led to some naïve views that nuclear-generated electricity would become so inexpensive it would be virtually free. A critical blow to the nuclear industry occurred with the partial meltdown of the Three Mile Island reactor in Pennsylvania in 1979. Despite no harm being done to humans by radiation and the recognition that human error was the cause, the American public was sufficiently frightened that every nuclear plant that was still in the planning process was cancelled. The nuclear power industry was dealt another serious blow when the Chernobyl power plant, near Kiev in the Ukraine, caught fire in 1986 and released massive amounts of radiation. Although the immediate death toll due to the Chernobyl accident was small, hundreds of thousands of people were irradiated and tens of thousands are expected to develop cancers in the several decades to follow.

Despite the fact that nuclear power currently accounts for approximately 20 percent of world electricity generation (18 percent in the United States, 20 percent in the United Kingdom, and around 70 percent in France) and the fact that there have been no major problems for more than 20 years, there remains much public distrust. Many countries have had to make difficult decisions in providing electricity for their growing and increasingly technologically oriented populations. Consequently, the future of nuclear power varies considerably from one country to another. For example, France has decided that nuclear plants are the best source for most of its electricity, but neighboring Germany has decided to phase out all existing nuclear plants. China, Japan, and India are constructing new plants to meet growing needs, but the United

States and Great Britain remain undecided regarding the construction of new nuclear plants, even as the aging original plants approach closure.

One of the most perplexing nuclear power plant problems has been what to do with the highly radioactive, but no longer usable ("depleted") fuel rods after they have been employed in a nuclear power plant. Countries variously store or reprocess the rods because some fuel remains in them, but the United States, for example, has not yet formally decided on either a reprocessing policy or a permanent storage facility for used fuel rods. Yucca Mountain, in the desert of southern Nevada, was selected as the most likely first storage site, but ongoing scientific studies as well as political disputes have delayed the opening and operation of the facility. Congressional agreements had mandated the receipt of nuclear waste at Yucca Mountain by 2010, but later discussions pushed the earliest opening date to 2017. Furthermore, in 2009, strong political groups announced that Yucca Mountain would never be used as a nuclear waste repository.

Consequently, the future of Yucca Mountain remains unknown and the United States will have spent at least its first 60 years of nuclear power generation with no firm solution to the problem of the disposal of its fuel wastes. In addition to concerns about the actual burial sites, there has long been opposition to the transport of the highly radioactive spent fuel rods; the moving of rods from the approximately 100 different power plant sites around the United States to Nevada will certainly continue to be a contentious issue. The transport of nuclear fuel materials by railroad has already led to nonviolent protests in Germany and threats of similar actions in other countries.

RENEWABLE ENERGY

Despite our current dependence on fossil fuels and the significant contribution of nuclear power toward electricity generation worldwide, Earth has other major potential sources of energy including several that can be considered "renewable," unlike fossil fuels. Renewable energy sources range from long-used sources such as wind and hydroelectric power to those power sources used only more recently, such as solar power and geothermal energies. The three major problems of fossil fuels—rising cost, environmental impact, and growing foreign dependency—have focused attention and investment on alternative sources of energy in countries such as the U.S. and the U.K. Worldwide, the use of alternative energy sources is small, but they are growing rapidly and promise to take on much greater importance in the twenty-first century. Many scientists and some politicians have suggested investment in these alternative energy sources, but a special push came in mid-2008 when former American vice president and Nobel prizewinner Al Gore proposed that the United States convert all of its electricity production to renewable sources by 2018. By far the most productive of these renewable sources today is hydroelectric power, which supplies nearly all of the electricity in Norway, although only about 6 percent of the electricity in the United States. While viewed as environmentally friendly by some, hydroelectric power is decried as environmentally disruptive by others because rivers must be dammed to construct a plant. Environmental considerations have resulted in the decision to remove many hydroelectric power dams across the United States, but at the same time, many large dams continue to be constructed in other parts of the world. Construction of the greatest hydroelectric power dam of all time, the Three Gorges Dam, was recently completed in 2007 on the Yangtse River in China: it has displaced more than a million people and flooded hundreds of square miles of the river plain that are now lost to agriculture. Even greater in scope is the Great Inga Dam, proposed to be built on the Congo River in Africa. If built, it would be three times larger than the Three Gorges Dam and would provide one-third of the electrical needs of all African countries. However, the high cost, lack of infrastructure, and political instability of the region threaten the success of the project even before it has begun.

Other renewable energy forms, such as power from the wind, the Sun, the tides, geothermal heat, and from the biomass are locally abundant and useful but make only small contributions. Despite each of these being called a"green energy source," each actually brings its own environmental impact. Solar, or photovoltaic, energy is commonly viewed as the most promising of the alternative energies and is growing rapidly, but it presently accounts for only a tiny fraction of electricity production. It has great promise, but it will have to grow by many thousand times to replace just one percent of the production currently coming from fossil fuel or nuclear sources.

Wind energy has seen great growth worldwide, ranging from the plains of Texas to the offshore areas of Denmark and Germany. New designs and improved efficiencies have lowered costs of production of windmills and their generators, but environmental concerns regarding their aesthetic impact on the landscape and the hazard presented to migrating birds remain problems. Like solar power, wind energy production is growing rapidly and is highly regarded because it does not increase the CO_2 content of the atmosphere.

The digestion of plant materials to make alcohol as a substitute for gasoline has long been known and used locally (especially in Brazil in recent years). Concerns over the price of fossil fuels and the dependence on foreign oil have led to farm subsidies and increased production of biofuels, especially in the United States. Farmers rapidly shifted from growing corn for food to growing corn for biofuel production. Unfortunately, this did little to reduce foreign dependence on oil, but it did rapidly push up the cost of grain. This had especially unfortunate consequences for food prices in poorer countries. Concerns that fuel production might reduce the world's food supplies has, of course, led to increasing experimentation with nonfood vegetation, such as switchgrass or inedible stalks of wheat, corn, and rice. Unquestionably, the production and use of biofuels will grow over the coming years, but its consequences must be carefully monitored.

Two alternative energy sources that are viewed by many experts as having the long-term potential to meet human needs are hydrogen fuels and nuclear fusion. Hydrogen can be burned much like natural gas and can be used in modified automobiles directly as fuel or in fuel cells, with the only waste product being water vapor. However, hydrogen is difficult and dangerous to handle, and no infrastructure yet exists for its distribution. "Fusion is the energy of the future and always will be" is the view of many. Fusion is the energy source for the Sun, but successful controlled nuclear fusion on Earth appears to be decades away at the earliest. The great advantages of limitless power and few waste products are presently offset by the difficulty of carrying out sustainable fusion at reasonably attainable temperatures.

It is likely that our descendants will wish to use energy at least at the rate that we do today. However, the fossil fuels that presently supply most of the world's energy cannot continue to be used in the long term, so our descendants will have to get their energy from other sources. Earth's environment is energy rich, but much of this energy cannot be readily captured and put to use. Our descendants will face a major challenge, not so much in having energy available as in the price to be paid for it in money and damage to the environment.

THE ENERGY DEBATE

Although the need for energy to power modern societies is clear, the choice as to the source of that energy is not. There are continuing debates as to the appropriate sources that would serve best in different places. The parameters that need to be considered include cost, safety, availability, import reliance, international politics, environmental impact on land, air, water, wildlife, waste disposal, and even aesthetic factors. The decision to develop an energy supply for a town, power company, state, or region requires analysis of a complex matrix, weighing each of those factors.

The rapid expansion of the oil industry in the early part of the twentieth century combined with the belief that oil supplies were nearly limitless led to an intense clustering of oil wells, as seen at Signal Hills, California, in the 1920s. (Courtesy of Shell Oil Company.)

CHAPTER 5

Energy from Fossil Fuels

Oil has helped to make possible mastery over the physical world. It has given us our daily life and, literally, through agricultural chemicals and transportation our daily bread. It has also fueled the global struggles for political and economic primacy. Much blood has been spilled in its name. The fierce and sometimes violent quest for oil—and for the riches and power it conveys—will surely continue as long as oil holds a central place. For ours is a century in which every facet of our civilization has been transformed by the modern and mesmerizing alchemy of petroleum. Ours truly remains the age of oil.

Daniel Yergin, The Prize, 1991

FOCAL POINTS

- The most familiar fossil fuels and largest present-day energy sources are coal, petroleum, and natural gas. Less familiar but of local or growing importance are peat, oil shales, tar sands, and heavy oils.
- Coal, which forms as a result of the accumulation, burial, and progressive compaction and heating of land plants, occurs in ranks of increasing carbon content. It begins as peat, but may be transformed into lignite, sub-bituminous coal, bituminous coal, or anthracite.
- Petroleum forms from marine organic matter that is trapped in sediment and transformed as a result of increasing temperature and pressure during burial.
- Natural gas, composed primarily of methane (CH_4), forms biogenically when bacteria decompose shallowly buried organic matter, and thermogenically when the temperature and pressure of deep burial cause some decomposition and reconstitution of the organic matter.
- Petroleum, although known in ancient times, traces its modern production history back to a well drilled by Edwin L. Drake in Titusville, northwestern Pennsylvania, in 1859.
- Accumulations of petroleum are valuable only where sufficiently large quantities of petroleum have migrated from their source rocks into porous and permeable rocks that serve as structural or stratigraphic traps.
- *Petroleum* rarely comes out of wells as gushers. Oil is either forced out by pumps or is extracted with the injection of chemicals or steam to release it from the rock pores. Even so, close to half of the original petroleum remains trapped in the rocks after the oil has been pumped out.
- World petroleum production rose substantially from 1950 to 2008 and is presently about 85 million barrels per day (31 billion barrels annually). The rate of consumption is more than a million times faster than the rate at which nature creates new oil.
- Oil production in the United States peaked in 1970 and has been slowly declining since then; this trend, coupled with an increase in U.S. demand, has resulted in increased imports of oil into America.
- OPEC (Organization of Petroleum Exporting Countries) controls about three-quarters of the world's total oil reserves of about 1000 billion barrels; Saudi Arabia, with 260 billion barrels, has the largest reserves.

- The world's liquid petroleum production is predicted to peak in the first quarter of the twenty-first century. Anticipation of the event is encouraging the search for alternative energy sources.
- Natural gas usage has increased more than fivefold since 1960 and much is now transported by pipelines and as liquefied natural gas. Coal-bed methane is being increasingly developed as a new source of natural gas.
- Tar sands, heavy oils, and oil shales contain very large amounts of potentially extractable oil; most remain uneconomic, but emerging technologies, particularly in the case of tar sands, are beginning to allow for increasing production.
- Beginning in about 2003, the combination of reduced Iraq oil production, the rapid growth of world demand for oil, and the fall in the value of the U.S. dollar, the price of oil began a steady rise from about $60 per barrel in 2003 to about $150 in 2008. By February 2009, the price had fallen again to below $40 per barrel. This appears to have ushered in a period of rapid variations in the price of oil.

INTRODUCTION

Every action we take, every procedure we design—even the process of reading these words and thinking about their meaning—requires **energy**, which is the actual or potential ability to do **work**. The primary function of the machines we build is to convert energy into useful work. We ourselves are machines that convert the chemical energy stored in the food we eat into the physical and mental activities we consider as useful work. Unfortunately, we are not very efficient when judged simply as machines, and even when we are at our most industrious, each of us is only capable, on a continuing basis, of working at a rate sufficient to keep a 75-watt lightbulb burning. Because of the limitations of human muscle-energy, far back in history our ancestors found it expedient to turn to supplementary sources to provide energy to do things that they did not want to do or of which they were not physically capable. The first of the supplementary sources were probably other humans (slaves) and animals. Because these sources proved inadequate to meet their needs, they turned to progressively more sophisticated means such as sails for ships, windmills, waterwheels, steam and internal combustion engines, electric motors, and eventually nuclear power plants and integrated electrical circuits.

The human appetite for energy has grown so large that supplemental energy expenditures now far exceed our individual muscle energy in every aspect of life. Whereas our earliest ancestors relied upon the energy from their own "one-manpower" bodies, we now augment our own body energy with that from a vast array of supplemental sources. We can envision this supplemental energy as "silent slaves" working continuously to feed, clothe, and maintain each and every one of us. The number of these energy slaves available varies widely from one country and culture to another. Hence, the supplementary energy used per person is equivalent to 15 energy slaves in India, 30 in South America, 75 in Japan, 120 in Russia, 150 in England and Europe, and about 300 in the United States and Canada. We see the work in the form of numerous machines that perform various jobs for us, but each of the machines has an Earth-supplied energy source such as petroleum, natural gas, coal, running water, or uranium. Just how dependent we have become on these energy slaves can be envisioned by considering the consequences of them going on strike (which would be the case if all of Earth's energy supplies ran out or if power supplies were suddenly not available, as was feared by some at the beginning of the twenty-first century). We would be reduced to our own muscle power to supply all of our needs. Our technological society would come to an abrupt halt and we would very soon find ourselves unable to feed and maintain the world's population. Muscle power alone could not hold back the inevitable starvation, famine, and pestilence that would rapidly reduce the world's population. Survivors would be forced to live much as our early ancestors did thousands of years ago, prior to the development of the resources that now provide most of our supplemental energy.

ENERGY UNITS

Because energy is the capacity to do work and because there are many types of work—mechanical, electrical, and thermal, for example—there are many different types of units by which energy is measured. The **joule** is an electrical unit that is defined as the energy needed to maintain a flow of current of 1 ampere for 1 second at a potential of 1 volt; the **calorie** is a heat unit and is the energy needed to raise the temperature of 1 gram of water 1° Celsius, while the **British thermal unit** (BTU), another heat unit, is the energy needed to raise the temperature of 1 pound of water 1° Fahrenheit. The variety and interchangeability of energy units is shown in Table 5.1. To be able to compare the energy that humans might be able to extract from various sources, we shall be using the joule as our standard unit of energy. However, to also permit us to relate to the energy production statistics we hear about on a daily basis, we shall also equate joules to the commonly employed units such as metric tons of coal, barrels of oil, and trillions of cubic feet of gas.

TABLE 5.1 Energy equivalences

1 btu	= 252 gram-calories = 1055 joules = 2.93×10^{-4} kwh
1 joule	= 0.239 gram-calorie = 0.00095 btu = 2.78×10^{-7} kwh
1 gram, calorie	= 4.189 joules = 0.00397 btu
1 watt	= 1 joule/sec = 0.239 cal/sec = 0.0569 btu/min = 0.00134 horsepower
1 Quad (btu)	= 10^{15} btu = 1.05×10^{18} joules = 2.93×10^{11} kwh
1 million (10^6) btu equals approximately:	
90	pounds of bituminous coal and lignite production (1982)
125	pounds of oven-dried wood
8	gallons of motor gasoline or enough to move the average passenger car about 124 miles (1981 rate)
10	therms of natural gas (dry)
11	gallons of propane
1.2	days of per capita energy consumption in the United States (1982 rate)
2	months of dietary intake of a laborer
20	cases (240 bottles) of table wine
1 million btu of fossil fuels burned at electric utilities can generate about 100 kilowatt-hours of electricity, while about 300 kilowatt-hours of electricity generated at electric utilities can produce about 1 million btu of heat.	
1 quadrillion (10^{15}) btu equals approximately:	
44	million short tons of bituminous coal and lignite production
63	million short tons of oven-dried wood
1	trillion cubic feet of natural gas (dry)
170	million barrels of crude oil
500	thousand barrels per day of crude oil for 1 year
35	days of petroleum imports into the United States (1982 rate)
30	days of United States motor gasoline usage (1982 rate)
1 barrel of crude oil equals approximately:	
5.7	thousand cubic feet of natural gas (dry)
0.26	short tons of bituminous coal and lignite production
1700	kilowatt-hours of electricity consumed
1 short ton of bituminous coal and lignite production equals about:	
3.9	barrels of crude oil
22	thousand cubic feet of natural gas (dry)
6600	kilowatt-hours of electricity consumed
1 thousand cubic feet of natural gas equals approximately:	
0.18	barrels (or 7.5 gallons) of crude oil
0.045	short tones (or 90 pounds) of bituminous coal and lignite production
300	kilowatt-hours of electricity consumed
1 thousand kilowatt-hours of electricity equals approximately:	
0.59	barrels of crude oil (although it takes about 1.7 barrels of oil to produce 1000 kWh)
0.15	short tons of bituminous coal and lignite production (although it takes about 0.5 short tons to produce 1000 kWh)
3300	cubic feet of natural gas (dry) (although it takes about 10,000 cubic feet to produce 1000 kWh)
27.2	gallons of gasoline

Although the total energy available from any source is important, we must also be concerned about the rate at which energy is used and the maximum rate at which it can be supplied. In the case of a windmill, for example, we cannot use energy faster than it is supplied by the wind pushing the blades. We must, therefore, also consider a time-dependent function called **power**, by which we mean the energy used per unit of time. The widely used term **horsepower**, familiar to most people, originated from the use of horse-drawn plows and wagons. It has persisted as a measure of the strength of engines, including those in automobiles because in 1766,

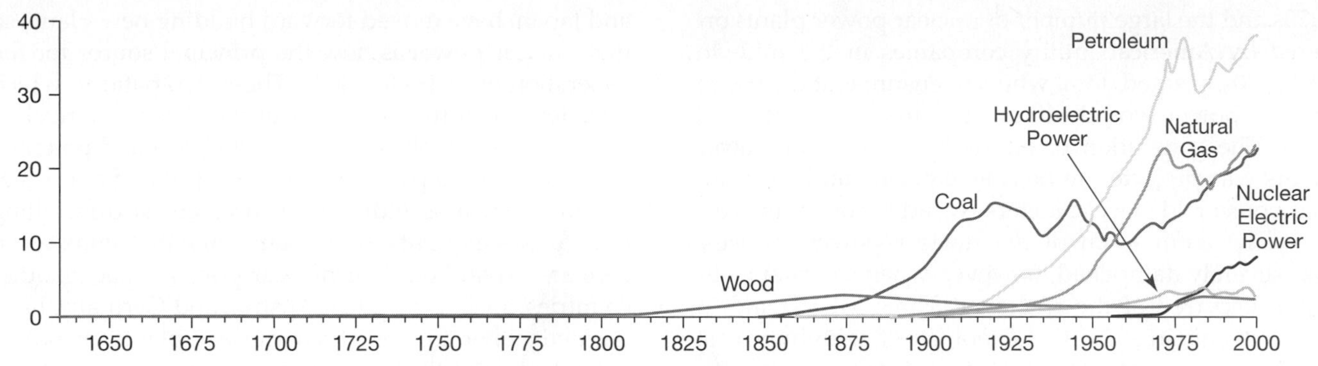

FIGURE 5.1 The changing nature of the United States principal energy supplies. Wood served as the dominant energy source until about 1880 when it was surpassed by coal. Coal then supplied the energy for the Industrial Revolution and remained the principal source until about 1950. The development of the automobile raised petroleum to the dominant position where it remains today. Natural gas gained in importance after the installation of transmission lines in the 1940s and 1950s; nuclear power only became available after the 1950s. It is clear that fossil fuels will continue to constitute the major sources of energy for the United States during the first half of the twenty-first century. (From the Energy Information Administation.)

James Watt measured the power of his steam engine against the power of a horse. In this text, for the sake of consistency, the power unit used is the **watt**, which is defined as the consumption of one joule of electrical energy per second. The watt is commonly used in every day discussion of household appliances and lightbulbs and will become even more familiar as we increasingly use energy from renewable sources, such as heat and light from the Sun. As with food and other renewable resources, it is not the total amount of solar energy that has or will reach Earth that is important, but rather the rate at which it continues to reach us.

THE CHANGING USE OF ENERGY

The human progression from simple hunter-gatherer to modern technological individual has been characterized both by vast increases in the amounts of supplemental energy consumed and by marked changes in the sources of the energy. Supplemental energy use remained relatively low until the Industrial Revolution swept across Europe and North America in the eighteenth and nineteenth centuries. Prior to that time, the primary sources of supplemental energy were wood, wind, running water, and animals.

The onset of the Industrial Revolution in England required unprecedented amounts of fuel to drive the newly invented steam engines. It was soon obvious that wood could not meet these needs and that the entire British Isles were in danger of being deforested. Consequently, the British turned to coal, which proved to be a superior heat source. Coal soon became the fuel for the Industrial Revolution worldwide; this led simultaneously to two significant changes in the energy-use pattern in the industrialized nations of the world: the total amount of energy used rose dramatically, and coal rapidly replaced wood as the major fuel. The scene was similar in each country and is especially well shown for the United States in Figure 5.1. From the time of the earliest settlers in the seventeenth century until the Civil War in the 1860s, wood constituted 90 percent, or more, of the American fuel source. With the introduction of the steam engine and other machines of the Industrial Revolution in the first half of the nineteenth century, coal use rose and wood dropped to 50 percent of energy source in 1880 and to only about 10 percent in 1900. Coal remained the dominant energy source, providing nearly 75 percent of the nation's energy in 1920, until the rapid expansion of oil and gas use surpassed it in the 1940s and 1950s. Total energy production in the United States and the world rose dramatically from the 1940s until the 1970s when the 1973 OPEC oil embargo and rapid price increases for petroleum shocked the world. These events led to a slower growth rate of world energy demand and a slight decrease in energy consumption of some major countries such as the United States (Figure 5.1). Nevertheless, the total energy used in the United States at the beginning of the twenty-first century was more than ten times the total amount used at the beginning of the twentieth century; in that same timespan, the U.S. population increased only about 4.5 times (from 76 million to about 275 millon).

Since about 1880, the fossil fuels—coal, oil, and natural gas—have served as the major sources of energy for the United States and all other industrialized nations. It has now become clear that supplies of these fuels, though large, are limited and that some day other fuels will be needed to replace them. Thus, it appears that the period from about 1880 to about 2200 will likely be known as the *Fossil Fuel Era* to historians in the future. This leads to an important question: "What will be the energy sources beyond the fossil fuel era?" The relatively rapid rise in nuclear power in the 1960s and

1970s and the large number of nuclear power plants ordered by American utility companies in the mid- to late-1970s, seemed, for a while, to ensure that the use of nuclear power would ultimately surpass that of fossil fuels. The generation of electricity by nuclear power plants was projected to be safe, efficient, and so cheap that household electric meters would become unnecessary. The enthusiasm of the nuclear power industry was severely dampened, however, when a partial meltdown occurred at the reactor at Three Mile Island in Pennsylvania in 1979. This event, along with huge cost overruns in the construction of nuclear power plants, reduced projection for electricity demand and resulted in the widespread realization that safe disposal of nuclear waste is an enormous problem. Ultimately, many projected nuclear power plants were cancelled, including many already partially constructed plants.

The magnitude of the nuclear power industry problems in the United States is perhaps best exemplified by the default of the Washington Public Power Supply System (known as WPPSS in Washington or "Woops" on Wall Street) in construction of power plants after the investment of $2.8 billion, much of which was in privately held bonds. All of these problems were compounded and magnified by the meltdown at the Chernobyl nuclear plant in the Ukraine region of the former Soviet Union in 1986. Accordingly, the long-term fate of the 110 nuclear power plants operating in the United States is uncertain. While the United States has continued to debate the potential and desirability of nuclear power, other countries, and in particular France and Japan, have moved forward building new plants so that nuclear power is now the principal source for the generation of their electricity. These two nations, which have few fossil fuel resources of their own, see nuclear energy as preferable to the increasing cost and potential unreliability of imported fossil fuel supplies. Some other countries, such as India and China, are also installing nuclear power plants. At the same time that some countries are expanding their nuclear power capacity, other countries, such as Sweden, Austria, and Germany, have decided either to close their nuclear plants or not to grant new operating permits.

In any discussion of energy, it is important to consider the usage as well as the source. This is shown in Figure 5.2, which, although drawn for the United States, is representative of most industrialized nations. This diagram illustrates the complex nature of energy generation and the use of energy in the transportation, industrial, residential, and commercial realms. Unfortunately, the laws of thermodynamics prevent us from building 100 percent efficient machines. Actual energy losses range widely, from 10 percent to 65 percent, depending on the means of generation and conversion, and are highest in electricity production (Figure 5.3). The rapid rise of energy costs in the 1970s and early 1980s led to much energy conservation and significant improvements in efficiency, but the overall efficiency will probably never rise above about 80 percent. Our discussion of energy begins with the fossil fuels because they are still the principal energy sources in the early twenty-first century.

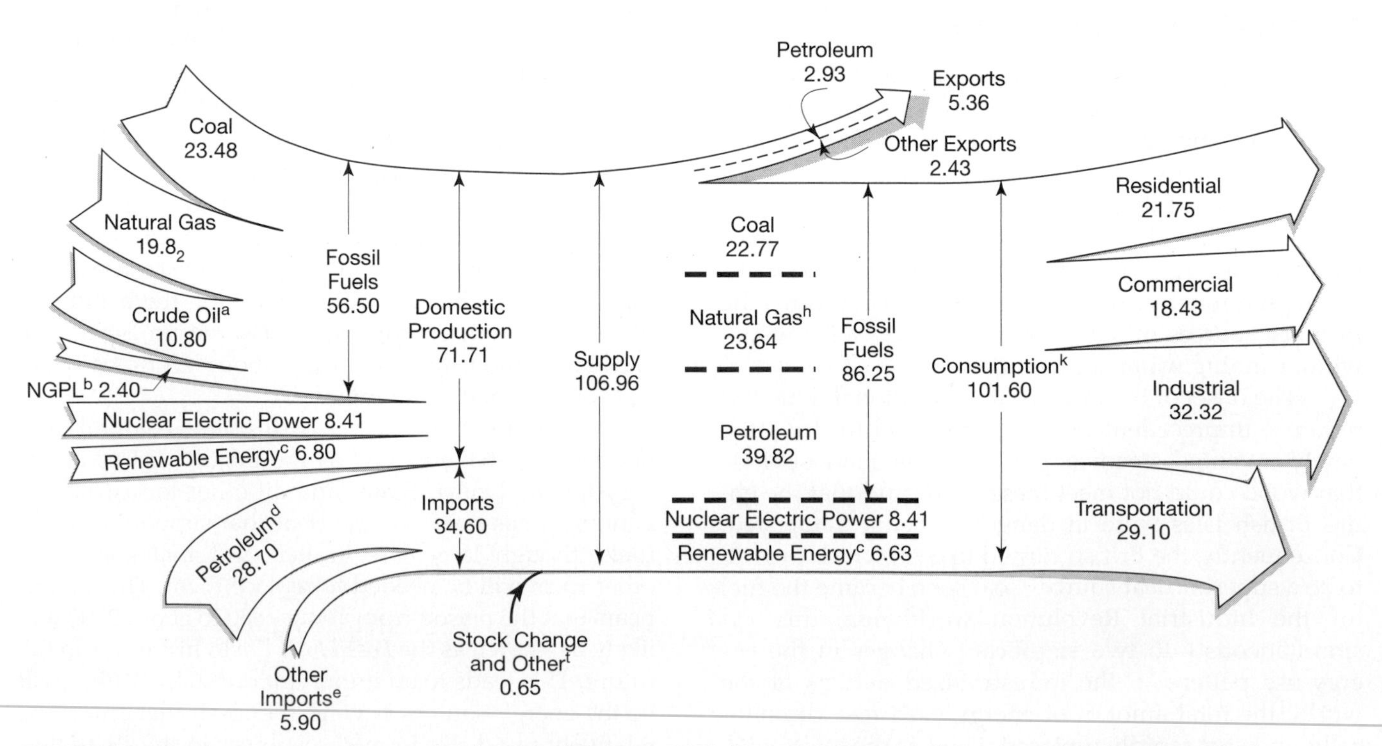

FIGURE 5.2 Total energy flow diagram for the United States in 2007. (From U.S. Energy Information Administration, 2009.)

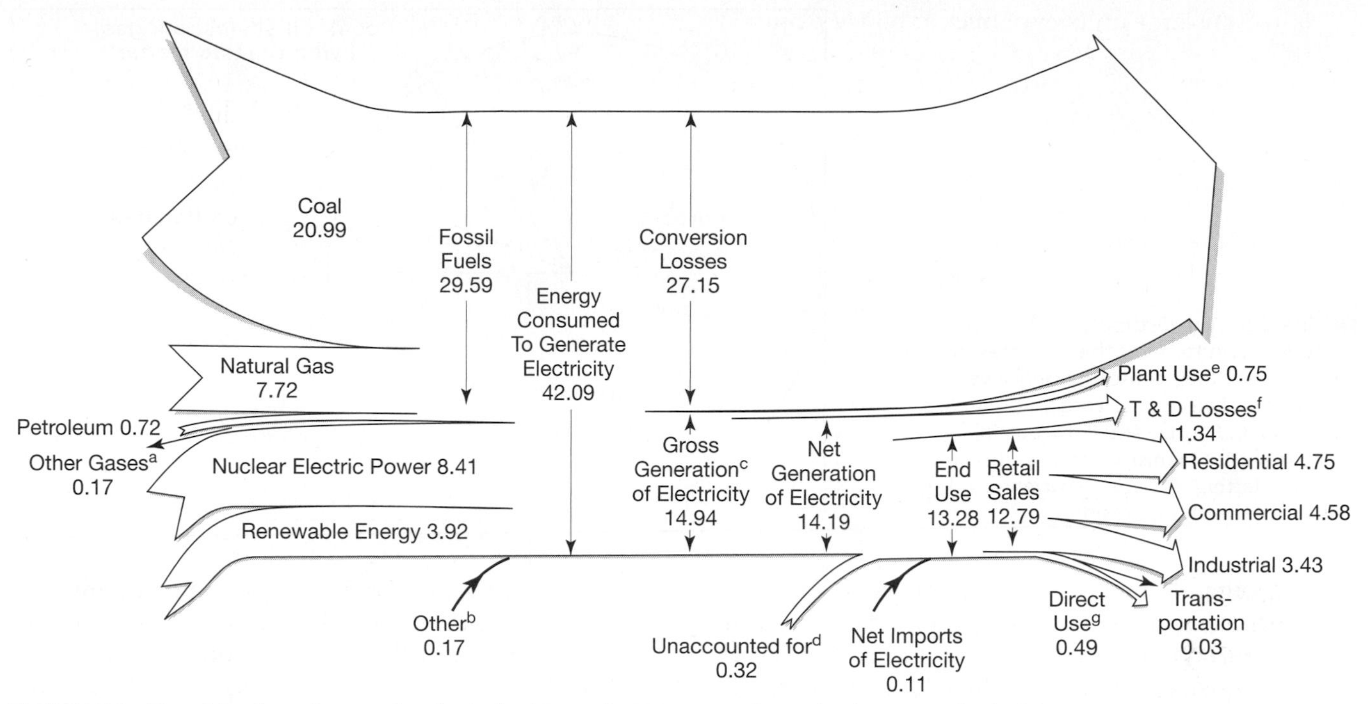

FIGURE 5.3 Electricity flow diagram for the United States in 2007. Note that only about one-third of the energy actually is converted into usable electricity. The large conversion losses result from the inefficiency of converting heat energy into the mechanical energy of turbines in conventional electricity plants. (From the U.S. Energy Information Administration, 2009.)

FOSSIL FUELS

All sedimentary rocks contain organic matter, with amounts ranging from traces in some sandstones to the major constituents in coals and oil shales. This organic matter, consisting of hydrocarbons in a great variety of different molecular species, is all part of the **carbon cycle** (Figure 5.4). The carbon cycle is a complex series of chemical reactions by which the carbon passes back and forth from solid rock to carbon dioxide in the air, to dissolved gas in waters, to plants and organisms. Most

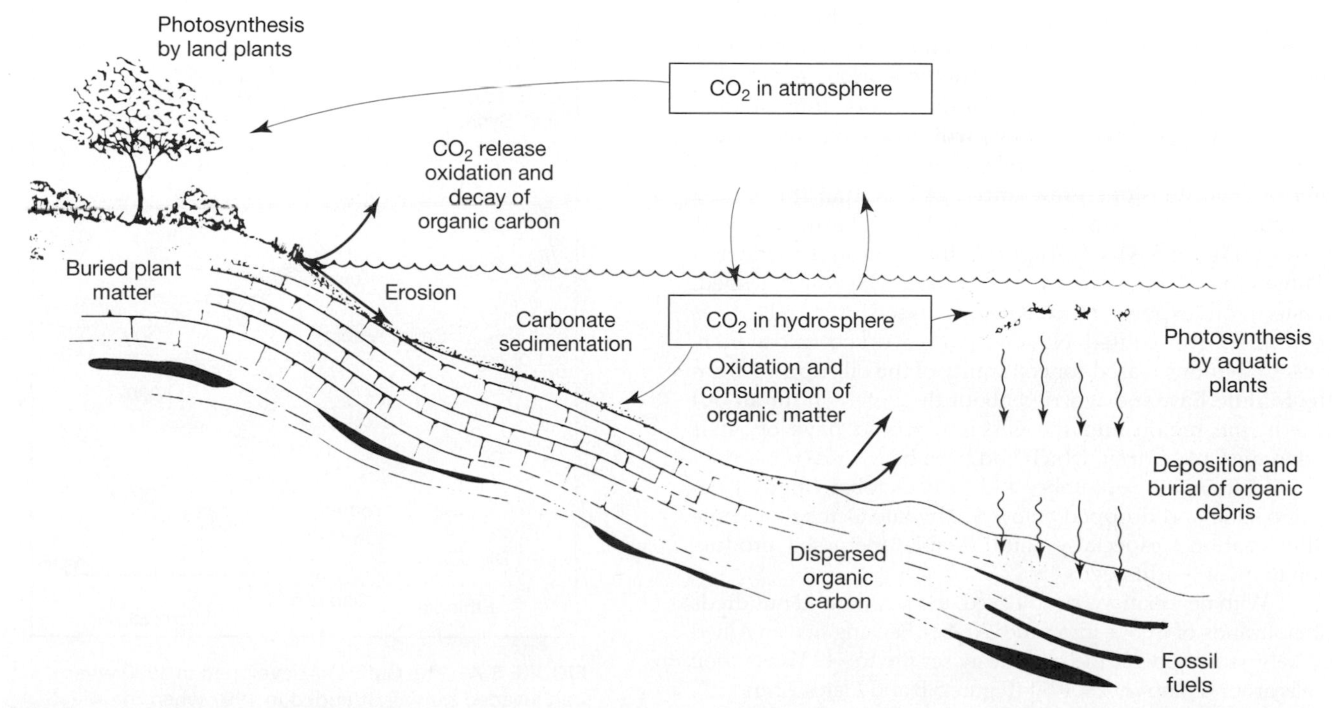

FIGURE 5.4 The main components of the carbon cycle. Fossil fuels are formed as a result of atmospheric carbon dioxide being converted by photosynthesis into organic matter. Subsequently, some of the organic matter is trapped in the sediments, and only a small portion of that is preserved as fossil fuels.

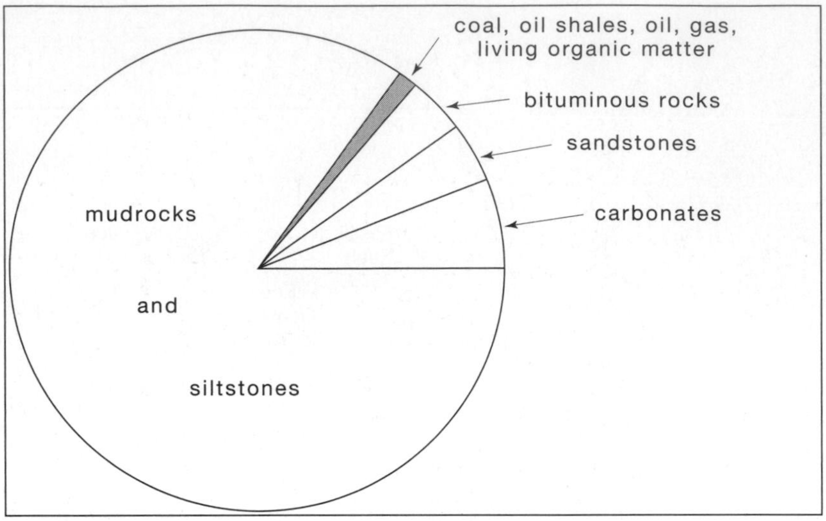

FIGURE 5.5 The distribution of organic matter in sediments in terms of the total mass of organic carbon. Fossil fuels constitute only a very small proportion of all organic carbon. (From M. A. Barnes, W. C. Barnes, and R. M. Bustin, "Diagenesis 8: Chemistry and Evolution of Organic Matter," *Canada Geoscience*, vol. 11 [1984]. Used with permission.)

of the organic matter, or *biomass,* that forms is ultimately consumed or destroyed by oxidation, returning the carbon to other parts of the carbon cycle. However, a small fraction of the organic matter, estimated to be no more than 1 percent of the total biomass, is preserved by burial in various types of sediments. Of this preserved material, most occurs as minor disseminated components of fine-grained sediments and some bituminous rocks (Figure 5.5). Lesser amounts are present in carbonate rocks and sandstones. Only a very small percentage of the preserved organic material is sufficiently concentrated to be valuable as a **fossil fuel**. The term *fossil fuel* is rather loosely defined, but is generally understood as the organically derived sedimentary rocks and rock products that can be burned for fuel.

BOX 5.1

The Persian Gulf War 1990–1991: Oil, Politics, Environment

The world focused its attention on the Middle East from the summer of 1990 through the fall of 1991 as a drama involving resources, politics, and the environment played out. This region, with more than 60 percent of the world's recoverable liquid petroleum reserves, has been a major international concern since 1973 when OPEC, with many member countries in the area, imposed an oil embargo. Through the summer of 1990, tensions grew once again as Iraq threatened occupation of its small but oil-rich southern neighbor, Kuwait (Figure 5.A). On August 2, Iraq overran Kuwait and claimed the country as a province of Iraq. The world gasped, western oil exports from Kuwait ceased, and oil prices soared. As the United Nations (UN) worked out a plan, western nations feared for continuity of the oil supplies from the Middle East and worried about the potential for all-out war in the region and the effects it would have on their economies. Oil prices, which had been below $20 per barrel, rose to $40 in late September and early October, but they began to slide and dropped below $30 by late October as some OPEC nations, especially Saudi Arabia, increased production to meet world needs.

With negotiations deadlocked, the UN poured hundreds of thousands of troops into Saudi Arabia. Sensing that an Allied strike was imminent, the Iraqi army set fire to 749 Kuwaiti oil wells around January 28, 1991 (Figure 5.B and Plates 22 and 23). Millions of barrels of oil burned every day; smoke from the

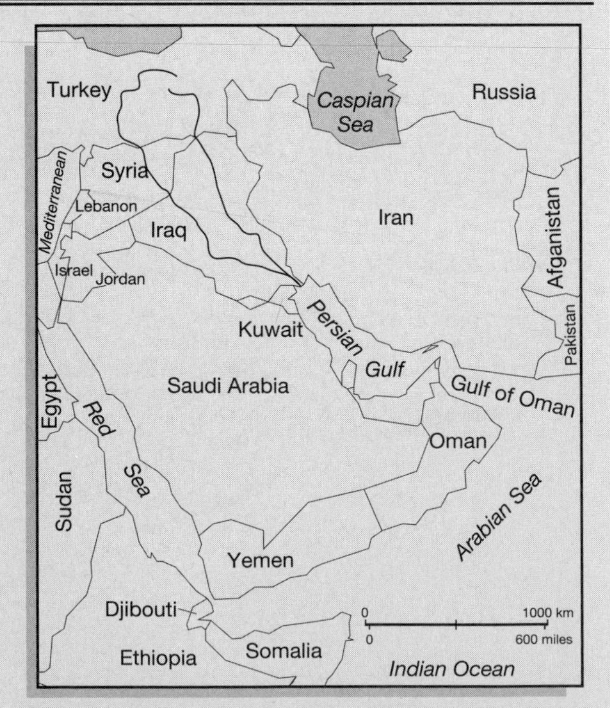

FIGURE 5.A The Gulf War developed in 1990 when Iraq invaded Kuwait. It ended in 1991 when the Allied forces liberated Kuwait.

FIGURE 5.B Damaged oil wells in Kuwait at the end of the Gulf War burned for up to six months, consuming large quantities of oil and creating extensive pollution. (Photograph by Jonas Jordan, U.S. Army Corps of Engineers.)

wells was so thick that it appeared to be nighttime in the middle of the day. The high gas pressure in the escaping oil resulted in roaring plumes of fire that rose scores of meters (nearly 100 feet) in the air and created trails of black soot and smoke that extended hundreds of kilometers (miles) downwind. Simultaneously, the Iraqi army began releasing crude oil into the Persian Gulf as a means of slowing any UN amphibious assault on the coasts of Kuwait and Iraq. These actions constituted an enormous loss of crude oil and were widely viewed as ecological terrorism. The UN forces attacked on February 24, 1991, raced rapidly through western Kuwait and southern Iraq and forced a surrender by Iraqi forces within five days.

The combatant side of the war was over, but the flames of 749 burning wells and the outflow of oil into the Persian Gulf continued. Initial estimates were that well fires might burn for as long as two years and that the oil spill into the Gulf was as large as 11 million barrels. Fortunately, both of these were wrong. Close to 9000 workers from 37 countries managed to completely extinguish the fires, one by one, with the last fire finally quenched on November 6, 1991.

The true volume of oil released into the Persian Gulf will probably never be known, but estimates ranged from 2 million to 11 million barrels; this represents one of the largest spills in history. At the time of the hostilities, individual oil slicks exceeded 130 kilometers (were nearly 100 miles) in length and killed countless birds, fish, turtles, and other types of aquatic life. One year after the war, as much as 600 kilometers (400 miles) of Saudi Arabian beaches remained paved with slabs of sticky crude oil–sand mixtures. This devastated the marine life of the intertidal zone; the best estimates for complete recovery are 80 to 100 years.

There were many fears that this story would be repeated when the United States and its allies invaded Iraq in 2003 in response to a perceived threat from the Iraqi dictator, Saddam Hussein, who was at that time believed to have acquired weapons of mass destruction. Despite the intensity and prolonged nature of the Second Gulf War, there was no major oil pollution. Iraq's oil production was curtailed but had nearly returned to pre-war levels by the end of 2008.

Fossil fuels occur in three major forms that are familiar to most people—**coal, oil,** and **natural gas**. Less well known, because of only limited and local use, are **peat, oil shale, tar sand,** and heavy oil. Although the various fossil fuels are quite different in appearance and are processed and utilized in different ways, they all share a common origin—they are trapped organic debris in sedimentary rocks (Figure 5.6). Their variety results from differences in the kinds of original organic matter (for example, leaves and stems in a freshwater swamp versus phytoplanktonic organic matter in a marine basin) and the degree of alteration that occurs as a result of bacterial decay, during and after trapping, and as a result of rising temperature and pressure due to increasing depth of burial.

The changes that occur in buried organic matter are both progressive and irreversible; they are shown

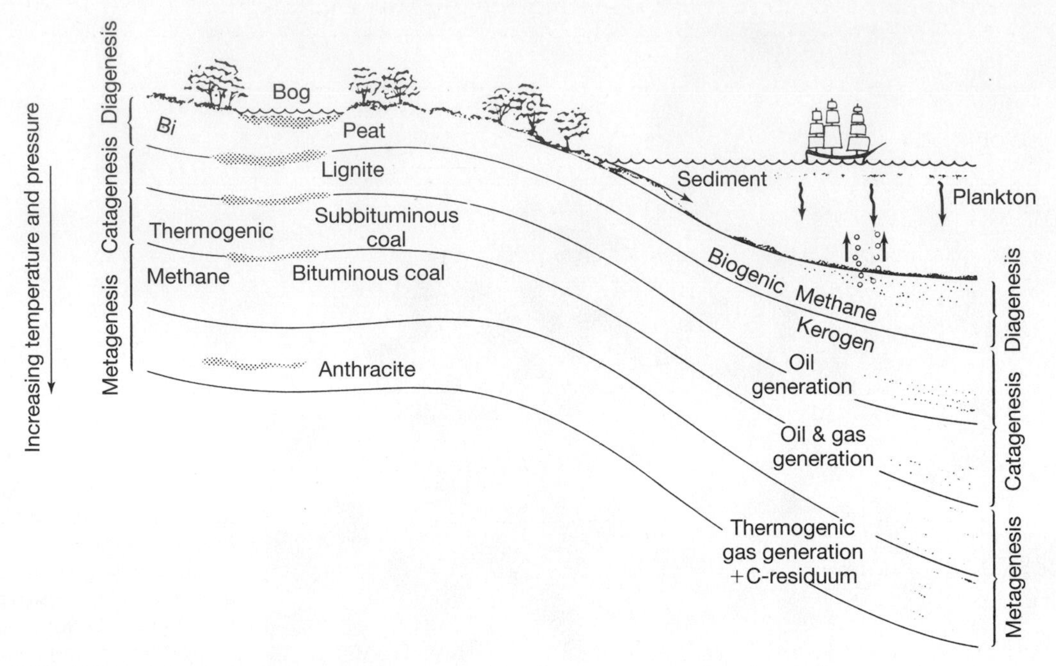

FIGURE 5.6 Schematic diagram of the generation of various ranks of coal, petroleum, and gas from terrigeneous and marine organic matter. See discussion in the text.

schematically in Figure 5.6. In general, the earliest changes, which take place at very shallow burial depths, are biochemical and occur as a result of the metabolism of bacteria, fungi, and other microorganisms. A primary product of this activity, regardless of the type or location of the organic matter, is **methane gas** (CH_4), sometimes known as **swamp gas** or **marsh gas**. As the sediment depth increases, microbial activity slows and increased temperature and pressure drive off water and bring about the **cracking**, or breakup, of complex hydrocarbon molecules.

The original type of organic matter plays an important role in the generation of fossil fuels. Prior to the Silurian Period (about 415 million years ago), there were no large land plants; hence, microscopic photosynthetic marine phytoplankton and bacteria were the principal sources of the organic matter found in the sediments. Similar microorganisms still constitute most of the organics found in modern marine sediments; they contribute mainly proteins, lipids, and carbohydrates.When plants emerged from the water and became established on the land, a new source of organic matter had arrived. In modern terrestrial environments, the higher plants contribute resins, waxes, lignins, and carbohydrates in the form of cellulose. Table 5.2 presents a comparison of some of the major organic constituents in the different types of living matter that serve as precursors to fossil fuels, plus the carbon, hydrogen, sulfur, nitrogen, and oxygen contents of these constituents and the fossil fuels.

The conversion from organic debris to fossil fuel involves the expulsion of water, a reduction in the contents of oxygen and nitrogen, and an increase in carbon and hydrogen. As a result of these changes, the fossil fuels are superior heat sources compared to fresh organic matter, such as wood and leaves. Organics in the marine realm generally are altered into natural gas and liquid petroleum, whereas those in the terrestrial rocks tend to form gas and coal. In some organic-rich shales—commonly called *oil shales*—the burial temperatures have never been high enough to completely break down the original organic molecules but only sufficient to alter them to large waxy molecules known as **kerogen**. These can be converted to oil and gas by various refining processes.

Although the fossil fuels represent parts of a broad spectrum of organic compounds, they are discussed individually here. The conventional fossil fuels—coal, oil, and natural gas—are discussed first, followed by oil shales, tar sands, and heavy oils. All of the fossil fuels are similar chemically in that they consist primarily of hydrocarbon molecules. Differences in the physical properties—from an invisible gas such as methane, to a yellow to brown syrupy liquid such as oil, to a black solid such as anthracite—reflect the vast differences in arrangements and sizes of the hundreds of types of hydrocarbon molecules and the differing ratios of hydrogen to carbon in the fuels. During burning, it is the combustion of carbon and hydrogen with oxygen from the atmosphere that produces the heat; hence, the higher the contents of these elements, the better is the fuel. At the same time, the higher the carbon content, the more CO_2 is generated, whereas the higher the hydrogen content, the more H_2O is generated.

TABLE 5.2 Representative compositions of living matter and fossil fuels

Part A: Living Matter

	Major Constituents (wt%)		
Substances	**Lipids**	**Proteins**	**Carbohydrates**
Green plants	2	7	75
Humus	6	10	77
Phytoplankton	11	15	66
Zooplankton	15	53	5
Bacteria (veg.)	20	60	20
Spores	50	8	42

Part B: Petroleum

	Elemental Composition (wt%)				
Substances	**C**	**H**	**S**	**N**	**O**
Lipids	80	10	—	—	10
Proteins	53	7	2	16	22
Carbohydrates	44	6	—	—	50
Lignin	63	5	0.1	0.3	31
Kerogen	79	6	5	2	8
Natural gas	75–80	20–25	trace–0.2	trace–minor	—
Asphalt	81–87	9–11	0.3–6	0.8–2.2	0–4
Petroleum	82–87	12–15	0.15	0.1–5	0.1–2

Part C: Coal

	Elemental Composition (wt%)				
Substances	**C**	**H**	**S**	**N**	**O**
Peat	21.0	8.3	—	1.1	62.9*
Lignite	42.4	6.6	1.7	0.6	42.1
Sub-bituminous	76.3	4.7	0.5	1.5	17.0
Bituminous	87.0	5.4	1.0	1.4	5.2
Semianthracite	92.2	3.8	0.6	1.2	2.2
Anthracite	94.4	1.8	1.0	0.7	2.1

*Remainder is ash and moisture.

(From Chilingarian and Yen. *Bitumens, Asphalts, and Tar Sands*, and from *Coal Development*, S. Bureau of Land Management, 1983.)

Coal

The fossil fuels that bear the greatest witness to the original organic matter from which they were derived are peat and coal. In most peats and coals, there are abundant imprints of leaves, stems, seeds, and spores of the plants that were compacted to form the fuels we now mine. Because peat is the clear precursor of coal, it is included in the following discussion under the general title of coal. Coal was a very common household fuel in the United States and Europe in the second half of the nineteenth century and first half of the twentieth century. In the second half of the twentieth century, its use in homes was largely replaced by oil, gas, or electric heat because these fuels were more readily available and much cleaner to use. Although coal is no longer in direct use in homes, it is nevertheless used because coal is the major heat source in the power plants that supply electricity to the homes (Figure 5.7). Cheap and convenient petroleum replaced coal in some power plants in the 1950s and 1960s, but there was a return to increased coal usage following the 1973 OPEC oil embargo and the consequent concerns over supplies triggered by sharp rises in the cost of petroleum. Today, the world's coal reserves contain more recoverable energy than do the known reserves of oil or natural gas; it seems clear that we shall be relying on coal as an energy source for many years to come. At the same time, coal has come under much criticism because of the environmental impact of coal mining and its significant contributions to acid rain and CO_2 in the atmosphere on combustion.

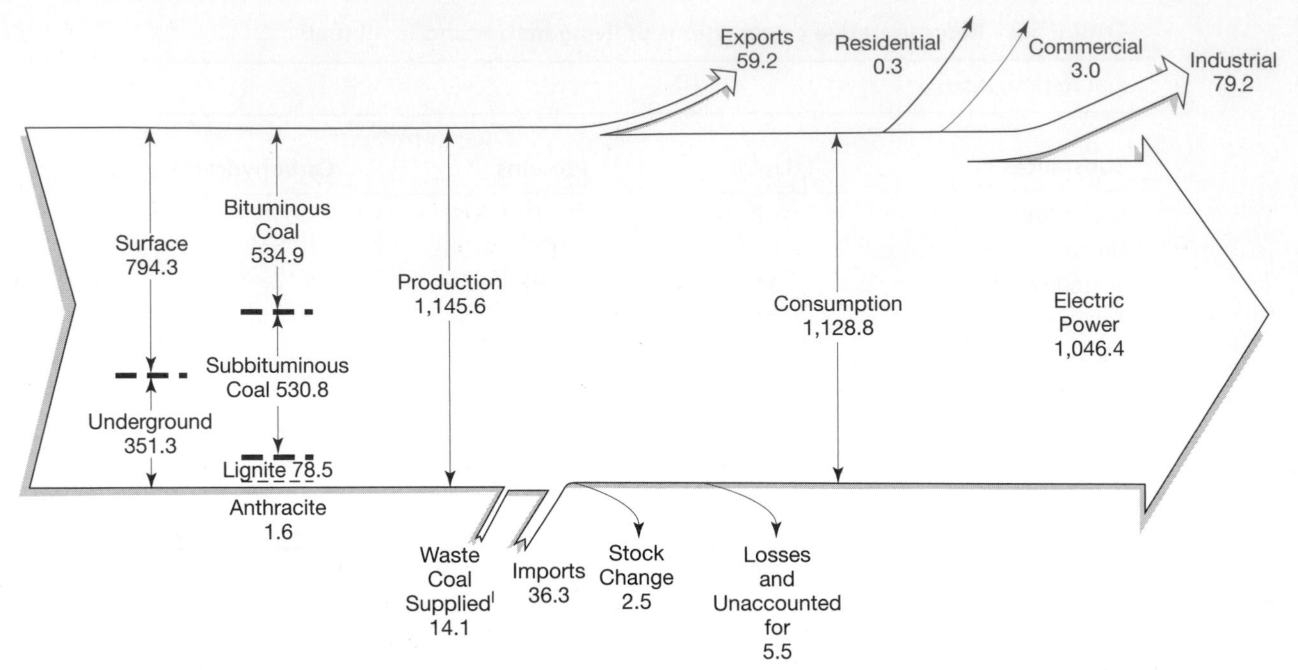

FIGURE 5.7 Coal flow diagram for the United States in 2007. More than 97 percent of the coal used is domestically produced and more than 90 percent of coal consumption is for the generation of electricity. (From the U.S. Energy Information Administration.)

Coal has been classified in several ways. We shall employ the simple and widely used terms of *peat, lignite, bituminous coal,* and *anthracite.*

HISTORY OF COAL USE. The origins of coal use are not known, but there is evidence that 3000 to 4000 years ago, Bronze Age tribes in Wales used coal in funeral pyres of their dead. Coal was probably also used by the Chinese as early as 1100 B.C. and by the Greeks in 200 to 300 B.C., but nearly a thousand years passed before coal had any lasting impact on civilization.

Widespread use of coal as a fuel began in the twelfth century when the inhabitants on the northeast coast of England found that black rocks weathering out of coastal cliffs would burn hotter than wood. The name, in fact, derives from the Anglo-Saxon **col** first used to refer to charcoal; this evolved to *cole,* the spelling used until about 300 years ago. Inefficient burning of impure coals released repugnant odors that caused Londoners to complain about air pollution in 1273 and ultimately led to an edict from King Edward I in 1306 banning the use of coal. However, by the onset of the Industrial Revolution in the late 1600s and 1700s, England was facing a crisis as its forests were being depleted to make charcoal, and the Admiralty feared for sufficient timber to maintain its fleets. The value of coal as a substitute for fuel wood was apparent, but two further developments brought coal into its own as a fuel. About 1710, Abraham Darby, a Shropshire iron maker, developed a method of using coke, made by heating coal in the absence of air, to smelt iron. The first commercial steam engine, produced in 1698, burned wood or charcoal; however, as these engines were perfected and came into wide usage during the 1700s, coal became the fuel to drive them.

Although the Pueblo Indians of the southwest United States used coal in pottery-making for many years, the first recorded discovery of coal in North America was by a French exploration party in 1679 along the Illinois River, about 130 kilometers (80 miles) southwest of Chicago. The first New World mining effort began in 1750 near Richmond, Virginia, where a Huguenot colony worked exposed seams of coal. Coal mining then began in western Pennsylvania in 1759 and soon spread throughout the area of the Appalachian coal fields.

The Industrial Revolution rapidly increased the demand for coal both in Europe and in North America. In England and in parts of the eastern United States, the most efficient way to transport the coal was by canal systems. The development of railroads in the early 1800s supplanted, in part, the canal systems but also provided another major market for the coal. Then in the 1890s, with the development of the steam-driven electric generator, coal became the principal fuel for electric power plants, a position it continues to hold today.

It was the discovery and widespread use of coal that probably saved the great forests of the eastern areas of North America. If coal had not been available when the Industrial Revolution reached North

America in the late 1800s, the only likely alternative fuel source would have been wood, a situation that would have resulted in a massive deforestation, as had happened earlier in Great Britain. The availability of coal and its superiority as a heat source (a ton of coal roughly equaling about one cord of dried hard wood) rapidly turned attention away from the forests toward the coal mines as industrial fuel sources. Hence, the forests were spared; in fact, as major agriculture moved westward and southward, the eastern forests have actually increased in area relative to the 1880s.

FORMATION OF COAL. Coals of all types are the compacted and variously preserved remains of land plants. Many plant remains such as leaves, stems, and tree trunks are visible to the naked eye, whereas others, such as spores, are visible under the microscope. Most plant matter today, as in the past, is not preserved but decomposes where it falls or breaks down during the early stages of burial. Only where plant growth is abundant and the conditions for preservation are optimum can there be an accumulation of a mass of organic matter that could ultimately become coal. The trees, with the cellulose-rich stems and leaves that are found in coals, did not evolve on the continents until late in the Devonian Period (about 360 million years ago). Hence, Precambrian and early Paleozoic rocks do not contain any coal beds. The Paleozoic coal beds of the Carboniferous Period in Europe and North America were dominated by ferns and scale-trees (Figure 5.8). In contrast, the Mesozoic and Cenozoic swamps consisted of angiosperm (flowering) plants much like those forming today.

Most of the world's coals are known as **humic coals** and consist of organic debris that has passed through a peat stage. Their major components are lustrous black to dark brown materials known as **macerals,** the organic equivalents of the minerals that constitute a rock (Figure 5.9). Much less common, but locally important, is another type of coal known as **sapropelic coal**. The two varieties of sapropelic coals, called **boghead** and **cannel coals,** consist primarily of fine-grained, featureless algal debris that collected in oxygen-deficient ponds, lakes, and lagoons. These coals have compositions similar to the kerogen that is the precursor of oil; indeed, when subject to higher temperatures and pressures, they yield oil and gas rather than the black vitreous macerals seen in humic coals.

The formation of the major coals (humic types) begins with the accumulation of organic debris in peat swamps where the stagnant waters prevent oxidation and destruction. It has been estimated that under average peat-forming conditions only about 10 percent of the plant matter is preserved. The highest rates of plant growth occur in tropical forest swamps, but these are also the sites of the greatest bacterial activity that destroys vegetable matter; hence, few peats develop in the tropics. Today, the major peat-forming areas occur in the temperate to cold regions such as Ireland, Scandinavia, Alaska, and Canada, where abundant rainfall promotes rapid plant growth, but where the cooler temperatures retard bacterial decay. If the rates of peat accumulation in the past were similar to the 1 millimeter per year we see now, the major coal basins must represent swamps that persisted for tens to hundreds of thousands of years.

FIGURE 5.8 Many coal beds contain imprints of the plants from which they were formed. Here the bark of a cycad tree from a Mississippian (Carboniferous)-age coal is visible. (Photograph courtesy of S. Scheckler.)

FIGURE 5.9 The horizontal layering in the coal represents the layering of compacted organic matter; differences in the reflectivity result from differences in the content of the macerals. Pyrite, (FeS_2) although not visible in this sample, occurs locally in coal beds and is the principal contaminant that results in environmental pollution. (Photograph by J. R. Craig.)

FIGURE 5.10 The cyclical development of coal-forming conditions resulted in the deposition of multiple coal beds in many areas, as are visible in this road cut in West Virginia. (Photograph courtesy of Kenneth A. Eriksson).

The geologic study of coal-bearing sequences reveals that many formed in areas of successive transgression and regression of shorelines. This is seen in the interbedding of marine sediments with the coals and with lacustrine and terrestrial beds. A common sequence would have been the development of a near-coastal swamp along the margin of a basin with the accumulation of thick masses of peat as the basin slowly subsided. If the rate of subsidence exceeded the slow buildup of peat and sediment, the sea advanced (or transgressed) over the swamp, covering and preserving the peat with sands and muds. Where the seawater was warm and clean, lime muds often accumulated. Frequently, the subsidence was followed by a reemergence of the coastal area (with a retreat or regression of the ocean) and a subsequent reestablishment of a swamp and the deposition of another layer of peat. As a result of such cyclical deposition, many of the world's major coal basins contain dozens of individual coal beds separated by sandstones, shales, and limestones (Figure 5.10). The individual coal beds range in thickness from only a few centimeters (inches) to dozens of meters (many tens of feet). The individual coal seams often split into two or more seams, indicating that deposition in the peat swamp was continuous in some areas but interrupted in others. This situation is seen today in swamps that lie in major deltas where the distributaries keep changing direction and where many differences exist in the rates of subsidence and sediment accumulation.

Peat formation has taken place continuously since the Devonian Period, but the sizes of the swamps and the degree of coal preservation have not been uniform. By far the greatest period of coal swamp formation took place during the last 70 million years or so of the Paleozoic Era (Figure 5.11). During this time, the great coal beds of Britain, Australia, the eastern United States, and elsewhere were laid down; the abundance of the coal in these

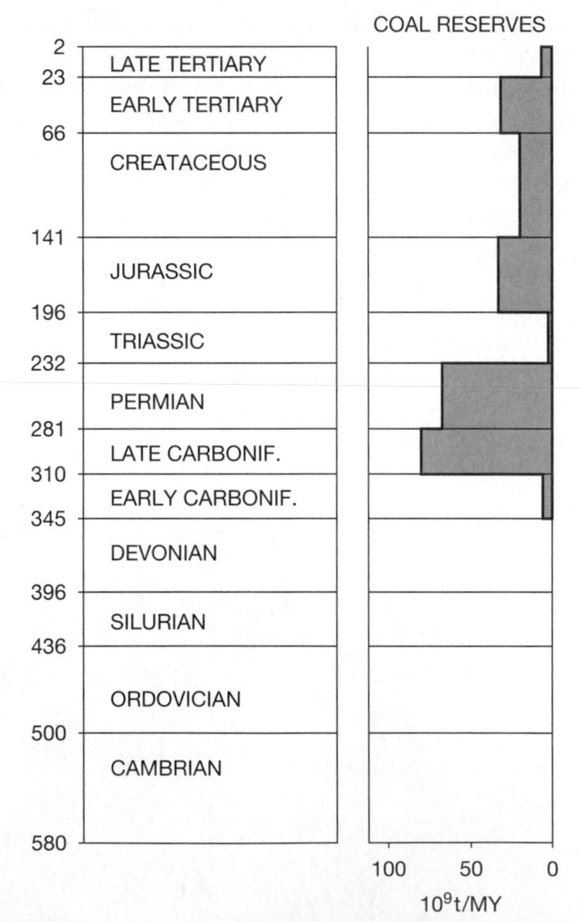

FIGURE 5.11 The right portion of the graph shows the worldwide distribution of coal reserves in terms of millions of tons of accumulation per million years during geologic time. The absence of land plants prior to the Devonian precludes earlier coals; the late Carboniferous (Mississippian) and Permian were the periods of most prolific deposition. (Figure after Demaison [1977] and Bestougeff [1980] and from Bois, Bouch, and Pelet, *American Association of Petroleum Geologists Bulletin*, vol. 66 [1982], p. 1264. Used with permission.)

Rank increasing ↓

		Rank stages	Characteristics	H_2O %	Heat content
		Peat	Large pores Details of original plant matter still recognizable Free cellulose		3000kcal/kg (5400 Btu/lb)
Brown coal and lignite		Soft brown coal	No free cellulose	~75	
	Hard brown coal	Dull brown coal	Marked compaction of plant structures	~35	4000kcal/kg (7200 Btu/lb)
		Bright brown coal	Plant structures partly recognizable	~25	5500kcal/kg (9900 Btu/lb)
Hard coal		Bituminous		~10	7000kcal/kg (12,000 Btu/lb)
		Anthracite	Plant structures no longer recognizable		8650kcal/kg (15,500 Btu/lb)

FIGURE 5.12 The ranks of coal, the calorific value, and some important physical characteristics. (From *The International Handbook of Coal Petrography* [1963].)

beds in Britain led to this span of geologic time being named the Carboniferous period. Another great period of coal deposition extended from the beginning of the Jurassic until the mid-Paleocene; it was during this period that the major coals of the western United States were deposited. Deposition continues today in localized areas such as the Everglades of Florida, the Dismal Swamp of Virginia and North Carolina in the United States, and in the coastal swamps of Canada, Scandinavia, and Ireland. Although these do not rival the peat swamps of the past, they do give us a firsthand opportunity to understand the conditions that must be present for peats and coals to form.

Immediately after accumulation of the dead plant tissue, compaction occurs and bacterial and fungal attacks begin. In the formation of peat, the cellulose and other original plant components are decomposed, resulting in the production of biogenic methane gas (CH_4), carbon dioxide (CO_2), and some ammonia (NH_3). The remnant is a mass of brown, hydrated gels, rich in large hydrocarbon molecules.

Coalification is the process by which the organic components that survived peat formation undergo further physical and chemical changes as a result of biochemical action and of rising temperature and pressure. As burial depth increases, temperature is by far the most important factor in coalification. The rank of coal progresses from peat, to lignite (or brown coal), to bituminous (or hard) coal, and then anthracite, as shown in Figure 5.12. The precise assignment of rank is determined by the carbon content, the calorific value (or heat given off during burning), the moisture content, and the content of volatile matter. The most significant chemical changes are the progressive decrease in the oxygen and hydrogen contents and the resultant increase in the carbon content, as shown in what is known as a van Krevelen diagram (Figure 5.13). As this process proceeds, the number of distinguishable plant remains decreases, and more and more of the shiny black macerals are formed. A progressive decrease occurs in moisture, and an increase in density, an

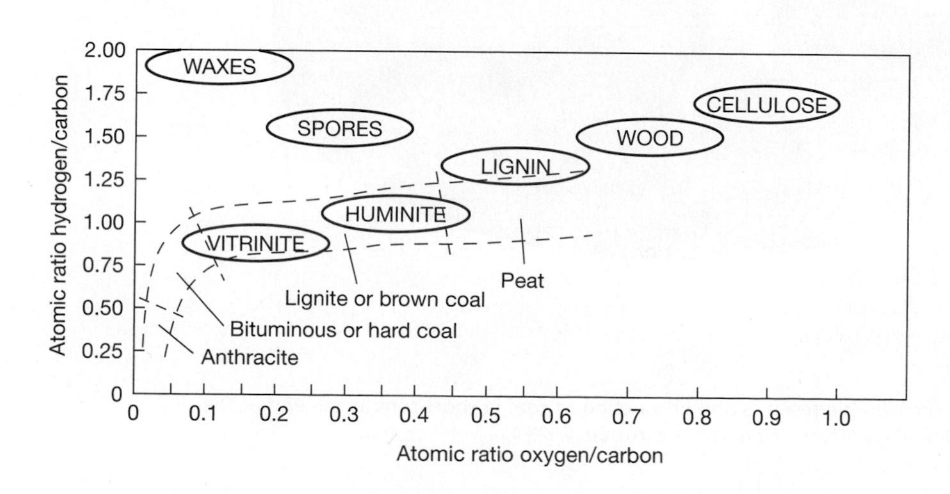

FIGURE 5.13 The van Krevelen diagram illustrating the evolution of the composition of organic matter as it is converted into coal. The carbon content and the calorific value rise as the organic matter progresses to lower hydrogen/carbon and oxygen/carbon ratios.

increase in calorific value (up to anthracite), and an increase in the degree of polymerization (chemical bonds between carbon atoms) take place.

THE WORLD'S COAL RESERVES AND COAL PRODUCTION. Earth's coal resources are large but very irregularly distributed. Of the nearly 1 trillion tons of recoverable coal reserves, more than two-thirds occur in the United States, Russia, China, and Australia [Figure 5.14(a)]. The other countries with significant reserves include Germany, India, and South Africa (Figure 5.15). In contrast, coal reserves are virtually absent in the entire continent of South America. The amount of coal production parallels the magnitude of the reserves, with the United States, Russia, and China accounting for two-thirds of the world total [Figure 5.14(b)]. The large size of the world's coal reserves and the existing infrastructure have led to the belief that coal will remain one of the major energy sources for the foreseeable future, especially for the generation of

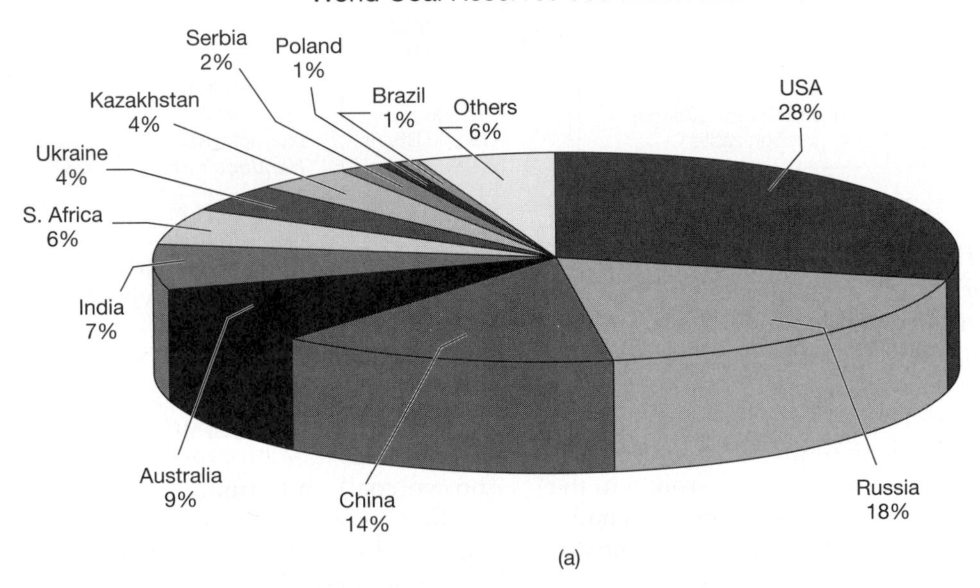

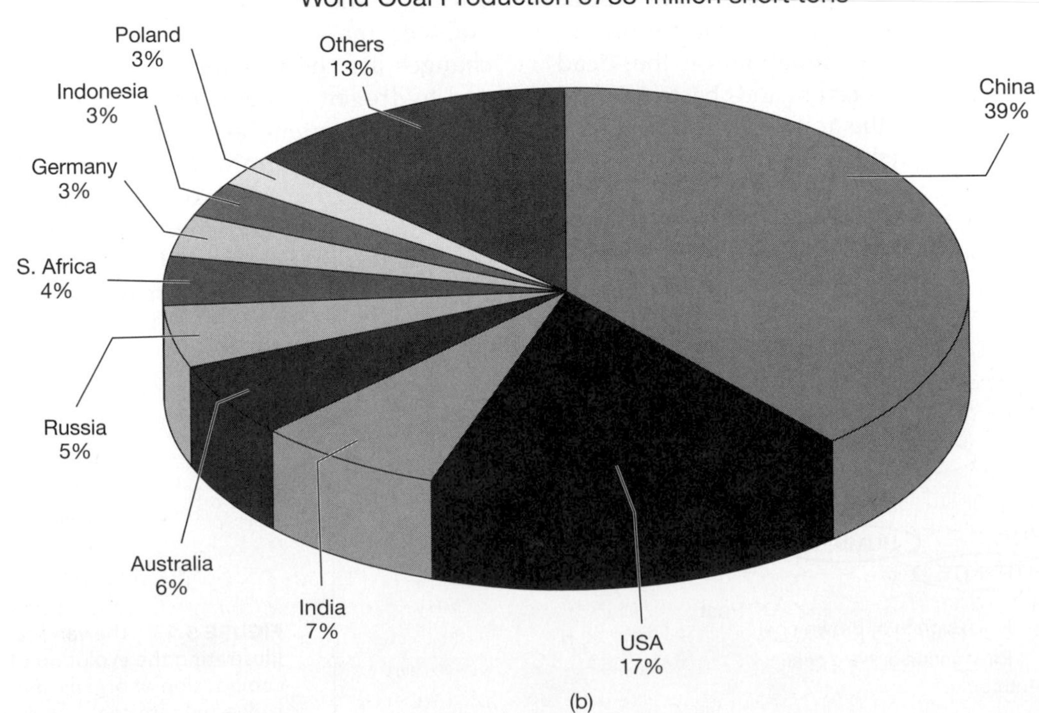

FIGURE 5.14 International recoverable reserves and production of coal in short tons at the end of the twentieth century. (From U.S. Energy Information Administration, 2009.)

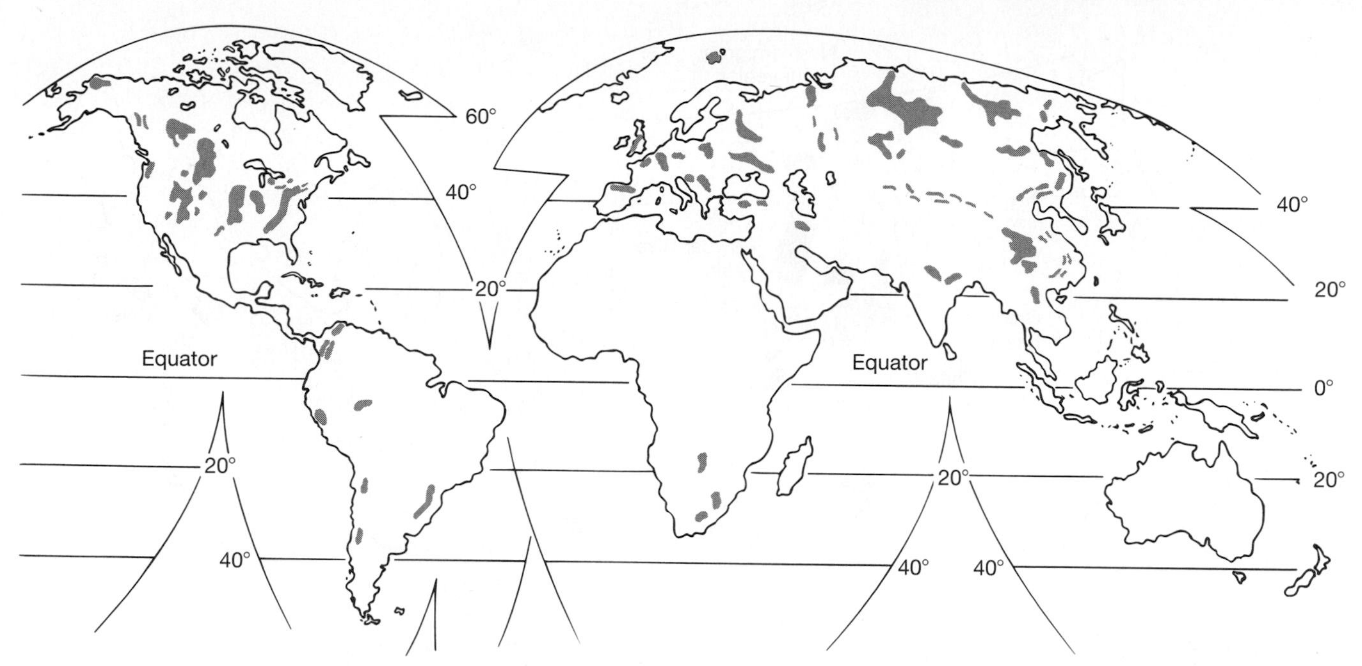

FIGURE 5.15 The geographic distribution of coal fields throughout the world. (After Fettweis, World Coal Resources, Elsevier, Amsterdam, 1979.)

electricity. A June 2009 report by the Energy Information Administration reported that the United States' coal reserves have been grossly overstated and are only about half the the previously stated values. Regardless of which estimate is correct, coal will continue as a major source of energy and carbon dioxide release for the first quarter or half of the twenty-first century.

Within the United States, coal fields are widespread, with four major provinces accounting for most of the reserves (Figure 5.16). In general, the eastern and interior provinces contain bituminous coal, whereas the Rocky Mountain and Northern Great Plains provinces are richest in sub-bituminous coal and lignite. Anthracite occurs locally in all provinces, but the country's major production and reserves lie in eastern Pennsylvania. Alaska contains significant amounts of sub-bituminous coal, but its remoteness leaves these deposits largely as resources rather than as reserves.

The mining of coal, like the mining of metallic ores, began as a very labor intensive industry. The coal beds were easier to follow than metallic ores because many were nearly horizontal, and they were easier to mine because they were safer and easier to break with a pick and shovel than were metal ores. In contrast, conditions in coal mines were more difficult and much more dangerous because many coal seams were less than 1 meter (3 feet) in thickness and the mine openings were only as high as the seam was thick; furthermore, mine fires and explosions resulting from seepage of methane gas into the workings were common events. In the 1800s, in Britain, the United States, and in many other countries, children worked in the coal mines because of their small size and because they could be paid so little. Prior to 1898, the workweek in the United States bituminous mines was 60 hours; it was then reduced to 52 hours where it remained until 1917, and finally to 40 hours in 1933. The problems of low pay and long hours led to unionization and brought about the establishment of very strong unions in the early years of the twentieth century.

Despite union efforts to maintain jobs, however, mechanization has markedly changed coal mining and resulted in a reduction in the number of miners in the United States. In 1923, some 700,000 miners produced 510 million metric tons of coal; by the mid-1990s, some 120,000 miners produced 900 million metric tons; by 2008, some 80,000 miners produced 1.1 billion metric tons. Similar trends in mining efficiency have been seen in all of the major coal-producing countries. In Great Britain, where the coal mining industry was for many years in public ownership, a return to private ownership combined with mechanization led to a drastic reduction in the size of the workforce; even so, increased competition from cheaper imported coal forced the closure of many mines during the 1980s and 1990s.

Similarly, the breakup of the former Soviet Union and the shift away from a heavily subsidized state-owned mining industry led to the closing of many mines because they could not compete economically with more efficient operations elsewhere in the world. World coal production rose slowly but steadily during the nineteenth century, reaching 1 billion metric tons

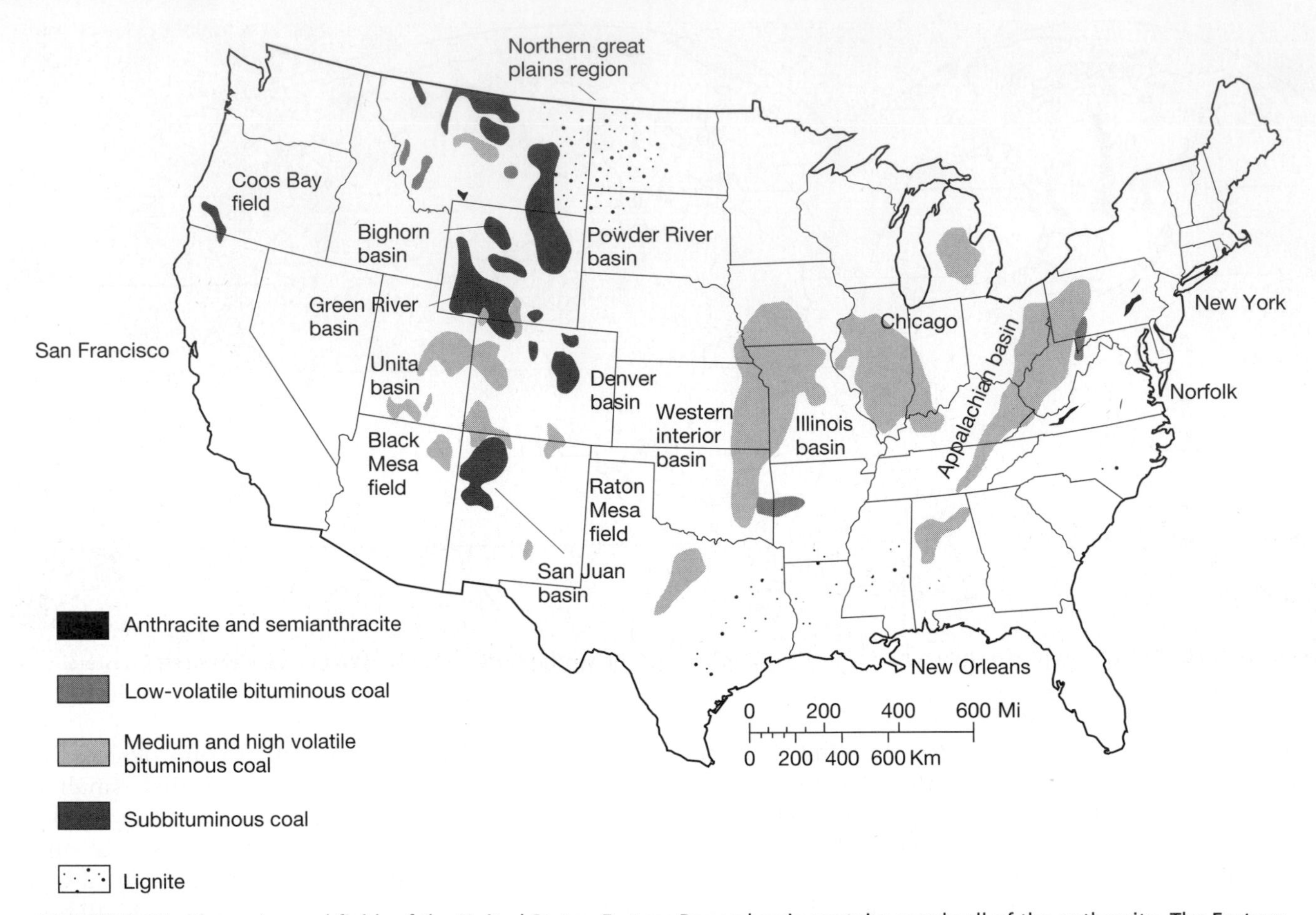

FIGURE 5.16 The major coal fields of the United States. Eastern Pennsylvania contains nearly all of the anthracite. The Eastern and Interior provinces are dominantly bituminous coal, whereas the Rocky Mountain province contains mostly sub-bituminous coal and lignite. (From the U.S. Department of Energy.)

in 1907 (Figure 5.17). Production continued to grow, except for disruptions due to World War II, until the 1950s, when production rates increased markedly. Annual production reached 2 billion metric tons in 1953, rose to 3 billion by 1970, 4 billion by 1982, 5 billion by the mid-1990s, and will likely reach 6 billion by 2010.

Modern coal mining is carried out by either underground (deep-mining) or by surface stripping or open cast methods, as described in Chapter 3. In underground mining, the pick and the shovel of the past have been replaced by drills and efficient cutting machines, which cut into the coal and dump it onto a conveyor belt for removal to a loading site or to the surface. The most common type of mining removes only about 50 percent of the coal while leaving the rest as pillars to support the overlying rock (see Figure 5.18). Both the spacing and the size of the pillars depend upon the depth below surface, the thickness of the coal, the stability of the roof rock, and

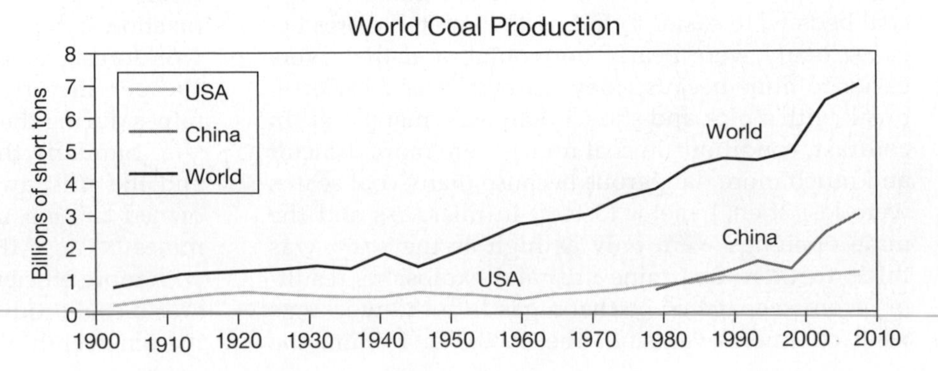

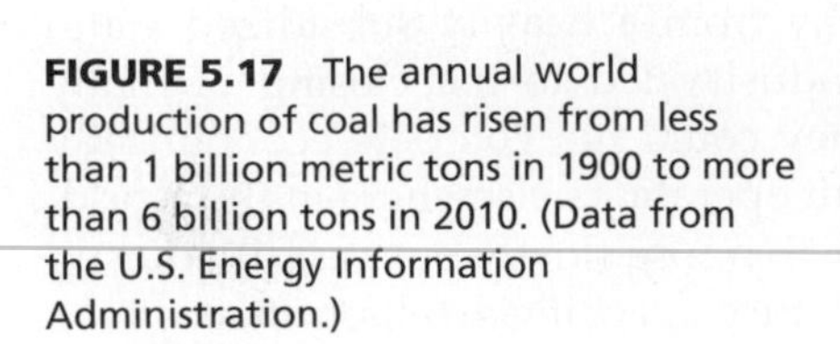

FIGURE 5.17 The annual world production of coal has risen from less than 1 billion metric tons in 1900 to more than 6 billion tons in 2010. (Data from the U.S. Energy Information Administration.)

FIGURE 5.18 Pillars of coal are often left on each side of the area where coal is removed in order to support the roof as mining progresses. These pillars, which may contain a large proportion of the coal, are commonly removed as miners "retreat" from a mine when its reserves are becoming exhausted. (Photograph courtesy of Bethlehem Steel Company.)

the number of individual coal seams being mined. After initial mining has been completed, some of the coal in the pillars is also recovered in a process called "robbing the pillars." This occurs during the retreat or final stages of mining. A newer and even more efficient mining method makes use of a continuous mining machine (Figure 5.19) that moves back and forth, removing nearly 100 percent of the coal by a rotary cutter. The machine and the conveyor belt that carries out the coal are protected by a steel canopy. The machine, the belt, and the canopy advance together as coal is cut and the overlying rock is allowed to subside in a continuous but controlled manner.

The thickness and the quality of a coal determine whether a given seam is economic to mine, but generally beds greater than about two feet in thickness are mineable. Underground mining is inherently dangerous, and tragically many miners are killed every year by roof collapses. Many coals are also gassy and yield large amounts of methane. Although methane is a great hazard, it can also prove a valuable resource, as discussed later in this chapter. Despite

FIGURE 5.19 A continuous mining machine in operation. The coal is cut by the rotating cutting drum at the front and is carried by conveyor belt to the back where it is loaded into rail cars or placed on another belt for removal from the mine. (Photograph courtesy of The A.T. Massey Coal Co., Inc., and Heffner and Cook, Inc.)

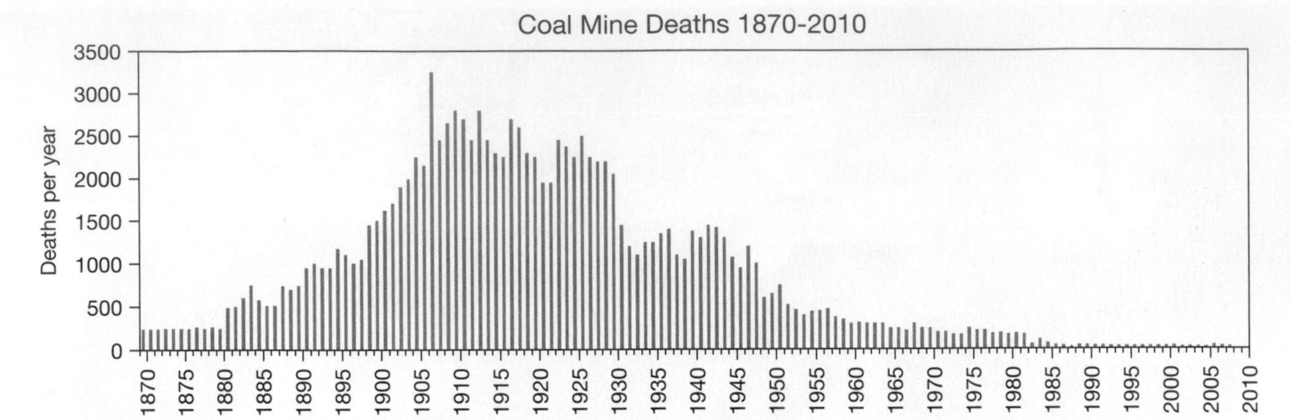

FIGURE 5.20 Since 1870, more than 120,000 miners have been killed in coal mine accidents in the United States. More than 3200 were killed in 1907 alone. Since the 1920s, union efforts and more stringent safety measures and codes have reduced the numbers of deaths. Nevertheless, as recently as 2008, more than 30 miners per year were still being killed in accidents.

the former use of canaries and the modern use of air-sensing equipment to test for methane, rapid gas buildup can still result in explosions (see Plate 24 and Figure 4.9). Perhaps the magnitude of the safety problem can be appreciated when one realizes that since 1870 more than 120,000 coal miners have been killed in the United States; more than 20,000 of them were killed in West Virginia mines alone. Even with all of today's precautions about thirty miners are killed in American coal mines annually (Figure 5.20). Terrible as these statistics are, they pale in comparison to the 3000 to 5000 miners killed annually in modern Chinese coal mines. Regardless of the measure, coal mining remains a difficult and dangerous occupation everywhere in the world.

Surface mining requires the removal of the overlying strata to expose the beds of coal; once exposed, coal is removed by heavy excavating equipment such as bulldozers, front-end loaders, power shovels, and large drag lines. To remove large tonnages of overburden rapidly and efficiently, excavating equipment has constantly increased in size. The largest power shovel in the United States, the "Gem," which operates in Ohio, weighs about 7000 metric tons, is about 60 meters (190 feet) high, has a shovel capacity of 130 metric tons; it is said to be the largest mobile land piece of equipment in the world (Figure 5.21). Economics generally dictate which coal beds can be extracted through surface techniques; the rule of thumb is that surface extraction is economic if the ratio of the depth of overburden to be

FIGURE 5.21 The "Gem," or giant earth mover, the largest power shovel in the United States, is shown here mining in Ohio. The shovel, with a capacity of 300 to 500 metric tons per shovel load, is used to strip off the overburden to expose the underlying coal beds. (Courtesy of Consolidation Coal Company.)

FIGURE 5.22 In the Powder River Basin of Wyoming, a modern surface mine lies beyond an area where underground mining in the early 1900s removed coal from the same bed. The holes in the foreground are the result of surface collapse into the old workings. (From U.S. Geological Survey Professional Paper 1164 [1983].)

removed to the coal thickness does not exceed 20:1. Where the overlying rock thickness is too great for surface removal, the mining may proceed as an underground mine, or coal may be removed by augering. Augers are drills, up to 1 meter (3 feet) or more in diameter, that can be driven into horizontal or gently dipping coal beds. As the auger turns, it cuts the coal and feeds the broken pieces out just as a hand drill or brace-and-bit feeds out wood chips. Although the augers can remove only about 50 percent of the coal in a seam, it is coal that could not be economically recovered by any other means.

Most coal production in the first half of the twentieth century came from underground mines because coals exposed at the surface were too deeply weathered to be useful or because there were no practical means of removing the large quantities of overlying rock to expose deeper coal beds. Most of the beds that were amenable to surface mining, such as those in the western United States, were so far removed from markets that transportation was either not available or not economic. Efficient transportation systems and development of very large power shovels and drag-lines made surface mining of beds as much as 30 meters (90 feet) deep economic in the second half of the twentieth century. The conversion of an area from underground to surface mining is visible in the Powder River Basin area of Wyoming (Figure 5.22) where subsidence of early twentieth-century underground workings occurs adjacent to a modern strip mine working the same coal bed. In the second half of the twentieth century and in the twenty-first, there were a series of progressive changes in American coal mining practices, as shown in Figure 5.23. The most important included a rapid increase in mining west of the Mississippi River, a significant increase in the mining of sub-bituminous coal and lignite, a sharp decline in the mining of anthracite, and a rapid increase in surface mining at the expense of deep mining. The main reasons for these changes included: (1) greater demand for the lower-sulfur, near-surface coals that are found in the western states (Figure 5.24); (2) the development of power plants near the western coals; (3) concern for increased safety; (4) the greater per man productivity in surface mines; and (5) the absence of unions in many of the surface mines in the West.

Coal mining and coal use have been the subject of many environmental concerns, especially in recent years. Most of these concerns have centered on the problems of acid mine drainage, acid rain (see discussion in Box 4.1), increased carbon dioxide levels in the atmosphere, and surface mine reclamation. The first two concerns are related to the sulfur, which is present in all coals in amounts between 0.2 and 7.0 percent. Generally, about one-half of the sulfur present is bound within organic macerals; the remaining sulfur occurs principally as the two iron sulfide minerals, pyrite and marcasite (both FeS_2). The sulfur was originally derived from the organic matter, or from sulfate in groundwaters, from which it was reduced by bacterial action. Once the coal is exposed to air and water by mining, the iron sulfide is oxidized to ferrous sulfate, $FeSO_4$, and sulfuric acid, H_2SO_4. The escape of these acid mine wastes into rivers and streams has left thousands of kilometers (miles) of waterways devoid of fish and other aquatic fauna and flora. The ferrous iron is readily soluble in the strongly acid waters that wash off exposed coal beds or waste piles, but is quickly oxidized to ferric hydroxide as the solutions are

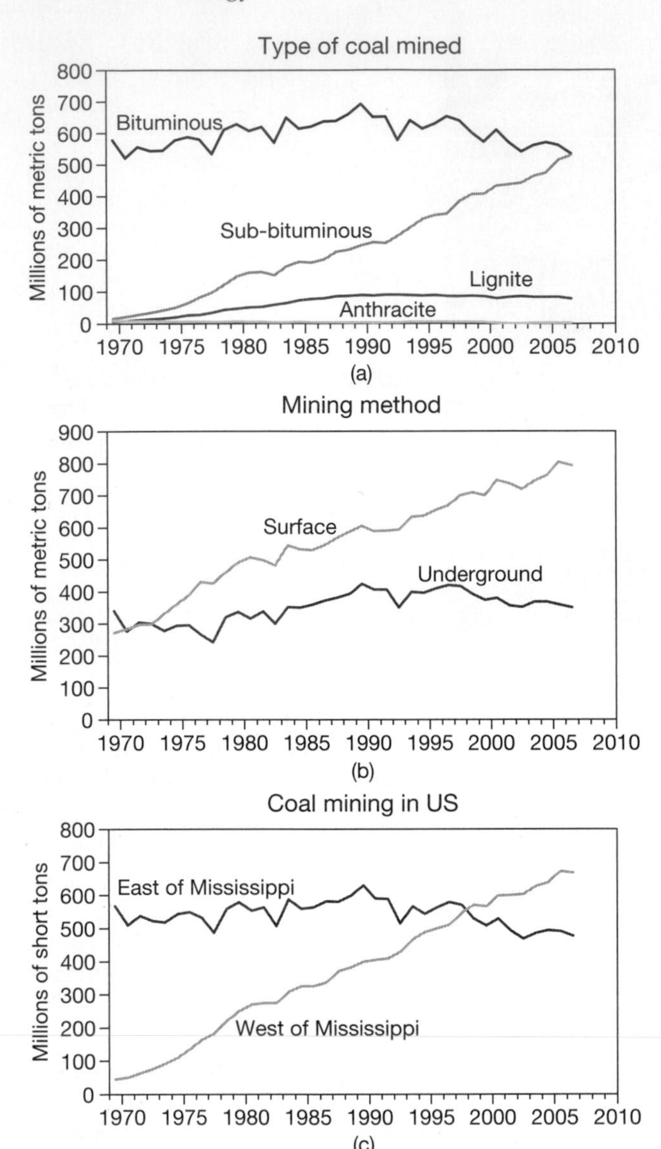

FIGURE 5.23 Changes in the United States coal industry since 1949. (a) Bituminous long dominated production but was surpassed by sub-bituminous and lignite mining. (b) Surface-mined coal has exceeded underground-mined coal since the early 1970s. (c) Mining west of the Mississippi River has risen sharply since 1970 and now exceeds mining east of the Mississippi. All of these changes reflect the rapid growth of coal mining in the western United States where the beds are closer to the surface but of lower rank. (Data from U.S. Energy Information Administration.)

diluted by stream water. The result is the precipitation of iron hydroxides as gelatinous, reddish-brown coatings on the rocks and plants. Our understanding of the processes involved in acid mine water pollution and the concern that no further abuses occur has led to much more stringent regulations regarding the discharge of mine waters and of the dumping of iron sulfide–rich mining debris. Much effort has been focused on producing "clean coal" by finely grinding the coal to remove impurities, especially iron sulfides, before burning it in power plants. This cleaning does not, however, reduce the amount of carbon dioxide that is produced during combustion.

Despite cleaning efforts to remove the pyrite from coals before it is burned, some pyrite always remains, and so does the organically bound sulfur in the coal. The burning of coal releases the sulfur as sulfur dioxide, SO_2, which is a significant contributor to acid rain.

On the third issue, there is no question that the burning of coal, or of any fossil fuel, raises the CO_2 level of the atmosphere. On average, the burning of a metric ton of bituminous coal releases about 2.8 metric tons of carbon dioxide to the atmosphere (Figure 5.25). For the same amount of energy produced, burning coal creates about 25 percent more carbon dioxide than does the burning of oil and about 2.5 times more than does the burning of natural gas. This disparity has led to an emphasis on replacing coal with natural gas at power plants and to the development of ways to sequester the carbon dioxide and so prevent it from entering the atmosphere. Both of these procedures increase the cost of power generation; thus, there is much incentive to maintain the *status quo*. As we have progressed into the fossil fuel era, the atmospheric CO_2 level has nearly doubled (see Figure 1.8). Because carbon dioxide is effective in keeping infrared radiation trapped in Earth's atmosphere, the carbon dioxide promotes general global warming. There remains, however, a great deal of uncertainty about how high the future carbon dioxide levels are likely to rise and what the magnitude of warming is likely to be.

All types of underground mining can lead to problems of subsidence after the mining is finished (see Figure 4.14). An example can be seen in the foreground of Figure 5.26, which shows an area in Wyoming where many of the mine openings, which were only a few dozen meters (several tens of feet) below the surface, have caved in. Similar openings underlie many mined areas in England, India, and the eastern and central United States. More spectacular than the subsidence of old coal mines are underground coal fires (see Plate 16). There are an estimated 300 underground coal fires still burning in the United States; in many other parts of the world, coal beds have been accidentally or spontaneously ignited. If there is an adequate oxygen supply, an underground fire can smolder for years and travel for considerable distances. The most famous mine fire was at Centralia, Pennsylvania, where a burning waste dump ignited an exposed coal seam in 1962. The fire, despite the expenditure of millions of dollars trying to extinguish it by water flooding, trenching, drilling, etc, continues to burn. The fire has progressed under much of the town of Centralia, causing local subsidence and giving rise to the sudden appearance of cracks that issue hot and noxious gases. Many residents left the town, and those who remained usually had household monitors for deadly gases such as

FIGURE 5.24 Some of the coal beds, such as this one in the Powder River Basin in Montana, are 80 to 100 feet thick and extend over thousands of square kilometers. The overburden visible on the right-hand side of the photo is stripped away to expose the coal seam in the bottom of the pit. The overburden is piled at the left-hand side of the photograph and is pushed back into the pit after the coal has been removed. (Photograph courtesy of K. Takahashi, U.S. Geological Survey.)

BOX 5.2

The United States Strategic Petroleum Reserve

The oil embargo of 1973 made OPEC (Organization of Petroleum Exporting Countries) a household word, and it made many Americans and western Europeans realize the degree to which their societies had become dependent upon oil imports. The United States, recognizing that any future cutoffs of imported oil would create many problems in the American

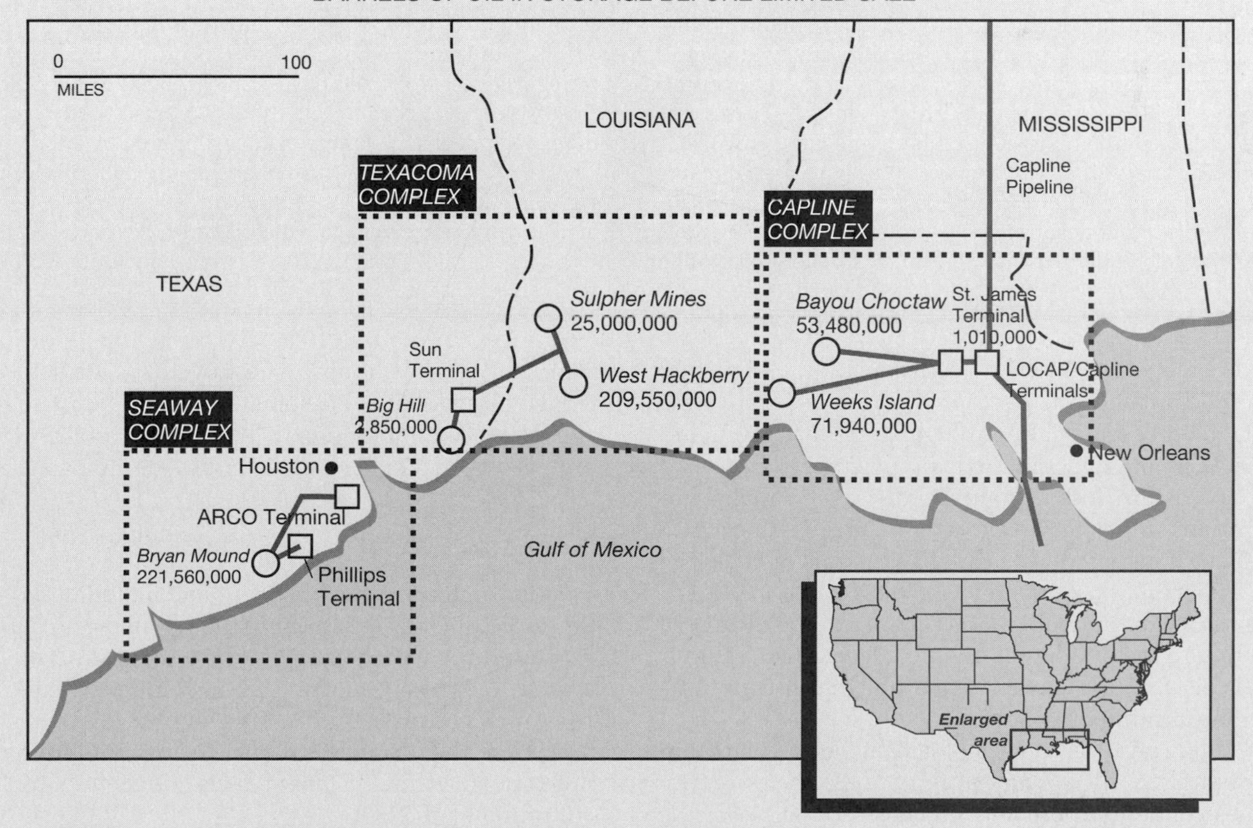

FIGURE 5.C Location map of the United States Strategic Petroleum Reserve.

(continued)

BOX 5.2

The United States Strategic Petroleum Reserve *(Continued)*

economy, decided to establish an oil reserve that could provide oil as needed. The Strategic Petroleum Reserve was created in 1975 for the express purpose of serving as a source of oil for the United States if there were any cutoff of imported oil. The goal was to place into storage 1 billion barrels of oil, enough to supply the country's import needs for more than 100 days.

Several alternatives were considered, but it was eventually decided that the best storage sites were cavities within salt domes in Louisiana and Texas. The domes formed over millions of years as salt slowly moved upward through the overlying sediments to create roughly cylindrical bodies up to 5 miles across and up to 10 miles in height. These salt domes have been the sites of much oil exploration because oil often migrates along upturned beds at the margins of the domes. Furthermore, tests in Germany showed that oil could be stored in salt for years with no deterioration in its quality. Ultimately, five salt domes were selected as sites for the development of underground storage cavities; a sixth site was a converted salt mine (see Figure 5.C). The Sulphur Mines site has subsequently been closed. The caverns, created by solution mining, are typically cylindrical in shape, being about 200 feet in diameter and 2000 feet in height. Each would easily hold the Empire State Building or three Washington Monuments on top of each other.

Solution mining began with the drilling of a standard well into the salt dome. Freshwater was pumped into the well to dissolve salt, and the resulting brine was pumped into deep injection wells or into the Gulf of Mexico. Once the cavity had expanded to the desired width, a small amount of oil was injected into the cavity to float on the water and to protect the salt roof of the cavern from any additional solution. Selective injection of the water was used to increase the size of the cavern and to control its shape (Figure 5.D).

The individual cavities were designed to ultimately hold between 25 million and 220 million barrels of petroleum. Filling of the Strategic Petroleum Reserve began in 1977. By late 1994, it contained approximately 600 million barrels of petroleum; that has now risen to about 750 million barrels. America hopes that it shall never again be faced with an embargo of imported oil, but if it is, the Strategic Petroleum Reserve will be available to help fill the needs.

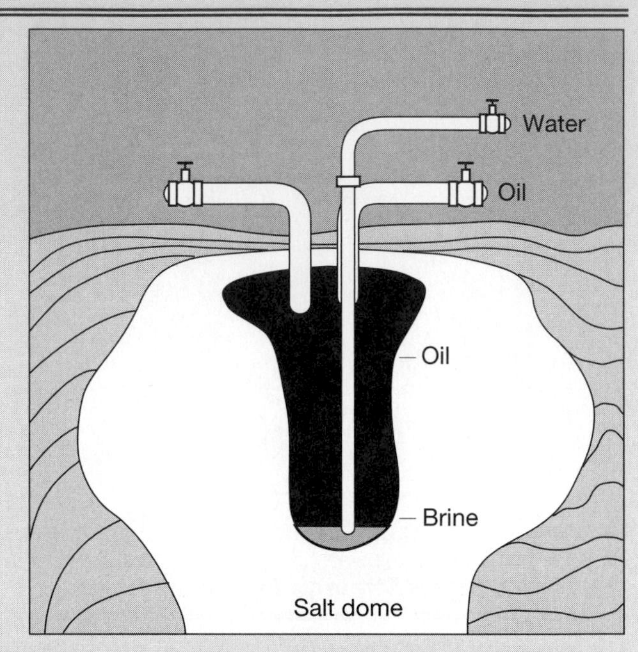

FIGURE 5.D Schematic cross section showing the shape and location of the oil-containing chambers in the salt domes.

An unfortunate footnote to the development of the Strategic Petroleum Reserve was the October 1995 report of a serious leak in one of the caverns. This forced costly repairs and extraction of some of the oil. Periodic concerns about the price of oil and national security have resulted in calls for use of the oil from the Strategic Petroleum Reserve. Relatively small tests have proven that the oil can be extracted safely, but no large-scale extractions had been carried out by 2009.

carbon monoxide, which occasionally seeped into homes. Ultimately, the decision was made to move the entire town because it became apparent that it was impractical, if not impossible, to extinguish the fire.

In eastern India, there are at least 65 major underground mine fires—some have burned since 1916 and have consumed at least 1 billion tons of coal. The area is pockmarked by crevasses and holes that belch steam, smoke, and poisonous gases, and the rate of burning appears to be increasing. Even where there is no visible fire, the ground is so hot that it cannot be used for farming or pasturing.

China has been described as having the worst underground coal fires of any country on Earth. More than 60 fires continue to burn as much as 20 million metric tons of coal annually along a zone 3000 miles long in northern China. These fires, which have so far proven impossible to put out, have forced local inhabitants to move, wrecked water supplies, destroyed vegetation, and driven off wildlife. It is estimated that the Chinese mine fires alone account for two to three percent of the world's annual carbon dioxide emissions into the atmosphere.

Peat Resources

Peat is a natural, renewable, organic material that covers approximately 4 percent of the world's land surface, mostly in the temperate regions of the northern hemisphere. Peat has been used as a fuel in European countries for centuries, but is little known as an energy source in the rest of the world. In many countries, including the United States, peat is widely used for agricultural purposes.

Peat is generally considered as a young coal because it consists of plant matter that has been only slightly compacted and decomposed. Peats are

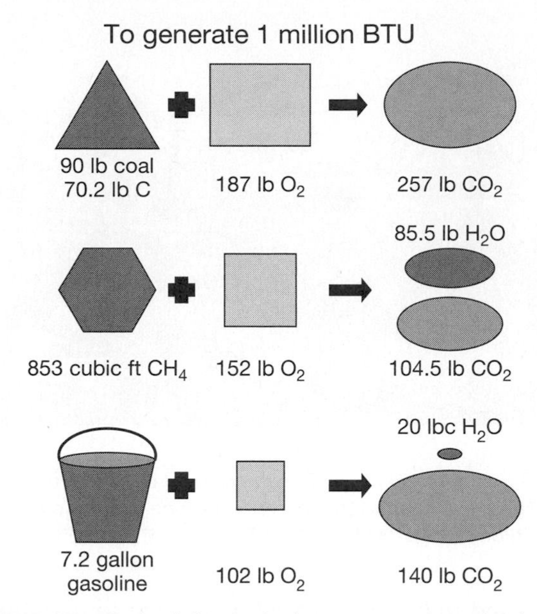

FIGURE 5.25 Much of the rise in the concentration of carbon dioxide in Earth's atmosphere has been blamed on burning of fossil fuels, but natural gas has been considered as a "clean" alternative to coal. Burning 90 pounds of coal, 853 cubic feet of natural gas, or 7.2 gallons of gasoline will produce 1 million BTU of heat. The coal burning will yield 257 pounds of carbon dioxide, gasoline will produce 140 pounds of carbon dioxide, but the burning of the natural gas will yield only 104.5 pounds of carbon dioxide (and 85.5 pounds of water vapor); hence, burning natural gas creates much less carbon dioxide than burning coal or gasoline to provide the same amount of energy.

categorized on the basis of the biological origin of the constituent plants and the state of decomposition. Peats that are least decomposed and richest in mosses are widely used for horticultural purposes as peat moss, but they have little value as fuel. Peats that are more decomposed and compressed have useful heat values and after air drying, they are only about 25 percent lower than lignite; such peats have considerable potential as a fuel. Peat used as a fuel, when air-dried to contain only about 35–40 weight percent moisture, contains more nitrogen and less sulfur than do higher ranks of coal. Direct combustion is the simplest way to derive heat from peat, but there has been increasing research into the conversion of peat into methane gas by bacterial digestion or by thermal breakdown at 400°– 500°C.

The world's largest peat producers by far are Finland and Ireland, where annual peat production averages more than nine million and four million tons, respectively. Several million tons have been used in the past in electrical power plants in Russia, but peat never amounted to more than 2 to 3 percent of Russia's total electrical production. Ireland is famous for its peat production and has a long history of hand cutting and drying peat for use in home heating [Figure 5.27(a)]. Although much of Ireland's production of peat in the past was used for power production, most today is used for horticultural purposes; its use in power plants has been largely replaced by natural gas.

There is no consensus as to the magnitude of the world's available peat resources, but the quantity

FIGURE 5.26 Subsidence resulting from collapse into old coal mine workings in Wyoming is a threat to cattle grazing. (Photograph by Gary Glass.)

(a)

(b)

FIGURE 5.27 (a) Mining peat in the traditional manner in Ireland. After digging, the peat is stacked and dried before burning. (Photograph courtesy of Irish Tourist Board.) (b) Logs of compressed peat being extruded as the sodding machine moves across a drained portion of a bog in North Carolina. (Photograph courtesy of First Colony Farms, Inc.)

is certainly in the trillions of metric tons (see Table 5.3 for a summary of the world peat reserves and production as of 2008). Furthermore, peat is constantly growing and can be reharvested in many areas every five to ten years. Several countries, including the United States and Canada, have considered large scale electricity generation using peat as fuel, but few of the plants have been economic in the long term. For example, Minnesota experimented by converting some heating plants in public buildings to peat-burning systems; there was a trial peat-to-energy project in 1984 when special sodding machines began scooping up wet peat and extruding it as log-shaped blocks [Figure 5.27(b)]. Similarly, a peat-fired power plant in Maine began supplying electricity in 1989 and operated successfully on a competitive and environmentally acceptable basis for several years before shutting down. Several additional plants were scheduled to be constructed in Florida to begin operation in the mid-1990s, but these were delayed because they would not have been cost competitive with plants using natural gas as a fuel. The peat occurrences of the United States are similar to those in many other parts of the world in that they are concentrated in shallow wetlands in coastal areas and in glacial terranes. Consequently, much of the peat in the United States lies in environmentally sensitive or restricted zones, such as protected wetlands and game refuges. The great peat resources of Canada commonly occur in remote and sometimes arctic regions where their extraction would not be economic. Nevertheless, peat does constitute a considerable fossil fuel resource, but its use in the near future will probably remain limited and local.

TABLE 5.3 World peat reserves and production as of 2008 (in millions of metric tons) (From Mineral Commodities Summaries 2009)

Country	Reserves	Resources
Finland	9.1	6000
Russia	1.3	1000
Canada	1.0	720
Belarus	0.3	400
Lithuania	0.3	190
Ireland	0.65	150
United States	4.3	160
Latvia	1.0	76
Estonia	2.0	60
Others	1.3	1400
World Total	19.15	10,000

Petroleum

Petroleum, long scorned by most of society as a sticky, foul-smelling material to be avoided, has now emerged as the principal fossil fuel of our times. In contrast to coal, the physical appearance of petroleum bears no evidence of its origin from marine planktonic organisms because a complete reconstitution of the tiny organisms occurs following burial. The discussion below considers the origin, occurrence, extraction, refining, and future potential of petroleum as a fuel.

HISTORY OF PETROLEUM USAGE. Although petroleum is the most important of the modern fuels, it has actually been used in a variety of ways since before recorded history. In the petroleum-rich areas of the Middle East, and especially near the Tigris and the Euphrates Rivers in today's Iraq, petroleum and **bitumen** occur in numerous natural seeps that have been exploited by many different peoples. The early

Mesopotamians did not use oil because they could not handle its flammability, but the later Akkadians, Babylonians, and Assyrians found numerous uses for the sticky bitumen as a glue for arrowheads, for setting inlays in tile designs, and as a mortar for holding building bricks together. The famous Tower of Babel, a seven-stage pyramid that reached a height of 295 feet above the roofs of Babylon, consisted of bricks cemented together with bitumen. Meanwhile, the peoples living along the rivers found that bitumen waterproofed their boats. This knowledge is evidenced in two early biblical narratives: (1) Noah, after building an ark of gopher wood, coated it with pitch [or bitumen] inside and out; (2) Moses' mother got a "papyrus basket for him, and coated it with tar and pitch." Natural floating masses of bitumen were harvested from the Dead Sea; in fact, Mark Anthony included the concession for the gathering of this material as one of his many love tokens given to Cleopatra. The Egyptians also found that the bitumen served very well as a preservative for mummies when they ran short of the resins originally used. In the Americas, the Indians used bitumen and oil from natural seeps to caulk canoes and to waterproof blankets; they had additional uses in medicines, in the gluing of Toltec mosaic tile designs, and probably as a fuel.

The use of bitumen probably changed little from the days of early Babylonia until about A.D. 1000 when Arab scientists discovered distillation; by the twelfth century, the Arabs were producing tons of kerosene. Unfortunately, this technological advance was lost with the decline of scientific progress in the Middle East after the twelfth century and was not rediscovered until the nineteenth century.

Through the 1600s and early 1700s, most Europeans and American settlers knew little, if anything, of petroleum. However, by about 1750, numerous oil seeps had been found in New York, Pennsylvania, and West Virginia; wells drilled for water and salt often produced small amounts of oil. This oil was generally considered a nuisance because of its smell and tendency to stick to everything. Some uses were discovered, and about 1847 Samuel M. Kier, who operated a salt business in Pittsburgh, began bottling oil as a sideline. Even the famed frontiersman Kit Carson collected oil and sold it as axle grease to pioneers moving west. Until the 1850s, the major sources of lubricants and illuminants were vegetable and animal oils, especially whale oil. A major step toward the petroleum industry occurred in 1852 when a Canadian geologist, Abraham Gesner, made the discovery that kerosene (called *coal oil*) for use in lamps could be produced from oil and coal by distillation. The usefulness of oil was rapidly realized in various parts of the world, and in 1857 James M. Wilson dug an oil well and built a refinery to produce lamp oil near Oil Springs, Ontario. In the same year, oil production from hand-dug pits reached 2000 barrels in Romania.

Despite these accomplishments, the modern oil industry generally traces its origins to the first American oil well, which was drilled in 1859 by Edwin L. Drake along Oil Creek near Titusville, Pennsylvania. Prompted by the potentialities of oil as a fuel, lubricant, and illuminant, George H. Bissel, a New Haven, Connecticut, businessman, found partners and in 1854, established the Pennsylvania Rock Oil Company to drill for oil near Titusville. Their hopes were indeed spurred when Professor Benjamin Silliman Jr. of Yale University analyzed a sample of crude oil skimmed from a Pennsylvania spring and reported in 1855: "In conclusion, gentlemen, it appears to me that there is much ground for encouragement in the belief that your company has in their possession a raw material from which, by simple and not expensive process, they may manufacture very valuable products." The initial effort failed and ended in bankruptcy, but the investors regrouped under a new name, Seneca Oil Company, and hired Drake, an unemployed railroad conductor, to direct its operation. Drilling began in June 1859 using a wooden rig and a steam-operated drill (Figure 5.28[a]). Because water and cave-ins threatened the well, Drake drove an iron pipe 39 feet into the ground and proceeded to drill inside the pipe. Oil-bearing strata were encountered at a depth of 69.5 feet on August 27, 1859. The oil rose to just below the ground surface; Drake mounted a pump on the well and began producing between 10 and 35 barrels of oil per day.

Initially, the oil sold for $20 per barrel, but the success resulted in the drilling of numerous other wells [Figure 5.28(b)], and the price of oil dropped to 10 cents per barrel within three years. Boomtowns of tents and shacks sprang up rapidly, and wagons and river barges carried the oil in wooden barrels to refineries built along the Atlantic Coast. Railroads soon built branch lines to the oil fields, and by 1865 the first oil pipeline was constructed to carry oil 8 kilometers (14 miles) to a railroad loading area. In 1874, a 97 kilometer pipeline (60 miles) was constructed to transport 3500 barrels a day from the oil fields to Pittsburgh. After Drake's success, oil discoveries spread rapidly—to West Virginia in 1860, Colorado in 1862, Texas in 1866, and California in 1875. In many of these areas, the initial discoveries led to the drilling of numerous very closely spaced wells, as shown in Figure 5.28(b) and in the photo at the beginning of this chapter. The first of the giant oil fields in the Gulf Coast area was opened when Spindletop gushed nearly 60 meters

(a)

(b)

FIGURE 5.28 (a) Edwin Drake (right) in front of his oil well on the banks of Oil Creek in Titusville, Pennsylvania. This well, drilled in 1859, represents the beginning of the modern extraction of oil. (b) The success of Drake's first well resulted in the drilling of large numbers of closely spaced wells, as shown here in 1861 on the Benninghoff Farm along Oil Creek. (Photographs courtesy of American Petroleum Institute Photographic and Film Services.)

FIGURE 5.29 Spindletop, in southeast Texas, was one of the most famous gushers. It began flowing on January 10, 1901, at a rate of 100,000 barrels per day in a gusher that reached a height of 175 feet. (Photograph courtesy of the American Petroleum Institute.)

(200 feet) into the air on January 10, 1901 (Figure 5.29) and yielded 100,000 barrels per day.

Commercial oil production spread rapidly throughout the world. Italy became a small producer in 1860 and was rapidly followed by Canada, Russia, Poland, Japan, Germany, India, Indonesia, Peru, Mexico, Argentina, and Trinidad. The knowledge that oil and tar seeps had been worked in the Middle East for thousands of years stimulated considerable exploration interest in the region in the late 1800s and early 1900s. Small but encouraging discoveries were made in Iran in 1908 and in Iraq in 1927. The true potential of the area finally became apparent when the first of the large fields was discovered in Saudi Arabia in 1938; subsequent drilling has shown that the Middle East contains more than half of the world's known oil reserves.

Petroleum exploration after World War II expanded throughout the world and led to discoveries on every continent and from the tropics to the polar regions. Although numerous small fields have been found, the two discoveries that have received the greatest publicity in recent years have been those of the North Sea in 1965 and the Alaskan North Slope in 1968 (Figure 5.30 and Plate 21).

THE FORMATION OF PETROLEUM AND NATURAL GAS. The most important fossil fuel in the modern industrial world is **petroleum**; it is the base for most lubricants and fuels and for more than 7000 organic compounds. Early discoverers applied the Latin terms

BOX 5.3

Coal Bed Methane

Methane gas in coal mines has long been feared because it causes explosions that have killed tens of thousands of miners around the world. People know of methane, a colorless, odorless gas, as the natural gas that is widely used for cooking, home heating, and industrial processing. Most of the gas we use today is derived from oil and gas fields after the gas has been cleaned to remove impurities, and a special odorant has been added so that it is easily detected by the human olfactory system.

Early coal miners carried canaries into the mines as their means of testing for the presence of methane. The canaries would pass out from the gas before the concentrations were great enough to harm workers. Sometimes, brave individuals were sent ahead of the miners to try to burn out small pockets of the gas before it created explosions (Figure 4.9). At concentrations below 5 percent, methane is inert; at concentrations above 15 percent, it burns with a steady, easily controlled flame. But between 5 and 15 percent, methane mixed with air is violently explosive (Plate 24). Although methane explosions still occasionally occur, modern detection devices can usually provide sufficient warning to prevent the concentrations from becoming deadly.

Today, coal-bed methane is being viewed in many places as an undeveloped valuable resource. After all, the same gas that explodes when accidentally released can be used as an energy source when systematically extracted. The methane in coal forms as the result of biogenic or thermogenic action. Biogenic gas can be generated by anaerobic bacterial decay either early in the burial history, or much later when groundwater flow introduces these bacteria into a coal bed. Thermogenic gas is generated during burial as hydrocarbons and other gases (such as water and nitrogen) are emitted by the organic matter as it is heated and compressed.

Whereas the methane gas associated with petroleum occurs as a free or dissolved phase, nearly all of the methane in coal beds is present as monomolecular layers adsorbed on the internal surfaces of the coal. Coal appears solid, but it is actually a microporous material with very large internal surface areas (tens to hundreds of square yards of surface per gram of coal). Hence, it can adsorb very large quantities of gas. The volumes of adsorbed gas can vary with many factors, but the highly volatile bituminous coals that dominate American coal production can readily contain 25 to 30 cubic centimeters (one and one-quarter cubic feet) of adsorbed gas per gram of coal. Economic production of the methane gas from coal beds is accomplished by drilling into the coal, lowering the pressure of water present (because that tends to prevent the release of methane from the coal), sometimes introducing extra fractures (called *cleats* in coal beds), and pumping out the gas. As pressure drops, the gas desorbs from micropores, diffuses through the coal, and escapes along the cleats (Figure 5.E).

The recoverable amounts of coal bed methane are not well known, but estimates of United States reserves and resources made by the U.S. Geological Survey in 2007 are shown in Table 5.A. The U.S. reserves are approximately equal to one year's total production and are about one-tenth of U.S. conventional gas reserves. The U.S. coal bed methane reserves are approximately nine times greater than the reserves. The world total of coal bed methane is not well known but was estimated a few years back as being similar in magnitude to the conventional natural gas reserves.

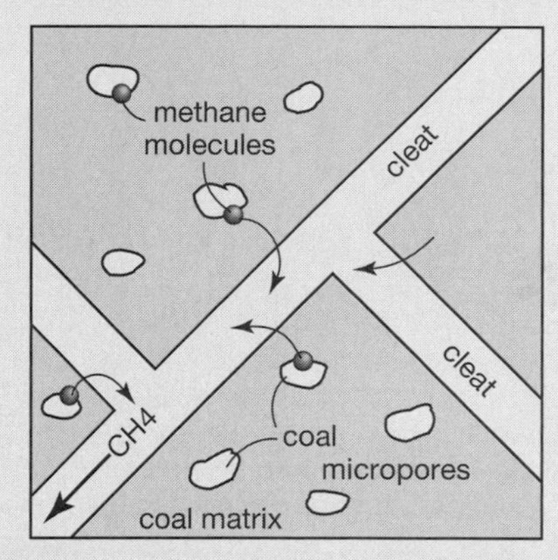

FIGURE 5.E Coal-bed methane (black sphere) desorbs from micropores in the coal matrix, diffuses through the coal, and flows along the cleats (fractures). It once only constituted a threat to mining, but now its recovery produces valuable resources and reduces dangers in some of the coal mines.

Coal-bed methane has been increasingly exploited in recent years and will likely see increased development in the years ahead. Some of the methane to be recovered will be extracted from coal beds that would not otherwise have yielded any energy product. However, some methane will be extracted from mineable seams so that a substance that could have caused a tragedy will instead serve as a valuable energy resource.

TABLE 5.A United States Coal Bed Methane Reserves and Resources in billions of cubic meters, 2008 (From U.S. Energy Information Administration)

Province	Reserves	Resources
Alaska	–	1614
Wyoming	71	697
Appalachians	117	490
New Mexico	318	360
Illinois	–	215
Missouri	–	173
Southern Rockies	–	326
Utah	21	156
Northern Rockies	–	142
Oklahoma	26	130
Others	–	255
World Total	553	4558

(a)

(b)

FIGURE 5.30 (a) Oil platforms such as these in the North Sea are used to drill for oil and then to pump it via submarine pipes to onshore facilities. The smoke results from the flaring or burning of some excess natural gas. (Photograph courtesy of Shell U.K.) (b) The Alaskan Pipeline, which extends for 960 miles and cost $9.5 billion to build, transports 1.5 million barrels of oil per day from Prudhoe Bay on the North Slope of Alaska to Valdez, where it is loaded onto tankers for transport to refineries. (Photography courtesy of Sohio Petroleum Company.)

petra, meaning rock, and *oleum*, meaning oil, because they found it seeping out of the rocks. Petroleum, now commonly called *oil, crude oil,* or *black gold,* has its origin, just as all fossil fuels do, in organic matter that has been trapped in sediments.

Petroleum rarely occurs without accompanying **natural gas**, which is a mixture of light molecular weight hydrocarbons that are gaseous under Earth's surface conditions. In contrast to the petroleum, which consists of at least scores, and sometimes hundreds, of different hydrocarbon compounds, natural gas is dominantly composed—often 99 percent or more—of methane (CH_4). Minor amounts of the

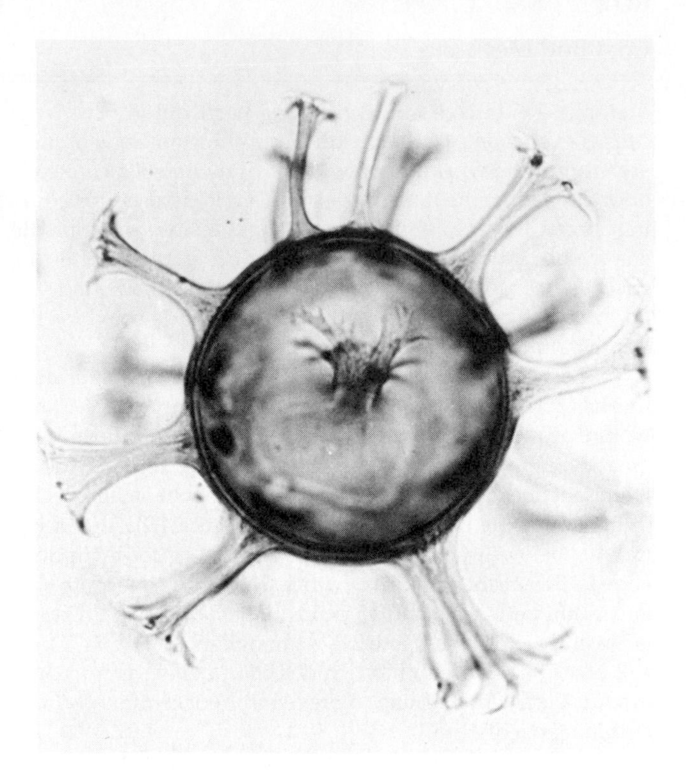

(a)

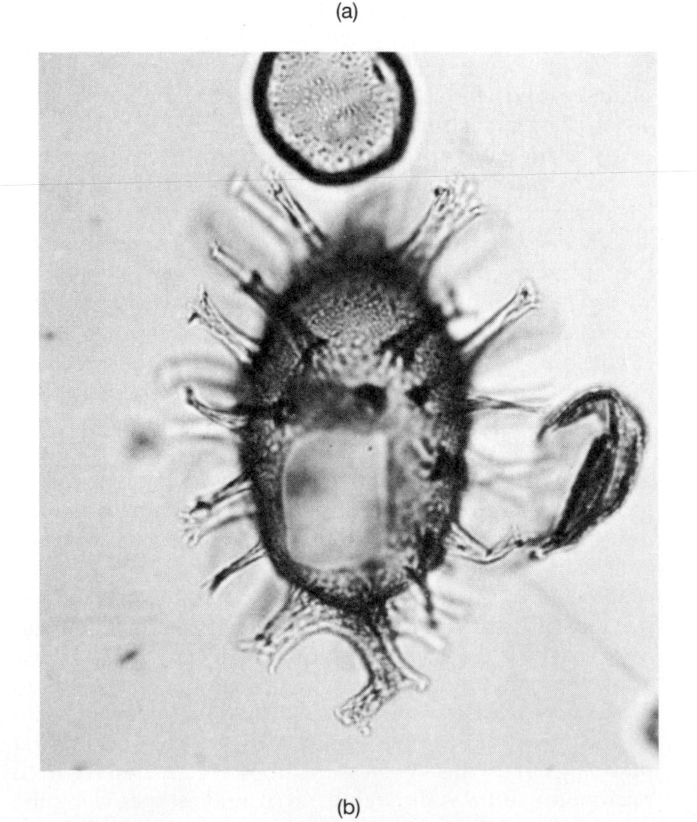

(b)

FIGURE 5.31 Petroleum forms from the accumulation of small floating plankton such as these modern dinoflagellates. The hydrocarbons of each of these, which average 0.55 mm in diameter, are converted into kerogen and ultimately into oil and gas. (Photograph courtesy of D. M. McLean.)

other hydrocarbon gases—such as ethane (C_2H_6), propane (C_3H_8), and butane (C_4H_{10})—may also be present (see Figure 5.49). In addition, natural gas can be admixed with variable small amounts of carbon dioxide (CO_2), hydrogen sulfide (H_2S), helium (He), nitrogen (N_2), hydrogen (H_2), water vapor, and ammonia (NH_3). Petroleum forms almost exclusively from the organic matter that has been trapped in marine sediments, whereas natural gas forms both in marine and terrestrial rocks. As early as 1781, the Abbey S. Volta of northern Italy provided insight into the formation of oil and gas when he noted, "Fermentation of buried animals and plants generates oil, which is transformed into naptha [a term for volatile colorless gasoline-like fluids], by distillation due to the underground heat, and in turn is elaborated into vapors."

The Abbey's views were quite accurate, for we now know that it is the modification of buried organic matter that leads to formation of oil and gas. As noted in Figure 5.6, natural gas forms with both oil and coal and by at least two processes. Most organic matter, even when buried, is totally decomposed by organisms or by oxygen in circulating waters. The portion that survives is still subject to attack in the oxygen-free environment by anaerobic bacteria. The product of the anaerobic bacterial action, which may occur any time from immediate burial to millions of years later, is **biogenic gas.** This gas is principally methane (CH_4), although variable amounts of other gases may also be present, and is constantly rising in small amounts from swamps, from soils, and from the sediments on the seafloor.

In the marine realm, the principal type of organic matter trapped in sediments is composed of the remains of free-floating, microscopic planktonic organisms that constantly rain onto the seafloor (Figure 5.31). This debris is rich in organic matter called **lipids,** but it also contains significant amounts of protein and carbohydrate; terrestrial plant debris is significantly different in that it contains large amounts of cellulose and lignin that make up woody tissue. The process of oil and gas formation in marine sediments is summarized in Figure 5.32. The depth scale, which also corresponds to a general increase in time and temperature, is only approximate and may vary with the nature of the original organic matter.

The depth of burial has been subdivided into three major zones in which the processes of **diagenesis, catagenesis,** and **metagenesis** are active. These processes, in which both minerals and organic matter are altered, constitute a continuum of increasing temperature and pressure in response to burial. *Diagenesis* occurs from the surface of the depositing sediment to depths of a few hundred meters (300 feet or so), where temperatures are generally less than 50°C. Minerals are dissolved and precipitated by groundwaters, and much organic matter is oxidized or consumed by burrowing organisms, or by bacteria. Anaerobic methanogenic bacteria are commonly very active in the upper parts of this zone and are responsible for the generation of considerable amounts of biogenic gas.

Catagenesis, which occurs in the temperature range of 50°C to about 150°C and pressures up to 1500 atmospheres at depths to about 3.5 to 5 kilometers (2 to 3 miles), brings about compaction of the rock and

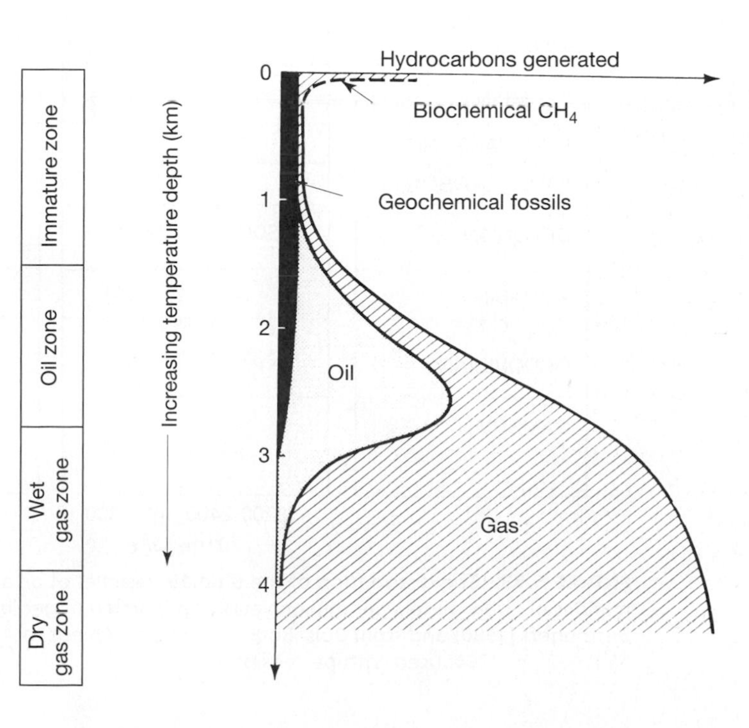

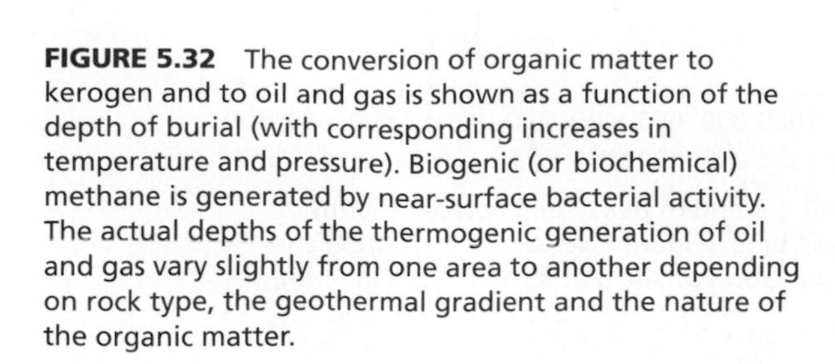

FIGURE 5.32 The conversion of organic matter to kerogen and to oil and gas is shown as a function of the depth of burial (with corresponding increases in temperature and pressure). Biogenic (or biochemical) methane is generated by near-surface bacterial activity. The actual depths of the thermogenic generation of oil and gas vary slightly from one area to another depending on rock type, the geothermal gradient and the nature of the organic matter.

the expulsion of water. The organic matter is progressively altered to kerogen and liquid petroleum. Biogenic gas-producing processes decrease in effectiveness with depth, but thermogenic gas processes become important and result in the formation of gas by thermal cracking of some of the kerogen. As depth continues to increase, petroleum-forming processes give way to **thermogenic gas** production. This gas is commonly referred to as "wet" gas because the dominant constituent, methane, is accompanied by minor amounts of ethane, propane, and butane, which are easily condensed into a fluid phase.

Below depths of 3500 to 4000 meters (11,000 to 13,000 feet) where temperatures exceed about 150°C and pressure rises above about 1500 atmospheres, the early stages of metamorphism occur, referred to as *metagenesis* when dealing with organic matter. At this stage, the remaining organic matter is either converted to dry gas, which is nearly pure thermogenic methane, or it remains as a carbon-rich solid residue. With deeper burial, metamorphic effects increase and the residue is converted to graphite.

From the previous discussion, it should be apparent that rocks containing different types of organic matter, or rocks that contain similar organic matter but which have been subjected to different burial conditions, can yield very different ratios of oil and gas. This is indeed true. Marine sediments with lipid-rich organic matter tend to yield oil and wet gas when subjected to catagenesis; terrestrial, cellulose-rich matter yields coal and dry gas when subjected to the same conditions. Organic-rich marine sediments buried to depths of 2000 to 3000 meters (6,400 to 9,500 feet) usually yield considerable oil and gas; but the same sediments, if buried an additional 1000 meters (3000 feet), usually yield much gas but little oil.

The formation of petroleum, including heavy oils and natural gas, depends on the availability and preservation of marine planktonic organisms. These creatures have been present in the world's oceans since the late Precambrian, but their abundance increased with the passage of time. Not surprisingly, the known reserves of liquid and gaseous hydrocarbons reflect this increase in plankton over time (Figure 5.33). The increased abundance of hydrocarbons in younger rocks no doubt also evidences the escape from, and destruction of, some oil and gas in the older rocks as a result of weathering and erosion. Methane gas can

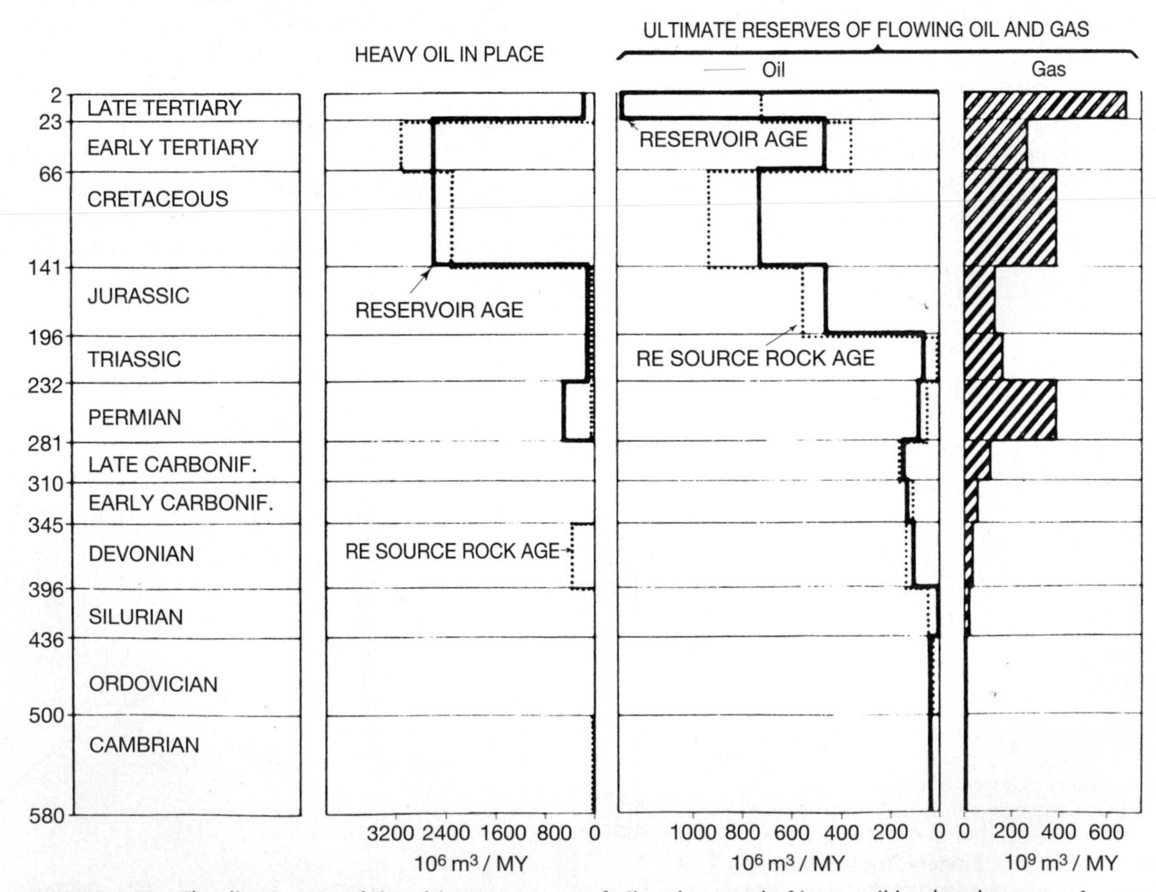

FIGURE 5.33 The distribution of the ultimate reserves of oil and gas and of heavy oil in place in terms of millions of cubic meters per million years as a function of geologic time. (Figure after Demaison [1977] and Bestougeff [1980] and from Bois, Bouch, and Pelet, *American Association of Petroleum Geologists Bulletin,* vol. 66 [1982], p. 1264; used with permission.)

Plate 1. Strip mining for coal can cause total destruction of arable farmland, but careful reclamation can restore the original productivity. The devastation of unreclaimed spoil piles is plainly visible at this coal mine in northern Texas, but when the spoil heaps are leveled and the topsoil is replaced, as in the foreground, a verdant pasture grows within 12 months. (Photograph by B. J. Skinner.)

Plate 2. Every scrap of arable land must be cultivated in order to feed the growing population of Nepal. When hill slopes are steep, terraces must be built to prevent topsoil from washing away, to reduce the possibility of landslides, and to increase the efficiency of irrigation systems. (Photograph by Howard Massey.)

Plate 3. The growing human population and rapid advances in technology create changes in the patterns and types of resources used. This is especially apparent in many parts of the developing world, such as Tongling, Anhui Province, China, where human-powered carts are pushed past buildings with satellite dishes on their roofs. (Photograph by J. R. Craig.)

Plate 4. Mabry Mill, on the Blue Ridge Parkway in Virginia, is an example of a nineteenth-century mill that used water to power saws, to cut lumber, and to grind grain. Water power was a major source of energy before the development of modern fossil fuel energy systems and the wide availability of electricity. (Photograph by J. R. Craig.)

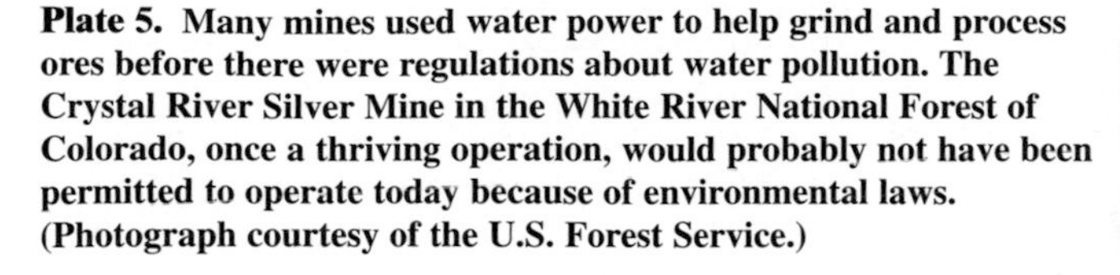

Plate 5. Many mines used water power to help grind and process ores before there were regulations about water pollution. The Crystal River Silver Mine in the White River National Forest of Colorado, once a thriving operation, would probably not have been permitted to operate today because of environmental laws. (Photograph courtesy of the U.S. Forest Service.)

Plate 6. Remains of engine-houses and shaft-towers of the Botallock Mine cling to the rugged cliffs of Cornwall, England. From such mines, some of which extended far out under the ocean, a rich stream of tin, copper, lead, and other metals flowed for over 2000 years. In the nineteenth century, the Cornish production was so great that England was a major world producer. (Photograph by J. R. Craig.)

Plate 7. Production of gold ore at the Bessi Mine, Japan, during the fourteenth century. When the miner had broken enough ore to fill a basket, his assistant would carry it to the surface for processing by climbing a system of ladders made by notching tree trunks. (Photograph by W. Sacco.)

Plate 8. Before miners had access to explosives or power equipment, tunnels were commonly cut by heating rock with fires to cause cracking. Cold water was thrown on the rock to cause more cracks and the broken rock was removed by hand. This nearly round fire cut tunnel was cut by this method at the famed silver deposits at Kongsberg, Norway. (Photograph by J. R. Craig.)

Plate 9. Room and pillar mining has commonly been used to remove areas of rich zinc and lead ore (the rooms) while leaving supporting columns (the pillars) of waste rock. In some areas, the rooms may reach heights of 60 meters and lengths up to 100 meters. (Photograph courtesy of ASARCO Inc.)

Plate 10. This collapse of Interstate 70 in northeastern Ohio occurred in March 1995, approximately 60 years after coal had been mined about 25 meters below. It exemplifies how some effects of resource exploitation may be long delayed in their appearance. (Photograph by Sandy Brandt; courtesy of Ohio Department of Transportation.)

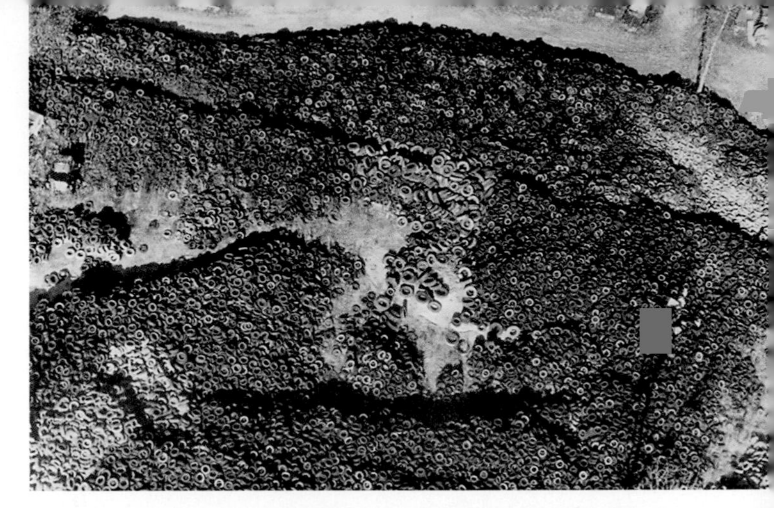

Plate 11. Approximately 250 million automobile and truck tires are discarded every year in the United States (and perhaps 500 million worldwide). This aerial view of a single illegal dump containing 5–8 million tires in King George County, Virginia, shows what has happened to all too many tires despite efforts to find constructive uses for them. (Photograph by U.S. Army Corps of Engineers; courtesy Virginia Department of Environmental Quality.)

Plate 12. Despite the efforts of many groups to prevent the pollution of waters and wetlands by solid and fluid waste, there are still many occurrences of indiscriminate dumping such as this one on Tangier Island, Virginia, in the center of Chesapeake Bay. (Photograph by J. R. Craig.)

Plate 13. Municipal solid waste, here being dumped into a modern lined landfill, contains a vast variety of materials, but most of it could potentially be recycled or converted into energy. Nevertheless, land filling is commonly chosen because it is usually the cheapest method of disposal. (Photograph courtesy of Draper-Aden Associates.)

Plate 14. Acid waters seeping from the base of a pile of waste rock left after coal mining in southwestern Pennsylvania. When iron sulfide minerals in the coal are exposed to the air as a consequence of mining, they become oxidized and produce soluble sulfates that cause the acidity. The yellow crust arises from sulfate compounds that precipitate from the seeping drainage water. (Photograph by J. R. Craig.)

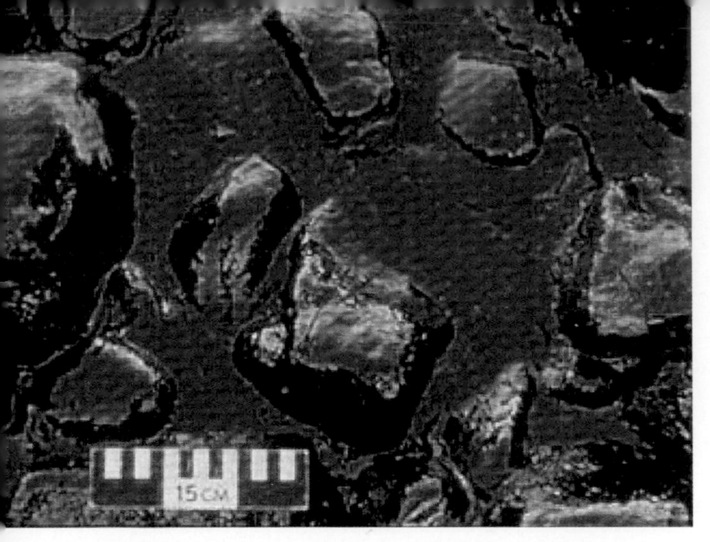

Plate 15. Oil spilled from the *Exxon Valdez* coated hundreds of miles of rocky beaches along Prince William Sound in Alaska. Oil from such spills can clog the feathers of sea birds and the fur of aquatic mammals. Prolonged exposure leads to the death of the animals. This thick crude oil was visible on the surface of the rocks; some of it was recovered or cleaned off the rocks. More than ten years after the spill, there was still much oil beneath the surface, an indication that full recovery will take considerably longer. (Photograph from the National Oceanographic and Atmospheric Administration.)

Plate 16. Underground coal mine fires, such as this one near Sheridan, Wyoming, represent waste of a potential resource and can create extreme local environmental hazards. Once ignited, underground coal fires have commonly proved to be very difficult to extinguish. (Photograph by C. R. Dunrud, U.S. Geological Survey.)

Plate 17. Surface collapse features above the old Hanna Number 3 Mine in Wyoming are environmental hazards; their presence suggests that the entire area underlain by the old mine is dangerous. (Photograph courtesy Gary B. Glass, Wyoming Geological Survey.)

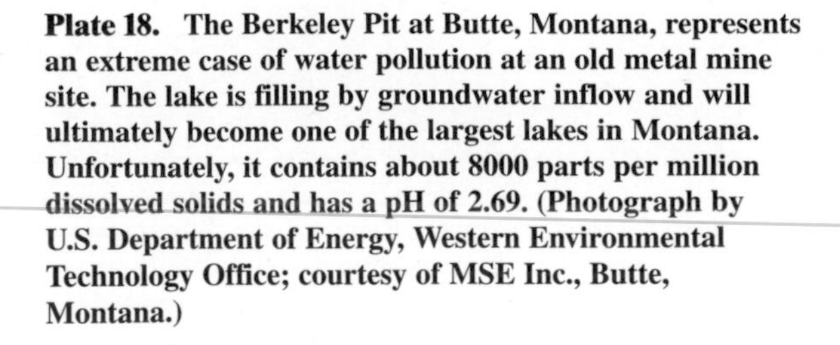

Plate 18. The Berkeley Pit at Butte, Montana, represents an extreme case of water pollution at an old metal mine site. The lake is filling by groundwater inflow and will ultimately become one of the largest lakes in Montana. Unfortunately, it contains about 8000 parts per million dissolved solids and has a pH of 2.69. (Photograph by U.S. Department of Energy, Western Environmental Technology Office; courtesy of MSE Inc., Butte, Montana.)

Plate 19. An increasing amount of the world's oil production is coming from offshore fields and requires the construction of sophisticated drilling platforms. (Left) Oil drilling platform in operation in the North Sea. (Above) Oil drilling platform being towed from port toward the North Sea. (Photographs courtesy of Elf Enterprises and U.K. Texaco.)

Plate 20. The Clinch River power plant in southwestern Virginia is a typical coal-fired electricity generation facility, burning approximately 2000 tons of coal per day. (Photograph courtesy of Appalachian Power Company.)

Plate 21. The Alaskan pipeline transmits more than one million barrels of oil per day from Prudhoe Bay on the Arctic Ocean to the shipping port at Valdez. Fortunately, concerns that the pipeline would disturb migration of the caribou have not come true; they routinely walk over or under the pipeline. (Photograph courtesy of BP America.)

Plate 22. During the 1991 Gulf War, the retreating Iraqi army set fire to more than 700 oil wells, some of which burned for nearly six months. The loss of oil from wells, such as those seen behind a damaged Iraqi tank, was more than a million barrels per day. (Photograph by Jonas Jordan, U.S. Army Corps of Engineers.)

Plate 23. As much as one-billion barrels of oil were spilled into the Persian Gulf, wasted onto the desert floor, or set ablaze as the result of Iraqi army sabotage during the Gulf War in 1991. The effects of the severe pollution were felt for many years. (Photograph by Jonas Jordan, U.S. Army Corps of Engineers.)

Plate 24. Methane gas and coal dust explosions, such as this one at an experimental mine in Pittsburgh, have been a major cause of deaths and injuries in coal mines. Sparks generated during mining can ignite methane gas if the gas is suddenly released from the coal. Once the methane ignites, the air pressure generated may raise much coal dust which then also becomes explosive. (Photograph courtesy of U.S. Bureau of Mines.)

Plate 25. The Rossing Mine, Namibia (in southwest Africa), is one of the world's largest and most mechanized uranium mines. A truckload of ore has just passed under a radioactive scanning device that measures the richness of the ore and sends the information ahead to the processing mill in order to recover the uranium as efficiently as possible. (Photograph by B. J. Skinner.)

Plate 26. Yucca Mountain, in southern Nevada, is the site chosen by the United States government as a permanent repository for the United States' high-level radioactive waste. (Photograph courtesy of U.S. Department of Energy.)

Plate 27. This entrance provides access into the Yucca Mountain nuclear waste storage facility in Nevada. Years of development, as well as many political decisions, remain before the facility is ready to accept nuclear waste (now scheduled for the year 2010). (Photograph courtesy of U.S. Department of Energy.)

Plate 28. A forest of windmills at Altamont Pass Windmill Farm, California. The windmills convert the kinetic energy of flowing air (wind) into electrical energy. The amount of electricity produced by each mill is small, but the total from all mills on the farm is large. (Photograph courtesy of U.S. Department of Energy.)

Plate 29. Modern waste incinerators reduce the volume of municipal solid waste by about 90 percent and often produce commercially salable steam and electricity during operation. Gas emissions are strictly controlled, but the disposal of the ash, which concentrates metals in the waste, remains the subject of considerable debate. (Photograph courtesy of Ogden-Martin Systems Inc.)

Plate 30. Solar cells, such as the onc utilized at the oil well in the photo above, provide electrical power at remote locations. The 240 watts produced by these solar cells are used to protect the steel oil well casings from corrosion. (Photograph courtesy of Solarex Corporation.)

Plate 31. Molten steel is held in a huge ladle prior to being poured into useful forms. At this point, impurities are removed and alloying metals added to provide the desired properties. (Photograph courtesy of Wheeling-Pittsburgh Steel Company.)

Plate 32. Nodular bauxite, the preferred ore of aluminum. This rich ore is the highest grade product mined at Weipa, in Queensland, Australia. Weipa also contains one of the world's largest resources of bauxite. (Photograph by H. Murray.)

Plate 33. Finely banded layers of cherty silica (white) and hematite (red) in a specimen from a banded iron formation known as the Negaunee Formation, Marquette District, Michigan. Banded iron formations are ancient chemical sediments, known to exist on every continent, and they contain the largest resources of mineable iron in the world. (Photograph by H. L. James, Economic Geology.)

Plate 34. Titanium mineral-bearing sands are mined by using high-pressure water cannons, at Eneabba, Australia. The titanium minerals, which make up less than 5 percent of the sands, are removed, and the remaining sand is then returned to the original sites. (Photograph courtesy of RGC, Inc.)

Plate 35. The weathering of iron sulfide-rich ores usually produces gossan, masses of porous iron oxides and hydroxides. Gossan have sometimes served as sources of iron but usually are of little value except for indicating the likely presence of ores beneath the surface. (Photograph of the Sulphur Mine, Louisa County, Virginia, by J. R. Craig.)

Plate 36. Rich ore from Almadén, Spain, one of the world's most famous mercury mines, is red because of the high content of the mineral cinnabar (HgS). Many parts of the ore body also contain free liquid mercury. Thus, the ore is rich, but the fumes pose health hazards to the workers. (Photograph by J. R. Craig.)

form through some types of metamorphic and igneous reactions and is found trapped in tiny inclusions in some minerals. Most petroleum geologists believe, however, that only organically produced gas will ever be found in sufficient quantities to be economically recovered.

The initial amounts of organic matter in nearly all sediments are too low and too dispersed to form commercial quantities of oil. Economic accumulations occur only where the petroleum has migrated from **source rocks** rich in organic matter, in which it was generated in small dispersed amounts, into porous and permeable carrier or **reservoir rocks** (usually sandstones or porous limestones). The final requirement for commercial oil and gas accumulations are **traps** (Figure 5.35), zones in which the migrating hydrocarbons become confined and prevented from further movement by an impermeable seal or cap rock. There are two general types of traps: **structural traps,** formed by folding or faulting, and **stratigraphic traps,** created when layers of porous, permeable rocks are sealed off by overlying impermeable beds. In the Gulf Coast region of the United States and a few other geologically similar areas, considerable amounts of petroleum occur in structural traps that have developed adjacent to salt domes, which have bowed

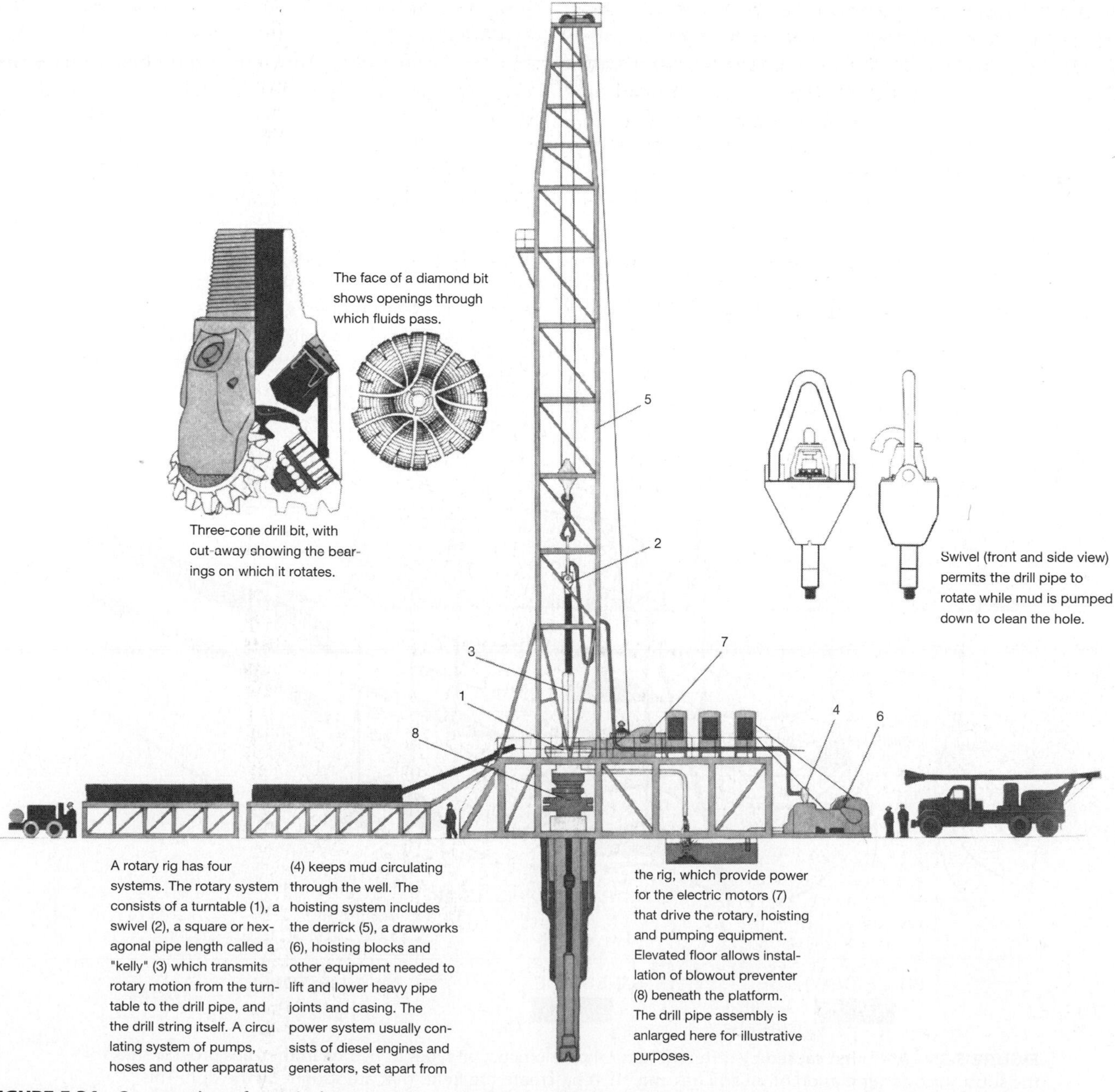

FIGURE 5.34 Cross section of a typical rotary drill rig used on land in the search for oil and gas. (From *The Upstream,* Exxon Background series. Reprinted with permission of Exxon Corporation.)

up and penetrated oil-bearing strata, as shown in Figure 5.35. The migration of water and natural gas along with the oil usually results in a layered configuration in the traps, with the more dense water below, and the less dense gas above the oil.

OIL RECOVERY. The earliest recovery of oil was from natural seeps where oil, and commonly natural gas, had migrated along faults or along bedding planes either to Earth's surface or into zones of moving groundwater that then carried the oil until it surfaced in a spring. At the surface, the natural gas dissipated in the air and the light fractions of the oil evaporated, leaving behind bitumen, or pitch. The pitch was gathered from the seashores or scooped off the surface, as early as thousands of years ago in the Middle East and in the Americas. Although oil still locally seeps out onto Earth's surface, virtually all of the oil produced in the world today is recovered through wells that employ the primary or secondary methods of extraction described next.

The earliest oil wells were drilled by driving rotating pieces of pipe, which had crudely formed cutting teeth, into the soil and rock. These simple drills were replaced by **cable-tool drills,** which consist of a heavy bit attached to a long steel cable. The cable raises the bit and drops it again and again so that it cuts little by little into the rock; periodically, the cable and bit are withdrawn from the hole and the loose fragments of rock are flushed out. The cable tool drills, which can effectively reach depths of several thousand meters (many thousand feet), have now been replaced by **rotary drills** that are more efficient and can operate at depths greater than 10,000 meters (30,000 feet). The rotary drill employs a complex bit with rotating teeth that cut into the rock as the bit is rotated; the bit is attached to the base of a series of hollow steel pipes that are rotated by motors on the drill platform. Drilling fluids are pumped down the center of the pipe to cool the bit and to flush the rock chips out of the hole. To change drill bits, the drill crew must raise the entire stem of drill pipes and disconnect them in 10- to 20-meter (30 to 60 foot) lengths; once the bit has been replaced, the entire stem must be reassembled pipe-by-pipe as it is lowered into the hole. On land, drilling is carried out by rigs of the type shown in Figure 5.34 In recent years, much oil well drilling has been carried out offshore on the continental shelves and has employed either floating, stationary, or boat-mounted rotary drills (Figure 5.36; see also Plate 19).

Once a series of holes have been drilled, they are linked together by a production manifold that allows oil from the individual wells to be safely transported to loading sites. A good example is the Strathspey oil field in the North Sea where fifteen wells, including water-injection wells, gas wells, and oil wells, are linked to a single manifold; oil then flows to the loading platform via a 36-inch diameter pipeline. Early oil drillers assumed that the holes they drilled were vertical beneath the drill rig, to a depth equal to the

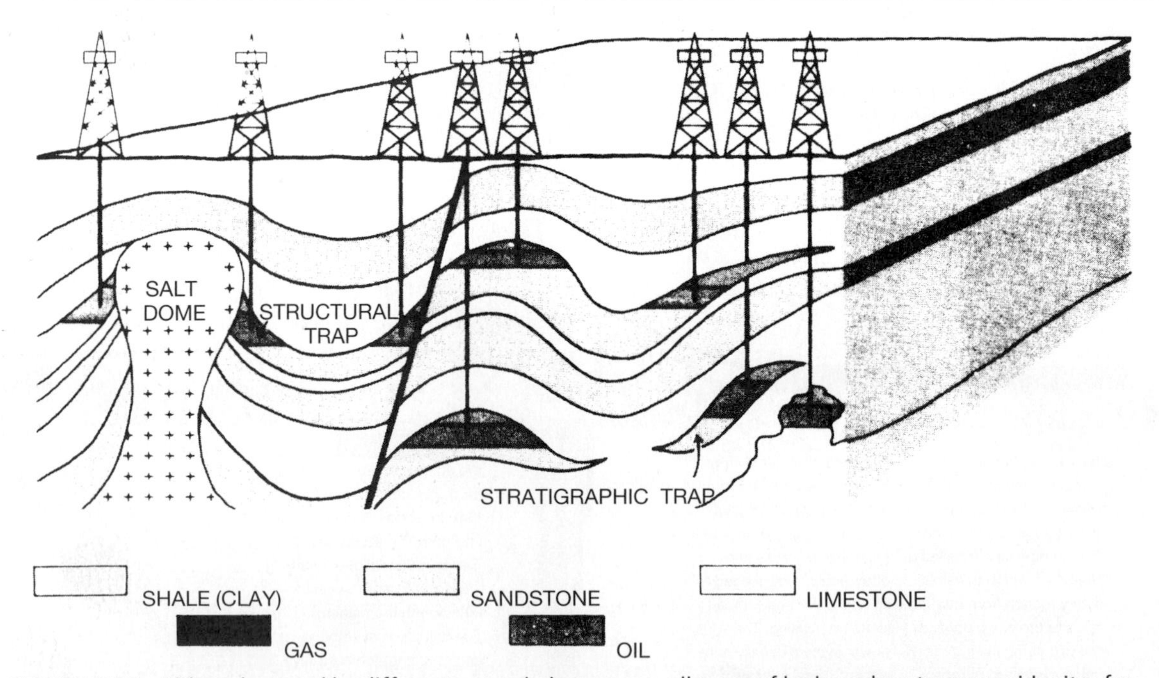

FIGURE 5.35 Although caused by different natural phenomena, all types of hydrocarbon traps provide sites for the subsurface accumulation of oil and gas and, thereby, create the fields that are sought by the drill. This simplified sketch shows examples of major traps. (From *How Much Oil and Gas*, Exxon Background Series; reprinted with permission of the Exxon Corporation.)

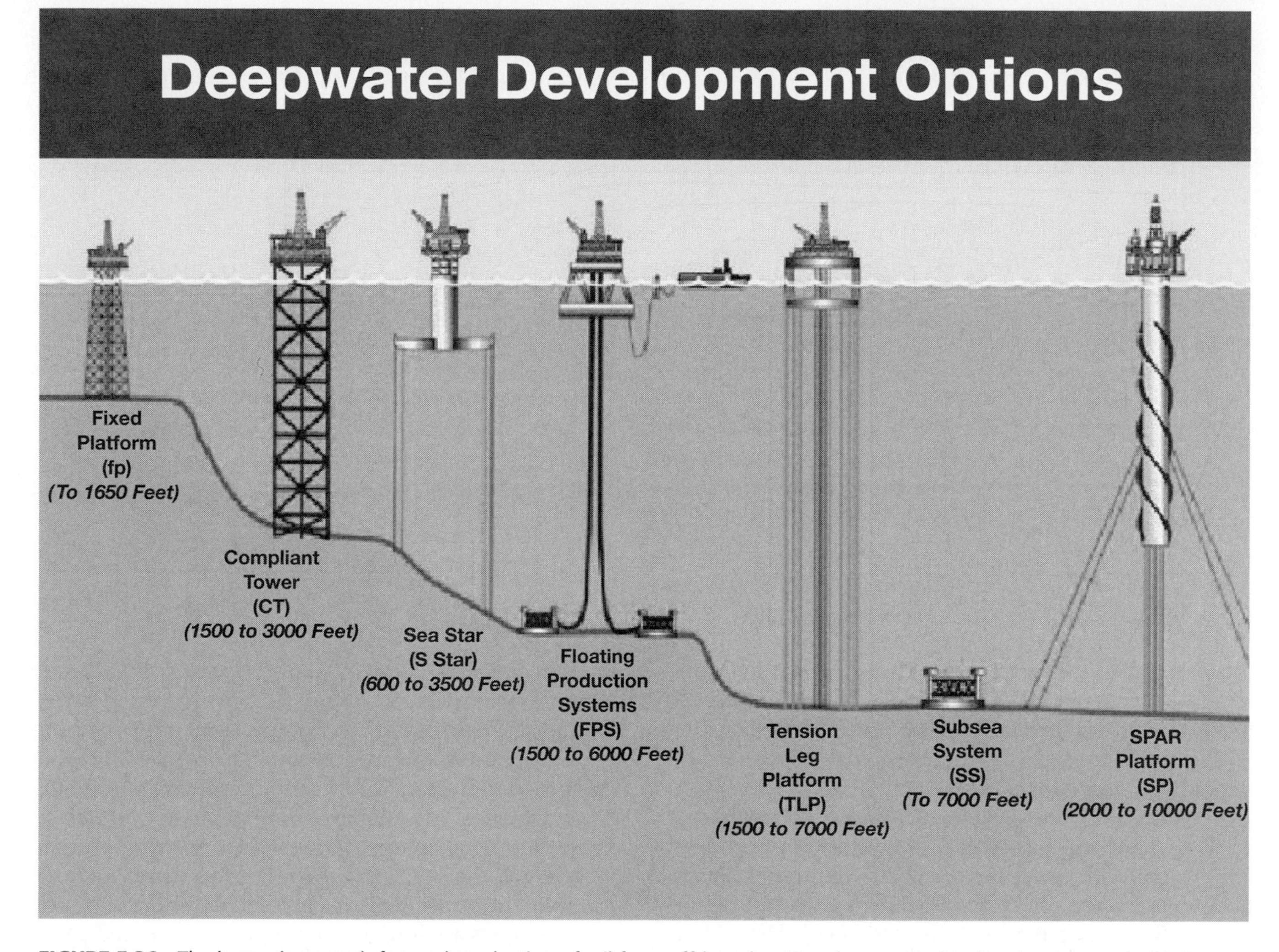

FIGURE 5.36 The increasing search for, and production of, oil from offshore locations has resulted in the development of taller fixed platforms and a variety of floating platforms that are held in place by being tied to the seafloor. In addition, once wells have been drilled and metal casings installed, production may be accomplished by seafloor systems operated by remote control.

length of the cable or drill pipe. This was often true, but in many instances the differences in the composition and hardness of rocks, especially when the strata were inclined, caused the drills to veer off course, missing their original targets. As the technology for drilling has improved through time, drills can now be kept on target, and *directional drilling* has also developed. With directional drilling holes can be drilled at any angle, including horizontally (Figure 5.37). An impressive horizontal well, drilled in Qatar in 2008, extended a total drilling length of more than 40,000 feet (7.6 miles or 16.7 kilometers) reaching out horizontally more than 36,000 feet. These techniques are also now widely used to reach previously inaccessible places, such as under lakes, rivers, or bays, to more effectively probe for oil accumulations occurring in vertical fractures, to more efficiently extract oil from broad horizontal zones, and to permit the drilling of several reservoirs from a single drill site. Directional drilling is accomplished by using drill motors that are lowered down into the holes and are guided by complex navigational and survey systems. Directional drilling is especially important when drilling from a single stationary site, such as the large rigs in the North Sea or the Gulf of Mexico, and in areas where vertical drilling may be considered environmentally unacceptable, such as in Chesapeake Bay.

One good example of the effectiveness of horizontal drilling can be seen in the Captain Field, which holds 1.5 billion barrels of oil 83 miles off the north coast of Scotland in the North Sea. The use of horizontal wells reduced the number of wells needed from 100 to 30, the drilling locations from six to two, and allowed single wells to produce from a 2000-meter-(6000 ft) wide zone across the oil field. A conventional vertical well could have produced oil from a zone only a few hundred meters (several hundreds of feet) in thickness. Subsequently, techniques have advanced and

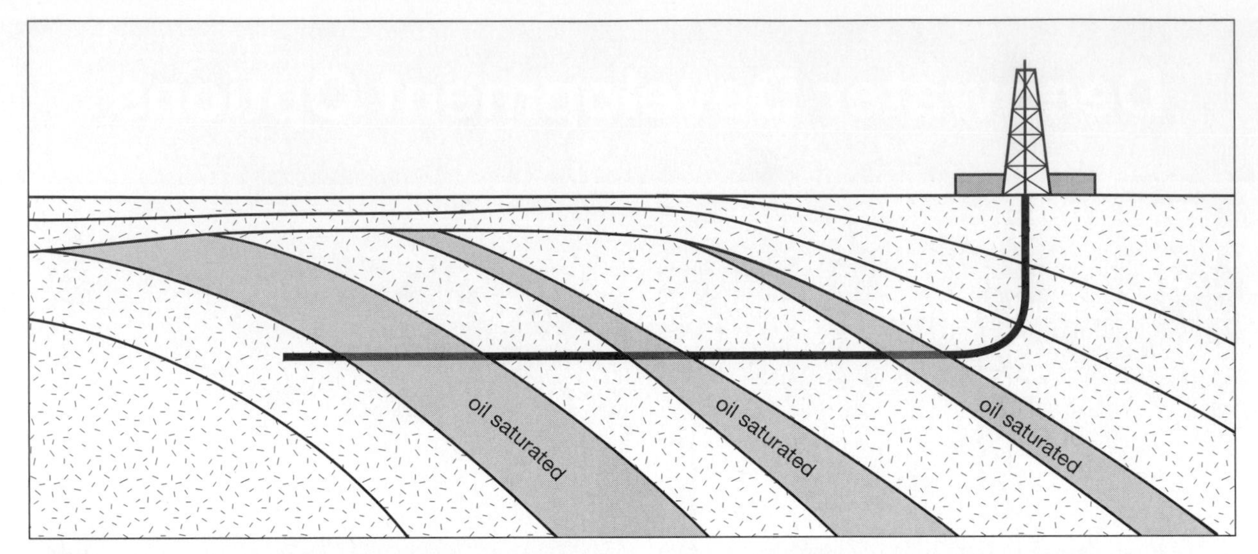

FIGURE 5.37 Directional drilling, including horizontal drilling, is now common in many oil fields. Inclined or horizontal holes can be useful in increasing recoverable oil or gas by cutting through wider producing zones or across multiple vertical or sloping fractures or sands. Furthermore, many separate directional wells may be directed from a single drill site, thus decreasing the environmental impact and the costs.

today there are wells extending as much as 10 kilometers (6 miles) laterally from the central well site. This has reduced costs, lessened environmental impacts, boosted discovery success rates, and increased the productivity of individual reservoirs.

Because the petroleum within rocks occurs as small droplets and films between grains, in pores, and along small fractures, as shown in Figure 5.38, its movement is usually very slow. To allow the petroleum to flow more readily, it is often necessary either to enlarge the natural flow channels in the rock or to make new ones. One common method is the injection of strong acid solutions that dissolve cementing agents holding mineral grains together, thereby increasing the permeability between pore spaces. In another commonly employed method to increase permeability, water and coarse sand are pumped down the well under a pressure high enough to fracture the oil-bearing rock; the sand grains prop open the fractures to allow oil movement. In other areas, high-caliber bullets or small explosive charges are used to fracture the rock adjacent to the well. All of these procedures increase the total amount of oil recovered because they open new channels and break open tightly constricted pores from which oil would not otherwise have been able to flow to the well.

FIGURE 5.38 Oil and gas occur in, and move through, the interstices between the grains in the reservoir rocks. The porosity rarely exceeds 30 percent of the total rock volume. Natural gas moves rapidly through interconnected pores, but oil sticks to the mineral surfaces and cracks so that the primary recovery is usually only about 30 to 35 percent. Additional oil can often be recovered by use of the secondary recovery methods.

Primary recovery is the simplest and least expensive method of recovering oil from wells because it takes advantage of natural pressures within the petroleum reservoir to push the oil to, and sometimes up, the well. If natural pressures are extreme, the initial drilling into a reservoir can produce a **gusher** (Figure 5.29), in which the oil is sent spouting out of the top of the hole. Gushers were actually quite rare in the past, and today they are nearly always prevented by the presence of special valves that will stop or control the flow of oil if high pressure is present. If the natural pressures are low, the oil must be pushed to the surface by pumps placed at the bottom of the well.

The natural pressures that aid in the concentration and extraction of oil are water drive, gas expansion, and the evolution of dissolved gas from the oil [Figure 5.39(a),(b),(c)]. Most petroleum contains dissolved gases held in solution by pressure, much as CO_2 is held in solution in soda water as long as a can or bottle is unopened. When petroleum removal begins, the confining pressure drops and some of the gas comes out of the solution. Just as the evolution of

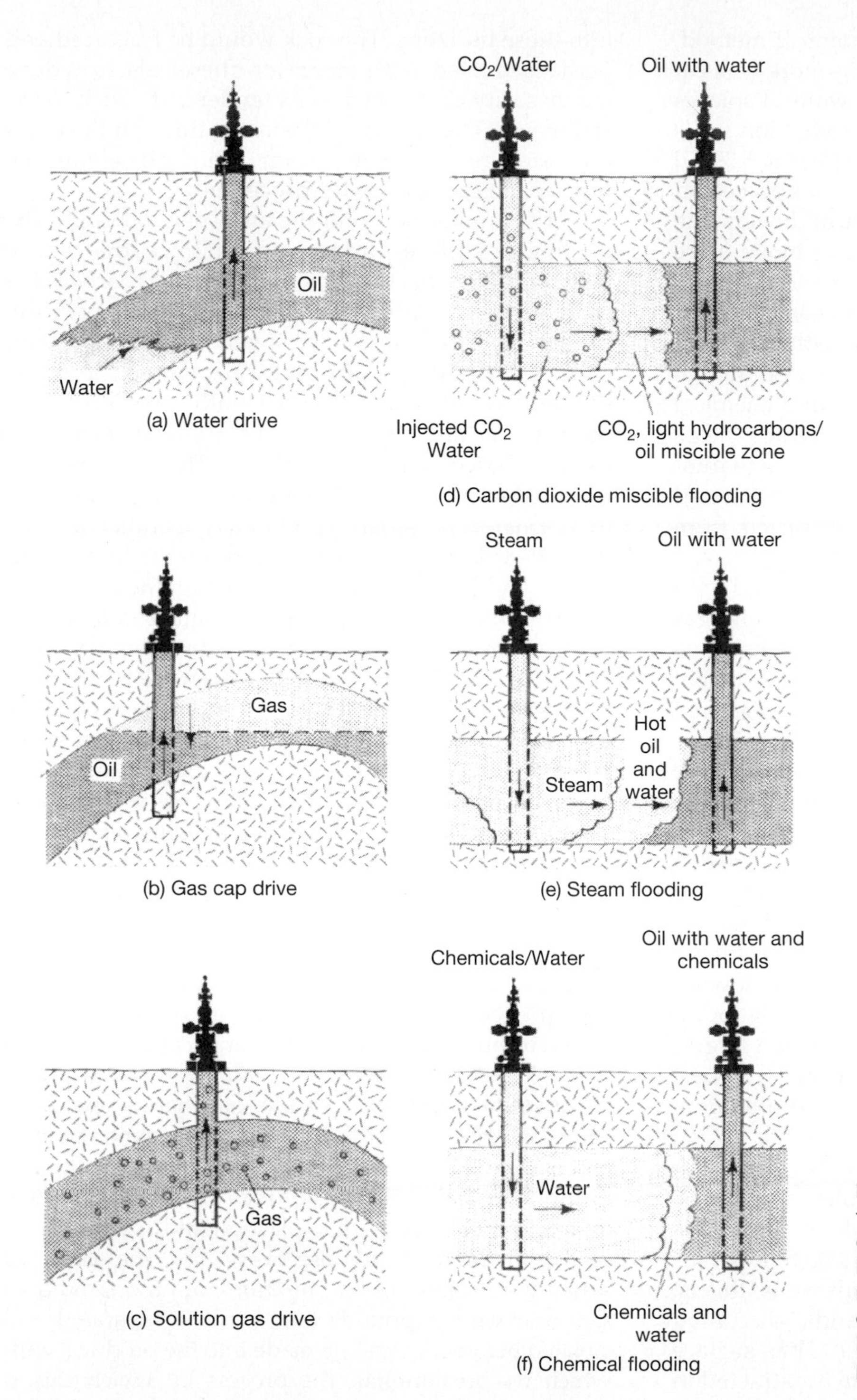

FIGURE 5.39 Recovery of oil and gas from traps may be by primary methods—(a) water drive, (b) gas drive, (c) solution gas drive; or by secondary methods—(d) carbon dioxide miscibility flooding, (e) steam flooding, (f) chemical flooding. (From *The Upstream*, Exxon Background Series; reprinted with permission of Exxon Corporation.)

CO_2 can cause some of the soda to overflow from the can or bottle, the evolution of the natural gases can force some of the oil from the rock into the reservoir and up the well.

In many traps, natural gas either in liquid or gas form lies above the petroleum, as shown in Figure 5.39(b), and exerts considerable pressure from above. As oil is withdrawn up the well the gas cap expands and thus continues to help push oil toward the well. Water drive [Figure 5.39(a)], usually the most efficient of the natural processes, occurs when the pressure in the water of the underlying portion of the reservoir is sufficient to push the oil ahead of it as oil is removed.

Although natural processes can help significantly in oil flow and recovery, the primary phase of oil recovery commonly recovers only about 20 to 30 percent of the total oil in the reservoir. To increase the recovery of the oil, artificial or **secondary recovery** techniques are employed. Secondary recovery procedures most commonly employ water, steam, or chemical solutions to flood, displace, or dissolve and mobilize the oil [Figure 5.39(d), (e), (f)]. **Water**

flooding, the most common and economical method, involves injection of water down one or more wells at the periphery of the trap or field. The water displaces some of the oil and drives it toward production wells where it is pumped out. **Steam flooding** [Figure 5.39(e)] is a variant of water flooding in which superheated steam is injected into a well. The heat lowers the viscosity of the oil, thus permitting it to move more easily; the condensed steam then serves as water flooding. **Chemical flooding** [Figure 5.39(f)] is similar to water flooding except that it employs either a chemical (such as a light hydrocarbon), which is soluble in the petroleum and reduces its viscosity, or a chemical (such as a polymer or a surfactant), which helps reduce the tendency for the petroleum to stick to mineral grains. Generally, chemical solutions are injected into one well and the petroleum is extracted from another.

Sometimes the natural gas that is recovered with oil is separated and reinjected into an expanding gas cap to help maintain reservoir pressure and improve oil recovery. This gas then also represents a potential future resource because much of it can be extracted later if economic conditions are favorable. Despite expensive secondary recovery procedures, commonly 50 percent or more of the original oil remains in the ground.

In recent years, especially after the rapid rise in oil prices between 1979 and 1981 (see Figure 3.14), there has been considerable speculation about **oil mining.** Interest in such procedures has been fueled by economics, by the declining amounts of flowing liquid petroleum reserves, and by the observation of the American Petroleum Institute that the ten largest oil fields in the United States will still contain 63 percent of their original oil in place after full production. In the United States alone, this amounts to an estimated 300 billion barrels, fifteen times the country's known recoverable liquid petroleum reserves. Worldwide, the potentially recoverable original oil is no doubt many times larger than the 2000 billion barrels of oil expected to flow out through wells; this is, in addition to oil shales and tar sands, discussed later. Oil mining could be carried out (1) as surface mining, much as low-grade copper ore is extracted in many parts of the world today; (2) as underground mining, using large tonnage extraction procedures as employed in many metal mines; or (3) as underground drainage systems, in which the oil drains into underground cavities in response to gravity. In the first two techniques, the rock, which may contain 0.1 to 0.5 of a barrel of oil per metric ton, is extracted, crushed, and then either treated with steam or chemical solvents to liberate the oil. Underground drainage systems would involve excavating tunnels under the oil-bearing horizons and then drilling up into those horizons. The rock would be fractured and perhaps treated with steam or chemicals to reduce the viscosity of the oil and its tendency to stick to the sediments. The oil would drain out through the holes into storage chambers, from which it would be pumped to the surface for processing.

Oil mining is by no means a new concept. The digging of surface pits to gather oil occurred in the Baku area of Persia (today's Azerbaijan) as early as the sixth century B.C.; in 1830, this area was reportedly producing 30,000 barrels of oil per year from pits and springs. Hand-dug shafts and wells were yielding oil that seeped into them in the sixteenth and seventeenth centuries in areas as diverse as Sumatra, Germany, Cuba, Mexico, and Switzerland. The only serious twentieth-century oil mining efforts were carried out by Germany and Japan with limited operations, when they were faced with wartime needs during World War II, and by the United States, when the need for a special grade of oil spurred a similar attempt in Pennsylvania. The reduction in petroleum prices in the 1980s and 1990s reduced the incentive for oil mining (as well as oil shale recovery), and most previously proposed operations were postponed or cancelled. However, the rapid rise in the price of oil after 2000 made oil mining of the tar sands of western Canada economic and in a very short time increased Canada's petroleum reserves from less than 20 billion to nearly 200 billion barrels. The vast deposits of oil sands transformed communities such as Fort McMurray into "boomtowns" as thousands of workers moved into the area. The mining proved to be economical but also brought about considerable criticism of the detrimental environmental effects on the landscape.

PETROLEUM REFINING. Crude oil that is extracted from an oil well is a black, sticky fluid with a consistency that varies from watery to syrupy. It often bears little resemblance to the gasoline, kerosene, lubricating oils, chemicals, plastics, and myriad other petroleum products we use every day. The crude oil actually consists of a mixture of thousands of hydrocarbon compounds that must be separated and isolated before they can be made into the products with which we are familiar; the process by which this is accomplished is called **refining.** A primitive method of refining was developed by Arab scientists in the Middle East in about A.D. 1000 when they boiled bitumen and condensed it on a hide or in a water-cooled glass column. Although knowledge of these technological innovations was lost with the decline of science during the Middle Ages, it was rediscovered in the nineteenth century.

The first step in refining is the **distillation** or **fractionation** of the crude oil into a series of fractions on the basis of their condensation temperatures. The

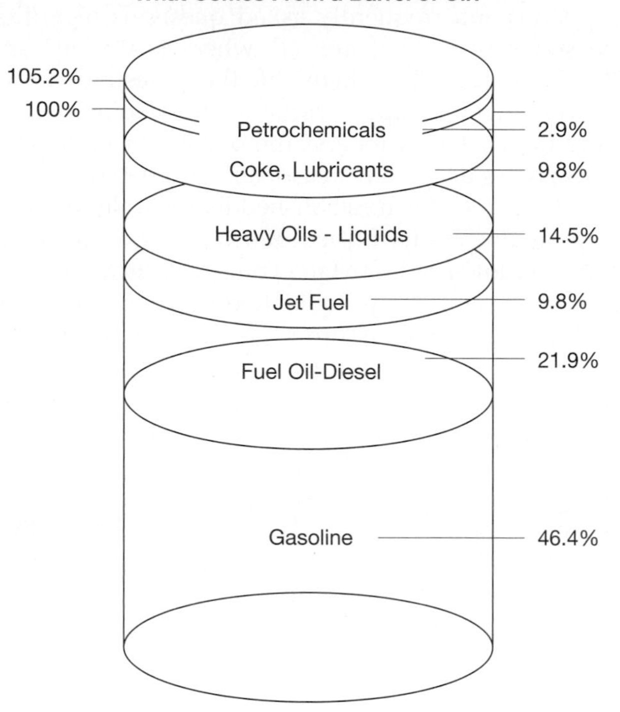

FIGURE 5.40 What comes from a barrel of oil? Oils vary significantly in terms of chemical and physical characteristics, but this diagram illustrates the percentages of products produced from an average barrel of oil today. Hydrogen is added to the oil during refining to upgrade the products that can be produced; hence, the final products produced are approximately 5 percent greater than the original amount of oil treated. (Data from U.S. Energy Information Administration.)

crude oil is first heated to nearly 500°C and then separated into a large number of different products, as shown in Figure 5.40. The lightest of the fractions, light gasoline, rises to the top of the tower; heavier fractions, such as diesel fuel and heating oil, condense at a lower level; the heaviest residuum, which is asphalt, is taken out at the base of the tower. The composition of a typical crude oil is given in Table 5.4.

TABLE 5.4 Composition of typical crude oil

Components (molecular size)	Volume %
Gasoline (C_4 to C_{10})	27
Kerosene (C_{11} to C_{13})	13
Diesel fuel (C_{14} to C_{18})	12
Heavy gas oil (C_{19} to C_{25})	10
Lubricating oil (C_{26} to C_{40})	20
Residuum (>C_{40})	18
Total	100

(From Hunt, *Petroleum Geochemistry and Geology*, 1979.) The physical nature of hydrocarbon compounds is dependent on the arrangement of the carbon atoms in the molecules as well as the number of those atoms present.

In the early days of the oil industry, distillation was the primary means of separating products. The most useful fractions were kerosene, heating oils, and lubrication oils; gasoline was too explosive for household use and was commonly discarded. With the age of the automobile, there was an increasing demand for a higher percentage of gasoline. This brought about the development of techniques for the conversion of less useful heavier fractions into lighter ones.

GASOLINE USAGE AND DEMAND—THE EXAMPLE OF THE UNITED STATES

Most of the petroleum produced today is converted into some sort of transportation fuel—gasoline, aviation fuel, diesel oil. Thus, more than 9 million barrels of gasoline are produced daily from the more than 20 million barrels of crude oil consumed in America every day. This equates to nearly 400 million gallons of gasoline per day.

Because the American economy and lifestyle are so dependent on gasoline and diesel fuel, and, hence, the petroleum from which these are made, there have been numerous political declarations about the United States becoming self-sufficient by increasing domestic oil production, substituting alternative fuels, or improving the mileage obtained from a gallon of gasoline in vehicles. Despite the drilling of more oil wells and new enhanced recovery methods, U.S. domestic oil production has continued to decline since its peak in 1970—and will likely contine to decline as the major fields are depleted.

After the oil embargo and the gasoline shortages in 1973, the U.S. Congress instituted the Corporate Average Fuel Economy (or CAFÉ) standards in 1975. These were intended to increase gasoline mileage performance in cars and thus reduce the reliance on imported oil. As a result, the U.S. average automobile mileage was expected to rise from 13.5 mpg (miles per gallon) in 1974, to 20 mpg in 1980, and to 27.5 mpg by 1985. It actually peaked at only 26.2, but has slowly declined since. Hence, actual average mileage has fallen consistently for the past twenty years despite much political posturing. However, what Congress failed to do, the economic environment brought about. The rapid rise in gasoline prices, reaching more than $4.00 per gallon across America in 2008, combined with increased concern about the rising levels of CO_2 in the atmosphere, created a market for electric vehicles and hybrid cars capable of operating much of the time on electricity. At the same time, there was an increasing, albeit controversial, push for the increased production of alcohol to be used in gasohol (see the discussion in Chapter 6).

Europeans have always paid higher prices for gasoline than Americans, in large measure because of higher taxes. Consequently, most European manufactured cars have been smaller than American made cars and they have achieved much better gas mileage.

Thermal cracking is the application of heat and pressure to heavy hydrocarbons to "crack" or break them into lighter ones. **Catalytic cracking** produces the same result through the use of a catalyst, usually a synthetic zeolite mineral, that speeds and facilitates the process. Zeolite catalysts not only reduce the energy requirements for cracking; they are also useful in adding hydrogen to the hydrocarbons in oil, a process called **hydrogenation,** thereby increasing the production of gasoline. Crude oils range widely in quality in terms of the distillation products and the amounts of contaminants (mostly sulfur, but also some metals such as nickel and vanadium), but modern refineries can produce nearly 50 percent gasoline, 30 percent fuel oil, and 7.5 percent jet fuel from the original oil. Separation of the sulfur is necessary to make the use of the fuel products environmentally acceptable and has resulted in petroleum refineries becoming major producers of sulfur. In fact, in the United States, petroleum refining now accounts for more than 55 percent of domestic sulfur production.

WHERE IS THE WORLD'S OIL FOUND? Two of the most important and frequently asked questions regarding the sufficiency of oil are: (1) where is the oil? and (2) how much oil is there? Neither question can be answered with complete certainty, but exploration for more than a 150 years and the drilling of millions of wells allows us to make reasonable estimates.

It has long been established that petroleum forms from organic matter buried in sedimentary rocks and that even low-grade metamorphism destroys, or converts to graphite, any liquid hydrocarbons originally held in sediments. Accordingly, the search for oil is confined to the approximately 600 sedimentary provinces known to exist around the world on the basis of geologic mapping (Figure 5.41). By the mid-1980s, more than 420 of these basins had been tested, at least by exploratory drilling. Virtually all of the sedimentary provinces contain some hydrocarbons, and more than 240 have oil or gas that is economically producible. However, the widespread occurrence of hydrocarbons should not mislead us from recognizing that significant

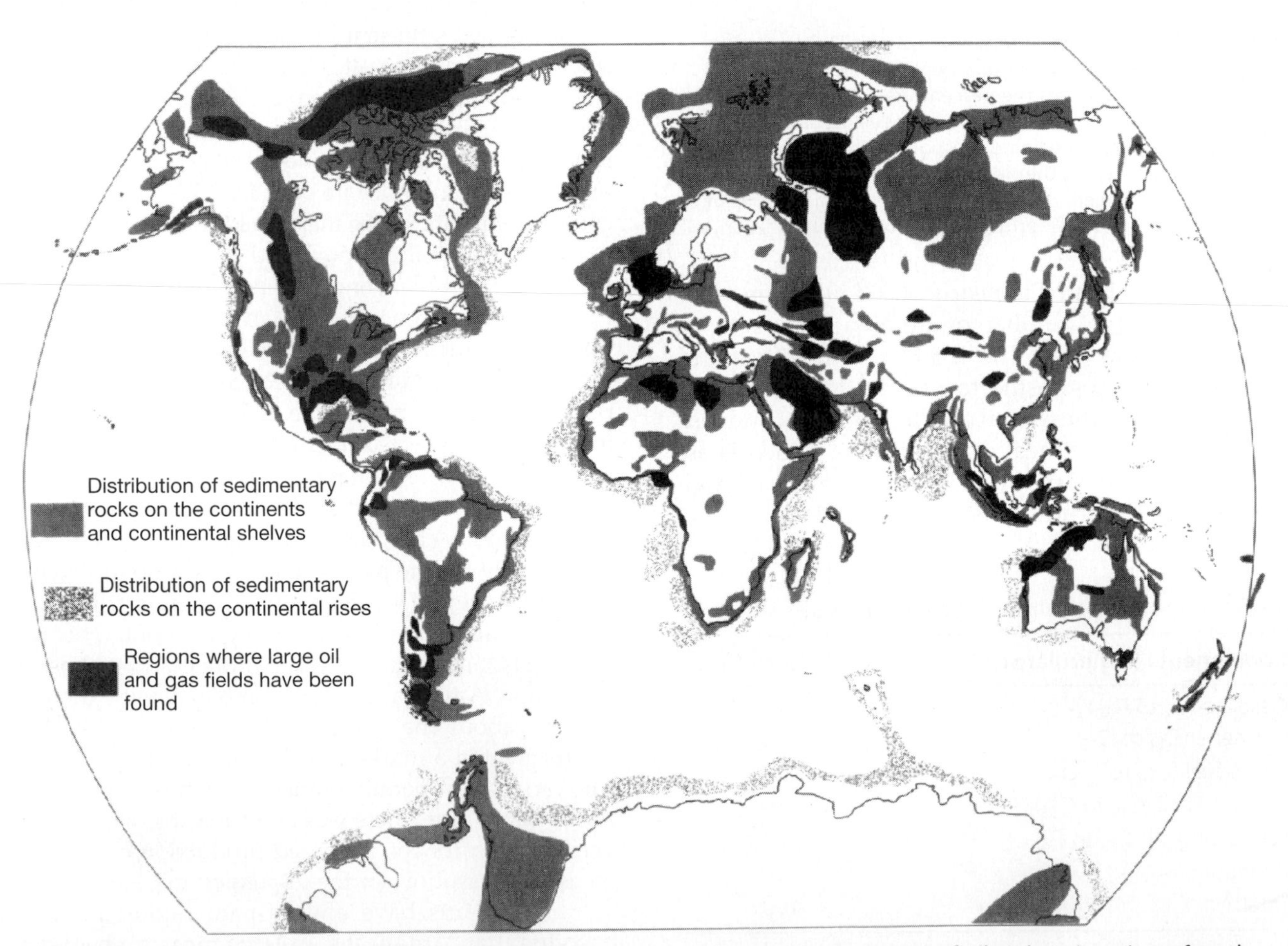

FIGURE 5.41 The major areas of sedimentary rocks and the regions where major occurences of oil and gas have been found. The largest field, by far, is that which occurs in the vicinity of the Persian Gulf (see Table 5.7). Future discoveries will most likely be made on the deeper parts of continental margins and in areas that are little known, such as the continental shelves around Antarctica. (From B. J. Skinner, *Earth Resources,* 3rd ed., Englewood Cliffs, NJ: Prentice Hall, 1986, p. 38.)

accumulations of oil or gas are very rare events, both geologically and statistically.

After more than 150 years of exploration in greater than 75 percent of the potential oil-bearing sedimentary areas, including all of the largest and most accessible ones, we have found only seven provinces that have contained more conventionally recoverable oil than the world uses today in a single year (Table 5.5). Seven provinces (each with more than 25 billion barrels of known recoverable oil) contained more than two-thirds of known world recoverable reserves. The one megaprovince, the Arabian-Iranian, more commonly just referred to as the Middle East, contained 626 billion barrels, nearly half of the world's presently known resources. The 25 major provinces, those each with at least 8.5 billion barrels of known recoverable liquid oil, contained 1118 billion barrels or more than 88 percent of the total produced and remaining world oil reserves. This means that only 6 percent of the world's explored sedimentary provinces and 10 percent of those with producible oil have contained almost 90 percent of the known recoverable world oil.

We do not know where, when, or how much additional conventionally recoverable oil will be found.

TABLE 5.5 The major oil provinces of the world (containing conventionally recoverable oil)

Province	Location	Original Recoverable (billion barrels)	Age of Major Source Rock(s)
	Megaprovinces (100 billion barrels plus)		
1. Arabian–Iranian	Arabian–Persian Gulf	626.3	Cretaceous, Jurassic
	Superprovinces (25–100 billion barrels)		
2. Maracaibo	Venezuela–Colombia	49.0	Cretaceous
3. West Siberian	Former Soviet Union	45.0	Jurassic, Cretaceous
4. Reforma–Campeche	Mexico	42.2	Jurassic, Cretaceous
5. Volga–Ural	Former Soviet Union	41.0	Devonian
6. Permian	United States	32.6	Permian, Pennsylvanian
7. Sirte	Libya	28.0	Cretaceous, Paleocene
Superprovinces Subtotal		237.8	
	Other Major Provinces (7.5–25 billion barrels)		
8. Mississippi Delta	United States	22.4	Miocene–Oligocene
9. Northern North Sea	United Kingdom-Norway–Denmark	22.4	Jurassic
10. Niger Delta	Nigeria–Cameroon	20.8	Oligocene–Miocene
11. Eastern Venezuela	Venezuela–Trinidad	19.5	Cretaceous
12. Texas Gulf Coast–Burgos	United States–Mexico	18.7	Oligocene–Miocene, Eocene
13. Alberta	Canada	17.0	Cretaceous, Devonian
14. East Texas–Arkansas–Louisiana	United States	15.2	Cretaceous
15. Triassic	Algeria–Tunisia	13.5	Silurian
16. San Joaquin	United States	13.0	Miocene
17. North Caucasus–Mangyshlak	Former Soviet Union	12.0	Oligocene–Miocene, Jurassic
18. South Caspian	Former Soviet Union	12.0	Miocene
19. Anadarko–Amarillo–Ardmore	United States	10.8	Pennsylvanian
20. Tampico–Misantla	Mexico	10.7	Jurassic, Eocene
21. Arctic Slope	United States	10.3	Cretaceous
22. Central Sumatra	Indonesia	10.0	Miocene
23. Los Angeles	United States	8.9	Miocene
24. Chautauqua	United States	8.5	Pennsylvanian
25. Sungliao	China	8.5	Cretaceous
Other Major Provinces Subtotal		254.2	
All Major Provinces Subtotal		1118.3	
All Other Provinces Subtotal		202.0	
World Total		1320.3	

However, the largest and most accessible sedimentary areas have already been extensively explored; hence, future discoveries will most likely be made in smaller and more remote areas where production will be more difficult and more costly. Furthermore, the rates of worldwide oil discovery have been decreasing; some areas such as the Atlantic continental shelf of North America, once thought to possess significant oil potential, have failed to yield any recoverable oil. Much of the Atlantic continental shelf along the east coast of the United States was long ruled as "off-limits" to oil and gas drilling because of fears of pollution. However, the rise in oil prices from 2002 to 2008 renewed calls for energy independence and for opening of the continental shelf to exploration and production. Prior to 1935, the rate of oil discovery worldwide (based on five-year running analyses) was never greater than 12 billion barrels annually; from 1935 to 1970, discovery averaged 25 billion to 30 billion barrels per year; and since 1970, it has averaged only 15 billion to 18 billion barrels.

The geologic and climatic conditions necessary for oil formation and preservation have been present in many parts of the world at different times in geologic history. Petroleum formation has been a continuous process since Precambrian time, but worldwide there were four intervals that seem to have been most productive in terms of oil source sediment generation (see Figure 5.33): (1) Devonian (396–345 million years ago); (2) late Carboniferous (310–281 million years ago); (3) late Triassic to late Cretaceous (200–65 million years ago); (4) Oligocene to middle Miocene (35–12 million years ago). Each of these intervals began with plate tectonic activity, which created rapidly subsiding tectonic basins that filled with organic-rich sediment. Petroleum formation continues today and no doubt will in the future as the same types of geologic processes operate, but the rate of its development is far too slow relative to the rates of consumption for petroleum to be considered as a renewable resource. Assuming a generous estimate of 3000 billion barrels of ultimately recoverable liquid petroleum having formed in the last 200 million years, the rate of formation averaged only about 15,000 barrels annually. The world's present rate of liquid oil consumption is more than 30 billion barrels per year, about 2 million times faster than the oil is forming!

INTERNATIONAL PETROLEUM PRODUCTION AND TRADE. Because petroleum is the principal energy source in the world today, it is the most important and most valuable commodity of international trade. A complex web of trade routes crisscrosses the world's oceans and continents linking the politics and economics of countries that would otherwise have little in common. Since the 1940s, international oil trade has more or less followed total world production. After peaking in 1980 when the international flow of oil exceeded 31 million barrels per day, oil production decreased to approximately 28 million barrels daily in 1984 and then slowly rose to about 38 million barrels per day by 2000. This represents only about 50 percent of total world production, largely because some major producers such as the United States, Russia, and the United Kingdom, domestically use much of the oil they produce. As might be expected, the oil-rich Middle East has been, and still is, the world's major oil exporter, with supplies going mainly to Western Europe, Japan, China, and the United States. Western Europe also obtains oil from the North Sea fields and from Africa, Japan gets oil from Indonesia, and the United States gets some of its oil from Mexico, Venezuela, Africa, and the North Sea.

To meet the large and constant demands for oil and to minimize transportation costs, modern oil tankers (Figure 5.42) are now the biggest ships afloat,

FIGURE 5.42 The tanker *Batillus* (550,000 tons = 499,000 metric tons, 1350 feet = 430 meters long, 206 feet = 66 meters wide), shown here loading oil in the Persian Gulf, is representative of the modern large oil transport ships. (Photograph courtesy of Arabian-American Oil Company and American Petroleum Institute Photographic and Film Services.) (b) As a result of the *Exxon Valdez* oil spill in Alaska, most modern oil tankers are being built with a double-hulled design, as shown here. The inner and outer hulls are separated by about 3 meters (10 feet) and have much reinforcing between hulls to minimize the potential for leaking if there are accidents. (Photograph courtesy of CONOCO, Inc.)

with the largest ones reaching lengths of more than 1300 feet, widths of more than 206 feet, and hauling capacities of more than 3.5 million barrels of oil (500,000 metric tons). Prior to the closing of the Suez Canal in the 1967 Middle East conflict, most tankers that carried Middle Eastern oil to western Europe or the United States passed through the canal; now, none of the largest tankers will fit, so they must sail around the Cape of Good Hope in order to deliver their cargo.

The United States was the world's principal petroleum producer from the days of the Drake well in 1859 until the mid-1970s, when it was surpassed by Saudi Arabia and the Soviet Union. The Soviet Union kept most of its oil for domestic use and for supplying the Soviet satellite nations in Eastern Europe. In contrast, Saudi Arabia and the other OPEC members (the history of this organization is discussed in Chapter 2) became the major suppliers for the non-communist world. The share of world oil produced by OPEC rose from about 41.5 percent in 1960 to more than 55 percent in 1973. The subsequent increased production in the Soviet Union and other countries, and the reduction in OPEC production after 1979, had reduced the OPEC share of world production to about 33 percent by 1983. Through the latter part of the 1980s and the first half of the 1990s, total world petroleum output gradually began to increase, even as U.S. production declined and as the Soviet Union's (Russia's after 1991) output peaked and then declined. By the mid-1990s, world production had again exceeded 60 million barrels per day, and the share produced by OPEC had grown to more than 40 percent of the total. In 2008, world oil production exceeded 80 million barrels per day, with OPEC production accounting for about 40 percent. OPEC nations are estimated to hold more than 60 percent of the conventially recoverable oil reserves.

The total annual world production of petroleum rose from a few thousand barrels per year in 1859 to more than 20 billion barrels per year (55 million barrels per day) in 1979, and to 30 billion barrels per year in 2008 (see Figure 5.43). The rise was gradual and irregular through the first half of the twentieth century

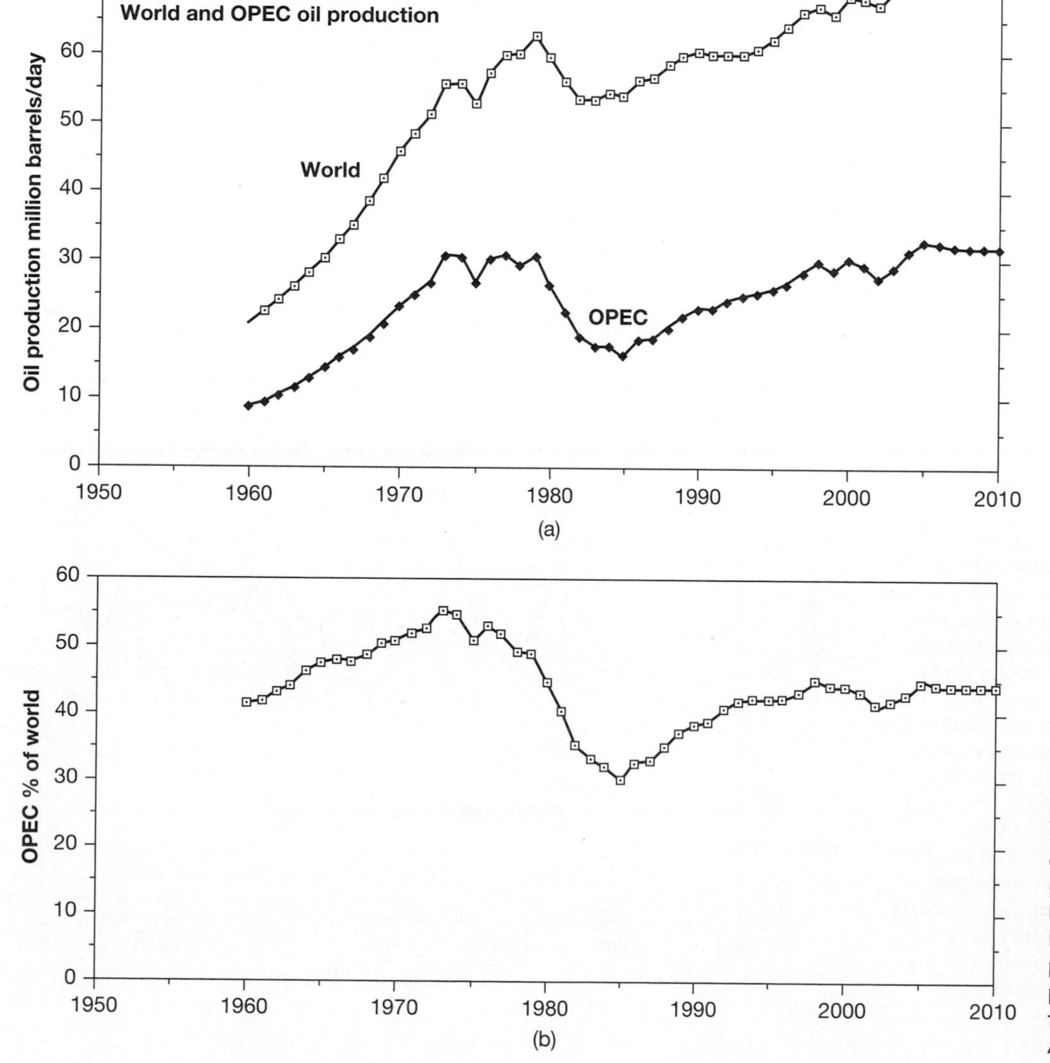

FIGURE 5.43 The international production of crude oil 1960–2008 and the portion of the total produced by the members of OPEC. The lower curve shows the proportion of total world oil production from OPEC. (Data from U.S. Energy Information Administration.)

as wars promoted petroleum use, and the Great Depression of the 1930s lessened demand. After World War II, production and demand rose steadily until 1973 when the OPEC oil embargo spread fear of oil shortages. After a slight dip in world production in 1975, production rose again until 1979 when energy conservation, stimulated by a tripling of oil prices, resulted in a reduction of approximately 15 percent by 1983 (Figure 5.43). This reduction in oil consumption, and especially in dependence on OPEC, is well demonstrated in the case of the United States (Figure 5.44). Total petroleum use by the United States rose from 6.1 million barrels per day in 1949 to 19.2 million barrels per day in 1978, but then dropped to less than 16 million barrels in 1983. However, by the mid-1990s, the rate of consumption had again moved back to over 17 million barrels per day and by 2008 exceeded 21 million barrels per day. The use of imported oil, and especially that from OPEC nations, followed similar trends. Total U.S. imports reached 8.5 million barrels per day (OPEC portion, 6.2 million barrels) in 1977, but were reduced to 5.3 million barrels per day (OPEC portion only 1.8 million barrels) by 1983. By the late 1990s, U.S. petroleum imports had grown back to nearly 10.0 million barrels per day, with OPEC supplying more than 5.3 million barrels of that total. By 2005, U.S. imports were more than 13 million barrels per day with about 25 percent of that coming from OPEC suppliers.

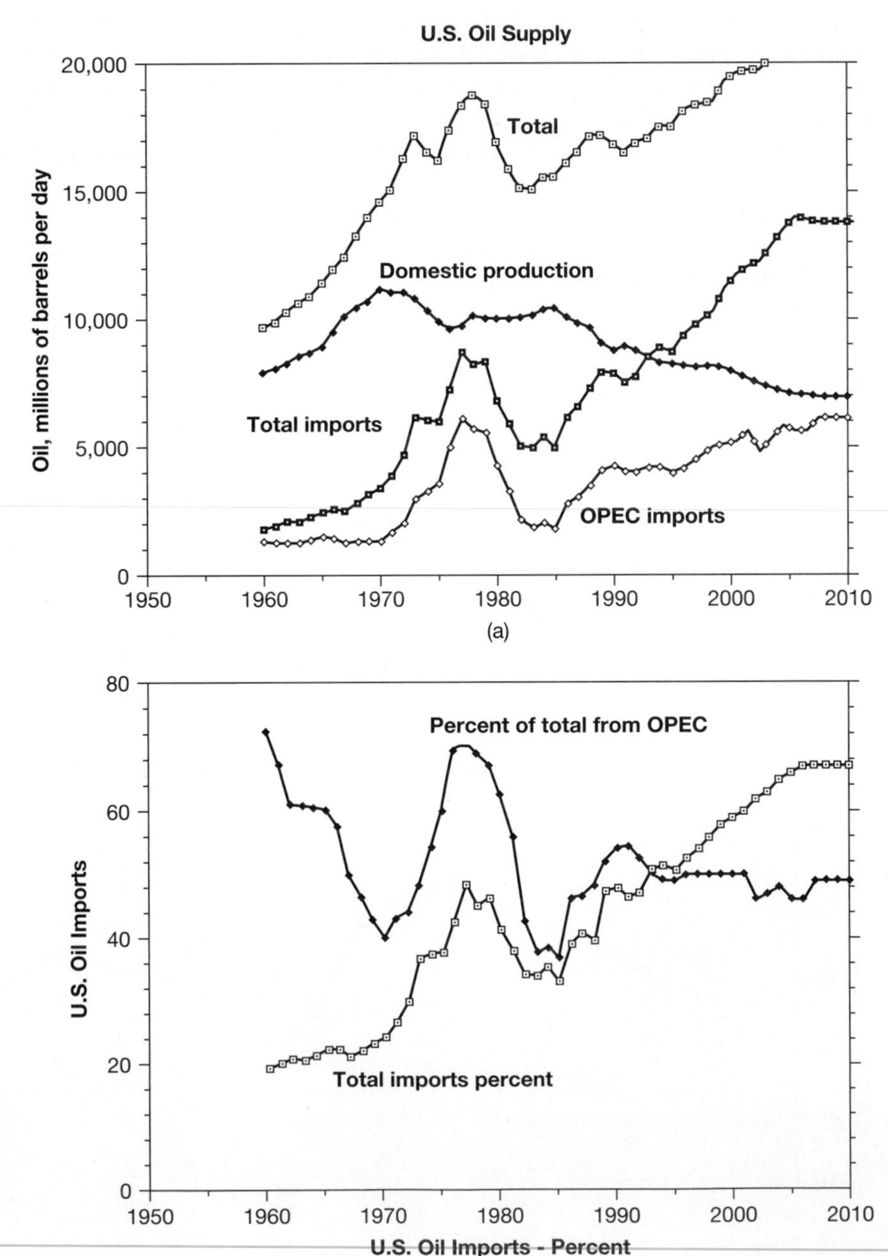

FIGURE 5.44 U.S. oil supply 1960–2008. The uppermost curve in the top half of the diagram shows the U.S. total oil supply. The next two curves show the domestic production and the total imports of oil. The lowest curve notes the amount of the total imports that are derived from the OPEC nations. Note that U.S. domestic production reached its maximum in 1970 and that total imports began to exceed domestic production in the early 1990s. The lower portion of the diagram shows the total percentage of imports of oil used by the United States; this rose from about 20 percent of oil used in 1960 to more than 50 percent in the 1990s. The percentage of the total imports coming from OPEc decreased from about 70 percent in 1960 to less than 50 percent in the 1990s. (Data from U.S. Energy Information Administration.)

The reduction in the rate of increase in world usage of oil that began in about 1980 was projected to extend the life of world oil supplies and to reduce the economic impact of OPEC. However, world oil demand began to rise steadily after 2000, expecially as a result of economic growth in China. Furthermore, despite U.S. and European requests that OPEC increase oil producion to meet demand and lower prices, OPEC kept production more-or-less constant and prices soared after 2005 and reached nearly $150 per barrel in 2008. The political importance of oil is greater by far than that of any other commodity. Consequently, there remains a concern for a continuous flow of oil supplies to meet world needs. Any disruption, or even a threat of disruption of oil supplies especially at major "choke points" (Figure 5.45), creates tension, rapidly boosts prices, and can bring military responses. Never was this more apparent than at the time of the Persian Gulf War of 1990–1991 and in the Iraq War since its inception in 2003.

HOW MUCH OIL IS THERE AND HOW LONG WILL IT LAST? Throughout the first 100 years of the modern petroleum industry, little thought was given to how much extractable oil was contained within Earth. The exploration of new sedimentary basins and the drilling of new wells on land and on the continental shelves led continuously to the discovery of new fields, and world oil reserves rose faster than we could use the oil. The availability, relative cleanliness of use, and low price of oil-based products, especially gasoline, promoted ever-increasing usage. It was not until 1973, when the OPEC embargo cut off a significant proportion of petroleum supplies to western Europe and the United States, that most of the general public began to realize there are limits to the world's oil stocks. The shock of the embargo, coupled with a subsequent increase in the price of petroleum from about $3 per barrel to about $10 per barrel in 1973–1974, and from about $10 to about $20 per barrel in 1979–1980, to about $35 per barrel in 1981, and to nearly $150 per barrel in 2008, led to the use of the terms *oil crisis* and *shortage.* These terms may have been overused, but they have helped focus attention on the very real fact that the world's oil supplies are indeed limited.

As early as 1948, Dr. M. King Hubbert of the United States Geological Survey (USGS) presented a diagram that indicated a severe decline in the world's oil reserves before the year 2000 if the trend in oil production were left unrestricted. His prediction, made at a time when the expanding postwar economies of Europe, Japan, and the United States were being fueled by what seemed to be endless supplies of oil, was little noticed by the general public and the oil industry. However, the massive amounts

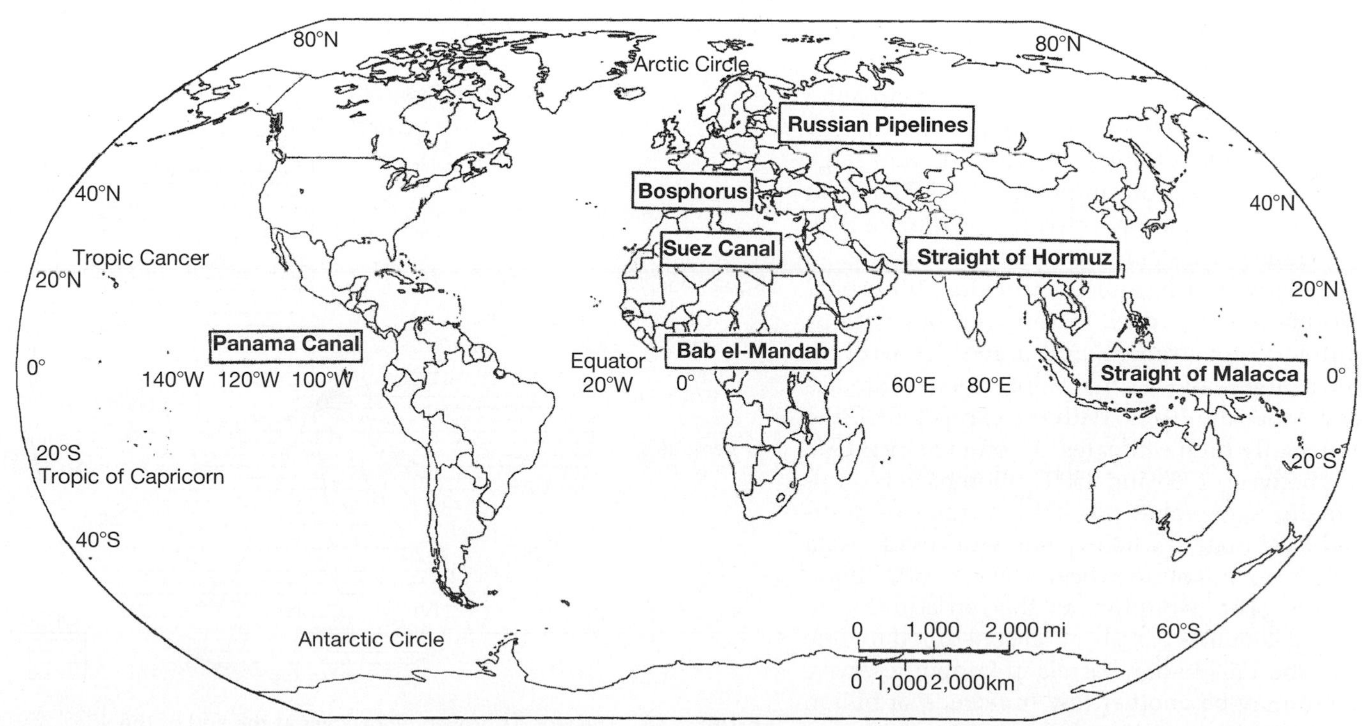

FIGURE 5.45 The world's major "choke points" for the flow of petroleum. The choke points are considered to be the critical points where disruption (due to natural disasters or political events) could severely constrict world oil supplies. The single most critical choke point is the Strait of Hormuz, where large oil tankers exit from the Persian Gulf. (Based on the views of the U.S. Energy Information Administration.)

of exploration, drilling, production, and computer modeling of world oil supplies in the intervening years have all confirmed Dr. Hubbert's predictions. His original curve assumed that unrestricted production would lead to a constant increase in world output until the mid-1990s and then result in an equally rapid decline.

The actual rate of world oil production is shown in Figure 5.43. The rapid rate of increase in the 1960s was slowed by the OPEC embargo and the sharp price rise in 1973. Usage began to rise again in the mid-1970s but was curtailed by the even bigger price increases from 1979 through 1981. The sharp price decline in the mid-1980s, followed by relatively stable prices through the 1990s except for a brief spike at the time of the Persian Gulf War in 1990–1991, saw another rise in oil consumption. World demand rose steadily after 2000, especially fueled by growth in China, until the worldwide recession cut demand in 2008. It is likely that demand will begin to rise again as there is recovery from the recession. Price and demand fluctuations, combined with new discoveries and recovery techniques, have made forecasting world oil production difficult and a subject of considerable debate.

It is clear from the examples of the past forty years that the oil price changes have a major impact on oil reserves in two ways: (1) a rise in the price of oil generally leads to a reduction in usage and to increases in the efficiency of usage; (2) a rise in the price of oil generally leads to increased exploration, increased rates of discovery, and increased percentages of recovery from known fields. Both of these result in an extension in the life of the world's liquid petroleum reserves. Nevertheless, liquid petroleum is a nonrenewable resource because of our large rates of annual consumption, about 31 billion barrels today and projected to reach more than 40 billion barrels per year in the next decade.

How much liquid petroleum still exists? No one really knows the answer, but various industry, government, and university workers have attempted to make informed estimates ever since the 1940s. With the exception of two unusually high estimates, all others since 1958 are between 1200 and 3500 billion barrels of ultimate recoverable oil. The average of post-1970 estimates, when better worldwide data became available, has been 2300 to 2500 billion barrels. This would mean that, in addition to the 1000 billion barrels of oil consumed to date and the 1300 billion barrels in known reserves, there may be another few hundreds of billion barrels to be discovered and recovered.

It is important to note that there is no proof that any of the undiscovered oil exists, and to also note that none of it is of any value or use until it has been discovered and exploited. Furthermore, it is highly probable that most future oil discoveries will be in small fields that will be harder to find, will occur in deeper rocks or under deeper water, and will reside in more remote and climatically hostile areas of the world. The statistical distribution of known oil fields and the degree to which the world's sedimentary basins have already been explored strongly suggest that we shall not find any more large oil fields such as those listed in Table 5.5. It is important to remember that the oil recoverable by simple pumping is only about one-third of the total amount present because most of the crude petroleum remains stuck on mineral surfaces and trapped in small pore spaces and fractures in the rock. Procedures such as water flooding and CO_2 injection, used to liberate more of the oil, generally raise the total to no more than 50 percent. The USGS has recently estimated that every 1 percent increase in the rate of oil recovery is equivalent to adding about 50 billion barrels to the world's total reserves.

The long-term availability of oil to countries like the United States and those in western Europe is a function of the total oil reserves, but also a function of the ownership of the oil as was demonstrated during the 1973 embargo. The present oil reserves are very irregularly distributed geographically, with more than 65 percent of the world total occurring in the Middle East (Figure 5.46 and Table 5.6). Saudi Arabia with 20 percent of the world reserves (260,000 million barrels)

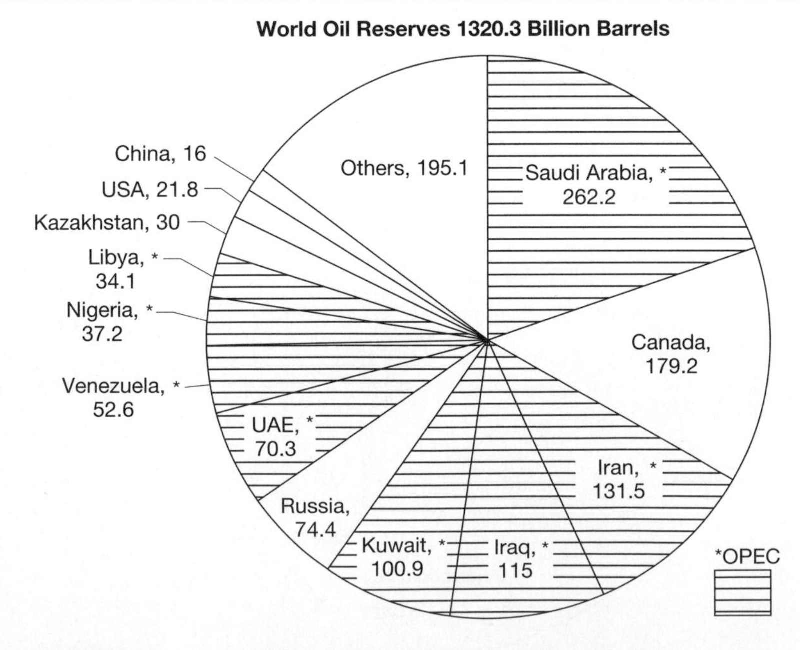

FIGURE 5.46 The world's proven oil reserves at the end of the twentieth century. Approximately 60 percent of the reserves are held by the OPEC nations (noted by an asterisk and lined on the diagram) with the largest quantity held by Saudi Arabia. (Data from U.S. Energy Information Administration, 2009.)

TABLE 5.6 International recoverable crude oil reserves as of 2007 (in billions of barrels)

North America		
Mexico	12.4	
United States	21.8	
Canada	179.2	
Total	213.3	(16.2%)
Central and South America		
Venezuela*	52.6	
Brazil	11.7	
Others	38.5	
Total	102.8	(7.8%)
Western Europe and Eastern Asia		
Russia	74.4	
United Kingdom	3.9	
Norway	7.8	
Kazakhstan	30.0	
Others	1.4	
Total	117.5	(8.9%)
Middle East		
Saudi Arabia*	262.2	
Kuwait*	100.9	
Iran*	131.5	
Iraq*	115.0	
United Arab Emirates*	70.3	
Others	44.1	
Total	739.2	(56.0%)
Africa		
Libya*	34.1	
Nigeria*	37.2	
Algeria*	11.4	
Angola	8.0	
Others	23.4	
Total	114.1	(8.6%)
Far East and Oceania		
China	16.0	
Indonesia*	4.3	
India	5.6	
Others	7.5	
Total	33.4	(2.5%)
World Total	1320.3	(100%)
OPEC Total	819.5	(62.1%)

*OPEC nations.

dwarfs all other nations; only Kuwait, United Arab Emirates, Iraq, and Iran have more than 70,000 million barrels. The United States, which consumes approximately 25 percent of world production, has only 22,000 million barrels or 1.7 percent of world reserves. At the current rates of production, world reserves would last about 43 years; U.S. reserves would last only about 8 years.

Today, and likely in the near future, the division of reserves will be seen in terms of OPEC and non-OPEC sources. The dilemma for many countries is that OPEC has the largest reserves, but non-OPEC sources supply most of the world's needs. A consequence is that non-OPEC reserves will be consumed faster leaving OPEC in an ever stronger negotiating position. Only when oil demand drops (unlikely to be soon) or alternative energy sources (such as renewables and tar sands) are developed on a massive scale, will OPEC's position of prominence be reduced. However, it is probable that the discovery of new conventional reserves will be in proportion to the reserves now held; consequently, the general trend of non-OPEC nations consuming their reserves more rapidly than the OPEC nations will continue. Thus, even with the discovery of new reserves, it is very likely that OPEC's share of remaining world liquid petroleum reserves will increase significantly during the first quarter of the twenty-first century. This will increase the political significance of the OPEC nations. Most investigators believe that the world's production of conventional liquid petroleum will actually peak before 2020, as shown in Figure 5.47. This projection is widely accepted but is subject to many uncertainties, especially that due to the price of oil because price will affect usage and the economic competitiveness of other forms of energy.

The shock of the 1973 oil embargo impressed upon the United States that disruption of international imports of oil could cripple the country. Therefore, following the idea of a strategic materials stockpile that evolved from wartime shortages of a variety of commodities, the United States has developed the Strategic Petroleum Reserve, as discussed in Box 5.2 on page 143 (Figure 5.C).

Natural Gas

THE HISTORY OF NATURAL GAS USAGE. The first recorded use of natural gas was in ancient China where the people had learned to pipe it through bamboo poles so that they could burn it to boil saline brines and obtain the residual salt. By about A.D. 600, temples in what is now the Baku area of Azerbaijan on the west coast of the Caspian Sea, contained eternal flames fueled with gas piped from fractures in the rocks.

The use of gas as an illuminant in the modern world began with gas distilled or manufactured from coal, wood, and peat in Belgium and England in the early 1600s. Several men experimented with gas lamps in homes, abbeys, and even university classrooms in the late 1700s. However, it was not until 1802 to 1804, when Scottish engineer William Murdock installed coal gaslights in cotton mills, that the gas industry be-

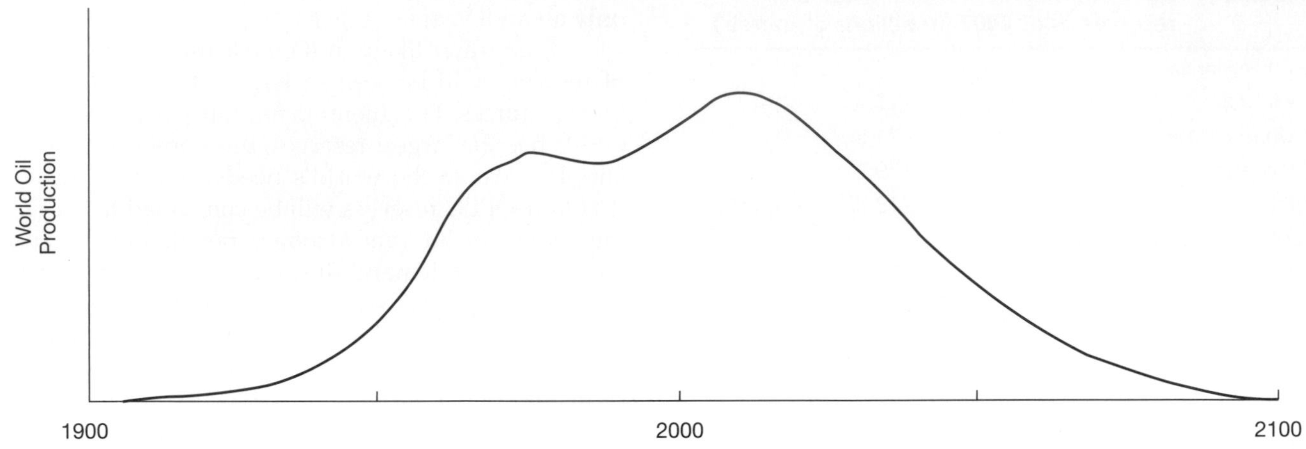

FIGURE 5.47 The world's oil production rates from 1900 to 2100. The curve represents an average of estimates made by several authorities of past and future oil production from conventional reserves. Production rates rose rapidly in the mid-twentieth century, decreased slightly as a result of price rises in the early 1980s, and are projected to peak in the first 15 to 20 years of the twenty-first century. Then, world production is projected to decline slowly; this will probably be accompanied by rising prices, which could stimulate the production of energy from alternative sources.

came industrially important. This led to the establishment of the commercial gaslight companies in London in 1812 and in Baltimore in 1816. Within a few years, gaslighting had spread throughout Europe, the Americas, and to the Orient. Manufactured gases are still important and are produced by a variety of methods involving heating coal and reacting water with calcium carbide, but their usage has been much overshadowed by that of natural gas.

The modern natural gas industry traces its origins to the United States where in 1775, French missionaries reported seeing "pillars of fire" caused by seeping gas that had accidentally been set ablaze. The same year, George Washington witnessed burning springs in which flaming gas rose from the water near Charleston, in what is now West Virginia. In 1821, mysterious bubbles were found rising from a well drilled for water at Fredonia, New York. After the driller abandoned the well, small boys accidentally ignited the escaping natural gas, creating a spectacular sight. Shortly thereafter a gunsmith named William Hart recognized the commercial potential of the gas and drilled a well 27 feet deep at the same site. The initial wooden pipes were made from hollowed logs but were later replaced by lead pipe, and the gas was piped to a local inn where it was used to fuel sixty-six gaslights. More gas wells were completed, and a natural gas distribution company was formed at Fredonia in 1865.

Natural gas was also found with oil in 1859 near Titusville, Pennsylvania, and in subsequent oil wells, but there was little market and no pipelines for its distribution; hence, its usage was much overshadowed by manufactured gas, which could be produced wherever needed. Long-distance pipelines to transport natural gas appeared in 1872 when a 25-mile wooden pipeline was constructed to supply hundreds of customers in Rochester, New York, and a 5.5-mile metal pipeline carried gas to Titusville, Pennsylvania. The fledgling industry was nearly killed when Thomas Edison's electric lightbulb first appeared in 1879, but grew slowly on the basis of its use as a heating fuel rather than as an illuminant. A major impetus came with the discovery of large gas fields in Texas, Oklahoma, and Louisiana; by 1925, there were 3.5 million gas consumers, but all were situated near the gas fields. The great expansion of the gas industry and its growth to the status of a major energy source came about in the late 1920s and 1930s with the introduction of strong, seamless, electrically welded pipes that could carry great quantities of gas under high pressure. In 1947, two pipelines, dubbed the Big Inch and the Little Inch, originally built during World War II to transport oil from east Texas to Pennsylvania, were converted to carry natural gas; these two pipelines opened up the U.S. east coast and brought about the dominance of natural gas over manufactured gas.

THE MODERN NATURAL GAS INDUSTRY. Prior to the 1940s, growth of the gas industry was slow (see Figure 5.48), and vast quantities of gas for which there was no market were allowed to escape or were burned off (flared). Seamless pipes that made possible the transmission of natural gas over long distances became available in the 1920s, but it was not until after World War II that the rebuilding of Europe and the growth of the American suburbs provided large new markets. The gas was found to be especially attractive as a fuel because it did not require refining and only minor processing was needed; it was easily handled, burned cleanly, and provided more heat per unit weight than any other

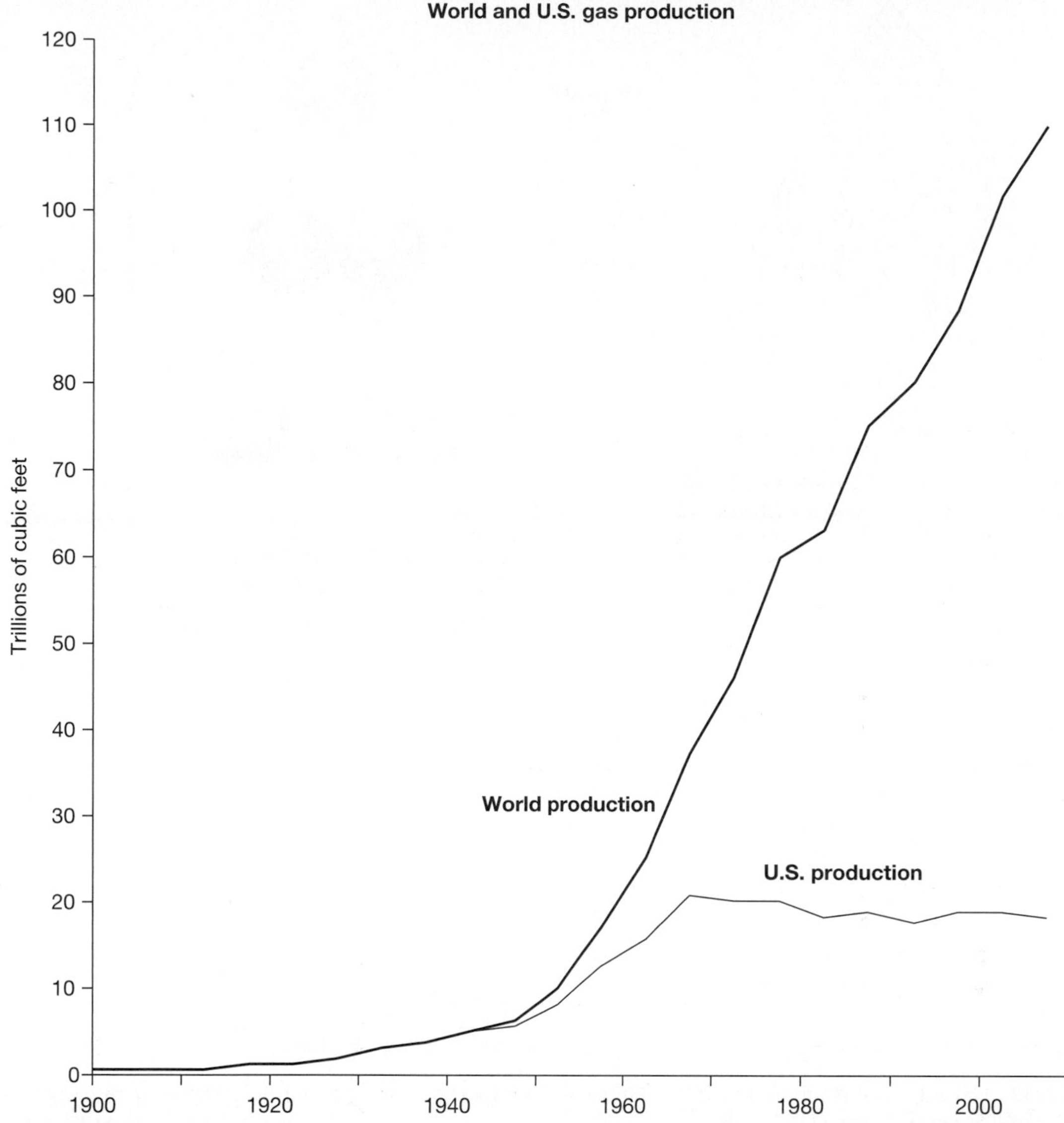

FIGURE 5.48 World production of natural gas has risen sharply since 1950 when pipelines and transport ships became available. Production in the United States peaked in the 1970s, but remains at nearly the same levels. (Data from the U.S. Energy Information Administration.)

fossil fuel. Today, natural gas is piped to tens of millions of homes, businesses, institutions, and industries, and provides approximately 25 percent of the total energy needs of the United States and Europe. In addition to its use as a heating fuel, natural gas is widely employed in the manufacture of thousands of chemicals, being the starting component for such diverse products as plastics, detergents, drugs, and fertilizers (see discussion of nitrogen fertilizers in Chapter 9)

The gas industry consists of three major sectors—production, transmission, and distribution. The search and drilling for gas is very similar to that for petroleum, but the extraction of the gas is often easier because it moves through pores and cracks in the rock more readily than oil and because it does not stick to the mineral grains. Generally, more than 99 percent of the useful gas consists of methane, but minor amounts of ethane, propane, butane, and some carbon dioxide, hydrogen sulfide, helium, hydrogen, nitrogen, and ammonia may also be present (Figure 5.49). Approximately 80 percent of the world's gas reserves are believed to be of thermogenic origin; this gas is recovered from wells that range from only a few hundred meters (several hundreds of feet) to about 10,000 meters in depth. The 20 percent of the natural gas that was biogenically produced usually is found at relatively shallow depths. More than 95 percent of the gas used today has been obtained in its present state; however, locally and for special purposes, gas is manufactured by heating coal or by reacting steam or hydrogen with coal or heavy oils. Probably the most widely known of the synthetic gases is acetylene (C_2H_2),

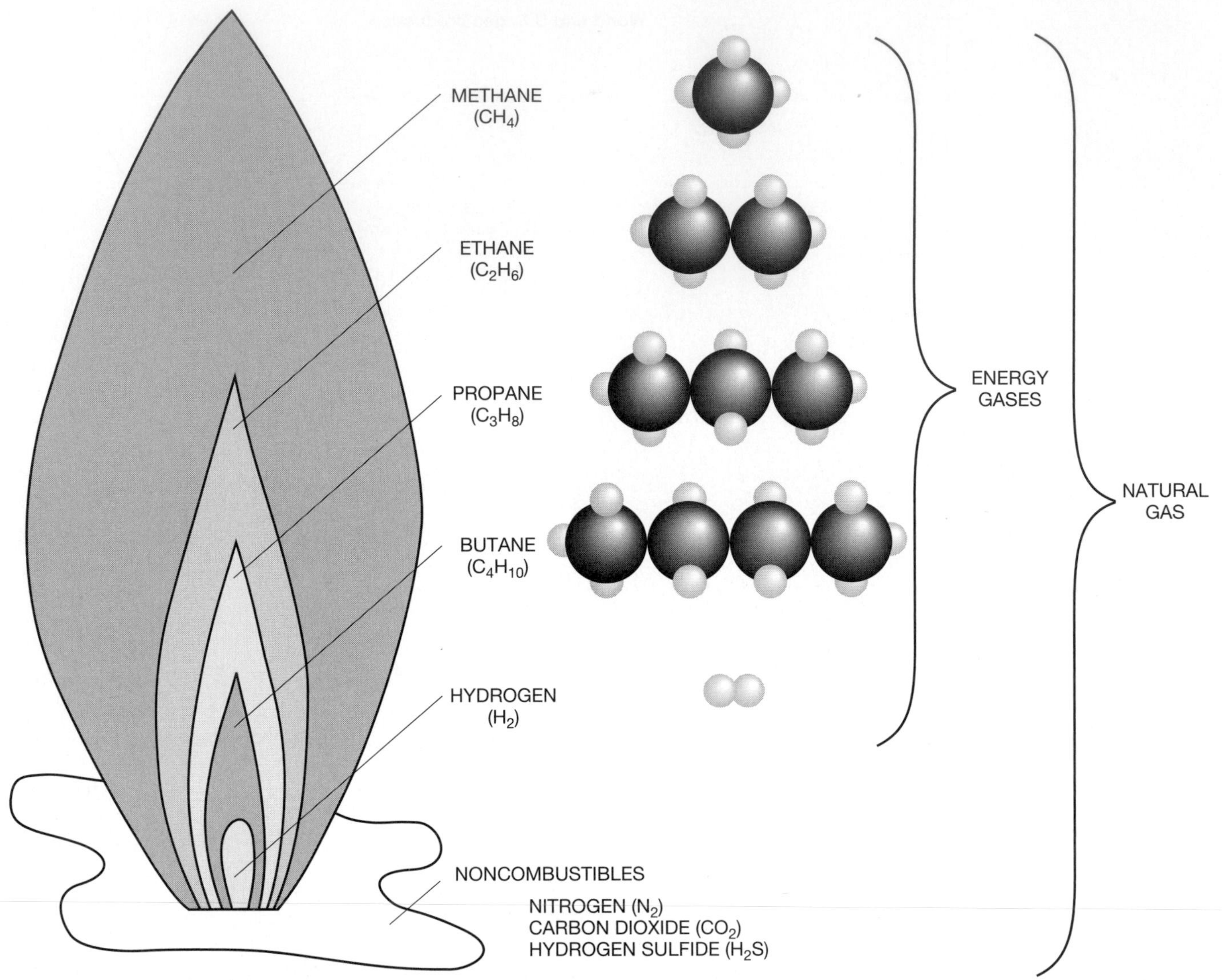

FIGURE 5.49 Natural gas is composed of 80 to 100 percent methane, but small quantities of other gases may also be present. (From U.S. Geological Survey Circular 1115.)

formed by the reaction of water with calcium carbide or by thermal cracking of methane. Acetylene is used in welding because it produces a hotter flame than do other gases.

After extraction, gas is piped to processing plants where contaminants such as water, sulfur, and other impurity gases are removed; at the same time, traces of the characteristic coal gas scent are added to enable human detection because natural gas otherwise has no aroma. The gas is then sent through transmission lines under high pressure and at velocities of about 24 kilometers (15 miles) per hour. Compression stations along the line restore the pressure as it drops due to friction and to tapping by communities. Within each community, the gas is then sent via the smaller distribution lines to individual homes and businesses. From the original 9-kilometer (6-mile) pipeline used to transport gas near Titusville, Pennsylvania, the pipeline system in the United States has grown to more than 1,800,000 kilometers, (1,100,000 miles) not counting the individual service lines to homes. Natural gas consumption for home heating is much greater in the winter months than in the summer months, and the long-range transmission system cannot handle sufficient gas to meet demand on the coldest days. Accordingly, much gas is pumped into underground storage facilities such as caverns and old gas fields along transmission lines during the summer months so that it may be extracted when needed during the winter.

Until recent years, it was impractical to bring natural gas from many oil fields scattered around the world to the major industrial consumers because there was no way to profitably transport it. The rising demand for gas has brought about the construction of many new pipelines through and between countries. Furthermore, much natural gas from the fields of the

BOX 5.4
Plastics

Plastics are now among the most widely used materials on Earth, serving in thousands of applications from the moment we are born to the day we die. Plastics are in the delivery rooms in the forms of medical supplies as we enter this world and in the flower holders and caskets as we depart. During our daily lives, plastics have become so commonplace that it is almost impossible to be in any room of any building and not see plastics serving some purpose. Furthermore, many individuals use plastics in normal daily life as the lenses in eyeglasses, in dental plates, heart valves, and replacement body sockets. Despite the fact that nearly everyone is familiar with and can recognize plastics, few people know what plastics actually are or from what source they are made. Their incredible diversity in form (from soft, flexible, transparent films to hard rigid automobile parts) give no evidence of their source materials.

Plastics are synthetic materials composed of giant chain-like molecules, known as *polymers,* which have relatively small repeating units linked in immensely long chains. The simplest and one of the best known of the polymers is *polyethylene,* which is built up from ethylene molecules as illustrated by:

$$nCH_2{=}CH_2 \longrightarrow \left(-\overset{H}{\underset{H}{\overset{|}{\underset{|}{C}}}} - \overset{H}{\underset{H}{\overset{|}{\underset{|}{C}}}} - \right)$$

The name *plastic* is derived from Greek *plastikos,* "to form," in recognition that they are synthetic materials capable of being formed into useful products by heating, milling, molding, or similar processes. All plastics are manufactured by some method of polymerization (forming of the long chains from simpler molecules) and fall into one of two very broad categories. **Thermosetting resins** are those that become insoluble and rigid upon heating. These include phenolics, furan resins, amino-plastics, alkyds, epoxy resins, polyurethanes, and silicones. In contrast, the **thermoplastic resins** are those that soften upon heating and include cellulose derivatives such as polyethylene, polypropylene, vinyl, nylon, and polycarbonates.

The modern age of plastics began in 1864 when a Swiss chemist soaked up nitric acid and sulfuric acid with a cotton cloth and accidentally created nitrated cellulose, a new type of material that could be molded when warm but that became tough but elastic when cooled. The first truly synthetic plastic, called Bakelite, was made in 1907 when phenol was reacted with formaldahyde. The resulting black thermoplastic proved useful as a moldable material to make telephones, electrical insulators, and billiard balls, and it paved the way for an explosive growth in the polymer chemistry industry. Thousands of new plastics, with an amazing array of properties, have been developed—none more famous than nylon which was developed in 1935. Today, the industry continues to expand.

The earliest materials were synthesized using cellulose-based derivatives. However, it soon became apparent that the tars and oils produced when coals were heated could lead to many more and different types of plastic materials. Even today, coal by-products contain most of the raw materials used to make plastics, but petroleum offers easier and more economic ways to manufacture these starting ingredients. Of particular importance are ethylene and propylene, which are produced on an enormous scale during the cracking of petroleum. Thus, today it is likely that the plastic materials we use every day, regardless of their physical properties, have been formed from petroleum pumped from a well not so long ago.

Despite the numerous contributions of plastics to modern life, they have also become one of Earth's greatest pollutants. As pointed out in Chapter 4, plastic waste is now found in every part of the continents and oceans. Dr. Anthony Andrady, in his book *Plastics in the Environment* notes: "Except for a small amount that's been incinerated, every bit of plastic manufactured in the world for the last 50 years or so still remains. It's somewhere in the environment." The most visible forms of plastic pollution in many parts of the world are shopping bags and beverage bottles.

Middle East, Africa, and South America that used to be burned off is now liquefied and transported in large, liquid natural gas (LNG) transport ships (Figure 5.50). The gas is cooled and held below -259°F where it becomes liquid and occupies only 1/600 of its gaseous volume. The refrigerated tanks on the ships permit large quantities of the LNG to be economically transported worldwide. Upon arrival at its destination, the LNG is allowed to warm and return to the gaseous state and is fed into the normal gas transmission lines.

Methane is also generated as a natural anaerobic decomposition product of organic materials. Consequently, most modern landfills are constructed so that the methane can be captured and used as an energy source. The total methane from such sites is small relative to that extracted from oil and gas fields, but can be locally significant as discussed in Chapter 6. It is also well-known that cattle generate significant amounts of methane during the digestion of grasses and other organic matter, but an efficient collection mechanism for this widely dispersed fuel has not yet been found.

FIGURE 5.50 Liquid natural gas transporting ship. The gas is liquefied by cooling, transported, and then allowed to convert back to a gas by warming so that it can be transmitted through pipes for industrial, commercial, and domestic use. (Photograph courtesy of American Gas Association.)

INTERNATIONAL GAS PRODUCTION AND RESERVES. The increasing demand for natural gas and its irregular global distribution has resulted in it becoming a major commodity of international trade. Western Europe is the recipient of piped gas, with major supplies coming from Russia; from the North Sea fields off England, Scotland, Norway, and the Netherlands; and from North Africa via pipelines across the Mediterranean. Large amounts of gas are also piped from the fields of western Canada and from Mexico to the United States. Japan, now a major consumer of gas, receives much of it via LNG carriers from Southeast Asia and the Middle East. The anomalous situation of the United States being a major gas importer but still sending the gas from Alaska and the Gulf Coast to Japan has developed because that gas can be sold on the international market for more than on the regulated domestic market.

International gas reserves and production are shown in Table 5.7. Russia is by far the world's leader in reserves and has, since the mid-1970s, become a major exporter, having built large pipelines to transport the gas to western Europe. Most of the former Soviet Union gas reserves lie in Russia, but other states, such as Turkmenistan, also have significant natural gas and are trying to develop the fields in order to export gas to Europe. The other European countries with major reserves and production were gas importers until the North Sea oil and gas fields were developed in the 1960s and 1970s. Norway and the Netherlands are now self-sufficient and are able to supply part of the needs of the rest of western Europe. Iran is second only to Russia in gas reserves, but its internal and international political problems, especially its long war with neighboring Iraq in the 1980s, resulted in little sale of gas through that period. If political stability can be maintained in the Middle East, it is likely that Iran and its neighbor countries will become major suppliers of gas on the world markets. The United States is well-endowed with natural gas, but it is also a very large gas consumer and imports approximately one-third of its needs. Since 1975, the rate of discovery of new gas

TABLE 5.7 World natural gas reserves, 2008 (at standard temperature and pressure) (Data from Energy Information Adminstration)

	Reserves	
	$ft^3 \times 10^{12}$	$m^3 \times 10^{12}$
North America		
United States	204	5.8
Canada	58	1.6
Mexico	15	0.4
South America		
Venezuela	152	4.3
Argentina	16	0.5
Others	73	2.1
Europe and Western Asia		
Russia	1680	47.6
Norway	82	2.3
Netherlands	49	1.4
United Kingdom	17	0.5
Turkmenistan	100	2.8
Others	87	2.5
Middle East		
Iran	974	27.6
Qatar	911	25.8
Saudi Arabia	240	6.8
United Arab Emirates	214	6.1
Iraq	112	3.2
Others	115	3.3
Africa		
Algeria	162	4.6
Nigeria	182	5.2
Libya	53	1.5
Egypt	59	1.7
Others	28	0.8
Far East and Oceania		
Malaysia	75	2.1
Indonesia	98	2.8
China	80	2.3
Australia	30	0.9
Others	137	3.9
World Total	6003	170.4

has been slightly less than the rate of domestic gas consumption; hence, the reserves have declined by about 10 percent in that time. This has left the United States with a reserve-to-production ratio of about 8 years; in contrast, the similar world ratio is approximately 65 years.

The world's ultimate resource potential of natural gas is not known with certainty, but the heat potential for worldwide gas reserves is presently equal to that of petroleum reserves. As with petroleum, most of the major sedimentary basins have been tested, and most of the major gas fields have probably been found. Future discoveries will be smaller than the large fields found in the past, and their development will be more costly. The relatively recent development of coal bed methane recovery systems has significantly added to potential gas reserves. (See the discussion in Box 5.3.)

Natural gas has become the energy source of choice at the beginning of the twenty-first century with usage expected nearly to double from 1996 through 2020. This demand results from relatively low costs, a new infrastructure of pipelines between countries, and the fact that natural gas generates smaller amounts of greenhouse gases than do other fossil fuels. Furthermore, gas generally contains fewer pollutants such as sulfur, is easier to clean, and burns with much higher efficiencies in power plants than do coal or oil. As a result, many countries, such as the United States and Germany, see natural gas as the primary energy source to make up for reductions in electricity generation by nuclear power over the next twenty years. The Kyoto Protocol calls for many nations to reduce the emissions of greenhouse gases, especially carbon dioxide. Switching from coal to natural gas in electric power plants would produce less than half as much carbon dioxide for the same amount of heat generated (Figure 5.25); this is seen by many countries as the easiest and cheapest way to reduce such emissions.

Natural gas reserves are more widespread geographically than are oil reserves, and thus estimation of the world's ultimately recoverable gas resources is difficult. Despite increasing rates of usage, new discoveries have kept the world's reserve to production ratio at 60 to 70 years. The discovery of vast quantities of natural gas hydrates buried in the sediments of the continental shelves in many parts of the world has led to much speculation about large-scale seabed gas recovery. The gas is held in an ice-like phase, which is a combination of water and methane called a *clathrate.* Methane clathrate remains stable in the sediments of the continental shelf because of the high pressure of the overlying water and the cold temperature, which is usually about 4°C. Warming of the sediment by some means would release large quantities of gas, which would rise upward and could be collected and piped to shore for use. As with other seabed mining proposals, the potential harvest is huge, but the difficulties and expense of recovery are very great. Furthermore, the environmental effects are unknown, and the uncontrolled release of the methane could have a significant impact on Earth's atmosphere. It has already been proposed that large natural releases of methane from seafloor clathrates contributed to past periods of rapid global warming because methane is a more effective greenhouse gas than is carbon dioxide. Thus, there is the fear that accidental triggering of massive methane release from gas hydrates could adversely affect global climate.

Heavy Oils and Tar Sands

The early history of petroleum centered on the use of bitumen, a black, viscous, and semisolid hydrocarbon material that is found where oil has lost its lightweight volatile components through exposure to air. The modern oil industry has concentrated on liquid petroleum, which flows readily and is much more easily extracted and processed, and is much more profitable to produce than bitumen. Nevertheless, there remain large quantities of natural bitumen-like hydrocarbons, which also are known by the names *heavy oils* and *tars* (also commonly referred to as *tar sands);* such materials will probably serve as important sources of petroleum in the future. Petroleum is a spectrum that ranges from the freely flowing crude that has a water-like viscosity, to heavy oil that is tar-like and will only flow if heated or altered by viscosity-lowering additives, to bitumen (tar sands) that cements sand grains together and which will not flow at all. The heavy oil and bitumens all are characterized by being (1) dark in color; (2) so viscous that they will neither flow naturally nor respond effectively to primary or secondary recovery techniques; (3) high in sulfur (3–6 percent), nickel, and vanadium (up to 500 ppm); and (4) rich in hydrocarbons known as **alphaltenes** (a primary constituent of asphalt). Heavy oils and tars, mainly in the form of tar sands, either occur alone or are found with liquid petroleum; they owe their origin to at least three processes that may have operated singly or jointly on petroleum. Some bitumens have formed, as did most of the early discovered bitumens, through **oxidation** of liquid petroleum and the loss of the lightweight volatile fractions, leaving behind the heavy organic molecules. The two other modes of origin involve **thermal maturation,** in which the light fractions have been driven off or converted to gas due to natural heating, or **biodegradation,** in which bacteria consume the lighter fractions, leaving heavier components behind.

Heavy oils and tar sands are known from several parts of the world [Figure 5.51(a)], but they remain relatively little publicized or exploited because both their recovery and their use are more difficult and expensive than liquid petroleum. They have received significant attention during times of high oil prices, such as the early 1980s and in 2008, but interest waned rapidly when liquid petroleum prices dropped. The largest-known deposits are in northern Alberta, Canada, and in the Orinoco district of Venezuela, but significant resources also exist in the United States, in the Middle East oil fields, and in Russia. Estimates of the oil in place and recoverable vary widely from one study to another, so the values given in Table 5.8 should only be taken as estimates.

In Canada, two commercial plants extract materials from tar sands in the Athabasca area [Figure 5.51(b)], but these two open-pit operations can only extract a few percent of the hundreds of billions of barrels of tar held there in their operating lifetimes of twenty-five years or so. The tar-rich sands are removed by large rotary-bucket wheel excavators or drag lines and then processed with hot water and chemicals to separate the oil from the sand. Once separated, the oil is processed in special refineries to remove sulfur and to produce a variety of usable petroleum products. Canadian pilot plants have also begun to experiment with a variety of *in situ* extraction techniques, including the so-called huff-and-puff procedure in which superheated steam is injected into an oil zone for about a month to heat and soften the oil; then the more fluid oil is pumped out for at least a month before the cycle is repeated.

The physical and chemical properties of heavy oils have severely limited their exploitation, and countless problems remain to be solved. Nevertheless, progress continues to be made in developing methods to convert the heavy oils and tar sands into liquid petroleum-type fluids. The rise in liquid oil prices to over $100 per barrel did, indeed, move some of the tar sand resources in western Canada into the status of reserves. As a consequence, Canada's recoverable crude oil reserves rose from about 5 billion barrels in 2000 to about 180 billion barrels in 2007. If petroleum prices again rise, it is likely that even more tar sands will also move from resource status to reserve status. However, because the hydrocarbons in these deposits are too viscous to be pumped, the recovery methods employ enhanced recovery techniques, or conventional mining, of the oil-bearing rock. The most commonly employed enhanced recovery techniques involve softening or liquefying the oil by heating it with injected high-pressure steam or by some sort of electrical device lowered down drill holes. Another, at present, experimental technique involves igniting some of the oil underground and then letting the heat generated by the fire melt or fractionate the lighter components so that they can be recovered from pumped wells. Other efforts use the injection of natural gas, which will dissolve some of the heavy oil and allow it to flow so that it can be pumped out.

TABLE 5.8 Regional distribution of estimated recoverable heavy oil and bitumen (tar sands)

In billions of barrels (From USGS Fact sheet 70-03, 2007)

Region	Heavy Oil	Natural Bitumen
North America	35.3	531
South America	266	0.1
Africa	7.2	43
Europe	4.9	0.2
Middle East	78.2	0
Asia	29.6	42.8
Other	22	34
World Total	443	651

Oil Shale

Oil shales are a diverse group of fine-grained rocks containing significant amounts of the waxy insoluble hydrocarbon substance known as *kerogen*. Kerogen is actually a mixture of a large number of complex, high molecular weight hydrocarbons, much of which can be converted to oil upon heating to temperature of 500°C or more. Oil shales are part of the spectrum of organic-bearing, fine-grained sediments that range from carbonaceous shales through oil shales to sapropelic coals. The amount of recoverable organic matter ranges from as little as about 5 percent to more than 25 percent, and the yield of petroleum during processing can be as much as 100 gallons per ton of shale.

Although oil shales are generally thought of as fuels of the future, the history of their exploitation extends at least back to 1694 when an English patent was granted for a process to make "oyle out of a kind of stone." Because the richest oil shales burn much like coal, it is probable that they had been used as solid fuels long before the potential for oil extraction was discovered. During the nineteenth century, small-scale oil shale industries developed in Europe, Africa, Asia, Australia, and North America, where petroleum was in short supply. A small oil shale industry flourished in France from about 1838 to about 1900, and a few deposits continued to be

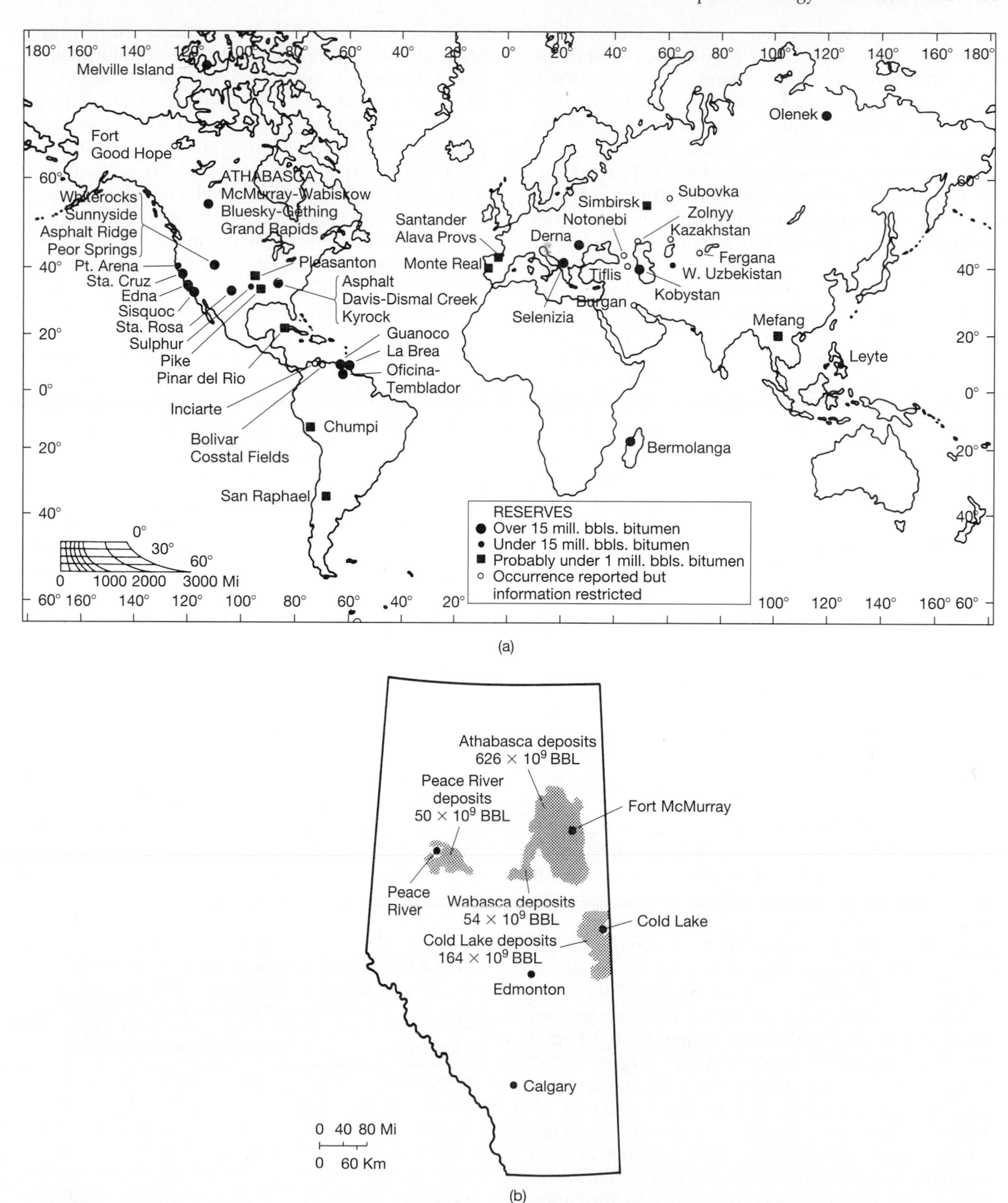

FIGURE 5.51 (a) The distribution of major tar sand deposits in the world. (b) The location and size of the heavy oil deposits in Alberta, Canada. (After P.H. Phizackerley and L.O. Scott and from F.K. Spragins, respectively, in G.V. Chilingarian and T.F. Yen, *Bitumens, Asphalts and Tar Sands,* Amsterdam, Elsevier Sci. Pub. Co. [1978] p. 57 and p. 94. Used with permission.)

mined through government subsidized operations until 1957. The largest and best-known operations were those in central Scotland, where oil-shale processing began about 1850 and continued until 1963. The average yields from the earliest Scottish operations were at least 30 gallons per ton; yields gradually decreased to about 25 gallons per ton, but the total production was about 100 million barrels. Less well documented but probably equally large production has been derived from both Estonia and the Fushun area of Manchuria where production continues. Modern production has also been reported from Spain, Sweden, Italy, Germany, Australia, and Switzerland.

Oil shales form where there is simultaneous deposition of fine-grained mineral debris and organic material in a nonoxidizing environment free of destructive organisms. The fine-grained nature of the sediments indicates that deposition must have occurred in quiet lakes, swamps, or marine basins that were rich in organic matter. No doubt many types of organic debris have contributed to the formation of the kerogen, but the principal precursor appears to have been the lipid fraction of blue-green algae species that can thrive in both freshwater and saltwater. Relatively rapid accumulation of the clays and organic debris under stagnant, reducing conditions protected them from destruction. Continued sedimentation of overlying rocks provided the compaction and burial depth so that temperatures probably rose to 100° to 150°C. This mild heating resulted in the loss of much of the most volatile fractions and left the heavier molecular weight and more refractory organic residue.

Although oil shales have been known and exploited on a small scale for many years, they began to receive a great deal of attention in the mid-1970s when the price of liquid petroleum rose sharply and the world's supply was uncertain. Oil shales are known from all the continents in Lower Paleozoic through Tertiary-age rocks; the major oil shale resources are listed by continent in Table 5.9. In the United States, the greatest attention has been focused on the Piceance Creek basin in eastern Colorado and the Uinta basin in western Utah (Figure 5.52), in which the Eocene Green River formation contains perhaps as much as 2 trillion barrels of oil (Table 5.9). Very large quantities of the hydrocarbon-rich Devonian Chattanooga shale underlie at least ten states in the eastern United States, as shown in Figure 5.53. Unfortunately, most of this formation and its equivalents will yield only 1 to 15 gallons of oil per metric ton by the conventional processing techniques discussed next; hence, their extraction for oil is not feasible in the foreseeable future. Several major projects were initiated in the western United States in the mid-1970s. However, a combination of rising production costs and declining liquid petroleum prices resulted in a suspension of all major operations in the early 1980s. The sharp rise in the price of oil after 2005 brought new interest in oil shales as it did for tar sands, and it is likely that some operations will be initiated. However, growing environmental concerns, especially regarding the use of large quantities of water in a dry climate, will make oil shale development in the western United States difficult.

Several production methods have been proposed for the recovery of the hydrocarbons from oil shales. The two principal ones are surface mining and processing, and *in situ* retorting (Figure 5.53). The first method involves open-pit mining or bulk underground mining of the shale, grinding it into fine particles and heating it to about 500°C in a large pressure cooker-like kiln called a **retort.** The volatilized hydrocarbons are condensed and can then be processed in the same manner as conventionally recovered oil and gas. To be viable, the refining process must yield more fuel than it consumes. The mining and processing facilities are very expensive to construct and create environmental problems, such as the generation of dust during mining, the need for very large quantities of water during processing, and the generation of huge quantities of waste rock. This last problem results from the expansion, or "popcorn effect," of the shale when it is heated; as a result, there is a larger volume of rock waste to be disposed of than there was when mined.

The second technique of handling oil shale, *in situ* retorting, is similar to processes used for enhanced oil and tar sand recovery. It involves the development of underground tunnels, known as *drifts*, either above and below or on both sides of a

TABLE 5.9 World oil shale resources in billions of barrels of recoverable oil (from USGS Science Investigations Report 2005–2006)

Country	Billions of BBL Recoverable
United States	2085
Russia	248
Congo	100
Brazil	82
Italy	73
Morocco	53
Jordan	34
Australia	32
World Total	2826

block of oil shales, as much as 10 meters (30 feet) or more on each side. After the tunnels are completed, the rock is shattered by explosives. The broken rock is then ignited and the rate of combustion is controlled by a flow of air, diesel fuel, and steam. As the fire burns downward in a vertical retort, or from one side to the other in a horizontal retort, the rock ahead of the combustion zone is heated to approximately 500°C; much of the kerogen is vaporized and driven ahead to be drawn out as oil and gas through wells or drains. The remaining bitumen serves as a fuel for the advancing fire. This method is lower in mining costs, greatly reduces the problem of disposal of processed rock, and requires much less water. The burned shale expands to fill the original chamber and leaves a relatively stable ground surface that reveals little evidence of the activity below. One potential problem, although less so in the arid oil shale–rich areas of the western United States than many other areas, is possible contamination of groundwater supplies by waters that leach through the burned out retorts.

The ultimate commercialization of oil shales will depend upon the costs of producing the oil; however, there will also be useful by-products such as ammonia and sulfur produced in the processing of the gases. Furthermore, some oil-shale deposits, such as those in the Piceance Creek Basin of Colorado, are rich in the minerals halite ($NaCl$), dawsonite [$NaAl(CO_3)(OH)_2$], and nahcolite ($NaHCO_3$), which could be recovered and be used as sources of chemicals and aluminum. Estimates have shown that a commercial, above-ground processing plant in the Piceance Creek Basin in Colorado, producing 13,000 barrels of oil per day, would also produce annually 220,000 tons of Al_2O_3, 550,000 tons of Na_2CO_3, and 1,600,000 tons of $NaHCO_3$ as by-products.

FIGURE 5.52 Oil shale deposits in the United States. The major oil shale deposits in the Green River Formation of Wyoming are considered the most likely to be processed for oil. (From C. F. Knutson and G. F. Dana, "Developments in Oil Shale in 1981," *AAPG Bulletin*, vol. 66, no. 11, 1982.)

(a)

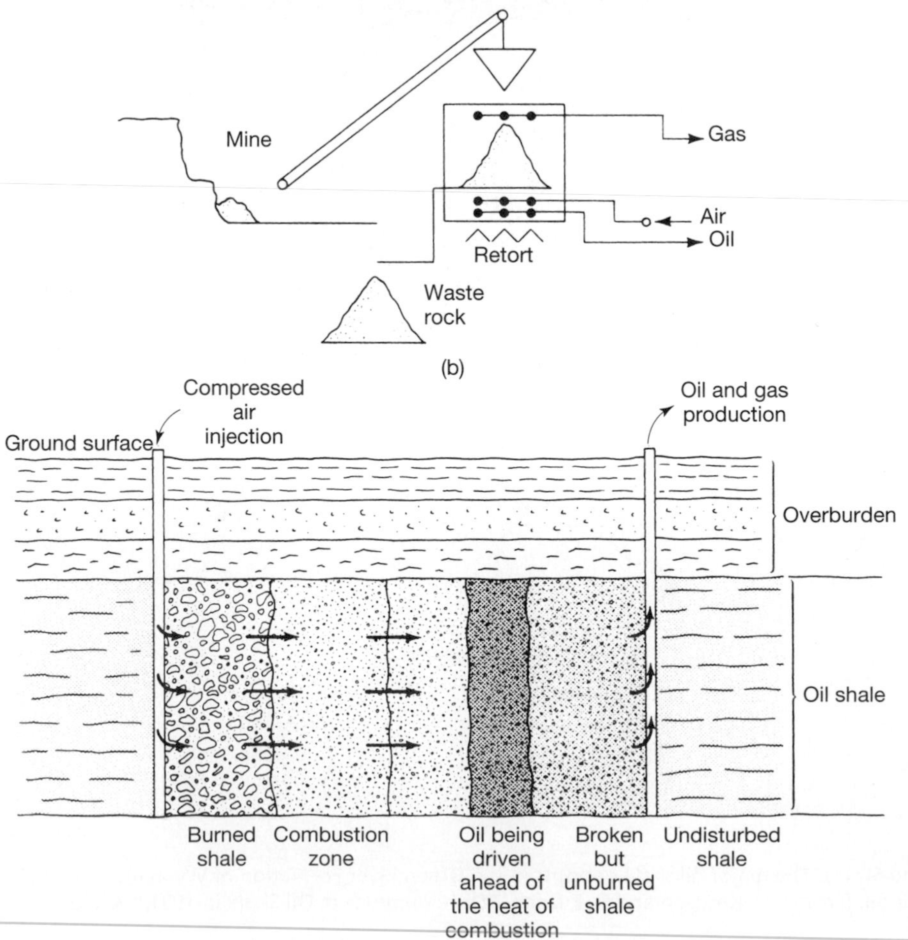

FIGURE 5.53 (a) A close-up view of a sample of rich oil shale. (b) Schematic diagram of the processing of oil shale after mining. The retort would be operated in 400° to 500°C to release the gas and oil from the crushed shale. (c) The *in situ* oil shale retorting process relies upon the movement of a combustion zone through the shale by injecting air into one well and extracting oil and gas from another well ahead of the combustion zone. The shale is first broken to permit the movement of air and gases.

FUTURE FOSSIL FUEL RESOURCES

We now live at the height of the fossil fuel era, and many aspects of our lifestyles are dependent upon a constant supply of fossil fuels. These fuels are not, however, evenly distributed geographically; hence, the fossil fuels have become important in the complex web of international economics and politics.

It is apparent that the world's present dependence on oil for more than 50 percent of energy supplies cannot continue much longer. Far greater energy potential exists in coal and in oil shale; it is to these fuels that we shall have to turn when the world's oil supplies dwindle. Massive, sudden changes in lifestyles are unlikely; however, when oil reserves decrease and oil prices rise in the years ahead, it is probable that increasing use will be made of the other fossil fuels and that fuels such as oil shale and some heavy oils, which are not now economic, will become cost-competitive.

Nuclear power plants such as the North Anna Power Station in Virginia have become very important sources of electrical power in many nations. Nuclear power is, however, a major political and environmental issue in some countries and its future is uncertain. The reactors are housed in the large dome-shaped concrete buildings. (Courtesy of Virginia Power.)

CHAPTER 6

Nuclear and Renewable Energy Sources

It is not too much to expect that our children will enjoy in their homes electrical energy too cheap to meter.

LEWIS L. STRAUSS, CHAIRMAN U.S. ATOMIC ENERGY COMMISSION, REMARKS AT THE FOUNDER'S DAY DINNER OF THE NATIONAL ASSOCIATION OF SCIENCE WRITERS, SEPTEMBER 16, 1954

FOCAL POINTS

- The natural isotopes of uranium and thorium spontaneously decay slowly with the emission of heat in a process called *radioactivity;* artificially induced breakdown in the process of *nuclear fission* accelerates the process and can be used to generate heat in nuclear power plants.
- ^{238}U is the most naturally abundant isotope of uranium (99.3 percent), but ^{235}U (0.7 percent) is the primary fuel of most nuclear reactors because its fission is easier to control.
- Uranium is extracted from ores commonly containing less than 1 percent U_3O_8 and is concentrated into oxide masses called *yellowcake;* this is processed to enrich it in the ^{235}U isotope and is prepared into pellets that are loaded into fuel rods.
- Nuclear reactors contain large numbers of fuel rods. The radiation from these rods creates a chain reaction in which large but controlled amounts of heat energy are released to create steam and drive turbines to generate electricity.
- The first commercial nuclear reactor started operation in England in 1956; subsequently, nearly 200 reactors have been built in Europe and worldwide, more than 400 reactors have been built in more than forty countries.
- A partial meltdown at the Three Mile Island nuclear plant in the United States in 1979 resulted in many design changes, much higher costs, and the cancellation of the building plans for a large number of nuclear power plants in the United States.
- An explosion and fire at the Chernobyl plant near Kiev in the Ukraine in April 1986 resulted in widespread radioactive fallout, particularly throughout Eastern Europe and Scandinavia.
- Worldwide, and especially in the United States and Great Britain, there is currently much renewed interest in nuclear energy because it does not contribute to global warming and climate change by emitting CO_2 to the atmosphere.
- Renewable energy is developing at a rapid pace around the world with the greatest emphasis on wind and solar because they are domestically available, new technologies have lowered the costs, and their operation does not contribute to carbon dioxide level of the atmosphere.

- Solar energy may be used passively to heat homes or water, or it may be used actively to generate electricity by means of photovoltaic cells.
- Hydroelectric power is generated by using flowing water to drive turbines; there is no resulting pollution, but there may be significant environmental impact resulting from the disruption of free-flowing streams.
- Wind energy, used since the earliest time to power ships and windmills, is seeing an increase in use to generate electricity on "wind farms" both on- and offshore; wind energy is only locally available and has been criticized for its impact on scenery, migrating birds, and bats.
- Wave power makes use of wind-generated wave action, and tidal power makes use of the regular rise and fall of tides in response to the gravitational pull of the Moon on the oceans. Both sources of energy can be used in suitable cases to generate electricity.
- Geothermal energy comes from heat in Earth's interior that is originally produced by the breakdown of naturally occurring radioactive elements. It is extracted from hot water and steam coming out of wells drilled in areas with unusually high geothermal gradients.
- Nuclear *fusion*, the process that occurs in the Sun, stars, and the hydrogen bomb, is the fusing together of light atoms to form heavier atoms, releasing vast amounts of energy. The process requires temperatures of millions of degrees and creates very little radioactive waste but will not likely be commercially available until well into the twenty-first century because of the great technical problems associated with controlled fusion.
- Hydrogen is viewed as a potentially important energy source for the future because it is abundant, burns without creating pollutants, and can be used in conventional combustion engines and in fuel cells. However, there are great challenges in the generation and distribution of hydrogen as a fuel.

INTRODUCTION

Although we now live in the fossil fuel era, as evidenced by the fact about 85 percent of the energy we use is derived from fossil fuels, it is clear that this heavy dependence on coal, oil, and related fuels is creating significant environmental problems and is now influencing the global climate. Furthermore, fossil fuels are nonrenewable energy sources and, because of the rate at which we are using them, they will be largely consumed by the end of the twenty-first century. Accordingly, it is important to consider nuclear and other alternative energy sources that might be used to meet our needs later in the twenty-first century and beyond.

Energy actually reaches Earth's surface from three sources (Figure 6.1). The first and most evident energy source is the Sun. As well as being a direct source of heat, the Sun warms the atmosphere and the ocean, producing wind, rain, and ocean currents. Eventually, all of the Sun's energy is radiated back into space so that Earth's surface remains in thermal balance. We have long made indirect use of **solar energy** by harnessing the power of running water and wind and by burning plant matter that has stored energy chemically through photosynthesis. Modern society relies dominantly on the fossil solar energy locked in oil, coal, and natural gas, but now, in the early years of the twenty-first century, we recognize the environmental impact of fossil fuel use and are harnessing solar energy more directly through solar collectors and photovoltaic devices, and indirectly through the wind.

The second source of energy comes from Earth's interior in the form of heat that is derived from the natural disintegration of radioactive elements such as uranium and thorium. Such processes occur naturally within Earth and provide a source of heat that can be exploited as **geothermal energy.** When radioactive processes are speeded up under artificial conditions, we have nuclear power plants.

The third source of energy comes from Earth's rotation and the gravitational interaction between Earth and Moon that produces the tides. **Tidal energy,** although small in magnitude compared with the two other sources, has great promise and is being increasingly exploited locally.

The natural disintegration (or **decay**) of radioactive elements takes place very slowly within Earth and involves a complex series of steps ultimately resulting in the formation of various stable isotopes. For example, radioactive isotopes of uranium decay to stable isotopes of lead, and the radioactive isotope of rubidium decays to a stable isotope of strontium. Each step in the decay produces a little bit of heat energy that we see as geothermal energy. Natural radioactive decay in Earth is so slow that we could not directly derive useful energy from it. However, the breakdown of uranium and thorium can be artificially accelerated by **nuclear fission,** a process in which the breakdown is very rapid and immediately releases very large amounts of energy. Today, the fission of nuclear fuels such as uranium takes place in nuclear power plants.

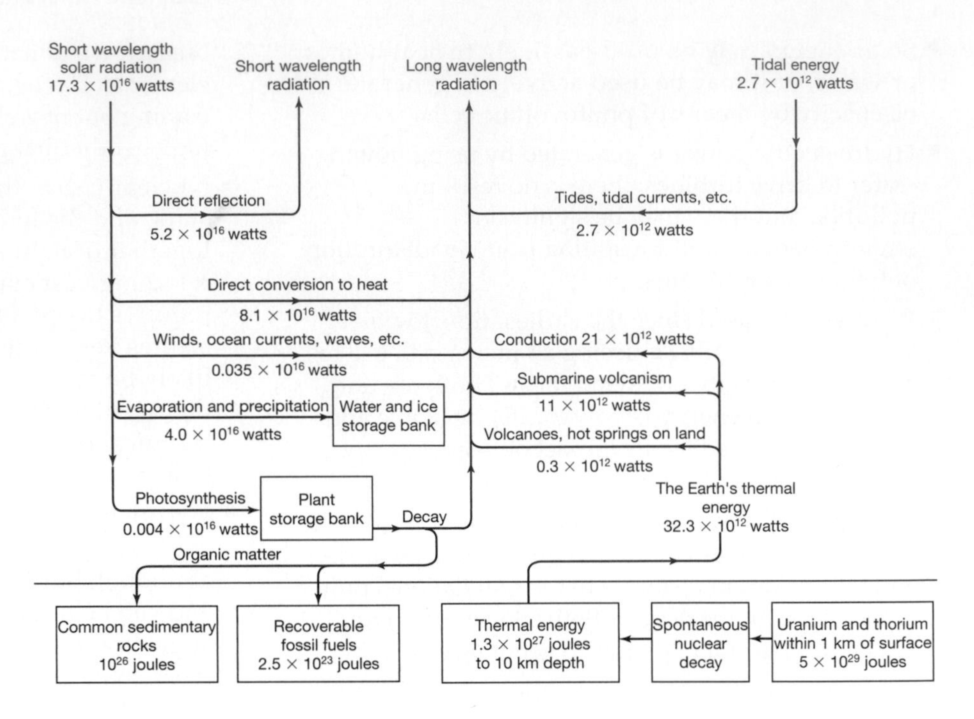

FIGURE 6.1 Energy flow diagram for Earth's surface. The principal source of energy reaching the surface is short-wavelength solar radiation; additional energy comes from tides and from heat flowing from Earth's interior. The nearly constant average temperature of Earth's surface indicates that the total energy radiated back into space must be just equal to the total energy reaching the surface. (After M. K. Hubbert, Energy Resources, Publication 1000-D, Committee on Natural Resources, National Academy of Sciences, National Research Council, Washington, DC, 1962.)

Uranium and thorium are, therefore, nonrenewable nuclear fuels mined in much the same ways as the fossil fuels.

NUCLEAR POWER—URANIUM AND NUCLEAR FISSION

The burning of a fossil fuel such as coal or oil is a *chemical* reaction that proceeds with the emission of heat. Energy is released as a result of changes in the bonds between the electron shells of the atoms. Chemical reactions involve only the transfer or sharing of electrons so that the atomic nuclei are unaffected and the chemical elements retain their identities. In contrast, nuclear energy is generated by changes in the forces that hold the nucleus together—forces that are roughly a million times greater than the energies of chemical bonds.

Early chemists believed that the chemical elements could never be created or destroyed and that the atom could not be "split." However, in 1896, the French scientist Henri Becquerel discovered that certain chemical elements (notably uranium) undergo a spontaneous disintegration with the emission of atomic particles, gamma rays, and heat energy, a process that was termed **radioactivity.** A few years later, at Manchester University in England, the New Zealand-born scientist Ernest Rutherford succeeded in splitting the atom. Subsequently, it was found that many naturally occurring chemical elements have isotopes that are radioactive. Radioactivity is determined by the ratio of protons to neutrons in the atomic nucleus. Some ratios are stable, others are not, and the unstable ones decay by spontaneous nuclear reactions to more stable combinations; the end result of which is the transformation of an isotope of the element concerned into one or more other elements. The energy emitted during radioactive decay is associated with that binding together the nucleus and, as can be seen from Figure 6.2, the binding energy of a nucleus varies as a function of the total number of protons plus neutrons making up that nucleus (the *mass number*). The curve of mass number versus binding energy is such that the lightest and heaviest elements have the highest nuclear binding energies.

The breakdown of a large nucleus such as uranium into two smaller nuclei, such as barium and krypton, is called **nuclear fission.** Another way of releasing energy is by the joining of the nuclei of very light elements, such as hydrogen or lithium, to form heavier elements, a process called **fusion.** In either case—fission or fusion—a small amount of matter is converted directly into a large amount of energy because whenever a nucleus is formed, its mass is slightly less than the sum of the masses of the individual protons and neutrons that comprise it. For example, helium (He) contains two protons and two neutrons in the nucleus and should weigh 4.03303 atomic mass units. In fact, helium only weighs 4.00260 units. The missing mass was converted into energy when the protons or neutrons joined to form the nucleus (i.e., it is the binding energy).

Until about a century ago, it was believed that matter could be neither created nor destroyed; however, in 1905, Albert Einstein postulated the equivalence

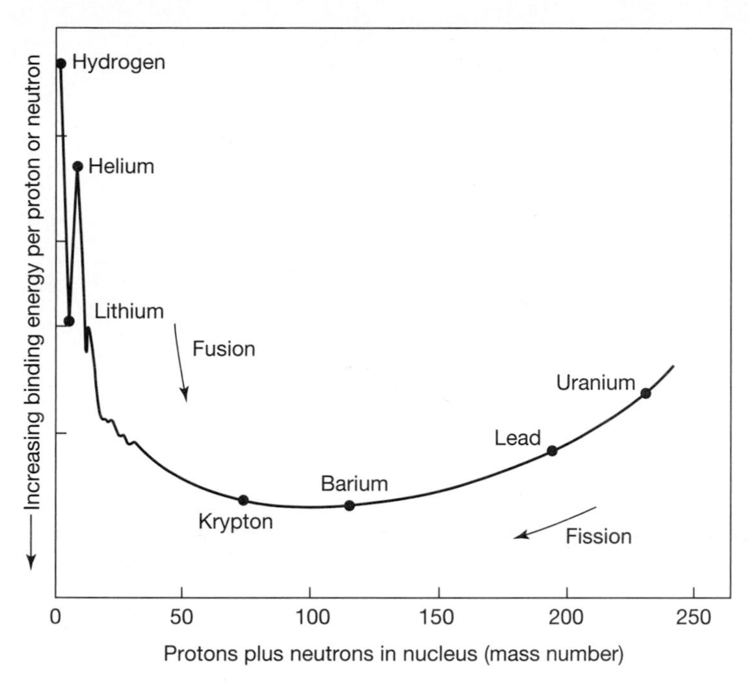

FIGURE 6.2 Binding energy, produced by conversion of some of the mass of an atomic nucleus, varies with the number of protons and neutrons. The higher an atom sits on the curve, the more the energy that will be given off when protons and neutrons combine to form its nucleus. Arrows indicate the directions of movement along the binding energy curve during fusion and fission.

of matter and energy. His famous equation, $E = mc^2$ (where E = energy, m = mass, and c = velocity of light), relates the conversion of a very small amount of matter into very large amounts of energy (because c^2 is a very large number).

Nuclear Fission

The decay of naturally occurring radioactive elements present in Earth today, such as uranium or thorium, takes place very slowly and over a great span of time (millions of years). We cannot slow the natural fission process, but we can accelerate it by bombarding radioactive nuclei with neutrons or by bringing together enough radioactive nuclei so that their natural rate of neutron emission is sufficient to cause an increase in the rate of decay. Uranium occurs in three naturally occurring isotopes, each with a markedly different abundance and half-life, as seen in Table 6.1; however, the only naturally occurring atom that is readily fissionable is the isotope of uranium with mass number 235 (^{235}U). Many reactions can occur during the fission of ^{235}U. For example, when a ^{235}U atom is bombarded with neutrons (n), it can break (undergo fission) into the isotopes of barium (^{141}Ba) and krypton (^{92}Kr) (Figure 6.3) with the release of further neutrons and energy:

$$^{235}_{92}U + n \rightarrow {}^{141}_{56}Ba + {}^{92}_{36}Kr + 3n + \text{energy (200 million electron volts)} \quad (6.1^*)$$

*1 million electron volts = 1.6×10^{-13} joules.

The released neutrons can penetrate the nuclei of adjacent atoms of ^{235}U so that the reaction can continue, provided enough uranium is present. Thus, a **chain reaction** is created if a sufficient amount (the **critical mass**) of uranium is present (Figure 6.3). If this chain reaction is allowed to proceed in an uncontrolled manner, the result is the explosive release of enormous amounts of energy, in other words, an atomic bomb. However, if the fissioning of uranium proceeds under carefully controlled conditions, the heat energy released can be kept at the levels needed for electric power generation.

Note that the reaction in equation 6.1 releases 200 million electron volts for every atom of uranium that decomposes. This means that just one gram of ^{235}U (a cube that is 4 mm on each side) releases the heat equivalent to burning 2.7 metric tons of coal or 13.7 barrels (12 metric tons) of crude oil (Figure 6.4). Not only that, in the fission process there is no smoke and no CO_2 released to pollute the environment—but there is some radiation. Obviously, uranium is a very attractive fuel to consider for power generation, and the technology necessary to exploit uranium fission developed quickly after World War II.

Thorium, which in nature is almost entirely made up of the isotope ^{232}Th, is an alternative source of fissionable material. Although ^{232}Th is not itself capable of sustaining a nuclear chain reaction, it can absorb neutrons from the controlled fission of ^{235}U and be converted to ^{233}U by passing through

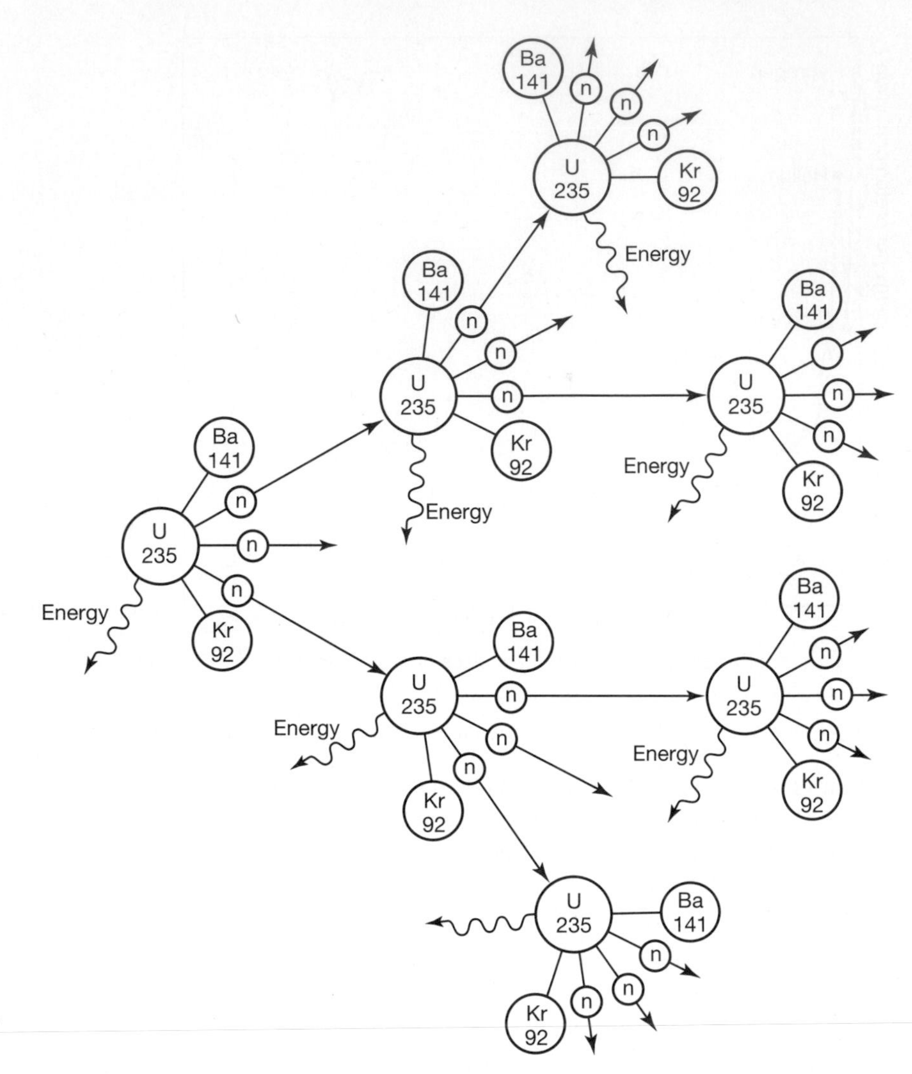

FIGURE 6.3 Schematic representation of a chain reaction in the fission of uranium. When a high-energy neutron (n) is absorbed by the nucleus of a uranium atom, it results in nearly spontaneous fission of the uranium nucleus, producing more neutrons and a large release of energy. The released neutrons cause the fission of other uranium nuclei, and the reaction proceeds in a rapidly expanding steplike, or "chain," reaction.

various fairly short-lived isotopes and emitting β particles:

$$^{232}_{90}\mathrm{Th} + \mathrm{n} \rightarrow {}^{233}_{90}\mathrm{Th};\ {}^{233}_{90}\mathrm{Th} - \beta^{-} \rightarrow {}^{233}_{90}\mathrm{Pa};$$
$$^{233}_{91}\mathrm{Pa} - \beta^{-} \rightarrow {}^{233}_{92}\mathrm{U} \qquad (6.2)$$

The isotope ^{233}U is fissionable and can be made, or "bred," from thorium in a nuclear reactor. However, the vast majority of nuclear power programs are based on uranium as a fuel, so the use of uranium as a source of energy will be emphasized in the following discussion.

Uranium and How It Is Used—The Nuclear Reactor

The controlled fissioning of uranium for power generation takes place in a **nuclear reactor.** However, uranium extracted in mining operations is dominantly made up of the ^{238}U isotope (99.3 percent); the fissionable ^{235}U isotope comprises only 0.7 percent of the total uranium (Table 6.1). Although reactors using unenriched natural uranium were among the first developed, particularly in England, most modern reactor designs use a fuel that has been enriched to about 4 percent ^{235}U. This necessitates a costly process of separation and concentration before the mined uranium can be used as a fuel. In fact, uranium differs from fossil fuels in that complex processing operations are often involved both before and after it is burned in the reactor, and these operations form a sequence known as the **nuclear fuel cycle** (Figure 6.5), which we shall discuss before looking at the reactors.

THE NUCLEAR FUEL CYCLE. Deposits of uranium minerals are discussed later in this chapter, but obviously the first stage in the nuclear fuel cycle (Figure 6.5) is the mining of uranium. Mining involves standard open-pit and underground operations employing essentially the same methods as those for the mining of other low-grade ores (see Chapter 4). Because many uranium ores contain less than 1

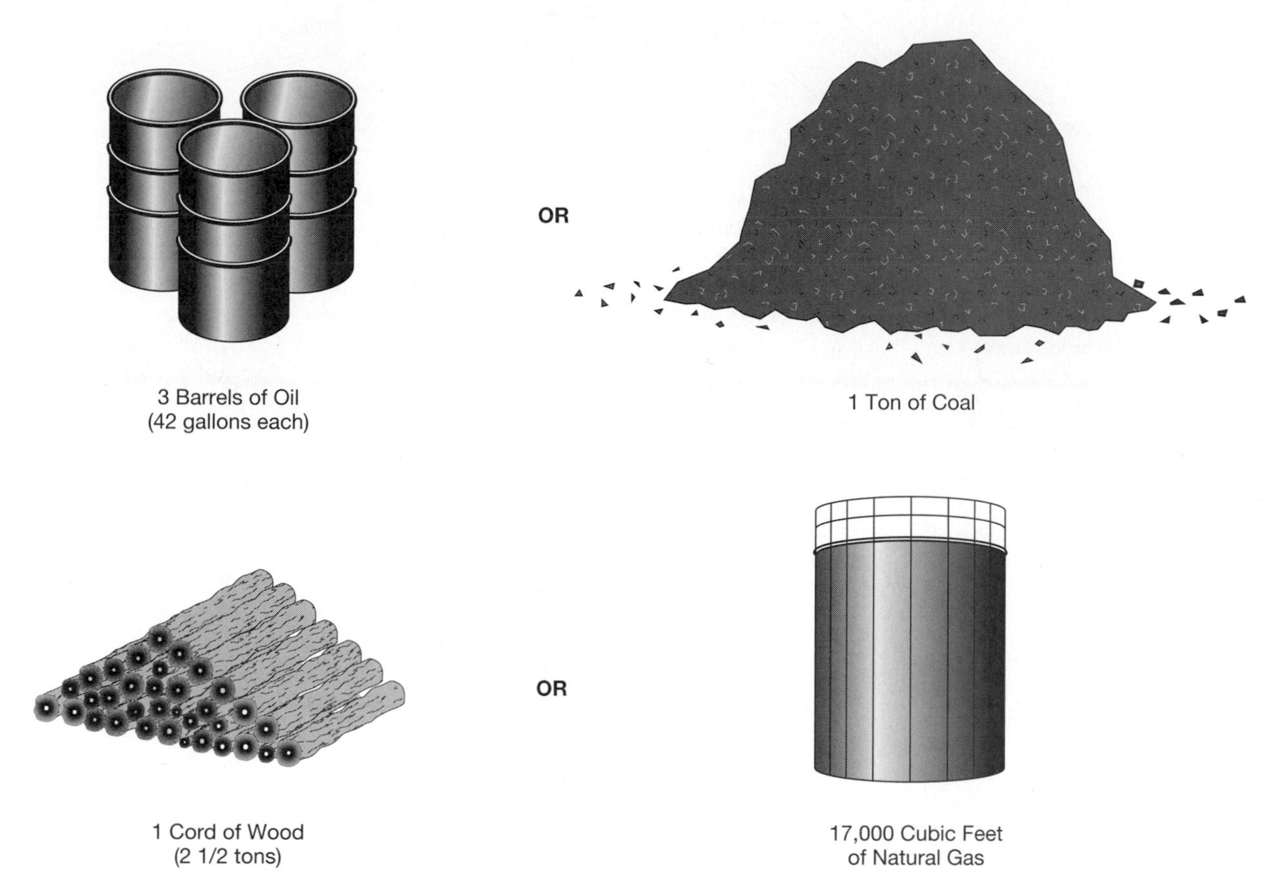

FIGURE 6.4 A uranium-bearing pellet, approximately 2 cm in length, that is placed in the fuel rod of a nuclear reactor contains only about one-third of a gram of ^{235}U, but releases as much energy as the burning of much larger amounts of fossil fuels or wood.

percent of U_3O_8 (conventionally, grades and production figures are expressed in terms of this oxide), extensive **beneficiation** is needed. This involves mechanical and chemical concentration to form an end product called **yellowcake,** a crude oxide containing 70–90 percent U_3O_8. After beneficiation, the yellowcake still only contains 0.7 percent ^{235}U, the isotope needed for conventional nuclear reactors (Figure 6.5).

The *enrichment* of the uranium to produce a fuel containing approximately 4 percent ^{235}U is a very difficult and expensive process because it involves separating two isotopes (^{235}U and ^{238}U) that behave the same chemically and have very little mass difference between them. The most common method of separation is *gaseous diffusion.* The method is based on the fact that a gas diffuses through a porous membrane at a rate inversely proportional to the square root of its mass. The U_3O_8 is converted to gaseous uranium hexafluoride (UF_6) before being passed through thousands of porous barriers to separate $^{235}UF_6$ from $^{238}UF_6$. The enrichment process accounts for roughly 30 percent of the nuclear fuel costs. Following enrichment, the UF_6 is converted to a ceramic powder such

TABLE 6.1 Naturally occurring isotopes of uranium

Uranium Isotope	Percentage of All Uranium	Half-life, Millions of Years
^{234}U	0.0054	0.247
^{235}U	0.7110	710
^{238}U	99.283	4510

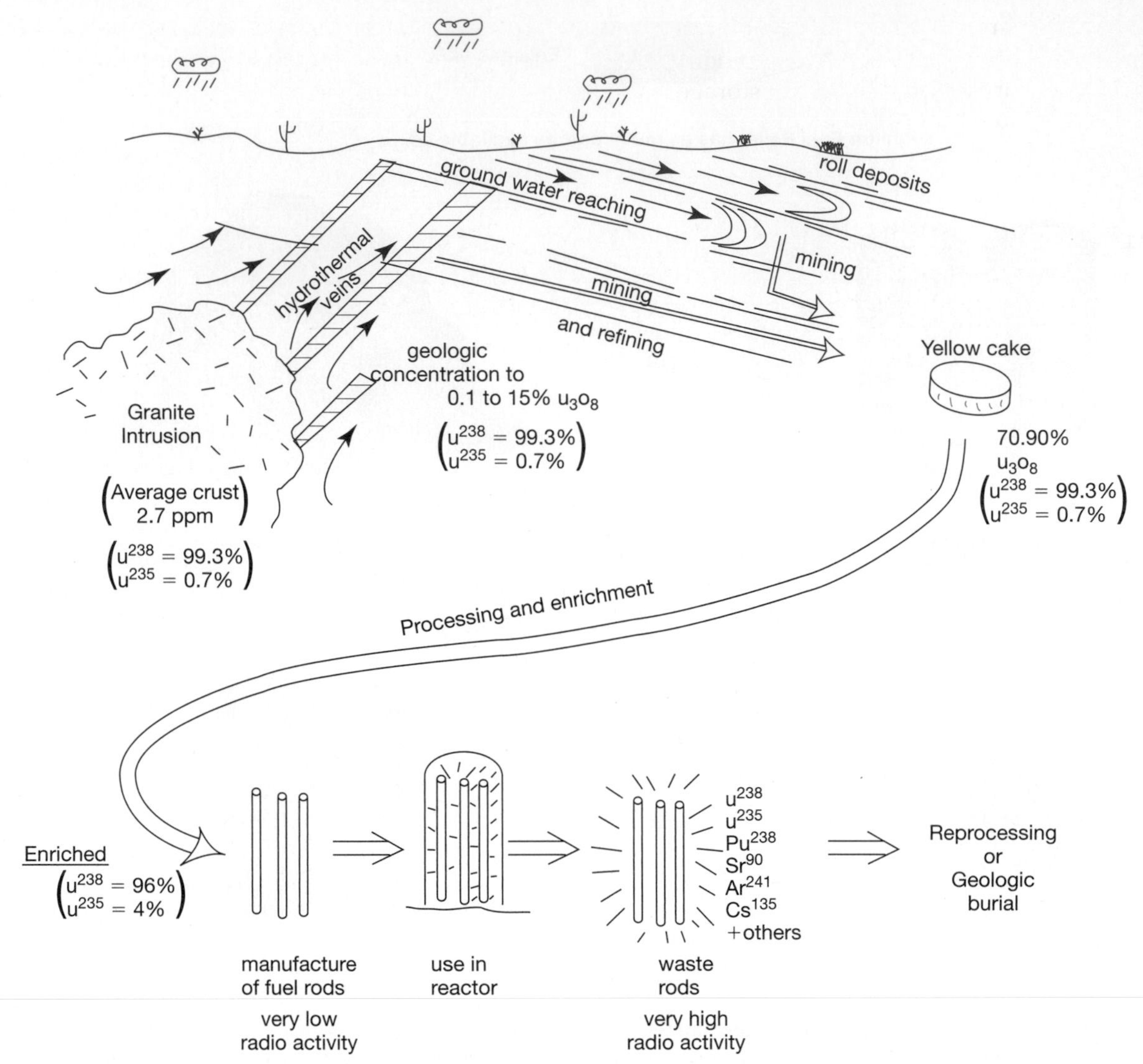

FIGURE 6.5 The nuclear fuel cycle schematically illustrates the steps involved in the conversion of uranium ore into nuclear fuel and ultimately into nuclear waste. Natural ores contain only about 0.7 percent ^{235}U, but this is enriched to about 4 percent in the preparation of the fuel rods that are used in nuclear reactors.

as an oxide (UO_2) and compacted into small pellets that can be loaded into metallic tubes to form **fuel rods** or **fuel elements.** The period after World War II, known as the Cold War era, saw the production of tens of thousands of weapons containing ^{235}U enriched to much higher concentrations than those of fuel rods. This politically charged period was one in which the former Soviet Union threatened to use nuclear weapons against the United States and Western Europe. Fortunately, a nuclear confrontation did not happen, and since the end of the Cold War, plans have evolved to dismantle many of the nuclear warheads and to remove and dilute the uranium down to about 4 percent ^{235}U so that it can be used peacefully in nuclear power plants.

The fuel elements, which are typically rods about 16 feet in length, are loaded into the reactor core where they are irradiated with neutrons that start the fissioning process; fuel rods can produce energy for three to five years. Eventually, the fissionable content of the rods drops to levels that will no longer sustain a chain reaction, and they are removed and replaced by new rods. Because the fuel rods at this point contain an abundance of dangerous but relatively short-lived radioactive isotopes, they are placed in tanks of water to allow some of these isotopes to decay. After several months, the fuel rods may be taken to a more permanent storage or may undergo reprocessing to recover the remaining uranium and some of the other isotopes. Long-term storage of nuclear waste has been the subject of intense debate ever since the first reactors began operating. Several European countries have made decisions on their storage facilities, but the United States will not have any permanent repository until at least the year 2017. For many years, it appeared that Yucca Mountain in southern Nevada would be the permanent United States repository, but political battles continue to forestall any final resolution. (See Box 6.1 for a discussion of the storage capabilties of Yucca Mountain.) The subject of radioactive waste disposal is discussed at greater length in Chapter 4.

THE NUCLEAR REACTOR. To understand the role of nuclear power at present and its future potential, it is necessary to know something about nuclear reactors. The different types and designs are many and varied, but certain major categories can be defined. The first experimental and commercial reactors employed natural (unenriched) uranium as a fuel.

A substantial fission chain reaction cannot be maintained in a simple block of natural uranium* because not enough of the neutrons emitted by the natural decay process would be captured by the small percentage of ^{235}U nuclei available. However, the neutrons released by natural decay travel at high velocities, and if they can be slowed to very low velocities, then the probability of a collision (more correctly termed **capture**) with further ^{235}U nuclei is greatly increased and a sustained fission chain reaction is possible. Neutrons can be slowed by allowing them to collide with the nuclei of light elements, a process that transfers some of the energy of a moving neutron in the same way that a moving billiard ball does on striking a stationary ball. The light elements used in this way are known as **moderators,** and two such elements have been widely used; one is carbon in the form of **graphite,** and the other is **deuterium** (the name given to the isotope of hydrogen with mass number 2) as deuterium oxide or *heavy water.*

Development of the graphite-moderated natural uranium reactor for power generation was pioneered in Britain where the world's first large-scale nuclear power station at Calder Hall, Cumbria, first produced electricity in 1956. From this prototype, the first generation of nuclear power stations in Britain was developed on the Magnox reactor (Figure 6.6). It is important to realize that in a nuclear power station, the reactor simply acts as a source of heat and so replaces the furnace in a conventional power station. Different rectors employ different methods of extracting the heat and, in this case, pressurized carbon dioxide gas is blown through the core and the hot gas is then passed through heat exchangers to heat water and produce steam that can drive turbines and generators, as in conventional power stations. The reactor must be operated under conditions in which a fission chain reaction can be maintained and at a constant power output—a condition known as **criticality.** In the Magnox reactor, this condition is maintained using boron steel control rods that very strongly absorb slow neutrons and are either lowered into the core to slow the fission reaction or raised to increase reactivity and power output. The other type of reactor that has been developed to use unenriched natural uranium as a fuel is a heavy-water-moderated system pioneered by the Canadians (in the so-called Candu program).

If natural uranium can be used as a fuel for power generation, one may ask, why bother with the costly process of fuel enrichment? The answer is that if natural uranium is enriched even to a modest degree by the addition of fissionable isotopes, many of the problems of reactor design found in unenriched uranium reactors are overcome, and there are important gains in manufacturing and operational efficiency. For example, ordinary water or organic liquids may be used to moderate neutrons, the moderator and fuel may be intimately mixed to form a *homogeneous system* (although serious corrosion problems arise), and, with high enrichment, it is even possible to dispense completely with the moderator. Other advantages include much greater freedom in the choice of constructional materials so that, for example, maximum operating temperatures can be increased.

The pressurized reactor [Figure 6.6(a)] allows the nuclear fuel to heat water, which is then circulated to a steam generator. There are three principal types of nuclear reactors that use enriched uranium for fuel. These are the *pressurized water reactor, the boiling water reactor,* and *the gas-cooled reactor.* In this generator, the heat is exchanged to a separate circuit in which steam is created to spin a turbine. The boiling water reactor [Figure 6.6(b)] allows the uranium-heated water to be converted directly into steam that then spins the turbine. The gas-cooled reactor [Figure 6.6(c)] employs CO_2 gas to convey the heat to a steam generator that spins the turbine. There are specific advantages and disadvantages to each type of reactor, but all share the common goal of efficiently converting the heat given off by the nuclear fuel into electricity by spinning a turbine.

One last category of reactors is the so-called **fast reactor.** When the uranium fuel is highly enriched, it is possible to maintain the chain reaction using high velocity (fast) neutrons and to dispense with the moderators entirely. In such a system, the reactor core can be quite small, and a high proportion of neutrons can be allowed to escape through the surface. If a blanket of the nonfissionable ^{238}U is wrapped around this core,

*A unique exception is the "fossil" natural fission reactor at a uranium mine at Oklo in Gabon, West Africa (*Scientific American,* 235, p. 36, 1976). Here, scattered pockets of rich ore achieved the necessary (**critical**) conditions in unusual geologic conditions at a time in Earth's history (about 2000 million years ago) when the natural relative abundance of ^{235}U was greater (about 3 percent of total uranium).

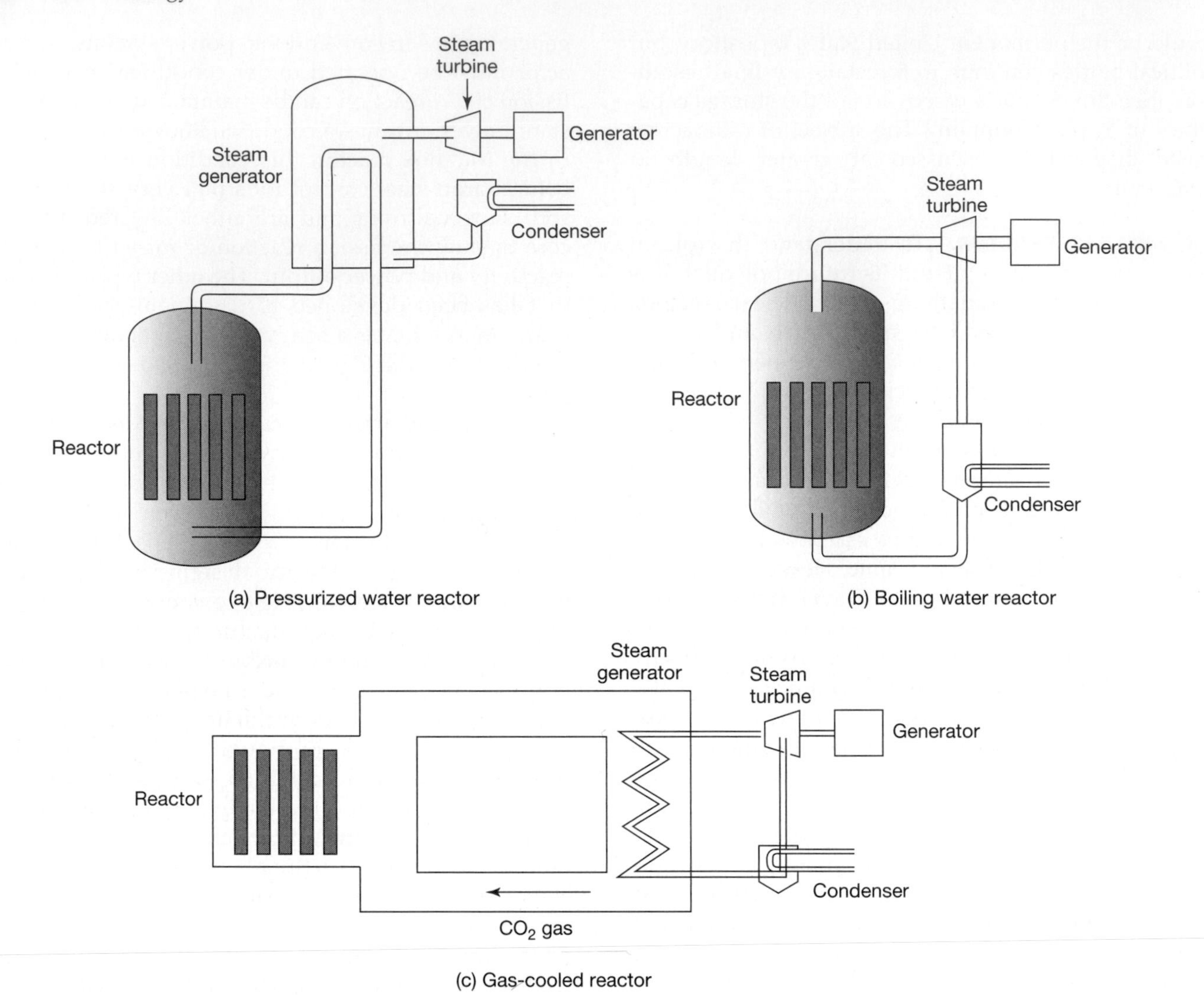

FIGURE 6.6 (a) The pressurized water reactor uses water under high pressure as both the moderator and the means of transferring heat from the reactor core to the electricity generating plant. (b) The boiling-water reactor is maintained at a pressure of only about 70 atmospheres, which allows boiling at about 300°C. The steam thus generated is used to directly drive turbines to generate electricity. (c) The gas-cooled reactor was an advanced form of reactor that used more enriched fuel and could operate more efficiently and at much higher temperatures.

then these emitted neutrons can be used to make or *breed* a fissile isotope of plutonium:

$$^{238}_{92}\text{U} + \text{n} \rightarrow {}^{239}_{92}\text{U};\ {}^{239}_{92}\text{U} - \beta^- \rightarrow {}^{239}_{92}\text{Np};$$
$$^{239}_{93}\text{Np} - \beta^- \rightarrow {}^{239}_{94}\text{Pu} \quad (6.3)$$

In this way, more fissile material can be made in the blanket than is consumed in the core. Such fast reactors are known as **fast-breeder reactors.** The implications of this development for nuclear power generation are considerable because the reactor is capable of breeding its own fuel by converting the vastly abundant uranium isotopes, ^{238}U, to usable isotopes (Figure 6.7).

A prototype fast-breeder reactor first became operational at Dounreay in Scotland in 1959 (Figure 6.8). The central core of 220 kilograms (kg) (480 lbs) was enriched to 46.5 percent ^{235}U, and the power output (60 megawatts) of a volume the size of a small garbage can was immense (comparable to that of 60,000 100-watt lightbulbs). The coolant used was molten sodium and potassium, a much more efficient coolant than water and gases used in other systems. A further 12 megawatts of power was produced from fission processes in the blanket of 20 tons of depleted uranium. The molten metal coolant, in turn, generated steam to drive the turbines and generators. This prototype reactor operated successfully until 1977 when it was replaced by another fast-breeder of improved design. The largest fast-breeder reactor in the world was the Super Phènix at

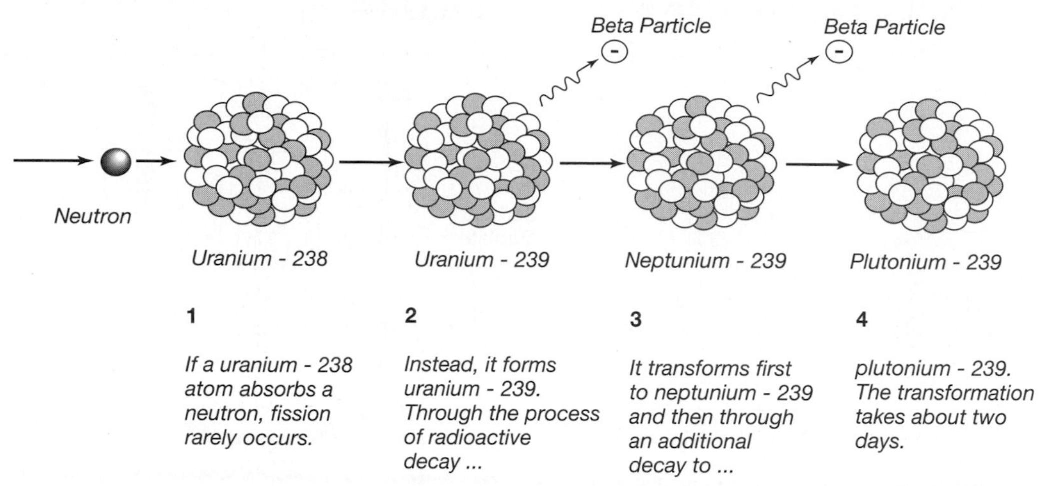

FIGURE 6.7 The ^{238}U used as fuel in a fast-breeder nuclear reactor is converted to ^{239}U, which then decays twice to form neptunium-239 and finally plutonium-239.

Creys-Malville in France, with an output of 1200 megawatts. The Super Phènix, which only operated about 50 percent of the time, and even then at low power, was closed in 1997. Other fast-breeder reactors have also been shut down and the last remaining operational plant is the Beloyarsk Unit 3 in Russia. Other developments in this field include work in the United States on a modified core for the PWR with a breeder blanket and molten salt. Here, features of the thermal reactor with its moderator are combined with a breeder cycle.

NUCLEAR REACTOR PROBLEMS AND SAFETY. Few subjects in the area of resources and energy cause greater public concern than the potential for accidents at nuclear power plants. Despite the very stringent safety precautions that are taken in the design, construction, and operation of nuclear plants, major accidents have occurred over the years. Therefore, it is important for us to examine the risks associated with nuclear power as objectively as possible when decisions concerning energy policy have to be made.

In all commercial nuclear reactors, a fission chain reaction is sustained at the level of criticality, which means constant power output. To maintain the balance in most reactor types, rods of material that capture neutrons very strongly and compete with the fission process can either be lowered into the reactor core to slow the reaction or raised to speed up the reaction. The control rods are usually made of cadmium or boron and are inserted or removed automatically in response to power fluctuations. Clearly, a first hazard is the failure of the control system, which could result in uncontrolled power output. To guard against this, reactors are equipped with ancillary control rods designed to respond automatically in such a situation (or other dangerous situations) and to automatically bring about a reactor shutdown, or *scram*, as it is commonly called. If these automatic safeguards fail, the reactor can rise above critically to a runaway state, in which case the rate of nuclear reaction would increase in an unchecked manner.

Fortunately, this situation could not produce a nuclear explosion like that of a nuclear weapon, but it could allow the temperature to rise to a level that would melt or even vaporize parts of the reactor. The damage to the reactor would probably render it no longer critical and power would eventually fall. The greatest hazard would be the escape of gaseous radioactive iodine (^{131}I), an isotope produced by fission. This dangerous isotope, with a half-life of eight days, is readily inhaled and absorbed in the human thyroid gland. Radioactive xenon and krypton gases, and solid isotopes of strontium (^{90}Sr) and cesium (^{137}Cs), are lesser dangers. To prevent the escape of dangerous fission products into the environment during a runaway event, most reactors are enclosed in various steel and concrete vessels.

In the heavy-water-moderated reactors of the Canadian Candu program, an additional safety feature is a dump tank into which the moderator can be readily emptied, should all else fail, to prevent a runaway. However, in the pressurized-water reactor, the loss of the moderator, which also acts as the coolant, is a further hazard. Loss of coolant or coolant flow is probably the most serious hazard in most reactor types, so coolant flow is always carefully monitored. Fast-breeder reactors present their own safety problems.

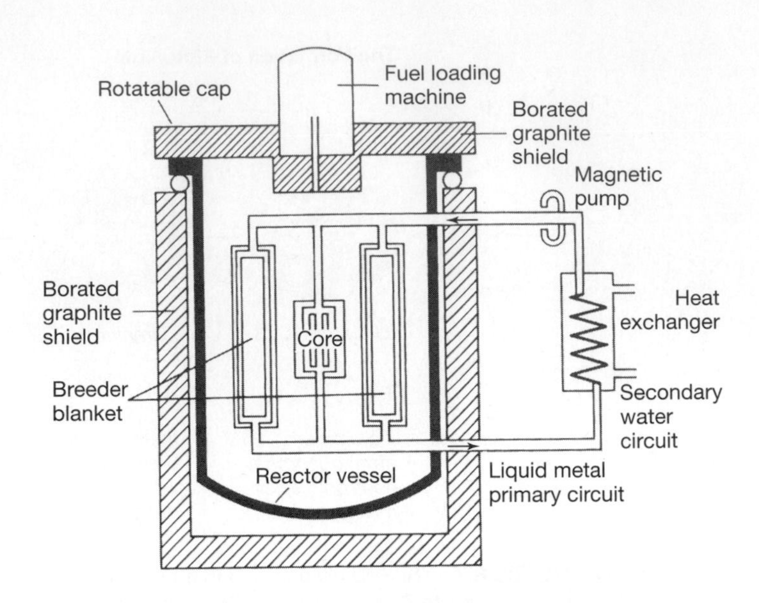

FIGURE 6.8 Schematic diagrams of fast-breeder nuclear reactors in which a blanket of non-fissionable ^{238}U is placed around the core. The ^{238}U is converted to ^{239}Pu, a fissionable isotope, at a rate faster than the core is consumed. (a) In the loop-type fast-breeder reactor, the heat is carried out of the reactor vessel by molten metal to a heat exchanger. (b) In the molten salt-breeder reactor, salt transports the heat to a series of heat exchangers so that electricity can be generated.

For example, if the core should melt, it could reform in such a way as to be even more reactive. Diverters positioned beneath the core ensure separation to prevent this should melting occur. Again, it must be emphasized that a nuclear reactor cannot explode like a nuclear bomb; the danger arises from the escape of highly poisonous or toxic fission products into the environment.

Up until 1979, the nuclear industry worldwide could claim an impressive safety record. The only significant release of radioactive material had been in Britain at Calder Hall (Windscale, Cumbria) in 1958; it involved a graphite-moderated, air-cooled reactor that was actually shut down at the time. The release, mainly of radioactive iodine (^{131}I) did not endanger life, but the milk produced by cows in an area around the reactor of ten by thirty miles was unusable for a time. Much more recently, two serious accidents that have significantly increased public concern and damaged the image of the nuclear power industry have been those at Three Mile Island (TMI), Pennsylvania, in March 1979 and at Chernobyl, in the Ukraine, in April 1986 (described in the discussion in Box 6.2).

The Three Mile Island facility, located on an island in the Susquehanna River near Harrisburg, Pennsylvania, operated as a pressurized water reactor. Such reactors have both a primary cooling system to carry the fission heat to the steam generator and a secondary cooling system to carry steam from the steam generator to the turbine. Each system has its own pumps, the reactor cooling system pumps and the feeder water pumps, respectively. Both systems carry heat from the reactor core to areas outside. At 4:00 a.m. on March 28, 1979, the main-feed water pump failed and, automatically, three reserve pumps (two electric and one steam-driven) went on. However, in the 15 seconds required for these pumps to build to normal pressure, the primary system heated and increased in pressure such that the automatic reactor shutdown procedure went into operation. A pressure relief valve in the primary system also opened to release pressure—all of these responses were quite proper and designed to occur. Unfortunately, unknown to the operators, valves connected to the reserve pumps, and supposed to be open at all times, were closed so that the steam generators soon boiled dry.

In addition to this, the pressure relief valve in the primary system failed to reset properly and began leaking. In the long, complex series of events that followed, the pump valves were opened and eventually the leaky valve blocked, but the operators were misled into thinking that there was too much water in the primary system instead of too little. As a result, the core lay uncovered for several hours and substantial damage occurred before the situation was finally brought under control.

Factors leading to the TMI accident involved three main ingredients: a temporary, abnormal situation aggravated by human error; a small loss-of-coolant accident; and misreading of the situation by the operators. Ultimately, although very costly damage was done to the reactor itself, the dangers to the public were very small. In the early stages of the accident, radioactive xenon (^{133}X) was released; actual exposure is calculated to have the potential to cause the death by cancer of less than one person in thirty to forty years

BOX 6.1

United States Nuclear Waste Storage—Yucca Mountain

Of the principal problems confronting the nuclear power industry, probably the single largest concern is the disposal of spent nuclear fuel rods. Although the typical nuclear fuel rod containing enriched ^{235}U (about 4 percent; the rest of the uranium is ^{238}U) has very low levels of natural radioactivity, the same fuel rod as extracted from the nuclear power plant at the end of its useful lifetime (3 to 4 years) exhibits very high levels of lethal radiation. Ever since the advent of commercial power generation by nuclear plants, first in England in 1956 and then in the United States in 1957, numerous studies and unending debates have occurred as to what to do with the spent fuel rods. The very slow rates of decay of the radioisotopes formed as a result of fission requires that the rods be effectively isolated from any human contact for about 10,000 years, when the radiation has declined to a safe level (see Figure 4.30).

Several countries have made decisions on storage sites—Germany decided on burial in salt beds, Sweden opted for burial in granite, but the decision has proven problematic in the United States. The major criteria for a storage site have been: (1) distance from population centers; (2) security; (3) prevention or minimization of any environmental impact, especially with regard to interaction with ground waters and surface waters. The first and third criteria effectively limited the choice to parts of the west of the country and led to detailed consideration of the Hanford facility in Washington State where the rocks are basalts, the WIPP (Waste Isolation Pilot Project) site in New Mexico using disposal in salt beds, and the Yucca Mountain site in Nevada, using disposal in volcanic tuffs. After years of debate, Yucca Mountain was chosen as the best site for development of a permanent facility. Yucca Mountain lies in an arid, unpopulated region of southern Nevada, about 100 miles northwest of Las Vegas (see Figures 4.33 and 4.34 and Plates 26 and 27) on a Bureau of Land Management facility.

Geologically, Yucca Mountain overlies a 1000-foot thick sequence of volcanic tuffs that form a solid mass as a result of the action of the volcanic heat and circulating aqueous fluids when the rocks were deposited more than 13 million years ago. Although there has been much faulting in southern Nevada, the block that constitutes Yucca Mountain is free of major faults. Several small volcanic cones occur in the area, but none appear to have shown any activity in the past 10,000 years. The groundwater table lies at a depth of 1500 to 2000 feet, which is 660 to 1300 feet below the level of the proposed nuclear storage site.

The proposed facility would be developed as a series of galleries tunneled into the body of the volcanic tuff. Here, waste nuclear fuel rods would be set into cylindrical storage cavities in the floors of the galleries. Heat buildup should be minimal and the relationship to the water table should prevent any groundwater contamination. Debate still occurs about the appropriateness of Yucca Mountain as the U.S. nuclear storage facility; the future of the entire project is murky at best. The government had promised to accept spent nuclear fuel by 1998, but construction delays pushed the opening date of the Yucca Mountain facility first to 2010 and then to 2017. The change in administration in 2009 was accompanied by strong governmental pressures to cancel the Yucca Mountain facility and to start a new search for a permanent repository. Furthermore, the administration cut off all funding for the licensing of Yucca Mountain in 2009, leading one Nevada congressman to refer to the site as "a failed $100 billion dinosaur in the Nevada desert." In the meantime, the waste builds up at the many nuclear power plants and the arguments and search for permanent storage sites continues. Even if the storage facility ever opens, there will be another major problem—how to transport the nuclear waste to Yucca Mountain. The general public does not like to see waste building up in temporary storage sites at the nation's reactors, but property owners also do not want to see dangerous wastes transported near their homes. At present, no final decision seems likely for several years.

after the incident. A footnote on the TMI incident is that, more than thirty years after the meltdown, no human being has entered the containment area because of the continuing high levels of radioactivity. It has been surveyed and some material has been removed by robots employing TV cameras, but most of the highly radioactive material remains in place.

Although there was no reported physical harm to any individual arising from the TMI incident, it did undermine confidence in the nuclear power industry. Prior to 1979, there was the expectation that nuclear power would ultimately serve as the principal energy source for the United States and many other nations. At the time of the TMI incident, there were about seventy reactors operating in the United States, producing about 11 percent of the electrical power. There were immediate calls for new reactor designs with fail-safe backup systems, which led to long construction delays and soaring costs. Plans for many of the hundred or so planned nuclear plants were immediately scrapped, and decisions on many others were delayed (only to be canceled later); only plants that were close to completion were allowed to continue with construction. Projected construction periods of five years stretched on to fifteen years, and original estimated costs of $500 million for plants ultimately became $5 billion. The power once prophesied to become so cheap (see the now famous quote of the Chairman of the Atomic Energy Commission at the beginning of this chapter) that electric meters would not even need be put into houses, turned into the most expensive form of energy.

Over the more than thirty years since TMI, hitherto unimaginable events unfolded—the Washington Power Supply System went bankrupt because of its investment in nuclear power plants; a nearly completed plant in Ohio was converted to burn fossil fuel; the Seabrook plant in New Hampshire was prevented from operating for many years because it was 8 miles from the Massachusetts border and that state refused to approve a 10-mile-radius emergency evacuation plan; and the Shoreham plant on Long Island, built at a cost of $6 billion, was sold to New York State for $1.00 and will never operate. The most tragic of nuclear accidents, however, occurred in April 1986 when a steam explosion at the Chernobyl reactor near Kiev in the Ukraine released large amounts of radioactivity into the environment, killed twenty to thirty workers, and left a legacy of cancer illnesses and deaths not yet fully counted (Figure 6.9). A more detailed description of this disaster is given in Box 6.2.

What can we say about the risks associated with nuclear power generation? It is still fair to say that the risk of death or serious injury to a member of the public as a result of a nuclear accident is very small, particularly when compared with the many risks taken by people in everyday life. Nevertheless, the lessons of TMI and Chernobyl must never be forgotten. To quote from the journal *Nature*,* "The difficulty is that what went wrong at Chernobyl on 26 April could have happened anywhere. That is the plain truth which no amount of technical comparison of different reactor types can possibly conceal. Moreover, there have been several occasions in the recent past when nuclear accidents, luckily smaller in scale, have been brought about, because operators have chosen to disregard the regulations they are supposed to live by, or

*Editorial, vol. 323, no. 6083

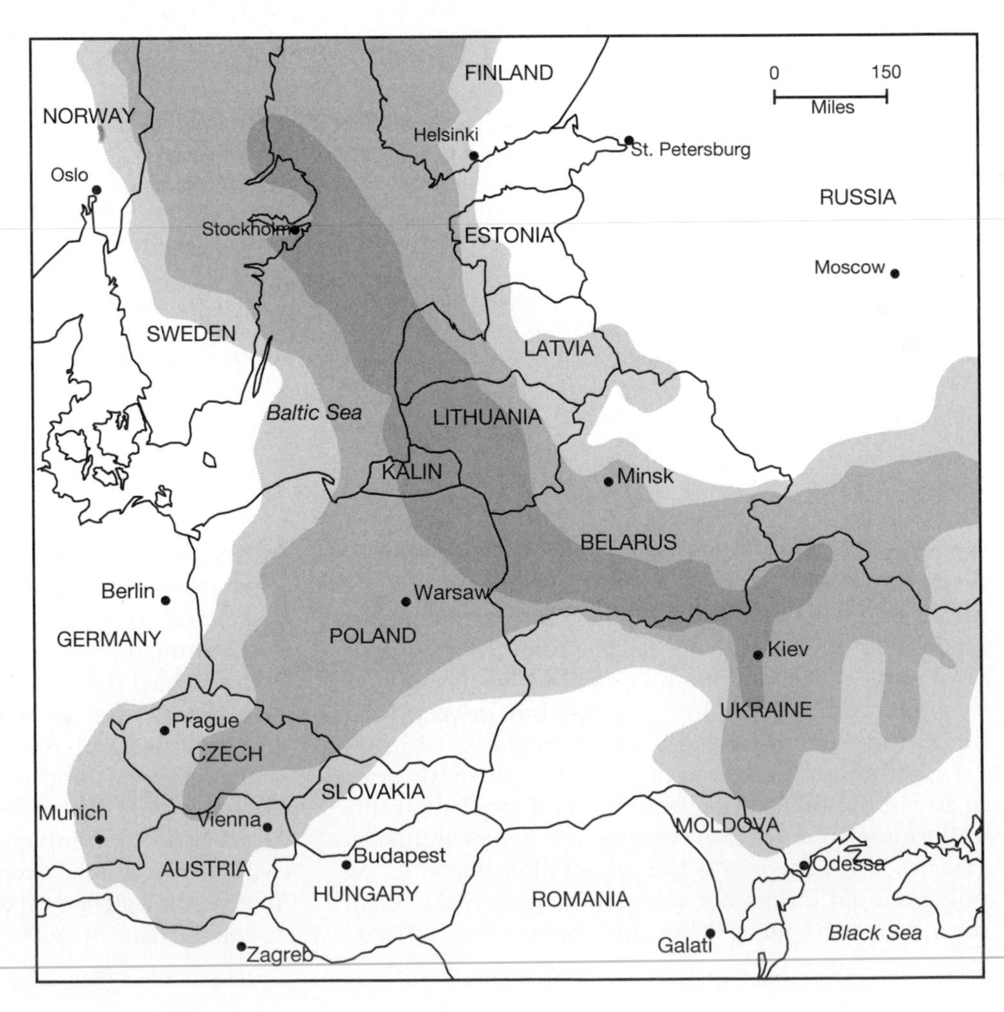

FIGURE 6.9 Map of the area most affected by the accident at the Chernobyl nuclear reactor near Kiev in the Ukraine in late April 1986. The shading shows simulation of the integrated dose of ^{131}I to adult thyroid glands accumulated from April 26 to May 1, 1986 (based on calculations done at the Lawrence Livermore National Laboratory in California). The central, darkest area had doses in excess of 1 rem, the intermediate zone had doses between 0.1 and 1 rem, and the outer zone had doses between 0.01 and 0.1 rem. (See Tables 3.3 and 3.5 for information on units of radioactivity and dosages experienced from other sources.)

have been deserted by common sense and elementary caution."

It is also important to remember that other aspects of safety and security, such as the hazards associated with nuclear waste disposal and the relationship that may exist between nuclear energy programs and the proliferation of nuclear weapons, must be considered in the overall formulation of policies regarding nuclear power. At the same time, it is important to realize that each of the competing energy sources creates its own problems and has its own costs. After all, most of the world's electricity is produced by burning coal and, worldwide, hundreds of coal miners are killed each year. However, because their deaths are not at the sites of power generation, we do not draw the same connection that we do about deaths at nuclear power plants. Interestingly, many environmental groups, initially much opposed to nuclear power, have now backed the building of new nuclear plants because they do not emit carbon dioxide, which environmentalists consider a greater global threat.

Nuclear Electric Power Production

Nuclear power moved from the bombs of war in 1945, to power plants in naval ships in 1954, to peaceful commercial electrical power generation in 1956. The opening of the first commercial nuclear power plant at Calder Hall (Windscale) in England in 1956 was soon followed by the opening of the first plant in the United States at Shippingport, Pennsylvania. This led to the speculation that there would be no more concerns about electrical power production and to the thought that power would be so abundant as to be free. For the next twenty years, everything went according to schedule, with approximately eighty nuclear power plants coming on line and hundreds more being planned in the United States alone (Figure 6.10). These plants were quiet, smoke-free, and were already producing about 12 percent of American electricity by 1977. All of this was dealt a serious blow and U.S. confidence was shattered when the Three Mile Island (TMI) meltdown occurred. The outcome was that scores of anticipated plants were canceled, and not one additional order or permit to build a new generating plant has ever been granted in the United States. Many partially built plants were abandoned or converted to burn fossil fuels and all plants that were nearly complete were delayed and modified with huge cost overruns; the last American plant was finally finished and became operational in 1996, about 15 years behind schedule.

Despite the effects of the TMI incident and Chernobyl disaster, nuclear power has become a major electrical energy source in many countries. In the United States, for example, as new plants came on line, the number of operating nuclear plants rose to a maximum of 110 in 1994. Subsequently, the number of operational plants dropped to 104; these produce about 20 percent of the United States' electricity. The nuclear power plants are widely distributed (Figure 6.11), but

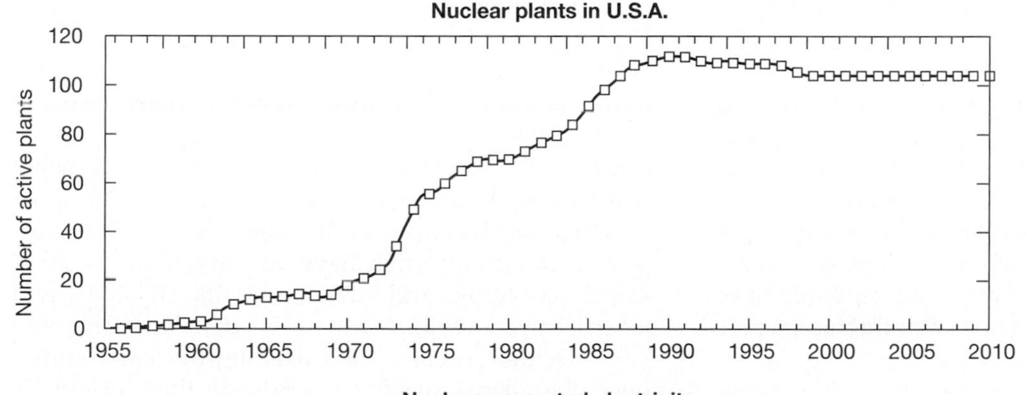

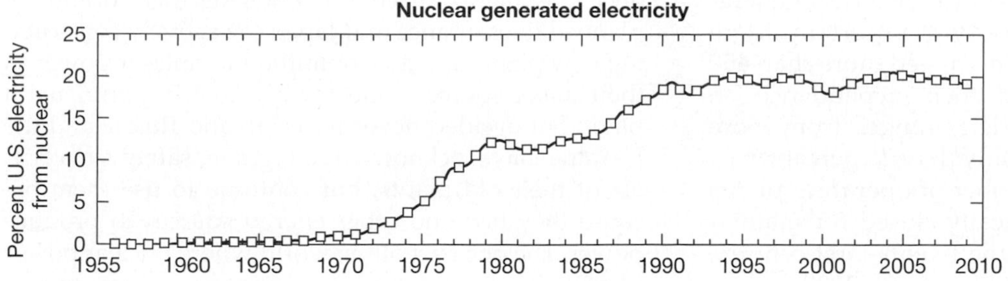

FIGURE 6.10 Nuclear power generation in the United States since the opening of the first nuclear power plant in 1957. (a) Number of operating nuclear power plants in the United States. (b) Percentage of U.S. electricity generated by nuclear plants. (Data from U.S. Energy Information Administration.)

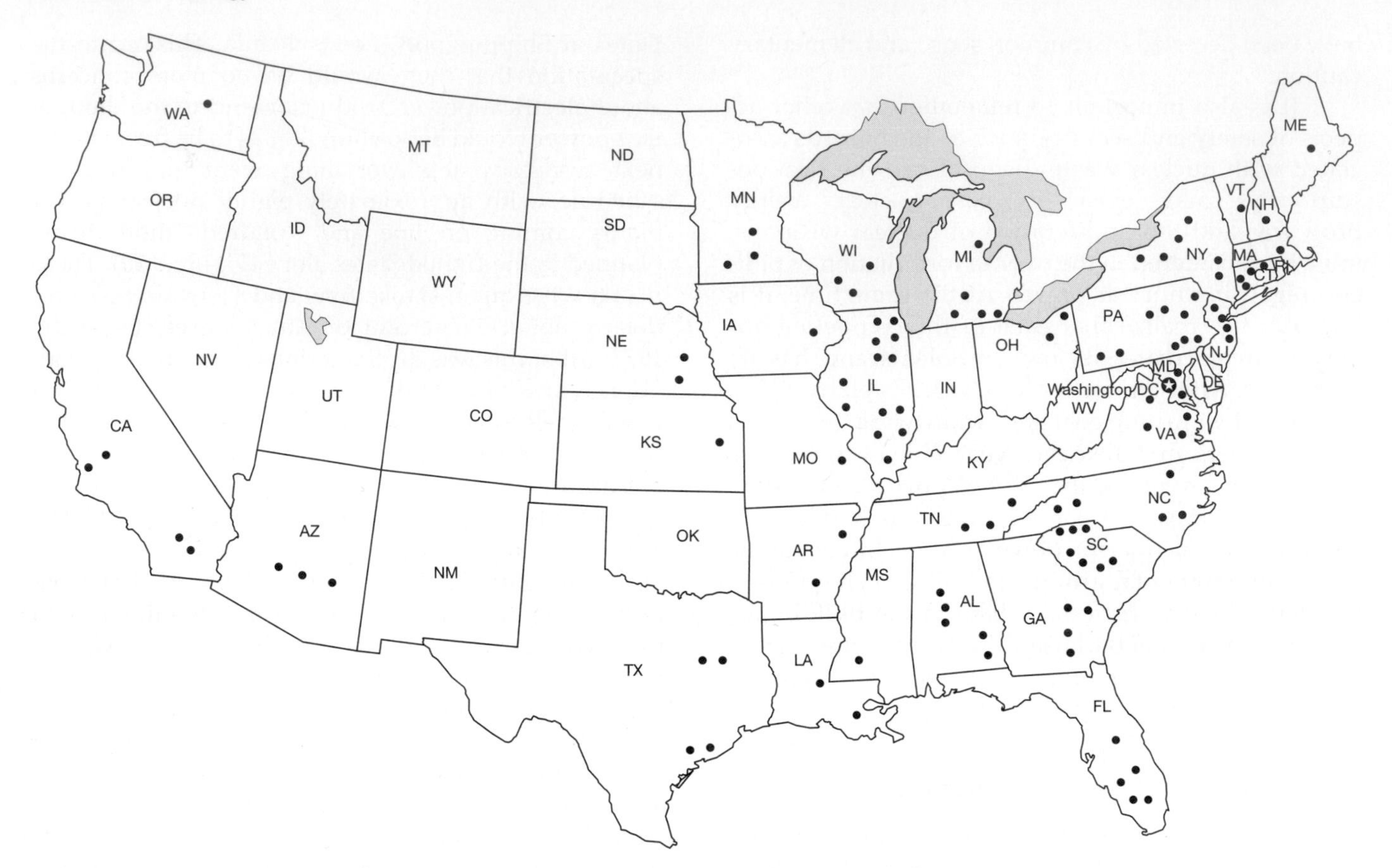

FIGURE 6.11 The distribution of nuclear power plants in the United States at the end of the twentieth century.

clusters of installations are located in the Northeast, near Chicago, and in California. Their electricity output reached about 23 percent of the U.S. total in the mid-1990s. Nuclear power plants have finite operational lives—usually about thirty to forty years—and many of the plants built in the 1960s are nearing the end of their original licensing periods and are subject to decisions on major refitting or shutdown. The first plant in the United States to face this was the Yankee Rowe in Massachusetts in 1994, and the decision was to shut it down. It is likely that the same choice will be made for several other older plants. This creates new problems because the dismantling of nuclear plants has never been attempted before, and methods have not been developed for the disposal of the radioactive components.

The use of nuclear power for electrical generation in the twenty-first century is uncertain. More than forty countries have constructed more than 450 nuclear power plants, and their dependence on these plants for their electricity ranges from more than 80 percent down to only 1 or 2 percent (see Figure 6.12). The actual number of operating plants varies as plants are periodically closed for maintanence. At the beginning of the twenty-first century, nuclear power plants provide about 2300 billion kilowatts of electricity, equal to about 19 percent of total world electricity; ten countries currently meet at least 40 percent of their electricity requirements from nuclear plants.

Although the United States and many other Western countries may have seen the peak of their nuclear power industries, nuclear power-generated electricity is going to be important for many years to come. Interestingly, nuclear power, once very much opposed by major environmental groups, is being reevaluated because it does not create acid mine drainage, acid rain, or contribute CO_2 to the atmosphere. Other countries have responded to citizens' safety concerns and the accidents at TMI and Chernobyl in various ways. Sweden and Germany have decided to close their existing nuclear facilities once the plants reach the ends of their originally planned lives; France and Japan are actively constructing new plants and are committed to nuclear power as their major source of electricity; Austria completed a plant, but decided never to use it; and Russia and the Ukraine have acknowledged grave safety concerns about their old plants, but continue to use them because they have no other energy sources to produce power. The one part of the world where nuclear power is still the most popular option for future energy

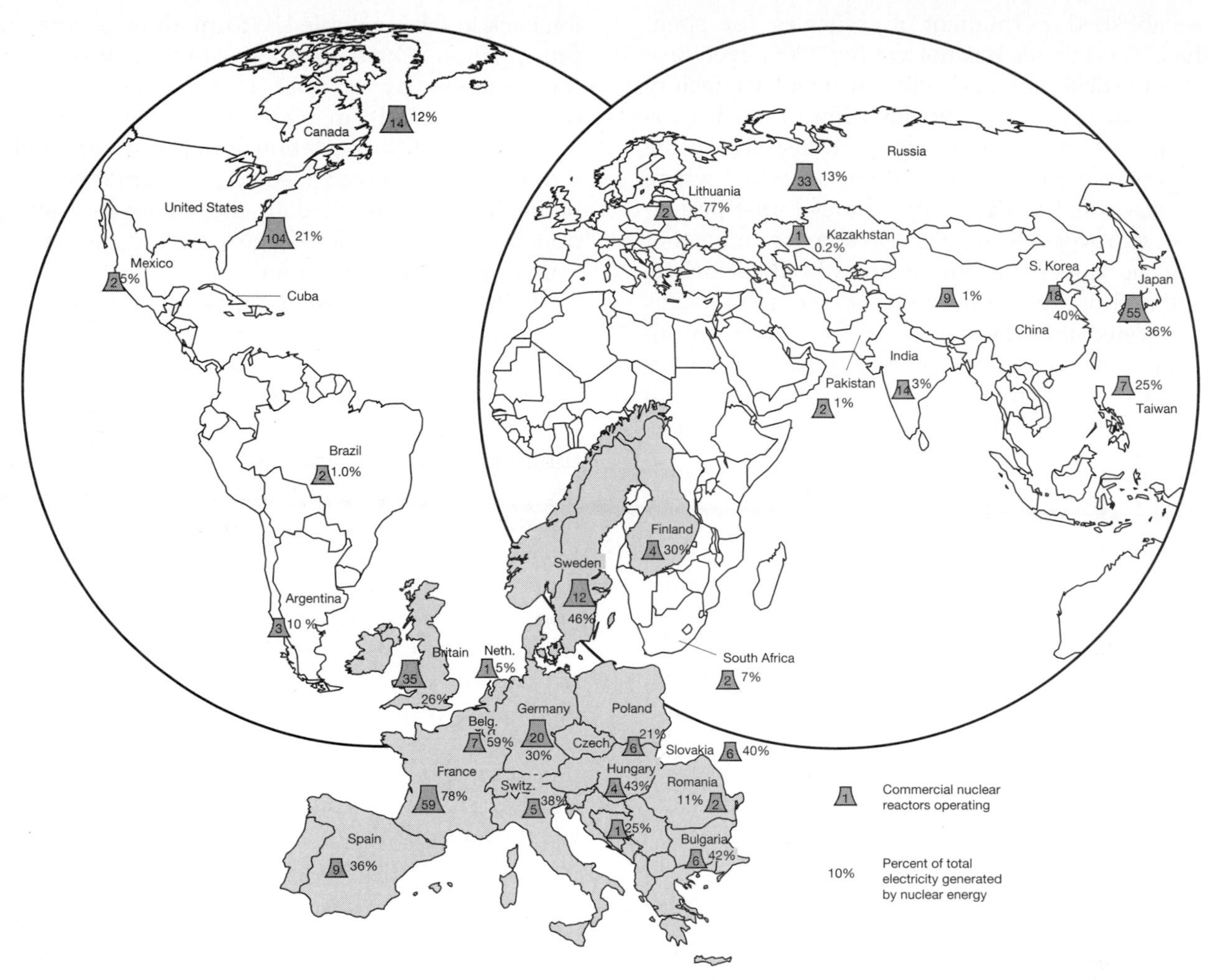

FIGURE 6.12 A map to illustrate the worldwide utilization of nuclear power at the end of the twentieth century. Countries involved in nuclear power generation are shown with the number of operating "commercial" nuclear reactors, reactors on order but not yet operating, and percentage of total electricity generated by nuclear energy.

generation is the Far East. Of the 100 reactors still under construction or in planning stages at the beginning of the twenty-first century, the majority are in Taiwan, Japan, China, South Korea, and other regions of Southeast Asia.

What is the future of nuclear power? The answer appears to depend upon where one lives. The most recent projections by the United States Energy Information Administration indicate that there will likely be a small increase (about 1 percent) in total world nuclear power until about 2010, and then there will be a significant decline (perhaps as much as 15 percent) by 2020. These figures actually disguise two opposing trends; the major industrialized nations will lose about 30 percent due to retiring older plants, while developing nations will increase their production by about 30 percent. Overall, this means that the world will have to find other means to produce the electricity that nuclear power might have generated. Furthermore, the loss of nuclear power, which produces no greenhouse gases, will make compliance with the Kyoto Protocol's call for lower carbon dioxide emissions more difficult.

Nuclear Waste Disposal

The solid and liquid wastes generated in nuclear power plants are among the most dangerous materials on Earth. Each year, every nuclear power plant must store or dispose of numerous highly radioactive, but no longer useful, fuel rods. An in-depth discussion of the options for handling these rods is presented in Chapter 4; suffice it to say that the issue of nuclear waste disposal has developed into one of the most contentious debates in the United States and several of the European countries over the past twenty years. In the United States, on-site storage capabilities have been exceeded at several operating plants, but there has been no general plan for the development of additional storage. Furthermore, even though some countries have established permanent repositories for spent fuel, the United States has not yet (by 2009) even designated an official site and will not have any facility ready until at least 2017; the situation in the U.K. is even more uncertain. Even then, transport of the wastes, for example from the 100 or so plants where they are now held in the United States to the permanent site at Yucca Mountain, Nevada (or any other site) will, no doubt, become a great political debate and legal nightmare. Numerous scientists and officials have suggested that the demise of the American nuclear industry will come about because plants will simply not have any more safe places left to store spent fuel rods and thus will not be able to use any more fuel.

Uranium in the Earth

Uranium has an average crustal abundance of 2.7 ppm. This is relatively low, but it is 600 times more abundant than gold (at 0.004 ppm). Although uranium can occur in trace amounts in a variety of minerals, the large size of the uranium atom tends to exclude it from early crystallizing minerals in magmas. Hence, uranium is commonly concentrated in the final (*residual*) melts and fluids and in silica-rich rocks such as granites that are rich in alkali elements such as sodium and potassium. In such rocks, uranium concentrations can range up to 100ppm or more. This uranium may be located in certain minerals that occur in such rocks as minor components (e.g., zircon, sphene, and apatite), or it may occur as the most important ore mineral of uranium, uraninite (UO_2, also called **pitchblende**). Here, the uranium is in the uranous (U^{4+}) state. Occasionally, within or close to such igneous rocks, very high concentrations of uranium minerals may occur either in veins or as more irregularly distributed disseminations. Such deposits were among the first uranium ores to be found and exploited and were the source of much of the uranium used by Pierre and Marie Curie for their pioneering work on radioactivity in the early years of the twentieth century.

Most of the uranium at Earth's surface probably formed in association with igneous rocks as described above. However, a very important feature of uranium in the U^{4+} state is that it is readily oxidized to the uranyl (U^{6+}) state. Whereas U^{4+} compounds are highly insoluble, U^{6+} combines with oxygen to form the uranyl ion $(UO_2)^{2+}$, which, in turn, can form soluble complex compounds with species such as carbonate, sulfate, and fluoride. Near-surface groundwaters are commonly oxidizing in nature and hence provide a ready means of leaching and carrying uranium in the U^{6+} state. Although much of this uranium may then be dispersed, groundwaters carrying the metal may pass into rocks in which reducing substances, commonly decaying organic matter, may reduce the soluble U^{6+} ion back to the insoluble U^{4+} form, thus resulting in its precipitation. Other reactions involve absorption of the uranium by another mineral such as apatite ($Ca_5[PO_4]_3[OH,F]$), in which U^{4+} replaces some Ca). The leaching, transportation, and precipitation of uranium in this way can lead to the formation of large bodies of rock enriched in the metal; deposits produced by these second-stage processes are the most important sources of uranium.

The most important types of uranium ore deposits are listed, with examples, in Table 6.2 and located on Figure 6.13. The igneous deposits include various types of ores disseminated in alkali rocks and granites, although these are not very substantial contributors to world uranium resources. Deposits of the metamorphic group include the concentrations (known as skarns) that have formed at the contact between molten igneous rocks and the rocks into which they have been emplaced. Also in this general category are the ores originally formed deeper in the crust when heating caused some melting and the migration of material.

Some of the most famous and important of all uranium deposits are detrital in origin—for example, those of the Witwatersrand in South Africa and Elliot Lake (also called Blind River) in Canada. Characteristically, ores of this type occur in very ancient (Precambrian) rocks made largely of quartz pebbles and regarded as representing former stream channels. The Witwatersrand area is better known as the world's greatest gold-producing region (see Chapter 8), but in recent years about 10 percent of the known world production of uranium has also come from this area. Although the processes by which these ores formed remain controversial, it is widely believed that the gold and uranium were carried along the bottoms of stream channels as sediment grains. These grains were eroded from their original sites of formation and washed into the sea; they were then concentrated by virtue of their high density (in much the same way as gold is concentrated in the prospector's pan). The persistence of the uranium minerals is taken as evidence of an early Earth atmosphere with little or no oxygen; if oxygen had been present, the uranium would have oxidized and been dissolved. The deposits at Elliot Lake in Ontario, Canada, exhibit all of the features shown by the Witwatersrand region, but a marked difference is that, although a major uranium producer, these ores do not contain significant quantities of gold.

The deposits categorized in Table 6.2 as hydrogenic include all those in which the uranium appears to have been deposited from water, either as a high-temperature fluid or a much lower temperature one such as groundwater. The boundaries between the different subcategories shown here are not always clear, but the divisions at least indicate the different

TABLE 6.2 Important types of uranium ore deposits

Deposit Type		Characteristic Elements	Examples
Igneous	In pegmatites, alkali igneous rocks, carbonatites, and related rocks (pegmatites)	U, Nb, Th, Cu, P, Ti, Zr, rare earths	Prairie Lake, Ontario, Canada; Pocos du Coldas, Brazil; Ilimaussaq, S. Greenland; Rossing, S.W. Africa
Metamorphic	In contact areas between igneous and host rocks (skarns) or from the partial melting of rocks deep in the earth	U, TH, Mo, rare earths, Nb, Ti	Rossing, S.W. Africa
Detrital	Deposited in the bottoms of ancient rivers and lakes (fossil placers)	U, Th, Ti, rare earths, Au, Zr, Co	Witwatersrand, S. Africa; Elliot Lake, Ontario, Canada
Unconformity	Occur close to a conspicuous Mid-Proterozoic unconformity	U ($\pm$ Ni, rare earths, Ti, etc.)	N. Saskatchewan, Canada (Rabbit Lake, Key Lake, etc.) Northern Territory, Australia (Jabiluka, Nabariek, etc.)
Hydrogenic (deposited from fluids and waters)	From high-temperature water or fluids (hydrothermal) forming disseminations or veins (vein type)	U, Th, rare earths, ($\pm$ Cu, F, Be, Nb, Zr)	Bokan Mt., Alaska; Rexspor, B.C., Canada
	U deposits formed at the same time as host shales, limestones, phosphate rocks, etc.	U, P, V, Cu, Co, Ni, As, Ag, C	Ronstad, Sweden; Kitts, Labrador, Canada
	U deposited from low-temperature waters introduced into sandstones, conglomerates, and forming disseminations or, sometimes, veins (sandstone type)	U, C ($\pm$ Cu, V, Mo, Ag, Ni, As, Co, Au, Se, Bi)	Colorado Plateau area and Wyoming; Cypress Hills, Saskatchewan, Canada; Beaverlodge, Saskatchewan, Canada; Port Radium, N.W.T., Canada
	As an encircled cap at the surface of other deposits	U, Cu, Ag, Ni, As	Eldorado, Saskatchewan, Canada; Rossing, S.W. Africa

processes at work. The veins formed by deposition from high-temperature fluids always show a fairly close spatial link with granites or similar rocks from which the fluids could have been derived. Such vein deposits are no longer a major world source of the metal. The richest uranium deposits in the United States are of the type found in Jurassic and Triassic sandstones in the Colorado Plateau area of western Colorado, eastern Utah, northeastern Arizona, and northwestern New Mexico. Similar deposits in younger rocks occur in Wyoming and are forming today in Texas (Figure 6.14). These deposits have formed through the precipitation of uranium from groundwaters carrying the metal in solution as uranyl complexes. The precipitation occurs as a result of reduction when the solutions encounter organic matter, as evidenced by the replacement of fossil logs by uranium minerals, or when the solution reacts with trapped H_2S. They owe their characteristic shape, as shown in the cross section in Figure 6.15, to the movement of fluid through the porous sandstones and the dissolution and reprecipitation of uranium and other minerals along a moving front. Uranium is also concentrated in organic-rich black shales (such as the Chattanooga Shale of Alabama and Kentucky), but generally at much lower concentration levels.

The Search for Uranium Deposits

From 1945 into the 1960s, uranium was the subject of the most intense mineral exploration activity ever undertaken for any metal. This activity involved both government agencies and private companies. The search employed all of the standard exploration techniques, but also made use of methods that detect radiation emitted during the slow, natural, radioactive decay of uranium.

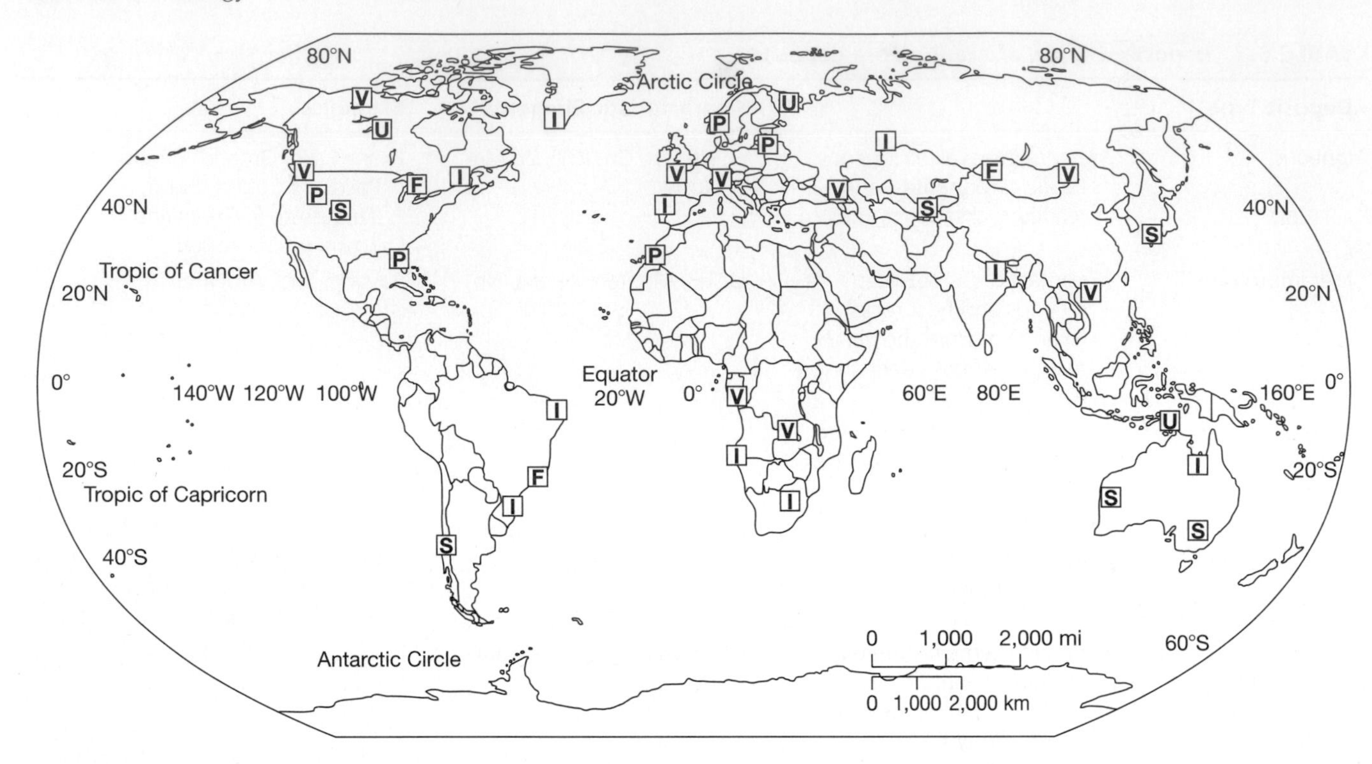

FIGURE 6.13 Major uranium deposits occur worldwide, as shown on this map. The deposit types are: V-veins; U-unconformity; S-sandstone; F-fossil placers; P-phosphate and black shale; I-igneous.

The two principal methods were simple Gross Count Surveys and Gamma Ray Spectrometry Surveys. The first method makes use of a simple Geiger-Muller detector, which detects all types of high-energy radiation, whereas the second method is able to separately detect the specific radiation emitted by each different radioactive element. Detectors range from handheld, to vehicle mounted, to airborne depending upon the desired sensitivity, size of area to be tested, and the allocated budget. Another method employed does not test for uranium directly, but for radon (^{222}Rn), a decay product of uranium. The radon (also discussed as a potential household contaminant in Chapter 4) is a gas that can migrate through rock

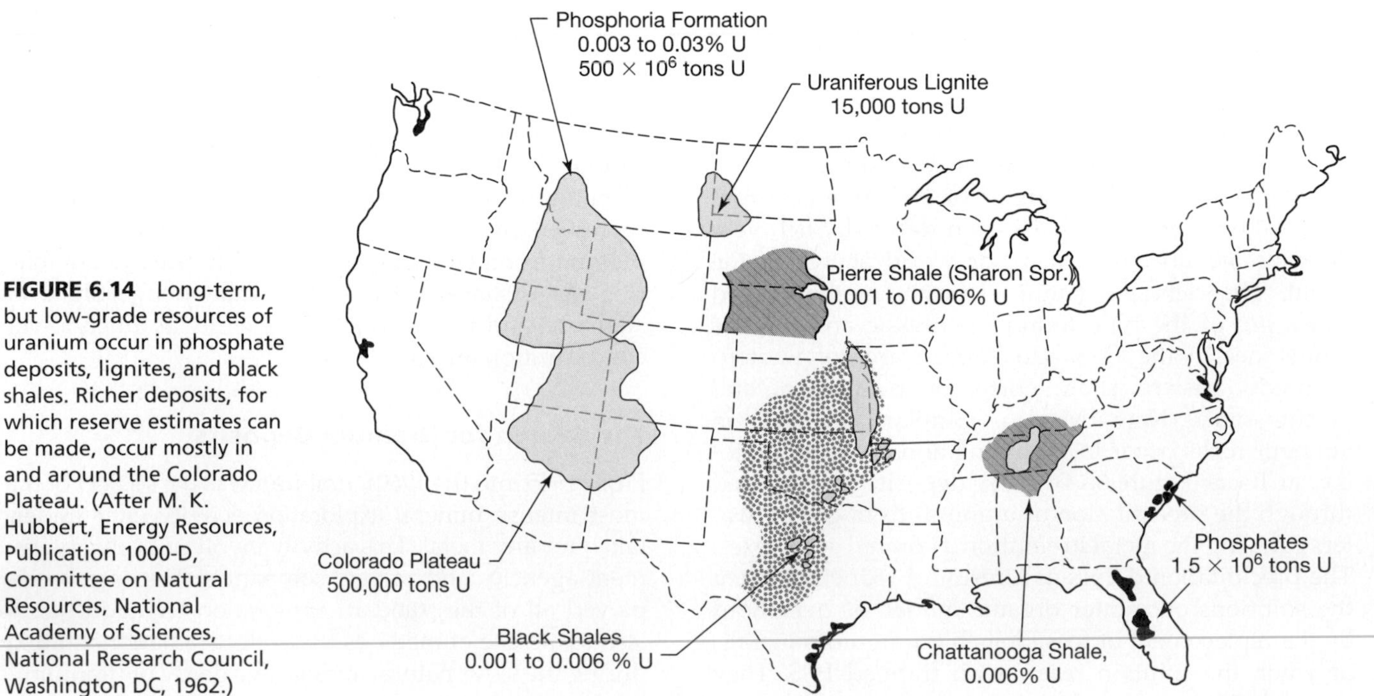

FIGURE 6.14 Long-term, but low-grade resources of uranium occur in phosphate deposits, lignites, and black shales. Richer deposits, for which reserve estimates can be made, occur mostly in and around the Colorado Plateau. (After M. K. Hubbert, Energy Resources, Publication 1000-D, Committee on Natural Resources, National Academy of Sciences, National Research Council, Washington DC, 1962.)

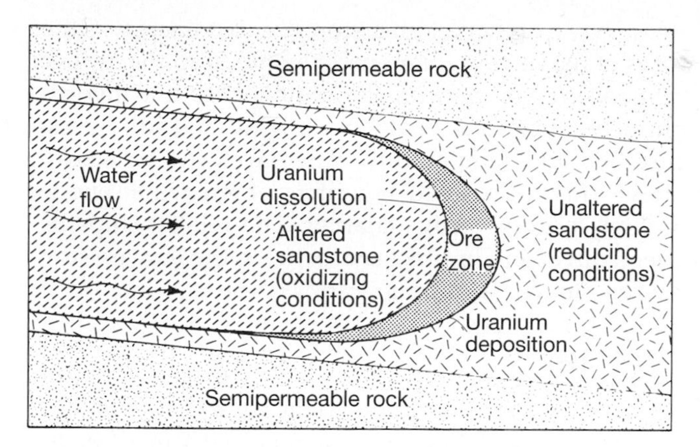

FIGURE 6.15 Many of the uranium deposits in the western United States are of the "roll-type" as shown in cross section above. Oxidizing groundwaters dissolve the low concentrations of uranium and reprecipitate it farther down an aquifer where reducing conditions are encountered. This results in a progressive accumulation of uranium and the movement of the arc-shaped deposit along the aquifer.

fractures and soil zones or become dissolved in water. Accordingly, radon can be dispersed around a uranium deposit, and its detection can be traced back to where the original uranium-bearing deposit is located.

Uranium Reserves and Resources and the Future of the Fission Reactor

Reserves of rich deposits of uranium are widespread but are not large (see Table 6.3), although the figures are probably conservative because an element of secrecy surrounds the subject of uranium. Substantial deposits that are economically recoverable now or in the near future are known on all the continents. Assessment of the potential of lower-grade deposits is difficult because the necessary data are not available. Within the United States, the kinds of source materials available have been broadly evaluated. It appears that the distribution of uranium in Earth's crust follows a log-normal abundance curve, with a 300-fold increase in recoverable uranium for each 10-fold decrease in ore grade. Such a trend would guarantee an ever-increasing uranium supply, as rising uranium prices would justify the mining of lower grades of ores.

Other views are less optimistic and suggest that uranium reserves are not large enough to support future extensive use of power stations burning ^{235}U. If all of the reserves and resources listed in Table 6.3 were used solely for their ^{235}U content and the conversion of heat energy to electricity was 40 percent efficient, the total energy produced would be only 8×10^{20} joules. Conversely, an equally efficient fast-breeder reactor would be capable of extracting 1140×10^{20} joules because it can utilize the much more abundant ^{238}U. Stockpiled ^{238}U and known reserves could supply energy through a fast-breeder reactor system for many hundreds of years.

The world's first commercial nuclear reactor at Calder Hall in Cumbria, England, started to supply electricity in 1956. About 55 years later, in 2011, nearly 450 reactors in more than 40 countries will be providing over 300,000 megawatts of the world's electricity. The global distribution of reactors shows the anticipated concentration of nuclear plants in the United

TABLE 6.3 Estimated uranium resources in ores rich enough for mining

Estimated uranium resources in ores rich enough to be mined for use in uranium-235 power plants, together with estimated rates of production for 2007. Data are reported as the oxide U_3O_8. No distinctions are drawn between reserves and resources, and no data are reported by the Communist countries (or the former USSR).

Country	Reasonably Assured Resources (m.t. of U_3O_8)	Estimated Production Rate, 1990* (m.t. of U_3O_8 per year)
Australia	1,240,00	8610
Kazakhstan	817,000	6640
Russia	550,000	3410
South Africa	435,000	540
Canada	3150	274
United States	342,000	1650
Brazil	278,000	300
All others	1,385,000	10,360

Production data from World Nuclear 2007.
Resource data from *Energy Information Administration*, 2005.

States, Western Europe, and Japan. The distribution of nuclear-generating capacity in terms of reactor type illustrates both this point and the dominance of the light-water reactors that are so extensively employed in the United States. However, the nuclear generating capacity is often only a relatively small proportion of the total demand for electricity of that estimated for the next few years. Also, in a number of countries (Sweden and Germany, for example), decisions have been taken to reduce or phase out nuclear power over the next few decades.

The world demand for nuclear electricity has not grown as much as anticipated in the 1970s. Safety fears after Three Mile Island and Chernobyl affected some customers; greater than expected costs affected other customers. However, since about 2000, there has been a resurgence of interest in nuclear power because of the rising costs of competing energy sources and because nuclear power does not contribute to the atmosphere's carbon dioxide content. Hence, it appears that the world may soon see significant expansion in nuclear-generated electricity.

RENEWABLE ENERGY SOURCES

Introduction

The Sun provides energy that is vital; it has been the major source of energy for most forms of life throughout geologic history (recall Figure 6.1). The Sun's direct role has been mainly through biological growth and the formation of biomass and fossil fuels such as wood, coal, and oil. These fuels (see Chapter 5) provide us with ways of harvesting stored solar energy. Other ways of harvesting solar energy come from the indirect influence of the Sun on the atmosphere and hydrosphere because solar heat causes wind and rain, ocean currents, and temperature differences in the oceans. These will be discussed as sources of renewable energy in later sections of this chapter. The relationships between direct and indirect energy from the Sun are summarized in the diagram in **Part 3.**

Solar Energy

The term solar energy refers to the direct conversion of the Sun's rays to energy in forms that can meet the needs of humankind. This energy can be best considered in two categories: **low-quality energy**—in which ordinary diffuse sunlight is used to produce low temperature forms of energy; and **high-quality energy**—which involves some form of solar concentrator or a physical or chemical process that produces electricity, or a chemical fuel such as hydrogen.

The Sun has a surface temperature of about 5500°C. From its distance of 1.5×10^8 kilometers, (93 million miles) a total of approximately 4×10^{24} joules per year of energy reach the surface of Earth. This energy is mainly infrared and visible light radiation with lesser amounts of ultraviolet radiation (Figure 6.16). The amount of each wavelength of radiation reaching Earth's surface depends on the distance that the Sun's rays have to travel through the atmosphere. This is so because direct radiation from the Sun is partly scattered and partly absorbed by molecules of various gases in the atmosphere, by water vapor, and

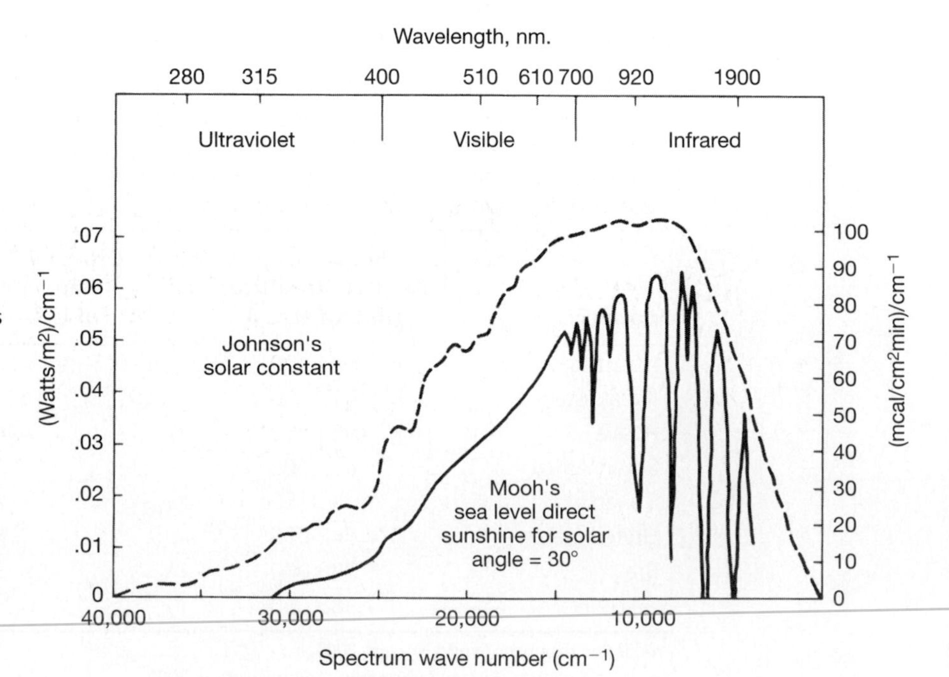

FIGURE 6.16 Energy from the Sun (shown both in watts and in microcalories on the vertical scales) as a function of the wavelength of the Sun's radiation expressed in both reciprocal centimeters (bottom scale) and in nanometers (nm, top scale). The dashed line shows the total flux of solar energy incident outside the atmosphere of Earth (Johnson's solar constant) and amounts of 178×10^{12} kilowatts continuous for the entire planet (1.5×10^{18} kilowatt-hours per year). The solid line represents the solar flux at sea level in direct sunshine for a solar attitude of 30°. The energy is depleted on passing through the atmosphere owing to absorption by water vapor, carbon dioxide, oxygen, nitrogen, ozone, and dust particles (in some cases at very specific wavelengths). The average solar energy is reduced in this way to 2.16×10^{17} kilowatt-hours per year. (Data from Task Force Report: Solar Energy, U.S. Federal Energy Administration, 1974.)

by dust suspended in the air. Absorption of infrared wavelengths (>700 nanometers) is due largely to water vapor and, to a lesser extent, to carbon dioxide. Absorption of smaller ultraviolet wavelengths (< 300 nanometers) is principally due to ozone (O_3). As noted in Chapter 4, depletion of the ozone in the atmosphere would result in more ultraviolet radiation passing through the atmosphere and in increases in the incidence of skin cancer. At around noon on a clear day in the mid-latitudes, direct radiation from the Sun is reduced about 30 percent by these processes. Very cloudy conditions may reduce direct radiation to less than 1 percent of the value above the atmosphere, but even under cloudy conditions, there is appreciable diffuse radiation derived from scattered direct radiation.

Expressed in units of power, solar energy arrives at Earth's surface at an average rate of 180 watts per square meter (m^2), but both the range and the distribution of the incident solar power in different parts of the globe vary greatly, primarily as a function of latitude (Figures 6.17 and 6.18). This energy constitutes a very large potential energy-resource. If only 20 percent of this average incident power were collected, a land area of 10,000 mi^2 (100 miles by 100 miles or about one-fourth the size of California) would be sufficient to supply the entire energy requirements of the United States.

LOW-QUALITY SOLAR ENERGY. Simple systems are used for the direct collection of solar energy in order to produce thermal energy (i.e., below 100°C). In the United States, the thermal energy available for an average day varies from about 5400 kilocalories/m^2 in the southwestern states such as New Mexico to about 2700 kilocalories/m^2 in the Northeast and Great Lakes states. This low-quality thermal energy is well suited to many applications, particularly the heating of water and interior space.

A south-facing window is the simplest possible type of solar collector. The sunlight that passes through the glass is absorbed by objects in the room and by the walls, from which it radiates to warm the air of the room. Provided that good insulation reduces heat losses as much as possible, windows exposed to direct sunlight can maintain comfortable temperatures in a room, even on a cold winter day. These simple principles are increasingly being used in the design of buildings—first, by arranging windows so as to capture sunlight, and second, by providing thermal storage facilities. Thermal storage can be accomplished by using massive objects—such as rocks, concrete, or containers filled with water—that are warmed by sunlight during the day and that then slowly release their heat after sunset. A combination of solar heating and thermal storage can substantially reduce the heating fuel needs of a building, but will rarely provide the sole means of heating.

All heating systems involve the transfer of heat energy, which can involve one or more of three mechanisms. Heat can be transferred (1) by **radiation,** in which waves (these can be visible light or infrared radiation) from a hot object such as the Sun are absorbed by matter and converted into heat; (2) by **convection,** in which heat is carried by the motion of heated masses of matter (e.g., hot water circulating through pipes in a building); and (3) by **conduction,** in which heat is transferred by contact between particles of matter (e.g., from hot water to a metal pipe through which it passes). More elaborate systems for the heating of buildings and water involve purpose-built solar collectors, a heat-transfer fluid such as air or water, and a heat-storage system such as a large mass of rock of water. The collector is normally a large panel with a collecting surface that has been blackened to absorb radiation. The air above the collector may be used as

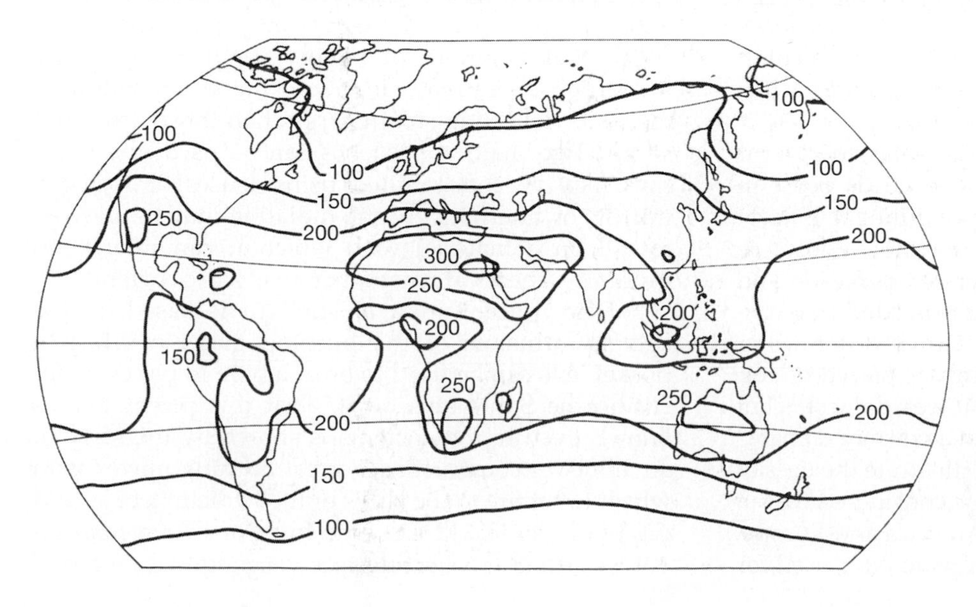

FIGURE 6.17 This world map shows the variation in annual mean solar energy flux (on a horizontal plane). Contours are in watts per square meter. The maximum amount of solar energy available occurs near the equator where the Sun's light is, on average, most vertical on Earth's surface.

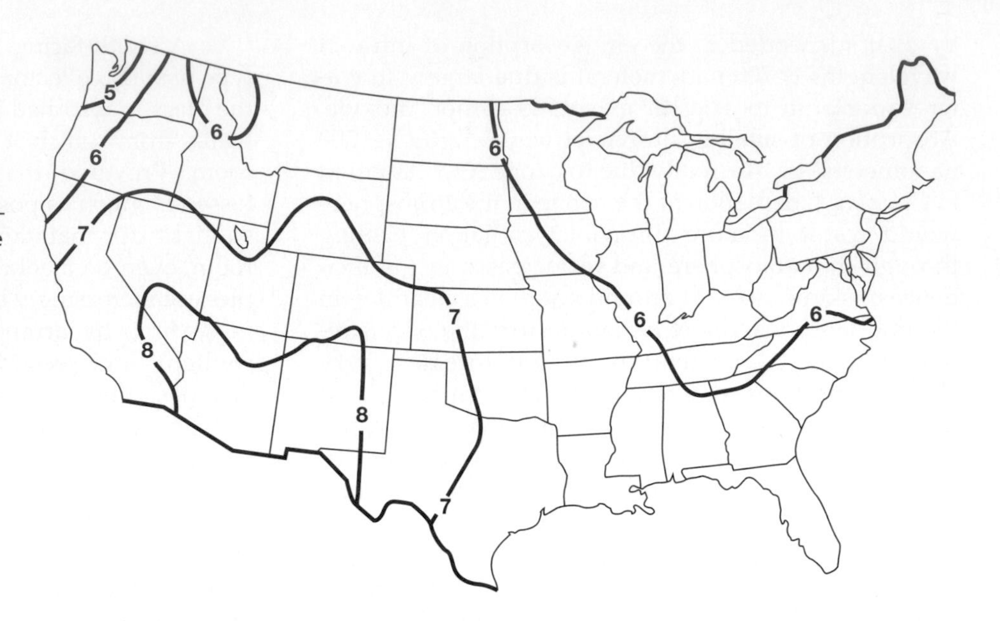

FIGURE 6.18 Solar energy, in kilowatt-hours per square meter available in the United States. In general, the amount of energy available is greatest in the southern parts of the country because the solar radiation is most nearly vertical there. The desert Southwest has much greater solar energy potential than the Southeast because there is very little cloud cover; in contrast, the Southeast has much greater cloud cover, which reflects much energy back into space. (From the U.S. Energy Information Administration.)

the heat-transfer fluid, or water may be circulated through tubes that form part of the collecting surface. An example of such a system is shown in Figure 6.19. Here, water that is heated in the solar panels is circulated through a heat exchanger to transfer this energy (by conduction) to a storage water tank that directly provides the supply to space-heating radiators or to the supply of running water. Such a system is nearly always coupled with conventional heaters and is controlled by a system of thermostats, valves, and timers to permit the most efficient combination of solar and conventional energy. Solar heating systems such as these are not technically complex; although operating costs are negligible, these systems tend to be costly to install. Also, the savings in conventional fuel costs may take many years to repay the investment in a system. Taking an average for the rate at which solar energy arrives at Earth's surface and assuming that the collector is about 50 percent efficient, its daily energy output per square meter will be roughly equivalent to burning one-tenth of a gallon of heating oil in a furnace that is 70 percent efficient.

One obvious problem in using solar energy for heating buildings is that the greatest needs occur in high latitudes when and where the sunlight is least available. This disadvantage does not apply so strongly to solar heating of water for domestic and commercial use because hot water is needed throughout the year regardless of climate. A solar water heater of the type shown in Figure 6.19 can be placed on top of flat-roofed buildings and can meet domestic hot water needs throughout the year in a warmer climate.

These few examples serve to illustrate the uses of low-quality solar energy. When it is considered that in countries such as those of North America and Europe, about one-third of all the energy consumed is used for space-heating and water-heating, the potential of low-quality solar energy becomes clear. Also, the limitations of climate are not as great as might be imagined. Even in countries such as Britain and Germany, which lie at relatively high latitudes and have cool and cloudy climates, it is possible to provide 50 percent of domestic heat by solar energy. Such systems are not more widely used because the conventional sources of energy are still readily and fairly cheaply available, and the initial cost of changing to any form of solar energy system is high. There is insufficient incentive for the individual homeowner to invest in such a system, even though it is both are very inexpensive to run and pollution free; but increases in the cost of conventional fuels in recent years have increased the use of solar heating.

HIGH-QUALITY SOLAR ENERGY. The generation of temperatures above 100°C, or the production of the substantial amounts of energy required by many industrial operations, involves more sophisticated means of collecting solar energy than those described so far. Two main approaches can be used: the Sun's rays may be concentrated using lenses or focusing mirrors or, alternatively, the radiation may be allowed to fall on a material with which it can interact to produce a chemical reaction or an electric current.

The concentration of sunlight is based on the age-old principle of the burning glass in which a pocket lens can be used to burn a hole in paper by focusing the Sun's rays on it. That this principle was known even to the ancients is shown by the story of Archimedes constructing a great burning mirror with which to set fire to the ships of the Persian fleet attacking Syracuse in 212 B.C. One form of this system involves a central collecting receiver mounted on top of

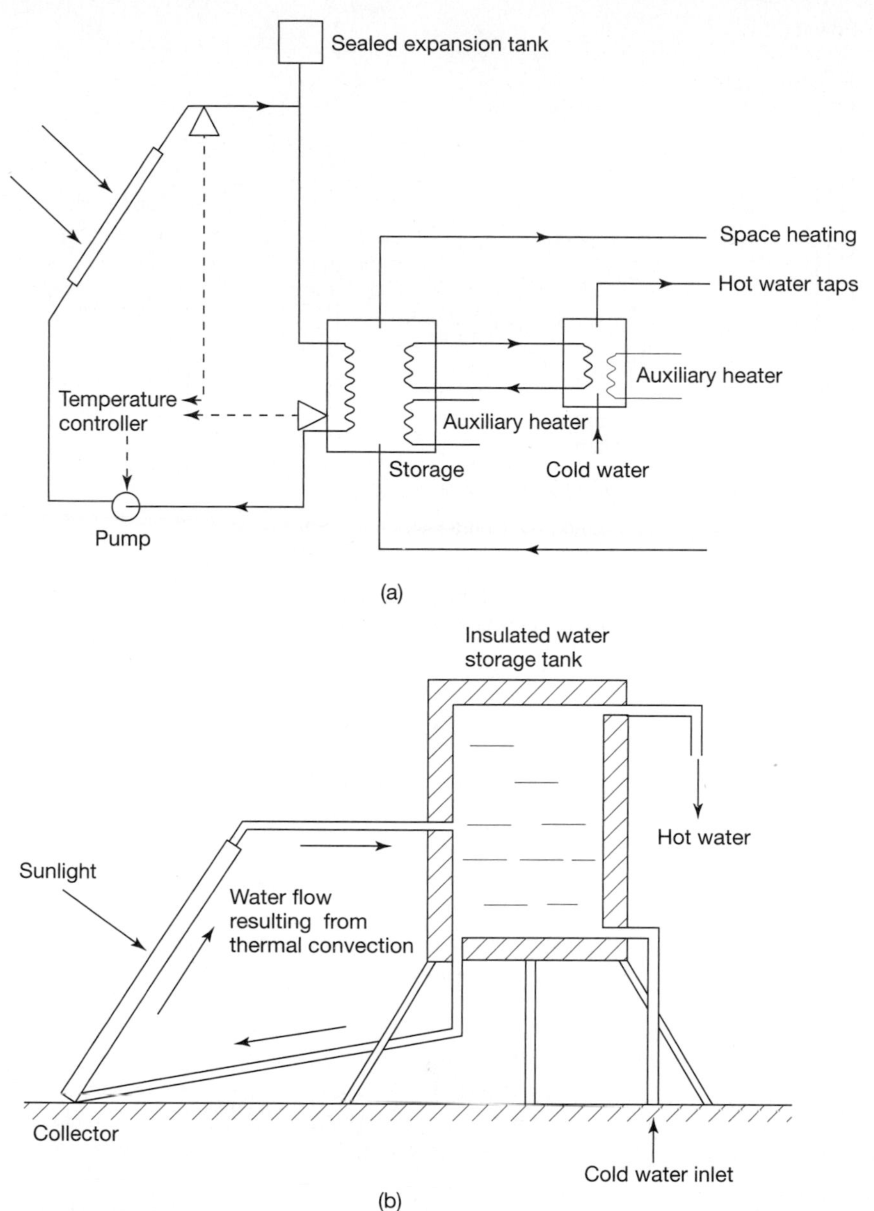

FIGURE 6.19 Domestic solar heating systems providing (a) both space heating and a supply of hot water, and (b) hot water only.

a high tower. As shown in Figure 6.20, the Sun's rays are reflected up to the receiver by a group of mirrors, called **heliostats,** which are programmed automatically to track the Sun and keep its rays focused on the receiver. Temperatures of approximately 1000°C can be generated at the receiver. This receiver can simply be a boiler that generates steam to drive a turbine, or a liquid metal such as sodium that is used to transfer the heat to a thermal storage system that, in turn, produces the steam for driving a turbine and generating electricity. Such a thermal storage facility, which might be tanks containing salts or hydrocarbon fluids, could maintain steam to drive the turbine during brief periods of cloud cover or it could extend the operating day for the plant.

A prototype facility with 2000 mirrors, each about 20 feet square, was constructed at Barstow, California (Figure 6.21) and can deliver 10 megawatts of electricity during daylight hours. A similar design plant, Solar Tres, with a 17 megawatt capacity, has been constructed in southern Spain; indeed in 2009, Spain was leading the world in the development of this technology with over 50 schemes approved and plans to generate more than 2 gigawatts of power from the Sun. Plants of this type convert the intercepted solar energy to electricity at about 20 percent efficiency. Full-scale plants would be ten or more times the size of the prototypes and would be located in desert regions where land costs are low and available solar energy is higher than for other regions. The costs

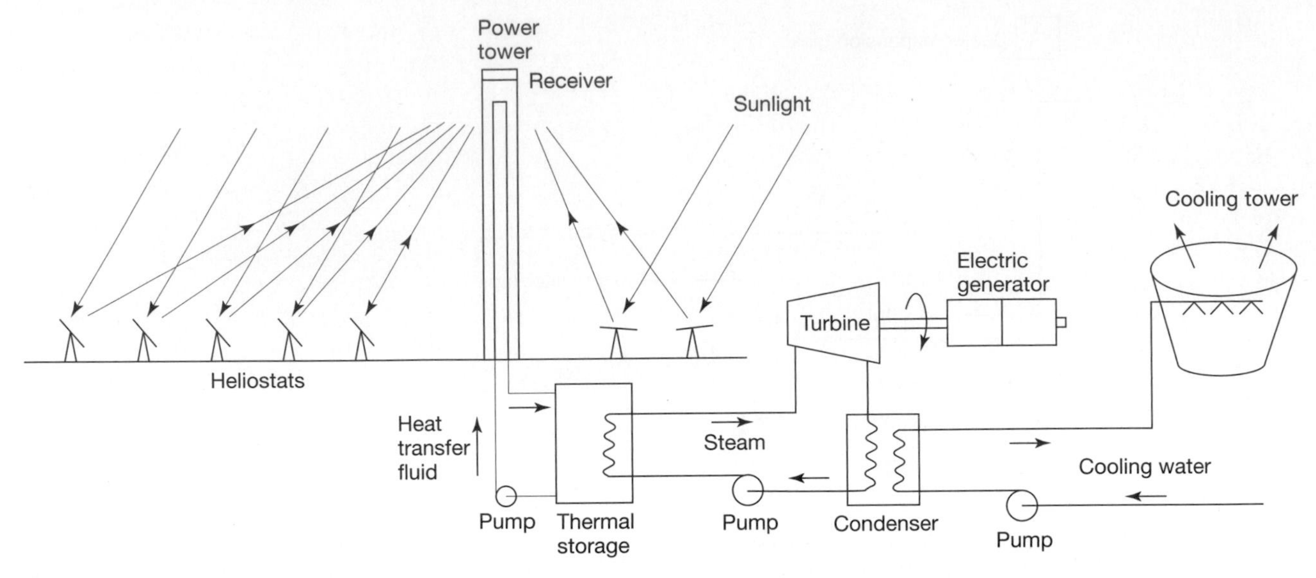

FIGURE 6.20 The basic design of a central tower solar electric power plant in which heliostats directs the Sun's rays onto a central receiver. The heat generated is transferred via a fluid to some form of thermal storage facility and used to raise steam and drive turbines.

of constructing such plants are much higher than for conventional fuel-burning power stations because large costs are involved in the making of mirrors and their control systems. However, as fossil fuel prices increase and the design techniques for large solar power stations improve, this type of solar power becomes commercially more attractive.

Another type of solar energy concentrator generally suitable for smaller power plants is the **parabolic reflector.** This employs a cylindrical reflector (Figure 6.22) to focus the sunlight onto a small-diameter collecting element, through which a heat-transporting medium such as a hydrocarbon or molten metal is passed. The transporting medium may circulate to a heat storage facility from which heat exchangers extract the energy to make steam and to drive turbines. Less complex movements are involved in tracking the Sun, and the whole system can be constructed on a smaller scale suitable for a community-sized power generation system. Parabolic reflectors and other types of mirror systems have also been used to construct solar furnaces such as the one in the French Pyrenees at Odeillo, where temperatures of

FIGURE 6.21 Solar furnace on the desert floor near Barstow, California, is an array of nearly 2000 mirrors, each with a surface area of 40 m^2 (430 ft^2). Computers aim the Sun-tracking mirrors to reflect sunlight on the central receiver 100 meters (300 feet) above ground. The receiver absorbs solar heat that converts water to steam that drives turbine to make electricity. (Courtesy of Southern California Edison Company.)

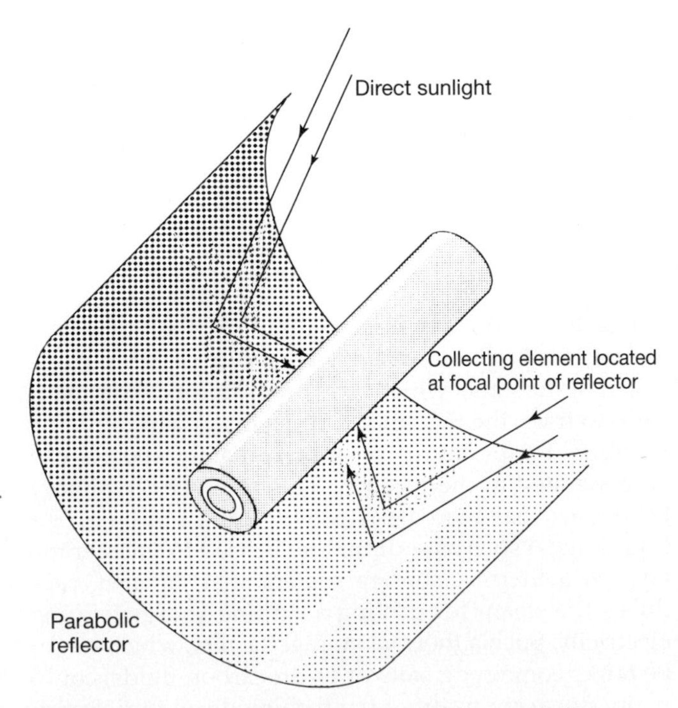

FIGURE 6.22 A focusing collector in which a parabolic reflector directs the Sun's rays onto a heat-transfer fluid located within an evacuated tube to minimize heat losses.

up to 400°C can be reached in a 50 cm^2 (8 square inch) hot spot.

Solar energy can also be converted directly into electrical energy by **photovoltaic cells.** In photovoltaic devices, light energy interacts directly with the electrons of a semiconductor to produce an electric current. These photovoltaic or solar cells are manufactured by processes similar to those employed in making transistors. A variety of materials can be used, but the most common one is silicon (see Figure 6.23 and Figure 7.27). The manufacture of solar cells is a complex and costly business because the compositions of the materials must be carefully controlled and very thin wafers of appreciable surface area (at least several square centimeters) must be fabricated.

The economics of large-scale electricity generation using photovoltaic cells has not been very attractive in the past, but increasing efficiencies and new materials, combined with governmental tax incentives, are changing the picture. Increasing numbers of buildings and homes are building solar panel arrays into their roofs; some new roofing shingles actually incorporate solar electric capacity in their design. The conversion efficiencies of the cells are fairly low: 12–15 percent for single crystal silicon cells and 4–6 percent for the cheaper but less reliable, cadmium sulfide/copper sulfide cells. Solar cells have the advantage of using the entire solar radiation, both direct and diffuse, and of converting solar energy directly to electricity. The direct current (DC) electricity produced may be used as is or it may be converted to alternating current (AC). It may be attached to some sort of storage systems so that power can be supplied even when there is insufficient solar radiation to produce electricity. A typical commercial silicon cell with a diameter of 7 centimeters (about 3 inches) would have an output of about 0.4 watts when operating in direct sunlight on a clear day. Obviously, very large numbers of these cells have to be mounted together to generate substantial power output, and the cost of manufacture and installation is very great. However, the cost of manufacturing photovoltaics has been dropping and their lifetimes have been lengthening. Recent calculations suggest that modern photovoltaics could generate the total energy needed to make them in three to four years and avoid carbon dioxide emissions at the same time. Over a useful lifetime of about thirty years, cells can produce roughly ten times the energy required to make them and can avoid large amounts of greenhouse gas emissions. Assuming a 10 percent overall conversion efficiency and a fixed flat-plate arrangement, it would be possible to produce all of the electrical energy used in the United States with a photovoltaic-covered square 100 miles on each side. The total area of the photovoltaics, 10,000 square miles (100 miles by 100 miles or 26,000 km^2) would be less than one-quarter of the area of paved roads in the United States.

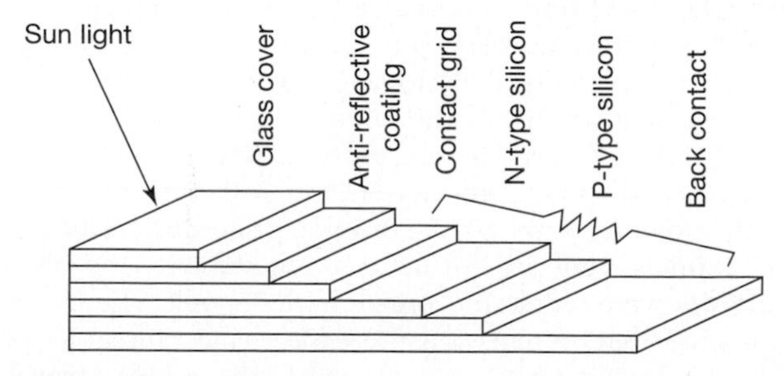

FIGURE 6.23 Cross section of a photovoltaic cell. The cells are composed of multiple layers of n- and p-type semiconductors that absorb solar photons and generate an electrical current.

The falling costs of photovoltaics and the growing interest in minimizing carbon dioxide emissions have resulted in significant growth in the use of photovoltaics. It is not surprising that these cells have mainly been employed as power sources in remote locations and in specialist applications such as solar-powered calculators. Perhaps the most spectacular success of the solar cell in energy generation has been in the powering of both manned and unmanned space vehicles. More widespread uses in routine power generation will only come as new manufacturing methods and designs reduce the cost of delivering power by comparison with more conventional methods. Nevertheless, major advances in efficiency and reliability of photovoltaic cells were made in the last decade of the twentieth century, and costs have been falling dramatically along with greatly increased production.

The first major photovoltaic project in the Third World was installed in Saudi Arabia in 1981 and generates 350×10^3 watts to meet the electricity needs of 3600 people in three villages. Indeed, it is in providing electricity for small isolated communities that this technology may make its greatest contribution. At the other end of the scale, there are plans for a 100 million watt power plant near Sacramento, California. Although only a very small fraction of the world's electricity demands is now met by photovoltaics, this is likely to change in the coming decades. In 2009, for example, the Chinese government announced plans for major investment in photovoltaics, with Chinese scientists noting that the country has many of the world's biggest deserts and that if 30–50 percent of those desert areas were covered with cells capturing only 10 percent of the incident solar energy, this would meet all of the current energy needs of the country. Such vast harvesting of solar energy is still a distant prospect; but it is interesting to note that in 2009, China was manufacturing more solar panels than any other country, although exporting 95 percent of their production. If vast scale collection of solar energy in desert areas seems futuristic, an even more futuristic proposal for massive power generation involves satellites in stationary orbit with very large arrays of photovoltaic cells that collect solar energy that is then transmitted to Earth by microwave beams (Figure 6.24).

An unfortunate footnote in the progression of solar energy occurred at the White House in Washington, DC in the 1980s. President Jimmy Carter recognized the need for the United States to advance in the realm of renewable energy, especially in light of the rising dependency on imported oil. As a gesture of that support, he had solar panels installed on the roof of the White House to supply some of the electricity. Unfortunately, his successor, Ronald Reagan, showed very little support for renewable energy and more support for large oil companies. Consequently, Reagan quietly had the solar panels removed and discarded. He also removed most of the federal support for research in solar energy.

Hydroelectric, Wind, Wave, Ocean, and Tidal Power

Nature transforms the solar energy reaching Earth into several other forms of energy. About 23 percent of incoming solar radiation is consumed by the evaporation of water that subsequently falls as rain and snow (Figure 6.1). In effect, the Sun acts as a

BOX 6.2
Chernobyl

The most tragic of nuclear accidents was that at Chernobyl in the Ukraine in April 1986; it resulted in loss of life and in significant contamination over a very large area. The accident centered around attempts to test a system for providing the necessary cooling water in the event of a reactor shutdown; these tests were to coincide with the actual closing of the reactor for its annual maintenance. In attempting to create the necessary conditions for this poorly planned test, automatic systems for operation of the control rods, emergency core-cooling systems, and various other fail-safe devices were overridden by the operators. When the test began to go wrong, actions taken by the operators were totally against regulations and led to a complete loss of control.

In the words of academician Legasov of the Russian Academy of Science, reporting to an international group of nuclear scientists in Vienna, the reactor was "free to do as it wished." Control rods were leaping up and down and water and steam were sloshing around uncontrollably. In less than a second, the power surged from 7 percent to several hundred times its normal level. The effect was like setting off half a ton of TNT in the core. Two explosions lifted the roof of the reactor, throwing red-hot lumps of graphite and pieces of uranium oxide fuel over the immediate area.

Over the next 10 days, while the Russians struggled to quench the fire, roughly 10 percent of the core material was dispersed into the atmosphere from which it fell out over Russia and Europe. Large areas of Russia, Poland, Sweden, and Finland (see Figure 6.9) were particularly affected, but significant increases in radiation levels were recorded as far away as Norway, Italy, and Britain, where fallout from Chernobyl led to restrictions on the sale of crops and livestock. In the immediate area of Chernobyl, one person was apparently killed within the reactor and about twenty were severely irradiated (of whom seventeen died within six weeks), but many thousands more face an increased risk of death from cancers associated with the radiation. The helicopter pilot who flew numerous missions dumping lead and sand on the burning reactor in the first few days allowed himself to be exposed to massive radiation, but his selflessness no doubt saved many others. He was praised for his heroism, but he died of leukemia resulting from the exposure.

The nightmare of Chernobyl has continued to unfold in the years since the event as more information has become available and as death counts have risen. In 1995, the government of the Ukraine blamed the accident for a nearly 16 percent rise in the death rate in the northern Ukraine and for a rapid rise in thyroid and organ cancers. More than 140,000 citizens had to abandon homes because of high levels of radioactivity and about 500,000 more continued to live in contaminated areas. The number of deaths resulting from the disaster is not known, but by 1995, death counts ranging from 8000 to 125,000 had been released. Twenty years after the incident, *Greenpeace* estimated that there had been 60,000 deaths since 1986 and there could be another 270,000 cancers with 93,000 fatalities.

The Soviet government continued to operate the two other nuclear reactors at the Chernobyl plant because of the need for electrical power. After the breakup of the Soviet Union, the Ukranian government also continued operation. A decision was finally made to shut down the other units by the end of 1993, but was reversed because there was no other adequate power supply for Kiev. In early 1995, the Ukranian government announced that it would close the two operational Chernobyl units. Subsequently, the government constructed two other nuclear plants and by 2008, the Ukraine was able to export electricity to neighboring countries.

Intensive studies of the wildlife in the contaminated area around the Chernobyl reactor ten years after the explosion yielded surprising results. Although the animals were slightly radioactive, they were thriving. Wild deer, bear, and other animals were present in large populations, and birds of all sorts were living and breeding quite well. The lesson learned was that humans constitute the greatest threat to the wildlife; when humans are removed from an area, even a radioactive one, wildlife prospers.

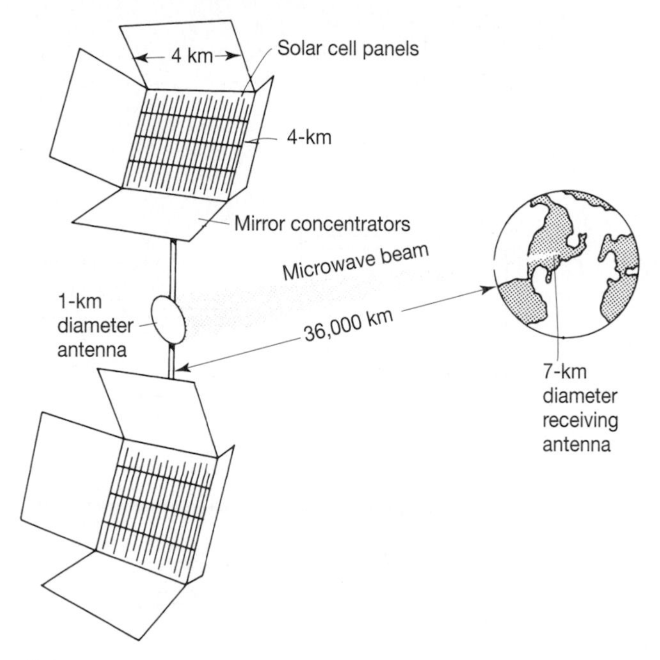

FIGURE 6.24 Schematic illustration of a satellite solar power station. Solar radiation is collected by large arrays of photovoltaic cells, and the electrical energy generated is transmitted to Earth by a microwave beam.

great pump drawing water from the sea and dropping it onto the land, from where it runs downward to the sea. Flowing water is therefore a renewable resource. A further 46 percent of incoming solar energy is absorbed by the oceans, the land, and the atmosphere. This energy warms the seas and produces ocean currents, winds, and waves. At least some of this can be considered a renewable, potential energy resource.

Water and wind power have both been used in small ways, such as waterwheels, boat sails, and windmills, for many thousands of years (Figure 6.25). Waterwheels were known to the ancient Greeks, but their capacity to generate power was very small. Toward the end of the eighteenth century, the largest waterwheels for industrial use did not exceed 10 horsepower. Nevertheless, they were a major source of power prior to development of the steam engine, which heralded the start of the Industrial Revolution in Europe. However, it was only at the beginning of the twentieth century that large-scale damming of rivers commenced for the generation of electricity.

HYDROELECTRIC POWER. **Hydroelectricity,** the electricity generated by the energy of flowing water, is usually produced at large dams. Dams are constructed to increase the height from which the water drops (or the "head of water") and to provide a constant flow of water through turbines that, in turn, drive electrical generators (Figure 6.26). Because falling water is a form of mechanical energy directly used to drive the turbines, the two-stage process involved is 80–90 percent efficient in converting that energy to electricity. In fuel powered generating stations, the heat produced by burning the fuel has first to be converted to mechanical energy by raising steam to drive turbines, which are then used to drive electrical generators; the efficiency of this three-stage process is much less (approximately 35–40 percent for fossil fuels and approximately 30 percent for nuclear fuels). Where the construction of a dam is not practical, water may be routed to turbines via canals or large pipes, as in the power plant at Niagara Falls, New York.

FIGURE 6.25 Mabry Mill, on the Blue Ridge Parkway in Virginia, is an example of a nineteenth-century waterwheel that provided power to turn a grinding mill and operate a large saw to cut trees into slab wood for construction. (Photograph by J. R. Craig.)

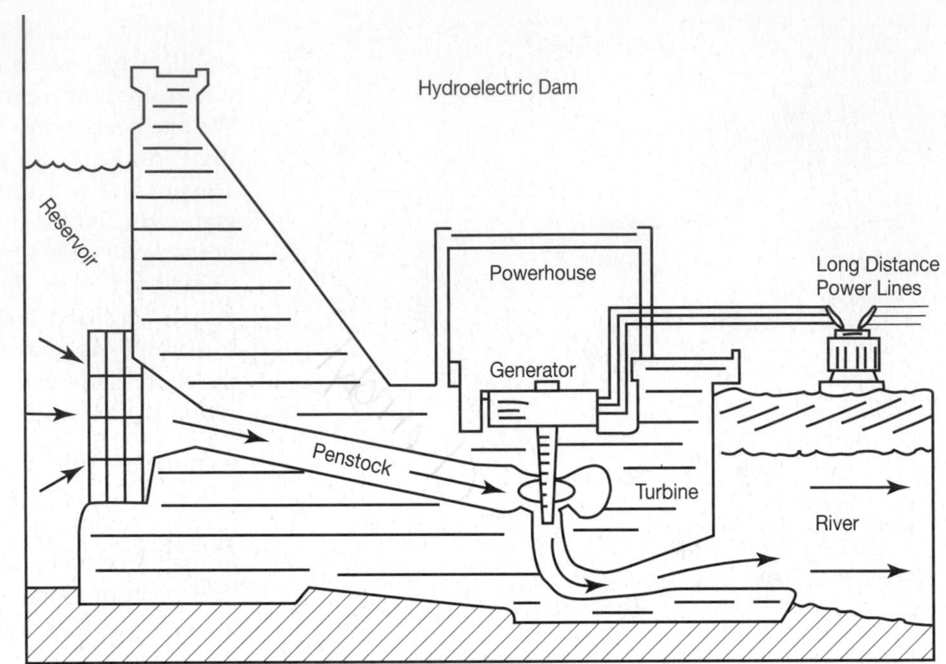

FIGURE 6.26 A schematic cross-section of a hydroelectric dam. The hydraulic head of the water behind the dam drives the water through the penstock with high velocity and spins the turbines that generate the electricity.

Hydroelectric power stations are generally very large installations, because vast quantities of water are needed to produce even modest amounts of energy. For example, for an elevation change of 50 meters (160 feet; approximately the height of Niagara Falls), 8 metric tons of water must flow through a turbine per second to produce 1 kilowatt of electricity.

If a reservoir is present as part of a hydroelectric system, the impounded water acts as a form of stored energy. Also, because hydroelectric systems can be started almost instantaneously, this energy can be made available as electricity at very short notice and can serve as a very effective backup system for plants using other sources of energy. An extension of this idea, now used in a number of countries, is the **pumped-water storage system** (Figure 6.27). When excess electricity is available within a linked network of power plants (often at night), water is pumped from a lower reservoir or storage area to a higher one and is then available to drive turbines and produce electricity when needed.

The installed hydroelectric-generating capacity around the world steadily increased through the twentieth century, and the quantities of hydroelectricity generated in the various continents and regions in 2005 are given in Table 6.4. The three principal countries generating hydroelectricity are China, Canada, and Brazil. The quantity and proportion of electrical energy provided from hydroelectric facilities varies widely from one country to another and from one continent to another, as shown in Figure 6.28. Thus, among larger countries, Norway produces more than 98 percent of its electricity by hydroelectric power; Brazil, more than 84 percent; and Canada, more than 59 percent. The United States, although generating nearly the same quantity of hydroelectricity as Canada, produces so much electricity from fossil fuel and nuclear sources that hydroelectricity is less than 7 percent of the U.S. total. Hydroelectricity in the United States is generated by passing three times more water than exists in all American rivers through power dams. This seeming impossibility results because much water passes through several dams each day and is counted each time; some water is also recycled through the same dam at pumped storage facilities. (See Box 6.3 for a discussion of hydroelectric power in the United States.)

Hydroelectricity has often been considered as environmentally friendly because there is no release of greenhouse gases, no acid rain emissions, and no radiation. Many countries see additional benefits from hydroelectricity because it does not depend upon imported fuels, and dams may be useful in mitigating floods and in storing water that can be used for agricultural, industrial, and household use. Damming of rivers is, however, not without environmental impact because the dams prevent fish migration, they flood bottom land, and they change parts of free-flowing rivers into lakes. Environmental pressures have led to the decision that no additional hydroelectric dams will be built in the United States and to the further decision to decommission and remove some of the installed dams in order to restore rivers to their original states so that fish, especially salmon, can migrate to their spawning grounds.

Many countries, however, view the beneficial aspects of dams as much more important than the negative aspects. Thus, China began construction of the

FIGURE 6.27 Hydroelectric water-pump storage facilities such as this one in Bath County, Virginia, generate electricity when water falling through large tunnels in the dam turns a turbine as shown in Figure 6.26. At times of low electricity demand, excess electricity generated in other fossil fuel or nuclear power plants is used to pump water back up into the reservoir so that it is available to generate power at peak demand times. (Courtesy of Virginia Power.)

world's largest hydroelectric project, the 18.2-gigawatt Three Gorges Dam, in 1995. When all parts are completed in about 2015, the $25 billion dam will be 2 kilometers (1.2 miles) long, 200 meters (600 feet) high, and will convert 600 kilometers of the Yangtze River into a lake. The dam will flood thousands of square kilometers (miles) of farmland and force the relocation of as many as 2 million people. Large hydroelectric dams are also planned for many countries in Southeast Asia, South America, and Africa. The Republic of the Congo has

TABLE 6.4 International hydroelectricity generation in billions of kilowatt-hours, 2005

Continent/region	Billion kilowatt-hours
North America	651
Central and South America	613
Europe	540
Eurasia	245
Middle East	21
Africa	89
Asia and Oceania	735
World	2,894

From Energy Information Administration, *Annual Energy Review,* 2005.

proposed construction of the Great Inge Dam on the Congo River. If this is built, it will be the largest hydroelectric dam in the world—but this project must first overcome many political and economic hurtles. Consequently, there will likely be a significant increase in hydroelectricity in many parts of the world while there is a decrease in the United States.

All dams, however well constructed, have finite and, sometimes, rather short lifetimes. All rivers carry suspended sediment, and the amount is a function of the velocity. When a river is dammed, the water stops flowing, so the sediment load is deposited. Depending on the sediment load, many reservoirs will be completely filled in periods ranging from 50 to 200 years. For example, the great Aswan High Dam on the Nile built in the 1960s will be at least half silted up by the year 2025. The world's largest undeveloped potential exists in South America and Africa; inasmuch as these continents have only small fossil fuel resources, it is fortunate that their water power is so plentiful. The long-term future for hydroelectric power in the Southern Hemisphere must be considered very promising.

WIND POWER. Wind has been tapped as an energy source for thousands of years. Today, this ancient energy source is receiving a great deal of attention as a potential major contribution to the world's growing energy needs. For many years, the largest windmills were those used in Holland [Figure 6.29(a)], which have become the picturesque symbol of that country. Smaller windmills were extensively used throughout the United States for pumping water before electricity became available; windmills are still used in many parts of the world [Figure 6.29(b)].

The building of the first electricity-producing windmill is attributed to Professor James Blythe who, in 1887, experimented with different designs,

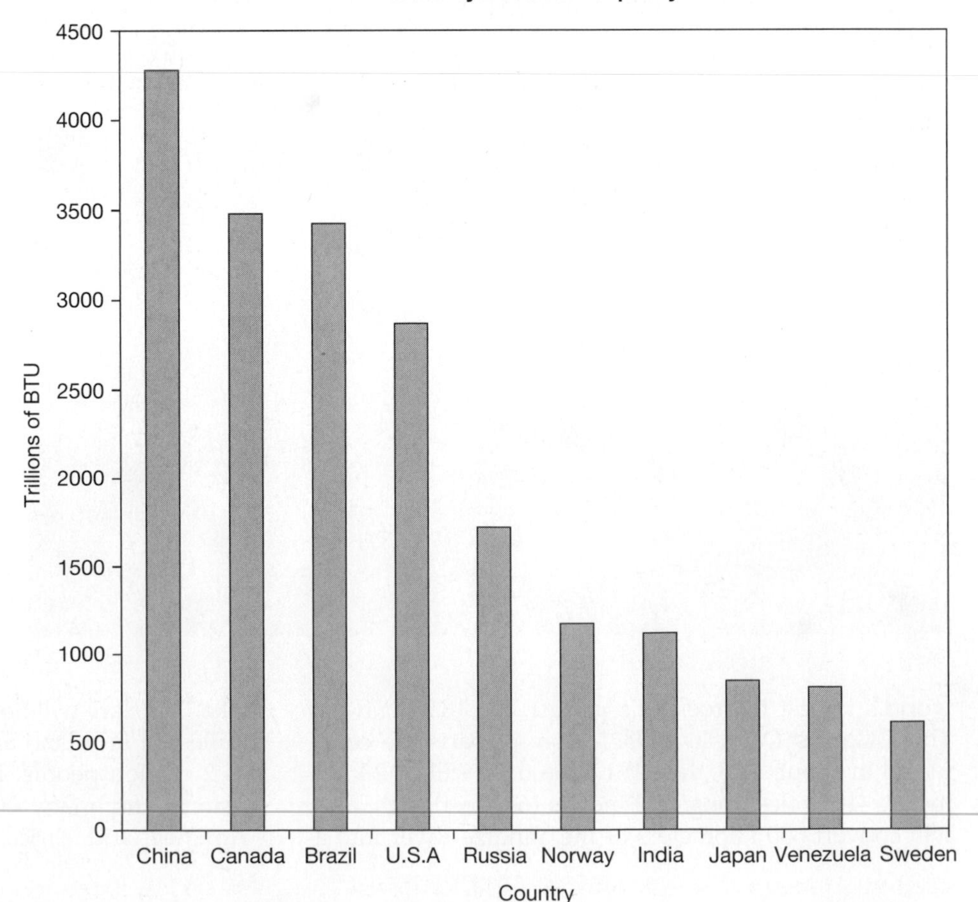

FIGURE 6.28 The principal countries that produce electricity by means of hydroelectric dams. (Based on data from the U.S. Energy Information Administration.)

(a)

(b)

(c)

FIGURE 6.29 (a) The traditional windmills of Holland have been used to harness wind energy for grinding grain and sawing wood since the middle of the thirteenth century. They also played an important role in shaping the Dutch landscape, because from 1414 to the present day they have been used to pump lakes dry and lower the water table to create the low-lying agricultural areas known as "polders." (Courtesy of Royal Netherlands Embassy.) (b) Simple windmills, such as this one in Patagonia in southern Argentina, are used worldwide to pump water for agricultural and domestic use. (Photograph by J. R. Craig.) (c) Modern "windmills" or wind turbines for use in generation of electricity. In the foreground, the so-called Darrieus wind turbine is a vertical axis machine, whereas the propeller-type machine in the background is a more conventional design. (Courtesy of Southern California Edison Company.)

one of which powered his Scottish home for twenty-five years. In 1895, a wind-electric system was built in Denmark, and by 1910, several hundred small wind-powered generators (5000 to 25,000 watts) were in operation there. It was not until 1931, however, when the Soviet Union built a 100,000-watt unit near Yalta that a really large wind turbine, expressly designed for producing electricity, was constructed. The United States built a two-bladed, 175-foot-diameter, propeller-like turbine at Grandpa's Knob, Vermont, that produced 1.25×10^6 watts in a 30-mile-an-hour (13.4 meters [42 feet] per second) wind, a machine that was tested between 1941 and 1945. From that time, through the second half of the twentieth century, there were numerous tests of windmill designs. An example is a machine with a blade nearly 70 meters (225 feet) in diameter, rated at 2×10^6 watts (at a wind velocity of 11.5 meters (35 feet per second), that was built in North Carolina in 1979. An unexpected problem developed, however, when this unit generated a low-frequency hum that kept many people in the nearby community awake at night; efforts to stop the sound were never successful, and the windmill was shut down and dismantled after a mechanical failure.

A very different design, sometimes called the Darrieus wind turbine, is a vertical axis machine [Figure 6.29(c)] that is less efficient than other designs but can be built to an even larger size. The small (17-meter diameter) machine is rated at 3×10^5 watts.

The United States, Canada, and many of the northern European countries have programs for the development and installation of larger machines. Indeed, in the United States, the Wind Energy Systems Act of 1980 initiated an eight-year, $900 million program to develop wind power systems. A consequence of this has been the sudden growth in **wind farms**—clusters of turbines connected to the electric grid—in parts of the United States. Although the world's first commercial wind farm began generating power in New Hampshire in 1981, the major developments since then have been in California, a state blessed with mountain passes that provide ideal wind farm sites. With farms like the one at Altamont Pass (Figure 6.30), where over 2000 machines generate up to 142×10^6 watts, the goal is to supply 8 percent of the state's electricity from wind power early in the twenty-first century. The largest single wind farm in the world is the Horse Hollow farm that is spread across 47,000 acres (19,000 hectares) of Texas. It has more than 420 turbines that generate 735 megawatts of electricity—enough to supply more than 200,000 homes. The installed capacity of wind energy in the United States is summarized in Table 6.5.

Worldwide, wind power grew only slowly until about 2000 although there were some countries, such as Germany and Denmark, which continued to develop wind farms. The combination of increasing costs of fossil fuels and recognition of their environmental impacts has, since 2000, spurred the rapid growth of wind power. It was announced in 2008 that the world had finally reached the milestone of producing more than one percent of its total electricity by wind power. Indeed, the future looks bright for wind power. It is currently being used to generate electricity in more than 70 countries and is producing nearly 100,000 megawatts of electricity. The current world leaders are Germany and the United States, each producing over 20,000 megawatts; Spain, India, and China are also large generators. Although Denmark's total production is only about 6000 megawatts, it now produces more than 20 percent of its total electricity needs from the many wind mills that dot the Jutland coast and countryside.

TABLE 6.5 U.S. wind energy capacity (2006)

State	Wind capacity in megawatts
Texas	2,738
California	2,255
Iowa	919
Minnesota	827
Washington	821
New Mexico	494
Oklahoma	480
Oregon	399
New York	370
Kansas	363
Colorado	288
Wyoming	287
Others	1088
Total	11,329

From Energy Information Administration, *Renewable Energy Annual,* 2006.

FIGURE 6.30 The U.S. Department of Energy operates an experimental wind farm with more than 2000 windmills at Altamont Pass, near San Francisco. (Courtesy of U.S. Department of Energy.)

Appropriately perhaps for the country where electricity was first generated in this way, Scotland now hosts Europe's largest onshore windfarm which is claimed to be powerful enough to meet the electricity needs of the City of Glasgow. There are also controversial plans to build an even larger scheme on the Shetland Islands. The commitment of the British to wind power is evidenced by around 200 operational windfarms, including offshore installations (the first of which was opened in 2003, approximately 8 km [5 miles] off the coast of north Wales) and advanced plans for several hundred more.

Future development of wind farms to provide significant amounts of electricity depend upon finding acceptable sites to locate large numbers of large windmills. The velocity of the wind generally increases, and is more constant, with increases in altitude. Hence, today most commercial windmills have two or three 230-foot long blades and reach as much as 660 feet above the surface. The U.S. Department of Energy has mapped the potential for commercial wind energy and found that the best zones lie along the coasts, in the Great Lakes area, and in a corridor that extends from western Texas northward into Montana and North Dakota. There are also numerous isolated sites on high peaks in the Appalachians and Rockies. Despite the potential for generating vast quantities of "green" electricity using windmills, there has been much opposition by those who view windmills and the necessary service roads and power lines as eyesores. Furthermore, there have been reports of night-flying migratory birds and bats colliding with the windmills. One problem arose when a large wind farm was proposed in the California mountains. The wind farm project was stopped because the windmills were considered a threat to the endangered California condors that live in the area and that tended to roost on such structures. Consequently, proponents of wind power have proposed the careful siting of windmills to avoid routes taken by birds and the use of audio or light signals to warn the birds.

In one of the most ambitious projects to date, Texas oilman T. Boone Pickins proposed, in 2008, the construction of 2700 windmills along the Texas to North Dakota corridor by 2014. This would provide more than 4000 megawatts of electricity to power one million homes. The main drawback is the construction cost of $6 to $10 billion. The problem of cost became a reality as soon as mid-2009 when the price of natural gas dropped significantly. This made the use of gas cheaper than the construction and operation of windmills in the short term. Furthermore, an infrastructure for the transport and use of gas already existed in many areas. In contrast, new giant wind farms would require not only the construction of windmills, but also the transmission grid to make the electricity available to markets. In the past several years, there has been increasing consideration of using "floaters", huge windmills 10 to 25 km (6 to 15 miles) offshore where the winds are generally strong and constant and where migratory birds would be a lesser problem. Such windmills would be mounted on floating barges much like those used for deepwater oil exploration and production.

What then is the likely future of wind power for large-scale energy generation? Clearly, wind energy will grow in importance in the future. The questions regarding this form of energy concern suitable sites offering appropriate conditions for efficient operation, sites that are acceptable to the public in terms of aesthetics and are environmentally benign. Also needed are improvements in efficiency and in the available capital to finance construction of the windmills and infrastructure. The United States government granted the first permits for offshore windmills near Atlantic City, New Jersey, in 2009 and more are likely in the near future. The future appears good for wind power, with experts predicting that wind farms will have an economic advantage over coal and nuclear power plants in many parts of the world during the twenty-first century.

The smaller-scale uses of wind power, as in transportation at sea, recreation, and farm use, have long had many applications, and experts also predict a resurgence of small-scale windmills and related devices in the future. We may see a rebirth of the large sailing ship in forms that use not only sails but wind turbines designed to drive propellers (Figure 6.31). Indeed, there is a Japanese prototype cargo ship which employs computer-controlled sails made of canvas on steel frames that can provide 58 percent of the power when the fully laden ship is traveling at 12 knots in a 30-knot wind.

WAVE POWER. Wave power is closely related to wind power because waves arise from winds blowing over the ocean. Waves contain much more energy than do winds of equal velocity because the mass of water involved is more than 800 times that of the same volume of air. A single wave that is 6 feet high, moving in water 29 feet deep, generates approximately 10^4 watts for each (three feet) of wavefront. Vast amounts of energy (estimated to be as much as 2.7×10^{12} kilowatts) are continuously being dissipated on the shorelines of the world. Although wave power has been used to ring bells and blow whistles for navigational aids for many years, large-scale energy recovery has only recently been considered. Many research groups have developed designs and tested

FIGURE 6.31 The *Minilace* is an experimental modern cargo ship that uses a sail as well as a conventional engine. The sail can increase speed and decrease fuel consumption and hence significantly increase efficiency. (Courtesy of the Wind Ship Development Company.)

small-scale prototypes that they claim could generate electricity at costs competitive with conventional power stations. One example is the Sea Energy Associates (SEA) Clam, which consists of a series of flexible air bags mounted along a long, hollow spine of reinforced concrete. Passing waves compress the bags and force the air into and out of the spine through a turbine. The self-rectifying turbine, which turns in the same direction whether the air is moving in or out, drives a conventional electric generator. Another device involves floating rafts or large tubes that transmit the mechanical energy of wave motion to hydraulic pumps that power a generator; another uses rigid concrete structures in which a column of air is trapped in such a way that the volume changes as waves pass, forcing the air through a turbine.

There is no shortage of ideas on how to harness wave power for the generation of electricity and to apply it on a small laboratory scale. However, none of the various schemes has yet received the very substantial funding needed to set up a large-scale commercial operation. It has been estimated, for example, that a 2×10^9 watts wave power SEA Clam system, comparable to a large conventional power station, would require 320 of these devices along 130 kilometers (80 miles) of coastline. The cost would be considerable, estimated at well over $10 billion in 2000. Because of the cost and because the ultimate potential of wave power is still unclear, many groups and companies that supported the relatively low cost of early development of wave power devices have not yet provided the funds to build more costly prototypes.

Wave power devices suffer from two problems of the marine environment: corrosion and storms. Seawater has proven to be incredibly corrosive; it is the home of organisms, such as barnacles and seaweeds, all of which are impediments to efficient operation. In addition, the violent and unpredictable storms in every ocean are likely to make operation difficult and repairs frequent.

OCEAN POWER. The term *ocean power* usually refers to a system of Ocean Thermal Energy Conversion (OTEC). The Sun warms the surface waters of the ocean, and this water, being less dense than the cold water at depth, remains near the surface. A temperature gradient is created, as shown in Figure 6.32. If the water at the two temperatures can be brought together, we have the basis for a heat engine that can generate electricity. The difficulty is the relatively small temperature difference involved, only about 20°C from top to bottom even in the tropics. Although the thermal efficiencies would only be 2–3 percent, the very large reservoir of heat in the oceans should make such a plant feasible. Small pilot plants that were set up and operated early in the twentieth century, in Cuba and on the West African coast, demonstrated that OTEC is possible in principle.

Most of the recent development efforts have centered on using a closed-cycle turbine employing a fluid such as ammonia that boils at a low temperature (25°C) but at a much higher pressure than water. As shown in Figure 6.32, the warm surface water is used to heat the fluid and vaporize it so that it expands

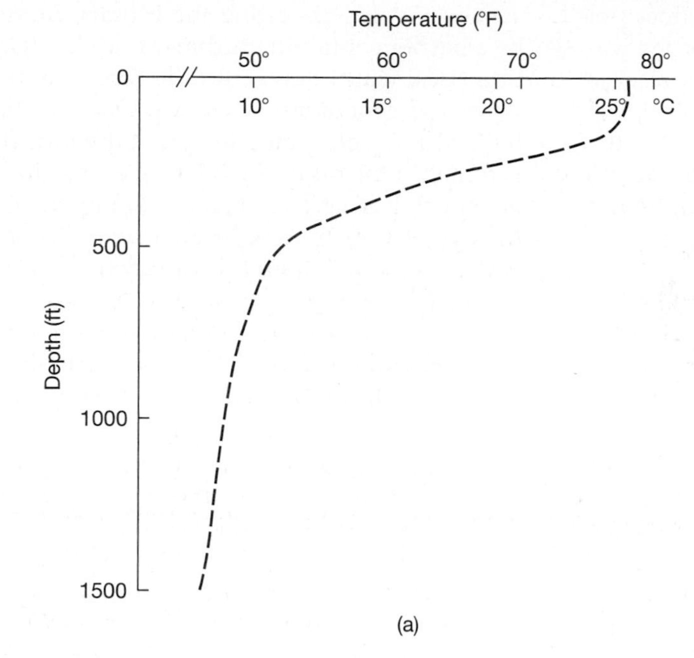

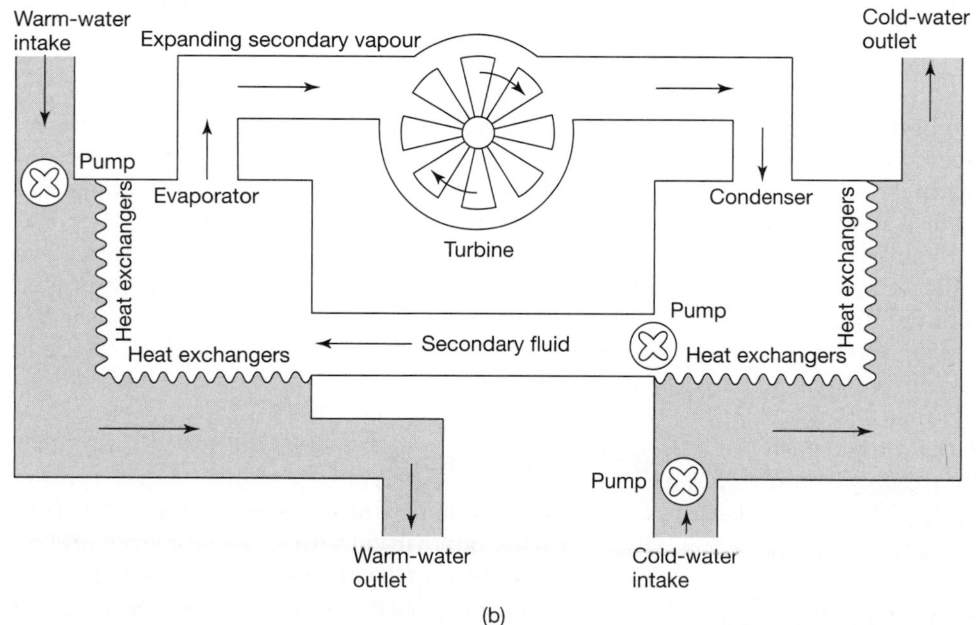

FIGURE 6.32 The OTEC approach to energy generation. (a) Typical temperature variation with depth in the ocean in equatorial regions. This difference is used as the basis for a heat engine cycle of the type illustrated in (b).

pushing a turbine; it is then condensed back to liquid by contact with cold water pumped from the ocean depths. A power plant would probably be enclosed in a submerged unit floating beneath the ocean surface. Electric power generated from the turbine could be transmitted ashore via submarine cables, and possibly used at the site of the plant to produce hydrogen by the electrolysis of water, the hydrogen then being shipped away in tankers.

Whether large-scale schemes using OTEC are economic remains uncertain and cost estimates vary widely. Unresolved problems include technical aspects of construction, problems of corrosion, encrustation of the machinery by marine life, and the environmental impact of OTEC plants. Supporters say that bringing cold nutrient-rich water to the surface would increase fish catches in the area around a plant. But surface sea temperatures play a major role in Earth's climate, and the effects of a large network of OTECs would need careful investigation. If all of these problems were to be solved, a very large source of energy would be made available. Just how large is difficult to estimate because it depends on the efficiency of the generating plant. Even if it were less than 1 percent efficient, the ocean's thermal energy potential exceeds the potential from fossil fuels. Another novel use of the ocean's deep cold waters (approximately 4°C; 39°F) does not involve electrical generation but makes use of the water's low temperatures as a refrigerant to preserve foods such as flour and corn and to serve as a tropical air-conditioning system. Of course, there is also the problem that deep waters with such low temperatures are usually far offshore from the cities where the use would be most beneficial.

TIDAL ENERGY. Tidal energy differs from the other energy sources discussed in not being derived ultimately from the heat of the Sun. The ocean's tides are the result of the gravitational pull of both the Moon and Sun on Earth and its oceans. The changes in ocean height resulting from the rhythmic rise and fall of tides can be used to drive a water turbine connected to an electric generator. However, only in certain parts of the world is the tidal rise and fall sufficient to justify constructing a power plant. The best areas, where tidal ranges exceed 10 meters, (30 feet) include the Bay of Fundy, the Bristol Channel in the U.K., the Patagonian coast of Argentina, the Murmansk coast (Barents Sea), and the coast of the Sea of Okhotsk (north of Japan). This is because both the effect of ocean bottom shape and contours of the shorelines enhance tidal rise and fall; elsewhere, the range is generally much less.

The harnessing of tides is not a new idea. For several centuries, beginning in 1580, 6.5 meter (20 foot) diameter waterwheels installed under London Bridge used the tidal rise and fall of the River Thames to pump water for London. At present, the only large-scale tidal power station in the world is at the Rance Estuary on the Brittany coast of France (Figure 6.33). This site has a peak electricity-generating capacity of 240×10^6 watts, but because of the rhythmic nature of tides, the average capacity is only 62×10^6 watts. The system involves isolating the estuary from the ocean by a barrier containing turbines and floodgates. As the tide rises, water flows into the reservoir, turning turbines and generating electricity. Once the tide reaches high, the water begins to flow outward, turning the turbines in the other direction. The system is set up to generate electricity as water flows in either direction, thus producing power at all times except when the water is briefly at high tide or low tide.

FIGURE 6.33 The dam built across the Rance Estuary on the French coast is equipped with floodgates and turbines, visible at the right side. As the tides rise and fall, the flow of water in and through the turbines generates electricity. (Photograph courtesy of Electricité de France.)

Tidal power is limited in the number of sites around the world that could be developed and in the total amount of energy potentially available. Development of all of the suitable sites would only generate about 16×10^9 watts, or less than 1 percent of the world's present total usage of electric power.

The oceans do contain other sources of energy in the form of the great surface currents. The Gulf Stream, for example, flows northward at a steady 5 mile per hour rate and thus could produce electricity 24 hours per day all year long. The Center for Ocean Energy Technology at Florida Atlantic University estimates that full implementation of a fleet of turbines anchored off the Florida Coast (Figure 6.34) could produce as much electricity as five or more nuclear power plants and could serve 5 million homes. Another but as yet totally unexploited source of energy in the oceans involves their salinity, or rather the salinity difference between fresh (river) water and salt (ocean) water. The difference in **osmotic pressure** between these two natural waters can produce a positive flow through a suitable membrane that could be used to raise saltwater that could then be discharged through a turbine to generate power.

Geothermal Energy

Earth's surface temperature is almost entirely controlled by the Sun through direct solar energy and by the redistribution of solar heat energy by wind and water. In contrast, Earth's subsurface temperature (at greater depths than 100 feet or so) is the result of the **geothermal gradient** which ranges from a 15° to 75°C per kilometer (25 to 120 per mile) increase, but averages 25° C per kilometer of depth below surface. The rate of temperature rise decreases at depths beyond 100 kilometers (60 miles), where rock strength decreases so much that convection is the main means of heat transfer. Estimates of the temperature at a depth of 100 kilometers are about 1000°C and at the core–mantle boundary about 5000°C. From this it is clear that a vast amount of heat energy is stored inside Earth.

The slow but continuous outward flow of heat from Earth averages 6.3×10^{-6} joules/cm^2 per second, or 32.3×10^{12} joules per second (32.3×10^{12} watts) over the entire Earth's surface. The total amount is vast but

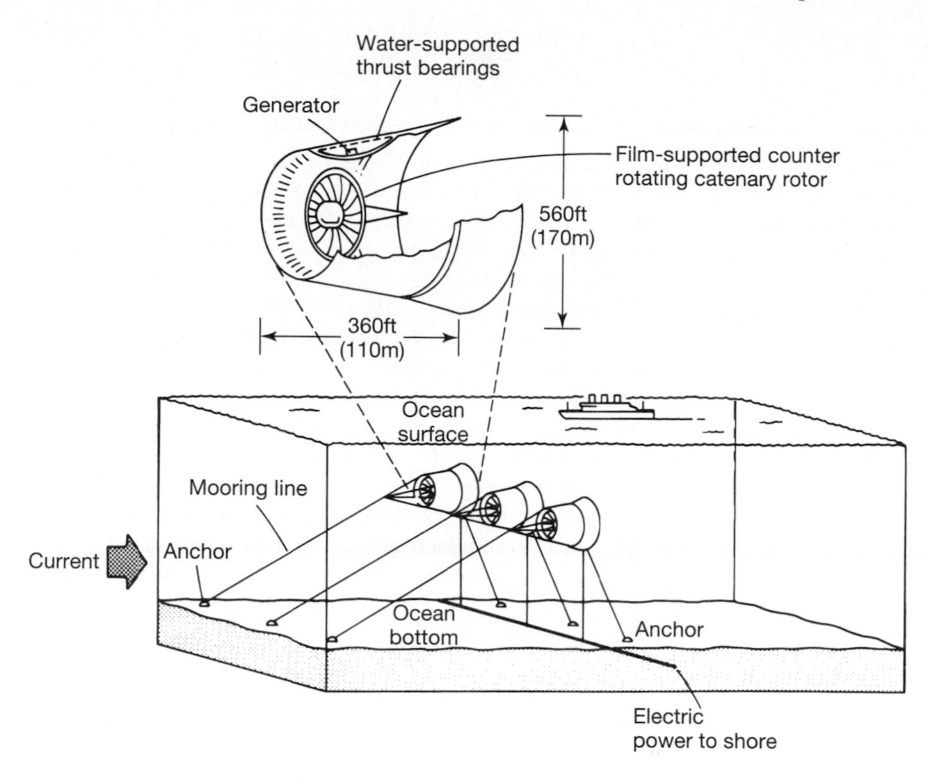

FIGURE 6.34 Proposed design and mooring arrangements of rim-driven turbines for use in generating electrical power from ocean currents.

is very diffuse, and the quantity of heat reaching the surface is equivalent to little more than 1/3000 of the heat received from the Sun. If all the heat escaping from 1 square meter (from 10 square feet) could somehow be gathered and used to heat a cup of water, it would take four days and nights to bring it to the boil.

Despite the heat loss, Earth is not cooling because new heat is constantly being added. Several naturally radioactive isotopes, principally uranium-238, uranium-235, thorium-232, and potassium-40, occur in trace amounts throughout Earth. Each time a radioactive atom disintegrates, a very small amount of heat is released. For example, atoms in an average igneous rock in the continental crust release 9.4×10^{-8} calories (3.93×10^{-7} joules) per gram per day. Although this is not much, summed over the entire Earth, it is enough to maintain a nearly constant average temperature distribution. New heat production is so low that we could never harness it, but the accumulated heat from millions of years can be used in particular places. This energy is not truly renewable because it will one day, in the very distant future, no longer be produced. However, its life span is similar to that of the Sun, so we consider it among the renewable energy sources.

How can geothermal energy be recovered? In certain special circumstances, nature has already provided the answer to this question. In some areas, such as in the vicinity of active volcanoes, abnormally hot rocks are found close to the surface. Groundwater slowly seeping downward is heated and may reemerge as geysers or hot springs [Figures 6.35(a) and (b)]. Regions where this occurs are known as **geothermal fields** and are of three types. The first is a field of low temperature water ($\leq 85°C$) that cannot be used efficiently in the generation of electrical power, but can be used for space-heating in homes, industry, and greenhouses. Resorts specializing in hot baths, as in the famous spa towns that were particularly popular in eighteenth- and nineteenth-century Europe, have long used geothermal hot springs. In Hungary and France, geothermal water is used to heat homes, and in Reykjavik, Iceland, the entire city is heated by geothermal hot water. Throughout the world more than 10^9 watts of low temperature thermal power is derived from geothermal wells, with nearly half of this in Iceland.

The geothermal energy used today as a source of power occurs in either **dry-steam** (vapor-dominated) or **wet-steam** (liquid-dominated) fields. Dry-steam fields occur in geothermal settings where the temperature is high and the water pressure is just a little more than atmospheric pressure. The water boils underground, making steam that fills fractures and pores in the rock and that can be tapped directly by wells drilled into the field. Examples of major dry-steam fields are Larderello in northern Italy, which has been producing power since 1904, and the Geysers field in California (about 90 miles north of San Francisco), where the first geothermal power plant in the United States was commissioned in 1960 (Figure 6.36). In the 1980s, seventeen power plants were providing over

BOX 6.3

Hydroelectric Power

Water, in addition to being one of the few resources that are literally vital for the existence of life, has served as a source of energy since before recorded time. For thousands of years, waterwheels have harnessed the power of flowing water to grind grain. Burgeoning technology permitted waterwheels to spin large saw blades to cut the wooden planks used in the construction of ships and buildings. The discovery of electricity and the understanding of its generation in the late eighteenth and nineteenth centuries led to the development of waterwheels that spun great turbines connected to large magnetic coils. When such coils are spun in a magnetic field, they produce a direct current, which can then be used or converted into the type of alternating current now used worldwide in homes and industries. Recognition that the force of the water, and hence the amount of electricity generated, increases if the water pressure increases led to the construction of dams to produce hydroelectric power. The general design, as shown in Figure 6.26 has changed little since the earliest dams were built. The increase in the water depth on the upstream side of the dam provides increasing pressure (the hydrostatic head, which is equal to the weight of the overlying water) that propels the blades of the turbine as the water is permitted to flow through a tunnel (called the *penstock*). Because water weighs approximately 1 metric ton/m^3, the pressure is 30 metric tons/m^2 at a depth of 30 meters, 100 metric tons/m^2 at 100 meters, and 600 metric tons/m^2 at 600 meters. Although modern dams vary greatly in size, many are 100 to 200 meters (300 to 600 feet) in height.

Although the earliest hydroelectric dams were built in Europe and the United States in the late 1800s, the greatest era of dam building was the early part of the 1900s. In the United States, the peak of dam building occurred in the Great Depression of the 1930s when the federal government employed thousands of workers to build large numbers of dams in areas such as the Tennessee Valley. The government established the TVA (Tennessee Valley Authority), which not only provided construction jobs, but was also the basis for electrification and development of the rural southeastern United States. The availability of cheap hydroelectricity became the springboard for long-term economic development; to this day TVA, a quasi-governmental agency, is a major power provider and employer in the southeastern United States.

Most hydroelectric dams have been constructed to the conventional design shown in Figure 6.26, but some have been built with a lake at both the base of the dam and behind the dam. These pumped-storage dams (see Figure 6.27) are designed to make use of excess electricity generated (especially at night) to pump water from the lower lake back up to the upper lake. This actually consumes more power than is returned when the water falls through the turbines, but it takes advantage of electricity that would otherwise go to waste. By pumping the water back to the upper lake, the water can be used again at times when extra power is needed. Thus, these operations may appear to actually be net energy losses (which they are), but they are employing otherwise unused energy to provide extra power at the times it is most needed.

Examination of the total amounts of water used in the generation of hydroelectricity reveals the apparent contradiction that countries such as the United States use more water in production of hydroelectric power than they actually have in their rivers. In a sense, this is true because the water is counted every time it passes through a hydroelectric dam; and many rivers have multiple dams so that the water passes through several each day, being counted each time.

The number of hydroelectric dams in the world today is probably in the hundreds of thousands; the United States alone has more than 50,000 of various sizes in operation. Worldwide, the total electricity produced by hydroelectric power is approximately 17 percent of the total. Within individual countries, the percentages range widely depending on the terrain, the rainfall, and the availability of other sources. The United States produces a large amount of electricity by hydroelectric systems (see Figure 6.28), but it accounts for only about 7 percent of its needs. Canada, with large rivers, produces about 60 percent of its electricity by hydro-power; Brazil and Norway produce 94 percent and 98 percent, respectively. Relatively flat or arid countries may produce little or no electricity by hydropower.

Concerns about the global impact of the burning of fossil fuels and the problems of disposal of nuclear waste have made hydroelectricity attractive as a clean or "green" energy source. It is sustainable in the long term, creates no emissions, and has no radioactive waste products. On the other hand, it has also come under strong criticism through being disruptive of rivers and streams, damaging to migratory fish stocks (especially salmon), and upsetting to ecologically sensitive areas. Hydropower, for all of its benefits, exacts a price environmentally, and its advantages must be weighed against its disadvantages.

1×10^9 watts capacity and by 2008, there were twenty-three operating fields in California alone. In the more common wet-steam fields, the reservoir of hot water (in effect, a **brine**) is under high pressure and can reach temperatures approaching 400°C without boiling. The most famous wet-steam field is at Wairakei in New Zealand.

The use of geothermal fields in electricity generation is shown in Table 6.6. Today, there are hundreds of geothermal power plants operating in 24 countries and producing 10,000 megawatts, which is sufficient to provide the electrical needs of 60 million people. This is a small amount of power when viewed in a global context, but it is most important in local areas

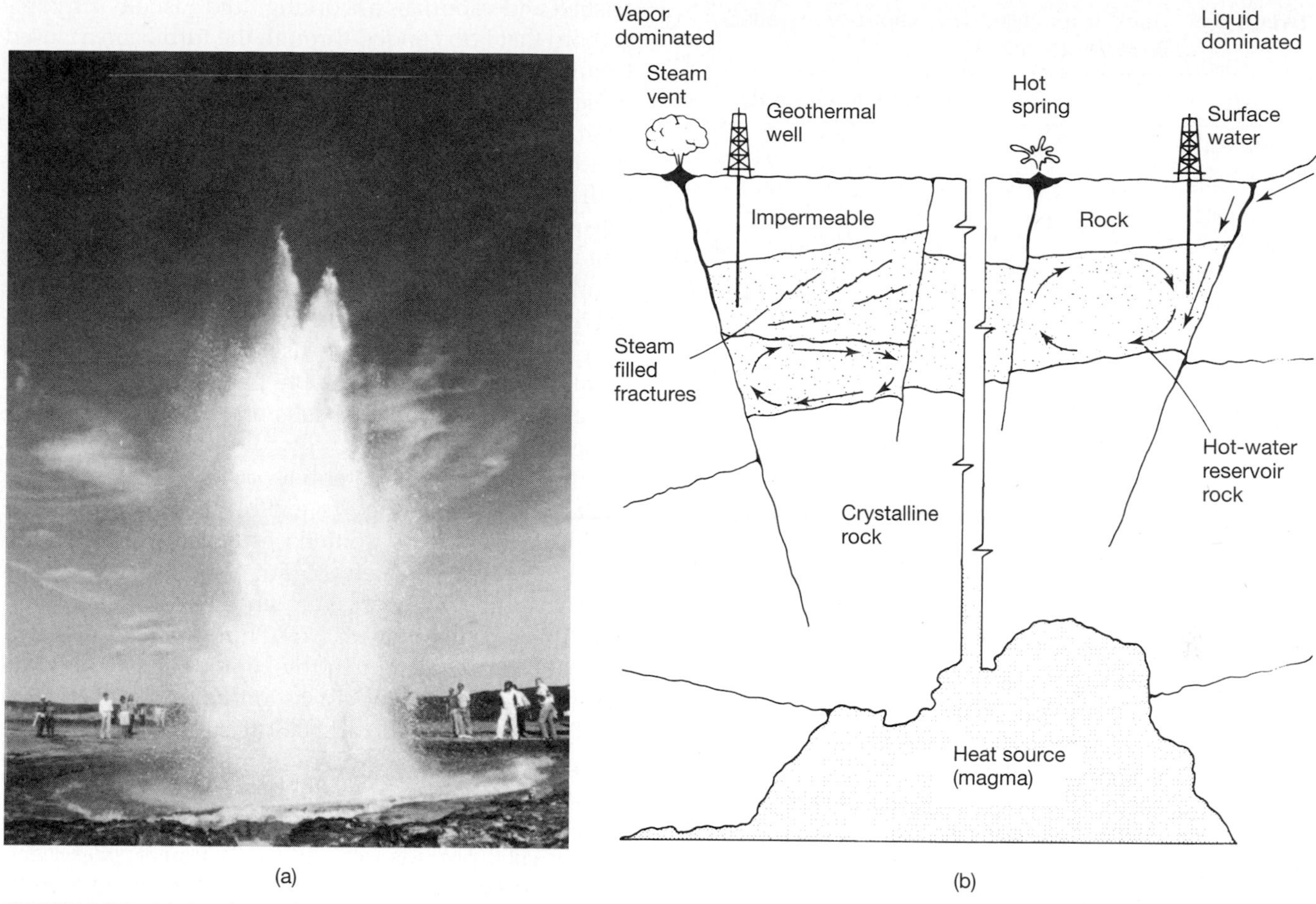

FIGURE 6.35 (a) Geothermal energy is locally evidenced by natural geysers such as those in Yellowstone Park, Wyoming. (Photograph courtesy of J. D. Rimstidt.) (b) Schematic diagram to show the geological features of dry-steam and wet-steam geothermal fields.

chiefly in areas ringing the Pacific, in Iceland, and in the central Mediterranean. No doubt, other sources will be developed, but they are likely to be localized in the areas of active volcanism.

A number of systems are being used, or have been proposed, for the conversion of geothermal energy into electricity. Two of them are illustrated in a simple way in Figure 6.37. The first is a *direct steam cycle* of the type used at dry-steam fields such as Larderello and the Geysers; steam brought out of the ground here is clean enough to go directly into the turbine, after which it is condensed and may simply be returned to the ground.

FIGURE 6.36 The Geysers power plant north of San Francisco, California, is the largest complex of geothermal power plants in the world. The steam rising from stainless steel-lined drill holes reaches the surface at more than 355°F and is used to spin turbines to generate electricity. (Courtesy of Pacific Gas and Electric Company.)

TABLE 6.6 Geothermal electricity capacity installed worldwide in 2007

Country	Total megawatts
United States	2924
Philippines	1970
Indonesia	992
Mexico	953
Japan	535
New Zealand	472
Iceland	421
El Salvador	204
Costa Rica	163
Kenya	129
Nicaragua	87
All Others	1118
World Total	9968

From Energy Information Administration, *Renewable Energy Annual*, 2006.

The *flash steam approach* is used in most of the wet-steam fields and here, the release of pressure in the flash chamber results in the spontaneous generation of the steam to drive the turbines, with the condensed steam again being returned to the ground. Two more advanced cycles have been developed for maximum efficiency at lower temperatures. In the *flash binary system,* the wet-steam is flashed and vaporizes a working fluid (usually a hydrocarbon) that is expanded through the turbine in a closed circuit, maintaining a clean, long-life turbine. In the *liquid-liquid binary system,* the brines are not allowed to flash, and heat is transferred to a working fluid to drive the turbine. This reduces some of the problems caused by flashing the brine and also isolates the hydrogen sulfide (H_2S), commonly found in such brines, from the atmosphere. In both cases, the condensed brine is returned to the ground. The practice of simply discharging the condensed brine is a wasteful one that is gradually being replaced by the use of flash distillation to desalinate and produce usable freshwater, or even by the extraction of valuable salts and minerals from the brines.

Several other potential geothermal resources exist but the technology to exploit them has yet to be developed. The first, sometimes called **geopressured zones,** involves pockets of hot water and methane trapped under high pressure and at fairly high temperatures (approximately 175°C) in deep sedimentary basins. Examples occur in the United States along the coasts of Louisiana and Texas at depths between 1200 to 8000 meters (4000 to 25000 feet). It is hoped that energy could be extracted from this resource via three routes: the geothermal heat of the water, the hydraulic energy of the water under high pressure (approximately 2000 pounds per square inch at surface), and the gas dissolved in the water. However, test wells

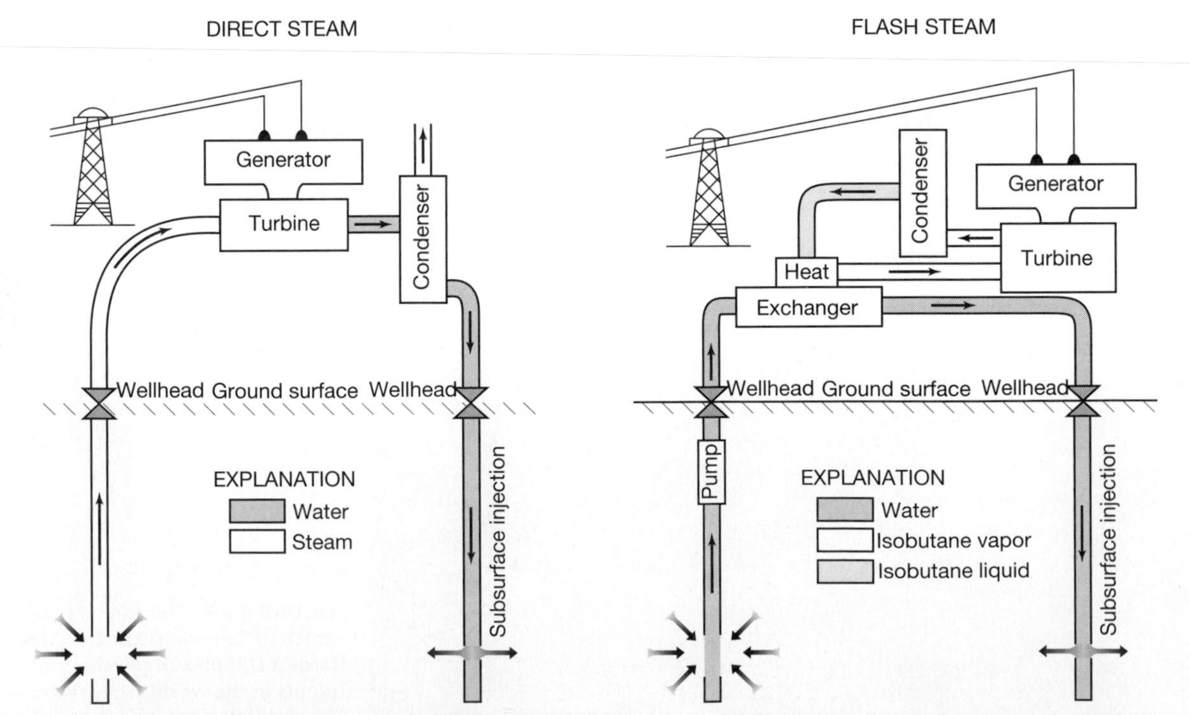

FIGURE 6.37 Systems for the conversion of geothermal energy into electricity. Steam from the geothermal well can be used directly to drive turbines or can be produced by release of pressure in a flash chamber and then used. Alernatively, heat from the geothermal fluid can be used to vaporize a working fluid that drives the turbine.

proved disappointing in terms of water temperatures, gas content, and the extent of this resource.

The second potential geothermal resource is usually called **hot dry rock** and, as the name implies, involves extracting the heat from dry rocks at depth. Unlike geopressured zones that are limited in extent, the hot dry rock resource is potentially very large worldwide and does not require the rather exceptional geological conditions associated with geothermal fields.

Many parts of Earth's surface are underlain by rocks such as granite that generate more heat than the average because of the relatively greater amounts of radioactive atoms they contain or by igneous intrusions that still retain excess heat from the time of their emplacement. The most commonly used means of exploiting the heat energy involves drilling two boreholes into a mass of rock such as granite and then explosively or hydraulically fracturing the zone between the bottoms of the two holes. Water is then pumped down the first hole, passing through the fractured rocks taking heat from them, and returned up the second hole to be used for electricity generation in the usual way on reaching the surface (Figure 6.38). This method of power generation is not as simple as it sounds. To reach regions with temperatures greater than 200°C generally requires wells about 5 to 7 kilometers (3 to 5 miles) deep that are expensive and difficult to drill. Furthermore, the transmission of fluids between the input and output wells has been plagued with problems and, at the low steam temperatures involved, present turbines are only about 10 percent efficient. Nevertheless, the technical problems can, in some cases, be overcome, as shown by a pilot plant at Los Alamos, New Mexico, that generated 6×10^4 watts from granite at 3000 meters (9500 feet). For the ultimate future of geothermal energy, it is also possible to consider drilling into molten magma chambers or exploiting dry rocks at greater depth in areas of only average geothermal gradient.

The extraction of energy from geothermal fields is well established, but it is a resource that is limited geographically and probably also in terms of worldwide reserves. The U.S. Geological Survey estimates that down to a depth of 3 kilometers (2 miles), which seems to be a limit for the occurrence of big geothermal fields, the worldwide reserves are 8×10^9 joules. Such a small amount suggests that this type of geothermal energy will be locally important, but globally insignificant. The total heat energy in the pools is, of course, much larger than 8×10^9 joules, but the estimate takes into account the low efficiency with which electricity can be generated from geothermal steam. Experience in Iceland, New Zealand, and Italy suggests that no more than 1 percent of the energy in a pool can be effectively recovered.

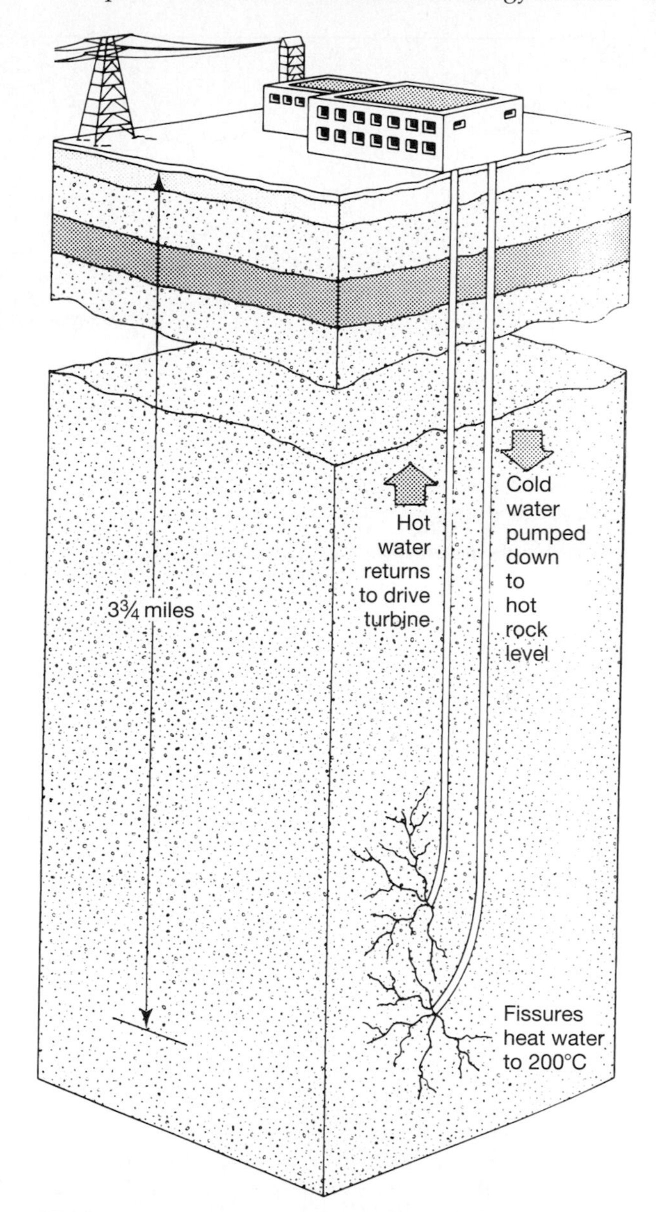

FIGURE 6.38 The generation of electricity using "hot dry rock" geothermal energy. Cold water is pumped down the well, is heated by passing through fissures in the fractured "hot" rocks at depth, and then is returned to the surface to be used to drive turbines.

Reserves of energy in geothermal fields represent only a tiny fraction of all geothermal heat. Experts cannot agree how much of the remainder should be considered a potential resource. A map of the United States showing areas and types of potential geothermal resources suggests a very large amount (Figure 6.39). However, the technological problems involved in exploiting even a very small fraction of this geothermal energy are likely to be considerable and the costs prohibitive. In the case of Britain, for example, it has been estimated that to meet between 1 percent and 2 percent of the national demand for electricity at the beginning

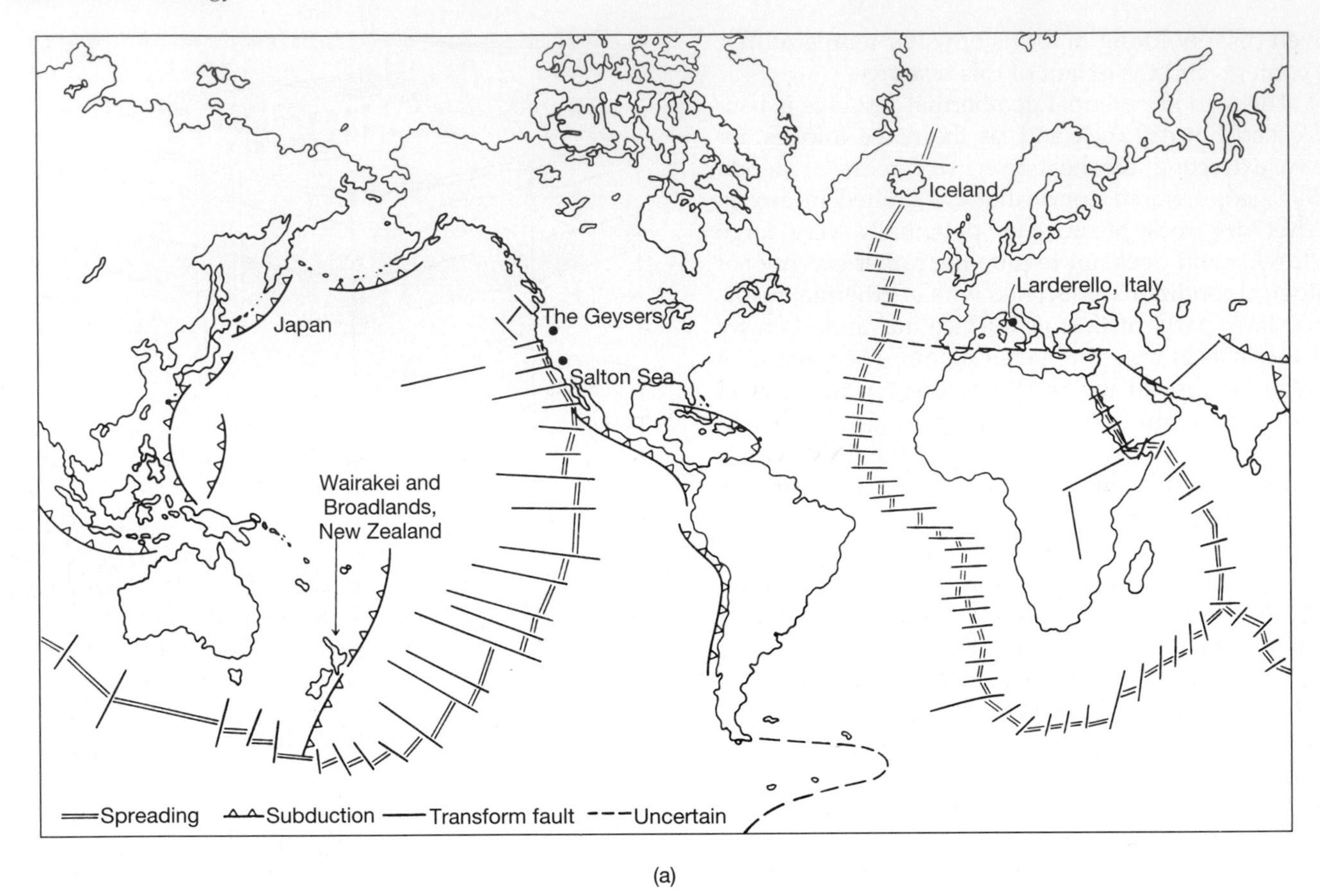

(a)

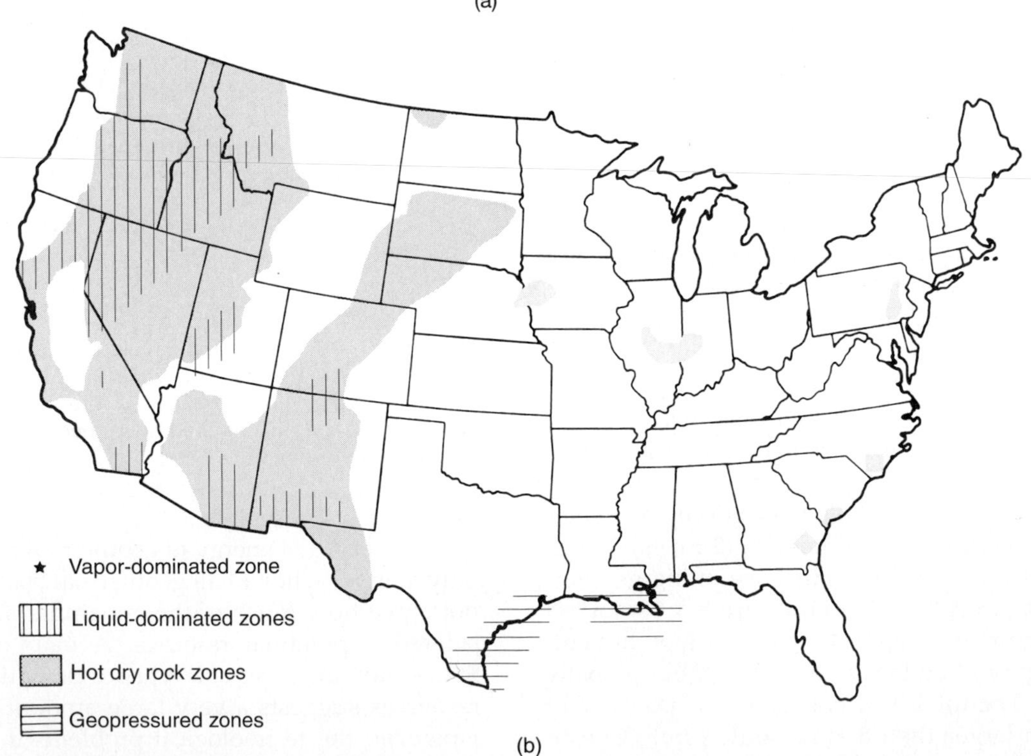

(b)

FIGURE 6.39 (a) The major geothermal areas of the world occur a long the major continental plate boundaries. (b) A map of the United States showing the location and types of geothermal resources.

of the twenty-first century would require 120 pairs of holes 6000 meters deep, costing nearly $2 billion for drilling alone.

Energy from Biofuels and Waste Products

The staple energy sources in many parts of the developing world are the simplest and most readily available combustible materials, such as wood, grain stalks, and even dried animal dung. In advanced societies, now primarily dependent on fossil fuels, interest in these same materials is growing because they are renewable, widespread, they lessen reliance on fossil fuels, and they reduce carbon dioxide emissions into the atmosphere. These are all examples of **biofuels**, which may be broadly defined as any solid, liquid, or gaseous fuel derived from recently living organisms.

The best known and widely used biofuel is wood. Many developing countries such as Haiti and Madagascar, with rapidly growing populations and rising energy needs, are experiencing rapid deforestation (see pg. 480 and Plate 69) following a pattern seen in Great Britain and the United States in the past. As they exhaust their forest resources, often with irreparable environmental damage, they too will be forced to turn to other resources for their energy needs. Unfortunately, the destruction of forests leads to erosion of precious soils, depletion of water supplies, and loss of wildlife.

Petroleum is the principal energy source used in the world today, and most petroleum is converted into one of the liquid fuels used in transportation—gasoline, diesel, or aviation fuel. Factors including rising petroleum costs, concerning countries such as the United States over import reliance, and worries over "greenhouse" gases and climate change have led to much interest in converting biomass into substitutes for gasoline and diesel fuel. Many different processes have been proposed; a generic example is presented in Figure 6.40. This example relies on normal biological photosynthesis by which plants extract carbon dioxide from the atmosphere and water from the soil to produce cellulose or other organic materials. These materials are converted into sugars through the use of enzymes. In turn, microbes convert the sugar into alcohol, which may be burned directly, or mixed with gasoline to produce a fuel called *gasohol* (see the following discussion). The burning of the alcohol produces water vapor and carbon dioxide that are released into the atmosphere. The two most obvious advantages of manufacturing biofuel are that there is a reduction on reliance on imported oil, and the overall amount of atmospheric carbon dioxide is not increased; burning a biofuel merely returns to the atmosphere the carbon dioxide that was originally removed by photosynthesis. Hence, it is a "closed carbon cycle." In contrast, the burning of a liquid fuel derived from petroleum raises the carbon dioxide content in the atmosphere by releasing carbon that had been sequestered in the subsurface.

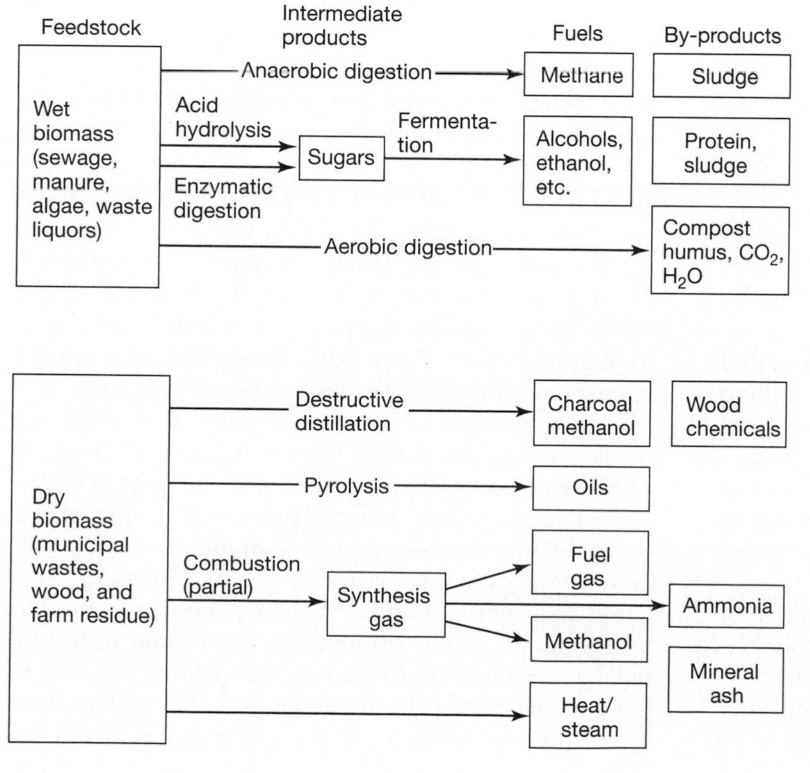

FIGURE 6.40 Block diagram showing how biomass can be converted into fuels, the processes involved, types of fuels produced, and by-products of the operation.

Wood was the fuel that sustained the early development of America. Harvesting of wood had little impact until the late 1800s when rapid growth of population and industry combined to demand larger and larger quantities of wood for construction and fuel. The rapid clearing of forests, had it continued, would certainly have stripped the American forests in the same manner as the British forests more than a hundred years earlier. Instead, industry, home heating, and the fledgling electricity generating utilities turned to coal, which provided far more energy per ton. This permitted the forests to recover and the U.S. forests of today, especially in the East, are larger than they were in 1900.

Various other processes are being developed for the conversion of biomass into fuels and Figure 6.40 illustrates several of the basic routes. For example, a fermentation process known as **anaerobic digestion** can be used to convert biomass into methane that can be directly substituted for natural gas. The technology is now available to produce methane gas in this way at costs similar to those involved in the gasification of coal. It is also possible to produce liquid fuels such as alcohols from biomass by a process called **pyrolysis,** which involves heating the organic material in an oxygen-deficient atmosphere. One use of this liquid fuel, much publicized in the United States, is as a partial substitute for gasoline in automobile engines. A mixture of roughly 10 percent ethyl alcohol derived from biomass and 90 percent unleaded gasoline produces a fuel termed **gasohol.** Although the energy savings in using gasohol appear to be marginal, if any, precious resources of petroleum are conserved. This is particularly important for countries such as Brazil, which in the past had to import large amounts of oil and which therefore focused on alternative energy sources. From the early 1980s until 1990, Brazil pushed for the operation of cars running on pure ethanol. Since 1990, all cars must be capable of operating on ethanol or on gasohol which today must contain 20–26 percent ethanol. In 2007, 15 million of Brazilian light vehicles use gasohol whereas 2.2 million additional vehicles use pure ethanol.

Brazil's effort to produce large amounts of ethyl alcohol (in 2007, 4.4 billion gallons or 16.5 billion liters) from sugar cane, sugar beet, cassava, and sorghum, as shown in Figure 6.41, has been so successful that today that Brazil not only meets its own needs, but also exports large quantities of ethanol to India, the United States, South Korea, Sweden, and lesser amounts to many other countries.

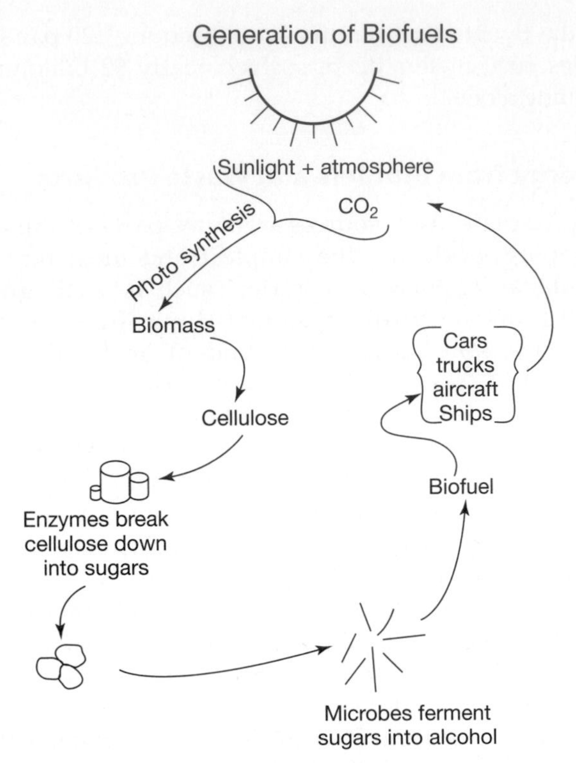

FIGURE 6.41 General schematic of the generation of biofuels using organic matter. The biomass is converted into sugars that are, in turn, converted into alcohol that can be used for fuel. This process returns carbon dioxide to the atmosphere but does not release extra carbon dioxide as does the burning of fossil fuels.

Interest in the use of gasohol in the United States and Europe has waxed and waned depending on the cost and availability of petroleum imports. After the OPEC embargo in 1973, in the early 1980s, and after 2005 as oil prices soared to over $100 per barrel, there was much discussion but little broad action. Gasohol, with about 10 percent alcohol, is available in many parts of the American midwest, but it is generally absent in the most populous areas of the east and west coasts. An additional stimulus for alcohol use appeared when the U.S. Environmental Protection Agency announced it would curtail the use of MTBE, an additive that reduces emissions from gasoline and increases octane, because MTBE was being found in many water supplies. The EPA suggested that ethyl alcohol could be used as an additive in place of the MTBE because it would present very little, if any, environmental hazard. Thus, ethyl alcohol might reappear in gasolines, this time for environmental reasons.

The example of Brazil draws attention to the use of biomass as an energy source not just as a waste product of other activities, but also through the biological harvesting of energy. It has been proposed, for example, that single-cell algae that are particularly efficient in producing hydrocarbons by photosynthesis could be grown in ponds, harvested, and fermented in a digester to produce methane. This could either be used as a substitute for natural gas or used on site to generate electricity. In the latter case, the carbon dioxide produced from methane combustion could be returned to the growing pond to promote the growth of the algae. Nobel laureate scientist Melvin Calvin had predicted that in the early parts of the twenty-first century, a significant contribution to world energy requirements would be made by plants. He envisaged the use of currently unproductive land (avoiding competition with food crops) to grow plants that could be mechanically harvested and dried. This material would be treated with chemical solvents to dissolve out sugars and other hydrocarbons; the sugars would be fermented to produce alcohol. Suitable plants for use as feedstock for such a "green factory" were identified and include the gopher plant (*Euphorbia lathyris*). From 1000 metric tons of plant material, Calvin suggested it would be possible to obtain 80 metric tons of hydrocarbons, 260 metric tons of sugars (that could yield 100 metric tons of alcohol on fermentation), and still have 200 metric tons of woody residue that itself could be used as a fuel. The important feature of the green factory is not its energy efficiency, but the fact that it produces liquid fuels and other chemicals (such as feedstocks) that currently depend heavily on oil supplies. The principles involved in the green factory could be extended to a wide range of farming activities, both on land and in the sea. The idea and plan were sound and may yet find wide application, but the low cost of fossil fuels in the later part of the twentieth century did not encourage the development of such alternative fuel systems.

In countries such as the United States, cultivated plants and grains are unlikely to provide a major future source of energy. For example, to satisfy the current U.S. consumption of natural gas using methane derived from plants would require a land area roughly equivalent to the total presently under cultivation (1.1 million square miles).

Diesel and Biodiesel

The diesel engine used in large trucks, farm equipment, and some types of cars, was invented by R.C.K. Diesel in 1892 and was originally fueled by oil extracted from peanuts. These engines were more efficient than many early gasoline engines, stronger in heavy vehicles, and used a fuel that was less flamable and less explosive than gasoline. However, within a few years the ready availability of cheap and abundant petroleum provided diesel fuel obtained from it as a distillation product, so the original vegetable oil fuels were discarded. Throughout the twentieth century, there were always novel diesel trucks or farm tractors running on some sort of animal fat or cooking oil, but 99.9 percent of diesel fuel was made by conventional petroleum refining. However, by 1990, interest had grown in a return to biologically based diesel production on a commercial scale because of the increasing cost of oil and concerns about the rise in atmospheric carbon dioxide. Production of biodiesel in the United States rose from 500,000 gallons in 1999 to more than 4,500,000 gallons in 2007. The U.S. Department of Energy notes that there is enough virgin soy oil, recycled greases, and other feedstocks readily available to produce about 1.7 billion gallons of biodiesel. Very optimistic projections place the rapidly growing production of biodiesel at 7.5 billion gallons (178 million barrels) annually by 2012.

Production of biodiesel uses a process called transesterfication in which a vegetable oil (such as soybean oil) or an animal fat is reacted with alcohol to remove glycerin. The result is biodiesel which can be simply mixed, at 5 to 20 percent, with regular diesel fuel in engines that require little or no modification. Proponents of biodiesel point to its leading to less use of imported oil, a reduction in unburned hydrocarbons, and zero sulfur emissions.

Hydrogen for Energy

Hydrogen has been used in transport as the gas that filled balloons and dirigibles more than a century ago and as one of the components in the fuel that today propels rockets into space. Hydrogen is now being viewed more generally as an energy source for aircraft and also for surface vehicles such as cars, trucks, and buses. Hydrogen is extremely abundant on Earth's surface, but it is always combined with other elements such as oxygen in water or in the vast array of hydrocarbon molecules that comprise biomaterials. When present in forms such as methane (CH_4), ethanol (C_2H_6O), or gasoline (a highly complex mixture of different molecules), hydrogen will burn, vigorously reacting with oxygen, producing heat and water vapor. In the search for alternative fuels, there are three very attractive aspects to hydrogen, or hydrogen-rich, fuels; first, hydrogen is abundant; second, the combustion of hydrogen produces large quantities of heat; and third, the combustion of hydrogen produces only water as a waste product.

As noted earlier, the burning of natural gas (chiefly comprised of methane, CH_4) yields much less carbon dioxide than does the burning of coal or oil; burning pure hydrogen yields no carbon dioxide. The problems are (1) how to produce large quantities of free hydrogen, and (2) how to transport and handle the hydrogen as a fuel. Hydrogen production from water requires a vast amount of energy. The gas itself is highly explosive and difficult to contain; cooling to condense it into a liquid form would require additional large quantities of energy. Hydrogen can be generated by a number of methods, but the most straightforward way is to pass an electrical current through water. Use of photovoltaic systems could provide the necessary electricity without burning fossil fuels. If hydrogen were to be used to power automobiles, it would be necessary to either develop large hydrogen production facilities and a new infrastructure to transport the hydrogen gas to service stations, or to develop many much smaller hydrogen production facilities that would be located near to each service station. The latter would eliminate the need for pipelines or tanker trucks to deliver the hydrogen. In either case, it would be necessary to keep the hydrogen under high pressure, or to liquefy it in order to have sufficient quantities to provide energy for significant travel distances. It is also important to recognize that highly compressed hydrogen gas, or liquid hydrogen, reacts more rapidly and much more violently with oxygen in the atmosphere than does gasoline and would therefore constitute a greater threat in accidents or fires.

Hydrogen-fueled engines could burn hydrogen much as automobile engines now burn gasoline, or they could make use of fuel cells. These devices combine the hydrogen with oxygen using catalysts and selective membranes such that they produce electricity. Thus, they do not generate gases to move pistons in cylinders but rather generate electrical currents that can propel a vehicle by using an electric motor. The hydrogen might be provided as a liquid or a compressed gas, or it might be generated on board a vehicle by using a reformer (converter) to separate the hydrogen from a fuel such as gasoline. In the twenty-first century, hydrogen fuel technology is expected to receive increasing attention in the hope that it may become a clean, and virtually inexhaustible, energy source for the future.

Nuclear Fusion—the Ultimate Energy Source?

The energy that is emitted by the Sun and by other stars throughout the universe results mainly from the process of **nuclear fusion,** in which nuclei of light atoms combine to form heavier atoms. The most promising candidates to provide fusion energy on Earth are heavy isotopes of hydrogen known as deuterium ($^{2}_{1}H$), and tritium ($^{3}_{1}H$), that can be fused to produce the heavier element helium (He), as $^{4}_{2}He$ or $^{3}_{2}He$. The fusion of two deuterium atoms can produce helium, a free neutron (n), and a great deal of energy:

$$^{2}_{1}H + ^{2}_{1}H \rightarrow ^{3}_{2}He + n + \text{energy (3.2 million electron volts)}. \quad (6.4)$$

A related reaction results in the formation of tritium by fusion of deuterium atoms:

$$^{2}_{1}H + ^{2}_{1}H \rightarrow ^{3}_{1}H + ^{1}_{1}H + n + \text{energy (4 million electron volts)}. \quad (6.5)$$

These two reactions are about equally probable when fusion of deuterium atoms occurs, but, whereas in the first case a stable product is formed, in the second case, the tritium atom reacts with another deuterium atom:

$$^{2}_{1}H + ^{3}_{1}H \rightarrow ^{4}_{2}H + n + \text{energy (17.6 million electron volts)}. \quad (6.6)$$

Therefore, the net result of these three reactions can be written as:

$$5\,^{2}_{1}H \rightarrow ^{4}_{2}He + ^{3}_{2}He + ^{1}_{1}H + 2n + \text{energy (24.8 million electron volts)}, \quad (6.7)$$

and the energy released per deuterium atom in these fusion reactions is 4.92 million electron volts. The fusion of deuterium and tritium to produce helium and release very large amounts of energy has already been achieved by humans, but only in the most uncontrolled and explosive manner imaginable. It is, in fact, the basis of the hydrogen (H), or thermonuclear, bomb. In the H-bomb, the conditions needed for fusion are created by first detonating an atomic bomb. The problem of using nuclear fusion as an energy source for the benefit of humankind consists entirely in the *controlled* and sustained production of such fusion reactions.

The technical problems involved in producing controlled fusion reactions are so great that, despite several decades of research already undertaken, it is most unlikely that commercial fusion reactors could be in operation within the next half century. To initiate a fusion reaction such as that involving two deuterium atoms, temperatures greater than 100 million degrees Celsius have to be reached. At such temperatures, a gas is so hot that its atoms have been torn apart by collisions into their component electrons and nuclei, and it is known as a **plasma.** This plasma has to be confined to allow collision and fusion, and one means of achieving this is to hold it within a strong magnetic field that is toroidal in shape (like a doughnut). The success of a Russian toroidal magnetic confinement machine called *Tokamak* has been followed by further work along these lines in the United States and England.

An alternative to this approach is called **inertial confinement** and involves firing a large amount of energy, which may be in the form of laser beams, electron beams, heavy ion beams, or even fragments of discrete matter, into a small blob of the mixture of hydrogen isotopes. Such beams, focused from different directions, can create shockwave compression and heating effects and, at the same time, confine the plasma. This method aims at producing fusion as a series of small explosions, perhaps several a second, whereas in magnetic confinement systems, a more continuous reaction is the objective. The plasma would have to be heated to the temperature of ignition when, like putting a match to a flammable material to start a fire, the reaction would begin and the energy released would maintain the temperature and hence, in turn, the reaction. Methods of heating the plasma, in addition to compression within an enormous magnetic field, have included shooting beams of neutral atoms into the plasma so as to collide with particles and raise the temperature, and the use of radio frequency heating, which is the principle employed in the domestic microwave oven. Whichever system is used to confine and heat the plasma, the system must be capable of controlling the three critical factors for fusion: temperature, plasma density, and time. A fusion reactor would, of course, generate energy in the form of large amounts of heat that could be extracted using conventional systems, probably based on those used in fission reactors.

Assuming that the great technical problems involved in controlled fusion can be overcome, what will be the fuel for such power plants and its cost and availability? If a reactor were to be constructed that utilizes the deuterium-deuterium reaction, then seawater would provide a vast supply of this fuel. Seawater contains one deuterium atom for every 6500 atoms of hydrogen, so that 1 m^3 of water contains 1.028×10^{25} atoms of deuterium that, if utilized in deuterium-deuterium fusion, has a potential fusion energy of 8.16×10^{12} joules. This is equivalent to the heat of combustion of 269 metric tons of coal or 1360 barrels of crude oil. If we extend this calculation to estimate the energy that would potentially be derived from 1 km^3 of seawater, this would be the equivalent to 269 *billion* metric tons of coal or 1360 *billion* barrels of crude oil. The latter figure is of the same order as some estimates of present and future world reserves of crude oil. Therefore, deuterium-deuterium fusion holds the possibility of generating the same amount of energy from a cubic kilometer or so of seawater as would be provided by all of the world's remaining oil; comparable estimates suggest that the world's remaining coal could be matched in energy by a few dozen cubic kilometers of seawater.

However, these calculations assume that the fusion reactor is based on the deuterium-deuterium reaction, whereas much current research is aimed at a controlled deuterium-tritium reaction (as in equation 6.6), which requires far less stringent experimental conditions. Tritium, however, is much rarer than deuterium and has to be produced by neutron bombardment of another light element, lithium (Li):

$$^{6}_{3}\text{Li} + \text{n} \rightarrow {}^{4}_{2}\text{He} + {}^{3}_{1}\text{H} + \text{energy (4.8 million electron volts)}, \quad (6.8)$$

and

$$^{7}_{3}\text{Li} + \text{n} \rightarrow {}^{4}_{2}\text{He} + {}^{3}_{1}\text{H} + \text{n} + \text{energy (2.5 million electron volts)}. \quad (6.9)$$

This tritium would then combine with deuterium as in reaction (6.10) and the net result would be equivalent to:

$$^{6}_{3}\text{Li} + {}^{2}_{1}\text{H} + \text{n} \rightarrow 2{}^{4}_{2}\text{He} + \text{n} + \text{energy (22.4 million electron volts)}. \quad (6.10)$$

Unfortunately, lithium is not available in great abundance (in seawater, for example, it occurs as only 1 part in 10 million), and the isotope ^{6}Li constitutes only 7.4 percent of natural lithium. It is an element that can be extracted from certain brines and is mined in the form of the mineral spodumene ($LiAlSi_2O_6$), which occurs in pegmatite deposits. It is much more limited as a resource; estimates suggest that if used as a major energy source, it may only last a few hundred years.

The ultimate fuel for fusion reactors would be hydrogen itself, the supply of which is virtually unlimited. It is hydrogen fusion that is chiefly responsible for energy production in the Sun and stars, but it requires still greater temperatures. The technical problems involved in harnessing hydrogen fusion for energy generation are even greater than for the other fusion reactions.

In 1989, two distinguished scientists working in the field of electrochemistry claimed to have achieved low-temperature nuclear fusion reaction at room temperature by passing a small electrical current through palladium electrodes immersed in a beaker containing heavy water (water in which the hydrogen atoms are deuterium). In their experiments, they claimed that helium and excess heat were generated by a fusion reaction involving deuterium. Unfortunately, this phenomenon, known as "cold fusion," could not be satisfactorily reproduced by other scientists and has been dismissed as misinterpreted observations.

Nuclear fusion can potentially provide humankind with an energy source that is almost limitless. A further advantage of nuclear fusion is that hazardous radioactive wastes are not produced as byproducts of reactor operation; thus, there is neither the problem of transport nor storage of dangerous fissionable materials. Neutron radiation is produced in fusion, however, as the above equations show, but some scientists believe that reactors could be developed employing reactions like that of deuterium –^{3}He reactions, which although requiring much higher temperatures than deuterium-deuterium or deuterium-tritium reactions (namely, about 300 million degrees) produce 10 to 50 times less neutron radiation. The potential of nuclear fusion as an energy source of the future is obvious, and its limitations lie not in the availability of fuel but in our ingenuity in developing the necessary technology.

THE FUTURE

There is not an energy shortage. Vast amounts of energy are available on Earth, more than we could ever use. Any energy crisis is of our making because we rely too heavily on relatively cheap and convenient fossil fuels. Many alternatives to the use of fossil fuels exist, as we have shown in this chapter, but each of the alternatives also has drawbacks.

Nuclear energy, once predicted to become the dominant energy source for the future, is now viewed as both promise and poison. Some countries are moving forward in making nuclear power their major electricity source, whereas others are abandoning it altogether. Although no new plants have been initiated in the United States for more than twenty years, there is renewed interest in nuclear power because of concerns over reliance on oil imports and the environmental impact of burning fossil fuels (especially CO_2 release into the atmosphere). Wastes from nuclear power plants remain lethally radioactive for periods of thousands to tens of thousands of years, and there is still no consensus on how to safely dispose of these wastes. Furthermore, breeder reactors produce more and more fissionable materials that could be used for nuclear weapons.

The other forms of energy, especially renewable forms discussed in this chapter, range from well-established sources such as hydroelectric power to the highly speculative sources such as ocean power and nuclear fusion. The two renewable energy sources seeing greatest expansion at present are wind and solar. Both share the problem of high initial capital cost, but their "green" operation which does not emit carbon dioxide into Earth's atmosphere positions them for continued growth. However, windmills do give rise to public concerns over aesthetics and their environmental impact on migrating birds. Solar cells will likely be more widely incorporated into construction materials. Even so, neither of these energy sources is likely to become the dominant source of electricity in the next 10–20 years.

Whatever sources ultimately replace the fossil fuels will depend on a whole range of economic, social, and political factors, some of which may be undreamed of as yet. Only two points about the future seem now to be certain. Energy needs will continue to rise for at least the near future, and energy sources will have to change. What is vital for many of the technologically advanced societies is to prepare for that change because it will surely involve changes in the need for many other resources and in the lifestyles of future generations.

Metals

Metals were rare and highly prized by our earliest ancestors, but they have become so commonplace that we use them today in myriad fashions, discard them in massive quantities, and rarely question where they come from. Metals vary so widely in their nature that we use them to build our homes and cars, to supply electricity, to make pots in which to cook our foods, to serve as our money, to be our jewelry, to become our weapons, to fill our teeth, and for countless other uses. It was metals, along with the fuels needed to refine and process them, that transformed the agrarian societies of our ancestors into the largely urban and technological societies of today.

HISTORY

Metal usage probably began with the discovery of small amounts of metals in their native (metallic) state, principally gold and copper. They were unique in appearance, required no refining, and could be easily melted, shaped, and sharpened. Their malleability meant that they rarely broke, and when they did, they could be reformed without loss of character, and used over and over again. From those earliest discoveries and uses more than 10,000 years ago, metals have transformed our society so greatly that we now name periods of human history for them—the *Copper Age,* the *Bronze Age,* and the *Iron Age.*

The rarity and softness of gold precluded its use in tools and relegated it to aesthetic applications. The greater abundance and hardness of copper, however, allowed it to serve in primitive technology—from pins and knives to spear tips and other weapons. The scarcity of native metals ensured that they were highly valued, widely traded, and continuously recycled. Scarcity also meant that objects made of metal were only owned by the ruling families. Prominent and widespread usage of metals in human society had to wait until the technology of metal smelting was developed approximately 6000 years ago; this allowed humans to tap the large supplies of metals held in ore deposits. This next step, however, required the additional technological developments involved in mining the ores from which the metals could be extracted and the development of transportation systems to move the ores and fuels to the smelters. Thus, one technological advance (smelting) spawned the need for others (mining and transportation).

Copper, the first metal to be used in technology, was joined by an alloy, bronze, about 5000 years ago, when it was discovered, probably accidentally, that tin enhances the durability of the copper. Iron was first discovered by people who picked up iron meteorites. The toughness of objects made from the iron in the meteorites brought recognition of the superiority of iron in many applications, but it was not until the development of iron smelting techniques, about 4000 years ago, that iron from terrestrial sources became available in large quantities. From that time to the present, iron has served as the primary metal for society and, throughout most of human history, iron and its alloys (the steels), have constituted more than 94 percent by weight of all the metal used.

The uses of all metals expanded vastly during the Industrial Revolution, and the diversity of metals used expanded again during the twentieth century. Mining has advanced from small labor-intensive operations that concentrated on the extraction of high-grade ores, to massive operations, where hundreds of thousands of tons of ore can be extracted in a single day from a single mine. Worldwide, more than 1.4 billion tons of refined metals are extracted annually from the more than 15 billion tons of ore and waste rock that are unearthed. The supply of metals today is maintained not only by this massive extraction effort but also by increasingly large quantities of recycled metals. Recycling not only provides an additional supply of a metal, but it saves the energy used in mining and refining, preserves landfill space, and lessens environmental contamination.

TYPES AND ABUNDANCES OF METALS

Metals can be separated into two categories—the **abundant metals,** whose presence in Earth's crust exceeds 0.1 weight percent, and the **scarce metals,** whose presence lies at or below 0.1 weight percent. The abundant metals—silicon, aluminum, iron, magnesium, manganese and titanium—occur in such quantities and are so widely distributed that their future availability will never be in doubt. Despite the abundance of these metals, only iron, which has served as the "backbone of civilization," has a long history of usage. Silicon, aluminum, magnesium, and titanium have only been identified in the past 200 years and have only found important uses in the past 100 years. These are also energy-intensive metals, meaning that they require much higher than average amounts of energy (usually electrical) to be won from their ores. This adds to their costs and tends to place some limits on their use.

The scarce metals are hundreds to thousands of times less abundant than the abundant metals, but they are vital in myriad applications because even in small amounts they keep industry efficient, effective, and healthy. They control the properties of the abundant metals in alloys, carry electrical currents, control the operation of computer chips, and keep automobiles and airplanes operational. The abundant metals, especially iron, may be likened to the flour used as the basic ingredient in baking, whereas the scarce metals would be likened to the sugar, salt, baking soda, and spices used to create a wide variety of tastes and textures in our foods. The scarce metals can be subdivided into groups based upon common characteristics and uses:

- The **ferro-alloy metals**—those used mainly to alter the properties of the iron-based alloys we call steels. The most important members of this group are nickel and chromium, which are incorporated in nearly all stainless steels.
- The **base metals**—a broad group originally viewed as having low value and inferior properties relative to precious metals, but now recognized as serving society in thousands of applications. Copper has the longest history of use, and today, it is the most important of the base metals because of its nearly universal use to transmit electricity. Lead was the most extensively used base metal before the rise of copper, and it remains important in automotive batteries, but its toxic properties are leading to its replacement in many other applications.
- The **precious metals**—those metals recognized historically as having great monetary and jewelry value and being resistant to corrosion (hence, also commonly called "noble metals") but also serving increasingly in many technological applications. Humans have had a long love affair with gold and have adorned the living and the dead with gold jewelry for thousands of years. No other metal has been so widely sought, nor has been the focus of such great mining rushes.

Although gold remains a metal of jewelry and finance, it also serves widely in areas ranging from electronics to medicine. Silver, also a metal of coinage and jewelry, came into its own as a technological metal in photography and even finds medical applications today. Platinum and its sister elements have only been recognized since the 1700s, and today, they are widely used for their catalytic properties which make them invaluable in chemical processing and in automotive catalytic converters.

- The **special metals**—a very broad group finding increasing importance in technological applications in the twenty-first century. Many of these metals (e.g., gallium, germanium, rare earths) find roles in modern society not for their native metallic properties but for the unique electronic properties they bring to devices such as computer chips, solar cells, integrated circuits, color phosphors, and lasers. Others find such diverse uses as flame retardants (antimony), wood preservatives (arsenic), shock resistant cookware (lithium), nuclear fuel rods (beryllium), and baseball bats (scandium).

It is difficult to envision a world without metals because we depend on them in thousands of applications. Modern automobiles contain as many as twenty-eight different kinds of metals, from the engine and frame to the computers and to the catalysts in the exhaust system. Were the metals denied us, we would return to the Stone Age and be largely reduced to a survival mode of existence. The roles of metals in our society are likely to increase as we continue to find new and novel applications for metals and new groups of metal-containing compounds.

The bridge at Ironbridge, constructed between 1775 and 1779 on the Severn River in England, was one of the first major structures constructed of iron. It remains today as a monument to the Industrial Revolution. (Courtesy of the Ironbridge Gorge Information Administration.)

CHAPTER 7

Abundant Metals

If we remove metals from the service of man, all methods of protecting and sustaining health and more carefully preserving the course of life are done away with. If there were no metals, men would pass a horrible and wretched existence in the midst of wild beasts;...will anyone be so foolish or obstinate as not to allow that metals are necessary for food and clothing and that they tend to preserve life?"

DE RE METALLICA, GEORGIUS AGRICOLA, 1556, TRANSLATION BY HERBERT C. AND LOU H. HOOVER, 1912

FOCAL POINTS

- Metals are extracted from ores obtained by the mining of *ore deposits*; the mineability of such deposits depends upon such factors as their mineralogy, grade, size, location, and the price of the metals extracted.
- The abundant metals, those with an average concentration in the Earth's crust of 0.1 percent by weight or greater, are silicon, aluminum, iron, magnesium, titanium, and manganese.
- Iron has accounted for about 94 percent of all the metal used throughout history.
- Iron deposits occur in all types of rocks; the primary minerals from which iron is extracted are hematite (Fe_2O_3), magnetite (Fe_3O_4), and goethite ($FeO{\cdot}OH$).
- The largest iron deposits in the world are the *banded iron formations;* these formed as chemical precipitates in the Precambrian oceans, possibly coinciding with the generation of oxygen by the first photosynthesizing plants.
- Iron smelting is carried out by reducing iron oxides to iron metal by reaction with carbon monoxide gas, usually derived from coke.
- World steel production has risen steadily since 1950, although production in the United States has remained approximately constant at about 100 million metric tons; thus, the U.S. proportion of world steel production dropped from nearly 50 percent in 1950 to only about 7 percent in 2007. Similarly, traditional steel-producing countries, such as Great Britain, have seen a decline in their relative positions, while newer producing countries, such as China and Japan, have greatly increased production.
- Manganese, a metal vital to steel production, is mined in many countries (although not the United States); it also occurs as ferromanganese nodules on many parts of the deep ocean floor.
- Aluminum, derived from bauxite (a soft heterogeneous mass of aluminum hydroxides occurring as a "soil" in some areas), has become the second most widely used metal because of its light weight, ability to conduct electricity, corrosion resistance, and workability; the major drawback of aluminum is the large amount of energy needed to extract it from its ores.

- Titanium, although used in many advanced technologies, is employed primarily in the preparation of white paint pigment.
- Magnesium, the lightest of the abundant metals, is used in the preparation of refractories and in lightweight alloys with aluminum.
- Silicon, the most abundant metal in Earth's crust, has long been used in steel manufacturing and is being increasingly used in new technologies such as solar cells and computer chips.
- Metals, recycled prehistorically because of their rarity, are recycled today for profit, to save energy, and to reduce environmental contamination. Iron recycling far exceeds the total of all other metals recycled.

METALS AND THEIR PROPERTIES

Metals are unique among the chemical elements in being opaque, tough, ductile, malleable, and fusible, and in possessing high thermal and electrical conductivities. Approximately half of the chemical elements possess some metallic properties, but all true metals have two or more of the special metallic properties. Our early ancestors were drawn to the use of metals because metals are tough (but not brittle like stone), malleable, and can be melted and cast. These same properties are important today, but we also rely heavily upon the special electrical and magnetic properties of metals and their machineable characteristics. Without metals, technology as we know it could not have come into being.

Today, approximately 35 metallic elements are available through mining and processing of their ores. The metals are generally separated into two groups, the abundant metals as listed in Table 7.1 and the scarce metals, discussed in Chapter 8. The abundant metals are those whose geochemical abundance exceeds 0.1 percent of the Earth's crust. Although many metallic elements are used in their pure forms because of unique properties, modern society commonly finds that chemical mixtures **(alloys)** of two or more metals, or metals and nonmetals, have superior characteristics. Alloys are also metallic, but they usually have properties of strength, durability, or corrosion resistance that improve upon the properties in the component-pure metals. Examples of common alloys are steel, brass, bronze and solder. **Steel** is an alloy in which the main constituent, iron, is combined with other metallic elements such as nickel, vanadium, or molybdenum, or a nonmetallic element such as carbon. Steels are tougher, less brittle, and more resistant to wear than iron alone. Brass is an alloy of copper and zinc. It melts at a lower temperature than copper, the metal it most nearly resembles, and it is much easier to cast. Bronze was the first alloy used by our ancestors and was developed about 3500 B.C.E. This alloy of copper and tin melts at a relatively low temperature, is very easily cast, is harder than pure copper, and is corrosion resistant. Common solder is an alloy of lead and tin that has an especially low melting temperature. The molten alloy has the property of combining with certain other metals. This property allows a solder, when cooled and solidified, to create a join between two pieces of the same metal or, in some cases, two fragments of different metals.

THE NATURE OF ORE DEPOSITS

Most metals are dispersed at low concentrations throughout the common rocks of the Earth's crust. Hence, economic extraction of the metals from most rocks is difficult to impossible and prohibitively expensive. Fortunately, the variety of geological processes discussed in Chapter 2 can be effective in extracting, transporting, and depositing metals into the concentatioins we call resources or reserves. Even so, it is important to remember that not all concentrations of metals actually constitute ore deposists. As noted in Chapter 1, a concentration must be legally, technically, and economically extractible to become an ore deposit.

Many factors determine whether or not a given rock is an ore, not only of the abundant metals but of

TABLE 7.1 The abundant metals

Element	Symbol	Atomic No.	Atomic Wt.	Crustal Abundance (%)	Density (g/cc)	Melting Point (°C)
Magnesium	Mg	12	24.31	2.3	1.74	649
Aluminum	Al	13	26.98	8.2	2.70	660
Silicon	Si	14	28.09	28.2	2.33	1410
Titanium	Ti	22	47.90	0.57	4.50	1660
Manganese	Mn	25	54.94	0.095	7.20	1244
Iron	Fe	26	55.85	5.6	7.86	1535

any metals; these have been discussed at length in Chapter 4, but it is worthwhile to recall that the most important factors are:

1. mineralogy
2. grade
3. grain size and texture
4. size of the deposit
5. depth of the deposit
6. geographic location
7. market value of metal
8. possible by-products

The mineralogy, or the chemical form in which a metal occurs, and the grade or proportion of metal content are the most important factors because they determine the extraction process and the energy that must be used to extract the metal and, thus, the cost of the extracted product. Each of the abundant metals occurs in a wide variety of minerals, but it is only the few listed in Table 7.2 that serve as **ore minerals** for these metals. Iron is present in large quantities in minerals such as fayalite (Fe_2SiO_4) and pyrite (FeS_2), but neither is mined primarily as an iron ore because of the large amounts of energy needed to extract the iron from fayalite and the pollution problems resulting in extraction of iron from pyrite. Similarly, aluminum is present in many silicate minerals, but the extraction is generally prohibitively expensive relative to that for the ore minerals listed.

The grain size of an ore mineral and the nature of its intergrowths with other minerals determine the method of processing and extraction and may, in the case of very fine grain size, actually make rich ores unworkable. The size and depth of the deposit control the mining method and, hence, much of the cost. Geographic location influences accessibility, environmental conditions, and even the tax or royalty charges incurred in mining.

The recovery of by-products commonly helps make the mining of mineral deposits profitable; in fact, some important metals such as cadmium, gallium, and germanium are only recovered as by-products. The abundant metals are generally recovered from ores in which they are the only metal refined, but many ores of the scarce metals yield a variety of useful by-product metals. The by-products are not present in sufficient grades to warrant their being mined alone, but they are recovered at little or no additional cost when mining the major metal, and their sales help pay for the total mining operation. In many instances, the by-products, which can also include waste rock products such as sand or agricultural limestone as well as metals, make mining operations profitable when major metal prices fall or when the grade of the major metal is insufficient alone to keep a mine profitable.

Note that most of the ore minerals listed in Table 7.2 are oxides, hydroxides, or carbonates. Even though silicate minerals are by far the most abundant in Earth's crust, they are rarely used as the sources for metals because they are difficult to handle and extremely expensive to smelt. However, it may be that the future will see many silicate minerals being used, but at present only two silicate minerals, kaolinite and anorthite, are viewed as potential sources for a metal, in this case,

TABLE 7.2 The abundant metals and their principal ore minerals

Metal	Important Ore Minerals	Amount of Metal in the Ore Mineral
Silicon	Quartz (SiO_2)	46.7
Aluminum	Boehmite ($AlO \cdot OH$)	45.0
	Diaspore ($AlO \cdot OH$)	45.0
	Gibbsite [$Al(OH)_3$]	34.6
	Kaolinite [$Al_2Si_2O_5(OH)_4$]	20.9
	Anorthite ($CaAl_2Si_2O_8$)	19.4
Iron	Magnetite(Fe_3O_4)	72.4
	Hematite (Fe_2O_3)	70.0
	Goethite ($FeO \cdot OH$)	62.9
	Siderite ($FeCO_3$)	62.1
	Chamosite [$Fe_3(Si,Al)_2O_5(OH)_4$]	45.7
Magnesium	Magnesite ($MgCO_3$)	28.7
	Dolomite [$CaMg(CO_3)_2$]	13.1
Titanium	Rutile (TiO_2)	60.0
	Ilmenite ($FeTiO_3$)	31.6
Manganese	Pyrolusite (MnO_2)	63.2
	Psilomelane ($BaMn_9O_{18} \cdot 2H_2O$)	46.0
	Rhodochrosite ($MnCO_3$)	39.0

aluminum. Where geochemically scarce metals are concerned, silicate minerals are sometimes used under special circumstances. Examples are beryl ($Be_3Al_2Si_6O_{18}$) used as a source of beryllium, zircon ($ZrSiO_4$) used as a source of zirconium, spodumene ($LiAlSi_2O_6$) used as a source for lithium, and willemite (Zn_2SiO_4), which has been mined for zinc in a few localities where the high grade of the ore warranted its recovery.

IRON: THE BACKBONE OF INDUSTRY

Iron is the third most abundant metal in Earth's crust, and historically, it has been the workhorse of industry. Today, iron and its alloys, steel, account for more than 94 percent by weight of all metals consumed, and a significant proportion of the remainder—most of the nickel, chromium, molybdenum, tungsten, vanadium, cobalt, and manganese—are mined principally for use as alloy components in the steel industry. The reasons for iron's dominance are not hard to find: the first is the abundance and ready accessibility of rich iron ores; the second is the relative ease with which the smelting process can be carried out; the third is low cost; and the fourth, and most important, is the special property of iron and its alloys which allow them to be tempered, shaped, sharpened and welded to give products that are exceptionally strong and durable. No other metal enjoys the same range of versatile properties. Rudyard Kipling in his brief poem "Cold Iron" captured the versatility of iron in a striking way:

> Gold is for the mistress—silver for the maid
> Copper for the craftsman, cunning at his trade.
> 'Good!' Said the Baron, sitting in his hall,
> but iron—cold iron—is master of them all.

Since 2004, the total annual world production of iron and steel has exceeded 1 billion (10^9) metric tons, and in industrial countries such as the United States, more than 600 kilograms (1300 pounds) per capita per annum are used. Although there are many end products of iron and steel production, the major categories today are construction (30 percent), transportation (25 percent), and machinery (20 percent). Other important uses include grocery cans, home appliances, and oil, gas, and water drilling equipment. Today, we take for granted the ready availability of iron and steel in a great variety of forms. It was not always so, and we need go back only about 250 years to recount the advances that have led iron and steel to their positions of prominence today.

Iron Minerals and Deposits

Iron, because of its crustal abundance and its chemical reactivity, occurs in hundreds of different mineral forms. However, like the other abundant metals, it is economically extracted from only a few minerals (Table 7.2). The iron oxides, **hematite** (Fe_2O_3) and **magnetite** (Fe_3O_4) are, by far, the most important ore minerals and will no doubt continue to be for many years. The other ore minerals were important in the past and are still locally significant today. Pyrite (FeS_2), although widespread and locally abundant, does not serve as an ore of iron because processing releases large quantities of noxious sulfur gases and produces inferior quality iron because of remaining traces of sulfur.

Iron deposits occur worldwide (Figure 7.1) and have been formed as a result of many different processes, and at many times throughout geologic time. The behavior of iron in geologic processes, especially at Earth's surface, is strongly influenced by its ability to exist in more than one oxidation state. Under the reducing conditions that occur beneath Earth's surface and in some deep and relatively oxygen-free waters, iron exists in the **ferrous** state (Fe^{2+}) and forms ferrous minerals such as siderite ($FeCO_3$), chamosite [$Fe_3(Si,Al)_2O_5(OH)_4$], and mixed ferrous-ferric minerals such as magnetite. Some ferrous iron compounds are relatively soluble. By comparison, at Earth's surface, or wherever oxygen is abundant, iron oxidizes to the **ferric** state (Fe^{3+}) and forms extremely insoluble compounds. Ferric iron is deposited in minerals such as hematite or **goethite** ($FeO{\cdot}OH$). Iron ore deposits form by **igneous**, **metamorphic**, and **sedimentary** processes. There are many distinct types of deposits, but nearly all of the major deposits mined today or in the recent past are of only a few types. These will likely continue to serve as our major sources of iron ore, and they are concisely described below.

DEPOSITS FORMED THROUGH IGNEOUS ACTIVITY. Three major types of iron ores arise from igneous activity: (1) accumulation in large **mafic** intrusions, (2) contact metamorphic deposits, and (3) ores formed through submarine volcanism.

(1) Large mafic igneous intrusions, such as the Bushveld Igneous Complex in the Republic of South Africa, commonly contain significant concentrations of magnetite. In these intrusions, the magnetite was precipitated and settled out of the magma in thick layers on the floor of the chamber as the magma crystallized. The total amount of magnetite in the igneous rock may be only a few percent, but its occurrence in nearly pure layers makes mining relatively easy. Although these accumulations in mafic igneous rocks are not presently mined for their iron content, in the Republic of South Africa they do serve as important ores of the vanadium that is also concentrated in the magnetite (these are discussed in Chapter 8 and shown in Figure 8.7).

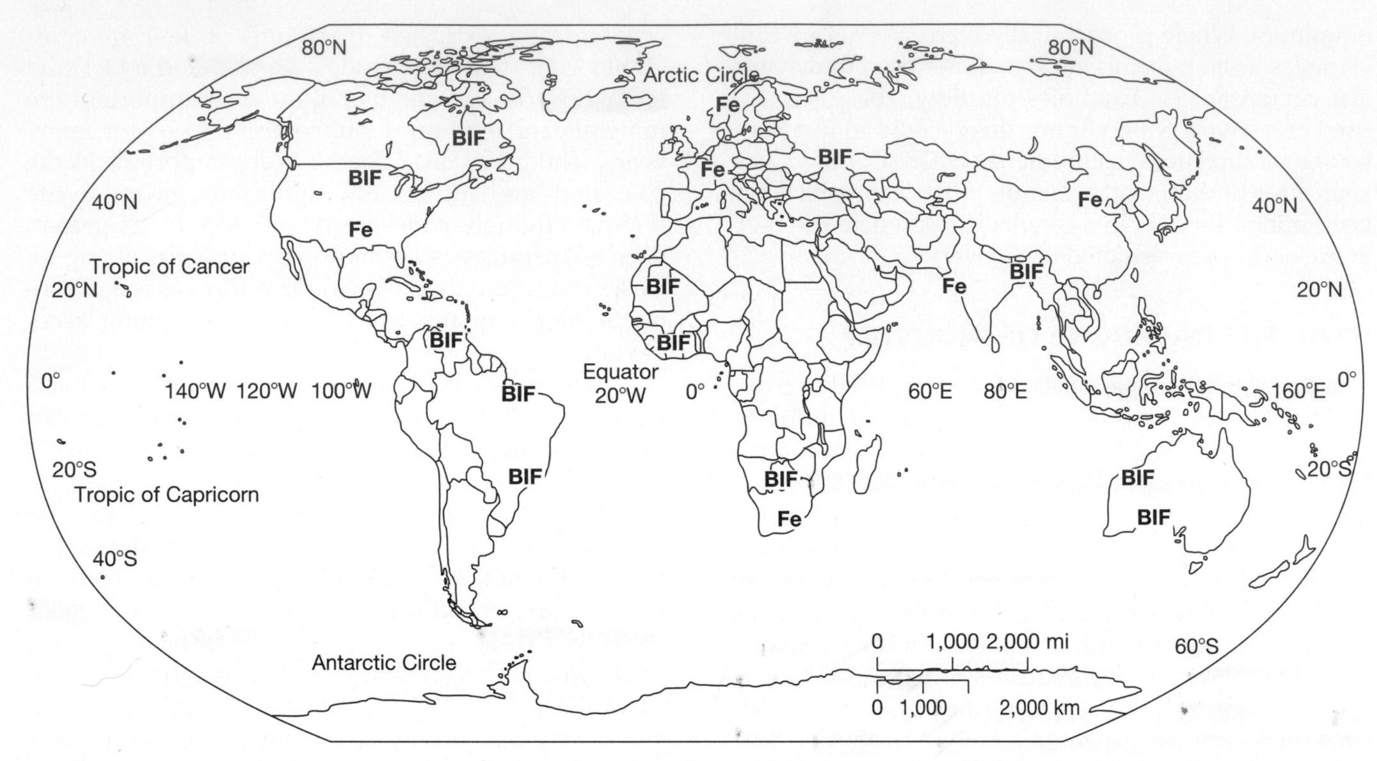

FIGURE 7.1 The locations of some of the major iron deposits in the world. The banded iron formations (BIFs) constitute the world's major producers today and are the major reserves that will provide iron in the twenty-first century.

(2) Contact metamorphic deposits form when iron-bearing fluids emitted by cooling igneous intrusions react with adjacent rocks, especially limestones (Figures 7.2 and 7.3). The hot fluids react with and sometimes completely replace the rocks that enclose the intruded magma, leaving a mixture of coarse-grained iron oxides and a host of unusual metamorphic minerals. Most contact metamorphic deposits are too small to be economically recovered, but in the United States at localities such as Cornwall and Morgantown, Pennsylvania, Iron Springs, Utah, and Pilot Knob, Missouri, massive contact metamorphic deposits were mined for many years.

(3) Seafloor volcanism is almost always accompanied by submarine hot springs that issue forth solutions rich in iron and silica. The rapid cooling and oxidation of the solutions as they mix with seawater results in the precipitation of iron oxide and silica known as **Algoma-type** deposits after the locality in the Canadian Shield where they were first recognized. A few of these deposits in Canada and elsewhere have proven rich enough to mine, but most are too small or too low in grade to be successfully worked.

RESIDUAL DEPOSITS. These deposits are formed where the weathering process leaches soluble minerals,

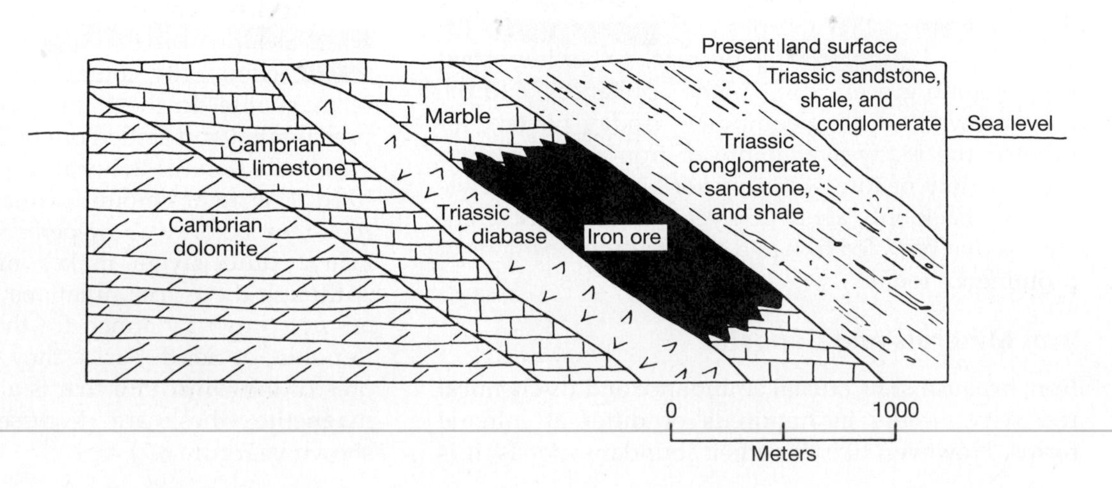

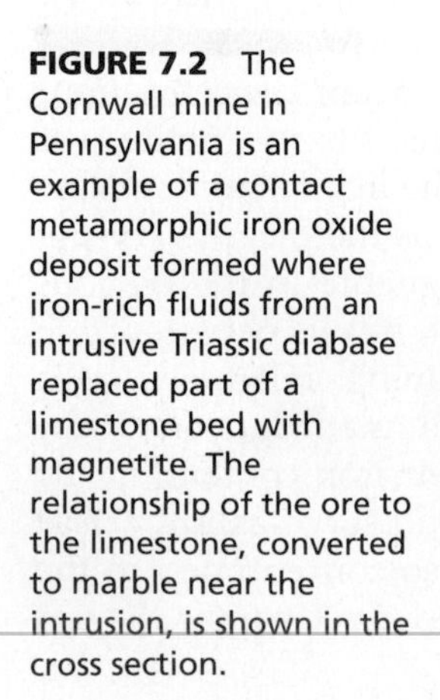
FIGURE 7.2 The Cornwall mine in Pennsylvania is an example of a contact metamorphic iron oxide deposit formed where iron-rich fluids from an intrusive Triassic diabase replaced part of a limestone bed with magnetite. The relationship of the ore to the limestone, converted to marble near the intrusion, is shown in the cross section.

FIGURE 7.3 A view of the open-pit mine at Cornwall, Pennsylvania, in which the massive magnetite ore (dark zone in the base of the pit) is visible below the white marble formed by metamorphism of the original limestone. (Photograph courtesy of Bethlehem Steel Corporation.)

oxidizes ferrous iron, and leaves behind a concentration of insoluble ferric iron minerals. This process accounts for the brown, yellow, and red colors that are familiar in most soils. Locally, especially in tropical regions where the chemical reactions are rapid because of higher temperatures and abundant rainfall, the weathering process removes the more soluble minerals such as feldspars and quartz, and leaves concentrated residues of iron oxides and hydroxides. These types of deposits, known as **brown ores**, have been forming since Precambrian times and are widespread. Most of the deposits are, however, small and although locally important in the past, they are not considered economic in terms of today's large-scale mining operations.

In tropical regions, intensive leaching has left large areas of hard red residual soils known as **laterites** (Figure 7.4), a name derived from the Latin *latere*, meaning *brick*. These soils are poor agriculturally and often become worse when they are used for farming because the exposure causes $Fe(OH)_3$ in the soil to irreversibly dehydrate to FeO·OH. This dehydration reaction is similar to that which occurs when clays are baked to form bricks and results in the soils becoming so hard that they do not absorb moisture and become unworkable. Where most intensely developed, laterites may contain 30 percent or more iron and may, thus, represent large future sources of iron. The tonnages of iron potentially available from the lateritic soils of the world probably exceed those from all other sources by a factor of at least ten, but their mining would cause massive environmental problems.

SEDIMENTARY DEPOSITS. Three important types of iron ores have formed through sedimentary processes: **bog iron deposits**, **ironstones**, and **banded iron formations.** The *bog iron deposits* are the smallest of these types of sedimentary deposits and, although not mined today, they provided ore for many early European operations, and they were the first iron ores to be exploited by the fledgling American industry in the seventeenth and eighteenth centuries. These deposits form locally in glaciated regions and in coastal plain sediments where ferrous iron, originally

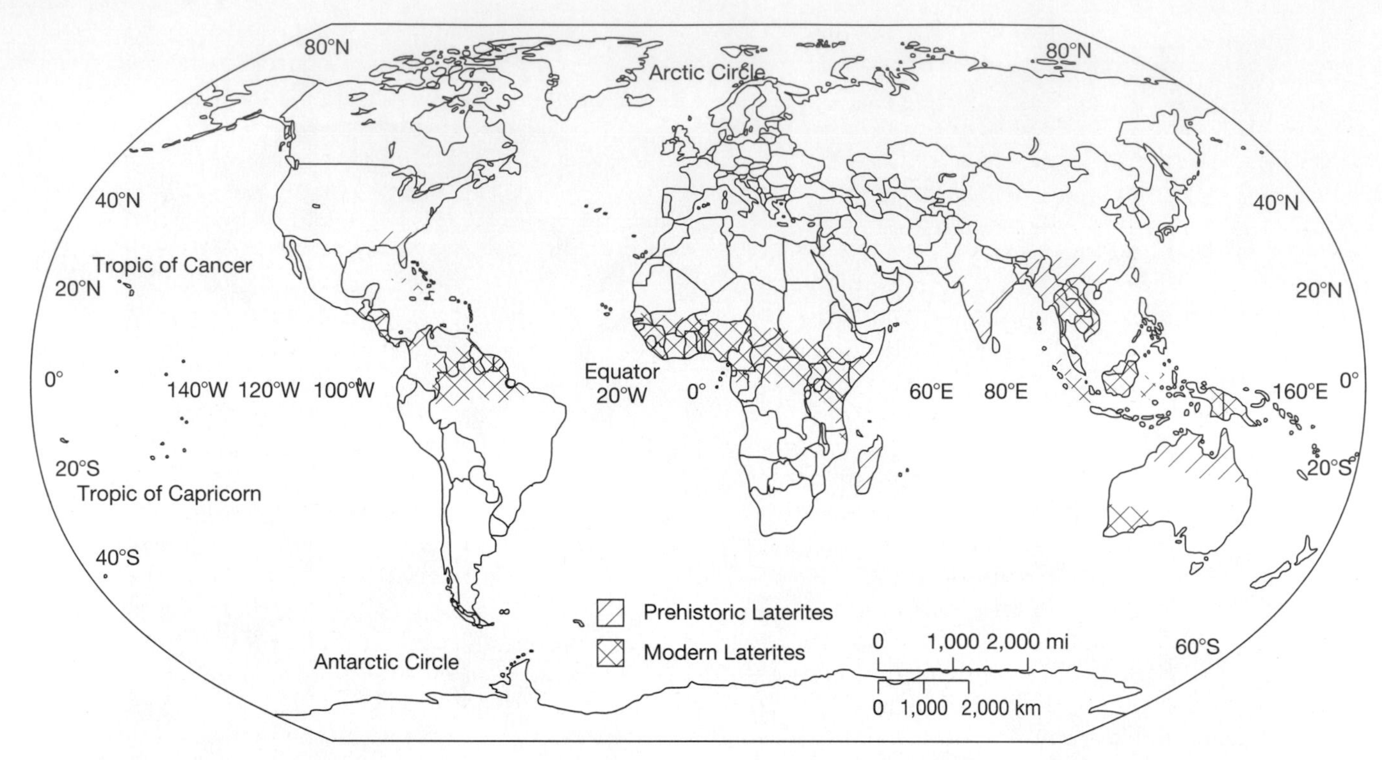

FIGURE 7.4 Laterites are hard, red, residual soils rich in iron minerals that formed in tropical to subtropical regions. The shaded regions on the map show where intense leaching has removed more soluble compounds and formed iron-rich laterite soils.

put into solution by the reducing conditions created by decaying organic matter, is oxidized and precipitated as lenses and sedimentary cements (Figure 7.5). The deposits are generally quite local in extent and are highly variable in grade.

Ironstones are iron-bearing formations that are much larger and more important than bog iron deposits. The ironstones are continuous sedimentary beds that can extend for hundreds of kilometers (miles) and are a few meters to dozens of meters (tens to hundreds of feet) in thickness. In many respects, they are similar to various other sedimentary rocks, containing cross-beds, abundant fossils, oolites, facies changes, and other sedimentary features. The main difference is that they contain significant amounts of goethite, hematite, siderite, or chamosite (a complex iron aluminum silicate) as coatings on mineral fragments, as oolites, and as replaced fossil fragments. The origins of the ironstones are not clear, but they appear to represent shallow, near-shore marine and, rarely, freshwater sediments where the formation of the iron minerals occurred both as direct sedimentation and as a diagenetic replacement. Distribution of the individual iron minerals was controlled by the nearness to shore, the water depth, and the amount of oxygen in the water. In the shallowest areas, the abundance of oxygen resulted in oxidation of the iron to the ferric state and the formation of goethite, whereas in the deeper areas, richer in CO_2 but poorer in oxygen, the iron precipitated in the form of siderite or chamosite.

The iron in ironstones must have been transported to the sites of deposition in the soluble ferrous state, either in groundwater or as deep basin fluids. The special conditions that led to the formation of the ironstones are believed to have resulted from generally warm and humid climates that permitted the generation of abundant organic matter. This resulted in significant amounts of organic acids and carbon dioxide in groundwater solutions that then readily reduced and dissolved iron minerals in the soils and rocks. The leached iron moved slowly via the groundwater system into lakes or shallow marine basins where the iron was precipitated. In some cases, the iron minerals precipitated directly on the sea floor or lake floor; in others, the iron minerals replaced calcium carbonate minerals, fossils, and oolites, or were precipitated by diagenetic processes interstitially among the other minerals.

The ironstones have formed since the beginning of the Cambrian Period, a time span inclusively called

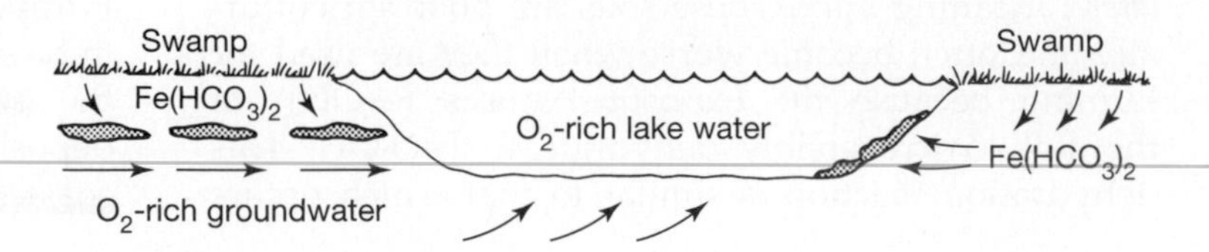

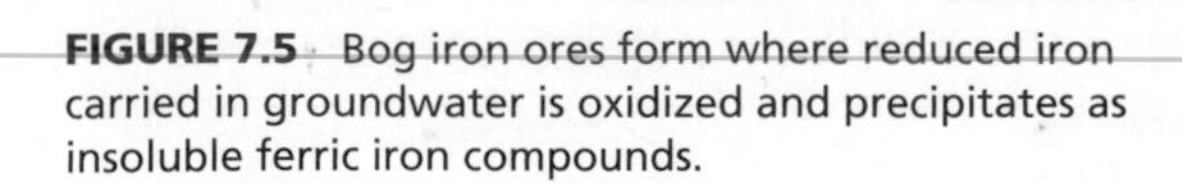
FIGURE 7.5 Bog iron ores form where reduced iron carried in groundwater is oxidized and precipitates as insoluble ferric iron compounds.

the Phanerozoic Eon. Consequently, these ores have sometimes been referred to as the Phanerozoic type. Iron ores of this type have been very important in Europe where they are collectively referred to as the **Minette type** and where they supplied much of the ore for the iron and steel industries of the United Kingdom, France, Germany, and Belgium. In North America, where Appalachian examples of these ores are called the **Clinton** (or **Wabana) type**, they were important resources from the late eighteenth century until the mid-twentieth century. Today, the importance of these ores is much diminished because the richest areas have been mined-out and because banded iron formations are richer and more economical to work.

The largest concentrations of iron oxides are found in *banded iron formations* (commonly called BIFs for short), which today supply most of the world's iron ores, and constitute the bulk of the world's iron ore reserves (Table 7.3). They are also known as **Lake Superior-type** ores after the large deposits of the kind that occur in the Lake Superior region. The BIFs occur in Precambrian rocks on all continents, and they are mined extensively in the United States, Canada, Brazil, Australia, South Africa, India, China, and Russia (see Figure 7.1). Various terms have been applied to deposits of this type in different parts of the world. Commonly encountered names for silica-rich ores are **taconite** (in the United States), **itabirite** (in Brazil), or **banded jaspilite** (in Australia). They are from 30 to 700 meters (100 to 2200 feet) in thickness and often extend over hundreds to thousands of square kilometers (miles). Many of these deposits, including all in the United States and Canada, have been metamorphosed to some degree so that they now consist of fine-grained magnetite and/or hematite in a matrix of quartz, iron silicate, and iron carbonate minerals in a very compact finely laminated rock (Figure 7.6). The BIFs, like the ironstones, commonly display strong facies development, indicating that at the same time differing conditions resulted in the formation of iron oxide-rich zones closest to shore and iron carbonate or iron silicate zones farther out into the basin. The iron contents of the BIFs vary widely, but the presently mined deposits typically have 20–40 percent iron.

The BIFs have produced billions of tons of ore and now are the world's major source of iron; however, their modes of origin remain somewhat enigmatic. Banded iron formations have several major characteristics. The most important is that most, if not all, formed during the period of 2.6 billion to 1.8 billion years ago. They all exhibit the typical banding visible in Figure 7.6, and they all are very low in aluminum content and are nearly free of common detrital sedimentary debris. They resemble the Algoma-type deposits in that they consist of fine layers of silica and iron oxide minerals, but the BIFs are much broader in extent and do not show any apparent relationship to submarine volcanism. Detailed studies of the BIFs indicate that they were formed in broad sedimentary basins following prolonged periods of continental weathering and erosion and the inundation of the land surface by shallow seas. The erosion had previously removed most detrital debris so that deposition in the basins was largely that of chemical precipitates. The origin of the iron is not clear, but it is likely that it was derived from several different sources including weathering of continental rocks, leaching of marine sediments, and submarine hydrothermal systems that discharged iron-rich fluids onto the seafloor.

TABLE 7.3 World iron ore reserves and production in 2008 in millions of metric tons

		Crude Ore		Iron Content	
Nation	Mine Production 2008	Reserves	Reserve Base	Reserves	Reserve Base
United States	54	6,900	15,000	2,100	4,600
Australia	330	16,000	45,000	10,000	28,000
Brazil	390	16,000	33,000	8,900	17,000
Canada	35	1,700	3,900	1,100	25,000
China	770	21,000	46,000	7,000	15,000
Russia	110	25,000	56,000	1,400	31,000
India	200	6,600	9,800	4,200	6,200
Kazakhstan	26	8,300	19,000	3,300	7,400
Mauritania	112	700	1,500	400	1,000
South Africa	42	1,000	2,300	650	1,500
Sweden	27	3,500	7,800	2,200	5,000
Ukraine	80	30,000	68,000	9,000	20,000
Other countries	50	11,000	30,000	6,200	17,000
World total (may be rounded)	2200	150,000	350,000	73,000	160,000

Source: From USGS, 2009.

FIGURE 7.6 Banded iron formations (BIFs) may have very regular layering (see Plate 33) or display irregular banding as shown here. The light colored bands consist of nearly pure iron oxides (hematite and magnetite), whereas the darker zones are nearly pure silica.

Today, it would be impossible to transport the huge quantities of iron from eroding land surfaces in rivers and streams or to disperse it widely from submarine vents because it would be quickly oxidized and precipitated in the insoluble ferric form. However, during the Precambrian when the BIFs formed, there was apparently little free oxygen in Earth's atmosphere or dissolved in surface waters. The carbon dioxide content of the atmosphere was probably higher, and gases such as methane, CH_4, might have been present. Under such conditions, rainwater, stream and lake waters, and ocean waters would have been slightly acidic and much less oxidizing, conditions that would have allowed for the ready transport of iron in solution in the soluble ferrous form. The iron accumulated in the broad shallow basins and gradually precipitated as iron oxides and hydroxides. The cause of the repetitive precipitation of layers of alternating iron oxides and silica has been a point of much debate, with suggestions including annual climatic changes, cyclical periods of evaporation, the effects of microorganisms altering silica availability and releasing oxygen, episodic volcanism, and many others. Whatever our final understanding of the origins of these important deposits, their formation was apparently directly controlled by the nature of the Precambrian atmosphere. As the atmosphere changed, so did the capacity of the ocean to serve as a transporter of iron. When photosynthesis started contributing large quantities of free oxygen into the atmosphere, banded iron deposits no longer formed; therefore, we have no analogous processes active in the world today.

Banded iron formations typically contain 20–40 percent iron, values that were long considered too low for economic recovery. In some areas, however, surficial chemical weathering has removed the associated siliceous or carbonate minerals and left enriched residual ores containing 55 percent or more iron. The initial mining efforts of the mid-to-late-1800s in the great Precambrian iron deposits of the world such as the Lake Superior district of the United States (see Figure 7.7 and Box 7.1) and in the 1950s and 1960s in the Labrador

FIGURE 7.7 Open-pit mining of the banded iron formations (locally called "taconites") in the Lake Superior District. The ore is blasted out in a series of benches and is then transferred by large power shovels into trains or large trucks for transport to the processing plant. (Photograph courtesy of U.S. Steel Corporation.)

BOX 7.1
The Iron Ranges

The principal metal of the modern era is iron, and the principal deposits of iron in North America are those of the Lake Superior District (Figure 7.A). These deposits have yielded more iron ore than any other district in the world, and they have provided the iron and steel on which the American Industrial Revolution was built.

The first European settlers in eastern North America had to rely totally upon iron shipped from Europe. Within a few decades, a fledgling American iron industry developed and drew upon a variety of small and often relatively low grade ores scattered on the coastal plain and along the Appalachians. Through the first half of the nineteenth century, explorers were opening up the central and western parts of what now constitute the United States and Canada. On September 19, 1844, a surveying party working on the Upper Peninsula of Michigan suddenly noted that their compasses displayed great variations, with compass needles commonly pointing west-southwest instead of north. Subsequent examination of the rocks in the area known as the Marquette Range, revealed that they consisted of high-grade iron ore containing magnetite (Fe_3O_4). Within a year, the Jackson Mining Company was incorporated and bought up 1 square mile (640 acres) of land for $2.50 an acre. Mining commenced, and the first loads of ore were shipped in 1848; however, the lack of any major transportation system delayed large-scale development until 1855 when completion of a shipping port at Sault Ste. Marie opened the way to Lake Superior.

The discovery of the Marquette iron ores led to a broad-scale search for additional deposits around Lake Superior and resulted in the discovery of the Menominee Range in 1849, the Gogebic Range in 1884, the Vermilion Range in 1885, the Mesabi Range in 1890, and the Cuyuna Range in 1903. Not only were the discoveries numerous but the amount of ore was almost beyond belief. The Marquette Range is 53 kilometers (33 miles) long, the Menoninee District is 80 kilometers (50 miles) long, and the Mesabi Range is 177 kilometers (110 miles) long. Furthermore, the thicknesses of the ore-bearing zones are high by any standard, 105–230 meters (340–750 feet) in the Mesabi Range and up to 760 meters (2500 feet) in the Marquette Range. Previously discovered ore deposits were considered large if they were measured in millions of tons, but these ores were gigantic and were measured in the billions of tons.

Despite the enormous size of the iron ore deposits of the Lake Superior District, a major problem remained—transportation. To get the ores from the isolated areas around the western side of Lake Superior, it was necessary to construct canals and railroads to provide the low-cost transportation of millions of tons of ores to shipping terminals on the Great Lakes. This allowed the Lake Superior mines to become the primary suppliers of ore for the great steel centers of Pittsburgh, Pennsylvania, and Gary, Indiana, and resulted in the closure of the smaller eastern mines. The large quantities of high-grade iron ore that became available provided the backbone for American industry as the Industrial

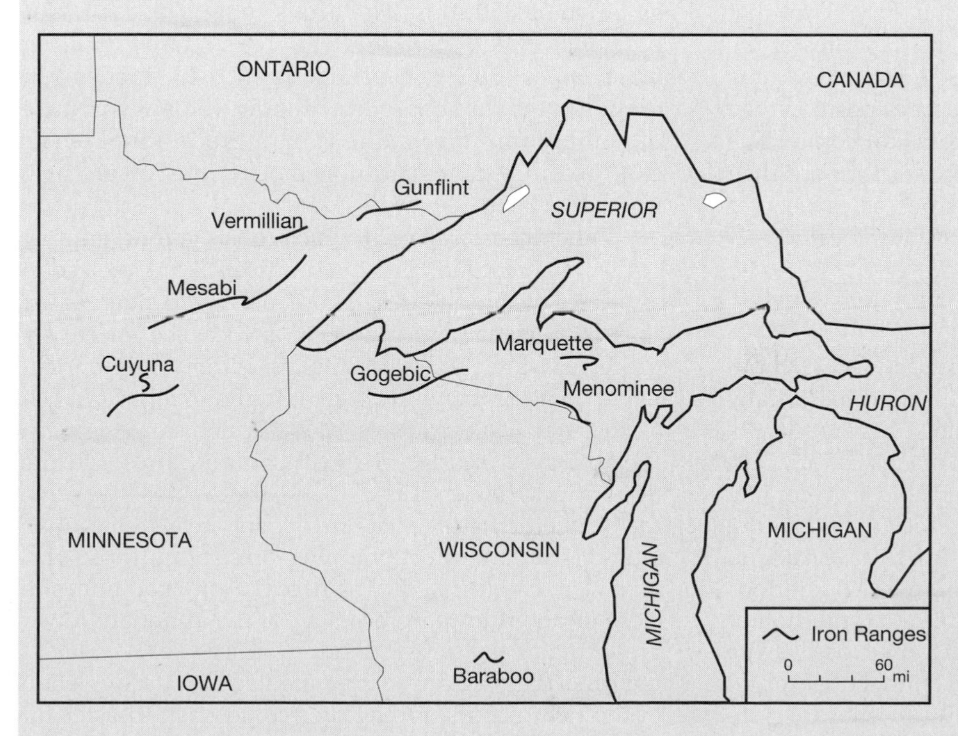

FIGURE 7.A The great banded iron formation-bearing ranges of the Lake Superior District occur as linear belts as much as 150 kilometers (90 miles) in length in Michigan, Wisconsin, Minnesota, and Ontario.

(continued)

BOX 7.1

The Iron Ranges (*Continued*)

Revolution transformed the United States from a rural country to an industrial giant. Iron ore production, which was only about a few hundred thousand tons in 1860, increased to more than 2.5 million metric tons by 1885, to 20 million metric tons in 1900, and ultimately to nearly 100 million metric tons in 1951. In all, nearly 600 individual mines produced ore from the iron ranges, with nearly 300 mines lying along the Mesabi Range.

By 2000 the total production of iron ore from the Lake Superior iron ranges had exceeded 5.5 billion tons, far surpassing any other district for any kind of ore in the world. Although the original high-grade ores, which contained as much as 65 percent iron, have been mined out and foreign production has reduced the market, the iron ranges are still the heart of the iron and steel industry in the United States, providing nearly 50 million metric tons of ore each year. Today's ores, referred to as "taconites," average only about 25 percent iron and must be processed to remove silica before being smelted, but the reserves exceed 50 billion tons and it appears that they will continue to serve as major sources of iron for the United States for many years to come.

Trough in Canada, Cerro Bolivar in Venezuela, Minas Gerais in Brazil, the Hamersley Range in Australia, and Krivoi Rog in Russia were all for the enriched ores. In the early years of the twentieth century, there was concern, echoed by the words of Andrew Carnegie and reproduced on page 243, that these rich ores were nearing depletion in places such as the Lake Superior District. Initial efforts in the early 1900s to concentrate the lower grade ores, the taconites, were only moderately successful, but renewed interest in the 1940s led to the development of economic concentration techniques. This new technology opened the way for exploitation of billions of tons of ores previously considered waste and led to the present iron ore mining industry. Since that time the technology for processing the lean ores of the BIFs has spread worldwide, and these ores have become the world's dominant source of iron ore, a position they will certainly hold for many years to come. The Lake Superior District alone had produced more than 5 billion tons of ore by 1990 and still possesses reserves several times greater than the ore thus far mined.

Mining and Beneficiation

Iron ore mining on a small scale dates back several thousand years. The earliest operations obtained ores locally by digging in a wide variety of shallow pits and underground tunnels. However, as better means of transport evolved, many smaller mines closed and iron mining became concentrated in larger, more efficient operations.

Today, more than 85 percent of the world iron ore production is mined from open-pit operations because the iron ore bodies have large lateral dimensions and many of them lie relatively close to the surface. Furthermore, open-pit mines have larger production capacities, are generally cheaper to operate per ton of ore, and are easier and safer to maintain than underground mines. As a result, the underground iron mines have found their ability to compete with open-pit mines continually reduced; this is evidenced by a drop in the number of underground iron mines in the United States from about thirty in 1951 to none today. The U.S. open-pit mines operate at a disadvantage relative to many foreign mines because they must mine 5–6 metric tons of rock (about 3 metric tons of ore and 2–3 metric tons of overburden waste) for each metric ton of iron oxide product produced. In contrast, the large Brazilian and Australian mines only have to mine 1.5–2 metric tons for each ton of iron oxide produced.

Open-pit mining is done primarily with large power shovels and with trucks having capacities of as much as 350 tons. The ore is removed in a series of steps, called **benches** (Figure 7.7), by drilling blast holes 30–38 centimeters (12–15 inches) in diameter that are charged with an explosive mixture of ammonium nitrate and fuel oil; individual blasts can break up to 1.5 million metric tons of ore at a time.

The richest iron ores, which consist almost entirely of the iron ore minerals listed in Table 7.2, are referred to as **direct shipping ores.** These usually contain more than 50 percent iron by weight and can be effectively processed directly at the smelter without any enrichment. Some of the world's great iron ore deposits, such as those at Hamersley in Australia and at Carajas, in Brazil, consist mostly of direct shipping ores. On the other hand, the deposits of the Lake Superior District in the United States contained only relatively small amounts of the direct shipping ores. Once those rich ores were exhausted, it was necessary for the mines to close or adapt to the processing of the abundant lower grade materials. In fact, Andrew Carnegie, pioneer of the American steel industry, lamented that the depletion of the rich direct shipping ores could spell disaster for the American iron mining industry. In his address to the Conference of Governors at the White House, May 13–15, 1908, he said:

> I have for many years been impressed with the steady depletion of our iron ore supply. It is staggering to learn that our once-supposed ample supply of rich ores can hardly outlast the generation now appearing, leaving only the leaner ores for the later years of the century. It is my judgment, as a practical man accustomed to dealing with those material factors on which our national prosperity is based, that it is time to take thought for the morrow.

The lower-grade materials of the Lake Superior District are called taconites and consist of a banded admixture of iron oxide and silicate minerals. They commonly contain only 50 percent iron oxide minerals and have grades of about 30 percent iron. Fortunately, the technology to effectively process these ores has now been developed, and Andrew Carnegie's fears for the exhaustion of iron ores are no longer relevant. To use taconites, they must be upgraded by beneficiation, which serves to concentrate the valuable iron minerals and remove valueless silicate minerals, including any problem impurities such as minerals containing phosphorus and sulfur that might be present. The final product consists of powdery or fine, sand-sized grains of iron oxides; the grains are difficult to handle and cannot be fed directly into blast furnaces without being blown out the top. Therefore, the iron oxide grains are formed into 1–2 centimeter pellets [Figure 7.8(a)] by adding a binder, usually bentonite, fine volcanic ash or clay, and then fired so that they are strong enough to be transported and easily handled. These pellets have proven very useful to the iron and steel industries because they are easy to handle, have uniform compositions of 63–65 percent iron, and because their natural porosity allows them to react rapidly with the carbon monoxide gas in the blast furnace during the smelting process.

(a)

(b)

FIGURE 7.8 (a) After the iron ores have been mined, they are finely ground to liberate the iron oxides from the gangue minerals. The powdery oxides are then formed into pellets which are suitable for loading into the blast furnaces. (b) The iron ore pellets are transported across the Great Lakes in large ships such as the *Burns Harbor*, shown here loading up for its voyage. (Photograph courtesy of Bethlehem Steel Corporation.)

Iron and Steel Smelting

The very first iron objects were made from iron meteorites that did not require smelting. But meteorites are rare, and the use of terrestrial iron ores consisting of iron combined with other elements requires **smelting** to obtain the pure metal from the ore mineral. The origins of iron smelting are unknown but probably lay in the Middle East or Asia Minor more than 3000 years ago. The Iron Age is usually dated as beginning about 1200 B.C.E. when the use of iron tools spread rapidly across the Middle East, and then across Asia to China. Ironmaking subsequently spread throughout Europe and possibly into Africa by the conquests of the Romans, who learned the techniques from the Greeks. European colonialization then disseminated iron smelting to other parts of the world, such as the Americas, where it was not previously known. The modern blast furnace had its origins in about A.D. 1340 and slowly evolved into its present form, shown in Figure 7.9. It consists of a refractory-lined cylindrical shaft into the top of which the charge can be introduced and from the bottom of which molten iron and slag can be drawn.

To produce 1 metric ton of iron, 1.6 metric tons of iron ore pellets must be mixed with 0.7 metric tons of coke and 0.2 metric tons of limestone. The mixture is introduced into the blast furnace where it reacts with 3.6 metric tons of air at temperatures of about 1600°C (3000°F). The chemical reactions are complex, but the two principal reactions are the controlled combustion of the coke to produce carbon monoxide and the reduction of the iron oxides to iron metal by the carbon monoxide, which is simultaneously oxidized to carbon dioxide.

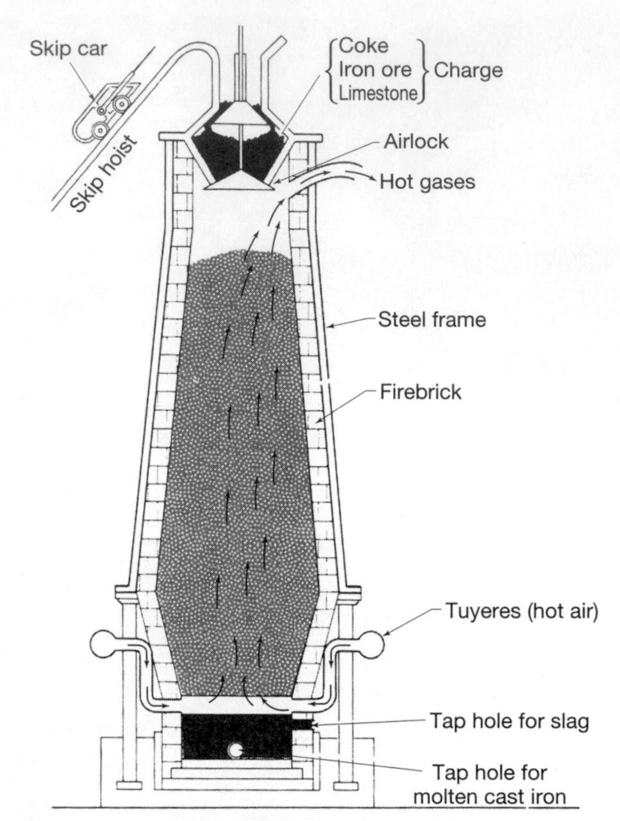

FIGURE 7.9 Diagrammatic representation of the interior of a blast furnace. The production of each metric ton of pig iron from pellets containing 60 percent iron requires approximately 250 kilograms of limestone as a flux and a metric ton of coking coal. Electric arc and basic oxygen furnaces require different mixes and are noteworthy for being more efficient in their use of coke, but the same three ingredients are used. Production of iron illustrates how interdependent are the uses of the different metals. (From Skinner, *Earth Resources*, Englewood Cliffs, NJ: Prentice Hall, 1986.)

$$\underset{\text{(coke)}}{C} + \underset{\text{(air)}}{1/2O_2} \rightarrow \underset{\text{(carbon monoxide gas)}}{CO}$$

$$\underset{\text{(carbon monoxide gas)}}{3CO} + \underset{\text{(iron ore)}}{Fe_2O_3} \rightarrow \underset{\text{(free iron)}}{2Fe} + \underset{\text{(carbon dioxide gas)}}{3CO_2}$$

The limestone aids the formation of a slag that floats on top of the molten iron and absorbs undesirable elements from the charge. Molten iron, referred to as "**hot metal**" is tapped from the bottom of the blast furnace into a transfer ladle that delivers it to the steel-making furnace (Figure 7.10). The slag, which is tapped separately, used to be discarded, but today may be used as concrete aggregate, railroad ballast, or soil conditioner.

Approximately two-thirds of all the iron produced is used in the manufacture of steel, an alloy of iron with one or more other elements that give the resulting metal specifically desired properties. Carbon steel, the most easily made and most widely used, is produced by adjusting the amount of carbon left after original smelting. Today, as shown in Table 7.4, many different metals are added to iron to produce steels with different properties for specific uses (Figure 7.11). Steel production takes place in a separate furnace where the alloying elements are carefully admixed. After the steel is refined, it is cast, rolled, or otherwise shaped into the useful forms we see around us.

Iron and Steel Production

Ever since the Industrial Revolution, the quantity of iron ore mined and the amount of iron and steel produced have far surpassed the production of all other

FIGURE 7.10 Iron, melted in a blast furnace, is transferred by a large ladle into a steel-making furnace where it will be converted to a specific steel alloy, prior to processing into finished products. (Photograph courtesy of U.S. Steel Corporation.)

TABLE 7.4 Elements added to iron to give desirable properties to steel

Element	Function in Steel
Aluminum	Remove oxygen; control grain size
Chromium	High temperature strength; corrosion resistance
Cobalt	High temperature hardness
Niobium	Strength
Copper	Corrosion resistance
Lead	Machinability
Manganese	Remove oxygen and sulfur; wear resistance
Molybdenum	High temperature hardness; brittleness control
Nickel	Low temperature toughness; corrosion resistance
Rare earths	Ductility; toughness
Silicon	Remove oxygen; electrical properties
Sulfur	Machinability
Tungsten	High temperature hardness
Vanadium	High temperature hardness; control grain size

metals combined. Indeed, iron and steel production has sometimes served as a general measure of the economic well-being of a nation.

The iron ores that fed the fledgling European iron industry at the beginning of the Industrial Revolution in the 1700s were taken from large deposits in England, in the Alsace-Lorraine area along the French–German border, and in Sweden. The first European settlers in eastern North America had to import iron products from Europe; however, as they explored the Atlantic coastal plain, they discovered local accumulations of bog iron deposits in the swamps and bogs. Such ores had been mined in earlier times in Europe but were used up by about 1700. As the American settlers moved westward, they found sedimentary ironstones in the high ridges of the Appalachians from New York to Alabama. Although these formations contain significant quantities of iron oxide minerals wherever they occur, iron concentrations only locally reached economic levels (40–50 percent Fe). The largest district was that near Birmingham, Alabama, which has subsequently closed down.

Westward expansion in the United States eventually resulted in the discovery of extremely large iron deposits near the western end of Lake Superior in 1845. Production from the vast deposits there gradually displaced all other sources. Furthermore, the discovery of similar deposits in many parts of the world and the development of techniques for bulk mining and processing these ores have resulted in the Lake Superior-type ores becoming a major source of iron in the world today, a position they are likely to retain for many years to come.

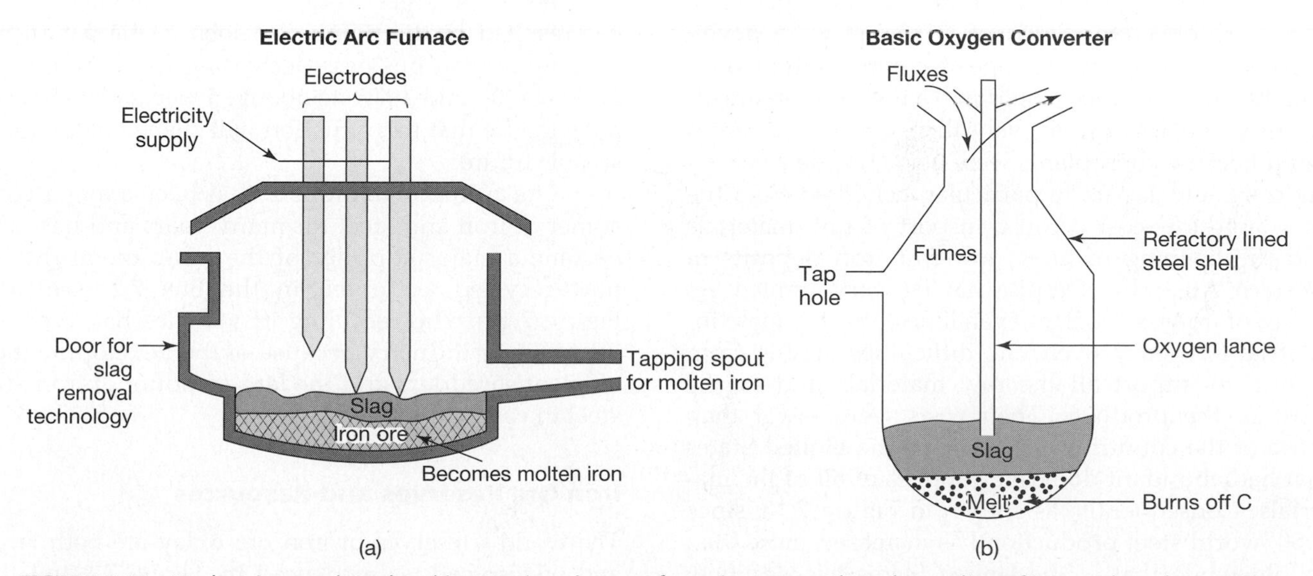

FIGURE 7.11 Modern iron and steel-making involves the use of new iron ore and much recycling of iron and steel scrap. The blast furnace still provides the primary means of making raw iron from new ore, but the basic oxygen and electric arc furnaces use varieties of scrap as their primary charges. Ferro-alloy metals are added to produce the specific types of steels required.

The modern steel industry traces its origins to the development of blast furnaces in central Europe in the fourteenth century. Growth was slow and production limited in quantity and quality until Abraham Darby found the way to use coke in iron smelting in 1709 in Shropshire, England. Europe, and England in particular, were the centers of iron and steel production and technology until the mid-nineteenth century when the combined effects of the discovery of the Lake Superior ores, the development of the Bessemer process for mass production of steel (in England), and the rapidly expanding American economy shifted the focus across the Atlantic. By the beginning of the twentieth century, the United States had become the world's dominant steel producer, a position it held until the 1970s. The United States was well supplied with high-grade ores, abundant coal and limestone, and effective waterway and rail transport systems were developed to bring the ores to places such as Bethlehem and Pittsburgh, Pennsylvania, and Gary, Indiana, where the major iron and steel centers flourished.

Iron and steel production surged worldwide in the first years of the twentieth century but then were reduced in continental Europe by the effects of World War I. Production began to expand in the 1920s, only to stagnate during the Great Depression of the early 1930s. By the late 1930s, production once again began to increase as economies recovered and as Germany and Japan began preparations for World War II. The war drastically reduced European and Japanese capacity for iron and steel production but left the United States with an intact industry and a large world market.

Thus, in 1950, the United States was the preeminent steel producer, supplying 47 percent of the world's total (Figure 7.12). Since that time, there has been a very marked change in the industry as Japan and Western Europe were rebuilt and as larger scale and more effective means of transport were developed. Japan and the European countries that once bought American steel could make their own more efficiently and less expensively than the United States could because their plants were new and their operating costs low. Japan, in particular, benefited from the new large low-cost ocean transport of raw materials and the opening of huge, low-cost iron deposits in Western Australia. Despite having very limited reserves of iron ore, coal, or even limestone, Japanese industrial efficiency overcame difficulties arising from having to import all the raw materials and export most of the products. Their costs were lower than those of the countries in Europe or the United States that had abundant domestic supplies of all of the materials. Consequently, as shown in Figure 7.13, since 1950, world steel production has increased more than sixfold while that of the United States has declined; U.S. production in 2008 was only 7 percent of the world's total, whereas China, with the newest furnaces, produced 38 percent. The decline in American iron and steel production, combined with cost-cutting efforts to try to remain competitive, has resulted in a reduction in the number of jobs in the blast furnaces and steel mills from more than 470,000 in the period 1976–1979, to less than 240,000 in 2008 (Figure 7.13). There are no indications that this trend will change in the years ahead. The situation is very similar in a number of European countries, especially Great Britain, where the iron and steel industry has experienced a large loss of markets and jobs because it cannot compete effectively with Japan and China.

While the American and British iron and steel industries struggle, other nations are expanding theirs—China and South Korea are particular examples, South Korea, which like Japan is resource-limited, has developed an extremely modern and efficient industry that competes very effectively with Japan in terms of iron and steel exports. China, which is large and has rich coal deposits to use as fuel in the blast furnaces, has made great strides in building state-of-the-art steel plants that can compete with any in the world.

The problems of the American iron and steel industry today are not those of resource adequacy, but rather are problems of economic competitiveness. Mines and mills that were operating far below capacity, and the massive steel mills of Pittsburgh and Bethlehem, Pennsylvania, are now closed. They have been replaced by much smaller custom mills that make specialty steels for specific needs. Since 1980, American iron ore imports have ranged from 19–37 percent of consumption and steel imports have ranged from 15–23 percent of consumption. Canada has been the largest foreign supplier of iron ore, whereas the European Economic Community, Brazil, and Japan have been the largest sources of imported steel. This has resulted in the number of jobs in the American iron ore mining business decreasing from more than 20,000 in the mid-1970s to about 7500 in 2000. There is no evidence that this situation will change in the foreseeable future.

The automotive industry has been a major consumer of iron and steel for many years and has also become a major supplier of the scrap metal that is now recycled, as noted in the Box 7.2 (see also Figure 7.14). The recycling of vehicles has evolved into a major industry because of the development of efficient shredders and the large amount of iron and steel in each vehicle.

Iron Ore Reserves and Resources

The world's reserves of iron ore today are both large and widespread, as evidenced by Figure 7.1 and the

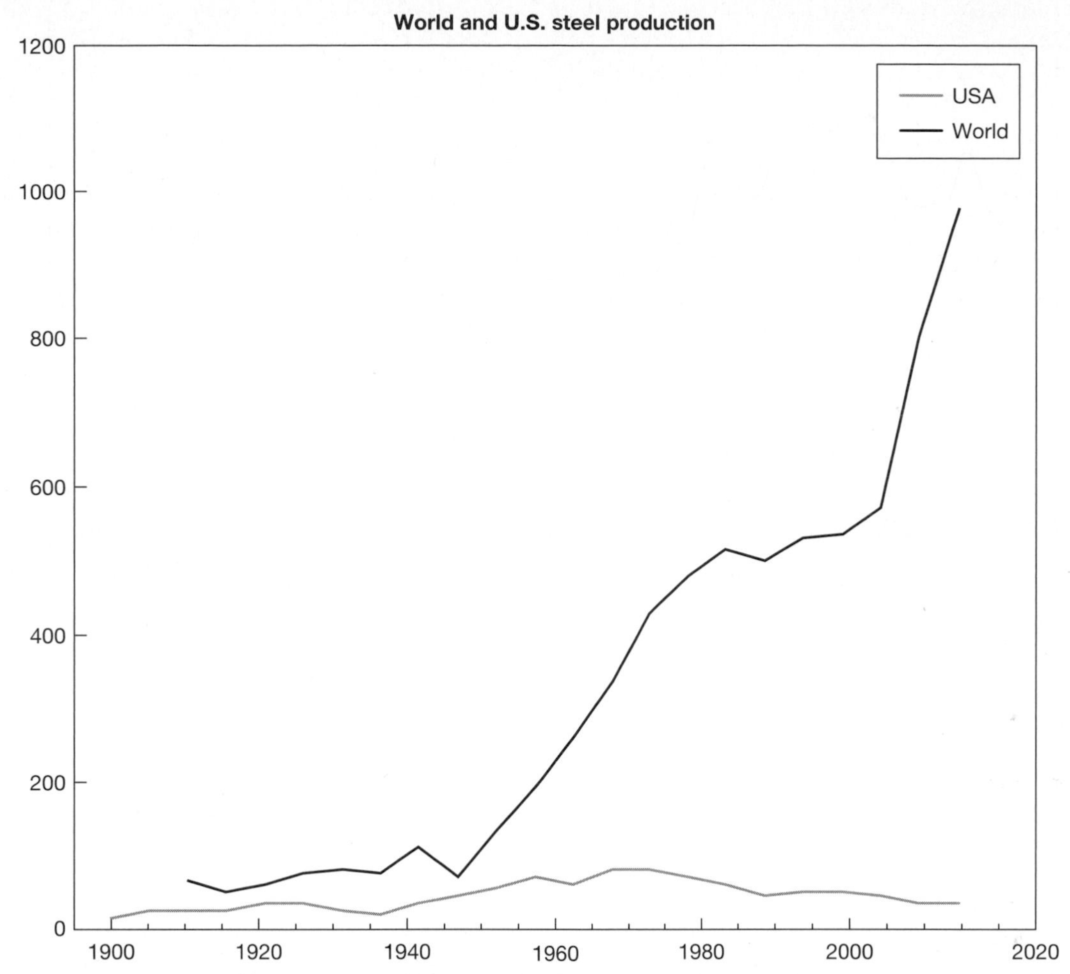

FIGURE 7.12 Raw steel production from 1950 through 2008 for the United States and the world. In 1950, the United States produced about 47 percent of the world total; by 2008, U.S. production, which was nearly the same tonnage as sixty years earlier, constituted only about 7 percent of the world's production. (Data from U.S. Geological Survey.)

data in Table 7.3. Considering that annual world production of iron ore is now about 2.2 billion metric tons, the known world reserves of about 150 billion metric tons will be sufficient to last for many years. Furthermore, the United States Geological Survey estimates world iron ore resources to be at least 350 billion tons, containing more than 160 billion tons of recoverable iron. It is apparent that sufficient iron ore is available to meet our needs far into the future.

MANGANESE

Manganese is a metal that is little known to the general public but which is very important to modern society because it is essential for the production of iron and steel. Manganese, like iron, exists in nature in more than one oxidation state; these oxidation states (Mn^{2+}, Mn^{3+}, Mn^{4+}) control its geological behavior and distribution. Manganese tends to be concentrated by chemical sedimentary processes, and important manganese resources in the world are sedimentary rocks, mixed sedimentary and volcanic rocks, or residual deposits formed by leaching of primary deposits. Like iron, manganese is very soluble in acidic or reducing solutions that carry it as Mn^{2+}, but when it becomes oxidized to Mn^{4+}, it is quite insoluble and precipitates as minerals such as pyrolusite (MnO_2) or psilomelane ($BaMn_9O_{18} \cdot 2H_2O$).

Manganese has little use on its own as a pure metal because of its brittleness, and it is little used even as an alloy in steels. However, there is no other substitute for its use as a scavenger of detrimental impurities such as sulfur and oxygen during the smelting of iron. Although the distinctive purple colors of manganese compounds and strong oxidizing properties of manganese had been utilized for centuries, manganese was not isolated as an element until 1774, and it did not become important industrially until its importance in steel-making was discovered in the mid-1800s. After that, it was soon found that the

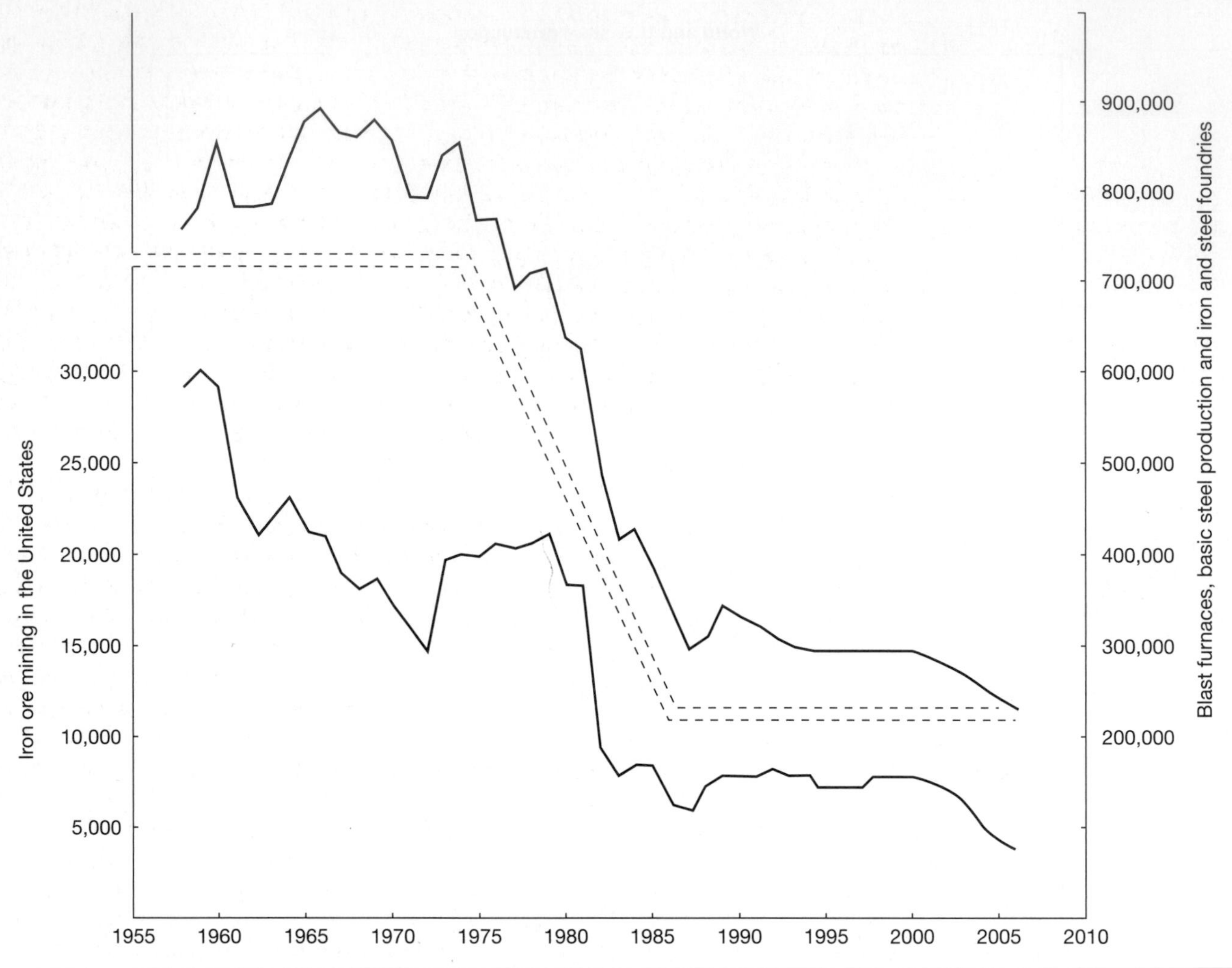

FIGURE 7.13 The U.S. labor force involved in iron ore mining (lower curve, left-hand scale) and in the steel industry (upper curve, right-hand scale) has dropped dramatically since the late 1950s, reflecting a small decline in production and a large increase in the efficiency of operation. (Data from U.S. Geological Survey.)

most useful form of manganese, still used today, is as **ferromanganese**, an alloy of iron and manganese containing 78–90 percent manganese. Up to 7 kilograms of manganese (added as ferromanganese during smelting) are necessary for the production of each metric ton of iron or steel. Although more than 90 percent of the world's manganese consumption is by the iron and steel industry, there are also important uses for manganese oxides in the chemical industry. Two of the best-known uses are as potassium permanganate, a powerful oxidizing agent used for water treatment and purification, and as manganese dioxide, a necessary component in dry cell batteries.

Manganese ores usually consist of dark brown to black oxides, especially pyrolusite (MnO_2) and romanechite [$BaMn_9O_{16}(OH)_4$], that range from hard and compact to friable and earthy; locally the carbonate, rhodochrosite ($MnCO_3$) or the silicate, rhodonite ($MnSiO_3$) are important ores. The largest and most important sedimentary deposits include Groote Eylandt off the north coast of Australia, the Molango District in Mexico, the Kalahari Field in the Republic of South Africa, and the Bol'shoy Tokmak, Chiatura, and Nikopol' deposits in Russia. Important residual deposits include the Sierra do Navio in Brazil, the Moanda in Gabon, and several deposits in India. At present, only deposits containing 35 percent or more manganese constitute reserves (see Figure 7.15). If deposits with manganese contents down to approximately 30 percent are also considered in the reserve base, the available manganese increases about fourfold. When world reserves are compared against total annual production, it is apparent that there will be sufficient manganese to meet human needs for many years to come. However, despite the relative abundance of manganese on a global scale, the distribution of ores is irregular and the United States, Japan and Western Europe all lack economic deposits of this important metal. Relatively small deposits were mined in the United States as early as the 1830s, and they

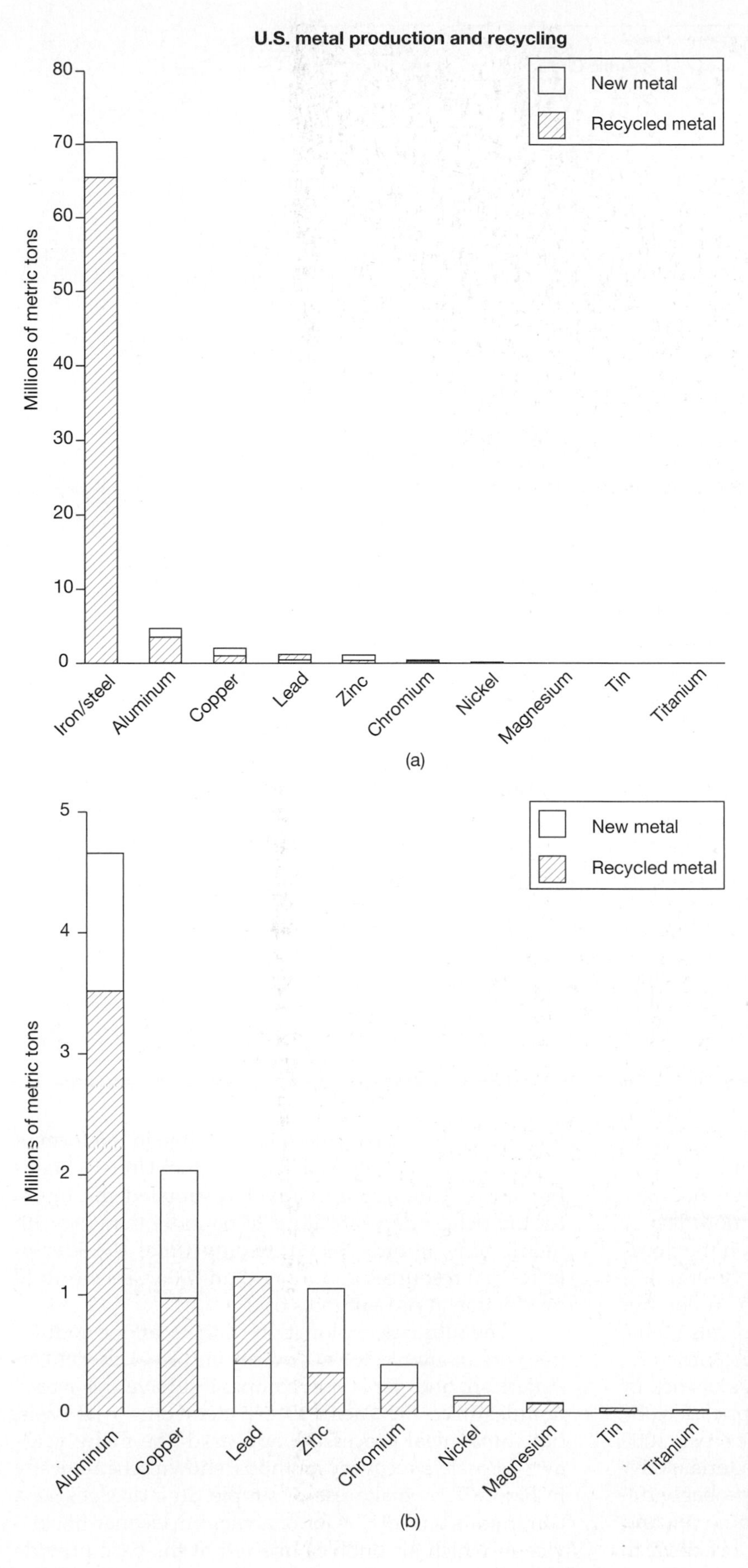

FIGURE 7.14 Metal recycling in the United States. (a) Recycled iron and steel constitute more than one-third of total iron and steel production; the tonnages of primary production and recycling far overshadow all other metals. (b) With iron removed from the graph, it is possible to see that recycling contributes significantly to copper, lead, and zinc supplies. Chromium, magnesium, and nickel are used in much smaller amounts but the recycled metals also contribute to their availability. (Data from the U.S. Geological Survey.)

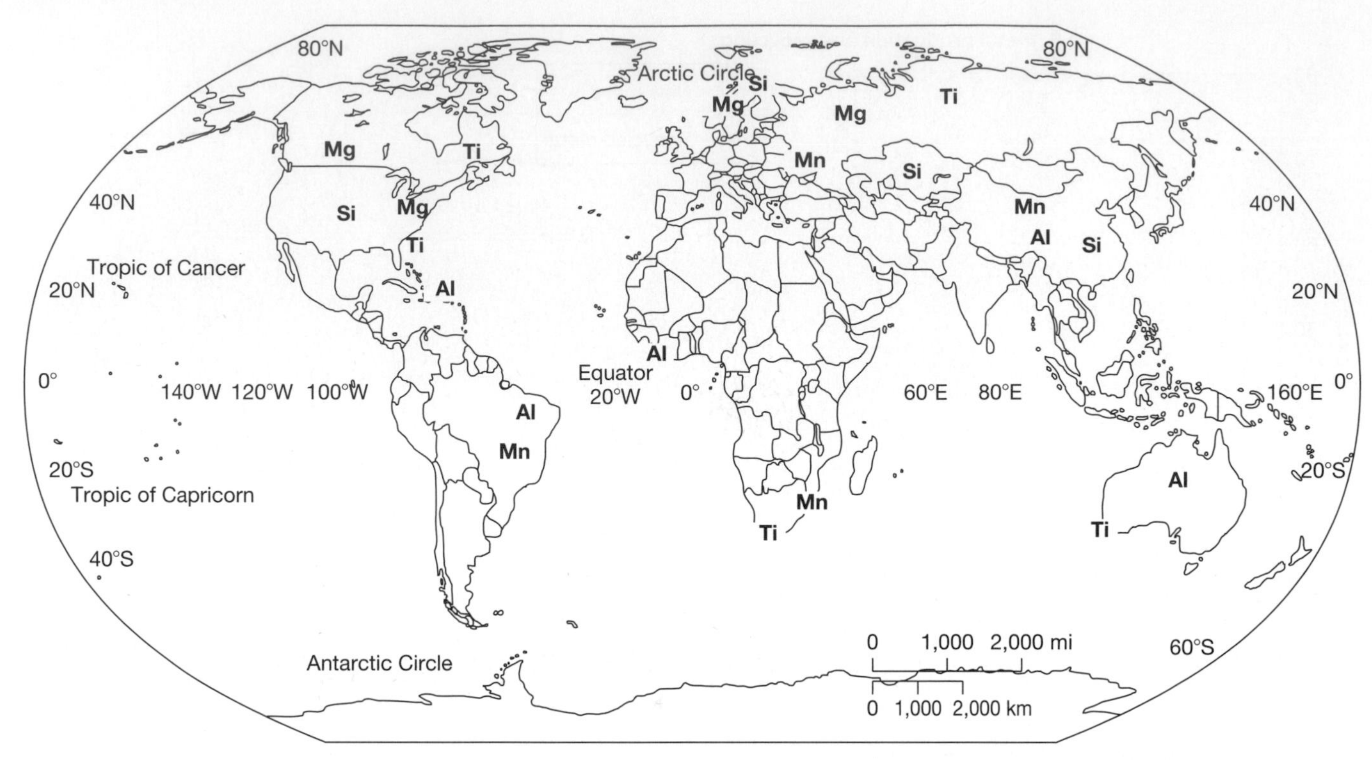

FIGURE 7.15 The major world reserves of aluminum (Al), titanium (Ti), and manganese (Mn) are shown. Also shown are the largest producers of silicon (Si) and magnesium (Mg); these latter two metals are extracted from sources that are virtually universal in occurrence and inexhaustible.

continued to be mined until the end of World War II, but the most significant remaining domestic deposits have average grades of less than 20 percent. This has led to a concern about the adequacy of stable long-term supplies from foreign producers, but the presence of large reserves in many countries will probably ensure adequacy of imports.

There is yet another potentially exploitable manganese resource—the deep ocean floor. The ship *Challenger*, sent around the world by the (London-based) Royal Society between 1873 and 1876 to gather data about the waters, rocks, plants, and animals of the oceans, found that some parts of the deep ocean floor are covered by quantities of black nodules up to several centimeters in diameter [Figure 7.16(a)]. Subsequent studies have found that these nodules, generally referred to as **ferromanganese nodules**, or just manganese nodules, are widespread on the floors of the oceans and are complex mixtures of iron and manganese oxides and hydroxides, with minor but potentially important amounts of other metals (Table 7.5). The nodules consist of onion-like concentric layers that have grown over a central nucleus of rock or shell material [Figure 7.16(b)]. Growth appears to be very slow—on the order of 1 millimeter per 1000 years—and may well be influenced by bacterial activity. The manganese and other metals are probably derived both from the land, as terrestrial weathering and erosion slowly liberate metals and transport them to the oceans, and from submarine hydrothermal and volcanic vents that occur along mid-ocean ridges. The total quantity of manganese recoverable in the form of nodules is not well known, but the United States Bureau of Mines conservatively estimated the figure for the richest deposits alone to be more than 16×10^9 metric tons, approximately twenty times the known terrestrial resources and more than 800 years worth of production at present rates of use.

TABLE 7.5 Average elemental compositions of ferromanganese nodules from the major oceans. Values in weight percent

Element	Atlantic	Indian	Pacific	Pacific*
Manganese	15.5	15.3	19.3	24.6
Iron	23.0	13.4	11.8	6.8
Nickel	0.3	0.5	0.9	1.1
Copper	0.1	0.3	0.7	1.1
Cobalt	0.2	0.3	0.3	0.2
Zinc	—	—	—	0.1

*A 230 km^2 area at 8°20′N and 153°W.

The ultimate exploitation of the seafloor nodules presents economic, technological, and legal challenges. American and Japanese companies have recovered nodules from the Pacific Ocean floor on a trial basis, but commercial processing appears to be many years away. Possible recovery methods, shown schematically in Figure 7.17, make use of simple drag dredges, of a continuous bucket line, or of a vacuum cleaner-like device in which air bubbles injected at the base provide the suction. Regardless of the technique employed, the recovery of nodules from depths of 4000 meters

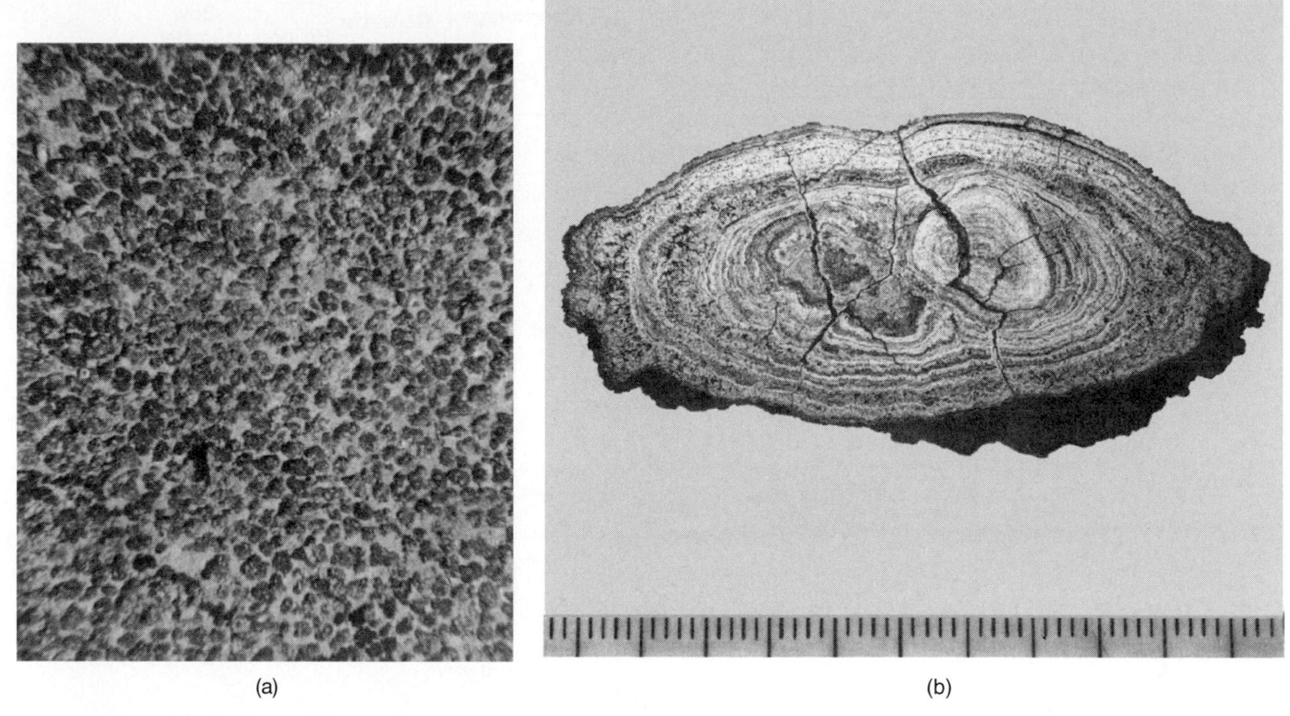

(a) (b)

FIGURE 7.16 (a) Manganese nodules are common on many parts of the deep ocean floor. These nodules in the Pacific Ocean are 5 to 10 centimeters (2 to 5 inches) in diameter. (Photograph by W. T. Allen, Deep Sea Ventures.) (b) A cross section cut through a typical manganese nodule shows the concentric nature of the manganese and iron oxides within the manganese nodule. (Photograph by B. J. Skinner.)

(12,000 feet) is difficult and expensive. The exploitation of nodules on the deep seafloor raises two additional questions. The first is: What are the possible environmental effects? Little is known of the deep-sea lifeforms and the extent, if any, to which they could be harmed by sediment disturbance caused by seafloor mining. The second question is: Who has the right to mine on the ocean floor? The International Law of the Sea conference of the United Nations worked for many years to try to define ownership of, and right of access to, mid-ocean resources. It resulted in the general recognition of exclusive economic zones covering the continental shelves but did not resolve the problems of mining manganese nodules and other deep-ocean resources. Serious international problems remain, and a legal framework for the recovery of manganese nodules has still to be worked out and to be accepted by all countries. There was much reluctance by the United States to accept any agreement because of concerns about deep-sea resources. When the United States finally signed the treaty in 1994, a consensus was reached that the industrial countries would have significant control over resources, such as manganese nodules, that they might ultimately wish to mine.

ALUMINUM, THE METAL OF THE TWENTY-FIRST CENTURY

Aluminum is the second most abundant metallic element (after silicon) in Earth's crust, where it occurs at an average concentration of 8.2 percent. However, it is so difficult to free the metal from its minerals that aluminum has been produced commercially for only about 125 years. Despite its relatively recent appearance on the industrial scene, aluminum has proven to be a remarkably useful metal. It weighs only about one-third as much as either iron or copper; it is malleable and ductile, easily machined and cast; it is corrosion resistant; and it is an excellent conductor of electricity. This versatility has resulted in such widespread use that today aluminum is the second most widely used metal after iron. Rubies and sapphires have been valued since biblical times, but it was not until the end of the eighteenth century that these gems (Figure 10.30), and corundum, were recognized as oxides of aluminum (Al_2O_3), and were collectively called *alumina*. From this, the metal was named *aluminum* in 1809, but it was not isolated in its free state until 1825.

Initially, and for a short time, because of the difficulty of producing aluminum and because of its novelty as a new metal, it was valued more highly than gold. Napoleon III, nephew of Napoleon Bonaparte and Emperor of France from 1852 until 1871, even had an aluminum baby rattle made for his infant son, and his most prized eating utensils were made from aluminum. The breakthrough that permitted commercial production, and hence the wide-scale use of aluminum, came in 1886 when Charles Hall in the United States and Paul Heroult in France developed an electrolytic process to release the metal from the oxide. At about the same time, in Austria, Karl Bayer developed a chemical process to produce alumina in large quantities from

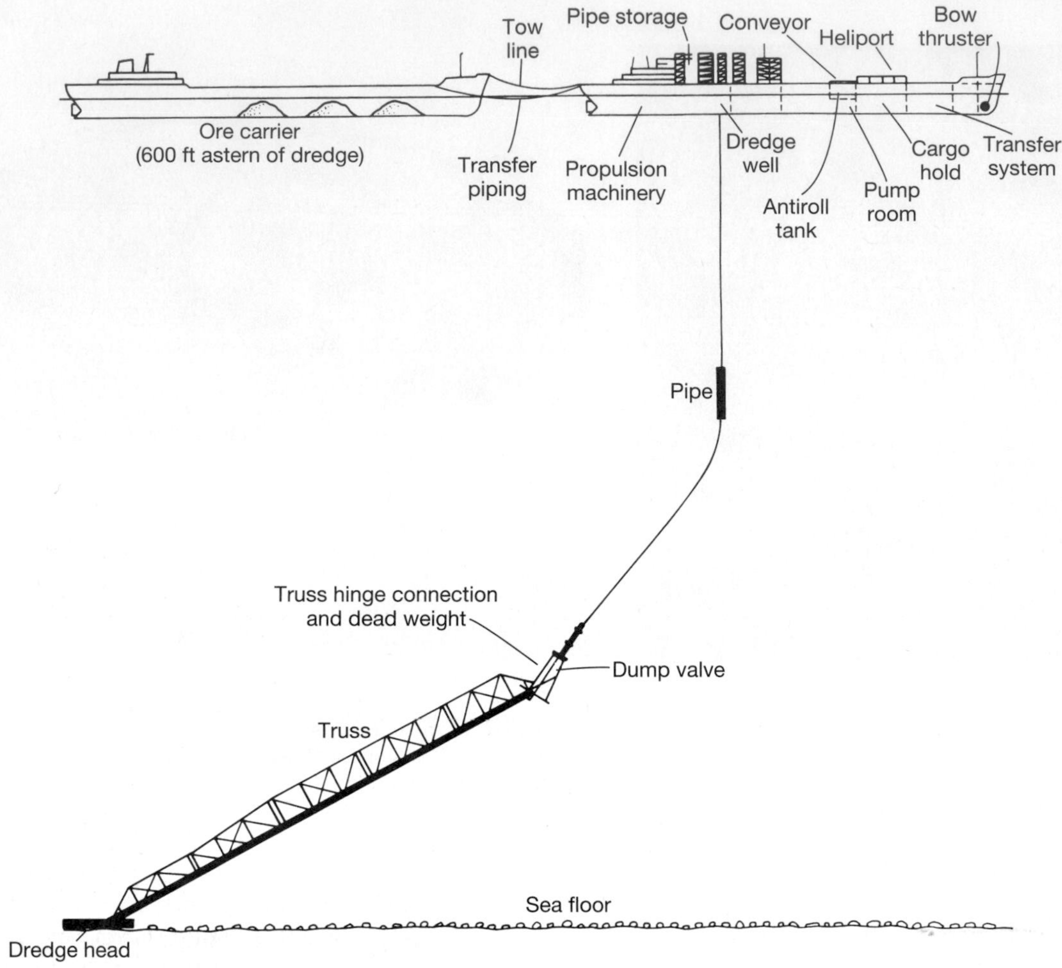

FIGURE 7.17 Recovery of manganese nodules might be accomplished by vacuum cleaner-like systems such as shown here or by bucket-line systems. The major problems would be in maintaining continuous and economic recovery in ocean depths of 3300 to 5000 meters (10,000 to 15,000 feet). (Courtesy of Deep Sea Ventures.)

bauxite. The Hall–Heroult and Bayer processes laid the foundations of the modern aluminum industry; commercial production began in 1888, and the processes continue to be used in only slightly modified forms today.

Aluminum Products and Usage

Because of its versatility, aluminum now finds a wide range of uses in our daily lives. The annual usage of aluminum metal in the United States is about 6 million metric tons, or about 20 kilograms (44 pounds) per person. No other country uses so much aluminum in total, nor so much per person, as the United States, but its use is increasing significantly in nearly every country. In the United States, the major uses are transportation (38 percent), packaging and containers (22 percent), and building products (16 percent); other important uses include electrical (7 percent) and consumer durable goods, such as refrigerators (7 percent). The principal packaging use is in aluminum cans, 100 billion of which were being produced in the United States by the early 1990s. Transportation uses continue to be important because the light weight of aluminum affords more efficient use of fuels and because aluminum is so resistant to corrosion. Approximately 70 kilograms (150 pounds) of aluminum are used in the average American-built car today; and the quantity is projected ultimately to reach 90 kilograms (200 pounds). Aluminum is widely used in construction because of its light weight and resistance to weathering, and it seems likely that there will be increased consumption in this area. Although copper is nearly always used for household electrical wiring, it is aluminum that serves as virtually all of the high-power transmission lines that extend across the countryside. It is the light weight and relatively high strength-to-weight ratio of aluminum that allow the construction of the long spans of wire between towers.

Less visible than the uses of aluminum metal are the uses of aluminum compounds. The most important of these are alumina and aluminum hydroxide, $Al(OH)_3$, which is also called *activated bauxite.* Both

BOX 7.2

Recycling Automobiles

There is no item more characteristic of modern developed societies than the automobile. In 2008, more than 250 million passenger vehicles were registered in the United States and the total around the world was approaching one billion. Thus, the approximately 4.5 percent of human population living in the United States operates approximately 25 percent of the world's total registered motor vehicles. There are, of course, many tens of millions of additional unregistered vehicles, some operating and many in new- and used-car inventories. Furthermore, developing countries such as China, India, and Indonesia are adding cars at a very fast pace.

The automotive industry is one of the major consumers of many kinds of metallic mineral resources, especially iron and aluminum, and once a vehicle is manufactured, the fuel for these vehicles constitutes the principal use of petroleum products. Up until the worldwide recession began in 2007, approximately 13 million motor vehicles (7 million cars and 6 million trucks and buses) were manufactured annually in the United States. At the same time, approximately 13 million vehicles were being retired from service and being recycled. The initial step is the recovery of the lead storage battery and the platinum metal-bearing catalytic converter, thus permitting complete recycling of those metals. Next, typically, the vehicle is crushed so that it is no more than about 18 inches thick, or squeezed into an approximate 3-foot cube, for ease in transporting to a recycling center. The crushed vehicle is fed into a powerful shredding machine where it is ripped into 4- or 5- inch fragments. Strong magnets and screens and air jets separate the pieces into three streams (1) iron and steel, (2) non-ferrous metals, and (3) "fluff" (plastics, fabric, glass, rubber, etc). Generally, all of the iron and non-ferrous metals are recovered for recycling, and much of the fluff is burned to recover the energy content.

The Steel Recycling Institute points out that the more than 14 million tons of steel recovered from the end of life vehicles provides enough material to build 45,000 steel-framed homes or saves the equivalent energy to power 18 million households annually. Although iron and steel, at more than a metric ton per unit, are the dominant metals recovered from vehicle recycling, there are many other metals covered as well. Typically, 2008 vehicles contain close to 300 pounds (135 kg) of aluminum and 55 pounds (25 kg) of copper wire, nearly all of which is recovered and recycled. Other metals recovered, in varying percentages, include zinc, tin, magnesium, titanium, chromium, mercury, cadmium, nickel, cobalt, vanadium, manganese, tungsten, and silver.

For many years old automobiles, their batteries and their tires were often discarded in junk yards, in fields, or along back roads (see also Plate 12). This resulted in unsightly litter from the automobile chassis and tires, and pollution from the lead batteries, engine fluids, and burning tires. Automotive recycling has progressed far from the old days and today, many manufacturers are moving toward designs which will make automotive recycling easier and even more complete. However, there still remains the problem of the discarding of 240 million worn tires every year and the backlog of 2 to 3 billion old ones dispersed about the countryside. They are a special concern because individual accumulations of many millions of tires are unsightly and pose great environmental and health hazards if they catch fire.

FIGURE 7.B Crushed automobiles being placed on a conveyor belt in preparation for shredding, which will allow separation of the various components for recycling. Approximately 10 million vehicles are shredded and recycled each year in the United States. (Photograph courtesy of the Institute of Scrap Recycling Industries, Inc.)

activated bauxite and alumina are widely employed in the petroleum industry as absorbents in oil and gas refining; other important uses are as fire retardants and as fillers in plastics and paper. Alumina has long served as a major component in **refractories** for the steel industry because it has a very high melting point and is relatively unreactive. Alumina has also long been used as an important grinding and polishing compound, but increasingly it has to compete with harder materials such as synthetic diamonds and silicon carbides. A quantitatively small but very important and growing use of alumina is in the production of synthetic rubies and sapphires used in the construction of lasers, as jewel bearings in precision mechanisms, and as synthetic gemstones.

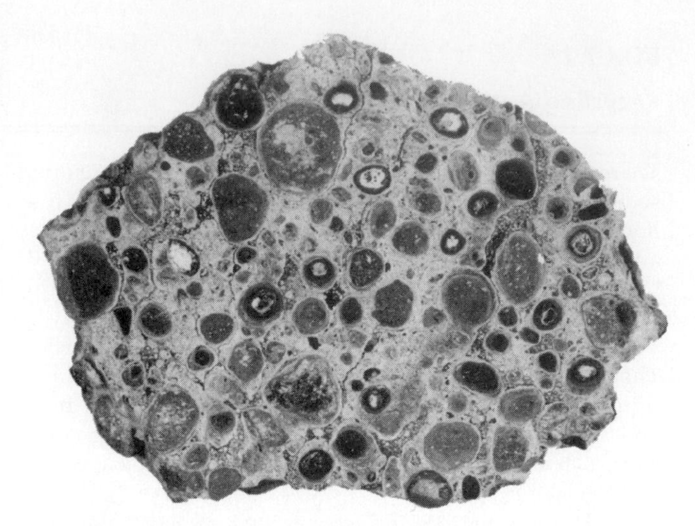

FIGURE 7.18 Bauxite, the ore of aluminum, commonly exhibits a characteristic pisolitic texture. (Photograph by B. J. Skinner.)

Aluminum Ores

Aluminum, like iron, is such an abundant element that it is a constituent of many common minerals. The most important are feldspar (the most abundant mineral in Earth's crust), mica, and clay. To date, and probably for the near future, most aluminum production has been from bauxite (a name derived from the southern French village of Les Baux where it was first recognized in 1821), a heterogeneous material composed chiefly of the aluminum hydroxides gibbsite [$Al(OH)_3$] and boehmite and diaspore (both $AlO{\cdot}OH$). These relatively uncommon minerals are formed by the breakdown of aluminum-bearing rocks under special conditions of lateritic weathering. These conditions occur most frequently in subtropical to tropical climates where there is abundant rainfall, the groundwater is neither too acid nor too alkaline, there are aluminous parent rocks, and there is subsurface drainage but low relief so that mechanical erosion is slow relative to chemical leaching.

During the intense chemical weathering typical of lateritic conditions, the three least soluble components are SiO_2, Al_2O_3, and Fe_2O_3. After the more soluble ions, such as Na^+, K^+, Mg^{2+}, and Ca^{2+}, have been removed in solution, the residue is mainly iron hydroxide plus clays such as kaolinite [$Al_2Si_2O_5(OH)_4$]. Percolating waters, made slightly acid by the decay of organic debris, slowly dissolve the clays and remove the silica and some of the iron. What remains are the hydroxides of aluminum and iron. Where the ratio of aluminum hydroxides to iron hydroxides is high, the resulting rock is an aluminum-rich laterite called *bauxite*. Successive solution and precipitation commonly result in a characteristic pisolitic texture of the type shown in Figure. 7.18 and Plate 32.

Aluminum-rich rocks at times serve as the parent rock for some deposits, but bauxites can form from the weathering of any rock that is aluminum-bearing. In fact, some important bauxites known as *terra rosa* types develop on limestones that contain very little aluminum. In these cases, the calcium carbonate of the limestones is rapidly dissolved in the acidic tropical groundwaters. This leaves a clay residue that can be altered to form discontinuous and localized, but rich, lenses of bauxite. Most major commercial bauxite deposits formed during the past 60 million years, and all of the largest have formed in the tropics over the past 25 million years. Deposits such as those in Arkansas or France, found in areas that are presently temperate, formed in earlier geologic times when the climates of those areas were tropical. No doubt many additional bauxites have formed during Earth's long history, but because they were surficial deposits, they were later destroyed by erosion. Bauxites are unknown in arctic regions because they would not form there today and because any deposits formed in the geologic past would likely have been removed by glaciation.

Aluminum Smelting and Production

The conversion of raw bauxite into alumina then into aluminum metal is a multistep process, as shown in Figure 7.19. Bauxite mining is relatively straightforward because the deposits lie at or near Earth's surface and because the ores are usually soft and easily removed. In contrast, the processing, and especially the production, of metallic aluminum is complex and extremely energy-intensive. Unfortunately, bauxite deposits, inexpensive energy supplies, and markets for the end product tend to be widely separated. Most mining of bauxite takes place in tropical regions that have neither abundant cheap electricity nor large markets for the aluminum products. Thus, to maximize the efficiency of shipping to processing sites, the bauxite is first crushed, washed to remove impurities, and then dried, as shown in Figure 7.19. The washed concentrate is then shipped to countries such as Norway,

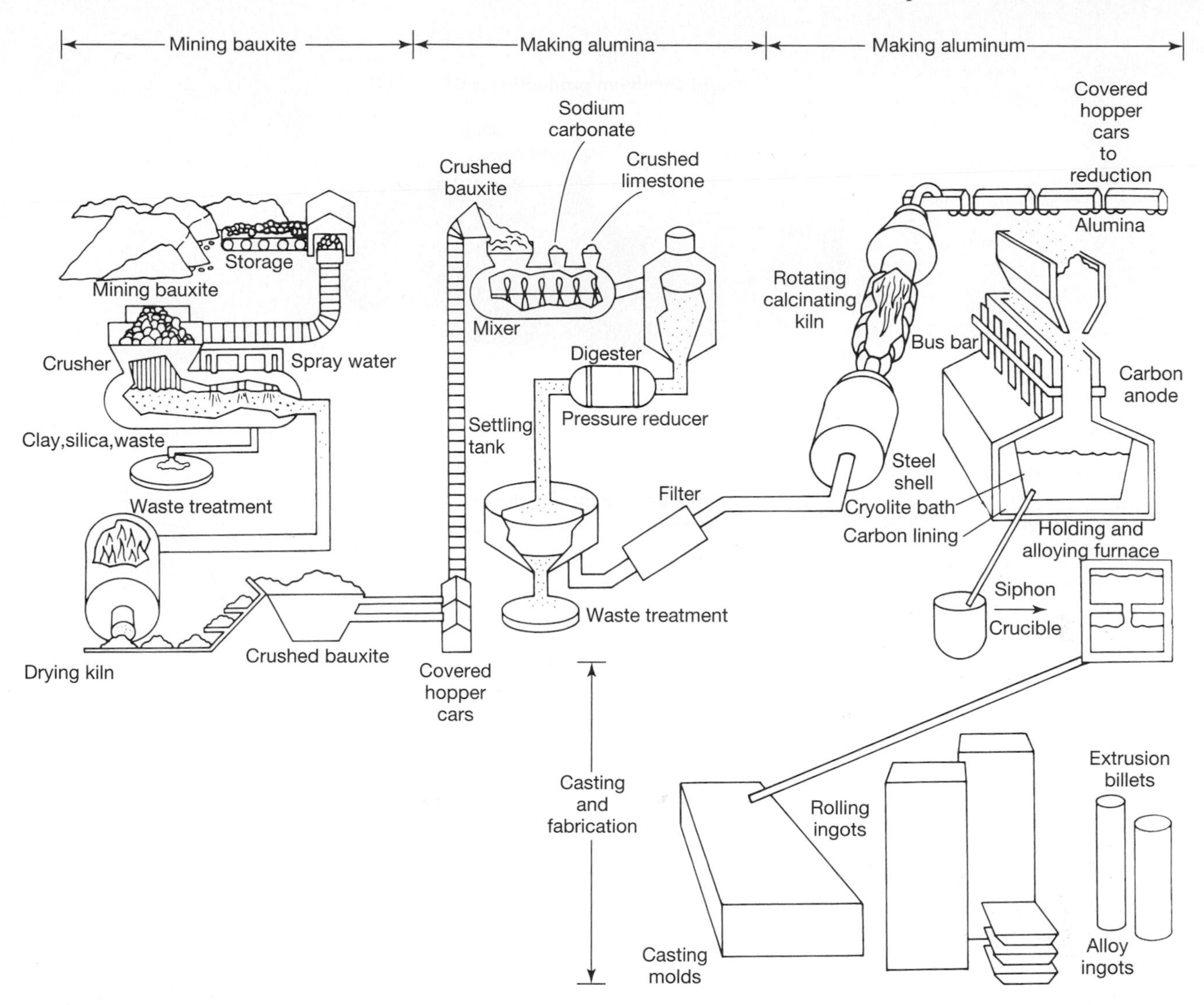

FIGURE 7.19 Schematic diagram showing the steps in mining, processing and smelting of aluminum. (Courtesy of The Aluminum Association.)

Canada, and the United States where there is now (or was when plants were established) abundant cheap electrical power. In each of these countries, the source of the power for processing is principally hydroelectric plants. There is also a large plant in Australia that uses cheap coal as its energy source and a plant that will use geothermal power is being built in Iceland.

The production of aluminum metal is accomplished by the electrolytic reduction of alumina in a molten bath of natural or synthetic cryolite (Na_3AlF_6) that serves both as an electrolyte and a solvent. The actual metal production takes place in a series of large bathtub-like vats called a **pot line,** where hundreds of aluminum ingots, each weighing up to a metric ton, are produced simultaneously. The metal reduction process uses very large amounts of electricity because the temperatures must be maintained at 950°C or more and because the electrical currents and voltages must reach as much as 150,000 amperes and 1000 volts, respectively. In the United States, the growth of the aluminum industry in the first half of the twentieth century coincided with the development of regional power networks and the building of major hydroelectric facilities. The availability of the large amounts of electricity generated in the Columbia River Basin and the Tennessee Valley led to the construction of major aluminum smelters in those areas. This situation worked well because the aluminum companies, while getting the cheap power they needed, provided a use for the surplus electricity generated by the hydroelectric dams. Consequently, aluminum metal production in the United States rose rapidly from the beginning of the twentieth century until 1973, when it exceeded 4100 million kilograms annually (Figure 7.20). At that time, the oil embargo raised energy prices and ended the period of growth. As a result of rising energy costs, American production leveled off between 3500 and 4000 metric tons per year in the 1980's and then declined below 2700 metric tons in the years since 2000. Countries that were dependent upon expensive imported fuel

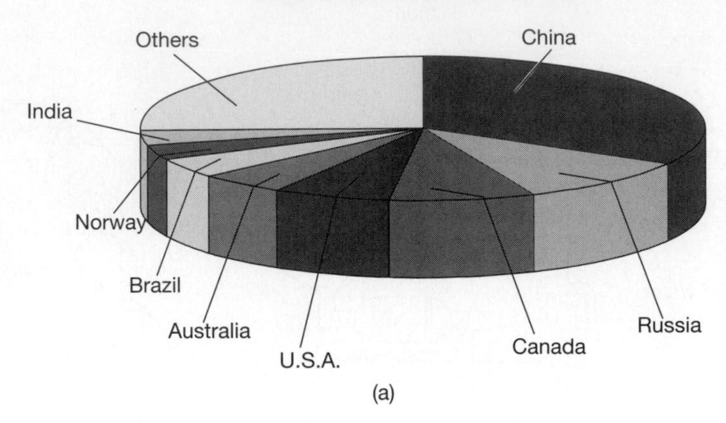

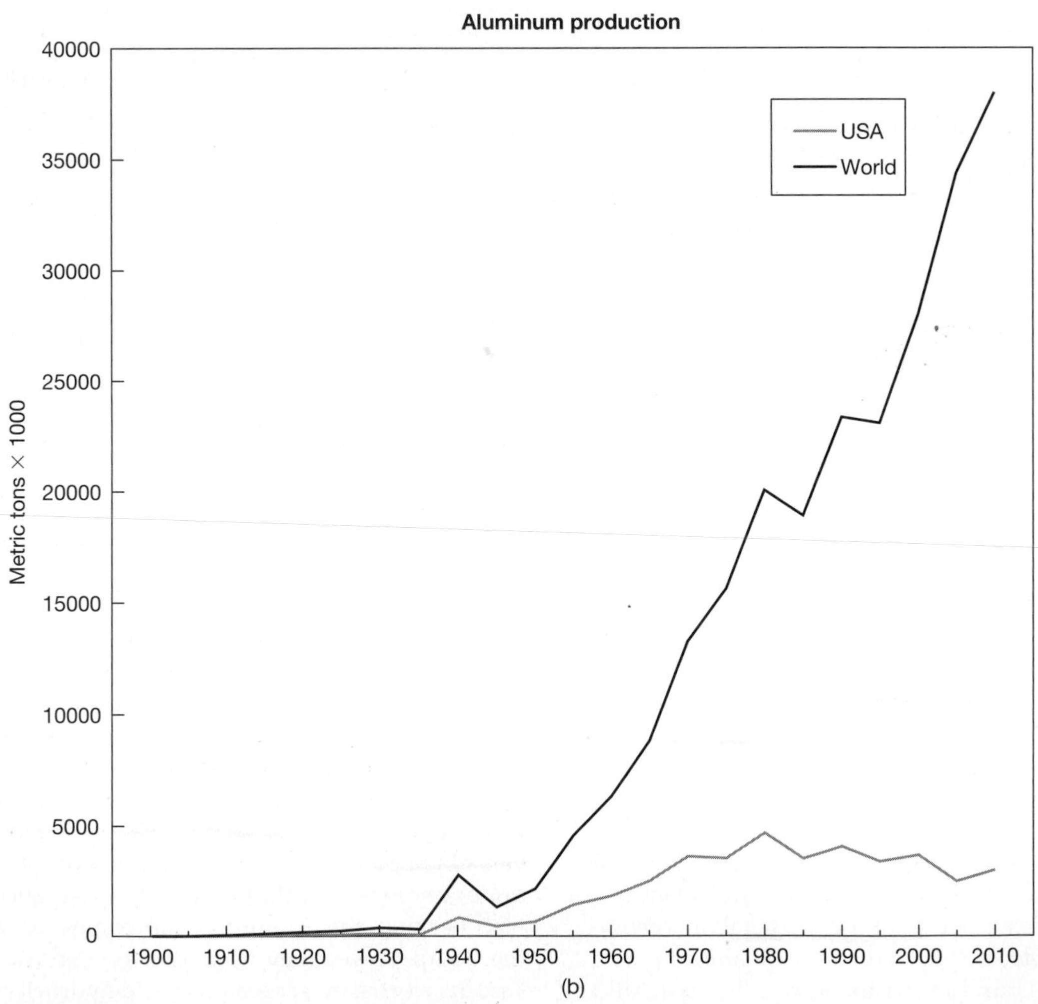

FIGURE 7.20 World aluminum production increased rapidly in the second half of the twentieth century. The decline in production in the early 1970s and early 1980s resulted from the high costs of energy at those times and reflect the energy-intensive nature of aluminum production. Subsequent demand has increased to about 40 million metric tons per year. World bauxite reserves are very large, so the greater concern for aluminum availability is the high cost of production relative to other metals.

supplies either sharply curtailed or, in the case of Japan, terminated primary aluminum production, deciding that it was more economic to import aluminum metal than to refine it. This underscores what is generally recognized as the major problem of aluminum refining—it is energy-intensive.

The amount of energy needed to refine aluminum relative to that for several other metals is shown in Figure 7.21. It is apparent that the energy required to extract a ton of aluminum from a typical bauxite is approximately twice that required per ton of iron extracted from taconite, and nearly three times that required to refine copper from sulfide ores. Furthermore, the energy that would be required if we were to refine aluminum from other common aluminum-bearing minerals such as clay or feldspar would be many times greater than that used in refining bauxite.

A concern for aluminum resources coupled with the energy-intensive extraction of the metal has led to a great deal of recycling of aluminum. By the 1990s, recycling of aluminum scrap in the United States had reached 3.6 million tons, equal to 30 percent of the American aluminum production. This quantity was second only to iron in terms of recycled metal. It consisted of new scrap, which is waste material generated in the production of new aluminum products, and old scrap, which are old aluminum products that have been discarded (cans, foil, engine parts, wire, etc.). In 1970, 3 billion (3×10^9) aluminum beverage cans were recycled in the United States; in 1983, that number had risen to 27.5 billion (27.5×10^9) cans, and in 2008, it was 63 billion (63×10^9). As noted in Chapter 4, recycling of these cans (and all aluminum scrap) is important because it represents not just a saving of resources, but also a large saving of energy. This is so because the remelting and reforming of a new aluminum can from an old one requires only 5 percent of the energy needed to make the can from bauxite in the first place. Thus, the recycling of aluminum represents a 95 percent energy saving. One way to picture the energy saved in recycling an aluminum can is to see it about one-third full of gasoline because this is the equivalent amount of energy lost if the can is discarded and needs to be replaced using newly produced metal. When Japan announced the closing of its primary aluminum production facilities in 1985, the country decided to step up the importing of aluminum scrap for reprocessing because it is economically more efficient and saves the large costs and problems involved in the importing of the fossil fuels and bauxite needed for primary production. In a very real sense, the importing of the aluminum scrap represents the importing of energy, the energy that some other country has supplied to refine the metal initially.

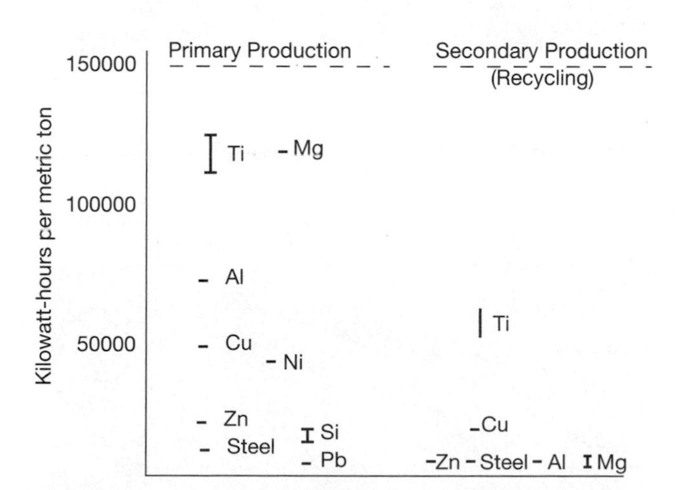

FIGURE 7.21 Energy in the primary production and recycling of metals. The lines in the left half of the diagram indicate the amount of energy required to produce a metric ton of each of the metals from their ores. In contrast, secondary production (commonly referred to as *recycling*) of several of these same metals is shown in the right-hand side of the diagram. Thus, it takes only 5 percent as much energy to recycle aluminum as it does to produce it from bauxite initially.

Bauxite Reserves

The specification of bauxite reserves is complicated because the quality of deposits is not judged solely on accessibility and aluminum content, but also on other chemical properties such as iron and silica contents, upon local factors such as energy costs, and on the proximity to markets. Typical bauxites mined in Jamaica contain 49 percent Al_2O_3 (on a dry basis) and high-grade ores from South America, Guinea, and Queensland, Australia, contain 50 to 60 percent Al_2O_3, whereas mineable reserves in Arkansas and Western Australia have only 40 percent Al_2O_3. Most European bauxites have 45 to 65 percent Al_2O_3.

World bauxite reserves (Figure 7.15) are concentrated in tropical and semitropical regions. Exploration over the past fifty years has greatly expanded reserves from 1×10^9 metric tons in 1945, to 3×10^9 metric tons in 1955, to 6×10^9 metric tons in 1965, to 21×10^9 metric tons in 1985, and to 27×10^9 metric tons in 2008. Guinea and Australia together have about one-half of world reserves; more than 25 percent occurs in the Western Hemisphere in Brazil, Jamaica, Guyana, and Surinam. The principal U.S. deposits are about 20 million metric tons in central Arkansas plus 2 to 3 million tons in Alabama and Georgia. At the present rate of world mining, approximately 200 million metric tons of bauxite annually, the reserves presently identified would last more than 100 years. Estimates of total world bauxite resources that may one day be mineable are 40 to 50×10^9 metric tons and would allow for bauxite mining for a much longer period of time. From this a quantity, of 9 to 11×10^9 metric tons of aluminum metal could be extracted. This quantity is indeed large, but very much smaller than the amount of iron that will ultimately be recoverable.

Concerns over the sufficiency of bauxite as the source of aluminum, especially in countries that have limited bauxite reserves, have led to the consideration of other types of materials as potential sources of aluminum. Russia, during the era of the Soviet Union, produced alumina from nepheline [$(Na,K)AlSiO_4$] and alunite [$KAl_3(SO_4)_2)(OH)_7$]. In addition to nepheline and alunite, particular interest has been focused upon clays (especially kaolinite), oil shales (the aluminum-rich waste left after processing them for oil), and anorthosite (a rock composed primarily of the **plagioclase feldspar**, anorthite, $CaAl_2Si_2O_8$). The potential resources of clays and anorthite are large and widespread; unfortunately, the processing of these materials for aluminum requires even more energy than needed to process bauxite. The potential use of oil shales would also be dependent upon the economic extraction of the oil, something that seems unlikely in the forseeable future.

In summary, total bauxite reserves appear to be adequate for many years, but several other mineral commodities will continue to be investigated as possible aluminum sources. This arises from the hope that their processing costs can be made competitive with that of bauxite and from the desire of many industrial nations to be less dependent upon foreign sources for the aluminum raw materials.

TITANIUM

Titanium, the least common of the abundant metals, comprises only 0.56 percent of Earth's crust. Although titanium was recognized as a chemical element in 1790, more than a century passed before any commercial potential was realized, and it has only been used on a large scale in the past fifty years. Today, it has two major applications. The first is in a variety of alloys where it imparts a high strength-to-weight ratio, a high melting point, and great resistance to corrosion. These properties have led to its widespread use in aircraft engines and air frames, electricity-generating plants, welding rods, and a wide variety of chemical processing and handling equipment, and to its designation as a strategic metal. The second use, which accounts for approximately 95 percent of the world's consumption of titanium minerals, is the preparation of white titanium oxide pigment. Because of its whiteness, opaqueness, permanence of color, and low toxicity, it is now the principal white pigment used in paint, paper, plastic, rubber, and many other materials. Formerly, white lead oxide had been widely used in paints, but this resulted in numerous cases of lead poisoning. In addition, white lead is more susceptible to changes in color when subject to air pollution. The use of titanium is relatively inconspicuous; nevertheless, the amounts used every year are very large. For example, in 2008, the United States annually produced approximately 10,000 metric tons of titanium metal and about 1.3 million metric tons of titanium dioxide pigment, with a total value of more than \$3.7 billion valued at more than \$3 billion.

Titanium occurs in minor amounts in most types of rocks as the oxide minerals **rutile** (TiO_2) and **ilmenite** ($FeTiO_3$); locally, **leucoxene**, an alteration product of ilmenite, is also present. Generally, these minerals are widely dispersed in igneous and metamorphic rocks as accessory minerals. Locally, however, the igneous processes involved in the formation of **mafic** rocks (**gabbro** to **anorthosite**) have concentrated large amounts of iron and titanium oxides, especially ilmenite, into lenses or thick layers. When the iron oxide present is magnetite (Fe_3O_4), separation of the titanium minerals into a relatively pure concentrate is usually possible because the minerals are coarse-grained and because the magnetite is magnetic. In contrast, mixtures of ilmenite with hematite are usually very fine-grained intergrowths, as shown in Figure 7.22(b), and are nearly

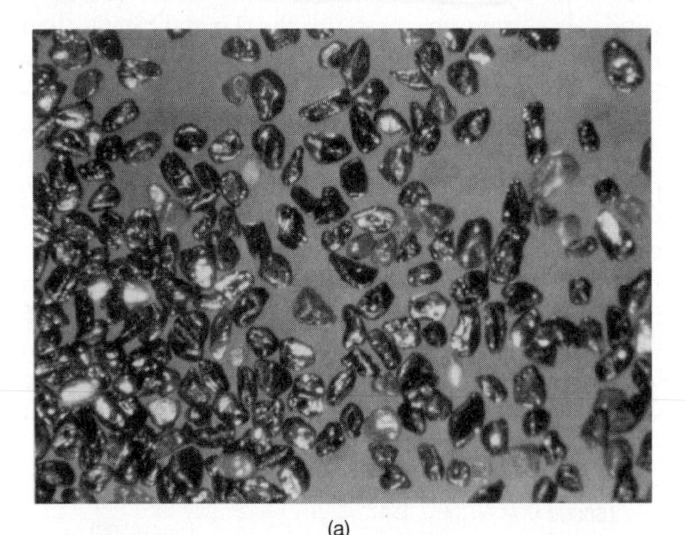

(a)

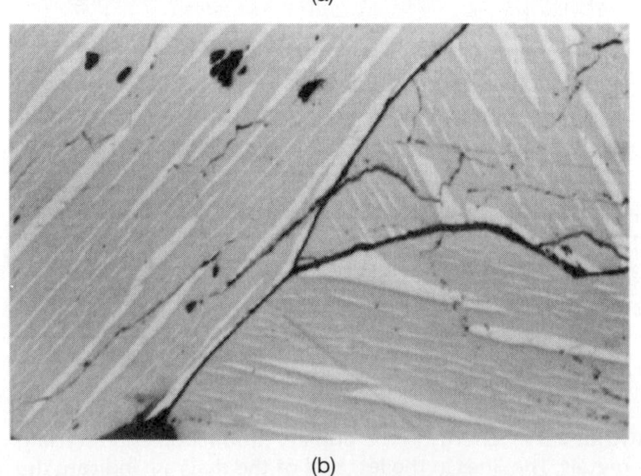

(b)

FIGURE 7.22 The principal mineralogical sources of titanium are rutile (TiO_2) and ilmenite ($FeTiO_3$) and commonly occur as the "black sands" in beaches and streams. (a) Placer rutile and ilmenite from a beach deposit. (b) Microscopic view of intimately intergrown ilmenite (darker phase) and hematite (white phase) in a hard rock deposit. (Photographs by J. R. Craig.)

impossible to separate into pure concentrates by mechanical means. The only way to thoroughly separate the titanium from these ores is by expensive chemical processes in which the ores are dissolved.

Most sedimentary rocks contain minor amounts of titanium oxide minerals derived from igneous or metamorphic rocks by weathering and erosion. The ilmenite and rutile are hard (5–6.5 on the **Mohs scale**) and resistant to solution or chemical attack, thus, they survive the weathering and erosional processes intact [Figure 7.22(a)]. Because their specific gravities (rutile = 4.25; ilmenite = 4.8) are higher than those of common constituents of river and beach sands (quartz = 2.65; feldspar = 2.6–2.7), the titanium minerals may be selectively concentrated into specific zones or sedimentary horizons (forming what are known as **placer deposits**). Gold panners make use of the same properties as they concentrate black sand and gold in the pan; in fact, the black sand is often largely ilmenite and rutile. The TiO_2 content of the ilmenite in placer deposits varies as a function of the initial composition of the ilmenite and the degree of weathering because the slow alteration can result in preferential leaching of iron. Thus, placer ilmenite concentrates from South Africa average only 48 percent TiO_2, whereas those from Florida and New Jersey have 61–65 percent TiO_2. Rutile concentrates are usually 93–96 percent TiO_2, and leucoxene concentrates contain up to 90 percent TiO_2. Even though the grade of the titanium may not be so rich as that in igneous deposits, the unconsolidated nature of the sediments makes processing both simple and economic.

Titanium minerals are mined today both from deposits in igneous rocks (*hard rock* mines) and from placer sand occurrences. Until about 1942, nearly all commercial ilmenite and rutile production came from placer deposits; today, rutile production still comes solely from placers, but nearly 40 percent of the ilmenite derives from hard rock mines. The worldwide production of titanium comes from the mining of about 4 million metric tons of placer ilmenite and 500,000 tons of rutile. The total world reserves are estimated at about 300 million tons of contained TiO_2, and world resources are at least 1.2 billion tons of TiO_2 content. Consequently, there will be sufficient titanium to meet our needs for many years to come. Furthermore, recent investigations of the detrital deposits of the continental shelves have revealed the presence of very much larger potential resources of placer titanium minerals that could be exploited if those on land are exhausted.

Titanium, like silicon, aluminum, and magnesium, requires large amounts of energy for processing from its source mineral forms to produce either pigment or metal. The energy required to produce titanium metal from ilmenite and rutile is shown relative to several other metals in Figure 7.21.

Titanium occurs in small amounts in all types of rocks worldwide, but economically viable deposits are not common. The most important hard rock ilmenite reserves are those at Allard Lake (Quebec), Tahawus (New York), Tellnes (Norway), and Otanmaki (Finland). Important placer deposits are worked in Australia, India, the Republic of South Africa, Russia, Sri Lanka, and Sierra Leone. The difference between the large production of the titanium minerals and the limited production of the metal reflects the fact that most of the minerals are converted into TiO_2.

The mining of titanium minerals does not create any unusual environmental problems because the mining is usually carried out in simple, shallow open-pit operations or on beaches (Figure 7.23). After scooping up the loose sands, the recovery is usually accomplished by the use of spiral separators or jigs that separate the titanium minerals on the basis of specific gravity. Hence, there is no blasting and no use of chemicals. The economic sedimentary deposits usually contain 6–8 percent titanium minerals; consequently, reclamation returns 92–94 percent of the material, and there is often no evidence of the previous mining activity. However, the processing of ilmenite to produce pigment generates up to 3.5 metric tons of toxic sulfate and sulfuric acid waste per ton of product. Previous methods of disposing of such wastes into streams and the ocean have now been replaced by acid neutralization plants; unfortunately, the runoff of sulfates from the waste sites of older plants has seriously polluted streams and continues to cause problems locally.

MAGNESIUM

Magnesium, the eighth most abundant element in Earth's crust, is the lightest of the abundant metals. Like most of the other abundant metals, its common occurrence and its important uses are little appreciated by most people. It finds its largest use not as the metal but as the oxide, MgO (called **magnesia**), and as the silicate mineral, **forsterite** (Mg_2SiO_4), both of which are used as refractories in the steel industry and some base-metal industries (see Chapter 10). As a metal, magnesium is commonly mixed with aluminum to produce lightweight, corrosion-resistant alloys that are widely employed in beverage cans, automobiles, aircraft, and machinery. In addition, magnesium compounds are used in such varied materials as cement, rubber, fertilizers, animal feed, paper, insulation, and pharmaceuticals.

Magnesium-bearing raw materials are abundant and geographically widespread. They consist of several minerals and of brines and seawater (Figure 7.24). The first magnesium resources, exploited in the mid-eighteenth century, were **magnesite** ($MgCO_3$) deposits in the former Czechoslovakia, Austria, and Greece. Similar deposits were subsequently discovered in

(a)

(b)

FIGURE 7.23 Titanium sand mining. (a) A small dredge, at the left, digs the sands and sends them into the concentrator where the 5 percent of titanium minerals present are separated and recovered. (b) Mining of beach sands for the titanium minerals is carried out along the coasts in parts of Australia. (Photograph courtesy of RGC [USA] Minerals Sands Inc.)

California, and mining began there in 1887. During World War I, a process was developed whereby magnesium-bearing refractory materials could be extracted from dolomite [$CaMg(CO_3)_2$], a common sedimentary rock, by intense baking (called **calcining**) to drive off the CO_2 (Figure 7.25).

Magnesium is unique in being the only metal to be extracted directly from brines and from seawater.

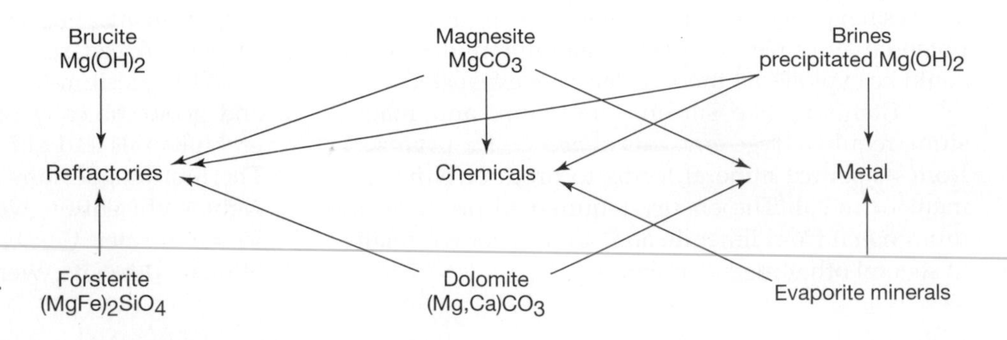

FIGURE 7.24 The uses of the principal raw materials of magnesium. (From the U.S. Geological Survey.)

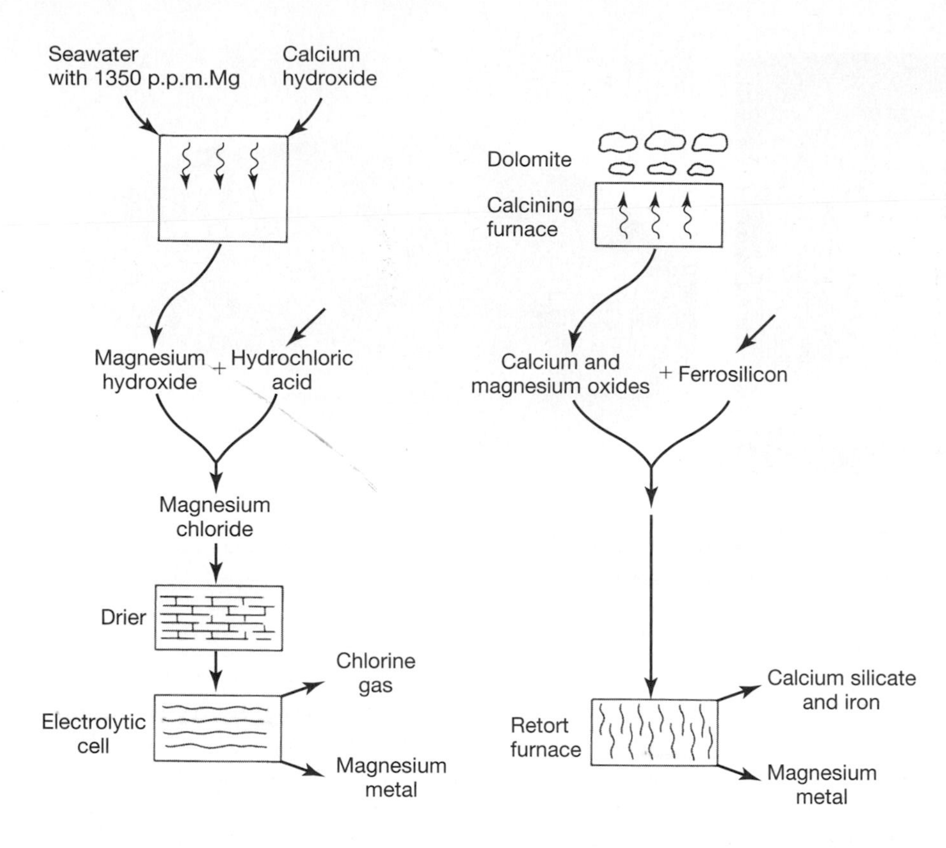

FIGURE 7.25 Magnesium metal is recovered from seawater by reaction with calcium hydroxide and hydrochloric acid, followed by electrolytic refining or by reacting calcined dolomite with ferrosilicon in a high-temperature retort furnace.

The recovery of magnesium metal from deep-well brines containing several thousand parts per million magnesium began in Michigan in 1917. There, brines are trapped in the thick sequence of evaporite minerals that underlie the Michigan Basin. Recovery of magnesium from seawater, in which it is the third most plentiful dissolved element [1350 parts per million (ppm)], began in 1940. The extraction process, shown schematically in Figure 7.24, is relatively simple chemically but requires large quantities of electricity. In 2009, magnesium metal and other compounds were being produced in the United States from seawater in California, Delaware, Florida, and Texas, from the lake brines of the Great Salt Lake in Utah and from well brines in Michigan.

Resources from which magnesium metal and its compounds can be recovered are globally widespread, and estimates range from very large to virtually unlimited. Reserves of the highest-quality raw materials, such as the magnesite mined in Nevada, exist in quantities sufficient to meet human needs for a very long time. Furthermore, the amounts of magnesium available in the form of dolomite, and in seawater and brines, are so vast that they will never be exhausted.

SILICON

Silicon is the second most abundant element in Earth's crust and is a major constituent of all the common silicate minerals that comprise many of the common rocks. Because of its very strong affinity for oxygen, with which it is combined in quartz (SiO_2) and all silicate minerals, free silicon only occurs in nature under the most unusual circumstances. The only confirmed occurrence is at a place in Michigan where an intense lightning strike fused some glacial debris, producing temperatures of about 2000°C and reducing some quartz to native silicon.

Despite its abundance in oxide forms all around us and its importance in modern technological applications, silicon metal is relatively unfamiliar to most people. Pure silicon is a lightweight, silvery substance that has a lustrous semimetallic appearance and is very brittle. Although the use of silicate minerals such as clays dates to prehistoric times, and the use of glass made from silicates began at least 12,000 years ago, free silicon was not prepared until 1824. In the late 1800s, the use of silicon as a deoxidizing (oxygen removing) agent for steels was discovered, and this led to a large demand in the growing steel industry. Today, silicon (or an iron--silicon alloy called **ferrosilicon**) is prepared by melting clean quartz, usually coarse-vein quartz or well-cemented quartzite [Figure 7.26(a)], with iron or steel scrap and coal, coke, or charcoal (as a reductant) in a large electric arc furnace. These furnaces, up to 13 meters (40 feet) in diameter and 13 meters (40 feet) high, can prepare 150 to 200 metric tons per day. Periodically, the furnaces are tapped and the molten silicon or ferrosilicon is cast into ingots. The ferrosilicon is mixed with the molten iron

(a)

(b)

FIGURE 7.26 Quartz in the form of (a) sand, quartzite, and quartz crystals serves as the source of silicon that now finds broad applications as (b) chips in many computers. (Photograph of silicon disc courtesy of ITT; photograph of silica sources by S. Llyn Sharp.)

or steel to remove oxygen and to serve as an alloying agent. The addition of up to 17 percent silicon in cast iron reduces scaling and corrosion at high temperatures. Silicon is also added to aluminum and copper alloys in amounts up to 25 percent because it improves the casting properties, adds strength, and reduces corrosion.

In recent years, silicon has found many additional uses outside of metallurgy. The best known of these began in 1949 when E.I. du Pont de Nemours and Co. produced the first silicon pure enough for use in transistors and other semiconductor devices. Today, the **silicon chip** [Figure 7.26(b)] is the basis for many electrical devices, being used in computers, calculators, and communications equipment. The chips are prepared by first producing ultrapure single crystals of silicon and then by introducing into these crystals specific amounts of certain chemical elements in order to produce desired electrical properties. Another important use is in photovoltaic devices, commonly called **photocells** or **solar cells** (Figure 7.27), which convert sunlight into electrical energy and contain thin layers of silicon, either as single crystals or as amorphous films. Today, photovoltaic cells are employed in uses ranging from pocket calculators to Earth circling satellites, and their application is continuing to expand rapidly.

Silicon is also used to produce compounds such as **silanes** (silicon–hydrogen compounds, for example, SiH_4) that are used in the manufacture of numerous kinds of silicone resins, rubbers, lubricants, adhesives, antifoaming agents, and water-repellent compounds. In 1891, E. G. Acheson failed in his attempts to synthesize diamonds, but he accidentally discovered silicon carbide (SiC), also known as *carborundum*. This substance, with a Mohs hardness of 9.5 (compared with 9 for corundum and 10 for diamond), is now one of

(a)

(b)

FIGURE 7.27 Modern solar cells produce electricity for an increasing variety of applications. These contain layers of crystalline or amorphous silicon in which the incident solar radiation produces electricity. (Photograph courtesy of Solarex Corporation.)

the most widely used commercial abrasives and is commonly found in hardware stores on some of the better grades of sandpaper.

The world's resources of silicon in the form of silica (SiO_2) in quartz and other silicates is virtually limitless. Constraints on the production of silicon or ferrosilicon are those of purity—which are reasonably met by quartz from many **quartzites**, pegmatite masses, and gravel deposits—and the availability of electrical power. The world output of about 3.5 million metric tons of silicon annually represents the efforts of many countries. However, the large increase in electrical power costs in recent years has seen a shift in silicon production from countries such as Japan and the United Kingdom to countries with lower power costs. Consequently, Norway, with abundant hydroelectric power and only a small domestic steel industry, has become a major exporter of ferrosilicon. China has increased its production of ferrosilicon greatly so that it is now the world's leading producer of total silicon forms. The growth in demand for silicon will be nearly totally dependent upon the world's steel industries because they account for most of the consumption. Semiconductor usage will likely continue to rise but accounts for less than 1 percent of the total output of silicon.

ABUNDANT METALS IN THE FUTURE

It is now apparent that Earth's crust contains vast quantities of the abundant metals in concentrations and forms that can be exploited by current technologies. The geochemical abundance of the metals, coupled with the increasingly diverse ways in which they are employed, ensures that the abundant metals will remain the principal metals used by society in the foreseeable future. Iron will no doubt remain the dominant metal because of its low cost, wide availability, and broad range of uses. Manganese will be needed because it is essential to the production of the steels made from iron. Aluminum, magnesium, titanium, and silicon will no doubt find new uses, especially in the construction and transportation industries, where they permit weight and energy savings. However, because these latter metals are energy intensive in terms of their extraction, their usage could be affected by the availability and costs of energy. The demand for silicon, especially in the form of solar panels will undoubtedly continue to rise as societies increasingly turn to renewable energy sources in the twenty-first century. We now turn our attention to the scarce metals, which though constituting only trace proportions of Earth's crust, serve modern society in a wide range of important roles.

The Quebrada Blanca mine in Chile is an example of a modern gold mine. The low-grade ore is extracted from a large open-pit mine (out of sight to the right of this scene), crushed, and then stacked in large rectangular heap leach piles visible in the background. Dilute cyanide-bearing solutions sprayed on these piles drain through them, dissolving the gold. The solutions drain onto large, impermeable plastic liners and are collected into the ponds visible in the center of the photo. Gold is finally extracted using chemical reactions and electrolytic processes conducted in the building in the central part of the operation. (Courtesy of Cominco.)

CHAPTER 8

The Geochemically Scarce Metals

The total volume of workable mineral deposits is an insignificant fraction of the earth's crust, and each deposit represents some geological accident in the remote past. Deposits must be mined where they occur—often far from centers of consumption. Each deposit has its limits; if worked . . . it must sooner or later be exhausted. No second crop will materialize. Rich mineral deposits are a nation's most valuable but ephemeral material possession—its quick assets.

T. S. LOVERING, 1969, "MINERAL RESOURCES FROM THE LAND" IN RESOURCES AND MAN, P. CLOUD, (EDITOR). SAN FRANCISCO: W. H. FREEMAN AND CO., PP. 109–134.

FOCAL POINTS

- There are more than thirty geochemically scarce elements that have metallic properties and that occur in Earth's crust at average abundances below 0.1 percent.
- The geochemically scarce metals commonly occur dispersed in common minerals; only when they are much concentrated (25 to 1000 or more times) do they form their own specific minerals and, in turn, mineable deposits.
- The four major groups of geochemically scarce metals are (1) ferro-alloy metals; (2) base metals; (3) precious metals; (4) special metals.
- The ferro-alloy metals—especially nickel (Ni), chromium (Cr), cobalt (Co), and molybdenum (Mo)—are used to alloy with iron to provide special steels (stainless, tool, high temperature steels, etc.).
- Nickel, chromium, and cobalt occur primarily associated with large mafic or ultramafic igneous rock bodies; molybdenum occurs in felsic, porphyritic rock bodies.
- The base metals—such as copper (Cu), lead (Pb), zinc (Zn), tin (Sn), and mercury (Hg)—occur primarily in deposits formed by precipitation from hydrothermal fluids; they are used in a broad range of technologies.
- Copper, the most widely used base metal, was one of the first metals known to humans and now serves as the most important metal for the electricity industry.

- Lead was the most extensively used base metal from Roman times until the twentieth century; it is used primarily for automotive batteries but it is also recognized as a dangerous pollutant. There are intensive efforts to remove lead from the environment and to limit human exposure.
- The precious metals—gold (Au), silver (Ag), and the platinum group—have long served as monetary standards and are used extensively in jewelry. They are finding increasing technological applications.
- The search for gold has been the driving force for much human exploration and colonization, and it continues to be the object of exploration activities.
- The special metals are a broad group that is playing an increasingly important role in new technologies; their use is likely to increase in the future.

PRODUCTION OF THE GEOCHEMICALLY SCARCE METALS

The backbone of industry is built from the geochemically abundant metals such as iron and aluminum. But it is the geochemically scarce metals that keep industry efficient, effective, and healthy because the scarce metals, used in small amounts, control the properties of alloys of the abundant metals, which carry electric currents and allow automobiles to run and planes to fly. Consider iron, which is so widely used that it accounts for about 95 percent by weight of all metals used. The properties of pure iron are so limited that iron alone could not possibly satisfy all the requirements of modern industry. For some uses iron must be hardened; for others, it must be made more flexible or more ductile, tougher to abrasion, or more resistant to corrosion. All such changes can be accomplished by the addition of small amounts of geochemically scarce metals as alloying agents. Consequently, several of the geochemically scarce metals—nickel, chromium, molybdenum, tungsten, vanadium, and cobalt—are mined and used principally as alloying components for special steels.

The geochemically **scarce metals** (Table 8.1) are those that are present in Earth's crust in such trace amounts that none exceeds 0.1 percent of the crust by weight. Indeed, some metals are so geochemically scarce that they make up a millionth of a percent or less of the mass of the crust. Examples are gold, which has a crustal abundance of 0.0000004 percent by weight, and ruthenium, which only has a crustal abundance of 0.00000001 percent! Despite their extreme geochemical scarcity, both gold and ruthenium have special properties that make them important, or even essential, commodities for industry. Approximately three dozen geochemically scarce metals are now mined and used for special industrial purposes. No other group of natural resources fills such a wide and varied range of needs. Because of the properties of the geochemically scarce metals, things happen more rapidly, more efficiently, and more effectively. In a sense, the geochemically scarce metals are like the enzymes that make our bodies work effectively by carrying out the complex chemical processes that keep us healthy. The geochemically scarce metals are the enzymes of industry. It is their special properties that have led to such technological marvels as the generation and distribution of electricity, and the inventions of the telephone, radio and television, automobiles, aircraft, rockets, computers, and the Internet. Yet it is in this same group of metals that many experts once believed shortages and restrictions of natural resources might appear. It now seems that such fears were unwarranted because new exploration techniques, better geologic understanding, and new recovery techniques have combined to increase the reserves of most of these metals. Furthermore, increasing quantities of many of these metals are being recycled, thus reducing the need for newly mined metals.

Many differences exist between the geochemically scarce and geochemically abundant metals in addition to the ways in which they are used. Where the world's annual production of iron has, in recent years, been more than a billion tons, only four of the geochemically scarce metals—chromium, copper, lead, and zinc—have ever been produced at rates that exceed a million tons a year. The production rate of many scarce metals is still less than a thousand tons per year (Table 8.1).

DISTRIBUTION OF SCARCE METALS IN THE CRUST

Ninety-two chemical elements have been identified in Earth's crust. The average concentrations of individual elements range from as low as 10^{-8} percent to as high as 45 percent by weight, but only nine elements account for 99 percent of the mass of the crust (Figure 8.1). The combined total of the remaining 83 elements accounts for only 1 percent of the mass of the crust. Most of the 83 minor elements fulfill the definition of geochemical scarcity in that their individual abundances are less than 0.1 percent by weight of Earth's crust.

The nine major elements are the ones that form the major minerals found in common rocks. Only the most common of the geochemically scarce metals—copper, zinc, and chromium—form minerals that can be found in common rocks, and even those minerals are of very limited occurrence. Nevertheless, careful analyses reveal that essentially all naturally occurring chemical elements are present in trace amounts in common rocks. However, most rocks contain at most three or four major minerals,

TABLE 8.1 The geochemically scarce metals in 2007 (USGS 2009)

Name	Chemical Symbol	Crustal Abundance (ppm)	Annual World Production (m.t.)	Major Producers
Ferro-Alloys				
Chromium	Cr	100	20×10^6	S. Africa, India, Kazakhstan
Cobalt	Co	25	70×10^3	Zambia, Canada, Congo, Australia
Molybdenum	Mo	1.7	140×10^3	U.S., China, Chile, Peru
Nickel	Ni	75	1.7×10^6	Russia, Canada, New Caledonia, Australia
Tungsten	W	1.5	90×10^3	China, Russia, Austria, Canada
Vanadium	V	135	60×10^3	S. Africa, Russia, China, U.S.
Base Metals				
Cadmium	Cd	0.2	20×10^3	Japan, Canada, Korea, China, Kazakhstan
Copper	Cu	55	15×10^6	Chile, U.S., Indonesia, Canada, Peru, China
Lead	Pb	12.5	3.5×10^6	China, Australia, U.S., Peru
Mercury	Hg	0.08	1.5×10^3	Spain, Kyrgyzstan, China
Tin	Sn	2	300×10^3	China, Indonesia, Peru, Brazil
Zinc	Zn	70	10×10^6	China, Canada, Australia, Peru
Precious Metals				
Gold	Au	0.004	2.5×10^3	S. Africa, U.S., Australia, Canada, China
Silver	Ag	0.07	20×10^3	Mexico, U.S., Peru, Canada
Palladium	Pd	0.01	230	S. Africa, Russia, Canada, U.S.
Platinum	Pt	0.005	230	S. Africa, Russia, U.S., Canada
Special Metals				
Antimony	Sb	0.2	135×10^3	China, Bolivia, Russia, S. Africa
Arsenic	As	1.8	60×10^3	China, Chile, Peru, Morocco
Beryllium	Be	2.8	200	U.S., China, Russia, Kazakhstan
Bismuth	Bi	0.17	6×10^3	Mexico, Peru, China, Bolivia
Gallium	Ga	15	80	China, Germany, Japan
Germanium	Ge	1.5	100	U.S., China, Russia
Indium	In	240	600	China, Korea, Japan, Canada
Niobium	Nb	20	45	Brazil, Canada, Australia
Rare-earth elements	REE	0.5–60	125×10^3	China, U.S., India, Brazil
Tantalum	Ta	2.4	420	Australia, Brazil, Canada
Zirconium	Zr	165	1200×10^3	Australia, S. Africa, China

plus an equal number of minor ones. Both the major and minor minerals are generally compounds of two to five of the nine major chemical elements. There is a simple explanation for the seemingly contradictory statement that all common rocks consist of minerals that contain the nine major elements, but that the same rocks also contain trace amounts of all of the geochemically scarce elements. The explanation is that the geochemically scarce elements are all present in the common minerals by **atomic substitution** or, as it is sometimes called, **solid solution.** Atoms of nickel, for example, can substitute for atoms of magnesium in the magnesium silicate mineral olivine (Mg_2SiO_4), and atoms of lead can substitute for atoms of potassium in orthoclase feldspar ($KAlSi_3O_8$). The properties that control solid solutions are like those that control liquid solutions. When the saturation limit of a liquid solution is exceeded, crystals of the solute start to grow. So, too, when a solid solution becomes saturated, a new mineral must form. When the solid solution limit is exceeded for lead (Pb) substituting for potassium (K) in potassium feldspar, a lead mineral must form. The reason that minerals of geochemically scarce metals do not occur in common rocks—or do so only in rare circumstances—is that the amounts of the scarce metals present in the crust do not exceed the conditions of saturation of the solid solutions in common minerals.

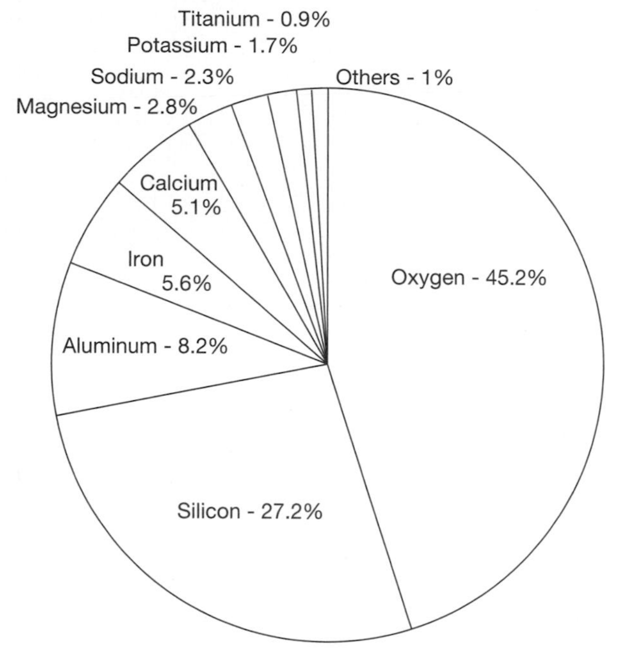

FIGURE 8.1 Nine chemical elements account for 99 percent of the mass of the continental crust of Earth. The geochemically scarce metals are nearly all present in abundances of less than 0.01 percent, as shown in Table 8.1.

TABLE 8.2 Calculated value of geochemically scarce metals in solid solution in a metric ton of average granite

Element	Crustal Conc. *	Grams per m.t.	\$ metal per kg†	\$ value in 1 m.t.
Thorium	9.6	9.6	200	1.92
Beryllium	2.8	2.8	308	0.86
Lithium	20	20	3.50	0.07
Niobium	20	20	22	0.22
Tantalum	2	2	80	0.16
Uranium	2.7	2.7	220	0.59
Zinc	70	70	3.30	0.23
Tungsten	1.5	1.5	4	0.01
Gold	0.004	0.004	29,000	0.12
Copper	55	55	6.60	0.36
Lead	12.5	12.5	2.90	0.04
Molybdenum	1.5	1.5	75	0.11
Silver	0.07	0.07	418	0.03
Tin	2	2	40	0.08
			Total =	\$ 4.80

*After Wedepohl (1978).

†Some values estimated because not quoted as metal.

Because of the differences between the properties of the various metals, there is no single rule concerning the concentration levels at which geochemically scarce metals form separate minerals. However, a rough rule of thumb is that at concentrations above about 0.1 percent, a mineral will form. Below that level, scarce metals occur only in solid solution. The rule is approximately correct for many metals including copper, lead, and zinc, but it is too low for a few metals such as gallium and germanium, and it is too high for a few others such as gold, molybdenum, and uranium.

Suppose we were to attempt to mine **granite,** a common igneous rock, and after crushing the rock to a powder, we attempted to break down the solid solutions to recover the most valuable of the geochemically scarce metals. The result, as shown in Table 8.2, would be economically ludicrous. Even large increases in the prices of the metals would not justify the extraction of the metals from the granite. It costs \$15 to \$50 a metric ton just to mine and crush granite to a fine powder suitable to start the necessary chemical treatments. The final cost of extraction would be hundreds of dollars a ton. Rather than attempting to mine metals from common rocks, therefore, we have always sought ore deposits, those localized geological circumstances in which ore minerals are found that carry unusually high contents of a desired scarce metal and that can be mined at a profit. Many of the geological processes active in ore deposit formation have been discussed in Chapter 2.

As described in Chapters 4 and 7, many factors determine whether a local concentration of such minerals can be considered an ore deposit. This is dependent upon grade, size, depth, location, and environmental laws. The minimum grade that a scarce metal ore deposit must reach ranges from as low as 0.0001 percent for gold, to 1 to 3 percent for zinc, to nearly 40 percent for chromium and manganese. This means that the natural metal concentrating processes operating in Earth's crust must create concentrations that are many times greater than those found in average rocks. Some of these concentration factors are enormous. Mercury, for example, must reach a local concentration 12,500 times greater than the crustal average in order for an ore deposit to be economic. The circumstances required for this happen very rarely indeed. Not surprisingly, ore deposits of geochemically scarce metals tend to be small and rare by comparison with ore deposits of geochemically abundant metals. The minimum concentration factors can change as the prices of metals change or as technological developments occur. Over the last 200 years, the minimum concentrations have tended to decline (Figure 8.2) because great advances have occurred in both mining and metallurgical technologies. It is not known how low concentration factors can be pushed, but if copper can be

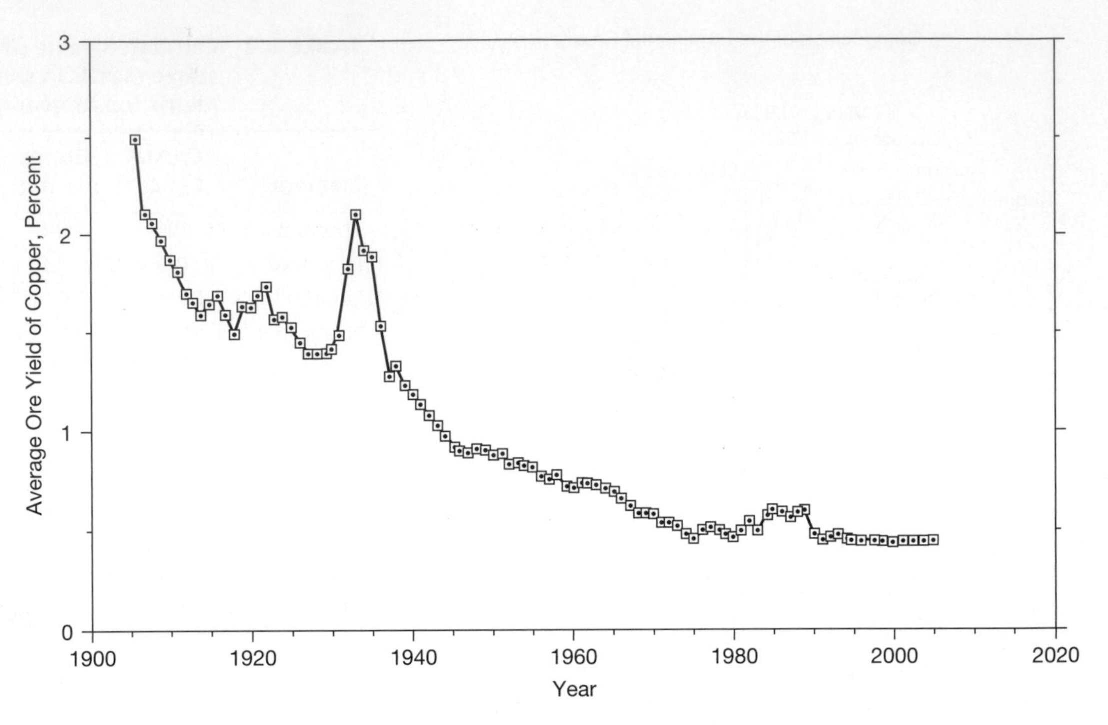

FIGURE 8.2 The average grade of copper ore mined in the United States has dropped from about 2.5 percent in the early part of the twentieth century to about 0.5 percent by the end of the century. This has been the result of mining away of the richer ores through time and the development of technologies that have permitted extraction of copper from lower-grade ores. (Data from the U.S. Geological Survey.)

taken as an example, it is likely that we are near the lower limits at which we can profitably mine the metal.

ORE MINERALS OF THE SCARCE METALS

Between 99.9 and 99.99 percent of the total amount of any given scarce metal is present in the crust in atomic substitution in common silicate minerals. Therefore, only a tiny fraction—between 0.01 and 0.1 percent—of a given metal occurs in ore minerals. Fortunately, the ore minerals tend to be found in localized concentrations rather than being scattered and disseminated. Fortunately, too, the geochemically scarce metals tend to form ore minerals that are sulfide or oxide compounds or, in a few cases, native metals; it is relatively easy to separate these ore minerals from the associated and valueless silicate or carbonate gangue minerals. Even so, some of the scarce metals (e.g. gallium, germanium, cadmium) never occur in sufficient concentrations to be the main objective of a mining operation. Instead, these elements, occurring in small but extractable concentrations as solid solutions in ore minerals, are only recovered as by-products (see the discussion in Box 8.3). Consequently, their availability for use in technology is controlled by mining primarily for other metals. If the mining for the other major metals were to cease, we would also lose the supply of the by-product metals. Thus, if our society no longer needed zinc, and all zinc mines were closed, we would have no major sources of cadmium or germanium.

The most important geochemically scarce metals and the most important ore minerals they form are listed in Table 8.3.

TABLE 8.3 Kinds of ore minerals formed by the geochemically scarce metals

	Examples of Ore Minerals
Sulfide Minerals	
Copper	Chalcocite (Cu_2S), Chalcopyrite ($CuFeS_2$)
Lead	Galena (PbS)
Zinc	Sphalerite (ZnS)
Mercury	Cinnabar (HgS)
Silver	Argentite (Ag_2S)
Cobalt	Linnaeite (Co_3S_4), Co-pyrite ($(Fe,Co)S_2$)
Molybdenum	Molybdenite (MoS_2)
Nickel	Pentlandite ($(Ni,Fe)_9S_8$)
Oxide Minerals	
Beryllium	Beryl ($Be_3Al_2Si_6O_{18}$)
Chromium	Chromite ($FeCr_2O_6$)
Niobium	Columbite ($FeNb_2O_6$)
Tantalum	Tantalite ($FeTa_2O_6$)
Tin	Cassiterite (SnO_2)
Tungsten	Wolframite ($FeWO_4$), Scheelite ($CaWO_4$)
Vanadium	V in solid solution in magnetite (Fe_3O_4)
Native Metals	
Gold	Native gold
Silver	Native silver
Platinum	Platinum-palladium alloy
Palladium	Platinum-palladium alloy
Iridium	Osmium-iridium alloy
Rhodium	Solid solution in osmium-iridium alloy
Ruthenium	Solid solution in osmium-iridium alloy
Osmium	Osmium-iridium alloy

CLASSIFICATION OF THE SCARCE METALS BY USAGE

All of the geochemically scarce metals can be divided into four major groups based on their properties and on the way the metals are used. The first group of scarce metals is known as the ferrous or ferro-alloy metals. It is a group of metals that are mined and used principally for their alloying properties, especially in the preparation of specialty steels. Examples of the ferrous metals are chromium, vanadium, nickel, and molybdenum.

The second group of scarce metals is variously called base metals or **nonferrous metals.** The term *base metal* is an old name that arose in the Middle Ages during the days of **alchemy.** Metals such as copper, lead, zinc, tin, and mercury were less valuable and less desirable than gold and silver—and, hence, base—so the ancient alchemists tried to convert them into the precious metals, gold and silver (Figure 8.3). The ancients were not successful in their efforts, and today we could argue that copper, zinc, and tin are actually more important than gold and silver because of their broad usage in industry and technology. Nevertheless, the terms *precious* and *base* remain with us. There is, however, an alternate designation for base metals. Because the principal uses of the base metals are for purposes other than alloying agents with iron, base metals are also known collectively by the term *nonferrous metals.* Neither base nor nonferrous is a completely correct description, but both terms are widely used.

The third group of scarce metals, known and used by our ancestors since the discovery of metals, is the precious metals. The precious metals of antiquity, gold and silver, were called **noble metals** because they were not readily debased by forming compounds with other chemical elements and, hence, rarely subject to corrosion. In more recent times, platinum, palladium, osmium, iridium, rhodium, and ruthenium (the so-called platinum group elements) have also come to be called precious or noble metals because they too exhibit nonreactive properties.

The fourth and final group of scarce metals, the special metals, does not fit into the previous three categories, but members of the group have unusual properties that make them important for industry. Tantalum, for example, is widely used in electronics because of its desirable electrical properties. Beryllium, in contrast, is a very useful metal in nuclear technology and high-speed aircraft. Production and use of the special metals are recent and most only came into use in the twentieth century. It is not surprising that the special metals do not readily fit into the traditional grouping of precious, base, and ferrous metals because the uses to which they are put have only been developed as a result of modern technology.

THE FERRO-ALLOY METALS

Although the individual ferro-alloy metals were all known and had been separated into their elemental forms well before the dawn of the twentieth century, their widespread use as alloying metals for special steels is a result of twentieth-century technology, and especially the technology that arose as a result of two world wars. Deposits of the ferro-alloy metals are widespread, as shown in Figure 8.4.

FIGURE 8.3 An alchemist (right) and his assistant testing formulas for the transmutation of base metals into gold. Even though the search for a successful transmutation was futile, many chemical processes were successfully developed by alchemists. (From a woodcut by Hans Weiditz, 1520.)

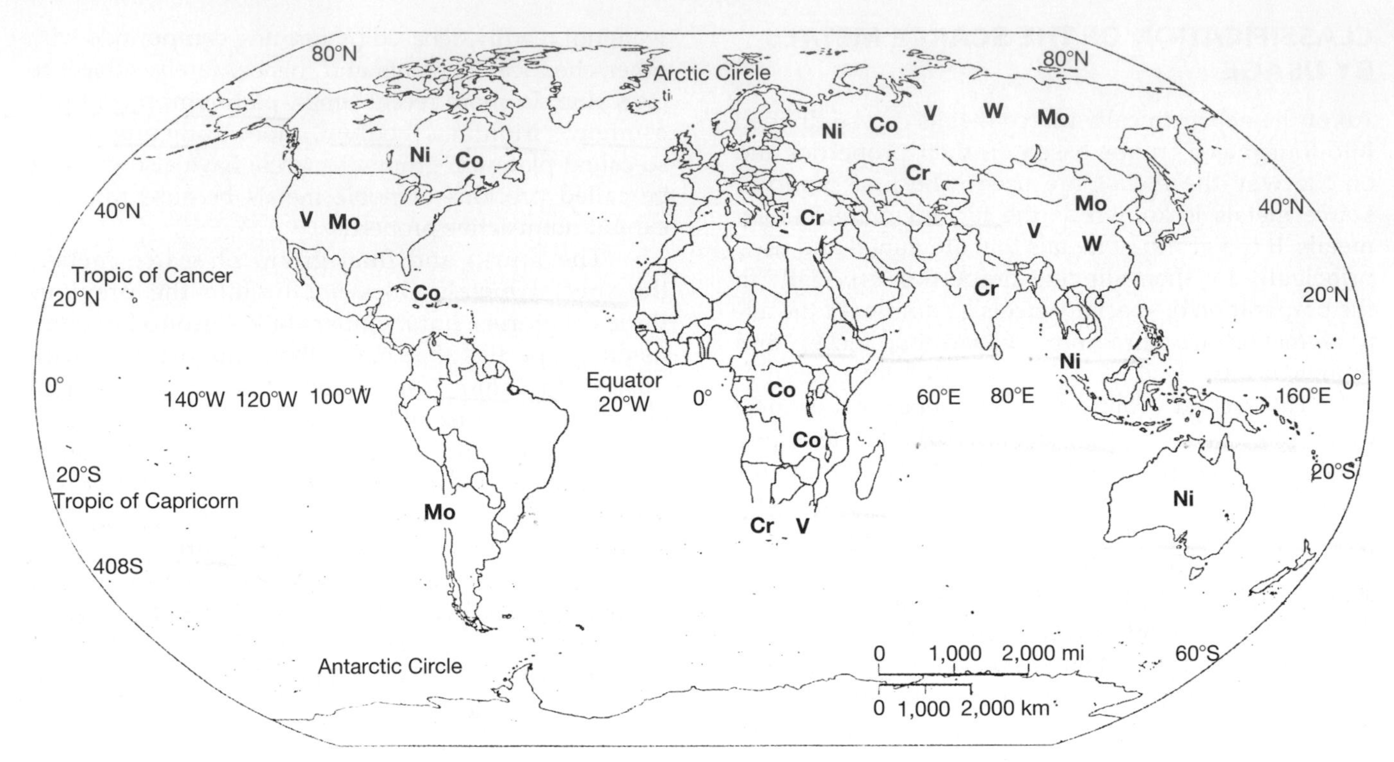

FIGURE 8.4 World map showing the locations of the largest reserves of the ferro-alloy metals.

Chromium

Chromium was first identified as a chemical element in 1765 through analysis of a chromate mineral found in Siberia. The metal was first separated as a pure chemical element in 1797. Chromate compounds have long been used in the tanning industry and in the manufacture of pigments for textiles, and they are still used for these purposes today. The properties of chromium as an alloying metal were discovered as early as 1820, but the widespread use of chromium alloys only started in 1899 when ferrochrome (an iron-chromium mixture produced by chemically reducing the mineral chromite [$FeCr_2O_4$]) was first produced in an electric furnace. Ferrochrome is still the way chromium is added to a batch of molten iron to make chromium-bearing steel.

Chromium is one of the most visible yet least widely recognized metals in our modern industrial society. Chromium plating on steel and chromium-containing alloys, such as **stainless steel**, are the shiny, noncorroding metal surfaces we find on automobiles, in kitchens, on faucets, and in the cutlery we use. Chromium is also a so-called strategic metal, which means it is a metal considered to be vital to national defense and the continued operation of industry. This designation results from the widespread use of chromium-steel alloys in aircraft engines, military vehicles, weapons, and the chemical industry.

Chromium finds important uses today in three broad fields—metallurgy, chemistry, and refractories. As discussed above, metallurgy is an essential usage because of chromium's unique alloying properties. Steel containing between 12 percent and 36 percent chromium by weight has a greatly reduced tendency to react with oxygen and water—that is, a chromium-bearing steel rusts very slowly. Chromium is also added to the steel used to make various machine tools because it increases hardness and resistance to wear. For refractories, chromium plays a vital role in the manufacture of high-temperature bricks used to construct furnaces.

The principal use of chromium as a chemical continues to lie in the production of pigments. Chromium pigments range in color from deep green and intense yellow to bright orange. Such pigments are widely used in paints, inks, roofing materials, and textile dyes. A lesser-known but still very important chemical use of chromium compounds is in the tanning of animal skins. The lightweight leathers used for furniture, clothing, shoes, wallets, and similar objects are produced by soaking raw animal hides in solutions of chromium sulfate under controlled conditions of temperature and acidity. Chromium in solution forms chemical bonds with the amino acids in the leather; this stabilizes the organic material by reducing its tendency for biological decay and, at the same time, increasing its resistance to heat.

In the metal industry, the ore mineral **chromite** has proven to be ideal for making the bricks used to line very high temperature smelter, blast, and gas furnaces.

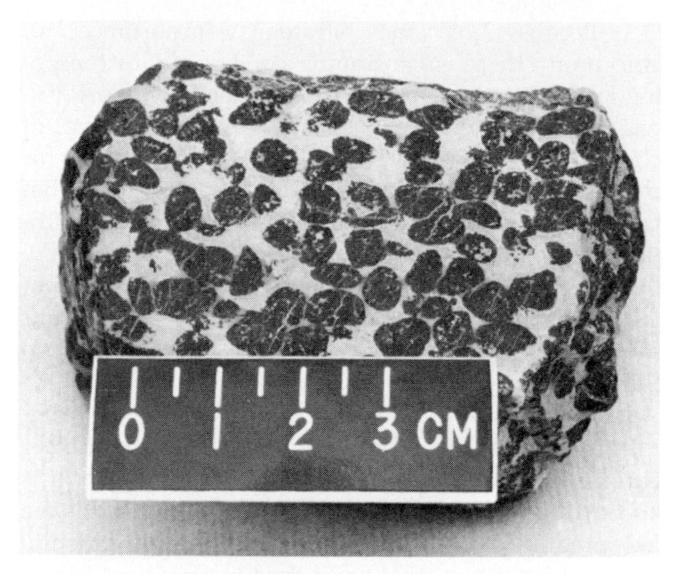

FIGURE 8.5 Pelletal grains of chromite in a podiform deposit from Greece. (Photograph by S. Llyn Sharp.)

Chromite is a member of the spinel family of minerals, which may contain variable percentages of iron, chromium, magnesium, manganese, titanium, and aluminum. Chromites of suitable composition for bricks are usually mixed with MgO in order to impart strength and to increase their resistance to thermal and chemical attack.

Geological Occurrence. Chromium is present in small amounts in all mafic and **ultramafic** rocks—that is, rocks rich in iron and magnesium but poor in silica. Chromite is the major ore mineral of chromium, and its occurrence in quantities sufficient to form ore deposits is essentially restricted to ultramafic rocks. Chromite occurs in two major types of ore bodies, **podiform** and **stratiform.** The podiform deposits appear as irregular pods or lenses that may range in mass from a few kilograms to several million tons and that nearly always occur in highly faulted and deformed portions of tectonically active zones. A common feature of these deposits is the appearance of the chromite as rounded or eye-like granules (Figure 8.5). The pods are enclosed in deformed masses of **dunites, serpentinites,** and related ultramafic rocks that are believed to be solid fragments from the upper mantle that were forced up during tectonic collisions between continents. Typical podiform deposits occur in the Ural Mountains of Russia and Kazakhstan; the Appalachian and Pacific Coast ranges of the United States; and also in Cuba, the Philippines, and the countries around the eastern end of the Mediterranean. The bodies, though small, contain chromites with very desirable compositions.

Stratiform chromite deposits are, as the name suggests, discrete, sharply bounded layers of essentially pure chromite (Figure 8.6) that occur in large, mafic intrusions where the layering developed as an artifact of the processes of cooling and crystallization (see Figure 2.7). Individual monomineralic layers of chromite are known to range up to several meters in thickness in the largest layered intrusions, but such thick layers are very rare. More commonly, chromite layers are a meter or less in thickness. Whether thick or thin, however, the chromite layers can extend laterally up to dozens of kilometers, and in the largest known

FIGURE 8.6 Stratiform layers of chromitite, a rock composed almost entirely of chromite, exposed along the banks of the Dwaal River, South Africa. The chromitite is interlayered with anorthositic norite (white) in the lower portion of the Bushveld Igneous Complex. (Photograph by B. J. Skinner.)

layered intrusion, the Bushveld Igneous Complex in South Africa, up to hundreds of kilometers. Stratiform chromite layers contain most of the world's known chromite resources, although in many cases the compositions of the stratiform chromites are not so desirable for metallurgical or refractory purposes as the chromites from podiform deposits.

The origin of stratiform layers of chromite remains in question. For many years, geologists favored the hypothesis that the layers formed when dense chromite grains crystallized from cooling bodies of **magma,** then settled to the floor of the magma chamber to form the monomineralic layers we see today. A dense mineral such as chromite was presumed to sink rapidly, whereas a less dense mineral such as pyroxene would sink more slowly, and as a result a separation would be effected. Recently, this simple picture based on crystal settling has been questioned and found wanting. The textures of the chromite grains, and indeed the sequence of minerals in the layers, suggest that some of the minerals actually grew on the bottom of the magma chamber. The thick, economic layers of chromite were formed during long periods of time when chromite was the only mineral crystallizing from the magma. We now hypothesize that the periods of chromite formation were brought about by relatively large-scale, but temporary, changes in the composition of the crystallizing magma as a result of contamination by overlying or underlying rocks. Such contamination would have altered the chemistry of the magma just enough that, for a period, only chromite crystallized; after that period, the normal sequence of igneous minerals would again crystallize.

By far the most important stratiform deposits of chromite in the world occur in the Bushveld Igneous Complex of South Africa. This enormous complex of layered intrusions covers 66,000 square kilometers, is 12 kilometers (7.5 miles) thick in places, and is also host to the world's largest known resources of vanadium and the platinum group metals.

PRODUCTION AND RESERVES. The production and reserves of chromite are dominated by Kazakhstan, India, and the Republic of South Africa, which together hold about 95 percent of the world's reserves; South Africa and Kazakhstan are the largest producers. The United States produces chromite from one small mine in Oregon, and all of Europe, except Finland, is without economically viable chromite deposits. The United States does have large, low-grade stratiform deposits in the Stillwater Complex, Montana, but the chromite has a composition that is difficult to process and expensive to use. As a result, the industrial nations of the world depend for their supplies on stratiform deposits in South Africa and Zimbabwe and podiform deposits in the Soviet Union for their chromite needs.

Because of the strategic importance of chromium, there is continuing concern about the potential for supplies being cut-off either due to civil unrest or for political reasons. Known world reserves of chromite are certainly adequate to meet needs for many years to come; furthermore, it seems likely that the Bushveld Igneous Complex probably has additional large resources that can also be exploited in the future. Chromium is thus one of the important commodities where future availability may well be far more dependent on political and social issues than on the physical limits of resources.

The widespread use of chromates as coloring agents in paints, in fabrics, and on paper means that chromium compounds are all around us. Concerns about the toxicity of some chromium compounds have led organizations such as the U.S. Environmental Protection Agency (EPA) and the Occupational Safety and Health Administration (OSHA) to introduce regulations on its use. Of particular concern has been the release of chromium into groundwater from leachates derived from old landfills and from old industrial sites where chromium-bearing wastes were disposed of by dumping or burial.

Vanadium

Vanadium was identified as a chemical element in 1830, and, like chromium, its salts soon found uses in the tanning of leather and in preparation of colored pigments for textiles, pottery, and ceramics. The use of vanadium in its metallic state as an alloying agent came much later. In 1896, French scientists found that vanadium so toughened steel that it could withstand the impact of bullets and, hence, could be used to make armor plate. It was soon discovered that vanadium steel also toughened the cutting edges of knives and swords and greatly increased the strength of certain constructional steels. This latter discovery led the American automobile industry to start using vanadium steel; by 1908, its use was advertised as a special feature of Ford motor vehicles.

Vanadium steels continue to be very widely used in automobiles and in industry in general because the incorporation of as little as 0.2 percent vanadium in an ordinary carbon steel greatly increases its strength, high temperature abrasion resistance, ductility, and even the ease with which steels can be welded. The use of high-strength vanadium steel allows a minimum weight of steel to be used in an automobile and this, in turn, leads to increased efficiency and to a reduction in the amount of fuel needed to run an automobile.

The ease and reliability with which vanadium steel can be welded has led to its widespread use in gas and oil transmission pipelines. The 1288-kilometer (800 mile) Alaskan pipeline that brings oil from

Prudhoe Bay to the port of Valdez incorporates 650 tons of vanadium.

GEOLOGICAL OCCURRENCE. Despite a crustal abundance of 0.014 percent, which makes vanadium one of the more common of the geochemically scarce elements, vanadium ore deposits are rare. The reason deposits are rare is that vanadium readily substitutes for ferric iron (Fe^{3+}) in common minerals such as magnetite (Fe_3O_4). The substitution of vanadium for iron is so extensive that the limits are rarely exceeded; hence, local concentrations of vanadium minerals are rare. The most important ore deposits of vanadium are thus vanadium-rich magnetites (containing approximately 2 percent V_2O_3). These are found as monomineralic stratiform layers of magnetite in certain layered intrusions of mafic igneous rock. By far the most important of the vanadiferous magnetite deposits discovered so far are in the Bushveld Igneous Complex, South Africa. The stratiform layers of vanadiferous magnetite, of which there are about ten, are near the top of the Bushveld Complex, whereas the stratiform chromite layers that they closely resemble (Figure 8.7) are near the base. The vanadium magnetite layers, like the chromite layers, appear to have formed as a result of monomineralic crystal growth on the floor of the magma chamber.

When separate deposits of vanadium minerals do occur, they apparently form as a result of weathering. When vanadium-bearing magnetites, or other vanadium-bearing minerals, are weathered in arid climates, the vanadium is oxidized from the trivalent V^{3+} state to the more soluble pentavalent V^{5+} state. Pentavalent vanadium can be transported long distances in solution. Precipitation of vanadium minerals can occur through evaporation or, as in the Colorado Plateau region of the United States, through contact with organic matter that serves as a reducing agent that causes the vanadium in solution to be converted back to the less soluble V^{3+} state. Uranium and copper also have more than one valency state, and they exhibit a behavior similar to vanadium. Uranium and vanadium are sometimes found concentrated together as a result. The region of Colorado, Wyoming, Utah, and New Mexico where deposits of this kind are found is known as the **Uravan District** because of the co-occurrence of uranium and vanadium. For many years, in the early part of the twentieth century, deposits in the Uravan District were worked for vanadium. From the time of World War II, which saw the development of the atom bomb and the rise of nuclear power, attention in the Uravan District has been focused almost entirely on uranium.

FIGURE 8.7 Dark-colored stratiform layer of magnetite in the upper portion of the Bushveld Igneous Complex, South Africa. The layers are almost entirely composed of magnetite and have the same origin as the chromitite layers shown in Figure 8.6. The magnetite contains vanadium in solid solution and is one of the world's major sources of this valuable alloy metal. (Photograph by Craig Schiffries.)

Vanadium tends to be concentrated, at least to a small degree, whenever concentrations of organic matter occur. The vanadium content of coal averages about 0.02 percent by weight; that of crude oil is about 0.005 percent. Certain very heavy oils (tars) have much higher vanadium contents. Thus, tar in the Athabasca Tar Sands of Canada contains up to 0.025 percent vanadium; tar from the Orinoco Tar Sands of Venezuela contains 0.05 percent. In Peru and Argentina, veins of solid bitumen, believed to have formed by distillation of petroleum, contain 0.1 percent and 0.85 percent vanadium, respectively.

The role of vanadium in fossil fuels is not well understood. It is clear that much of the vanadium must enter the deposits after sedimentation because the vanadium contents of living plants and animals are not high. The vanadium appears to enter during degradation of the original organic matter in the sedimentary pile, perhaps brought in by groundwater, and to be locked up in compounds called **porphyrins.** The atomic structures of the vanadium porphyrins found in crude oils resemble cages, with the vanadium atoms at the center. The cages are nearly identical to the structures of the chlorophylls (magnesium-centered porphyrins) of green plants and the hemoglobins (iron-centered porphyrins) of blood. Vanadium probably changes place in the structures with magnesium and iron during diagenesis.

Many vanadium compounds are considered to be toxic to humans; hence, there is much care taken in working with ores or products rich in vanadium. In all practical applications, the concentrations are quite low, and the metals are found in inert alloy forms; hence, there have been no reports of health problems.

PRODUCTION AND RESERVES. The bulk of the vanadium produced today (about 95 percent) comes from the vanadiferous magnetite deposits of South Africa, Russia, and China. Lesser amounts are recovered as by-products from the slags of iron smelters and from smelters that produce elemental phosphorous. In the western United States, the ores of the Uravan District were mined for uranium and vanadium was recovered as a by-product; here, the economic viability of the deposits was primarily dependent upon the price of uranium. The decline in the demand for uranium has severely curtailed production of vanadium from the major American deposits.

The recovery of vanadium by-products from the ash of burned fuel oils and from spent catalytic converters used in oil refining, now accounts for about 1510 metric tons of metal, or about 5 percent of the world's annual production. In the United States, this type of recovery accounts for more than 26 percent of the output. The distillation of petroleum during the petroleum refining process concentrates the tiny quantities of metals in the oil into a solid residue referred to as "petroleum coke." Increasingly, this coke is being used to generate so-called synthesis gas (a mixture of CO, CH_4, and H_2) that becomes a chemical feedstock, or that is burned to generate electricity. In the process, the metals, especially vanadium, are again concentrated and deposited in a residual slag where they may reach concentrations of up to 40 percent by weight.

It is apparent that known world reserves of vanadium are sufficient for at least another century. Furthermore, the increase in mining of rich oil sands for petroleum will also produce increasing amounts of by-product vanadium. Those sands in the Athabasca region of Alberta alone could supply well over 2 million tons of vanadium, sufficient to meet all needs for an exceedingly long time into the future.

Nickel

Small amounts of nickel are present in some of the ancient copper coins dug up in the Middle East, although it is probable that the addition of nickel was accidental and came about because nickel was present in small amounts as an unrecognized contaminant in the copper ore used to make the coins. Nickel was first recognized in the Western world during the seventeenth and eighteenth centuries when it was encountered by copper miners in Saxony, in what is now eastern Germany. Certain nickel minerals so resemble copper minerals in their color and in other properties that the Saxon miners attempted to smelt the ore to recover copper. What they obtained were specks of a shiny white metal that could not be worked into useful objects, so they named the material *kupfernickel*, or "Old Nick's" copper. They believed that the devil, Old Nick, and his mischievous gnomes had bewitched the copper ore. The frustrations of those old miners lives on in the name "nickel." Early in the eighteenth century, a Swedish chemist, Axel Cronstedt, showed that nickel is actually a separate chemical element, but it was not until 1781 that pure, metallic nickel was prepared.

The first practical use of metallic nickel was in a nickel-silver alloy, the so-called German silver that has been used for trays, teapots, and other household utensils. Extensive demand for nickel arose as a result of discoveries by the English scientist and inventor Michael Faraday, who developed the process of **electroplating.** This is a process by which metal is dissolved into solution from a metal plate connected to one terminal of a battery, and then deposited on another metal object connected to the other terminal. Nickel dissolves when connected to the anode (the positive terminal of the battery) then moves through the solution as a result of the electrical current and is deposited as a thin layer on the metal object connected to the cathode (the negative terminal). Because nickel resists corrosion and can be polished to a high luster, nickel plating soon became very popular, and by 1844 a plating industry was firmly established in England. The durability of nickel-copper alloys led to their use in coinage, first in Belgium in about 1860, and then in the United States in 1865. The use of the term "nickel" for the American and Canadian five-cent coins soon followed because they contained nickel and looked like nickel, even though they actually contained 75 percent copper and only 25 percent nickel. (See Box 8.1 for a discussion of metals in U.S. coins today.)

The use of nickel as an alloying agent with iron came about in the twentieth century. Nickel and chromium steels do not corrode or rust—hence, they are stainless. Used alone, or in combination with chromium and other alloying agents, nickel-bearing steels find many uses in the manufacture of aircraft, trucks, railroad cars, and other structures where great reliability and high strength are required. It is probably not an overstatement to say that nickel has proved to be the most versatile of all the ferro-alloy metals.

Nickel is still widely used as a plating metal, is still used in coinage, and is still used to harden copper and make versatile alloys with metals other than iron. But the major use of nickel today, accounting for about half of the world's total production, is as an alloying agent with iron.

GEOLOGICAL OCCURRENCE. All of the important nickel deposits in the world are found in, or adjacent to, mafic and ultramafic igneous rocks. Nickel is one of the chemical elements that is depleted in Earth's crust but enriched in the **mantle.** Thus, mafic and ultramafic igneous rocks derived from magmas generated in the mantle tend to have high nickel contents. Whereas the average nickel concentration in rocks of the continental crust is only 0.0072 percent by weight, many mafic and ultramafic igneous rocks derived from the mantle contain as much as 0.1 percent nickel. The nickel is present in mafic igneous rocks in solid solution in **pyroxenes** and **olivines,** where it substitutes in the structures for magnesium and iron.

Ordinary mafic and ultramafic igneous rocks are not sufficiently enriched in nickel to be considered ores. Further concentration occurs in two entirely different ways. The first is a form of magmatic segregation involving **liquid immiscibility.** When a magma cools and starts to crystallize, minerals such as olivine and pyroxene form and grow in the liquid. The growth of crystals of pyroxene in the magma indicates that the liquid has become saturated in pyroxene. Suppose, however, that the cooling magma becomes saturated in a compound but that the temperature is still above the melting temperature of that compound. That is what sometimes happens in the case of the iron sulfide mineral **pyrrhotite.** Instead of a crystal of pyrrhotite forming in the cooling magma, tiny drops of a molten iron sulfide liquid form. The silicate magma and the iron sulfide liquid do not mix and are said to be *immiscibile,* like oil and water. The immiscible liquid is not pure iron sulfide, but tends to scavenge atoms of nickel, cobalt, copper, platinum, and certain other chemical elements from the magma, so it is really an iron-nickel-copper-sulfide liquid. The immiscible sulfide drops are denser than the silicate magma, so they tend to sink and form sulfide-rich zones near the base of the magma chamber. When the sulfide liquid eventually crystallizes, the main mineral that forms is pyrrhotite ($Fe_{1-x}S$), but intergrown with the pyrrhotite are grains of the only important sulfide ore mineral of nickel, pentlandite ($(Fe, Ni)_9S_8$) (Figure 8.8), together with chalcopyrite ($CuFeS_2$), and tiny grains of metallic platinum and other platinum group minerals. By the processes of concentration through liquid immiscibility and magmatic segregation, nickel contents as high as 3–4 percent can be reached.

For many years, the world's richest and most important nickel ore bodies were sulfide ores that formed as a result of magmatic segregation. The most famous nickel ore deposits are at Sudbury, Ontario, where intrusive mafic igneous rocks form an elliptical ring, 56 kilometers (35 miles) on the long axis and 26 kilometers (16 miles) on the short axis. Around the outer edge of the basin, near the base of the intrusion, a number of large rich deposits occur (Figure 8.9). One of the strange features about Sudbury is that the intrusion has the shape of a cone rather than a flat sheet, as is usual for most dikes and sills. The probable reason for the conical shape was realized in the

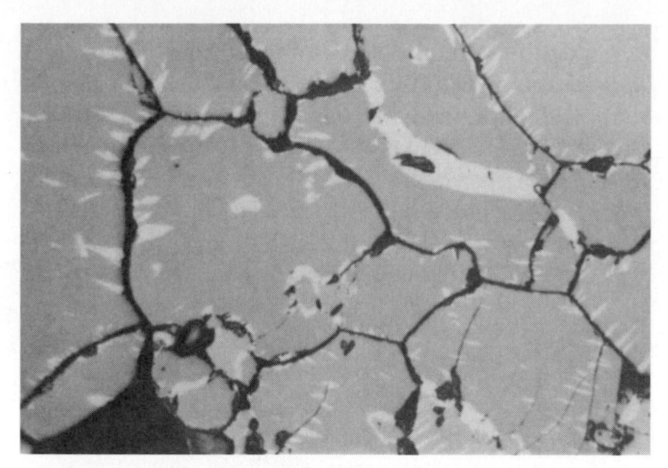

FIGURE 8.8 The primary ore mineral source for nickel is the mineral pentlandite ($(Fe,Ni)_9S_8$), which typically occurs as fine lamellae ("white" in this photo) exsolved from pyrrhotite ($Fe_{1-x}S$). This photomicrograph shows an area approximately 0.5 millimeter across. (Photograph by J. R. Craig.)

1960s as a result of investigations associated with space research. Many features on the surface of the Moon and other bodies in the solar system result from large meteorite impacts. As the properties and characteristics of impact structures on Earth were studied, it became apparent that the Sudbury Basin might be an ancient impact structure, as much as 1.9 billion years old; the magma carrying the immiscibile sulfide droplets probably rose from the mantle along fractures created by the impact event. Other rich and important sulfide ores of nickel are found in Canada in the Thompson Lake District of Manitoba, at Kambalda in Western Australia, in Botswana, Zimbabwe, and at Norils'k in Russia. The discovery of a huge, rich nickel sulfide deposit at Voisey's Bay, Labrador, in eastern Canada in the 1990s was hailed as one of the mining industry's greatest finds of the twentieth century. Unfortunately, it also became an example of the complications that arise from politics and economics. Disputes involving the mining company, the provincial government, and Native Americans over the siting of a processing mill and refinery have left the multibillion dollar deposits unmined and have cost all parties millions of dollars in revenues. Mining was finally begun after 2000, but has been hampered by various labor disputes.

The second important way that nickel can be concentrated in a mafic or ultramafic rock is through weathering. When mafic igneous rocks are subject to chemical weathering under tropical or semitropical conditions, the silicate minerals (pyroxene, olivine, and plagioclase) break down to form hydrous compounds, and iron is oxidized to the ferric state. The small amount of nickel present in solid solution in the olivines and pyroxenes is released in the process; it either forms nickel silicate minerals, or it is incorporated into the structure of other

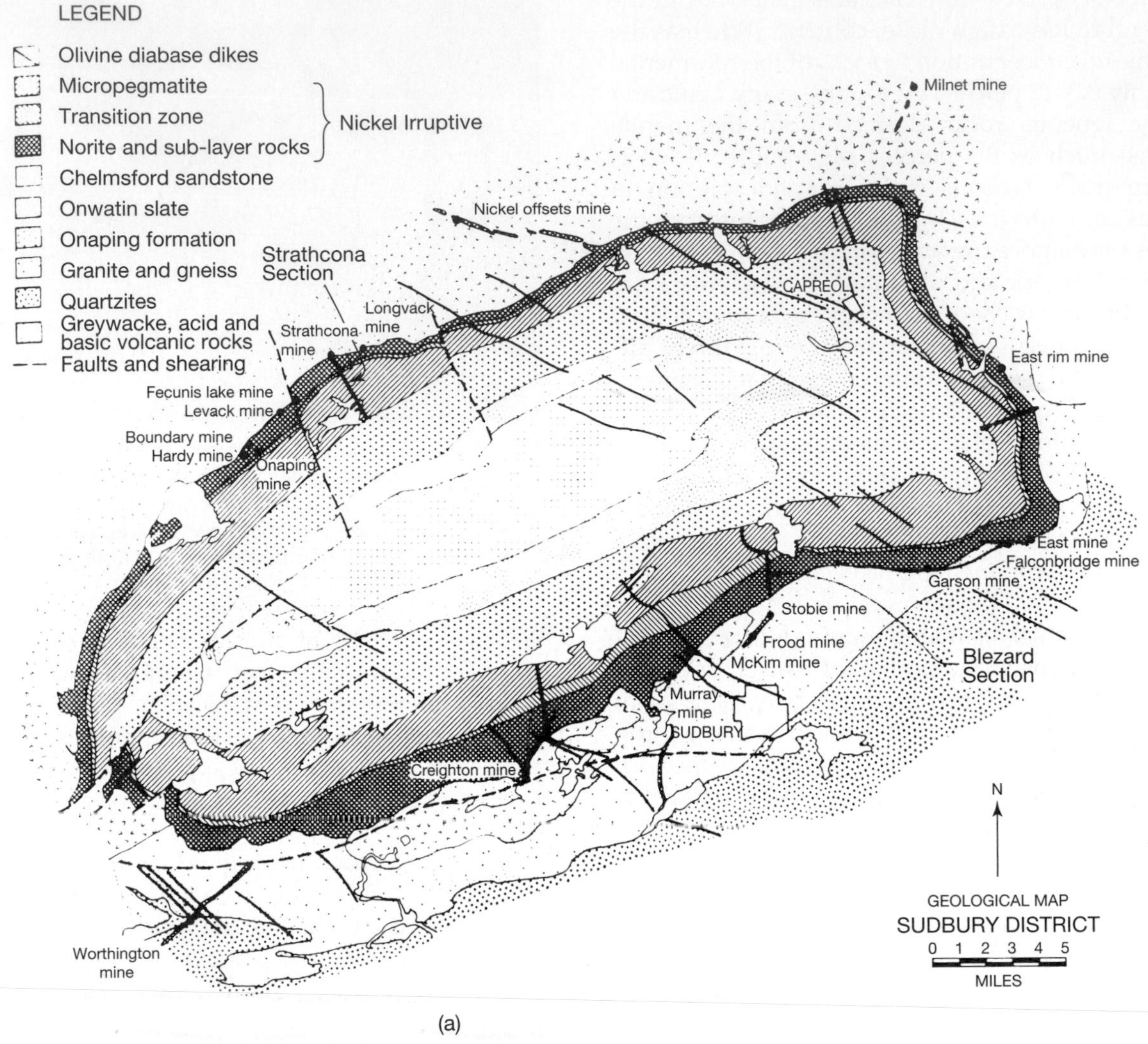

FIGURE 8.9 Map of the Sudbury intrusive in Canada, the largest nickel district in the world. (From E. Gasparini and A. J. Naldrett, *Economic Geology*, vol. 67, p. 605 [1972]. Used with permission.)

minerals formed during weathering. The weathering minerals, such as chlorite and serpentine, sometimes contain 1 or 2 percent of nickel; under certain circumstances the nickel silicate minerals, collectively called garnierite, may form and produce ores as rich as 4–5 percent nickel. Residual weathering ores are referred to as **laterite ores.** The most famous laterite ore is in New Caledonia, in the Pacific, where French interests have been mining garnierite-rich bodies for over a century. There, a nickeliferous **peridotite** has weathered to form lateritic garnierite ore (Figure 8.10). The development of a new technology, the pressurized acid-leach process, has made the recovery of nickel from the lateritic ores much more efficient and profitable. It is possible that production from the rich laterite deposits in Western Australia and New Caledonia may one day rival production at Sudbury. The surface mining of laterite ores and the high rates of recovery using hot acids under pressure will allow laterite ores to yield nickel and the by-product cobalt at about one-half the cost of mining and processing the deep sulfide ores of Sudbury.

PRODUCTION AND RESERVES. Nickel is one of the strategic minerals for which the United States has very limited resources of high-grade deposits. It does have very large low-grade deposits, however. These sulfide ores are in the Duluth Gabbro, Minnesota, and average about 0.21 percent nickel.

A large fraction of the world's nickel has, for many years, come from the Sudbury and the Thompson Lake Districts of Canada. A steady increase in the mining of nickel from the Norils'k Deposit in Russia has now brought the production there to levels higher than in Canada. Important production of sulfide ore also comes from Australia, Indonesia, Cuba, the Philippines, the Republic of South Africa, and

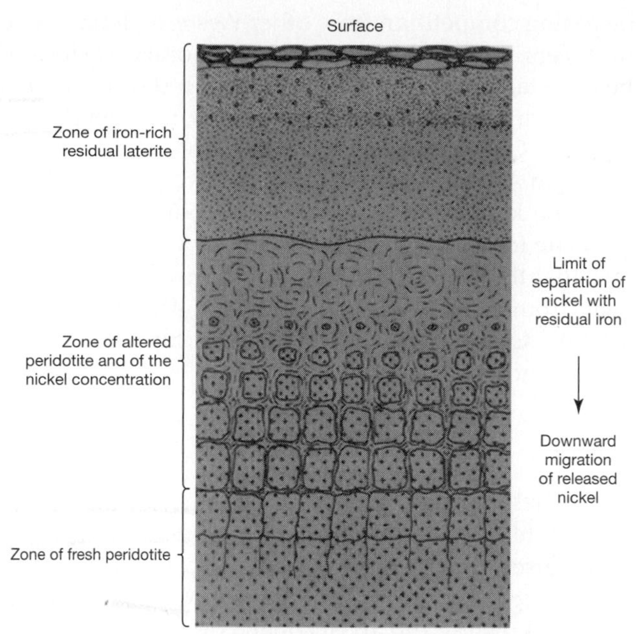

FIGURE 8.10 Chemical weathering of nickeliferous rocks, such as peridotite, release nickel trapped in the mineral olivine. The nickel is then reprecipitated in a new mineral called garnierite. Residual deposits of this type are becoming more important as nickel sources. (From E. de Chetalat, *Bull. Soc. Geol.* France, Ser. 63, Vol. SVII, 129, Fig. 4, [1967].)

China, while mining of lateritic ore is expanding in Australia, Indonesia, and New Caledonia.

Molybdenum

Molybdenum is a versatile and highly important member of the ferro-alloy family of metals, but it is relatively unknown outside of metallurgical circles. This is because molybdenum, like several other metals, is scarcely ever used in its pure form. Nevertheless, it is an extremely versatile metal employed in a wide assortment of products.

The name "molybdenum" comes from the Latin word *molybdaena*, and from the older Greek word *molybdos*, which actually refer to the lead mineral galena and to metallic lead, respectively. There are a number of soft, easily deformed, gray minerals that look like lead and galena. Among those confusing minerals are graphite (C) and the mineral we now call **molybdenite** (MoS_2). From the time of the Greeks and Romans, all of the lead-like minerals were called *molybdos*. The confusion was only resolved when the element was separated as the metal in 1782. The chemistry required to prepare the metal is so difficult that a pure metal was not produced until 1893.

As soon as pure molybdenum was available, its alloying properties were tested by French scientists. It proved to make a tough and resilient steel that was ideal for armor plating. By 1898, molybdenum tool steel had also been developed. However, neither of these uses created much demand. A few small mines were opened in the closing decade of the nineteenth century, but those in the United States had all ceased to operate by 1906. Production in those days was satisfied by molybdenum ores worked in Norway, Australia, and Canada.

Molybdenum finally found a market in World War I, when it was discovered that molybdenum steels could be substituted for the widely used tungsten steels in high-speed cutting tools and in armaments. As a result, large deposits were opened in Climax, Colorado, and Questa, New Mexico. When World War I ended, military demand for molybdenum slumped and production at Climax and Questa was stopped. Starting in about 1920, new uses were found for molybdenum steels—molybdenum alloys make desirable heating elements in the electrical industry. These new markets led to a resumption of full-scale mining by 1924.

About 80 percent of the molybdenum mined today is used as an alloying element in steels, cast irons, and **superalloys** where it imparts hardness, toughness, resistance to corrosion and abrasion, and adds strength at high temperatures. The molybdenum content in steel ranges from 0.1 percent to 10 percent by weight, and it is usually employed in combination with other ferro-alloy elements. The steels produced are now used in all areas of industry, but they find special demand in cutting tools, in transportation, and in the manufacture of oil and gas production equipment. The anticorrosion properties imparted by molybdenum have led to the increasing application of its steels in severe chemical environments and in seawater.

Nonmetallic applications for molybdenum include those as lubricants, catalysts, and pigments. Molybdenum disulfide (MoS_2) has a well-defined layer structure, somewhat like the structure of mica. The layered structure makes molybdenite soft and slippery to the touch and, because the compound resists breakdown even at high temperature and high pressure, it is an ideal lubricant under demanding conditions of temperature and pressure. Molybdenite is widely used as an additive to oils and greases, where it helps to reduce friction and wear significantly in automobile engines. Molybdenum catalysts are also used in the production of petroleum-based chemicals and alcohols. Molybdenum orange (MoO_2) is an important pigment in paints, dyes, and inks.

GEOLOGICAL OCCURRENCE. Nearly all of the known molybdenum ores consist of molybdenite, MoS_2, a lead-gray, metallic-looking mineral that occurs in *porphyry-type* igneous intrusions of the type described under copper. The large porphyry-type deposits, closely related to subduction zones at the edges of continental plate boundaries (see Figure 8.21) and commonly rich in copper minerals, usually contain minor amounts of molybdenite as well. A significant fraction of molybdenum production comes as a by-product from porphyry copper mining. A few large porphyry-type deposits, such as those at Climax and Urad in Colorado and Questa in New Mexico, contain molybdenite together with tin and tungsten minerals, almost to the exclusion of copper sulfides. The molybdenite occurs as disseminated mineral grains and in thin coatings in cross-cutting fractures in the porphyritic igneous rocks, and also in the surrounding rocks, as is evident in Figure 8.11. The molybdenite was deposited, along with large amounts of quartz, by **hydrothermal solutions** that were episodically released from the crystallizing magma.

Molybdenite also occurs in much lesser, and usually uneconomic, quantities in **contact metamorphic** zones adjacent to silica-rich igneous rocks, in quartz vein deposits, and in pegmatites. Molybdenite is usually recovered as a by-product from these deposits.

PRODUCTION AND RESERVES. The United States dominated the world molybdenum market for many years, with the bulk of production coming from the large deposits at Climax, Colorado, and Questa, New Mexico. In recent years, however, production has increased in Chile, Canada, China, and Peru. The decline in the U.S. steel industry, long the principal market for American production, combined with increasing competition from other Western Hemisphere producers, has resulted in at least temporary closings of the Questa, Climax, and Urad mines. Today, most U.S. molybdenum production comes from the Henderson Mine in Colorado, and as a by-product of copper mining in Arizona and Utah.

The known reserves of molybdenum are clearly adequate for many years to come. It does appear, however, that there will be a continuing shift in production away from the United States. Instead, the dominance of Chile, China, Mexico, and Peru as suppliers of the world's molybdenum is growing.

FIGURE 8.11 Specimen of ore from the molybdenum-bearing porphyry deposit at Climax, Colorado. This sample shows the typical intense fracturing along which the ore minerals have been deposited. The specimen is about 15 centimeters (6 inches) wide. (Photograph by M. Fortney.)

Cobalt

The earliest uses of cobalt as a brilliant blue coloring agent date from antiquity. Egyptian and Babylonian potters used cobalt oxide, known as "cobalt blue," as a coloring agent in glass and ceramics. Chinese artisans became skilled in the art of cobalt coloring of ceramics during the Ming Dynasty (fourteenth to seventeenth centuries); in Europe, Venetian artisans of the fifteenth and later centuries became renowned for their cobalt-colored glassware.

The name *cobalt*, like that of nickel, has an association with old German miners. During the sixteenth century, arsenic-bearing silver-cobalt ores were mined in the Harz Mountains of eastern Germany. The roasting of these ores released poisonous arsenical fumes that caused ulcers on the bodies of the miners tending the furnaces. The miners believed the source to be silver-stealing goblins called *Kobolds*, who replaced good silver minerals with useless cobalt arsenides that looked somewhat like silver.

Cobalt was shown to be a separate chemical element in 1780, and the modern history of cobalt dates from this time. The principal uses of cobalt continued to be in coloring agents and chemical compounds used in various industrial processes. In 1910, the use of cobalt as an alloying compound was finally demonstrated by an American, Elwood Haynes, who showed that the addition of about 5 percent cobalt to a steel containing chromium and tungsten greatly improved its qualities as a tool-steel. This alloy and other chromium-cobalt alloys were the forerunners of today's superalloys, which retain their mechanical strength at high temperatures and are resistant to corrosion by hot gases. The recognition of the importance of cobalt and its increasing use in superalloys employed in jet engines, rocket nozzles, and gas turbines after World War II has led to cobalt being designated a strategic metal. Cobalt alloys also have remarkable magnetic properties—the magnetism is very strong and it is retained forever. The alnico magnets used in the loudspeakers of high-fidelity sound systems and many other technological applications are alloys of aluminum, nickel, and cobalt.

GEOLOGICAL OCCURRENCE. Most of the world's cobalt is produced as a by-product of the mining and metallurgical treatment of the ores of copper, nickel, and silver. The most important cobalt ores are stratiform copper sulfide ores, as are found in the "copperbelt" of the Congo and Zambia in central Africa. This type of deposit consists of copper sulfide minerals, such as chalcopyrite ($CuFeS_2$), together with cobalt sulfide minerals, such as linnaeite (Co_3S_4) and cobaltiferous pyrite $(Fe,Co)S_2$, enclosed in fine-grained **clastic** sedimentary strata. The origin of the stratiform deposits is enigmatic, but the sulfide minerals appear to have been introduced into the sedimentary strata soon after deposition by warm saline solutions circulating in subsurface aquifers.

Cobalt also tends to be concentrated wherever nickel is concentrated. Thus, cobalt is an important by-product from the exploitation of both magmatic segregation nickel sulfide ores and lateritic nickel ores.

PRODUCTION AND RESERVES. One country, the Congo, produces about half of the cobalt mined in the world. The only other countries with reasonably large productions are Canada, Australia, Russia, and Zambia.

The world's largest cobalt reserves lie in the stratiform ores of the Congo and Zambia. The large Canadian reserves nearly all occur with the rich nickel ores in the Sudbury Basin; the cobalt is produced as a major by-product of the nickel mining. In addition to the known reserves, there are probably at least another 4.5 million metric tons of cobalt as identified resources. Large low-grade resources have also been identified in Cuba, Australia, and the United States, principally associated with the low-grade nickel deposits of the Duluth Gabbro in Minnesota. Additional very large deposits are believed to lie on the ocean floor, where some of the ferromanganese nodules, described in Chapter 7, contain up to 1 percent cobalt (Table 7.5). Thus, even though definite reserve figures cannot be quoted for cobalt, the long-term future for the metal appears to be assured.

Tungsten

Tungsten is a grayish-white metal that has many alloying properties similar to those of chromium and molybdenum. It has the distinction of being the metal with the highest melting temperature (3400°C) and the further distinction of being the metal with the highest tensile strength. Alloys of tungsten are therefore extremely hard and extremely stable at high temperatures.

Metallic tungsten and the ore minerals of tungsten are all very dense. The name *tungsten* recognizes this property; it originates from two Swedish words, *tung* meaning heavy, and *sten* meaning stone. Until the middle of the eighteenth century, the mineral we know today as scheelite ($CaWO_4$) but then called tungsten, was thought to be an ore mineral of tin. In 1781, the Swedish chemist K.W. Scheele demonstrated that tin was not present, and that a new chemical element was probably involved. Two years later, two Spanish chemists, the brothers d'Ellhuyar, separated metallic tungsten for the first time. The name *tungsten* was thereafter reserved for the element, and the mineral was named *scheelite* in honor of Scheele. The first use of tungsten as an additive to steel occurred in France in 1855. Within a few years, the toughness of tungsten steels came to be appreciated and, by 1868, small amounts of tungsten were added to the iron rails that carried French trains. Today tungsten steels are used in many circumstances where toughness, durability, and resistance to impact are needed.

The single most important use of tungsten is not for its alloying properties—although these uses are certainly very important—but for the manufacture of tungsten carbide (WC), a compound with a hardness approaching that of diamond. Tungsten carbide can be sintered (heated) to form intricate shapes, and it is widely used in the preparation of tungsten carbide tools, drill bits, cutting edges, and even armor-piercing projectiles.

The high melting temperature of tungsten facilitates its use in such places as the filaments of incandescent electric light bulbs (Figure 8.12) and in various heating elements. When the overall uses of tungsten are considered, more than 50 percent of the annual production is used to make tungsten carbide, 20 percent for ferro-alloys, 18 percent for tungsten metal and alloys in which tungsten is the major metal, and 11 percent for nonferrous alloys. All other uses total only 1 percent.

FIGURE 8.12 The filaments of incandescent lamps are prepared from tungsten because it has the highest melting point of any metal. Visible light is emitted when an electrical current heats the filament to temperatures of at least 800°C (1500°F.) (Photograph by J. R. Craig.)

GEOLOGICAL OCCURRENCE. Tungsten has a geochemical abundance in the crust of 0.001 percent by weight and is therefore one of the scarcer of the geochemically scarce metals. It forms only two minerals of economic importance, scheelite ($CaWO_4$) and wolframite ($(Fe,Mn)WO_4$). The most important tungsten deposits were formed by hydrothermal solutions. These fluids are widespread in Earth's crust; indeed, more ore deposits are formed by hydrothermal processes than by any other deposit-forming process. (The origin and chemistry of hydrothermal solutions are discussed more fully in this chapter in the section on copper.) Such solutions deposit small amounts of wolframite or scheelite, together with tin minerals, in quartz-rich **veins** that are usually closely related to intrusive igneous rocks that are granitic in character. Much of the world's tungsten production comes from quartz veins and from closely spaced quartz stringers called **stockworks.** A small amount of production arises as a by-product of gold, tin, or copper mining. Tungsten deposits also occur where igneous rocks that have been intruded into limestones of marbles form *contact metamorphic* deposits (also sometimes called **skarn ores**). The ore mineral in these deposits is usually scheelite and the deposits, although often rich, are generally small. Several deposits of this kind are worked in the United States in Nevada, Utah, and California.

PRODUCTION AND RESERVES. Global production of tungsten is about 54,000 metric tons per year. By far the largest producer in the world is China, responsible for about 80 percent of the world production in 2008. Other important producing countries are Russia, Canada, and the United States. Most Chinese ores are quartz vein deposits, and resources are believed to be huge, although they are not yet completely explored and tested. Tungsten production comes, too, from North Korea, Burma, and Thailand, three countries adjacent to China, and each country is endowed with the same kind of mineralization as that of China, though the reserves are much smaller. World reserves of tungsten are 3 million tons, which means that known deposits will be sufficient to meet needs far into the twenty-first century.

THE BASE METALS

A base metal is defined as a metal of comparatively low value and one that is relatively inferior in certain properties such as corrosion resistance. Such a definition seems to suggest that the base metals are second-class citizens in the family of metals. Nothing could be further from the truth, for the base metals are endowed with many important and unique properties. It was the base metals that our ancestors first learned to shape into useful objects. We still recognize those achievements in the terms *Copper Age* and *Bronze Age*. The Copper Age designates that first great step humankind took when it advanced from the Stone Age by learning to work and employ metals. The next step, into the Bronze Age, took our ancestors into the field of alloy metallurgy. Bronze is an alloy of copper and tin, but it may contain other metals as well. From the time our ancestors first learned how to work copper into needles, axes, arrowheads, and other useful objects, the practical uses found for the base metals have expanded continually. When the first indoor plumbing was installed by the Romans, they used lead pipes. When the ordinary working person wanted something better than leather or wooden plates to eat from, it was pewter, an alloy of tin that came into common use. Although copper was the first base metal our ancestors learned to use, lead eventually became the dominant base metal used by society until the widespread electrification of cities resulted in a great demand for copper to make the transmission lines. Is it possible to imagine how today's world of more than 6.7 billion people could operate without the widespread use of electricity? Nothing has changed our lives so much as this one great technological advance, and it is a base metal that has been the workhorse that brought about that advance.

The base metals of antiquity were copper, lead, tin, and mercury. Several hundred years ago, zinc was added to the list. Zinc was actually used by the Chinese and Romans over 2000 years ago, as is evidenced by its presence in some bronze dating from those times. But the Chinese and Romans apparently could not smelt and recover zinc metal, so the presence of the zinc in the alloy probably means that it was added as a sulfide or oxide mineral during smelting of the copper ore. The last of the base metals, cadmium, has properties similar to those of zinc. Like the ferro-alloy metals, cadmium use is a product of twentieth-century technology. Base metal deposits occur in a variety of geologic settings; the major occurrences are shown in Figure 8.13.

Copper

No one knows whether use of gold preceded that of copper or whether both were used at about the same time. Wherever and whenever the first use of metallic copper occurred, it was so far back in time that we will probably never find out who used it or for what

BOX 8.1
Metals in Modern Coins

Ever since coins were introduced as a medium of exchange in the sixth century B.C.E., they have served as major forms of money. Paper currency was introduced to lighten the burden of carrying coinage, and there has been much discussion in recent years that the use of credit cards, debit cards, and electronic banking will make metal coinage unnecessary. At the same time, there are suggestions that some coins (e.g., the Canadian dollar coin) work better and last longer than their paper equivalents. Regardless of the view taken, the mints in nearly every country on Earth are busier than ever stamping out coins at record rates. The earliest coins were minted from gold, silver, and copper and their alloys. The coins used in daily commerce today are generally made from copper, nickel, and zinc, with some use of aluminum and steel. Gold, silver, and platinum are used in the preparation of commemorative and bullion coins (those minted and held for the value of the metals they contain instead of their face value).

Minting of modern coinage is a big business that consumes vast quantities of metals and that has often made large profits for the governments. American coins (Figure 8.A) are used as examples because of the country's large economy and coinage demand, but other nations mint coins in similar ways and with similar results. We all use coins on a daily basis, but relatively few individuals recognize the metals that they are handling. Figure 8.B summarizes the structure and composition of modern U.S. coinage. Early coins were stamped from single metals or homogeneous alloys, but many modern coins have layered structures, as shown for the U.S. penny and dime. Coins in some countries even have center inserts that are different in composition and appearance from the outer parts. The U.S. one-cent coin, or penny, although much maligned as being of too little value to be useful, is now the coin minted in the largest quantity in the world. By 2008, more than 8.5 billion pennies (valued at more than $85 million) were being minted every year. In 2009, U.S. coin production was reduced by more than 50 percent due to the recession, but the Mint expects that production will again approach earlier levels within a year or two. The Mint's annual production in 2008 consumed more than 13,000 metric tons of zinc to constitute the cores of the coins and more than 340 metric tons of copper to make the coatings. It is hard to fully grasp the quantities involved in this coinage; perhaps two comparisons will help. If all of the pennies minted in just one year were piled one on top of the other, the stack would be more than 8000 miles high! If these coins were laid side by side at the equator, they would circle Earth four times! And this is just one year's production of U.S. pennies!

FIGURE 8.A The changing nature of precious metal coinage used in the United States. The Spanish 8 real coin (upper left, commonly known as the Spanish Piece of Eight) was minted in Spanish colonies in the New World, but was widely used in the United States until the mid-1800s. The $10 gold coin (upper right, commonly known as the Eagle) was minted as early as 1795 and was the most widely used gold coin until usage was stopped in the 1930s at the time of the Great Depression. The silver dollar (lower right) was as widely used as the gold Eagle, but was replaced by paper currency in the 1930s; it has been reissued in the later part of the twentieth century as a silver bullion coin, where its value is actually based on the price of silver rather than its nominal $1 value. The Sacagawea dollar (lower left) contains no precious metals, but is prepared from a manganese bronze with a golden color, to give it the appearance of value and to distinguish it from the quarter, which is of similar size.

(continued)

BOX 8.1

Metals in Modern Coins (*Continued*)

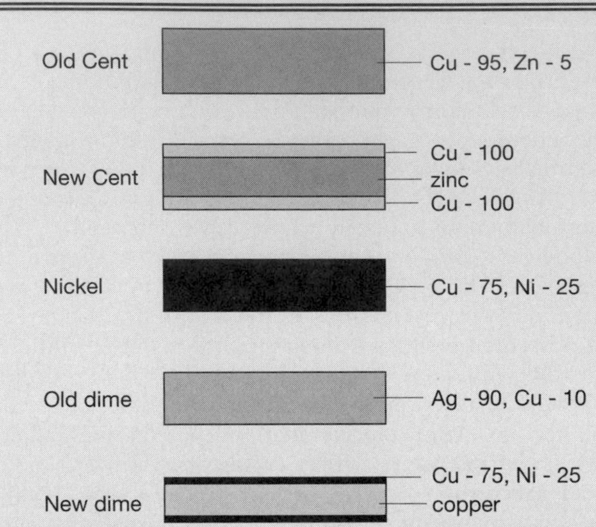

FIGURE 8.B The compositions and structures of modern U.S. coinage. One-cent pieces (pennies) prior to 1983 were uniformly composed of 95 percent copper and 5 percent zinc. In contrast, pennies made since 1983 have a thin coating of pure copper encasing a core of pure zinc. Composition of the five-cent piece (nickel) has remained constant since it was originally issued in 1866; it consists of 75 percent copper and 25 percent nickel. Ten-cent pieces (dimes) prior to 1965 were uniformly 90 percent silver and 10 percent copper. Since 1965, the dimes have had a pure copper core sandwiched between layers of 75 percent copper and 25 percent nickel. Both the twenty-five-cent pieces (quarters) and fifty-cent pieces (half-dollars) had compositions similar to dimes before 1965; after that date, quarters changed like dimes in composition and structure, and half-dollars had the silver phased out over a span of a few years.

The United States Mint consumes approximately 75,000 metric tons of metal annually to make all of its coins (13,000 tons of zinc; 40,000 tons of copper; and 23,000 tons of nickel). In an attempt to raise interest in coins and coin collecting, the United States mint issued tens of millions of state quarters from 1999 through 2008. These are the same size and composition as previous quarters, but they bear symbols representative of each of the 50 states. Subsequently, the Mint has issued newly designed quarters, nickels, and pennies. For many years, the minting of coins made large profits on the minting of all coins. In fact, profit was the primary motive in 1982 when the U.S. Mint changed the penny from a nearly pure copper coin to the present coin which is 97.5 percent zinc. The zinc was much cheaper than copper and more coins could be minted from a pound of zinc than could be minted from a pound of copper. Through the 1990s, it cost only about 0.6 cent to manufacture and distribute each penny. However, the rise in operational costs and the sharp increases in the prices of copper, zinc, and nickel changed the situation. Hence, the Mint announced in 2007 that the cost of producing and distributing each penny was 1.2 cent and the cost of producing each nickel was 5.7 cent. Dimes and quarters were still considerably more valued than their production costs. This situation led to much debate about the merits of eliminating the lower value coins or to minting them with cheaper metals, especially steel and aluminum. It is good to remember that the 1943 pennies were made of steel, with a thin zinc coating, because copper was in short supply. Nearly all of those steel pennies ultimately rusted away.

Most countries will continue to mint coins in very large numbers and many of them will be forced into decisions regarding the best metals to use.

purpose. It is not surprising that the use of copper should go so far back in time because copper, like gold, sometimes occurs as a native metal, so our ancestors did not have to first learn how to smelt ores in order to obtain the metal. What they had to learn was how to hammer native copper into a useful shape, rather than grinding or chipping it as they had been used to doing with stones. They had obviously learned to do so by 8000 years ago because excavations of the remains of civilizations that existed at that time yield finely wrought copper artifacts (Figure 8.14). The use of copper probably dates even further back in time because the smelting of copper ores seems to have started between 5500 and 6000 years ago; it is reasonable to believe that the use of native copper metal started much earlier.

The earliest underground mining activities, remains of which can still be seen today in several places in Europe, date back to the Stone Age and seem to have been for flints. The earliest underground mining for metal seems to have been for copper and to have been carried out at least 6000 years ago.

The value of copper, from antiquity until the latter years of the nineteenth century, was related to its **malleability** and the ease with which both copper and its alloys can be worked and cast. Copper metal, and the main copper alloys, *bronze* (copper with tin) and *brass* (copper with zinc), are attractive to look at, durable, and relatively corrosion resistant. Copper and its alloys found innumerable uses in weapons, utensils, tools, jewelry, statuary, pipes, roofs, and architectural features.

The nineteenth century brought about a great expansion in the use of copper because one of copper's most desirable properties is its high electrical conductivity. Not only can copper transmit electricity with a minimum loss of power, but the transmission wires are flexible and malleable, and they can be easily joined

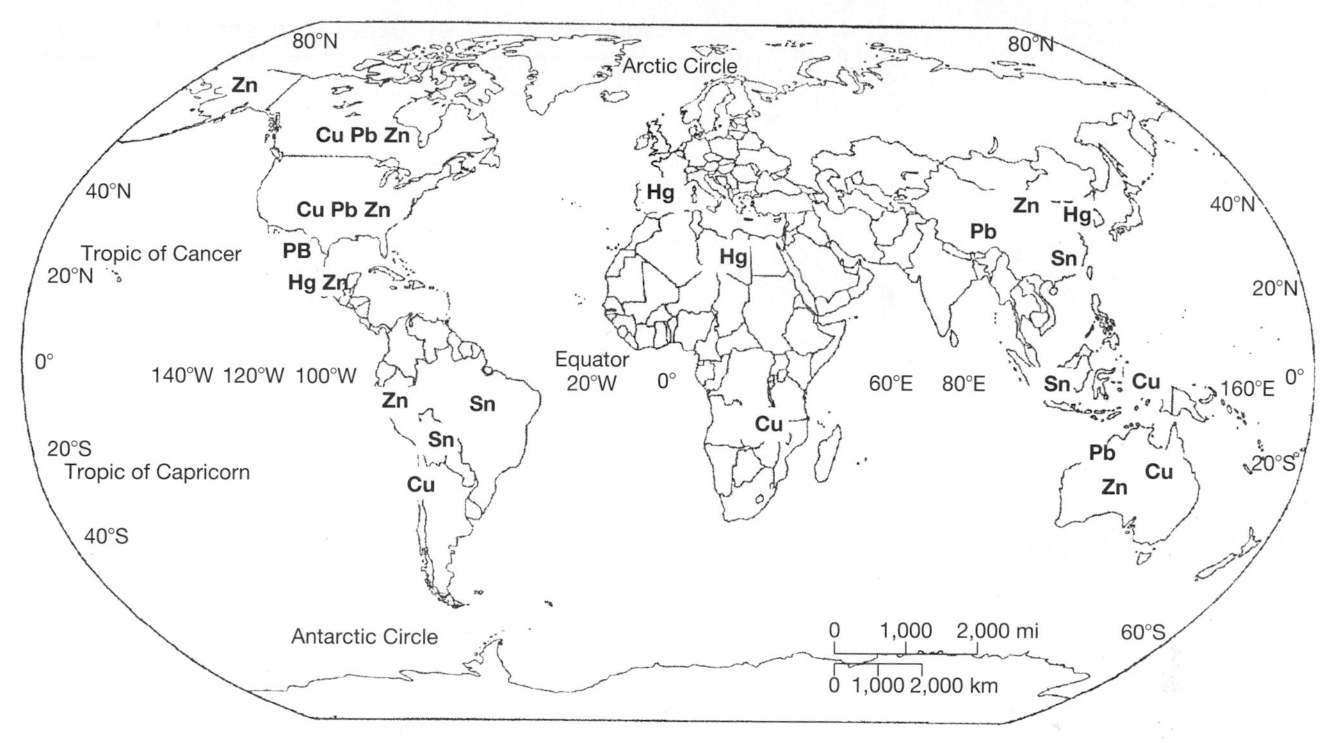

FIGURE 8.13 World map showing the locations of the largest reserves of the base metals.

FIGURE 8.14 A 20-centimeter (8 inch)-high copper figurine of Gudea of Lagash, made approximately 4200 years ago. (Specimen from the Yale Babylonian Collection. Photograph by W. Sacco.)

and soldered. The spectacular growth of the electrical power industry, plus all the industries using electricity, would certainly have been hampered without copper. Not surprisingly, the production of copper rose rapidly during the nineteenth century (Figure 8.15). In the early decades of the nineteenth century, average world annual production was less than 10,000 metric tons. We now produce more than double that amount each day! By the 1850s, production had risen to about 50,000 metric tons a year, and by the 1890s when electric power and telegraph needs started growing rapidly, production reached about 370,000 tons annually. By the beginning of the twenty-first century, the world's annual copper production had climbed to 15 million metric tons and is continuing to rise.

GEOLOGICAL OCCURRENCE. The geochemical abundance of copper is high—among the highest of the geochemically scarce metals. The average content of copper in the continental crust is 0.0055 percent, and in the oceanic crust it is even higher. Not surprisingly, for a metal that has been used for so long and in such large quantities, literally hundreds of thousands of copper deposits, large and small, have been discovered around the world. As a result, a great deal is known about copper deposits but, even so, many questions remain to be answered.

Most of the important copper ore minerals mined today are sulfides; they are chalcopyrite ($CuFeS_2$),

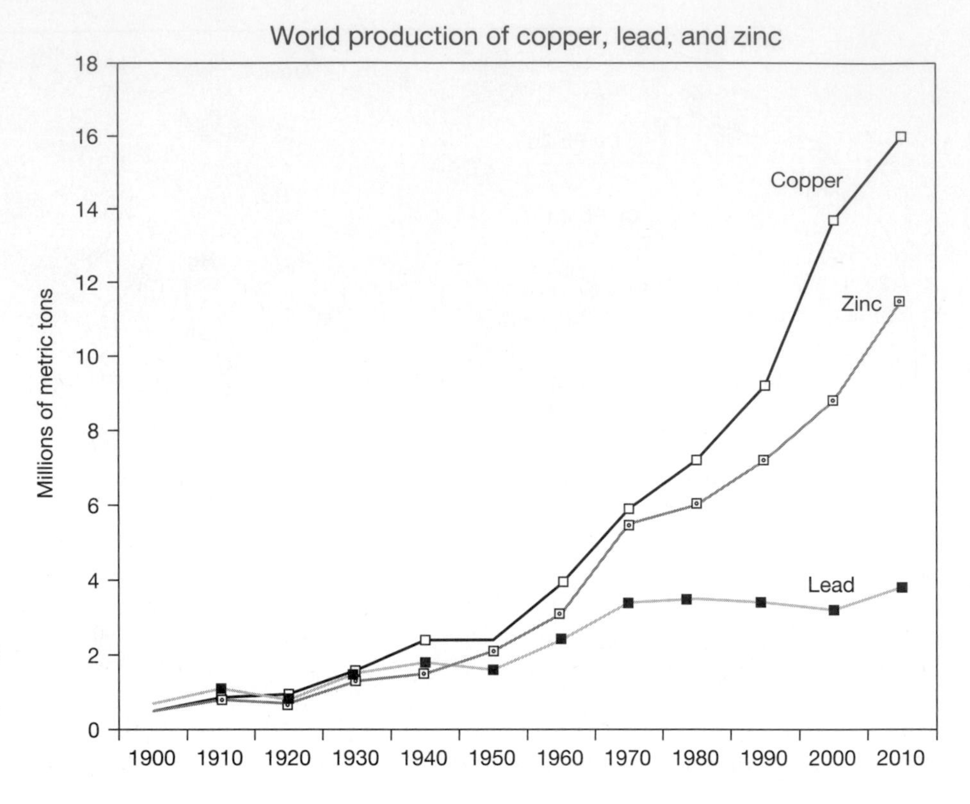

FIGURE 8.15 World production of copper, lead, and zinc since 1900. Lead was the dominant base metal produced before the beginning of the twentieth century when the spread of electrification required large quantities of copper to make wire. As with many other resources, growth in demand rose dramatically through the twentieth century.

digenite (Cu_9S_5), chalcocite (Cu_2S), bornite (Cu_5FeS_4), enargite (Cu_3AsS_4), and tetrahedrite ($Cu_{12}Sb_4S_{13}$); in addition, there are many other less important minerals. Historically, ores containing native copper, the two carbonate minerals azurite ($Cu(CO_3)_2(OH)_2$) and malachite ($Cu_2(CO_3)(OH)_2$), and the two copper oxide minerals tenorite (CuO) and cuprite (Cu_2O), have also been important. Native copper deposits, such as the famous ores of the Keweenaw District on Michigan's Northern Peninsula are infrequent; none is currently a large producer on the world scene. The oxide and carbonate minerals that copper form are a result of weathering interactions between the atmosphere and groundwater and a sulfide ore body. They are found in secondary, oxidized cappings above primary sulfide ores. However, beneath the oxidized cappings, there is sometimes a layer of very high-grade sulfide ore, called a zone of secondary enrichment, that has formed where copper has been leached out and reprecipitated at the water table, as shown in Figure 8.16. Secondary enrichment zones were highly prized by miners because the richness of the copper there usually earned great profits. Once the rich secondary ores were exhausted, mining turned to the lower grade disseminated sulfide ores. The minimum grade of ore that could be profitably mined has dropped over the past 100 years as the result of improved technology in mining and ore processing (for example, copper as shown in Figure 8.2). Today, most of the mines process sulfide ores if the copper contents are in the range of 0.5 to 1.0 weight percent. The oxidized zones, although commonly containing spectacular mineral specimens, were usually left unmined because the flotation techniques used with the sulfide minerals were not very effective in recovering the oxide and carbonate minerals.

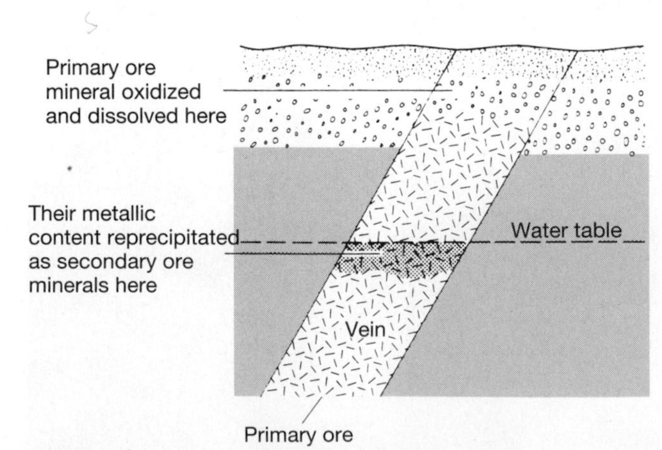

FIGURE 8.16 Descending groundwater oxidizes and impoverishes copper ore above the water table, forming an acid solution that removes soluble copper compounds and leaves a residue of limonite. The descending acid solution deposits copper at and below the water table, producing secondary enrichment below.

In recent years, the development of a new process called "solvent extraction–electrowinning" has permitted high rates of recovery of copper from the oxidized ores and has thereby much expanded the reserves of copper. In the solvent extraction–electrowinning process, acids are allowed to percolate through large piles of crushed oxide ore; the acids dissolve out the copper and are then treated by electrolysis to precipitate out the pure copper. The ores are very inexpensive to mine because they are at the surface; the only processing needed is some crushing to permit easy access by the acid. Consequently, many of the major mines are now extracting large amounts of copper from ores that were considered waste only a few years ago.

Primary copper sulfide deposits can be conveniently separated into three classes—magmatic segregation deposits, hydrothermal deposits, and sediment-hosted stratiform deposits. Each of the deposit classes deserves a separate discussion.

MAGMATIC SEGREGATION DEPOSITS. The least important class of copper deposits so far as current production is concerned is the *magmatic segregation* class. Such deposits have the same origin as the ores of nickel formed by magmatic segregation of immiscible sulfide liquids. The ore bodies are therefore associated with large bodies of mafic or ultramafic igneous rocks. Sudbury, Ontario, described in the discussion of nickel, is an example. Indeed, all of the magmatic segregation ores that are mined for nickel also produce significant amounts of copper.

HYDROTHERMAL DEPOSITS. Copper deposits (and deposits of many other metals) that are formed through the actions of hydrothermal solutions—hot aqueous solutions circulating through the crust (see Chapter 2)—are numerous. The most common hydrothermal deposits are small veins, usually found close to bodies of intrusive igneous rock but also found in metamorphic and sedimentary rocks far removed from known igneous intrusions. Such veins are usually quartz-rich and are almost always secondary features formed long after the host rocks became lithified. They tend to be confined in what were once open fractures through which the hydrothermal fluids flowed. Most vein deposits are small, containing only a few tons of ore and hence not of much economic interest. When they are large and rich, as were the great veins that surrounded the granitic intrusions in Cornwall, England (Figure 8.17), and at Butte, Montana, they constitute some of the richest and most profitable deposits ever found. Such rich bonanza veins were very important historically; today, many of the richest veins have been worked out. Nevertheless, investigations of vein deposits have been particularly informative in unraveling the complex questions that surround hydrothermal solutions.

Hydrothermal solutions are, as their name indicates, hot-water-based solutions. The water can come from two quite separate sources. First, the water can start as rainwater or seawater at Earth's surface. Such water trickles down the innumerable openings and fractures in surface rocks; below the water table, every opening is filled. At sufficient depth—a few thousand meters (several thousand feet)—the surrounding rocks are hot enough so that the buried waters become effective solvents. Small amounts of material are dissolved from the enclosing rocks; the hot water becomes a solution containing such constituents as NaCl, $MgCl_2$, $CaSO_4$, SiO_2, and, sometimes, small amounts of one or more of the geochemically scarce metals (see Table 2.1).

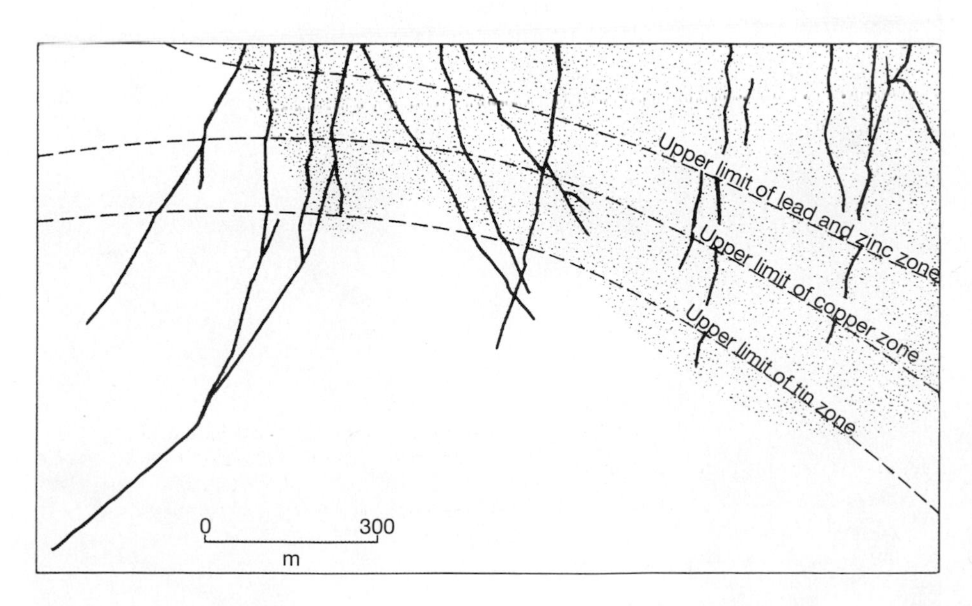

FIGURE 8.17 Mineral zoning developed in a system of veins formed around a granite stock in Cornwall, England. Each of the veins (as shown in Figure 2.3 and Plate 39) is mineralized and contains the tin mineral cassiterite (SnO_2), copper minerals of which chalcopyrite ($CuFeS_2$) is the most important, and zinc and lead minerals sphalerite (ZnS) and galena (PbS). The zonation is believed to arise from a drop in temperature of the hydrothermal solutions as they flowed through the fractures in which the veins now occur.

The second way a hydrothermal solution can arise is from a cooling magma. When a rock is melted and magma is formed deep in the crust or upper mantle, the magma assimilates any water that is present. Magmas formed in the crust, such as granites and diorites, generally have several weight percent water in them. When such magmas cool and crystallize, much of the dissolved water is released as a hot, aqueous solution that carries with it the same soluble constituents that deeply buried rainwater and seawater pick-up.

Because hydrothermal solutions from the two different sources are so similar in chemical composition, it is exceedingly difficult to determine the origin of a given fluid. The problem is complicated still further because the major driving force that causes hydrothermal solutions to flow is heat. When heated, a solution expands and rises convectively, thereby creating fluid flow. The principal heat sources in the crust that cause convective flow are shallow bodies of cooling magma and piles of hot lava, the very bodies that release magmatic hydrothermal solutions. But the rocks into which magmas are intruded also contain rainwater- and seawater-generated hydrothermal solutions in the pore spaces. A magmatic heat source will start these solutions flowing too. Once flow starts, both kinds of hydrothermal fluids became inextricably mixed. Hot springs in areas of active magmatic activity, such as those in Yellowstone National Park (Figure 8.18), are the tops of convectively driven hydrothermal systems. At Yellowstone, scientists have been able to show that more than 90 percent of the hot water started as rainfall.

When a hydrothermal solution starts to rise upward, a number of things can happen: the solution may start to cool; the solution may react with rocks such as limestone because such a solution tends to be slightly acidic; the pressure may drop; or boiling may occur. Each of these changes, or more likely a combination of them, can cause the solution to reach saturation and start precipitating the dissolved constituents. If it is the sulfide minerals of the geochemically scarce metals that precipitate, usually with associated quartz, an ore deposit may result (Figure 8.19). A rising hydrothermal solution tends to change progressively as the dissolved ore and waste minerals precipitate, thus developing a sequential or zonal pattern of metals.

The major kind of hydrothermal copper deposit being worked today is closely related to vein deposits. Known as **porphyry copper deposits** because the intrusive igneous rocks with which they are always associated have porphyritic textures, the deposits consist of innumerable tiny fractures, usually no more than a millimeter or so thick, spaced every few centimeters through a body of intensely fractured rock (see Figure 8.11). The geological setting of porphyry copper deposits, and the rocks with which they are associated, indicate that they represent subsurface conduits and chambers that once lay beneath volcanos. The formation process began with the intrusion of magma. As nonhydrous minerals such as feldspars began to crystallize around the outer edges of the intrusion, the water content of the remaining magma increased until the pressure was so great that

FIGURE 8.18 A hot spring and geyser, Yellowstone National Park, Wyoming. Such hot springs are believed to be the surface expressions of large, deeply circulating hydrothermal systems of the kind that formed the veins in Cornwall (Figures 2.3, 8.17 and Plates 38, 39). (Photograph by J. D. Rimstidt.)

FIGURE 8.19 Photograph of a tungsten-bearing vein at Panasqueira, Portugal, showing large white quartz crystals (on the lower side), dark wolframite crystals, and arsenopyrite (upper half of the vein). (Photograph by A. Arribas.)

steam explosions shattered the crystallized rock. The escaping hydrothermal solutions carrying silica, potassium, sodium salts, and also dissolved metal sulfides moved outward, precipitating minerals and sealing the fractures. As cooling continued, this process was repeated several times, leaving many generations of small fractures, several of which contain small but significant amounts of copper ore minerals. When the volume of shattered rock is large, a porphyry copper deposit can also be very large—over 1 billion (10^9) tons of ore in some of the largest. The deposits have distinctive zonal patterns that can be related to the fluid circulation system (Figure 8.20).

When porphyry copper deposits are mined, it is impractical to dig out the individual tiny veinlets. Instead, the entire body of shattered rock is mined. As a result, porphyry copper deposits tend to have lower grades than the richer bonanza vein deposits; grades range from 0.25 percent to 2 percent copper, plus small amounts of molybdenum and gold. What makes these porphyry-type deposits profitable at such low grades is their very large size and shape. Most are relatively near-surface, approximately cylindrical bodies of mineralized rock that lend themselves to very large mining schemes. Indeed, it was at the porphyry copper deposit at Bingham Canyon, Utah (see Figure 4.3), where D. C. Jackling and R. C. Gemmell first demonstrated that large-scale bulk mining is more profitable than small-scale, labor-intensive, selective mining. They made their proposal in 1899. By 1907, they had installed an open-pit mining scheme capable of producing 6000 metric tons of low-grade copper ore per day. Several modern mines are extracting more than 200,000 metric tons of ore per day; the maximum production in a 24-hour period was at the Morenci deposit in Arizona when 1.3 million tons of ore were removed.

Literally hundreds of porphyry copper deposits have been found; today, they account for more than half of the world's copper production. The deposits are concentrated around the rim of the Pacific Ocean Basin, where they were formed as a result of volcanism associated with the subduction of oceanic crust (Figures 8.20, and 8.21; see also Figure 2.4). The porphyry copper deposits around the Pacific region are geologically young and related to the present phase of plate tectonics. Older deposits are known and, in all cases, they seem to be related to ancient plate edges where subduction once occurred.

Two other kinds of hydrothermal copper deposits also deserve discussion because each is a significant producer of copper. The first is a *contact metamorphic* or *skarn* class of deposit. When granitic magma intrudes a pile of rock that contains limestone or marble, the acid

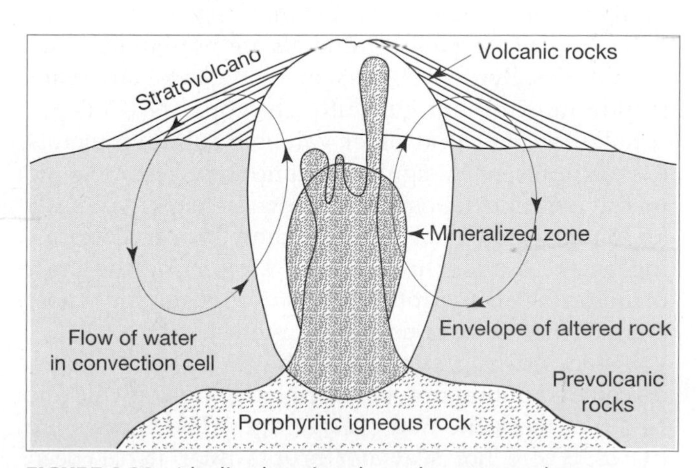

FIGURE 8.20 Idealized section through a strato-volcano showing the location of the mineralized zones and the nature of the groundwater convection established in the surrounding rocks.

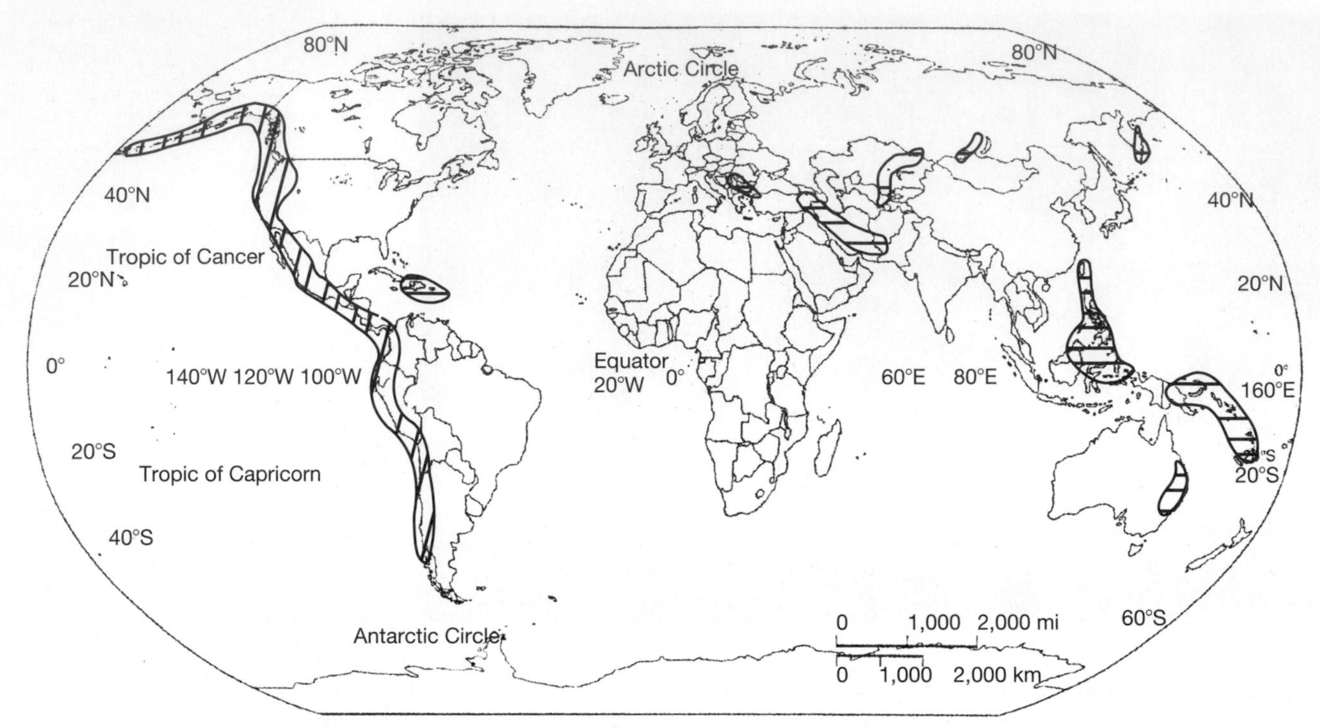

FIGURE 8.21 The world's greatest concentration of porphyry copper deposits exists along the western edge of the Americas where the oceanic tectonic plates of the Pacific are subducted beneath the continental plates of the American continents. The belt extends around the entire Pacific Basin and includes the Grasberg Deposit shown in Figure 13.5 and Plate 37.

solutions released by the cooling magma react with the $CaCO_3$ in the limestone. The reactions cause rapid precipitation of copper minerals and lead to very rich ores. Commonly, skarn ores form rich pockets associated with larger masses of porphyry-type ore. Examples are the Gaspé Copper Deposit in Quebec and the Yerington Deposit in Nevada.

The final class of hydrothermal copper deposit is known as a **volcanogenic** massive sulfide deposit. The enclosing rocks are invariably volcanic in origin and have features indicating that they were erupted on the seafloor. The ore minerals are always sulfides, and the term "massive" reflects the fact that very little volcanic debris or silicate gangue minerals are present to dilute the sulfides. Pyrite (FeS_2) is usually the most abundant sulfide mineral, but generally chalcopyrite ($CuFeS_2$), sphalerite (ZnS), and other valuable sulfide minerals are also present in significant amounts. Volcanogenic massive sulfide deposits are known in rocks as old as 3 billion years, but such deposits can be observed forming on the seafloor today. They are among the most common of all hydrothermal ore deposits, but most are too small to be of economic interest.

Deep-diving submarines have, in recent years, discovered submarine hot springs along the volcanic rifts that mark the mid-ocean ridges. The springs are places where hot seawater erupts after it has been heated at depth in the piles of volcanic rock that make up the upper layers of the oceanic crust. The hot circulating seawater is apparently heated by the magma chambers that underlie the mid-ocean ridge. It reacts with, and alters, the basaltic lavas it passes through; by the time it rises again to the ocean floor, it is a rapidly flowing jet of brine as hot as 350°C. When such a jet erupts into cold ocean water, it is quickly cooled; as a result, its dissolved load precipitates, sometimes as very fine soot-like sulfide minerals (thus, giving the name *black smokers* to some vents, Figure 8.22) and sometimes forms a blanket of massive sulfide ore around the vent. Only two of the modern mid-ocean ridge deposits so far discovered, in the Red Sea, and off the west coast of Canada, are large enough to be of potential interest for mining. More than a hundred others have been found in the eastern Pacific and mid-Atlantic, but all are small. Furthermore, the known deposits are all at water depths of 1500 meters (4500 feet) or more.

It may well be that other large, rich, modern, mid-ocean ridge deposits may someday be found on the ocean floor. For the present, the only massive sulfide deposits of this type that are being mined are in ancient fragments of oceanic crust found on the continents. Such massive sulfide deposits have been mined since ancient times in Cyprus; indeed, the word *copper* comes from the old Greek word *cyprus*.

Most recently, exploration of the sea floor has revealed that another class of modern massive sulfide deposit is being formed. These are deposits associated

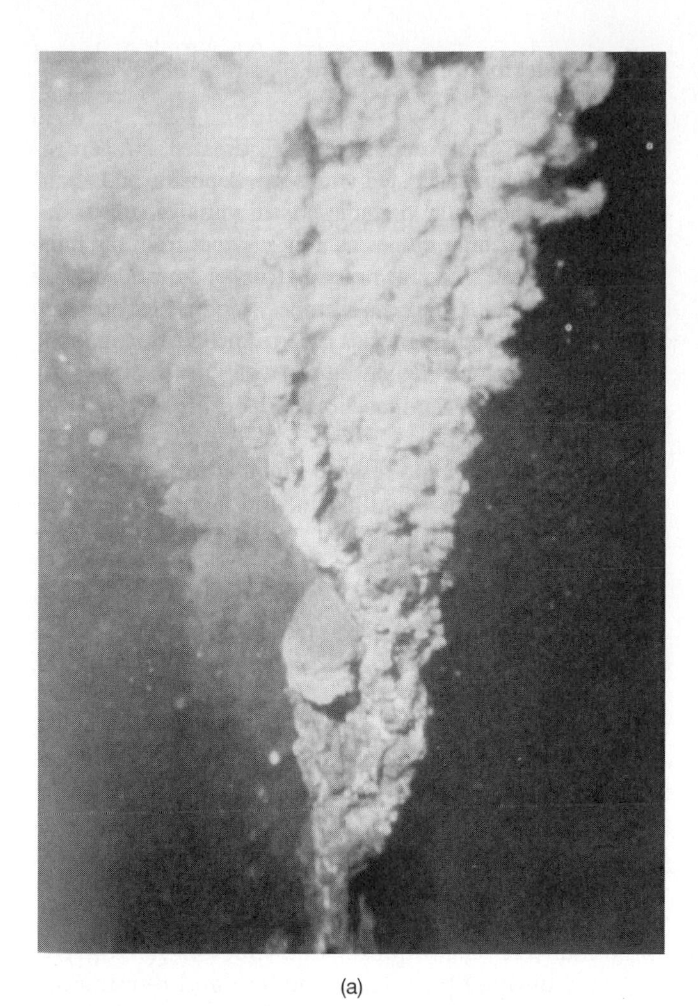

(a)

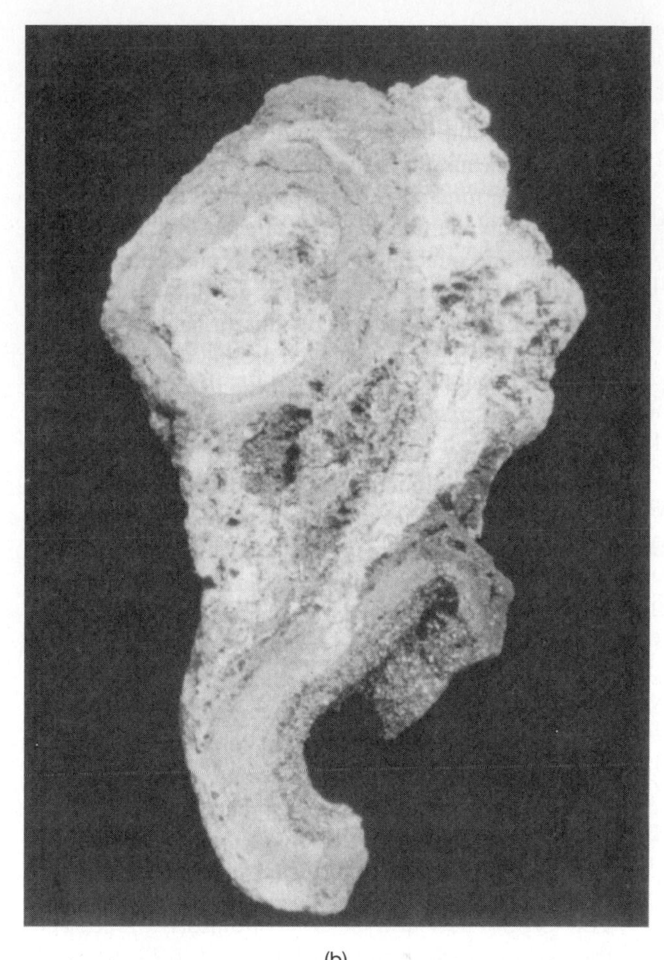

(b)

FIGURE 8.22 (a) A chimney-like structure of sulfide minerals built up around a vent from which 350°C hydrothermal solutions are being emitted into the cold waters above the East Pacific Rise. The solution is colorless, but when it mixes with the cold seawater, a dense cloud of exceedingly fine-grained sulfide minerals is precipitated. Such structures and precipitates are colloquially referred to as "chimneys" and "black smokers." (Photograph by R. Ballard.) (b) Section of a chimney recovered by a deep-diving submarine from a depth of 2500 meters (7700 feet) at 20°N latitude on the East Pacific Rise. The principal sulfide mineral present is pyrite, but small amounts of chalcopyrite and sphalerite are also intermixed. (The specimen is about 22 centimeters (9 inches) across.) (Photograph by B. J. Skinner.)

with the subduction edges of tectonic plates. Such deposits tend to be at water depths of 4000 meters (12,000 feet) or more, but in the ocean just north of New Guinea, the deposits are in shallower water; it is likely that these will be the first to be mined. On land, however, literally hundreds of old subduction related massive sulfide deposits have been exploited.

SEDIMENT-HOSTED STRATIFORM DEPOSITS. The final group of copper deposits all share distinctive and puzzling features. They are always in clastic, marine sedimentary rocks, usually shales, that contain a certain amount of organic matter and calcium carbonate, and they are usually stratiform, which means the mineralization occurs in strata-like layers. The ore minerals are usually sulfides, but native copper may also be present. Because neither copper metal nor sulfide minerals are known to precipitate directly from modern seawater, nor is it likely that ancient seawater had the composition for this to happen. Hence, the origin of such ores remains a mystery.

One group of theories concerning the origin of stratiform deposits suggests that the deposits are related to volcanogenic massive sulfide deposits and that the sulfide minerals were precipitated from submarine hot springs. The trouble with such theories is that hot springs usually produce massive ores, but the sediment-hosted stratiform ores always contain very large proportions of clastic silicate mineral grains. A second group of theories, which the authors of this volume prefer, ascribes the origin to hydrothermal solutions that circulated in coarse, clastic sedimentary layers beneath the now mineralized shales. Such theories ascribe the origin of the ores to reactions between the solutions and mineral constituents in the shales. The reactions seem to have occurred soon after the sediments were deposited and, in certain cases, before they were consolidated to solid rock. The most famous of the sediment-hosted stratiform copper deposits are enclosed in the **Kupferschiefer**, a Permian-aged shale found through much of northern Europe. The Kupferschiefer deposits, which average only about 20 centimeters (8 inches) in thickness but extend laterally over more than 6000 square kilometers (2500 square miles), have been mined continually since the fourteenth century. Larger and richer stratiform deposits, discovered in the twentieth century, are now worked in Zambia and the Congo, in central Africa. In the United States, there are deposits at White Pine, Michigan, and at Creta, Oklahoma, that are of this class.

PRODUCTION AND RESERVES. About 70 percent of the world's copper production comes from porphyry copper and associated skarn deposits. An estimated 15 percent comes from sediment-hosted stratiform deposits and about 10 percent derives from volcanic-hosted massive sulfide deposits. The remainder is a by-product from the mining of nickel, lead, and zinc ores, and from the chemical leaching of old mine dumps. Copper is so widely produced around the world that, in 2008, some sixty countries reported production. Not surprisingly for such a widely used and widely exploited metal, both the reserves and resources of copper are very large.

The production and reserve situation for copper since the end of World War II is illustrated in Figure 8.23 and is representative of most of the base metals. This diagram demonstrates that cumulative production of copper since 1946 would long ago have exhausted the copper reserves known in the world at that time. However, the figure shows that copper ores have been discovered more rapidly than they have been mined, even though mining rates have markedly increased and are now at their highest levels ever. This situation has developed because an increased understanding of the geology of copper deposits has resulted in successful exploration strategies and because new technologies have permitted the mining and processing of lower-grade ores. Similar situations exist for lead, zinc, and several other base metals.

Resources of copper in large deposits that are below today's minimum mining grades are quite large—several times larger than the reserves. In the United States, vast quantities of copper exist in magmatic segregation deposits in the Duluth Gabbro, Minnesota. Very large stratiform resources also exist in the late Proterozoic-aged Belt Series rocks of Montana. But the largest copper resources of all do not belong to any country. They are in the ferromanganese nodules that lie on the deep seafloor. In areas of slow sedimentation, such as the central Pacific Ocean, ferromanganese nodules average more than 1 percent copper (Table 7.5), and in some places the nodules are known to contain as much as 2 percent copper. The same nodules also carry important amounts of nickel and cobalt. Some estimates put the total copper resource in recoverable manganese nodules in excess of 1 billion (10^9) tons of copper. This suggests that the seafloor resources of copper are about the same size as the land resources. Despite the fact that copper is mined at such a high annual rate, the world's reserves and resources appear adequate for a century or more. The big question to be answered in the future concerns the cost of working present-day uneconomic resources.

Lead and Zinc

Lead and zinc are discussed together because their ore minerals so commonly occur together. Their uses as metals are rather different, however. Lead is another metal with a history that stretches back to antiquity. It rarely occurs in the native state, so the use of lead metal only commenced when our ancestors learned how to smelt lead ores. Lead was used in weights, sheet metal, solders, ceramic glazes, and glassware by the ancient Egyptians, the Phoenicians, the Greeks, and the Romans. The Phoenicians worked lead mines in Cyprus, Sardinia, and Spain, and they traded the metal around the Mediterranean. The famed mines at Laurium in Greece produced both silver and lead and supplied the ancient Greeks with much revenue. The great quantities of silver that enriched Rome and paid for so many of its conquests, and much of its high living, were derived from lead smelting and refining operations carried out in Spain, Sardinia, and Britain. In recent years, we have come to realize that the Roman smelting of lead and silver ores resulted in widespread lead pollution in several parts of Europe. See Box 8.2 for a look at the problems associated with lead use.

An unusual combination of properties has given lead a wide range of industrial uses. The metal is soft and easily worked, it is very dense, it has a low melting temperature, and it possesses very desirable alloying properties; it also resists corrosion, and it is an excellent shield against harmful radiation. The principal use of lead today is for automobile storage batteries; it is also used in crystal glass, for flashings in building construction, in ammunition, in various alloys (especially solder), in bearings, and in printer's type. But lead also has unpleasant properties. Many lead compounds are toxic; this has resulted in a reduction in the use of lead in situations where humans might ingest lead or lead compounds. Thus, the use of lead in paints and gasoline has been banned. For many years now, all new automobiles sold in North America, Europe, and most other countries have been designed to use unleaded gasoline; thus, the use of leaded gasoline has nearly ceased and it is used today only in antique cars.

Zinc is a relatively soft, bluish-white metal. It is more difficult to produce zinc metal by standard

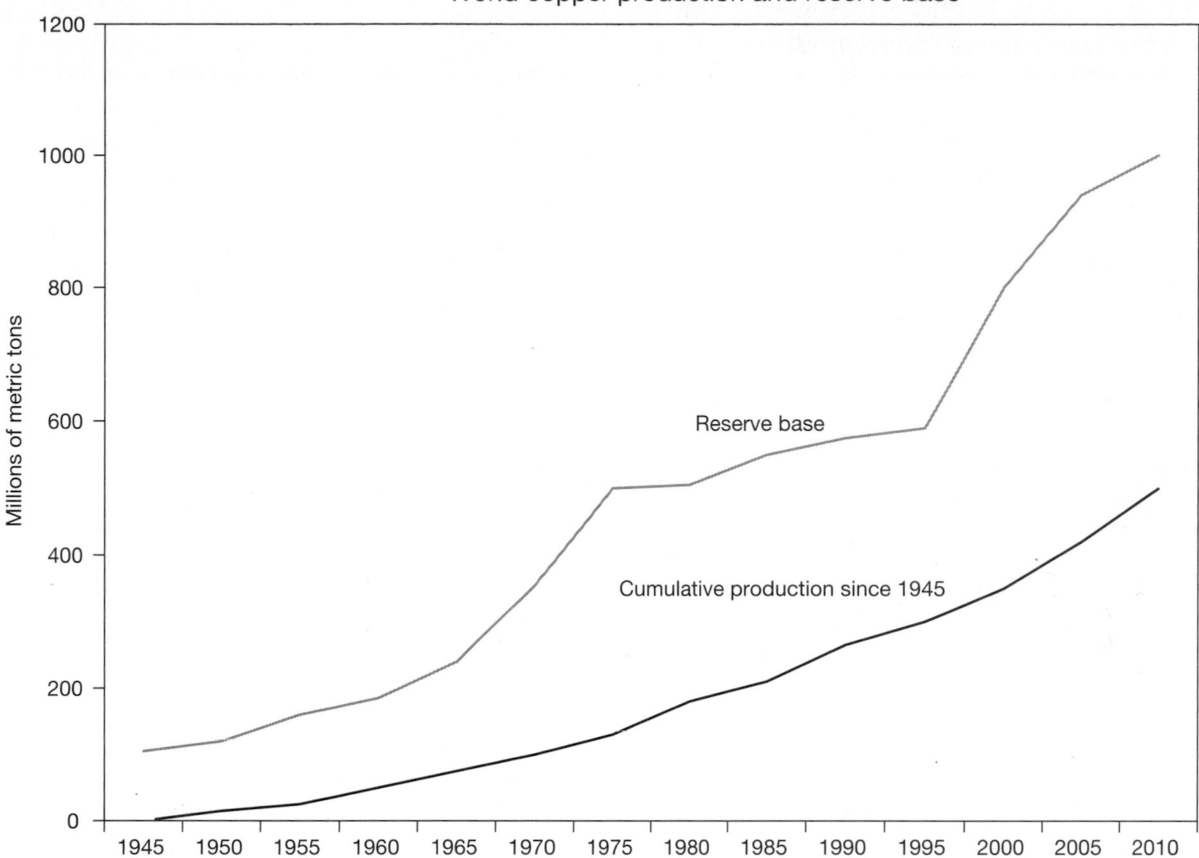

FIGURE 8.23 The cumulative world production of copper metal from 1946 through 2000 was about 450 million metric tons, but the reserves have increased dramatically since the 1960s when the plate tectonic theory established the relationship of porphyry deposits to subducting plate boundaries. (Data from the U.S. Geological Survey.)

BOX 8.2
Lead in the Environment

Lead is an ancient metal that is creating modern problems. Its usage dates back to antiquity when early metallurgists, or possibly potters, learned to refine the metal from galena (PbS); they soon found that the lead could be easily cast, molded, and otherwise shaped into pipes and storage vessels. Lead also worked well in ceramic glazes, solders, and even glassware. It was heavy but resisted corrosion reasonably well, so it was widely traded and extensively used. Lead became the common metal of water pipes, and lead alloys were used for printer's type, for munitions (especially small bullets and shot), and paints. With the advent of the automobile, lead storage batteries became the primary means of providing power to start engines. In addition, the additive *tetraethyl lead* became the primary means of increasing octane and preventing engine preignition, or *knocking*.

Until the second half of the twentieth century, lead seemed to be a useful and benign metal; only a few precautions were taken when it was used. Gradually, however, evidence began to accumulate that although lead metal is not poisonous (many people have lived for long periods with lead pellets or bullets in their bodies), many lead compounds are definitely poisonous. Lead, like many metals, does not necessarily remain in the form in which it was used; it is readily dissolved in acidic solutions and it slowly oxidizes to form lead oxides that have properties and solubilities very different from the properties of lead metal. Furthermore, the body eliminates lead only very slowly; hence, prolonged exposure can result in a gradual buildup. This lead buildup can be especially harmful in children because lead inhibits production of hemoglobin, interferes with normal growth and with the development of the central nervous system, and it locks on to essential enzymes in the brain. The effects of lead poisoning, or *plumbism*, include anemia, reduced hearing abilities, lack of coordination, drowsiness, cramps, and paralysis.

Mounting evidence of lead poisoning and the recognition that lead was widespread in our environment in fumes from leaded gasoline, in paints used in tens of millions of homes, and in lead shot used for hunting, resulted in a series of actions designed to limit lead exposure. In the 1970s, the U.S. government began to require that automobiles operate on

(continued)

BOX 8.2

Lead in the Environment (*Continued*)

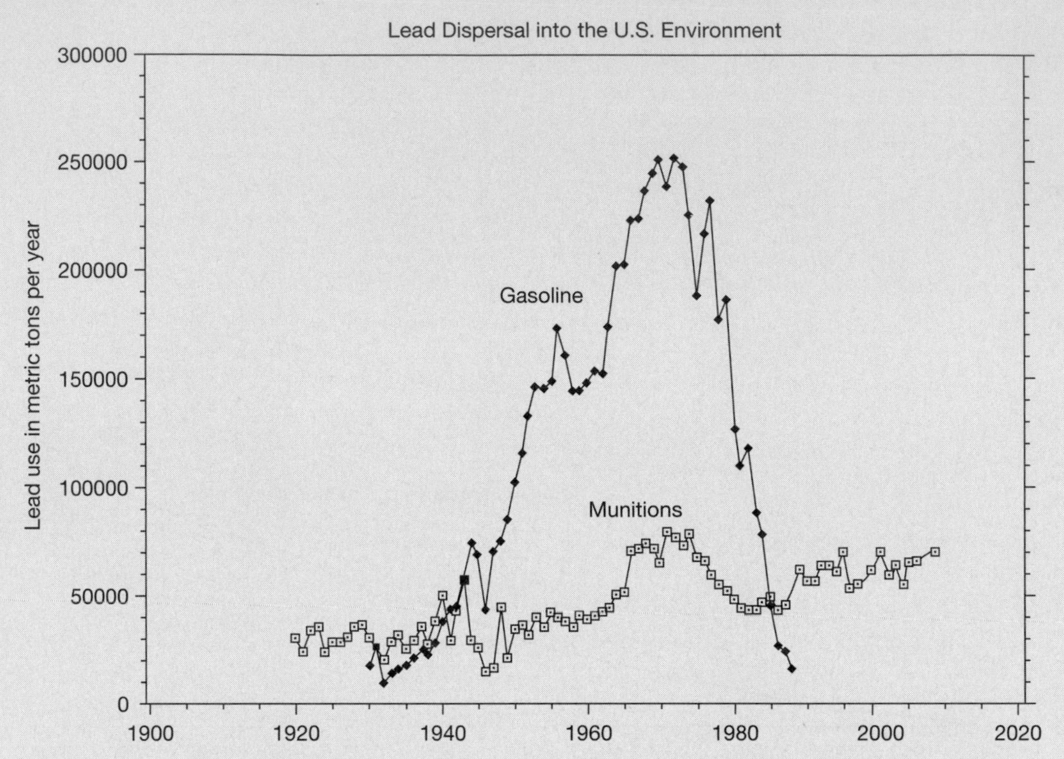

FIGURE 8.C Lead was added to nearly all gasoline in the United States from about 1930 until the early 1980s. Munitions used in recreational shooting now constitute the primary source of lead introduced into the American environment.

unleaded gasoline (Figure 8.C). By the end of the 1980s, the United States had virtually eliminated leaded gasoline; most European countries did likewise by the 1990s. These actions have sharply reduced exposure to lead, but the government estimates that three quarters of all homes built before 1978 (tens of millions) still contain old lead-based paints, and that more than 3 million children have been affected. The principal problem is the ingestion of dust and fragments of flaking lead-containing paints. A second source is drinking water, which may contain low levels of lead gradually dissolved (especially by acid waters) from lead pipes or solders used to join copper pipes. In 1992, the U.S. Environmental Protection Agency initiated a long-term program to remove lead from all public drinking water supplies.

Lead compounds have long been used in ceramic glazes on tableware and in lead glass sometimes used as decanters for wine. Only recently has it been learned that the contact with food and wine (which is acidic) can actually extract enough lead from the dishes or glassware to create a potential health hazard. Consequently, some types of tableware have been banned from sale and warnings have been issued not to store wines for long periods in lead glass containers.

During the 1970s, it also became apparent that there was widespread poisoning of water fowl which had ingested some of the hundreds of thousands of tons of lead shot that lay on the bottoms of lakes and wetlands. This led to the banning of lead shot in favor of steel shot for hunting. Unfortunately, there is no means of recovering the massive amounts of lead that remain in the wetlands after 200 years of hunting (Figure 8.D). Accordingly, lead exposure has been much reduced but not eliminated.

FIGURE 8.D Spent bullets from a typical shooting range can represent high local concentrations of lead. However, the development of corrosion layers on the lead greatly reduces its solubility in surface and groundwaters.

smelting practices than it is to produce copper or lead. Zinc has appeared in bronze and brass artifacts from Europe and Asia for several thousand years. The first commercial production of the metal is claimed to have been in China some 600 years ago, but details are poorly recorded. The first reliable record of zinc smelting was from Bristol, England, in 1740. As demand grew, other smelters were opened in Belgium, Germany, Russia, and in 1860, in the United States.

The largest part of the world's zinc production is used for *galvanizing*, a process by which protective coatings are put on steel and iron, mainly by a hot-dipping process in which the object to be coated is dipped into a bath of molten zinc. Because zinc resists corrosion, even a thin coating prevents rust from forming on iron and steel. Zinc also has desirable alloying properties; many die-cast objects that are not subject to abrasive wear are made from zinc-based alloys. Zinc is a constituent of brass, and in its oxide form (ZnO), it is used in paint pigments, ointments, lotions, and creams to prevent sunburn. Zinc became more prominent, if not more conspicuous, to Americans when the copper penny was replaced in 1982 by a zinc penny that is copper coated.

GEOLOGICAL OCCURRENCE. Lead and zinc minerals are formed in deposits that resemble copper deposits in many ways. There are four important kinds of deposits, and in all of them, the same two minerals occur—galena (PbS) and sphalerite (ZnS).

The first kind of deposit is a **hydrothermal vein type.** As with copper, lead and zinc minerals are very common in veins, but also as with copper, most of the veins are too small to be of much interest. Nevertheless, in many countries, including Peru and the United States, vein deposits still contribute to the production total.

Volcanogenic massive sulfide deposits associated with ancient subduction plate edges, the second kind of deposit, are often rich in zinc or in zinc plus lead, as

BOX 8.3

More Than Zinc from a Zinc Mine

What do you get from a zinc mine? The question seems straightforward enough, and most respondents would say "zinc." That answer is certainly correct, but it is far from complete. In fact, zinc mines, like most metal mines today, yield a variety of resources, many or most of which are by-products of the primary resource for which the mine is operated. By-products are defined as *secondary* or *additional* products, but they often play a critical role in the viability of a mine. Furthermore, several important resources are derived almost entirely as by-products; that is, we would lose access to some metals if we were to stop mining for others.

Zinc mines exemplify this relationship between primary products and by-products very well. The world's principal source of zinc is the mineral *sphalerite,* which has the idealized composition of ZnS. Sphalerite can form in a variety of types of ore deposits, but it is especially well-known in *carbonate hosted zinc* or *Mississippi Valley-type* deposits. The deposits, named for their occurrence in limestone or dolomite beds in the Mississippi River drainage basin, contain crusts of sphalerite that have precipitated in fractures and around the edges of fragments of carbonate (see Figure 8.25).

Natural sphalerite is never pure and always contains one or more of several other metals that can substitute for zinc atoms in the crystal structure. When the zinc, which was originally distributed in sediments and rocks in just a few parts per million, was dissolved in slowly moving, heated groundwaters, trace quantities of other metals were also dissolved. Iron is nearly always present in sphalerites and often contributes to the darkening of its color but is of no economic value in these ores. In contrast, some sphalerites, especially those from the Mississippi Valley-type deposits, contain a few tenths of one percent of gallium, germanium, cadmium, and sometimes indium. Although the concentrations are small relative to the zinc (which is over 60 percent in pure ZnS), the selling price of each of these metals is greater than zinc. The price of cadmium is about twice as much as zinc, indium about 250 times as much, gallium about 500 times as much, and germanium about 1000 times as much. Hence, once the ore has been mined and the sphalerite concentrated, most of the effort and the cost have been expended; relatively little added cost is required to extract the other metals and they may add significantly to the profitability. Gallium and germanium are used in components in light-emitting diodes and lasers; indium is important in solders, solar cells, and electronic coatings; and cadmium is widely used in anticorrosion coatings and in nickel-cadmium batteries. There are no mines in the world where cadmium, indium, and the other trace metals are the principal metals mined; accordingly, if the zinc mines closed, we would lose not only a supply of zinc, but also supplies of gallium, germanium, cadmium, and indium.

One other important by-product not to be overlooked is limestone or dolomite. Usually the sphalerite, with its by-product metals, constitutes less than 5 percent of what is actually mined. As a result, once the mill has processed the ore, 95 percent or more of the rock is left over as *waste.* Nearly all mines will sell this limestone or dolomite for building or agricultural purposes. It is low in value, but it is a "free leftover" and selling it brings in some revenue and means that it does not have to be stored. In fact, there have been times when the price of zinc has been depressed, and it has been the extra revenue generated by selling the limestone or dolomite that has kept the mines profitable. Indeed, there is often much more than zinc extracted from a zinc mine.

well as copper. The famous Kuroko deposits of northern Honshu, Japan, are of this type, and they have a long production history of both lead and zinc. In New Brunswick, Canada, large massive sulfide ores have been known and mined for many years. Many of the New Brunswick ores are so fine-grained that it is not possible to separate the galena and sphalerite in order to prepare concentrates suitable for smelting, so even though the deposits are large and easily mined, they cannot all be exploited profitably.

The third important class of hydrothermal lead-zinc deposits is known as the **Mississippi Valley type** (MVT) deposits, after the remarkable metallogenic province stretching from Oklahoma and Missouri to Kentucky and Wisconsin, a region that includes most of the drainage basin of the Mississippi River (Figure 8.24). Deposits with similar affinities have been discovered in Canada, northern Africa, Australia, Russia, and Europe. The MVT deposits always occur in limestones where the ore minerals have either replaced the limestone beds or have been precipitated in between rock fragments in a limestone **breccia.** The solutions that form the MVT deposits develop in sedimentary basins and flow laterally outward until they come in contact with limestones around the margins of the basins. There, the solutions may dissolve the limestone and precipitate the galena and/or sphalerite. The rich zinc deposits in Tennessee have formed where groundwaters had first dissolved much limestone, leaving a well-developed cave system. The ore-bearing solutions then deposited the ore minerals, sometimes as large spectacular crystals, in-between *breccia* blocks that developed either from cave collapse or tectonic fracturing (Figure 8.25). Other important MVT deposits are the lead deposits in southeast Missouri and the great zinc deposits of Pine Point in Canada.

An increasingly large production of lead and zinc now comes from deposits of the fourth class, namely sediment-hosted stratiform deposits. As with stratiform copper deposits, the origin of these deposits is enigmatic. The clearest example is the Kupferschiefer, where a zonal pattern of lead and zinc mineralization occurs around the copper deposits. The origin of the

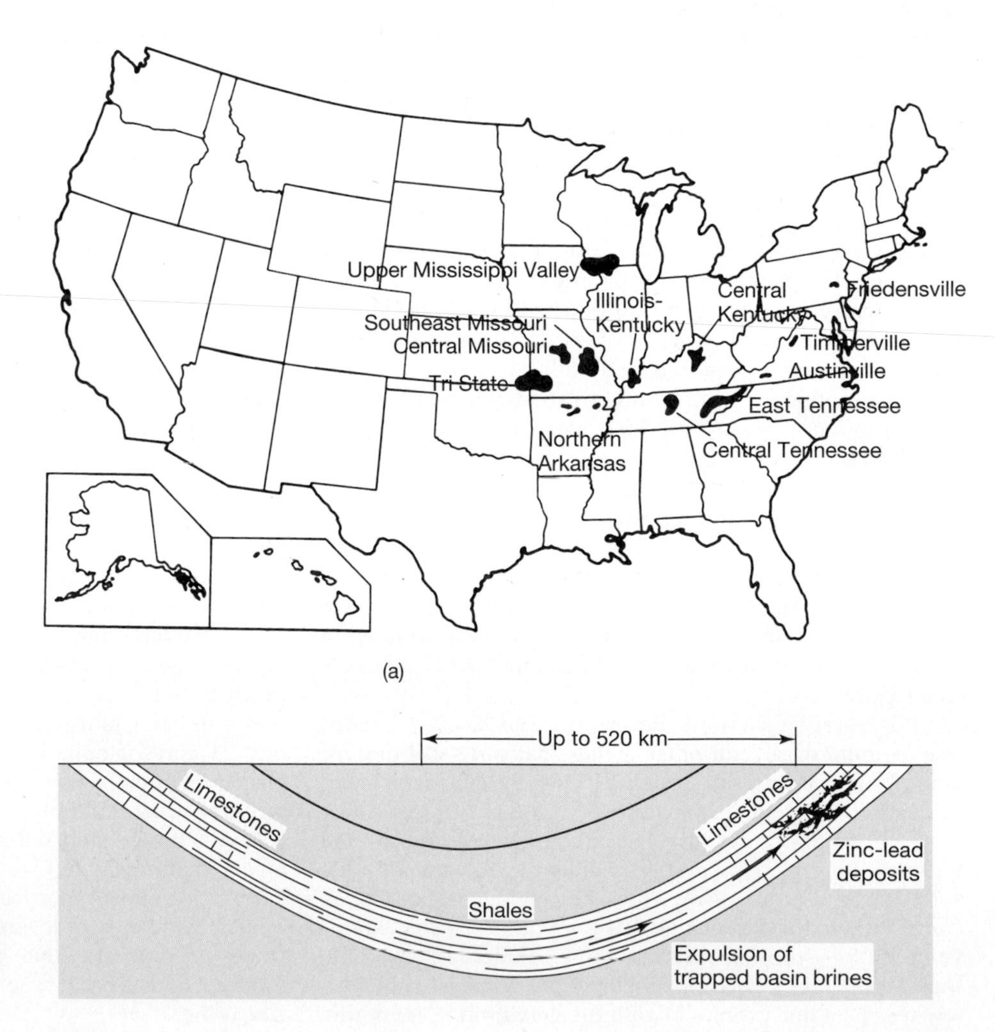

FIGURE 8.24 (a) Map showing the principal Mississippi Valley-type (carbonate-hosted) lead-zinc deposits in the United States. (b) Cross section of a large sedimentary basin showing expulsion of metal-bearing fluids from shales with subsequent deposition in the limestones on the flanks of the basin.

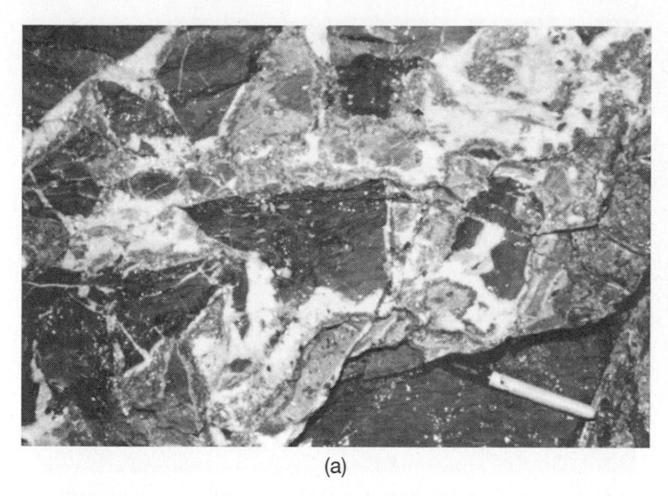

(a)

(b)

FIGURE 8.25 (a) Typical example of Mississippi Valley-type breccia ore from east Tennessee. The sphalerite was precipitated around the edges of the irregular and disoriented limestone and dolomite blocks; the white is late-stage white dolomite that filled the remaining voids after the deposition of the sphalerite. (Photograph by J. R. Craig.) (b) Example of well-developed crystals of sphalerite, fluorite, and barite from a cavity in the central Tennessee zinc deposits. (Photograph by S. Llyn Sharp.)

ores is, most likely, deposition in the Kupferschiefer by hydrothermal solutions that circulated in coarse, clastic sediments below the shales.

Most of the large stratiform lead-zinc deposits are Precambrian in age. One very large and very rich deposit, the Sullivan body, occurs at Kimberley in British Columbia. But the largest bodies of this kind have been found in Australia. One, at Broken Hill in New South Wales, is one of the richest ore bodies ever found. A second, at Mount Isa in Queensland, is also very large, and besides lead and zinc, it also has produced large quantities of copper. In the early 1990s, the Red Dog Deposit in Alaska was brought into production and has become the world's largest zinc producer. The deposits at Broken Hill, Mount Isa, Sullivan, and Red Dog are so rich that the ores contain combined values of lead plus zinc in excess of 20 percent—in fact, the ores are so rich, they closely resemble massive sulfide deposits.

PRODUCTION AND RESERVES. The world's production of lead and zinc is dominated by China, Australia, and Peru, but significant production is reported from about fifty countries.

World reserves of lead and zinc present a situation similar to that of copper; that is, increased geological understanding has led to increased discoveries of reserves. Furthermore, world demand for zinc is expected to only increase slowly, and world demand for lead, because of environmental concerns, could stabilize or decrease. As long as lead storage batteries are needed for automobiles and trucks, the demand for lead will continue to grow. This dominant use of lead also contributes to the large rate of recycling compared with other metals (see Figure 7.14). The increased use of electric vehicles to reduce pollution in cities would also increase the need for lead. However, if the efforts to develop alternative, lower weight batteries are successful, the world's use of lead would be dramatically reduced.

Tin

Tin shares the distinction with copper and lead of having been used by humans for more than 5000 years. Where and when tin was first used is not known, but it was probably either in the Middle East or in Southeast Asia. Most likely the earliest use was as an alloying agent. Two tin-bearing alloys have been used since very ancient times. The first is bronze, a copper-based alloy in which tin serves as a hardening agent for the copper. Bronze is tougher and more durable than pure copper; thus, bronze tools last longer and are more efficient than are copper tools. The discovery of bronze was one of the great milestones in the technological history of the human race. The second alloy of antiquity is pewter. Pure tin is too soft to be worked into spoons, knives, plates, and similar utensils (Figure 8.26), but when copper (and sometimes lead) is added to tin, an inexpensive, harder, and very useful alloy—pewter—is the result. With the addition of small amounts of antimony or copper, pewter can be made even stronger and more durable than an alloy of straight tin plus lead.

Tin is a soft white metal with a low melting temperature. It does not corrode or rust, and so is an effective coating agent to cover metals that do corrode. The method for plating tin on copper was developed at least 2000 years ago, and tin-plated iron was first manufactured in the sixteenth century. Today, the use of steel, aluminum, and plastic cans has largely supplanted tin-plated cans, though for some specialty purposes, tin-plating is still an important process.

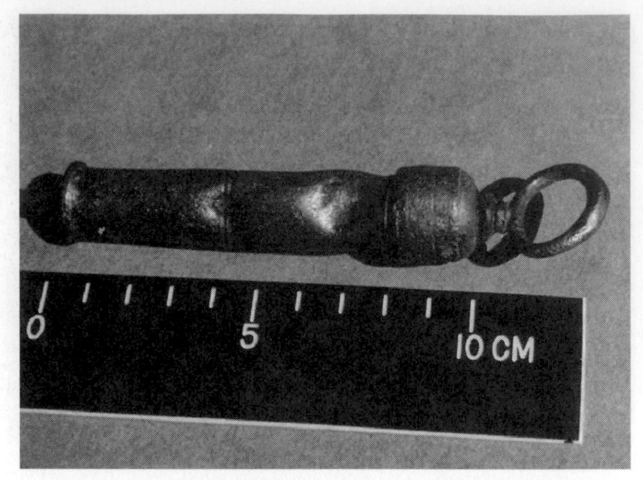

FIGURE 8.26 Pewter, made of tin or tin-rich alloys, has long been used to make a variety of decorative and utilitarian objects such as this pewter serving plate and the medical syringe. These objects were recovered from the wreck site of what is believed to be the *Queen Anne's Revenge*, the flagship of Blackbeard the pirate, which sank near Beaufort, North Carolina, in 1718. (Photographs by J. R. Craig.)

The low melting temperatures of tin, and of alloys of tin and lead, make tin the prime constituent of solders. Thus, the principal use of tin today is in solders, consuming approximately 35 percent of all tin mined. Bronze is used in certain soft-metal bearings, in ornamentation, in automobiles, and in aircraft.

GEOLOGICAL OCCURRENCE. There are many tin minerals, but almost all production comes from cassiterite (SnO_2). A small amount of tin is produced as a by-product from the mining of other base metal; in such cases, tin is usually present in the ore as the sulfide mineral stannite (Cu_2FeSnS_4). Cassiterite is sometimes found in pegmatites associated with granitic rocks, but more commonly it is found in hydrothermal deposits related to andesitic or rhyolitic volcanism. The hydrothermal deposits may be veins, disseminations in altered rocks, or even **skarns.**

Cassiterite is a dense, chemically stable mineral that does not alter or dissolve in stream waters. As a result, cassiterite is readily concentrated in **placers;** worldwide, almost as much tin is produced from placers as from hard-rock mines because placer mining is inexpensive compared to hard rock mining.

PRODUCTION AND RESERVES. Three countries dominate the tin market, China, Indonesia, and Peru. Tin reserves are reasonably large—more than ten times the annual production rate of 330,000 metric tons—and resources are even larger. The United States Geological Survey has estimated that tin resources might be as high as 30 million metric tons. Whether or not all the tin in the estimated resources can actually be found and mined is an open question. Present reserves plus resources will last for at least forty years and could last much longer.

Mercury

Mercury is the only metal that is a liquid at room temperature. It combines readily with other metals to form alloys—in the case of mercury, such alloys are called amalgams. One common use for a silver-mercury *amalgam* is filling cavities in teeth.

There is only one important ore mineral of mercury, cinnabar (HgS). It is a soft, blood-red-colored mineral found in hydrothermal vein deposits at a few places around the world. Small quantities of metallic mercury are often found with cinnabar. Mercury is produced both by primary mining and as a by-product from zinc and copper mining. The principal producing countries, in order of importance, are China, Kyrgyzstan, Algeria, and Spain (Figure 8.27). Among them, they account for about 75 percent of the world's production. Italy, although not presently a producer, has large reserves of mercury. The best estimates of mercury resources suggest that the world's needs will be met for at least another century. The principal uses of mercury are in batteries (30 percent), for chemical production (30 percent), and in scientific measuring devices (40 percent).

Cadmium

Cadmium is a soft, malleable, silver-white metal that was discovered and first separated as a pure metal in 1817 by a German scientist who was investigating impurities in zinc minerals. Cadmium minerals are rare, but cadmium itself is widespread in trace amounts in the principal zinc ore mineral sphalerite (ZnS). Cadmium replaces zinc by atomic substitution,

FIGURE 8.27 The mercury collection system used at Almaden in Spain for more than 200 years, from 1720 to 1928. The cinnabar-bearing ore (see Plate 36) was roasted in the chambers at the left and the mercury vapor condensed as it passed through the terra cotta pipes. A hole in the pipe at the low point allowed the liquid mercury to drip into a channel and flow to a collection sited at the far end. Modern systems operate in the same manner, but are enclosed to prevent exposure to the mercury vapor. (Photograph by J. R. Craig.)

and in the smelting of zinc, the cadmium can be recovered. All cadmium production now comes as a by-product of zinc mining (see discussion in Box 8.3).

For about sixty years after discovery of the chemical element cadmium, its only use was in chemical compounds as pigments for paints. Cadmium compounds have very bright and intense colors—yellows, reds, blues, and greens. By the 1830s, they were being used by French artists and, in the 1840s, by English watercolorists. Starting about 1890, cadmium found new uses as a metal in low-melting alloys and in chemical reagents. In 1919, a cadmium-electroplating process was developed; cadmium plating is similar to zinc plating, and this accounts for the bulk of the cadmium used today. Other uses are in batteries, pigments, and alloys.

Because cadmium is produced as a by-product of smelting zinc ores, it is the large zinc-smelting countries—Canada, China, South Korea, and Japan—that are the main producers. In 2008, the world's total production of cadmium was about 20,000 metric tons per year. It is somewhat meaningless to talk about cadmium reserves, which are estimated to be about 500,000 metric tons, because the production of cadmium is dependent upon the mining of zinc.

In recent years, the U.S. Occupational Safety and Health Administration (OSHA) has determined that cadmium can be toxic to humans. This has increased workplace safety regulations and brought about special concerns over cadmium in paints and in the leachates from landfills.

THE PRECIOUS METALS

The precious metals owe their name to their high values. Two of them, gold and silver, have been known and prized since antiquity. The other precious metals, the six platinum group metals, joined this select list in more recent times. The metals are precious (and expensive) for several reasons, but two are paramount. First, they are all rare and costly to find and recover. Second, to varying degrees, the precious metals resist corrosion and have great lasting qualities. Their resistance to chemical reaction is said to be due to their nobleness, hence, their alternate name—noble metals. The precious metals are found widely in small quantities; the locations of the major deposits are indicated on Figure 8.28.

Gold

Metallic gold is soft and malleable, but also extremely resistant to chemical attack, and it is corrosion-free. Indeed, gold is so stable and has so little tendency to corrode that most of the gold that has ever been mined is still in use. The gold that was present in the bracelets that adorned Cleopatra's arms could possibly now be residing in someone's teeth, a wedding band, or a gold coin. Once gold has been recovered from the ground, it can be melted down repeatedly and used again and again.

Gold was certainly one of the first metals used by our ancestors; in all likelihood, the first uses were for ornamentation. Gold's rarity and durability soon made it a medium of exchange, and eventually it became an accepted measure of value around the world

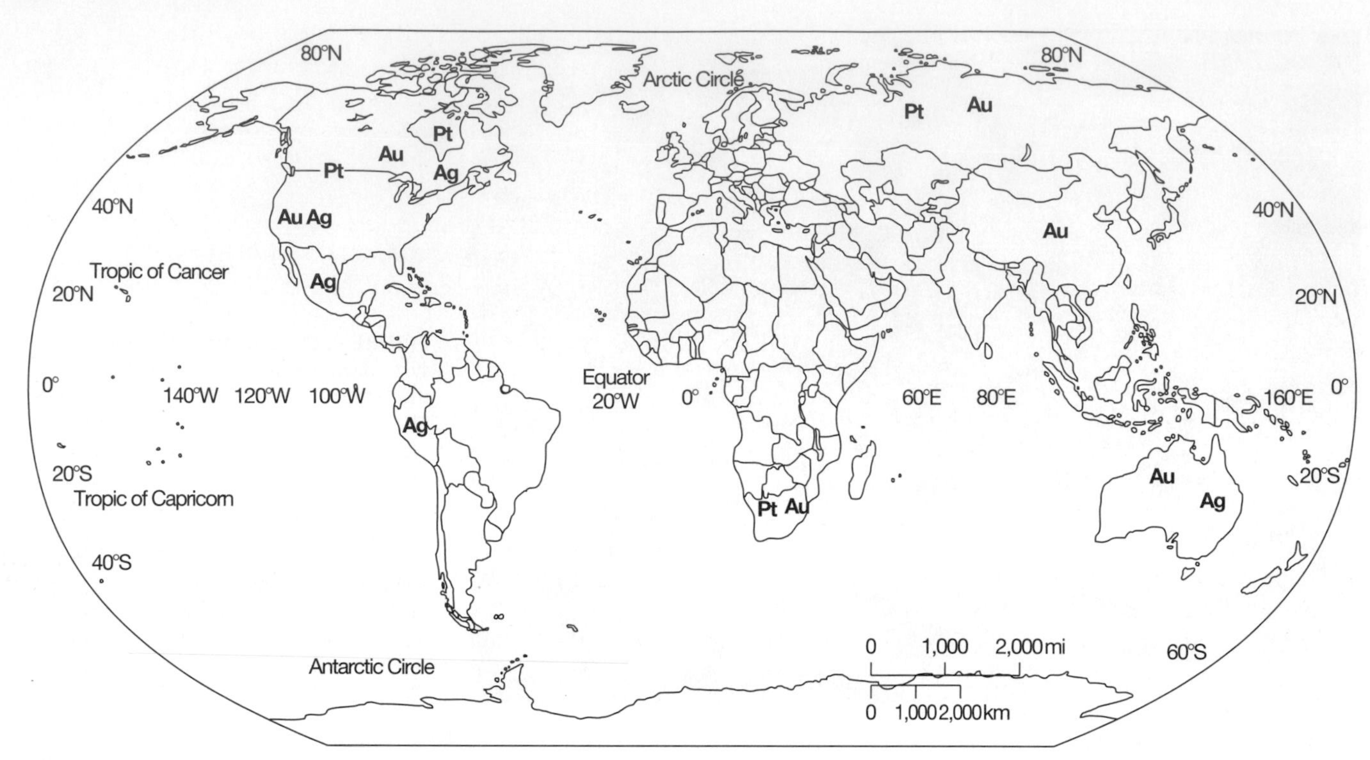

FIGURE 8.28 World map showing locations of the largest reserves of the precious metals—gold (Au), silver (Ag), and Platinum (Pt). South Africa contains the largest reserves of both gold and platinum.

(Figure 8.29). To a certain extent, it still fills this role, although few countries still back their issue of paper currency with gold. In addition, as the use of gold for monetary purposes has declined, its use for industrial purposes has grown so that today more than half the world's annual gold production is used in electronic products, aerospace applications, special alloys, and dentistry.

FIGURE 8.29 Stacks of gold bullion in the Federal Reserve Bank vault. Gold has often served as a monetary standard and remains a measure of wealth. (Photograph courtesy of the Federal Reserve Bank of New York.)

Throughout history, gold has been considered a safe investment in times of economic uncertainty. When people have feared that other goods or products could lose value, they have invested in gold, driving up its value. This accounts for much of the sharp rise in the price of gold since 2005; after years in the $300 to $400 range, it exceeded $1000 per troy ounce in March 2008 for the first time. Where the price will go in the next 10 years is difficult to predict, but it will certainly be much affected by world stability and the overall economic conditions.

GEOLOGICAL OCCURRENCE. Gold deposits can be formed by hydrothermal solutions. By far, the most important mineral in such deposits is native gold, but in a few instances, telluride minerals such as calaverite ($AuTe_2$) and sylvanite ($(AuAg)Te_2$) are also important. Hydrothermal gold deposits are veins in which the associated gangue mineral is generally quartz. Many of the highest-grade gold deposits are hydrothermal veins that formed in close proximity to granitic or dioritic igneous intrusions. Such deposits are highly variable in grade (up to several ounces of gold per ton of ore) and in thickness, but consist mostly of quartz with minor amounts of pyrite and other sulfides. The discovery of rich veins led to the explosive growth of boom towns in the Rocky Mountains. Unfortunately, many of the deposits, though very rich, were also very small and were rapidly mined out; exhaustion of the ores resulted in a mass exodus of the miners to new discoveries, leaving ghost towns in their wake.

Two events in the 1960s dramatically changed the world of gold mining. The first was the development of bulk mining and processing techniques that permitted economic recovery of gold from ores with grades below 1/10 of a troy ounce per ton (below 3 ppm). The techniques were first developed and demonstrated at the Carlin Mine in Nevada, but then spread rapidly through the western United States and to other countries. These new techniques immediately converted large quantities of low-grade resources into economic reserves. (See Box 8.4 for a discussion of other gold extraction techniques.) The second event in the 1960s was removal of the fixed price for gold (long held at $35 per troy ounce) so it could fluctuate according to market forces. The price actually changed very little at first, but it increased more in the 1970s and skyrocketed to over $800 per ounce in 1980. It fell back but remained above $400 per ounce for much of the 1980s. This significant price rise converted more noneconomic resources into economic reserves. The combination of new technology and higher price stimulated gold production (Figure 8.30) and exploration, with the result that more deposits were found; world gold production and reserves are at all-time record levels at the beginning of the twenty-first century and

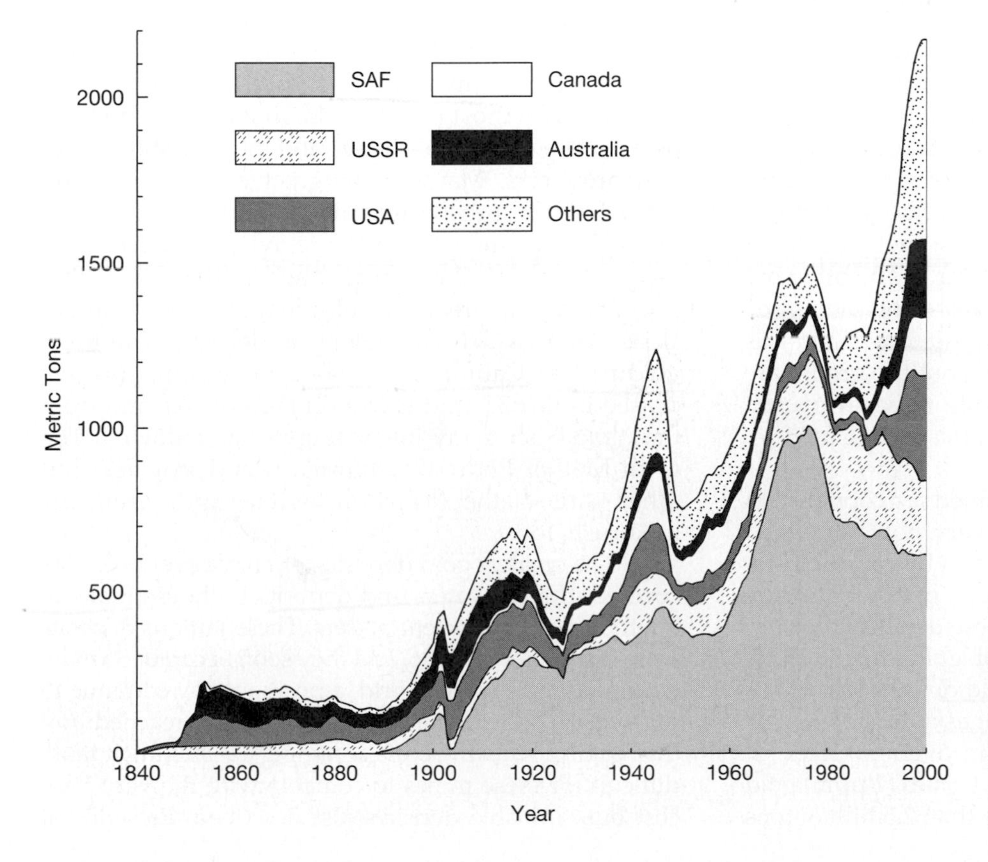

FIGURE 8.30 Annual world gold production since 1840. The discovery of gold in California in 1848 dramatically increased world production rates; this was followed by Australian discoveries in 1851 and the great discoveries in South Africa in 1886. South Africa has dominated world production since shortly after 1900.

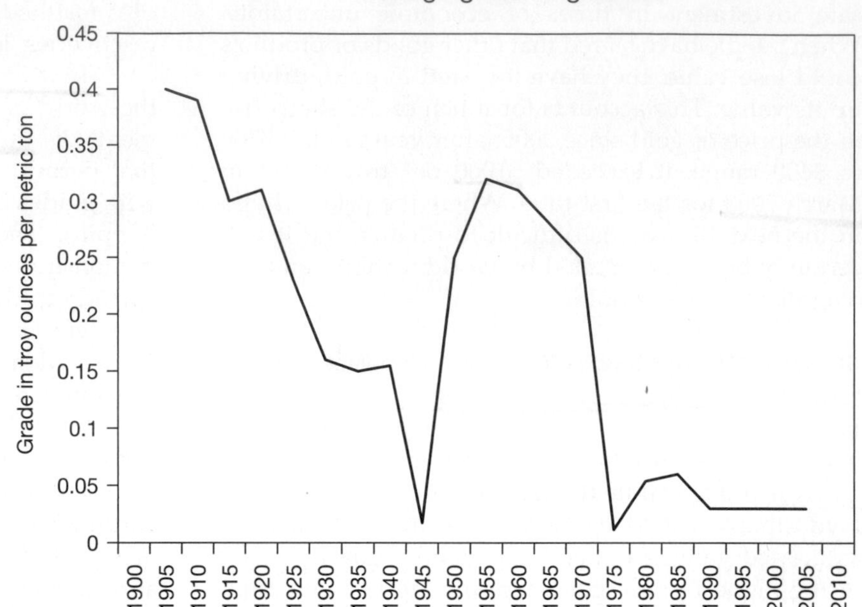

FIGURE 8.31 The changing grade of gold ores mined in the United States during the twentieth century. The average grade was about 0.4 troy ounces per ton (12 ppm) in the very early 1900s, but dropped to about 0.03 troy ounces per ton (1 ppm) by 2000. The primary factor in lowering the grade was the use of cyanide leaching, which permitted simple and inexpensive extraction of gold from very low grade ores.

grades of ore being treated are much lower than those in the past (Figure 8.31). The other major consequence has been that most U.S. and world production now comes from disseminated deposits, some with invisible gold that was not even considered an ore a few years ago.

There are two types of large low-grade gold deposits. One is represented by the Carlin Deposit, formed where hot spring-type waters circulated deeply in Earth's crust along major faults dissolving out trace amounts of gold and then redepositing them near the surface. It is believed that many modern hot springs and geysers represent deposits of this type in the making. A great bonus for miners of these ores is that they are at or very near Earth's surface and can be mined in large open pits by low-cost methods. They are processed as described in Chapter 4. The other ore type is the large porphyry copper-bearing intrusions formed near subducting plate boundaries. Gold has long been known in these deposits, but the grades are low and were marginally economic to recover until the new technologies and higher prices made gold recovery very profitable.

Many deposits, originally mined for copper with some minor by-product gold recovery, now are mined for gold, with copper considered as a by-product. The gold grades are usually low (about 1 ppm or 0.03 troy ounces per ton) but the sizes of these deposits are very large; hence, the total amount of gold can be quite large. In fact, the largest single gold deposit known today is the Grasberg mine in Indonesia. It is a copper-bearing porphyry deposit with an average grade of about 1 percent copper and about 1 gram (1 ppm) gold per ton; the deposit contains more than 2.5 billion tons of ore, with a total of about 70 million ounces of recoverable gold. The Busang gold swindle, discussed on page 14, was based on a deposit described by its promoters as of this same type, adding to its credibility until the "scam" was exposed.

Gold is a dense, almost indestructible mineral. It is therefore readily concentrated in streams where the action of flowing water washes away the less dense sand grains and leaves the gold concentrated behind barriers, and near the base of the stream channel (Figure 8.32). Alluvial placers, as these deposits are called, were almost certainly the first deposits to have been worked by our ancestors and they are still important producers. Many of the great gold rushes were started when prospectors discovered alluvial gold. Not only was the gold in the alluvial deposits worth recovering, but there was always the hope that prospecting upstream would lead to the "Mother Lode" from which the gold was derived. The great gold rush to California in 1849 is an example (as discussed in Box 3.1 and shown in Figure 3.A), and from that gold rush many interesting things followed. The great Mother Lode of California was discovered, but other parts of the American heritage grew from the gold rush, too.

The greatest gold deposits that have ever been discovered, the Witwatersrand deposits in the Republic of South Africa, are ancient placers. These famous deposits were discovered in 1886, and they soon became the main gold producers in the world, a position they continue to fill today. The extreme depths now being reached and the rapidly escalating costs of mining are making it more difficult for these mines to compete with the very low-cost bulk minable deposits just described. Presently, it

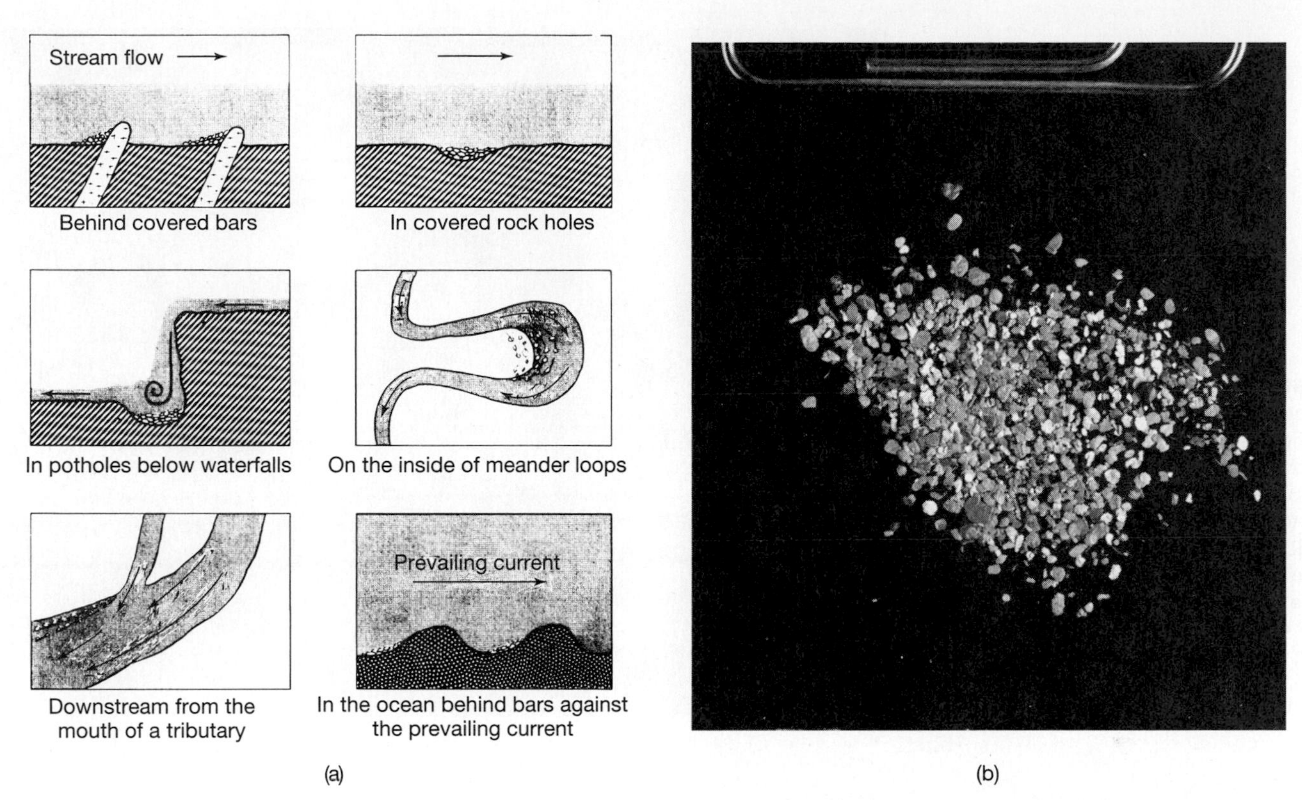

FIGURE 8.32 (a) Typical sites of placer gold accumulations where obstructing or deflecting barriers allow faster-moving waters to carry away the suspended load of light and fine-grained material while trapping the denser and coarser particles that are moving along the bottom by rolling or partial suspension. Placers can form whenever there is moving water, although they are most commonly associated with streams. (b) Grains of placer gold from California. By continual pounding in fast-moving streams, malleable gold is freed from brittle quartz and other valueless minerals. The tiny but dense gold grains accumulate in placers. It was gold such as this that led to the California Gold Rush and the "forty-niners," who mined the streams of California. The grains are 1 to 2 millimeters in diameter. (Photograph by W. Sacco.)

costs many of the deepest South African mines twice as much to produce an ounce of gold as it does to produce gold using the bulk minable operations at places such as Grasberg and Carlin, despite the fact that the ore grades in the Witwatersrand deposits are four to eight times richer than the deposits in Grasberg and Carlin.

The Witwatersrand deposits are ancient conglomerates that were laid down in a shallow marine basin (Figure 8.33). Into the basin ran ancient rivers, and at the mouth of each river a delta slowly built up. The gold and other heavy minerals are found in the coarse clastic sediments—the **conglomerates**—that make up the deltas (Figure 8.34); they are presumed to have been brought into the basin as clastic particles by the same streams that brought the pebbles into the conglomerates. The deposits are quite old, about 2.8 billion years, and they are enormous—more than twenty times larger than any other single gold district in the world. The source of the gold and how it came to be in the ancient conglomerates have been the subjects of fierce debate for over a century. The primary source of the gold is not known and is presumed to have been eroded away. The presence of large amounts of rounded placer pyrite grains and significant quantities of placer uranium minerals have led to the suggestion that the erosion of the original source rocks, and the subsequent deposition in the Witwatersrand deltas, took place when Earth's atmosphere contained very little free oxygen. If oxygen were present, the pyrite would have oxidized (as it does today) and the uranium would have dissolved. Hence, the pyrite and uranium in these ores, like the great banded-iron formations discussed in the previous chapter, are consistent with views that Earth's early atmosphere contained very little free oxygen. The Witwatersrand deposits are certainly unique in their size; similar but smaller deposits occur in rocks of about the same age in the Elliott Lake region of Canada and in the Jacobina region of Brazil, but none have been found to be as rich as South Africa's basin.

The gold mines of South Africa are the world's deepest. By 2008, the deepest mining activity was being carried out 12,100 feet below ground level; plans are being laid to carry mining activities to depths of more than 13,100 feet). Whether it will be possible to mine safely and profitably at these extreme depths remains to be seen, but on that question rides South Africa's hope of continuing as the world's major producer of gold.

FIGURE 8.33 The place where the fabulous Witwatersrand gold conglomerates were discovered by two prospectors, George Harrison and George Walker, in 1886. The conglomerates were laid down as deltaic deposits about 2 billion years ago and are now somewhat weathered due to oxidation of pyrite (compare Plate 40). The man in the photograph is the late Dr. Desmond Pretorius, one of the greatest authorities on the geology of these remarkable deposits.

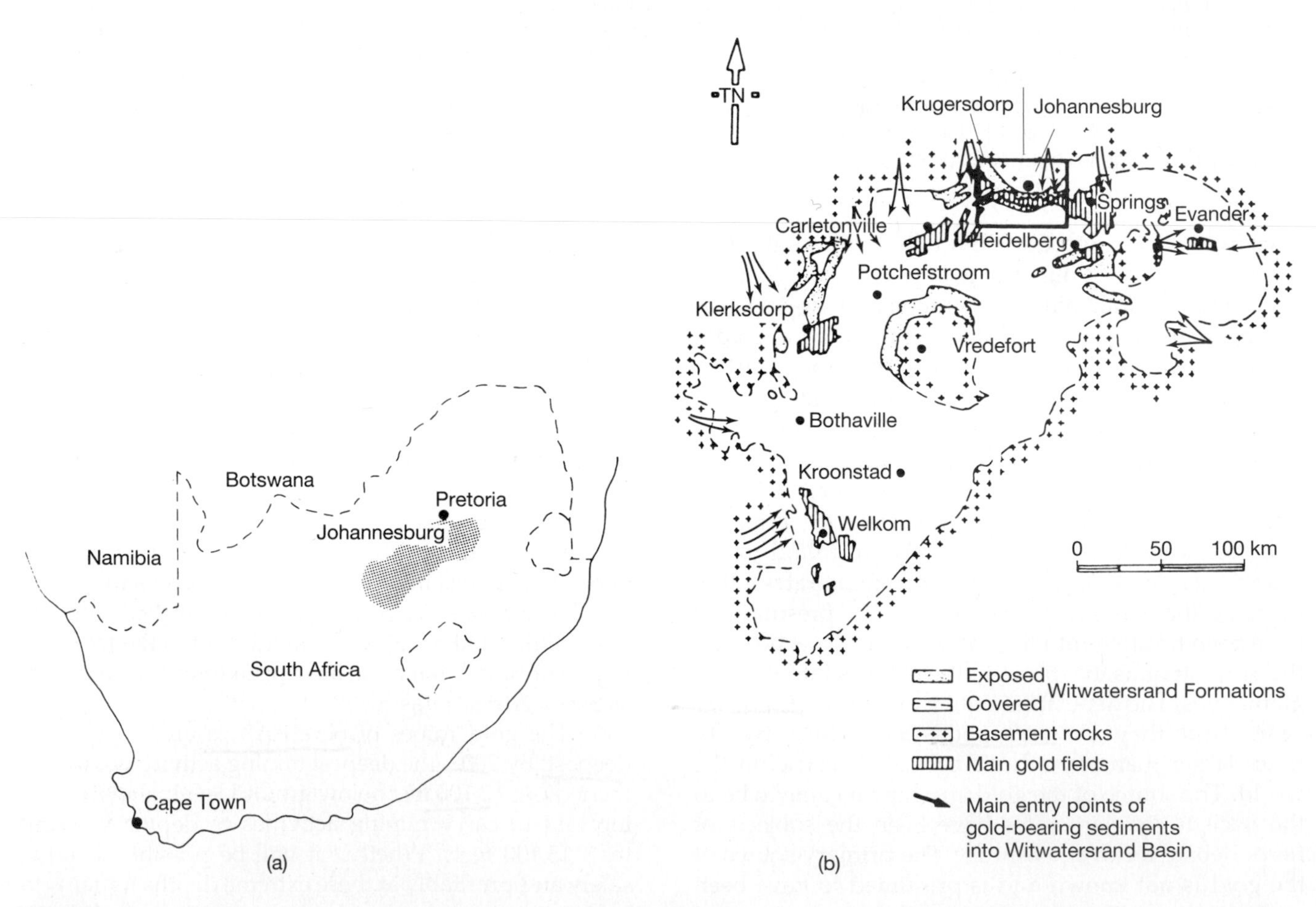

FIGURE 8.34 The Witwatersrand gold deposits constitute the world's largest gold reserves. These maps of the Witwatersrand gold fields show (a) the general location of the gold fields in South Africa and (b) the basin with the major gold fields, with arrows indicating the direction of transport of the conglomerates into the basin.

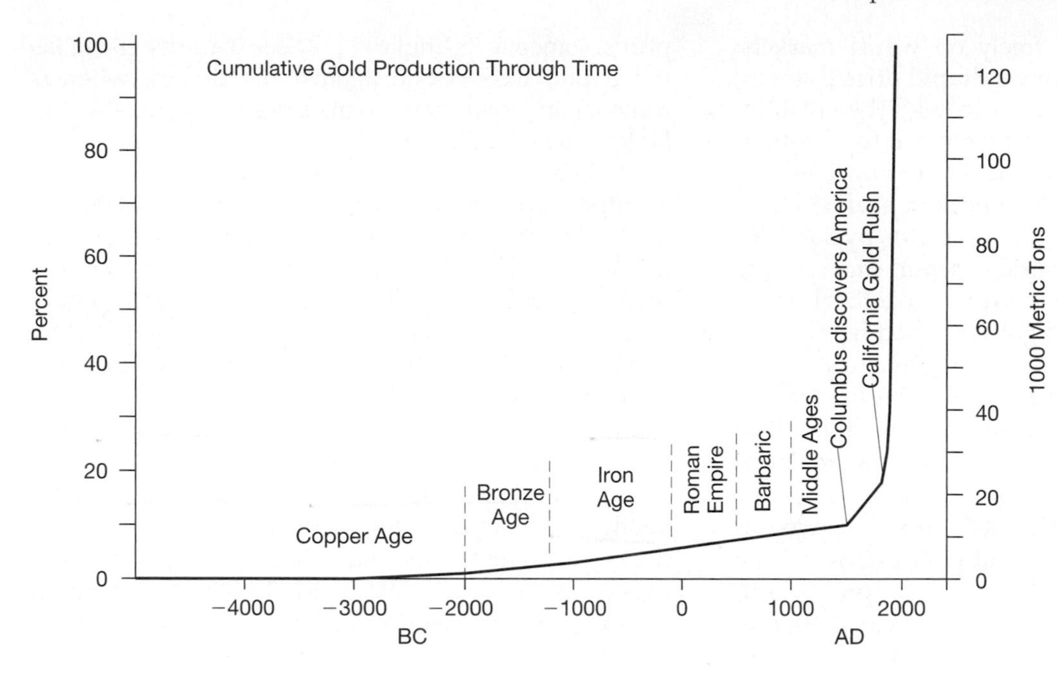

FIGURE 8.35 Gold was one of the first metals found and mined by humans. Rates of production were low through most of human history, but rose significantly after three major events: (1) Columbus's voyage that opened the New World to the European colonists and explorers; (2) the California Gold Rush, which tripled the rates of world gold production in a period of a few years; and (3) the release in the price of gold, in 1968, from fixed values to values that were controlled by market economics. Total world gold production throughout all of human history until the year 2010 is estimated to be about 150,000 metric tons.

PRODUCTION AND RESERVES. There are two widely used units in discussions of the production of precious metals—the *troy ounce* and the *gram.* One ounce (troy) is equal to 31.104 grams. The total world production annually during the beginning years of the twenty-first century was about 2500 metric tons per year (2.5 billion grams or 80 million troy ounces). The total amount of gold produced through all of history to the year 2008 is estimated at approximately 150,000 metric tons, more than 50 percent of which has been in the last fifty years and about 85 percent since 1900.

We have better records for the production of gold than for any other metal because of its long-held status of high value and, consequently, we can construct a reasonable estimate of the production of gold over the past 8000 years (Figure 8.35). There have been three times when worldwide production sharply increased: (1) the discovery of the New World in 1492, (2) the California Gold Rush in 1849, and (3) the end of regulated gold prices in 1968. The California gold fields increased world production manyfold, but the most important gold source for the past 100 years has been South Africa, as is evident in Figure 8.30. The price of gold was held fixed at various levels for many years, most notably at $35 (U.S.) from 1934 until 1968. This may have stabilized currency markets, but it lessened the value of gold because the value of everything else rose with inflation while gold remained fixed. In 1968, as a result of growing world pressures, the price of

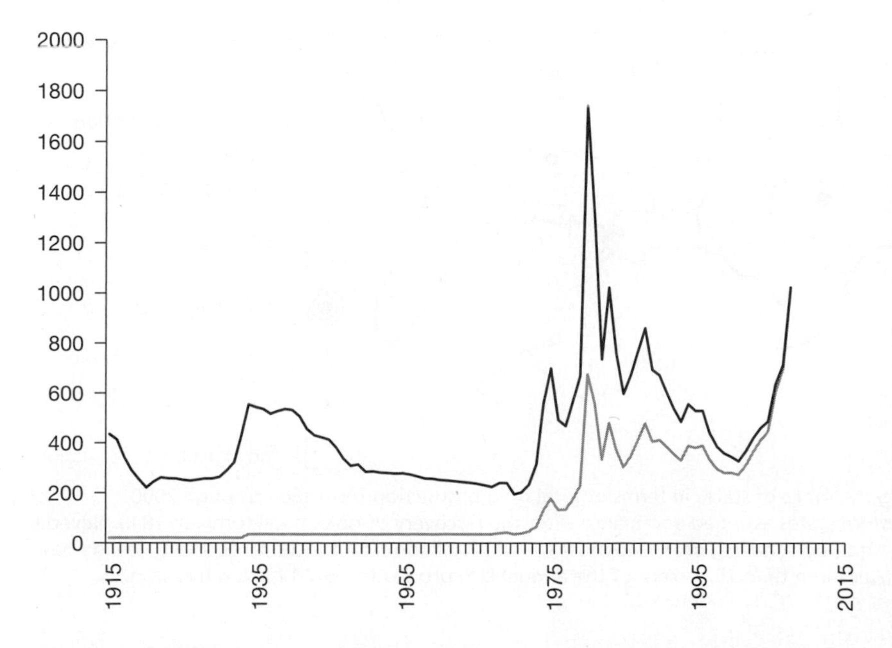

FIGURE 8.36 The price of gold from 1900 to 2008 (lower gray curve). The actual price of gold was fixed at $21.67 per troy ounce from 1900 through 1934, when it was raised to $35.00 per troy ounce, where it remained until 1968. The price was then allowed to float freely and rose to a maximum of over $850 per troy ounce briefly in 1980 and to $1000 in 2008. The value of gold relative to all other commodities is actually seen by considering how it was affected by inflation (using the Consumer Price Index); this value is depicted by the upper curve. The effect is clearly shown during the time interval when the price of gold was held constant (1935 to 1968) and inflation reduced the purchasing value of the gold by more than 50 percent.

gold was allowed to float freely on world markets. Despite much expectation for very rapid price changes, the value of gold at first only rose slowly. The volatility of economic markets has subsequently led to an irregular pattern, as shown through 2008 in Figure 8.36. The large price surge in 1980, which peaked near $850 per troy ounce, resulted primarily from efforts by some individuals to corner the world's silver market (the prices of gold and silver often move together). From 1980 until 2005, the value of gold declined nearly continuously as investment was placed in real estate and other commodities. However, the wars in the Middle East, especially Iraq and Afghanistan, and the fall in the value of the U.S. dollar, have led investors to return to the "safe haven" of gold. This drove prices to over $1000 per troy ounce by early 2008. There are two important points to remember about gold prices: no one has been able to make accurate long-term predictions, and not everyone in the world sees the price of gold remaining the same or even changing in the same direction. This is because gold is always quoted in U.S. dollar values, but there are constant changes in the value of the dollar relative to the currencies of other countries. Hence, at a time when Americans see steady gold prices, someone in England may see the price rising (as the pound loses value against the dollar), whereas someone in Japan may see the price falling (as the dollar loses against the yen).

Gold is so widely produced that more than sixty countries reported production between the 1980s and 2008. Almost certainly there were small amounts produced in many other countries, too, but they were not reported. Despite the broad extent of gold mining around the world, modern production is dominated by eight nations that each normally extract more than 100 metric tons of the metal each year (Figure 8.30). South Africa has remained the world's largest producer, but both its production and its percentage of the total have been declining in recent years. The Vaal Reefs mine of South Africa remains the world's largest single producer, but its importance has waned as total world production has reached 2500 metric tons per year. South Africa has led the world in producing a total of about 60,000 metric tons of gold to 2008; the United States is second in all time production at about 18,000 metric tons. The documented history of gold production in the United States began with Thomas Jefferson's description of a discovery in Virginia in 1782. Production began

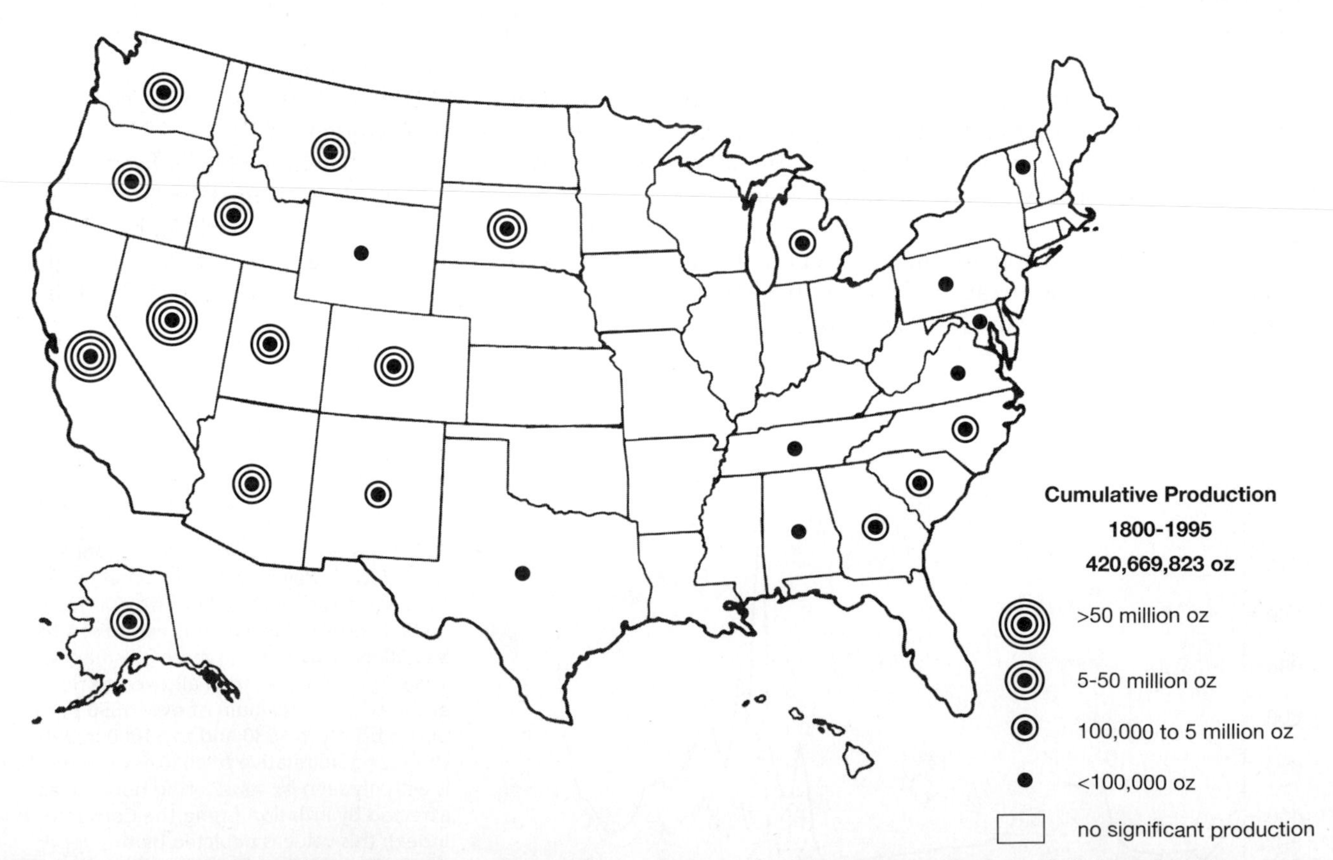

FIGURE 8.37 Map of the United States illustrating the ranks of states in terms of total gold production from 1800 through 2000. Production began in the eastern states, but the western states assumed dominance after the discovery of gold in California in 1848. Nevada and California have been the two most dominant states, both having produced more than 100 million troy ounces. Since 1990, Nevada has been the most important gold source; it now produces more than 75 percent of the annual U.S. production of 11 million troy ounces.

FIGURE 8.E The Chilean Mill, such as this one at Gold Hill, North Carolina, was pulled by a horse so that it rotated about a central post. The large wheels, cut from quartzite, crushed gold-bearing ores so that the ore could be passed over mercury-coated copper plates where the gold would stick to the mercury. (Photograph by J. R. Craig.)

extraction of gold by mercury is no longer practiced, although small scale miners still use mercury.

The accidental loss of mercury through inefficient methods and by heating amalgams has left small amounts of mercury in many areas that were prospected for gold in the 1800s and early 1900s. The mercury occurs in the sediments as small droplets that are slowly converted into a variety of compounds, some of which (like methyl mercury) are poisonous. In addition, many *weekend panners,* who pan for gold as a hobby and who continue to use mercury amalgamation, lose small amounts of mercury into streams. Despite government regulations, mercury amalgamation remains widely used in several areas of the Amazon River Basin where Brazilian panners known as *garimpeiros* still suffer widespread mercury poisoning. Brazilian officials estimate that the panners have released at least a kilogram of mercury into the environment for every kilogram of gold recovered—an incredible total of nearly 2000 tons of mercury—in the past decade.

Today, commercial gold extraction is carried out using cyanide because cyanide is cheaper, more efficient, and, if used properly, is safer than mercury. Cyanide is very poisonous and must be handled with great care to protect workers and wildlife. Gold, long ago referred to as a *noble metal* because it is so unreactive and insoluble in most solutions, even strong acids, is extremely soluble in cyanide solutions, forming compounds such as $NaAu(CN)_2$. As a consequence, in gold mines where ore grades may be as low as about 1 ppm (0.03 troy ounces per ton) or 0.0001 percent, gold can be efficiently extracted. The gold-bearing rock, once mined, is crushed to a size just small enough to allow the tiny gold grains (often no larger than 0.01 millimeter across) to be exposed; the crushed rocks are then piled on large impermeable sheets of plastic. Water containing only about 0.05 percent sodium cyanide, and maintained at a pH above 10 to prevent generation of poisonous gases, is allowed to percolate down through the crushed rock to dissolve the gold

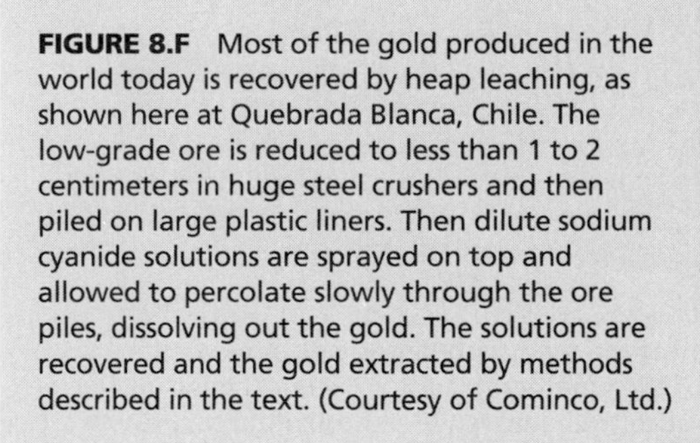

FIGURE 8.F Most of the gold produced in the world today is recovered by heap leaching, as shown here at Quebrada Blanca, Chile. The low-grade ore is reduced to less than 1 to 2 centimeters in huge steel crushers and then piled on large plastic liners. Then dilute sodium cyanide solutions are sprayed on top and allowed to percolate slowly through the ore piles, dissolving out the gold. The solutions are recovered and the gold extracted by methods described in the text. (Courtesy of Cominco, Ltd.)

(continued)

BOX 8.4

Gold Extraction Using Mercury and Cyanide (*Continued*)

(Figure 8.F). When the water flows out of the bottom of the pile, it is collected and allowed to pass through filters of activated charcoal onto which the gold is adsorbed. The gold is then redissolved and precipitated electrolytically, and the cyanide solutions are reused or rendered harmless so they do not damage the environment.

One commonly used variation to this *heap leaching* is called *carbon-in-pulp*. In this method, the activated charcoal is mixed directly with very finely ground ore in the dilute cyanide solution in large tanks. The mixture is agitated to promote gold solution and its absorption by the charcoal; then the charcoal and fine ore particles are separated, the spent ore is discarded, and the charcoal is taken for processing to recover the gold. The carbon-in-pulp treatment is more expensive to operate but allows for much faster treatment of the ore (hours instead of months) and can be used where it is not possible to construct large leach pads. The efficiency of the cyanide technique is so great that much of the gold recovered today comes from rocks where the gold is not even visible to the unaided eye and where the gold panners of old never dreamed that gold occurred and could be recovered.

FIGURE 8.40 The rich silver ores discovered at Eureka, Nevada, were one of the many rich deposits that made the United States the world's leading silver producer during the second half of the nineteenth century. The KC mine, as it looked in 1873, was one of several operating mines at Eureka. The entrance to the mine was a horizontal tunnel covered by the shed in the center of the photograph. Mined ore was tipped down the covered chutes to the processing plant just visible to the lower right. (From the collections of Beinecke Library, Yale University.)

The United States held that position until 1900, when Mexico became the world's leading producer. Today, China and Peru have joined Mexico as the principal producers.

The major uses of silver include coinage, medals, industrial applications, jewelry, silverware, and photography. Industrial usage is increasing, including medical (where it helps reduce infection), batteries, solders, catalytic converters, cell phones, and dental amalgams. The discovery that silver compounds are light sensitive led to its very important use in photography. This usage grew steadily through the twentieth century, but began to decline in about 2000 as digital photography replaced much conventional silver-coated film photography. Film usage is slowly decreasing but industrial applications are increasing, so overall demand is still rising slowly.

GEOLOGICAL OCCURRENCE. Silver is a chemical element that resembles copper in many ways. As a result, silver can readily proxy for copper, by atomic substitution, in most copper minerals. Not surprisingly, perhaps, much of the silver now mined comes as a by-product from copper mining. Silver also has an affinity for lead; a great deal of the world's silver is also produced as a by-product of lead mining. Today, more than 75 percent of all the silver produced is a by-product of copper and lead mining. Inasmuch as both copper and lead deposits are largely of hydrothermal origin, the by-product silver is also of hydrothermal origin.

A number of deposits around the world are still worked principally for silver. The minerals recovered are mainly argentite (Ag_2S) and tetrahedrite ($(CuAg)_{12}Sb_4S_{13}$), but a number of other silver minerals are also recovered. The deposits in which all of these minerals are found are hydrothermal veins, and the geological settings in which the veins are located are principally volcanic rocks of andesitic and rhyolitic affinity. Because the great mountain chain that runs down the western edge of the Americas is made up of andesitic and rhyolitic volcanoes, it is from here that most of the world's silver has been won. More than three-quarters of all the silver that has ever been produced has been mined in the Americas.

PRODUCTION AND RESERVES. Silver resembles gold in that a large number of countries (more than sixty at the beginning of the twenty-first century) report production. No single country dominates silver production, but seven countries do produce about 80 percent of the newly mined silver each year: Mexico, Peru, United States, Australia, China, Poland, and Chile.

It is apparent from recent studies that the annual world production of silver, at about 20,000 metric tons, vastly exceeds the world's production of gold. There is, nevertheless, a real problem concerning silver. In recent years, the consumption of silver has often exceeded its production. Because silver is subject to a certain amount of corrosion, and in many of the uses to which it is put it is not recoverable, there is a small but steady loss of silver. To make up the difference between the silver consumed and that produced, silver is withdrawn from inventory, from coinage, and from private hoards. Eventually, if consumption continues to exceed production, the hoards of silver will be depleted. The production of silver is not likely to rise very much, however, because so much of the production is as a by-product. If copper and lead production rises, silver production does too. If copper and lead production falls, silver output falls. Because the amount of copper in porphyry copper deposits and in volcanogenic massive sulfide deposits is very large, silver output will not cease, but someday it may well decline. A drop would produce a rise in price and this, in turn, would make some of the uses to which silver is put uneconomic. Reduced consumption would then follow. The development of digital photography is considered by some to be the first step in the decline of the photographic applications that have consumed so much silver. Based on this shift, the future demand for silver will significantly decrease.

Platinum Group Metals

The six platinum group metals—platinum, palladium, rhodium, iridium, ruthenium, and osmium—always occur together. They occur in the same geological settings and, to a certain degree, they can replace each other in minerals by atomic substitution. Many of their chemical and physical properties are also similar, so it is convenient to discuss these metals as a group rather than individually.

The group takes its name from its most abundant member, platinum. Each of the metals is silvery white in color, and although the metals are all malleable to a certain degree, only platinum and palladium are sufficiently malleable that they can sometimes be mistaken for silver. Indeed, the first uses of platinum probably occurred as a result of such a misidentification. Platinum metal occurs in the native form in placers, and grains of platinum, intermixed with silver grains, were used by ancient Egyptian artisans in certain works of ornamentation. The artisans must have realized platinum was more difficult to work than silver, but because it can be soldered they probably regarded the platinum as just an impure form of silver.

When the Spanish conquered South America, they discovered finely wrought objects of a strange white metal among the Indian treasures they looted. The Indian metal-smiths had learned to shape, solder, and even alloy platinum. The Spaniards lacked these skills and did not attempt to learn from the

Indians. They considered platinum to be valueless because they could not work it as they could silver. Consequently, they called the metal *platina*, a degrading term meaning "little silver." Furthermore, the importing of platinum into Europe from the New World was banned so that it would not degrade gold and silver, and much platinum was thrown into rivers and the sea.

The first pure separation of platinum as a chemical element, and the first clear statement of its properties, arose from the work of an English scientist, William Lewis. He published the results of his studies in 1763; soon, other chemists in Germany and France learned how to purify the metal and, as a result, they discovered many of platinum's alloying properties. During the purification of platinum, it was discovered that other platinum-like metals were also present in many of the ores. In 1803, the English scientist William Wollaston separated and identified palladium. In 1804, Wollaston described rhodium, while his countryman, Smithson Tennant, isolated and named osmium and iridium. Forty years were to pass until a German named Karl Klaus, working in Russia, discovered the last of the platinum group metals. He named it *ruthenium* in honor of *Ruthenia,* the Latinized name for his adopted home country of Russia.

All of the platinum group metals are resistant to corrosion, each has a high melting temperature, and each has interesting properties as a **catalyst**. Speeding chemical reactions through catalysis, plus the use of the metals in highly corrosive and very high temperature environments, account for the main uses of all of the platinum group metals.

Of particular importance is the use of platinum group elements in automobile catalytic converters (Figure 8.41), where platinum or palladium deposited on the inner surfaces of the sieve-like openings of a ceramic base convert harmful exhaust gases into inert ones. These converters have long been mandatory on automobiles and trucks in the United States and are being increasingly required on such vehicles around the world as concerns for air quality rise.

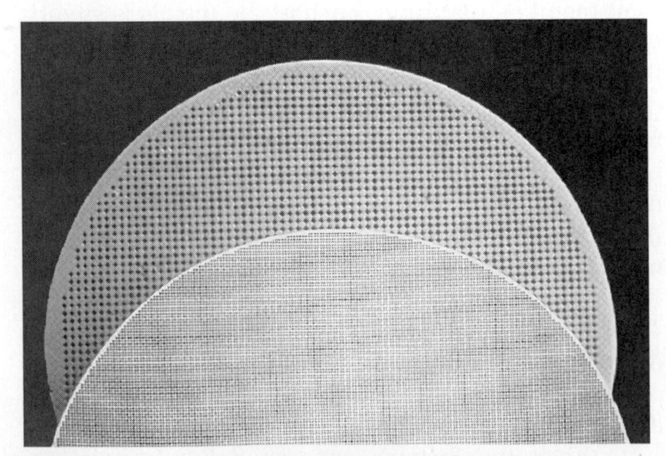

FIGURE 8.41 Automotive catalytic converters are prepared of cordierite ($Mg_2Al_4Si_5O_{18}$) in the form of narrow square tunnels and are coated with platinum-group metals, which convert hot noxious gases into harmless compounds as the exhaust passes through. (Courtesy of Corning, Inc.)

GEOLOGICAL OCCURRENCE. The platinum group metals are found as native metals and as sulfide and arsenide minerals. There are only two important geological settings in which these minerals are found in economic quantities. The first is in mafic and ultramafic rocks where the platinum group metals are concentrated in both chromite horizons and nickel-rich sulfide layers in layered intrusions. As a result, production of platinum group metals comes from the same deposits that produce nickel, copper, and chromium from magmatic segregation ores—Sudbury in Ontario, the Bushveld Igneous Complex in South Africa, and the great Norils'k deposits in Russia.

The second geological environment in which the platinum group metals are concentrated is in placers. Like gold, the metals are dense and very resistant to corrosion. They concentrate readily in alluvial sediments. Commonly, the source of the metals in placers is ultramafic rocks such as serpentine and peridotites. The great placer deposits of Russia have this origin. Serpentine rocks in the Ural Mountains contain small amounts of platinum group metals; thus, streams that carry debris from the weathering of serpentine from the Urals have valuable placers associated with them. These provided sufficient platinum that Russia minted and circulated large numbers of platinum coins from 1828 to 1841.

PRODUCTION AND RESERVES. Separate production figures are not reported for all of the platinum group metals, but it is possible to estimate the approximate percentages from the amounts of metals consumed. Platinum and palladium each account for about 40 percent of the production. Rhodium accounts for 9 percent, iridium 6 percent, ruthenium 4 percent, and osmium 1 percent.

The world's output of platinum group metals is overwhelmingly dominated by two countries, Russia and the Republic of South Africa. Between them, they produce about 90 percent of all the platinum group metals mined. Between them, too, they own most of the world's reserves. Of particular importance are two horizons within the Bushveld Complex in South Africa. One horizon is the 0.5 meter (1.5 foot)-thick Merensky Reef that has served as the major source for platinum for many years and that contains reserves estimated at 500 million to 600 million troy ounces (15 to 19 × 10^9 grams). The other horizon, a meter (three foot)-thick layer of massive chromite known as the

UG-2, lies below the Merensky Reef and is estimated to contain reserves of 800 million to 1350 million troy ounces (25 to 41 × 10^9 grams). In the United States, large resources of platinum group metals are present in the Stillwater Complex, Montana. These deposits occur in a Merensky Reef-like zone in a layered intrusion that is smaller, but that resembles the Bushveld Complex. Mining began on this zone in the late 1980s, and U.S. production has been steadily increasing.

THE SPECIAL METALS

The special metals earned their name because of their unique properties and the special roles they fill in modern-day technology. None of them are mined in large quantities, but each one fills one or more special roles that make their continued availability important to society.

Niobium and Tantalum

Niobium (also called columbium) and tantalum have similar properties and commonly occur together in nature. They were identified as separate chemical elements in 1801, but for more than 125 years no use was found for them. Eventually, in the 1920s, tantalum came to be used in the chemical and electrical industries, but it was only at the time of World War II that it found uses in electrical capacitors and in alloys for armaments. Niobium was first used about 1930 when it was added to steel to make very high-temperature alloys.

Niobium is still used for high-strength alloys needed in such demanding environments as gas turbines and the engines of jet aircraft. It is also used to make superconducting magnets. Most of the tantalum used is employed in the electronics industry for capacitors and rectifiers. It has a very important, relatively new market in the essential components of cell phones, DVD players, and computers. The tantalum exported to European and American markets from the Democratic Republic of Congo has been cited by experts as a key factor in financing civil wars in that region (where "coltan" is the colloquial name for the columbite-tantalite ores). Other applications of tantalum include tantalum carbide for high-temperature cutting tools and, because it is so corrosion resistant, insertion of tantalum mesh and pins in the human body during surgical repairs.

The main minerals that contain niobium and tantalum are columbite ($(FeMn)Nb_2O_6$) and tantalite ($(FeMn)Ta_2O_6$). Columbite always contains some tantalum in solid solution, and tantalite always contains niobium, so the two metals are always produced together. Tantalum and niobium minerals are found in several kinds of igneous rock. The first is an alkali-rich rock called a **nepheline syenite.** Such rocks are rarely rich enough to be mined, but in the Kola Peninsula of the Soviet Union, they do reach mining grade and are being exploited. The second kind of igneous rock is a **carbonatite.** This rare and unusual rock consists largely of calcium carbonate, and it is known both as an intrusive and extrusive igneous rock. Carbonatites have been worked for their niobium and tantalum contents at Oka in Quebec and at Axana in Brazil. The third type of igneous rock that often contains interesting amounts of niobium and tantalum is pegmatite. One of the major sources of tantalum in North America is the Bernic Lake pegmatite in Manitoba, Canada.

Both columbite and tantalite are chemically stable, hard, and dense. Therefore, the minerals are concentrated in placers. Such deposits have been worked in Brazil, Western Australia, and West Africa. The total world production in 2008 was about 60,000 metric tons of niobium per year and about 800 metric tons of tantalum per year. The major producing countries are Brazil, Australia, and Canada.

Arsenic, Antimony, and Bismuth

The three elements arsenic, antimony, and bismuth have similar properties and tend to occur in the same kinds of geological environments. Arsenic is little used in its elemental form; it is a brittle, grayish-colored substance that lacks the malleability usually associated with metals. For this reason, arsenic is often called a semimetal. The main use for arsenic is in chemical compounds, mainly as arsenates used in wood preservation, fungicides, insecticides, and pesticides, although the notorious toxicity of arsenic has led to it being used far less today than in the past. All of the arsenic produced comes as a by-product from base metal mining. The world's annual production since the 1980s has been around 50,000 metric tons of arsenic. The main producers have been China, Chile, Morocco, and Peru.

Antimony, like arsenic, is a semimetal but it has useful alloying properties; thus, much of the production finds a use in the metallic form. Antimony metal is added to the lead in batteries to toughen the lead. It is also used as an alloying agent with other base metals to harden the alloys and to make them more resistant to corrosion. Antimony compounds find uses as pigments in paints and plastics, as fire-retarding agents, as stabilizers in glass, and in many other applications. Annual world production of antimony in 2008 was about 165,000 metric tons. The major producers, in order of importance, were China, Bolivia, Russia, and the Republic of South Africa.

Bismuth, the heaviest of this trio of semimetals, is also the most metallic in its properties. Most people know of bismuth in its various chemical compounds that are used for medicinal purposes. The well-known antacid Pepto-Bismol is a proprietary form of a medicinal bismuth compound. The two largest uses of bismuth are in medicines and in cosmetic compounds. The third major use of bismuth is as an additive to low-melting alloys. Annual world production of bismuth is small, amounting to only about 6000 metric tons; China and Mexico are the main producers.

Arsenic, antimony, and bismuth are all produced as by-products of the mining for more abundant metals such as lead, zinc, and copper. All of the producing deposits, whether direct or by-product, are hydrothermal and all of the ore minerals are sulfides.

FIGURE 8.42 Beryl crystal from a pegmatite at Portland, Connecticut. Note the hexagonal shape of the crystal, which is 4 centimeters (2 inches) in diameter. Beryl, when occurring in gem-quality crystals, is known as emerald, as shown in the example in Plate 55. (Photograph by W. Sacco.)

Germanium, Gallium, and Indium

Germanium, gallium, and indium are very much metals of the twenty-first century. In each case, production is a by-product of the processing of major metals, principally aluminum and zinc but, to a lesser extent, lead and copper.

The total quantity of the three metals produced is relatively small, but the uses to which they are put are wide-ranging. Germanium has a high electrical conductivity, and it is the material from which some of the semiconductors in computers are made. Gallium has the unusual property of expanding on crystallizing, and this leads to some of the most important alloying effects. Indium has a very low melting temperature and is very soft and highly malleable, so it too has very interesting, although highly specialized, alloy properties. The principal uses for all three metals are in the electronics industry.

World production of the three metals in 2008 was 105 metric tons of germanium, 19 metric tons of gallium, and 570 metric tons of indium. Because the uses to which the metals are put are so politically sensitive, individual country productions are usually not published. The reserves of germanium and indium are held almost entirely in the form of solid solutions in the zinc sulfide, sphalerite. Hence, the availability of these elements is almost entirely dependent upon the mining of zinc. Potential resources of germanium would increase to millions of metric tons if it were to be recovered from fly ash and flue dusts that result from the burning of coal. Gallium also occurs in solid solution in some sphalerites, but mostly is present in concentrations of about 50 ppm in bauxites.

Beryllium

Beryllium metal production is a relatively young industry because many of its uses have arisen as a result of the nuclear and space programs. But one beryllium compound has been known and used since antiquity—the two gemstones emerald and aquamarine are green and blue forms of the beryllium silicate mineral, beryl (Figure 8.42).

The chemical element beryllium was first identified in 1797, but only in 1828 was the metal itself separated. It is a light reddish-colored metal that is strong and has a high melting temperature. Mixed with copper, beryllium produces a very hard but elastic alloy. The low atomic weight of beryllium makes it almost transparent to X-rays and to thermal neutrons, so the metal is widely used for the windows in X-ray tubes.

There are two major sources of beryllium. The first is the mineral beryl that is found in many pegmatites around the world. Unfortunately, none of the deposits is particularly large or rich. The second important mineral is bertrandite ($Be_4Si_2O_7(OH)_2$), which occurs in certain hydrothermal deposits associated with rhyolitic volcanic rocks. The deposits at Spor Mountain, Utah, are the largest known deposits of this kind.

The world's annual production of beryllium metal was only about 180 metric tons in 2008. Production had been a little larger in earlier times, but at no time has beryllium ever been produced in very large quantities. The restriction on production is twofold. First, deposits are small and quite expensive to work. Second, the separation of beryllium from its ores is a very expensive and exacting process.

Rare-Earth Elements

The **rare-earth elements (REE)** really do not deserve their name because they are much more abundant than many other geochemically scarce metals. The fifteen REE (Table 8.4), starting with lanthanum (atomic number 57) and ending with lutetium (atomic number 71), all have very similar chemical properties. When first discovered, the REE were known only in their oxide form; because they resembled the oxides of the alkaline earths such as CaO and BaO and did not seem to form common minerals, they were labeled *rare-earth elements.*

The REE are the basis of a small industry that started more than a century ago. The oxides are stable at high temperatures and they were added to the thorium oxide gas lamp mantles used by our great-grandparents. In more recent times, they have found such diverse applications as lasers, petroleum-cracking catalysts, opacifiers, coloring agents in the glass and ceramics industry, inhibitors of radiation in television tubes, and special optical glasses.

The REE are produced mainly from two minerals, monazite ($CeYPO_4$) and bastnaesite ($CeFCO_3$). Although the formulae of both minerals are written for cerium compounds, all of the REE replace cerium in the structures by atomic substitution.

The REE minerals are found in small amounts in many igneous rocks but principally in pegmatites, carbonatites, and granites. Some of these deposits are mined, as at Mountain Pass in California, but monazite is also concentrated in placers. It is recovered as a by-product from the mining of rutile, ilmenite, cassiterite, and other placer minerals. About 90 percent of world production in 2008 came from a huge carbonatite deposit in Inner Mongolia, called Bayan Obo.

Annual world production of REE, reported as rare-earth oxides, was about 125,000 metric tons in 2008. In the past, much data regarding rare earths was considered proprietary; however, in recent years, the data have been more freely available, and the U.S. Geological Survey has made reasonable estimates of production and reserves.

TABLE 8.4 The rare-earth elements. The relative amount of each REE produced is in proportion to its geochemical abundance in the crust.

Name	Chemical Symbol		Atomic Number	Geochemical Abundance (wt %)
Yttrium*	Y	The light REE	39	0.0035
Lanthanum	La		57	0.005
Cerium	Ce		58	0.0083
Praseodymium	Pr		59	0.013
Neodymium	Nd		60	0.0044
Promethium	Pm		61	Human-made
Samarium	Sm		62	0.00077
Europium	Eu		63	0.00022
Gadolinium	Gd	The heavy REE	64	0.00063
Terbium	Tb		65	0.0001
Dysprosium	Dy		66	0.00085
Holmium	Ho		67	0.00016
Erbium	Er		68	0.00036
Thulium	Tm		69	0.000052
Ytterbium	Yb		70	0.00034
Lutetium	Lu		71	0.00008

*Yttrium is commonly classed with the REE because its properties are so similar.

Fertilizer, Chemical, Construction, and Industrial Resources

Two basic human needs—food and shelter—require the kinds of resources discussed in this section. Many of these resources are overlooked by average citizens because they tend to come from faraway places. Nevertheless, these behind-the-scenes resources are important to our survival and will become ever more so as world population rises through the twenty-first century. The earliest resources used by humans were rocks of convenient size and shape, selected to perform some common task or to fulfill some everyday function. The basic need for shelter was first satisfied by living in caves and other naturally protected places. When our ancestors moved to places that lacked natural shelters, it was necessary to construct artificial caves; they did so by stacking rocks and logs to provide protection from both the weather and predators. Even though we have grown vastly in numbers and in technological innovation, we still make our shelters from stones and wood, much the same as our ancestors did. We may now shape the rocks more than our ancestors did, and we may even make artificial rocks using cement to make concrete, but we still construct our dwellings primarily from resources derived from rocks and trees. The durability of rocks is evidenced by the persistence of ancient temples, castles, walled cities, and statuary on every continent. The deterioration of such monuments results not so much from a failure of the materials used for construction, but from neglect or deliberate destruction.

CONSTRUCTION AND INDUSTRIAL MINERALS

Natural rock materials and their derivatives are quarried in vast quantities and provide most of what is used in the infrastructure we have constructed to support our modern technological societies. The greatest component of this infrastructure is the world's highway systems—in the United States alone there are more than 4 million miles of paved roads. The highway system provides the transportation network required to move the rock needed by modern society; in the United States, the system itself has consumed more than 60 billion metric tons of rock in its own construction. An estimated additional 40 billion metric tons of rock have been used in the dwellings in major U.S. cities. In nearly every country, regardless of the degree of technological advancement, construction materials are the largest tonnage resources in use today.

Ordinary rock has little intrinsic value because it is abundant and relatively easy to extract. However, every activity needed to handle, transport, shape, refine, and alter rock adds value. Local rock supplies in the United States commonly are valued at only a few dollars per ton, but the same rock when cut, polished, or otherwise processed and transported significant distances may be valued at hundreds of dollars per ton. For example, raw talc rock sells for about $10 per ton, but the same material that has been finely ground and sized to make cosmetic powder is valued at up to $550 per ton.

The largest single type of building material used by modern societies is crushed rock, which serves as the foundation material for roads, airports, construction sites, and is also the basic granular component of concrete. The cement used to bond the concrete together is prepared from limestone and shale and serves as an instant rock, which may be shaped to fit nearly any need. Other basic construction and industrial materials serve us widely but commonly in rather inconspicuous ways; examples are the sand used in making window glass, the gypsum in wallboard, the clays used to make bricks and pottery, and the clays and other minerals employed to coat the paper of this book.

We humans have long had a love affair with the special group of industrial minerals we call *gems.* Most gems are just beautiful varieties of fairly common minerals, but because of their color or sparkle, we have sought and valued them. Whereas common construction rock is among the least valuable of resources, gems constitute the most valuable; even a conservative estimate would place the value of highest-quality gem diamonds at $10 billion per kilogram. We extract natural gems from an array of geologic settings, but today we also manufacture greater quantities of artificial gems than we mine.

FERTILIZER AND CHEMICAL MINERALS

Earth's carrying capacity, by which we mean the maximum number of humans who can safely inhabit this globe, is usually estimated from the amount of food it is thought possible to produce. Despite some earlier estimates that world population could not be sustained above 5 billion, the world continues to produce enough food to support the present population of over 6.7 billion. Although there are local food shortages, these usually result from regional politically based conflicts or from droughts. There is no indication that worldwide food production could not feed in excess of seven billion people. The three main reasons that food production has kept pace with increasing population are: (1) the breeding and genetic engineering of plants that are more productive; (2) increased irrigation that minimizes drought problems and expands the area of arable land; and (3) increased use of fertilizers that promote plant growth and food production. The agriculture of our ancestors depended upon the availability of naturally supplied soil nutrients. Farmers long ago noted that animal wastes, decaying organic matter, and even ashes from the fireplace would stimulate plant growth. With the rise of modern chemistry, it was recognized that three basic components in addition to soil itself, light, and water are necessary for plants to thrive; these are nitrogen, phosphorus, and potassium. Animal wastes and bones are rich in nitrogen-bearing compounds of ammonia and compounds of phosphorous, and the ashes left from burning hardwood trees are rich in potassium. By the mid-twentieth century, animal and plant wastes were insufficient to meet fertilizer needs and were superseded by nitrogen compounds prepared artificially from the nitrogen of the atmosphere, by large marine sedimentary phosphate deposits, and by brines and salt beds rich in potassium compounds. Resources of nitrogen from the atmosphere and phosphorus and potassium from Earth's crust are large, but the demands of a population that will greatly increase in the century ahead may well tax the complex system of supply and distribution needed to get fertilizer compounds to all farmers.

The raw materials for fertilizers also serve in a wide range of chemical applications and, thus, are also members of the broad group of resources called *chemical minerals.* Other common minerals with chemical applications are salt, sulfur, and borax, but there are many others that contribute to our general well-being in diverse ways, from food additives to toothpastes.

CHAPTER 9

Fertilizer and Chemical Minerals

The ocean is used as a source of salt, one of the most widely used chemical minerals. This photograph of salt recovery operations in Bonaire in the Netherlands Antilles shows piles of salt that have been recovered by evaporating seawater. Ocean water with about 3.5 percent dissolved salts is pumped into large lagoons where it is allowed to evaporate using only the energy of the Sun. Once the water has evaporated, the salt is harvested and shipped for use in a myriad of industries. (Photograph courtesy of AKZO NOBEL.)

Higher fertilizer use will continue to have both positive and negative impacts on the environment. Organic and mineral fertilizers will help to replace the nutrients removed by crops, and build up soil organic matter. However, groundwater nitrate contamination will continue to be an issue in most developed countries and will increasingly become a problem in developing countries.

AGRICULTURE: TOWARDS 2015/30, FOOD AND AGRICULTURE ORGANIZATION OF THE UNITED NATIONS, 2000 (FAO)

FOCAL POINTS

- Three principal fertilizer components—nitrogen, phosphorous, and potassium—are necessary for plant growth, and their application substantially increases crop yield.
- The use of animal wastes and ashes from cooking fires as fertilizers to increase food production began in prehistoric times.
- World fertilizer production is presently over 200 million tons per year, and will have to continue to rise to meet the food needs of a rising world population.
- Nitrogen-bearing fertilizers first became important in world trade in the early 1800s, with the shipment of large quantities of bird guano from western South America to Europe.
- The supply of natural nitrates, shipped from what is now northern Chile, was the focus of the War of the Pacific, 1878–1883; the conflict resulted in Bolivia losing its coastline.
- Nitrogen fertilizers today are synthesized from atmospheric nitrogen by reacting it with natural gas.
- Phosphate fertilizers, originally made from bones, are now processed from phosphate rock using sulfuric acid to make soluble superphosphates.
- Potassium fertilizers, long derived from the "pot-ash" of hardwoods, are today extracted from evaporite deposits.
- The runoff of excess fertilizer has led to the development of many "dead zones" on the ocean floor where algal growth has depleted the oxygen levels in the ocean water such that the normal marine organisms cannot survive.
- Sulfur, sometimes included in fertilizers, is a widely used chemical product. Free natural sulfur is extracted from underground deposits (associated with "salt domes") by pumping hot

water into the deposits to melt out the sulfur. Today, most sulfur is extracted from fossil fuels as they are prepared for energy production.

- Halite, or "common salt," is the most widely used chemical mineral; it is extracted in underground mines from large evaporite beds, from natural brines, and from seawater.
- Many different minerals are used in foods and food supplements to alter taste or to change properties. Others are used in personal products such as medicines, toothpastes, and cosmetics.
- A long list of mineral-derived chemicals serves modern society in the manufacture of many thousands of materials.

INTRODUCTION

The importance of the mineral resources used in our fertilizer and chemical industries often goes unappreciated because the public usually only sees the products of their use, not the minerals themselves. Fertilizer and chemical minerals are two of the major types referred to as **nonmetallic minerals**—a general term used to describe resources that are not processed for the metals they contain and that are not used as fuels. Instead, they are mined and processed either for the nonmetallic elements within them or for the physical or chemical properties they possess. Nearly everyone is familiar with fertilizers because they are widely used on home gardens and lawns, as well as on large commercial farms. In contrast, many of the chemical minerals are little known because it is the end products that we see, rather than the minerals themselves; for example, we see the copies emerge from a photocopier but not the selenium compounds that actually transfer the image, and we use soaps daily but never see the boron minerals from which they are made. The vast number of minerals used in chemical processes and the very specialized uses of many of them precludes a comprehensive discussion; the following account provides an overview and shows the diversity of the important chemical minerals.

MINERALS FOR FERTILIZERS

The growth of the world's population requires that there be a continued expansion in the production of food. The growth of food crops requires the availability of ten chemical elements—hydrogen, oxygen, carbon, nitrogen, phosphorus, potassium, sulfur, calcium, iron, and magnesium. The first three elements, which constitute 98 percent of the bulk of a living plant, are supplied in water drawn up from the soil and in carbon dioxide absorbed from the atmosphere (Figure 9.1). The other elements, although constituting only 2 percent of plant matter by weight, are vital to the growth process (Figure 9.2); these elements are extracted by the plants directly from mineral or organic matter in the soil or groundwaters. It is not so much the absolute concentration of these elements in the soil that is important, but rather the concentrations that are available in a water-soluble form that the plant can absorb. Natural soil-forming processes lead to slow decomposition of many of the primary rock-forming minerals into clays, oxides, and soluble salts from which the plants can extract these necessary elements. Long before our ancestors understood anything about soil formation, or knew anything of chemistry, they had discovered that many kinds of organic wastes, such as animal dung and fish heads, were helpful in increasing the yields of crops. Before considering the resources we now use for

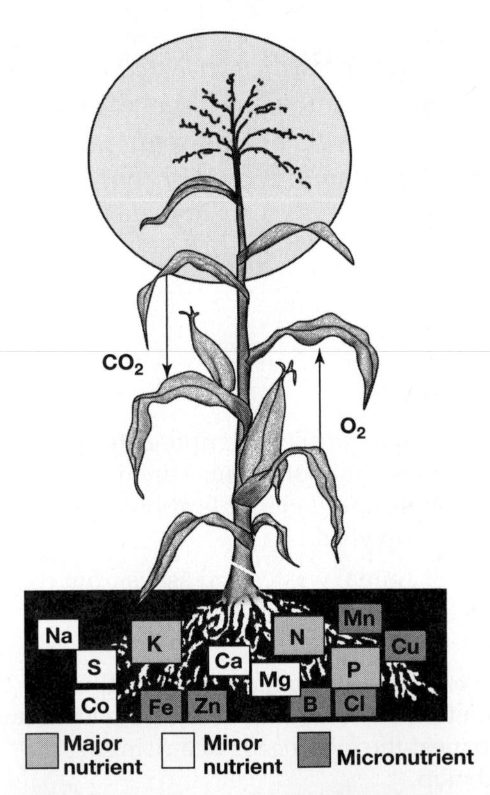

FIGURE 9.1 Sketch showing how plants require many different elements for optimum growth. The three nutrients needed in greatest concentration are nitrogen, phosphorus, and potassium; the bulk of the plant is composed of complex hydrocarbon molecules manufactured by photosynthesis from atmospheric carbon dioxide and water drawn up from the soil. Oxygen is released into the atmosphere during photosynthesis. (Graphic courtesy of the International Fertilizer Industry Association; www.fertilizer.org.)

FIGURE 9.2 The effects of adding or withholding major fertilizer components are demonstrated in these photographs of soybeans and corn that have received differing amounts of potassium. (Photograph courtesy of the Potash & Phosphate Institute.)

fertilizers, it is informative to consider the events leading to our present state of understanding.

HISTORICAL OVERVIEW OF FERTILIZERS

Our food supply depends upon three types of resources—**soils**, **water**, and **fertilizers.** Our earliest hunter-gatherer ancestors needed water directly for survival, but they gave little thought to soils and had no concept of fertilizers. However, when they became agriculturalists and began to plant and tend crops, they noted that the yields in different areas were not equal but were dependent on the availability of water and on some unseen characteristics of the soils.

We shall never know who first had the idea to fertilize crops, or when, but it is apparent that the earliest fertilizers were manures—animal and possibly human. Our ancestors did not know that these manures contain the three most important elements for plant growth—**nitrogen**, **phosphorus**, and **potassium**—but they did recognize the increased growth and yield that the applications of manure brought about for their crops. By Greek and Roman times, manures were classified according to their richness, with that of birds and fowl being rated the best. Xenophon, in about 400 B.C.E., observed: "The estate has gone to ruin [because] someone did not know it was well to manure the land." He perhaps gave the greatest compliment to this so often unappreciated product when he wrote, "There is nothing so good as manure."

Written records are sparse, but it is evident that the use of natural-waste organic materials such as fertilizers became a worldwide practice, nearly always on a local scale. An exception was the exploitation of the large Peruvian coastal **guano** (bird droppings) deposits that developed into a major article of commerce between Peru and Europe from 1808 until after 1880. Guano accumulated on easily accessible coastal islands where there was hardly any rainfall and where mining was inexpensive. Indigenous peoples had exploited the guano islands for at least sixty years but had been careful not to disturb or dislodge the large bird colonies that generated the deposits. England, seeking fertilizer for the newly introduced and important turnip crop, was thus a ready market, and English merchants rapidly took advantage of the decline of Spanish influence in the region. Shipments were sent to Germany and England before 1810, but significant commercial development did not occur until 1840. From 1840 until 1880, more than 4,350,000 metric tons of guano were shipped to England, with a peak of 274,000 tons in 1858; in the decade from 1855 to 1864, the value of the Peruvian guano cargoes to Britain exceeded £20 million (about £20 billion today). A Peruvian exporter of that period, who first believed that there could be no viable trade in guano, later noted: "The base manure could well be transformed into the purest gold."

The demise of the guano trade came in the period 1878–1885 as two other fertilizer industries—nitrates in South America and phosphates in Europe—were rapidly expanding. The nitrate compounds, commonly referred to as **saltpeter** or **niter** and consisting of potassium nitrate (KNO_3), sodium nitrate ($NaNO_3$), and calcium nitrate [$Ca(NO_3)_2$], occur in the very driest parts of the coastal regions of southern Peru (which was at that time western Bolivia) and northern Chile. They were mined on a small scale as early as 1810, and major shipments to Europe began about 1830. The nitrate exports rose to 21,300 metric tons in 1850, to 106,000 tons in 1867, to 535,000 tons in 1883, and peaked at 3.1 million tons in 1928. The total production of the area from 1830 has been estimated at 23.4 million metric tons of contained nitrogen from about 140 million metric tons of raw ore. The natural nitrates, all controlled by Chile after the War of the Pacific from 1879 to 1883, met with competition from ammonia produced by a coal-coking process beginning in 1892 and with nitrates prepared by the fixation of atmospheric nitrogen beginning about 1900. The rise in world nitrate demand and the rapid expansion of the by-product nitrate industries quickly relegated the natural Chilean nitrate mines to a minor role in world production; from 67 percent of the world's nitrogen production in 1900, to 22 percent in 1929, and to only 0.14 percent by 1980.

The earliest uses of **phosphates** parallel those of nitrates because many of the earliest fertilizer compounds contained both elements (Table 9.1). The earliest usage of phosphatic mineral substances was in about 1650, when English farmers applied ground bones to their fields. Usage gradually expanded, but it was not until 1835 that the phosphate in the bones was recognized as the valuable component. In 1840, during attempts to make bones more soluble, the German chemist Justus von Liebig dissolved some bones in sulfuric acid. His experiment provided the basis for modern phosphate fertilizer when he fused the phosphate with lime and produced a water insoluble product that seemed to be of no value. Within two years, John Lawes, in England, put Liebig's ideas to use and mixed sulfuric acid with bones and natural phosphate rock and produced what he called **superphosphate**, a name still used today. The value of this new fertilizer led to phosphate rock mining in France in 1846, England in 1847, Canada in 1863, and the United States in 1867.

Recognition of the value of potassium as a fertilizer component developed in the early 1800s and was finally confirmed by John Lawes and his co-worker, J. H. Gilbert, in 1855. Prior to this, many farmers had found that the application of wood ash, a material that can contain significant amounts of potassium, was beneficial to their crops. Although this **potash**, consisting of a mixture of potassium–sodium carbonates and hydroxides served as a fertilizer, it was primarily valued for its use in making soaps and glass. Significant world production of potassium salts began with the discovery of evaporite deposits at Strassfurt in eastern Germany in 1857, and in France in the early 1900s. These sources provided the relatively small world needs until World War I when the major German supplies were cut off from the rest of the world. This stimulated exploration for alternative sources and led to potassium production commencing in the United States in 1917 from subsurface brines at Searles Lake in California, and to the discovery of large bedded deposits near Carlsbad, New Mexico, in the 1930s. During the 1920s and early 1930s, discoveries were also made and production was begun in Poland, Palestine, Spain, and Russia. Rich deposits were found in Saskatchewan, Canada, in the late 1940s, but commercial production did not begin until 1958. Canada is today the world's major producer of potassium fertilizers.

TABLE 9.1 Compositions of fertilizer materials

	(%)		
	Nitrogen (N)	Phosphate (P_2O_5)	Potash (K_2O)
Natural Materials			
Saltpeter ($NaNO_3$)	15.6–16	—	—
Fish scrap	8	5–8	—
Sewage sludge	5–7	2–3.5	—
Urea	46	—	—
Bones	3.5	20–25	—
Kainite [$KMg(Cl,SO_4)\cdot 3H_2O$]	—	—	12–14
Seaweed ash	—	—	up to 30
Carnalite ($KMgCl_3.6H_2O$)	—	—	8–10
Wood ashes	—	2	up to 6
Peruvian guano	13	12.5	2.5
Synthetic Materials			
Superphosphate [$CaH_4(PO_4)_2\cdot H_2O$]	—	16–22	—
Triple superphosphate [($3CaH_4(PO_4)_2\cdot H_2O$]	—	44–52	—
Ammonium nitrate (NH_4NO_3)	33–35	—	—
Ammonium phosphate [$(NH_4)_3PO_4$]	11	60	—
Diammonium phosphate ($NH_4H_2PO_4$)	21	53	—
Potassium chloride (KCl)	-	—	48–62
10–10–10	10	10	10

Plate 37. The huge open pit at the Grasberg Mine in Irian Jaya, New Guinea, is one of the largest mining operations in the world and represents the largest single gold deposit, with more than 70 million ounces of recoverable gold. Its development in such a remote location and at high altitude (4100 meters or 13,500 feet) is an indication of the greater efforts that will have to be made to continue to provide the resources for a developing world. (Photograph courtesy of Freeport-McMoRan Copper and Gold, Inc.)

Plate 38. The M-Vein, Casapalca, Peru, is a rich mass of sulfide minerals containing silver, lead, copper, zinc, and other minerals. The ore minerals were deposited by a hydrothermal solution that flowed through a preexisting fracture in volcanic rocks. (Photograph by B. J. Skinner.)

Plate 39. In contrast to the single vein at Casapalca, Peru, shown in Plate 38, many deposits contain multiple veins such as these at Cligga Head in southwest England. Alteration zones occur along the sides of the narrow tin- and tungsten-bearing veins as a result of the ore fluids reacting with the wall rocks. (Photograph by J. R. Craig.)

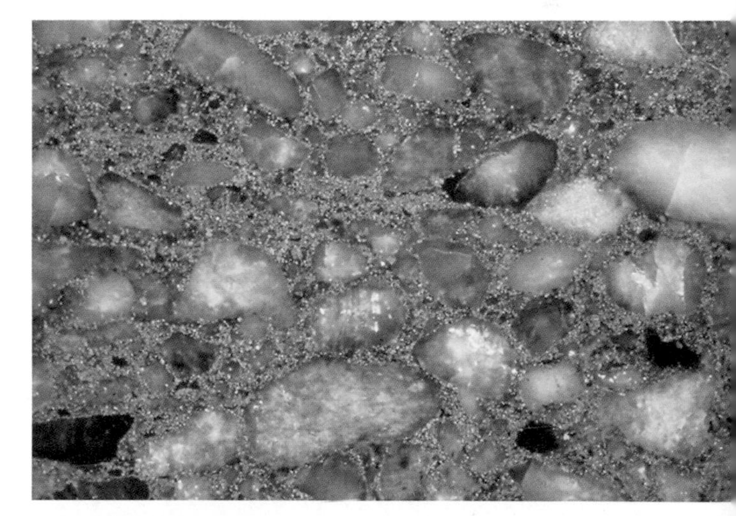

Plate 40. The Main Reef Leader, one of the rich, gold-bearing conglomerates mined in the Witwatersrand Basin, near Johannesburg, South Africa. The rounded quartz pebbles are set in a matrix of sand grains and rounded pyrite (FeS_2) grains, here seen as gold-colored particles. Gold is present but too fine-grained to be seen. (Photograph by Carlos Pais; courtesy of Geological Society of South Africa.)

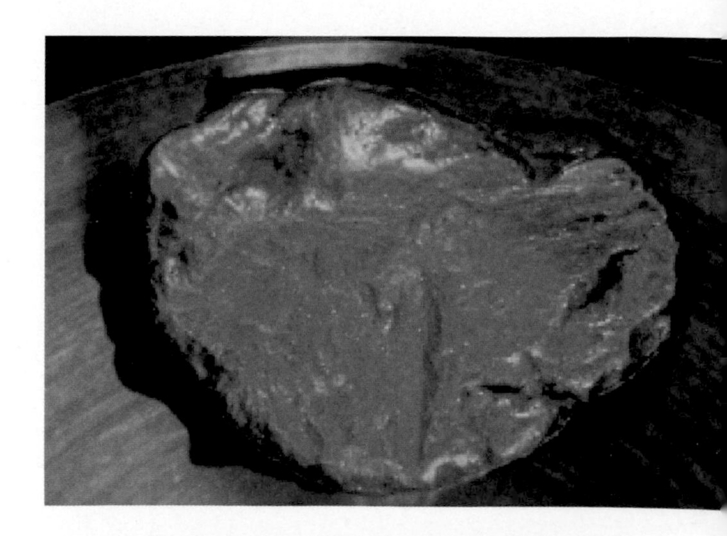

Plate 41. Small gold nuggets led to the discovery of most of the world's major gold districts, prior to the development of geochemical exploration techniques developed in the past few decades. This 3.5 gram nugget, 1 cm in length, was recovered from a stream in Cabarrus County, North Carolina, where the first gold mine in the United States was located. (Photograph by J. R. Craig.)

Plate 42. The Barney's Canyon Mine in Utah is typical of modern operations. Low-grade gold ores are mined from the open pit in the background, crushed, and placed in large heap leach pads visible in the foreground. Cyanide solutions that soak through the piles of crushed rock dissolving out the gold are then treated to recover the gold. (Photograph courtesy of Kennecott Corporation.)

Plate 43. A giant dragline scoops up rich phosphate ore in Florida. The ore is then pumped via a slurry pipeline to the processing plant. The very fine-grained waste material that remains after processing is allowed to settle and dry in huge settling ponds such as the one visible in the upper left-hand corner of the photograph. (Photograph courtesy of IMC Corporation.)

Plate 44. The reclamation of the land mined for phosphates, as shown in Plate 43, can produce productive wetlands and wildlife habitat within a few years of the mining operation. (Photograph courtesy of the Florida Institute of Phosphate Research.)

Plate 45. A line of fumaroles along a fault at Alae, on the southwest flank of Kilauea volcano, Hawaii. At the time this photograph was taken, volcanic vapors were depositing sulfur crystals in the fumarolic vents and laying down a yellow blanket of fine-grained sulfur on the countryside. (Photograph by B. J. Skinner.)

Plate 46. River sediments, such as these deposits laid down by the Pee Dee River in North Carolina, often serve as important sources of sand and gravel used for construction. Similar sediments can contain significant placer accumulations of gold or titanium minerals. (Photograph by J. R. Craig.)

Plate 47. Cranberry glass incorporated into beautiful ornamental vases makes use of gold to produce the rose colors. Cameo glass pieces such as these are made by first creating many layers of glass on a single shape. Then the outer layers are carefully etched away using acids or sandblasting to reveal the desired patterns and colors in the underlying layers. (Courtesy of Pilgrim Glass Corporation, Ceredo, W. Va)

Plate 48. The Taj Mahal, in Agra, India, is a breathtaking example of a building constructed with cut stone. The principal building stone is marble, but it is adorned with intricate insets of semi-precious stones, many in the form of passages from the Koran. The Taj Mahal is a mausoleum, completed about 1650 AD by Shah Jahan following the death of his favorite wife, Mumtaz Mahal. The Shah and his wife are both buried here. (Photograph courtesy of the Government of India Tourist Office.)

Plate 49. The Yule marble quarry in Colorado produces some of the finest marble for statuary. Blocks are carefully cut and shipped around the world. (Photograph by J. Groeneboer; courtesy of the U.S. Bureau of Mines.)

Plate 50. The volcanic ash deposits in the Cappadocia region of Turkey have weathered into unique conical shapes and have been soft enough for thousands of years of inhabitants to carve complexes of rooms. The rock is sturdy, and the insulation properties of the rock are effective in keeping the dwellings warm in winter and cool in summer. Göreme, Uçhisar. (Photograph courtesy of Embassy of Turkey.)

Plates 51. Concrete, an "instant rock," has become the most widely used construction material in the building of the infrastructure for modern Western societies. The intersection of two interstate highways in Tennessee required huge amounts of concrete in the construction of the four levels of crossing road surfaces. (Photograph courtesy of U.S. Federal Highway Administration.)

Plate 52. Concrete and steel have become superior building materials for the construction of beautiful and functional structures such as the Sunshine Skyway Bridge across Tampa Bay in Florida. (Photograph courtesy of U.S. Federal Highway Administration.)

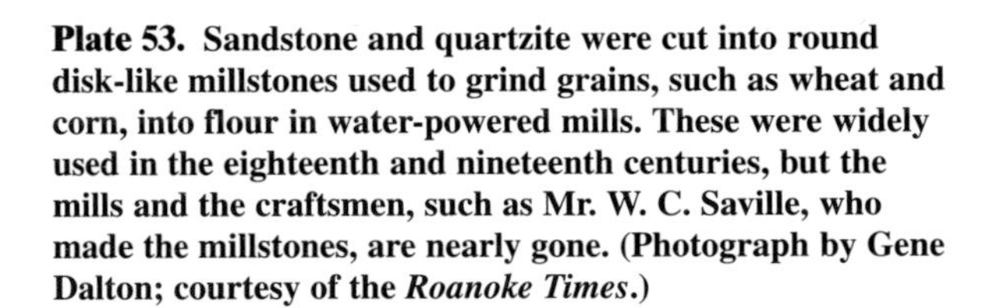

Plate 53. Sandstone and quartzite were cut into round disk-like millstones used to grind grains, such as wheat and corn, into flour in water-powered mills. These were widely used in the eighteenth and nineteenth centuries, but the mills and the craftsmen, such as Mr. W. C. Saville, who made the millstones, are nearly gone. (Photograph by Gene Dalton; courtesy of the *Roanoke Times*.)

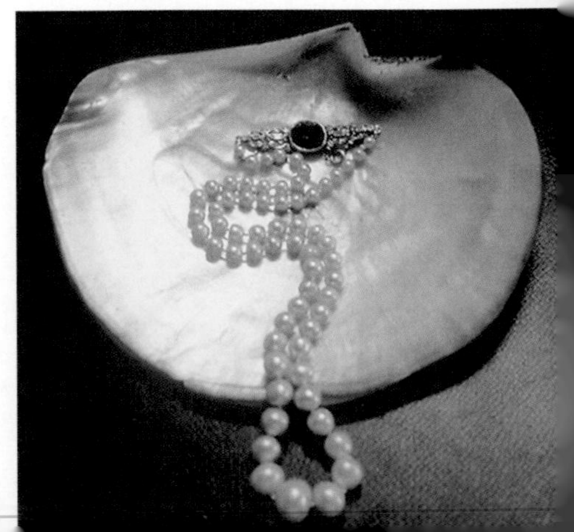

Plate 54. Gems are among the most highly valued resources and are usually beautiful crystalline forms of minerals. (left) The 128-carat yellow "Tiffany" Diamond. (right) Natural pearls and mother of pearl, the lustrous material on the inside of seashells that is identical to the material that forms the pearls. (Photographs courtesy of Tiffany Inc. [left] and CISGEM of the Chamber of Commerce of Milan, Italy [right].)

Plate 55. Some gems reveal their natural beauty even before they are cut and polished. Natural and cut emeralds from Colombia. (Photograph by Bart Curren; courtesy of ICA Gem Bureau.)

Plate 56. An intricate pattern of intersecting circles produced by a spray irrigation system used in the Midwest of the United States. The sprinkler system rotates around the water supply well at the center of the circle. (Photograph courtesy of Valmont Industries.)

Plate 57. The Central Arizona Project, a giant system of canals and pump stations, is designed to bring water from the sparsely populated drainage basin of the Colorado River system in northern Arizona to the heavily populated cities and irrigated farmlands of southern Arizona. (Photograph courtesy of U.S. Bureau of Reclamation.)

Plate 58. Subsidence in the Baytown area of Houston, Texas, has resulted from the subsurface extraction of oil and water and has resulted in the regular flooding of residential areas by high tides; this has led to the abandonment of many homes. (Photograph courtesy of Orin Pilkey.)

Plate 59. The Aral Sea, on the boundary between Uzbekistan and Kazakhstan, has lost 70 percent of its volume and has dropped 16.5 meters (54 feet), leaving ships and ports high and dry because the water that normally feeds the sea has been diverted for agricultural use. (Photograph courtesy of P. P. Micklin.)

Plate 60. Prolonged drought conditions can leave public water supplies, like the Gibralter Reservoir that serves Santa Barbara, California, dry. (Photograph courtesy of Santa Barbara County Public Works Department, Water Resources Division.)

(a)

(b)

Plate 61. (a) The Flood of 1993 inundated 11 million acres (17,000 square miles) of farmland for several months as heavy rains forced rivers out of their banks. (b) People in scores of towns, some more than 5 miles from the major rivers, were forced to evacuate. (Photographs courtesy of Federal Emergency Management Administration.)

Plate 62. Heavy rains in January 1995 flooded large parts of northern Europe. Areas such as Itteren (Province of Limburg) in the Netherlands were in great danger because much of the country lies below sea level. (Photograph by Aerophoto Schiphol BV; courtesy of The Royal Netherlands Embassy.)

Plate 63. Aqueducts, such as this one at Xiangao in China, have been used for thousands of years to transport water to major cities. Simple aqueducts are constructed so that the flow of water from source to users is controlled by gravity. (Photograph by J. R. Craig.)

Plate 64. Supplying the large quantities of water needed by large cities like New York requires a complex infrastructure of reservoirs, aqueducts, pump stations, and tunnels. The new Third Water Tunnel, presently under development, is 8–10 meters in diameter and lies 200–250 meters beneath the city. (Photograph by Ted Davey; courtesy of the New York City Department of Environmental Protection.)

Plate 65. The 216-meter (710-foot) high Glen Canyon Dam, on the Colorado River in northern Arizona, has controlled the river flow, creating Lake Powell. The water of the Colorado River is controlled by a series of dams and is distributed for use among the adjacent states by the Colorado River Compact. (Photograph courtesy of the U.S. Bureau of Land Management.)

(a)

(b)

Plate 66. Two soil profiles. (a) Layering is visible in highly weathered soils, such as those that underlie the stable land surface of the upper coastal plain of North Carolina. Technically, such soils are called plinthic paleudults. Iron has been leached from the upper layers, leaving them bleached and light colored. The leached iron is deposited about a meter below the surface in a pronounced reddish-color layer. (b) The soil profile at Poor Mountain, Virginia, formed above a base of sandstone and sandstone colluvium (brown). The uppermost, or A, zone is an organic-rich forest soil. Below it lies a thin, very dark Bh zone, rich in both iron and humus, and below that lies a bleached and leached E zone, where organic chelates have removed the iron oxides in solution. (Photographs by James C. Baker.)

Plate 67. Uncontrolled erosion at Twin Ponies, Iowa, has cut a deep furrow and is rapidly removing top soil. The severe loss of good top soil will reduce the productivity of the area. (Photograph courtesy of the U.S. Department of Agriculture, Soil Conservation Service.)

Plate 68. Contour farming, as shown at Red Rock, Iowa, is conducted to minimize soil erosion by plowing across the slopes and planting crops in sequences that help retain soil and mature at different times. (Photograph by Lynn Betts; courtesy of U.S. Department of Agriculture, Soil Conservation Service.)

Plate 69. Deforestation of several islands in the Caribbean, such as that shown on Nevis, has resulted from growing population and shortage of fuels. The results are often barren and unproductive fields that will not support farming, offer no fuel wood, have no wildlife, and retain little water. Efforts to reverse these effects have met with limited success. (Photograph by and courtesy of Bonham Richardson.)

Plate 70. Miners swarm over the working face of the fabulously rich gold deposit at Serra Pelada, Brazil. The Brazilian government did not allow mechanized mining operations, instead letting the small-time miner have a chance to make a fortune. The scene shown in the photograph, taken in 1985, has been repeated many times around the world in recent years as poor peasants have sought a way out of poverty. Unfortunately, few have been successful and the environmental damages have been significant. (Photograph by Glenn Allcott.)

Plate 71. The Polaris Mine, located on Little Cornwallis Island of the Nunavut Territory of the high Canadian arctic, is an example of the efforts now undertaken to find and extract ores. At 75 degrees north, this mine is well above the Arctic Circle and is ice-bound during parts of the year; the concentrates of zinc and lead produced must be stockpiled until ships (such as the one shown) can enter the harbor to remove them. (Photograph courtesy of Cominco, Ltd.)

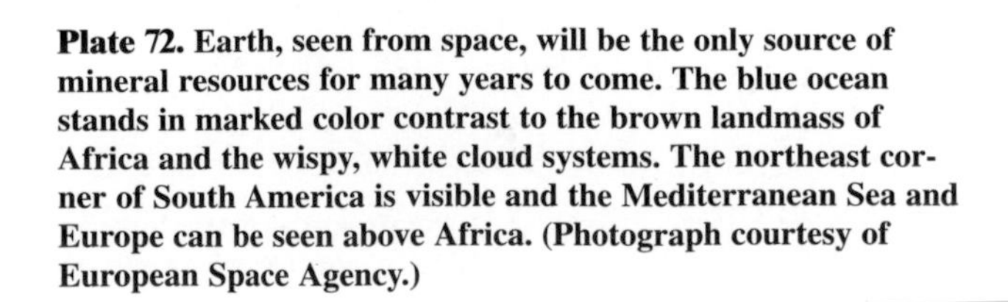

Plate 72. Earth, seen from space, will be the only source of mineral resources for many years to come. The blue ocean stands in marked color contrast to the brown landmass of Africa and the wispy, white cloud systems. The northeast corner of South America is visible and the Mediterranean Sea and Europe can be seen above Africa. (Photograph courtesy of European Space Agency.)

BOX 9.1

The Early Potash Industry and U.S. Patent Number 1

Samuel Hopkins of Pittsford, Vermont, was only twenty-five years old when he received the very first U.S. Patent on July 31, 1790, for an improvement "in the making of Potash and Pearl ash by a new apparatus and process." It was signed by President George Washington, Secretary of State Thomas Jefferson, and Attorney General Edmund Randolph. The term *potash* aptly describes the mixture of potassium hydroxides and carbonates that accumulate in the ashes when hardwood trees are burned; *pearl ash* is a purer variety that slightly resembles the color of a pearl. Potassium was already valuable, but Hopkins improved the quality of the product, thus contributing significantly to what has been called America's first industrial chemical.

Farmers in many parts of the world had long recognized that the ashes of some trees, especially the deciduous *hardwoods*, are rich in potash and serve well as fertilizers to promote the growth of crops. The colonial world did not, however, seek the potash so much for its use as a fertilizer, but rather for its value in making soap and glass, in dyeing fabrics, in baking, and in making saltpeter (KNO_3) used in gunpowder. By 1750, England had developed a great demand for soaps to wash wool before it was woven. Having largely exhausted its own forests in order to provide charcoal for steam engines and to make iron and steel, England turned to its colonies for new sources of potash. It was prepared by simply taking the black sooty ash that accumulated in the fireplace and leaching it with water in large pots (Figure 9.A). The soluble potassium salts dissolved and could then be reprecipitated when the water was evaporated off. The first operations were of small scale and relied on ashes purchased from local residents, but this was inefficient because both the amount and the quality of the supply were irregular. To meet the demand, commercial ventures developed across New England and southern Canada in the vast hardwood forests. The best potash yields came from elm, maple, ash, hickory, beech, and basswood; softwoods such as pine were of little value. Large steel pots were specially made for the leaching and boiling down of the residue. The American revolutionary figure Ethan Allen worked with his brothers in the manufacture of potash kettles, before he became the leader of the Green Mountain Boys.

During the American Revolution, the need for gunpowder increased the demand for potash and further spurred development of the industry. The potash was reacted with nitrogen-rich materials, such as bat guano, to make saltpeter (potassium nitrate), an essential constituent of gunpowder. A single large tree when burned and the ashes leached could yield about forty pounds (18 kilograms) of potash worth four dollars.

FIGURE 9.A Potash, a mixture of potassium and sodium carbonates and hydroxides, was extracted from the ashes of hardwood trees by dissolving the salts and then evaporating off the water. The name is derived from the ashes and the large pots used to extract the salts. (From Earl Palmer Collection at Virginia Tech.)

Samuel Hopkins's innovation was a technique of reburning the raw ashes to remove much of the unburned black wood char. This concentrated the potash and allowed for the production of higher quality potassium salts that commanded a much higher price. The United States became the world's leading producer of potash and remained as such until the discovery of the potassium-bearing evaporite beds in Germany in the 1860s. Although we no longer rely on Hopkins's methods, potash pots, or even on ashes for a supply of potassium salts, potash is still widely used in making soaps, glasses, and a large array of chemicals, including fertilizers.

Today, the world's fertilizer industry is focused on the extraction, manufacture, and distribution of nitrogen, phosphorus, and potassium either in bags for home use or in bulk liquid or solid forms for commercial farming. The numbers such as 5-10-10 or 5-10-20 seen on fertilizer bags refer to the percentages of nitrogen as nitrate (NO_3) or ammonia (NH_3), of phosphorus as phosphate (P_2O_5), and of potassium as potash (K_2O) present in the fertilizer. The remaining 65–75 percent is mostly inert material such as clay, although 1 or 2 percent of sulfur, calcium, and magnesium are also commonly present. The total percentage of major fertilizer components may seem low—only 25–35 percent—but greater concentrations could lead to damage of the delicate growth hairs on plant roots or upset the delicate levels of dissolved substances in plant fluids. Fertilizer demand has been climbing rapidly (Figure 9.3) so that the total world consumption has been doubling about every fifteen years. Unfortunately, the effectiveness of increased fertilizer usage appears to be diminishing. The **Global 2000 Report to the President** notes that a 200 million ton increase in grain production in the early 1960s was associated with a 20 million ton increase in fertilizer consumption. This 10-to-1 increase in grain production relative to fertilizer use in the early 1960s had dropped to about 8.5-to-1 in the 1970s, down to about 7-to-1 in the mid-1980s, and to about 5-to-1 in 2000. Clearly, world demand for fertilizers will continue to rise because they will be needed in the production of foodstuffs to feed an ever-increasing world population. Fortunately, reserves are large; however, except for nitrogen, the reserves suffer from the same problems that beset many of the scarce metals—they are geographically restricted, and we have already used the richest and most accessible deposits.

NITROGEN

Nitrogen is vital to plant growth because it is an essential component of all proteins, of more than 100 different amino acids, and is an integral part of the chlorophyll molecule responsible for **photosynthesis**, the food-making process in green plants. Most rocks and soils contain little or no nitrogen as discrete minerals, but many kinds of plants provide organic nitrogen, as ammonia or nitrate, into soils where it can be readily bound to the surfaces of clay particles.

The earliest nitrate fertilizers were the animal wastes that our ancestors spread on fields to stimulate plant growth. Locally derived farm manures were a sufficient source of nitrogen for crops until about the middle of the nineteenth century. At that time, the expansion of agriculture outpaced these local sources, and Europe turned to the importing of guano from Peru to provide its nitrogen fertilizer needs. The guano deposits accumulated on very arid coastal islands over thousands of years, from huge colonies of nesting seabirds. The local Indians had long made use of the guano (which is rich in both nitrogen and phosphate; Table 9.1), but it did not become known to Europeans until reports of it were

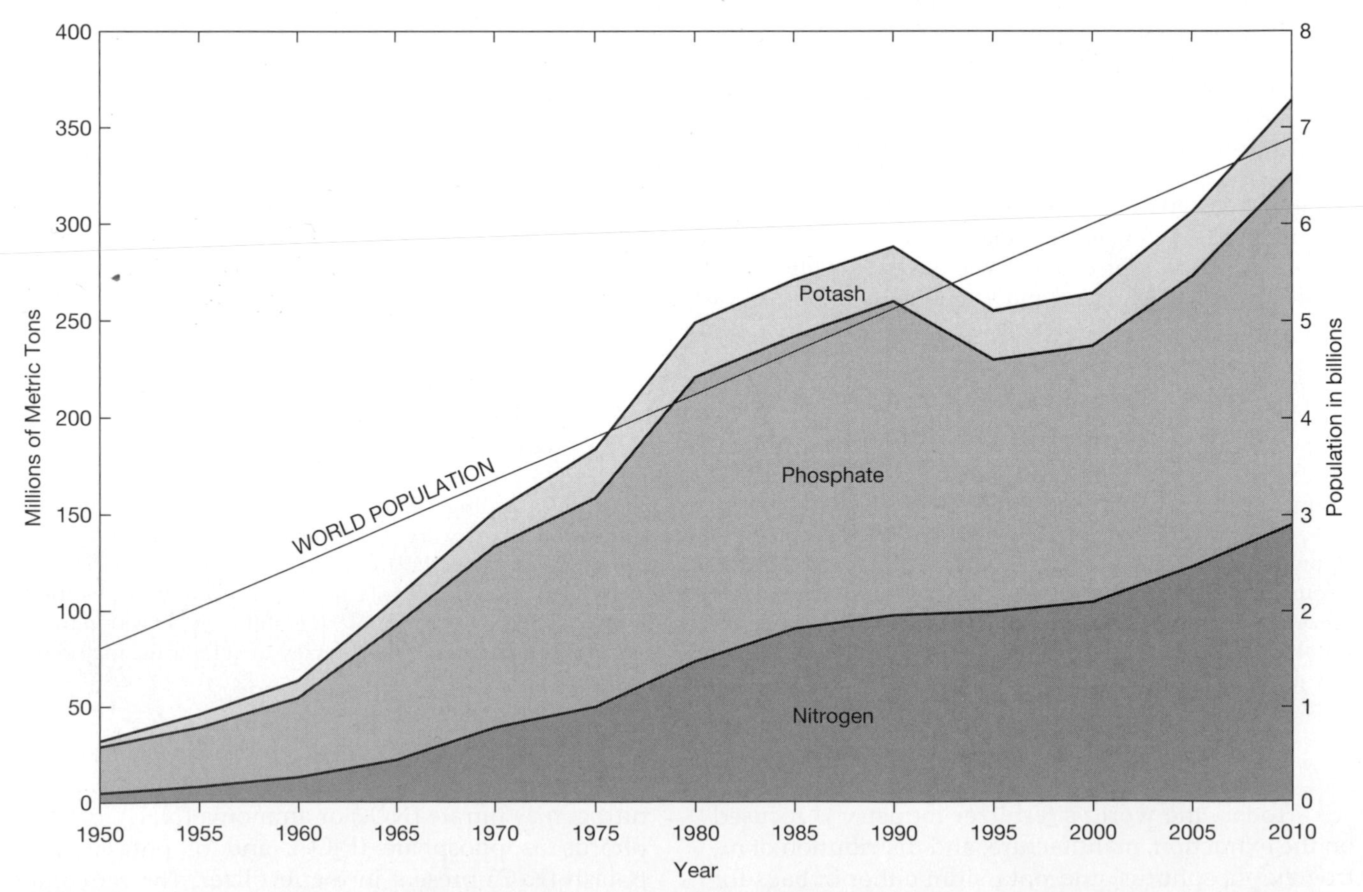

FIGURE 9.3 Changes in the world's population and the use of the major fertilizer components since 1950. The world's population has about doubled since 1950 (right-hand scale), but the use of fertilizer components has risen about ten times in the same time span in order to provide food for the growing population. Nitrogen is the dominant fertilizer component used, but the needs for phosphorus and potassium are also rising. (Data from the FAO of the United Nations Annual Reports).

brought back by Alexander von Humboldt after his visits there in 1802. Some small shipments were made by 1810, but the major trade did not develop until 1840, after which it flourished for forty years, then being displaced by the production of nitrate minerals from what is now northern Chile.

Natural nitrate mineral deposits occur only rarely because nearly all nitrate compounds are very soluble and hence are easily washed away by rains or percolating groundwaters. They generally occur as efflorescences or crusts resulting from the oxidation of nitrogen-bearing substances in the presence of other salts. Potassium nitrate is the most widespread of these minerals. It occurs as crusts on the walls and in the soils of some caves, such as those in Kentucky and Virginia, where it has apparently accumulated as a result of the evaporation of groundwaters that dissolved nitrogen compounds derived from organic matter in the overlying soils. In the eighteenth and nineteenth centuries, the demand for saltpeter as an ingredient in gunpowder led to the development of saltpeter plantations, or nitriaries, in France and Germany. In these plantations, the natural conditions of saltpeter formation were simulated by exposing heaps of decaying organic matter mixed with potash or lime to the atmosphere; the crusts of saltpeter were episodically gathered and processed. Calcium nitrate was also produced in the saltpeter plantations when lime was added to the organic matter. In addition, it commonly appeared as a crust on the walls of stables where it formed from reaction of the nitrogen compounds in horse wastes with lime that was used to reduce odors and insects.

Sodium nitrate, or ordinary niter, though much more restricted geographically than the potassium or calcium forms, has proven much more important as a source of nitrogen. This mineral occurs in very large deposits in the northern part of the Atacama Desert in what are now the two northernmost provinces of Chile (Figures 9.4 and 9.5). This is one of the driest places on Earth, with no precipitation at all for many

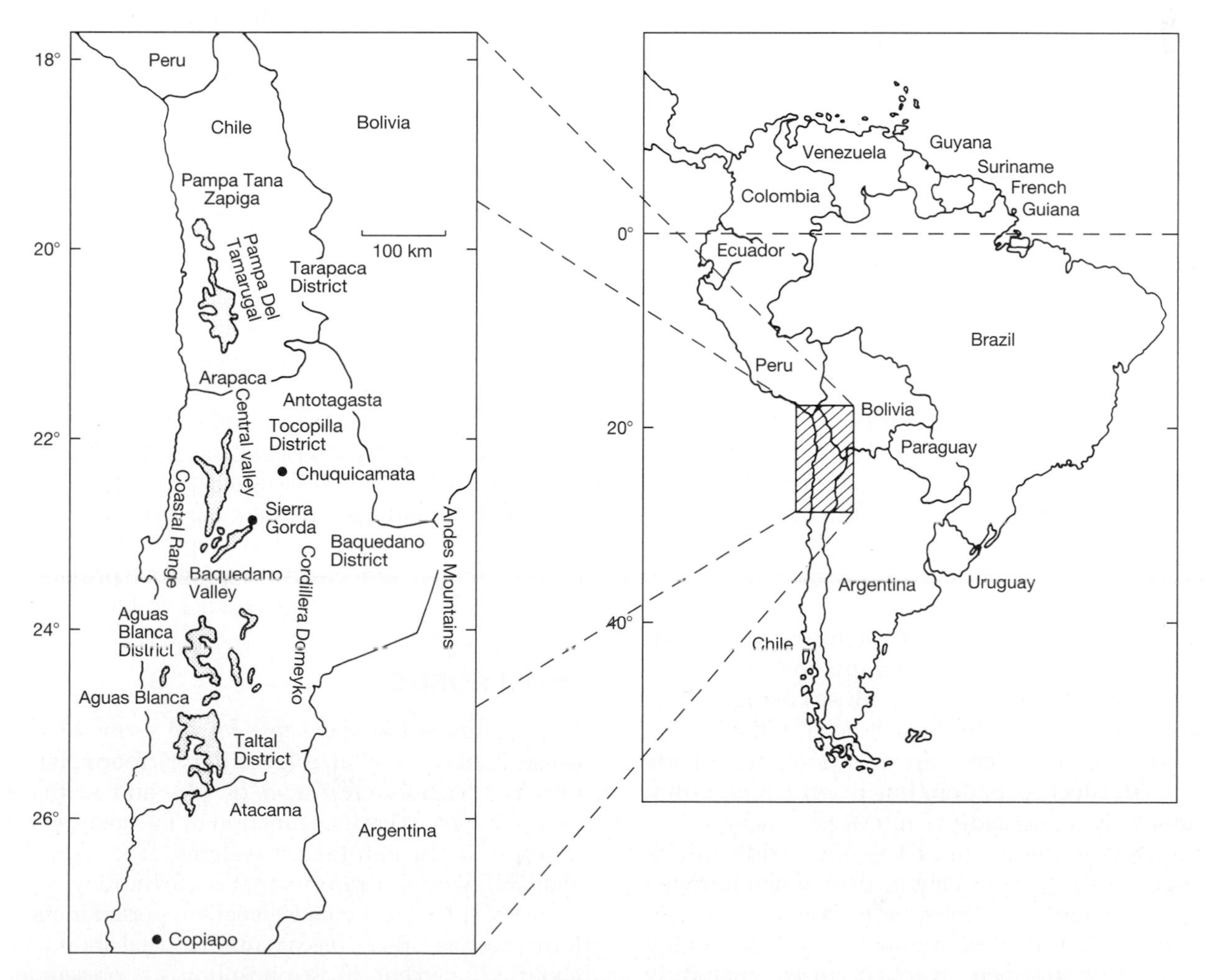

FIGURE 9.4 The world's largest nitrate deposits (shaded areas on the left-hand diagram) lie in the Tarapaca and Antofagasta provinces of what is now northern Chile. The extreme aridity of this area, which allows for the preservation of these deposits, results from the cold Humboldt Current that flows northward along the western coast of South America. Tarapaca and Antofagasta were parts of Peru and Bolivia before Chile annexed them in the War of the Pacific, 1878–1883; see additional discussion in the text.

FIGURE 9.5 The nitrate deposits of northern Chile are shown here being drilled before being processed and shipped to North America and Europe in the early part of the twentieth century. The nitrates formed slowly from nitrate-bearing mists that blew in from the Pacific Ocean and gradually cemented the desert soils. The very infrequent rains only served to allow the nitrates to percolate into the soils and form more solid rocks before the waters evaporated. There are few drier places on Earth—and none where nitrates have accumulated to this extent. The mining of these natural nitrates has declined dramatically, and nearly all nitrate fertilizer is now prepared from atmospheric nitrogen.

years at a time and an annual average rainfall of less than 2.5 centimeters. The nitrates are believed to have their origin in sea spray that precipitates in the soils from frequent fogs; the rare rain dissolves the very soluble nitrates and concentrates them as cements, or **caliches**, in the soils as a result of evaporation. Mining of the nitrate deposits began early in the nineteenth century, and by 1812 there were at least seven locations in which the caliches were boiled in water to dissolve the nitrates; the saltpeter settled out as the liquids were cooled. By the 1830s, the increased use of chemical fertilizers in European agriculture made the nitrates a useful return cargo for ships sailing to Europe. Although the mines were in the Tarapaca region of Peru and the Antofagasta region of Bolivia, both the labor and the investment were dominantly Chilean. In response to Peruvian attempts to expropriate the mines and to Bolivian attempts to increase taxation, Chile took over the two provinces using a small army and a naval force; this was the War of the Pacific, which crippled the Peruvian economy and made Bolivia a land-locked nation, but it left Chile with a near monopoly on the world's nitrate production.

Chilean nitrates continued as the world's principal source of nitrogen until about 1915, when nitrogen from coking ovens and atmospheric fixation processes became dominant. In 1892, a new type of coal coking oven—one that trapped expelled gases, including ammonia (NH_3)—was introduced. The ammonia immediately found use in the fertilizer and chemical industries. About 1900, it was also discovered that ammonia could be prepared from atmospheric nitrogen. This developed into the **Haber-Bosch process**, used in only a slightly modified form today. Fritz Haber, a famed German chemist, found that controlled combustion of a fossil fuel, coke, or gas with steam would yield carbon monoxide and hydrogen and that, with the aid of a catalyst, he could react the hydrogen with atmospheric nitrogen to form ammonia. (Today, the hydrogen is usually supplied by natural gas.) The ammonia can be used directly to make fertilizers or be oxidized to make explosives such as glyceryl trinitrate (formerly known as nitroglycerin) and trinitrotoluene (or TNT). Haber's discoveries were thus very important to the German military, which had been cut off from Chilean nitrate supplies during World War I.

Today, the synthetic nitrate industry satisfies more than 99.8 percent of world nitrogen needs. Chile still mines nitrates at rates of more than 500,000 metric tons per year, but the production constitutes only about 0.14 percent of world demand; even the total Chilean nitrogen reserves of 2.5×10^9 metric tons (containing more than 7 percent $NaNO_3$) would not equal one-half of the world's present yearly consumption. The probable resources at grades of less than 7 percent are thought to be no more than 22×10^9 metric tons.

Nitrogen remains primarily a fertilizer component, with more than 85 percent of world production for that use. The United States, India, and Canada are the world's primary fixed nitrogen producers, but the process is widely used in many countries; world production was 136 million metric tons of fixed nitrogen in 2008 (Figure 9.5). Nitrogen compounds have a variety of other important uses—plastics, fibers, resins, refrigerants, detonating agents for explosives, and the nitric acid widely used in the chemical industry. The availability of nitrogen from Earth's atmosphere for the production of fixed nitrogen (ammonia, NH_3) is unlimited. Because the hydrogen used to prepare the ammonia is derived from natural gas, there is a strong price dependence of fixed nitrogen on the cost of fossil fuels.

PHOSPHORUS

Phosphorus is indispensable for all forms of life because it plays a vital role in **deoxyribonucleic acid** (DNA), in **ribonucleic acid** (RNA), and in the ADP and ATP molecules that function in the energy cycle of living cells. In natural ecosystems, it is usually the availability of phosphorus that is life limiting.

Phosphorus, which is usually reported in its oxide form *phosphate*, P_2O_5, has a natural crustal abundance of about 0.23 percent. Most phosphorus is present as the mineral *apatite*, $Ca_5(PO_4)_3(F,Cl,OH)_2$. Apatite is a disseminated accessory mineral in many types of rocks, but it can occur in mineable concentrations in igneous rocks and in marine sedimentary rocks. Phosphorus is also present in the guano deposits left by birds and bats described in the discussion of nitrogen.

Apatite is the main ingredient in bones and teeth, and the first phosphate used as a fertilizer was in bone. English farmers, as early as the mid-1600s, had observed that applications of ground bone increased crop yield, and the Pilgrims learned from the Indians that buried fish carcasses and bones would help corn to grow. Although some nitrogen compounds are also present in bones, it is primarily the effect of the phosphorus that promotes growth. By the middle of the nineteenth century, European countries were importing bones, possibly some human, from every available source. The problem with apatite is its low solubility in water. Attempts to find a more soluble phosphate fertilizer were unsuccessful until 1842, when, as previously described, John Lawes developed **superphosphate** by treating bones with sulfuric acid. As a result of Lawes's success, there were fourteen fertilizer plants in Great Britain by 1853, and production had reached 100 metric tons per day by 1862. Lawes's techniques found rapid use with the development of phosphate-rock mining in France and England by 1850. In the United States, the first phosphate mining took place in South Carolina in 1867, but the major development of the industry dates from 1888 when the great deposits in Florida were discovered. Today, phosphate deposits are known in several states (Figure 9.6), but nearly 90 percent of American production comes from the large deposits in Florida and North Carolina.

The techniques used today to convert natural phosphate minerals into usable fertilizer are still based on John Lawes's experiments; the apatite in an ore is reacted with sulfuric acid to make superphosphate $[Ca_3(H_2PO_4)_2]$. Much is used in this form, but some is used as liquid fertilizers and some is transformed into triple superphosphates that have much higher P_2O_5 contents. Historically, the sedimentary phosphorites have been the world's dominant sources, but igneous bodies that contain large concentrations of apatite are slowly increasing in production, especially in Russia. The guano deposits of Chile and Peru, once major producers, have diminished in importance but still find demand among organic gardeners.

Phosphate rocks are found in many parts of the world (Figure 9.7), and major reserves are held and mined by many nations, but the major world suppliers of phosphate rock are China, the United States, Morocco, and Russia. The marine phosphorites, which constitute the principal reserves, presently supply approximately 80 percent of the world phosphate rock production and 100 percent of U.S. production. Although minor amounts of phosphate occur in nearly all marine sediments, major accumulations appear to have developed only where upwelling of cool phosphate-saturated seawater moved across shallow platforms and into near-coastal environments. Here

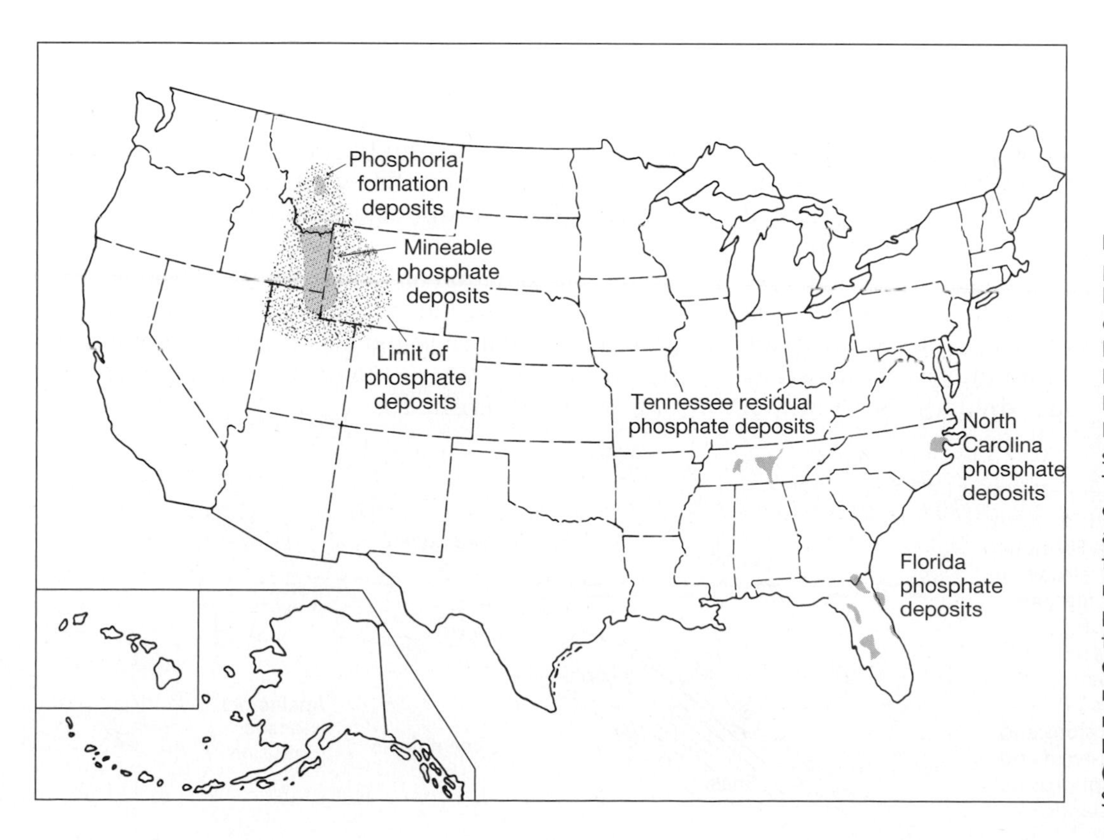

FIGURE 9.6 Major phosphate deposits have been worked in four areas of the United States. The beds of the Phosphoria Formation and those in North Carolina and part of Florida are primary marine sediments. The deposits in Tennessee and other parts of Florida are residual accumulations that result from weathering. Although nearly 90 percent of present production comes from Florida and North Carolina, the largest recoverable phosphate resources occur in the the Phosphoria Formation. (From U.S. Geological Survey Circular 888, 1984.)

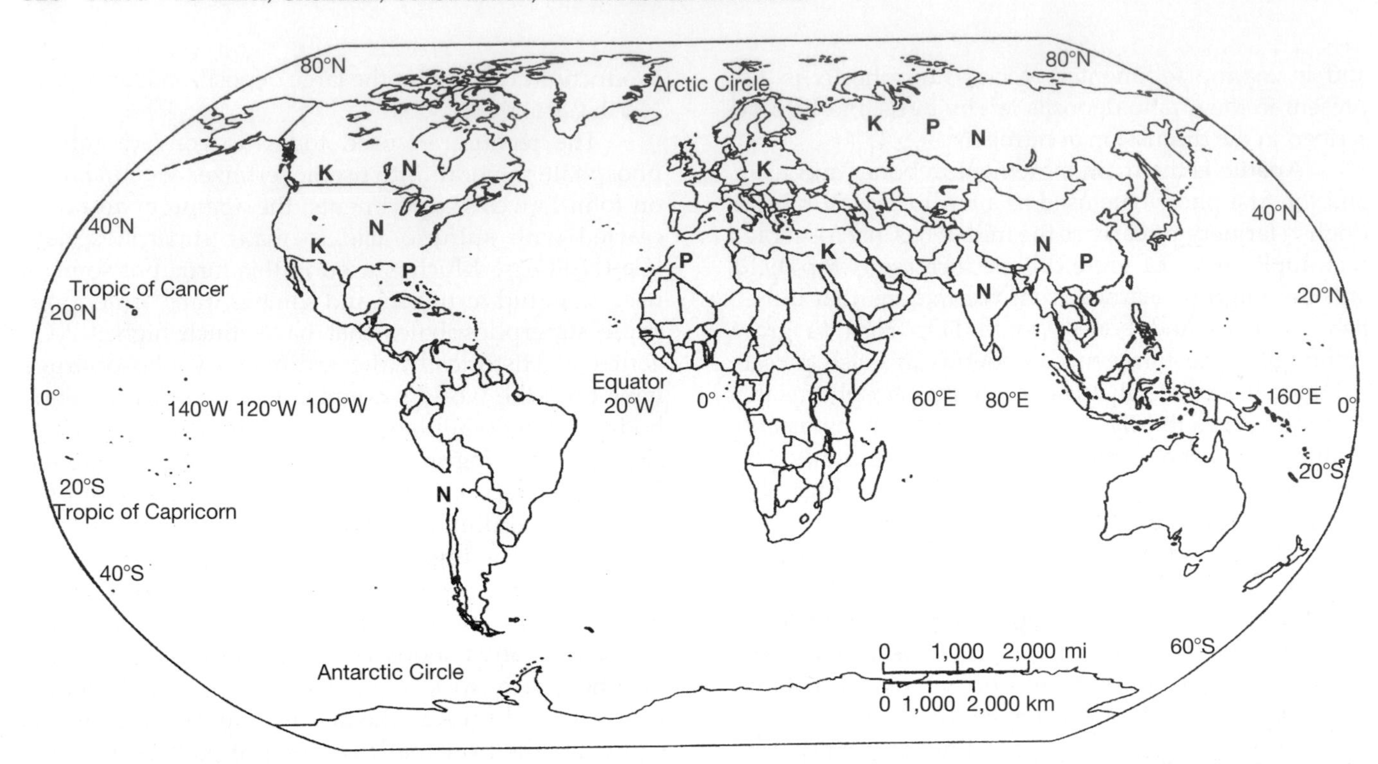

FIGURE 9.7 World map showing the locations of the major reserves of phospates and potassium salts extractable for fertilizers and the major producers of nitrogen-bearing fertilizers. The nitrogen-bearing fertilizers are now nearly all prepared using nitrogen from Earth's atmosphere and natural gas, so the only reserve of natural nitrates is in northern Chile.

the phosphate was precipitated, probably by complex microbiological processes, as microcrystalline muds, as nodules, and as hard crusts (Figure 9.8). In some areas, there are also vast accumulations of fish bones and teeth (Figure 9.9) that further contribute phosphate. Partial replacement of some calcite shells by apatite suggests that phosphate-bearing solutions also percolated through the sediments after deposition.

The largest of the marine phosphorite deposits occur in the Miocene-aged sediments of the North Carolina and Florida coastal plains of the United States, and in Morocco. These phosphorites are relatively thin (2 to 10 meters) beds that extend over wide areas (more than 2500 km^2 in Florida and at least 1200 km^2 in North Carolina). The sediments are mostly unconsolidated and are thus easily mined by using large drag lines and dredges as shown in Figure 9.10 and Plate 43. The North American deposits, which are one of the world's major producers, will continue to be major sources of phosphate rock, but they are the object of increasing environmental concerns. The three main problems are the vast amounts of groundwater pumped out of the mines and the underlying rock formations to permit mining at greater depths, the release of trace amounts of radioactive elements from the ores, and the generation of mountains of very fine-grained gypsum as a by-product of the sulfuric acid treatment. Another example of environmental problems is the small south Pacific island nation of Nauru, where phosphate rock was extracted without any ecological remediation plan. The once lush island was strip mined of its only resource, phosphate rock, and is now a barren, mined out, and impoverished state.

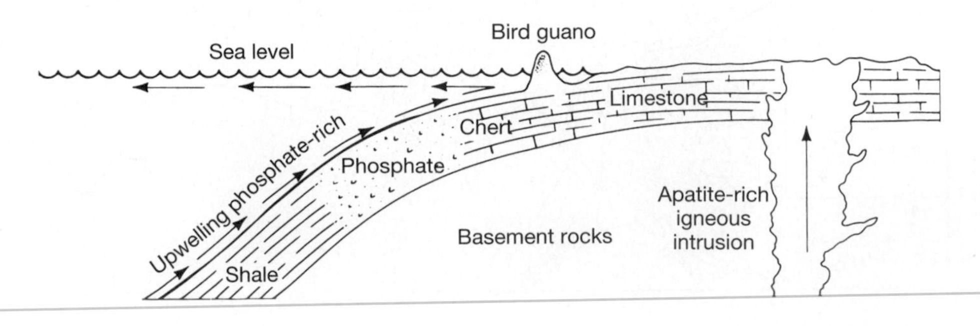

FIGURE 9.8 Marine sedimentary deposits such as those in Florida, North Carolina, and the Phosphoria Formation in the northern Rocky Mountains were deposited along continental margins where there was upwelling of phosphate-bearing ocean waters. The other major types of phosphate occurrences are in some unusual phosphate-rich igneous intrusions and where there have been long-term and large accumulations of phosphorus-rich bird guano on arid coastal islands.

FIGURE 9.9 The marine sedimentary phosphate deposits of North Carolina and Florida consist of unconsolidated pebbles and granules and contain abundant remains of fish, reptile, and mammal teeth and bones. (Photograph by J. R. Craig.)

Another particularly large (and unique), rock phosphate deposit was formed during Permian times in a shallow marine basin covering what are now parts of Idaho, Nevada, Utah, Colorado, Wyoming, and Montana. The phosphate-rich sediments, called the *Phosphoria Formation*, cover more than 160,000 km^2 and reach thicknesses of as much as 140 meters. However, over most of the area the thickness of the phosphatic bed is only 1 meter or less, and at best it can be considered only a potential resource. The tonnage, however, is enormous and is estimated at more than 2 billion metric tons.

There is no substitute for phosphate fertilizers; hence, the need for phosphate rock will continue to grow for at least the next 100 years in response to the need to feed the increasing world population. However, the world's phosphate reserves are large and, with the likelihood of new discoveries and of the technological advances to permit mining of lower-grade deposits, probably adequate for the twenty-first century. There will, however, be considerable change in world supply patterns because the relatively rapid depletion of the richest U.S. mines will force the United States, now a major exporter, to become an importer sometime in the twenty-first century. Unless there are major new discoveries, phosphate production in Florida, the principal U.S. producer, will begin dropping by about 2010. North Carolina will continue to produce significant amounts of phosphate rock, and minor production will be available from the western states.

Potential resources of phosphorus are large. Prospecting for phosphate deposits has not been thorough enough to be certain that all large deposits of phosphate have been discovered. In part, this is recognition of the economic difficulties entailed in opening new deposits in competition with existing mines. But in part, it also stems from the difficulty of recognizing

FIGURE 9.10 The phosphate deposits of Florida and North Carolina occur as flat-lying, unconsolidated beds that are mined by the use of large mobile drag-lines. After stripping off the overburden (the upper 3 meters), the phosphate-bearing ores (the 4 meters above the floor of the pit) are removed, and the land is returned to its original condition (see Plate 44). (Photograph by J. R. Craig.)

a phosphorus-rich rock. Even to an expert, many phosphate rocks look like ordinary shales and limestones. We can, therefore, probably anticipate discoveries of new deposits in the future.

One vast potential resource has already been discovered. The U.S. Geological Survey announced the finding of phosphatic crusts and nodules in the offshore continental shelf extension of the rich phosphate beds in Florida. Unfortunately, the deposits are lower grade than their landward equivalents, but the tonnages are large, probably as large as the present land reserves. We must conclude that the availability of phosphorus, like nitrogen and potassium, will not become a limitation to food production in the near future.

POTASSIUM

Potassium, the third of the important fertilizer elements, is the eighth most abundant element in the rocks of Earth's crust. Potassium occurs in nearly all rocks and soils, although its quantities vary widely. It occurs in igneous and metamorphic rocks primarily as potassium feldspar ($KAlSi_3O_8$); weathering releases the potassium, which is then incorporated in clay minerals such as illite [$KAl_2(Al,Si)_4O_{10}(OH)_2$]. The K^+ ion of the clay minerals is exchanged with plants by substitution of a hydrogen ion. Unlike nitrogen and phosphorus, the potassium does not form an integral part of plant components, but it is vital as a catalytic agent in numerous plant functions such as nitrogen metabolism, the synthesis of proteins, activation of enzymes, and maintenance of water content.

Although potassium occurs in most rocks, the only occurrences that can be economically extracted and processed into fertilizers are those in evaporite sequences. These formed by the evaporation of large amounts of seawater in broad basins (Figure 9.11) and are described in detail in the discussion of halite. Nearly complete evaporation results in the deposition of large amounts of halite (NaCl) and smaller amounts of a variety of potassium and potassium-magnesium salts, among which the most important are sylvite (KCl), langbeinite ($2MgSO_4 \cdot K_2SO_4$), kainite ($KCl \cdot MgSO_4 \cdot 3H_2O$), and carnallite ($KCl \cdot MgCl_2 \cdot 6H_2O$).

Because the potassium salts are very soluble, they are only preserved in arid regions or in salt beds that are buried below zones of groundwater flow. The evaporites generally occur as flat-lying beds that are now mined by rubber-tired diesel or electric mining machines (Figure 9.12). The salts are blasted or cut from walls in the mining areas and then brought to a surface refining facility where the ores are crushed and then separated into different minerals by a complex flotation system. The concentrates of potash minerals are then processed into a wide variety of solid and liquid fertilizers. Potassium salts may also be extracted from evaporites by solution mining, a process in which water is pumped down a drill hole, the salt dissolved, and the solution pumped back to the surface for processing.

Potassium, in the form of carbonates and hydroxides in wood ashes, was used as a fertilizer long before it was recognized in the 1840s as an element vital for plant growth. The name *potash* was derived from the custom of leaching wood ashes and then boiling the solutions in large iron pots to crystallize the soluble potassium salts, which were then used in soaps and in glass-making (see Figure 9.A). The first important mining of potash minerals began in 1857

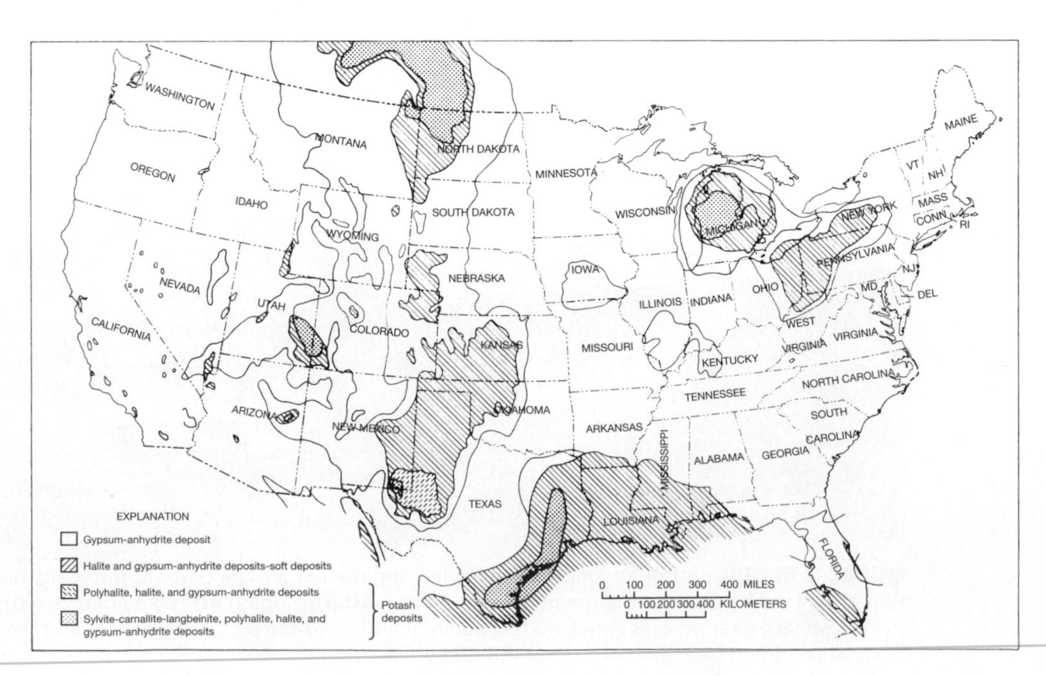

FIGURE 9.11 Large areas of the United States and southern Canada are underlain by major marine evaporite deposits that contain gypsum and anhydrite, halite, and potassium salts. These formed in the manner shown in Figures 9.16 and 9.17 and contain large quantities of potassium salts that are mined for the preparation of fertilizers. (After U.S. Geological Survey Bulletin 1019-J and U.S. Geological Survey Professional Paper 820.)

FIGURE 9.12 Salt mining from evaporite beds underlying the city of Cleveland, Ohio. Note the layering that probably represents annual cycles of salt deposition. The thicker white beds formed during the summers when there was more evaporation. The thinner beds represent periods of slower evaporation in winter and are dark because of the presence of minor amounts of silt and weathered organic matter. (Photograph by Burke and Smith Studios; courtesy of International Salt Company.)

when potassium chloride-bearing evaporite deposits were found at Strassfurt, Germany. These, and additional deposits discovered in the Alsace-Lorraine area, were controlled by a German cartel that had a virtual monopoly over the international trade in potash until 1915. In January of that year, Germany imposed an embargo on potash exports, and prices in the United States rose from their pre-embargo level of about $45 per ton to more than $480 per ton in 1916. Spurred by the shortage and the high prices, the United States stepped up production from wood ashes and discovered potassium-rich subsurface brines near Searles Lake in California. Following World War I, exports from Europe again became available, but the shortages during the war period had stimulated exploration and resulted in significant discoveries in Poland, Palestine, Spain, and Russia. In 1925, potash deposits were found near Carlsbad, New Mexico, by oil prospectors. Mines were opened in the 1930s, and soon these not only supplied the United States but became major exporters. High-grade deposits of potash were discovered in Saskatchewan, Canada, in the 1940s, again by oil drillers. These were not brought into production until 1958, but they now constitute the Western world's largest supplies and reserves.

The U.S. deposits are part of an extensive evaporite sequence underlying parts of New Mexico, Texas, Oklahoma, and Kansas where, in Permian times, a large and shallow sea deposited thick beds of evaporite salts over an area of at least 160,000 km^2. In a 4800-km^2 portion of the basin near Carlsbad, New Mexico, the sequence contains potassium salts in beds

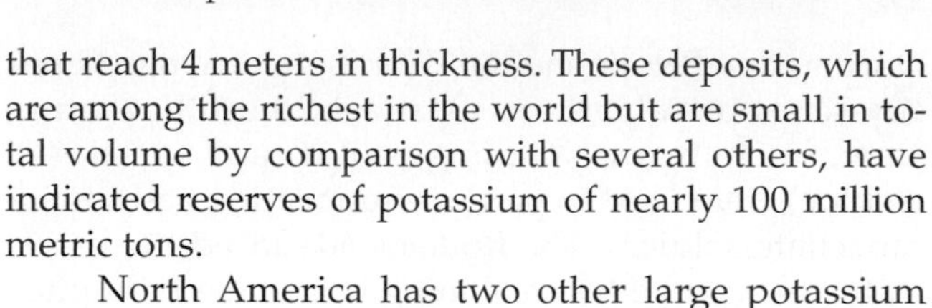

that reach 4 meters in thickness. These deposits, which are among the richest in the world but are small in total volume by comparison with several others, have indicated reserves of potassium of nearly 100 million metric tons.

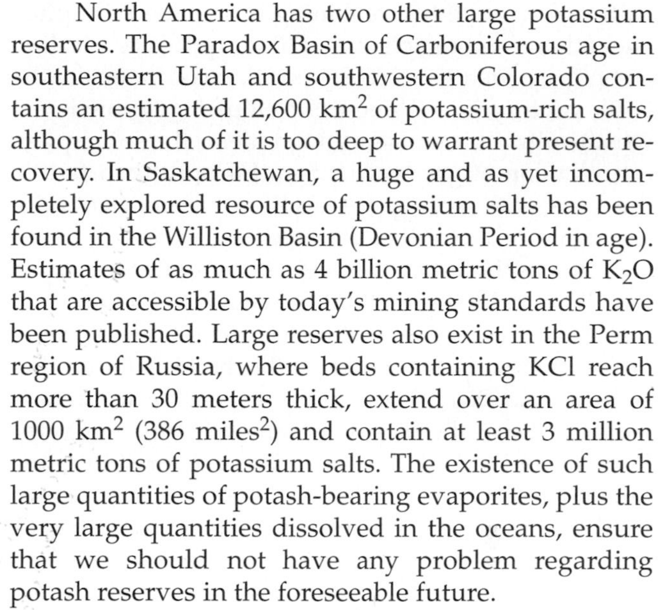

North America has two other large potassium reserves. The Paradox Basin of Carboniferous age in southeastern Utah and southwestern Colorado contains an estimated 12,600 km^2 of potassium-rich salts, although much of it is too deep to warrant present recovery. In Saskatchewan, a huge and as yet incompletely explored resource of potassium salts has been found in the Williston Basin (Devonian Period in age). Estimates of as much as 4 billion metric tons of K_2O that are accessible by today's mining standards have been published. Large reserves also exist in the Perm region of Russia, where beds containing KCl reach more than 30 meters thick, extend over an area of 1000 km^2 (386 miles^2) and contain at least 3 million metric tons of potassium salts. The existence of such large quantities of potash-bearing evaporites, plus the very large quantities dissolved in the oceans, ensure that we should not have any problem regarding potash reserves in the foreseeable future.

The application of fertilizer, intended to stimulate the growth of food crops or other desired plants has, unfortunately, had some very negative side effects. Over fertilization or excessive rainfall can result in significant amounts of fertilizer entering streams and rivers, lakes, seas, and oceans. There, the fertilizer stimulates the growth of algae, which, when they die, depletes the water of oxygen. This results in a "dead zone" in which normal lacustrine or marine biota cannot survive. There are many dead zones around the world, but the greatest is in the Gulf of Mexico off the Mississippi River delta. That zone has already severely damaged the Gulf fishing industry and is expected to increase in size over the next several years.

SULFUR

Sulfur, one of the first nonmetallic chemical elements discovered, is important because of its diverse uses. It is the fourth major fertilizer element, and the U.S. Bureau of Mines notes that "most products produced by industry require sulfur in one form or another during some stage of their manufacture." Sulfur is abundant on Earth's surface as native sulfur, in metal sulfides (especially pyrite, FeS_2), in mineral sulfates (primarily **gypsum**, $CaSO_4 \cdot 2H_2O$), as sulfate dissolved in the oceans, as hydrogen sulfide (sour gas) in natural gas, and as organic sulfur in petroleum and coal.

Sulfur was known in the ancient world as **brimstone**, the stone that burns, and has been used for thousands of years as a fumigant, a medicine, a bleaching agent, and as incense in exotic religious ceremonies.

During the Peloponnesian War between the Greek city-states of Athens and Sparta in the fifth century B.C.E., mixtures of burning sulfur and pitch (oil residue) were used to produce suffocating gases to incapacitate soldiers. The Romans advanced the use of sulfur in warfare by combining brimstone with pitch and other combustible materials to produce the first incendiary weapons. A thousand years later, in the tenth century, the Chinese developed gunpowder in which sulfur is a necessary ingredient; the subsequent introduction of gunpowder into European warfare in the fourteenth century made sulfur an important mineral commodity for the first time.

It was, however, the development of chemistry in the 1700s and the growth of the chemical industries in the 1800s that brought sulfur to its position of prominence in the modern world. Early chemists found that sulfuric acid was simple and inexpensive to prepare and that it was the most versatile of the mineral acids. Prior to the mid-1800s, world demand for sulfur was satisfied primarily by the native sulfur deposits in Sicily. A rise in demand and a large price increase as a result of the Sicilian monopoly resulted in a shift to pyrite (FeS_2) as a major sulfur source. Pyrite, when roasted in air, yields sulfur oxide gases that readily react with water to form sulfuric acid. In 1894, the **Frasch process** for mining subsurface native sulfur deposits associated with Gulf Coast salt domes was introduced. The Frasch process employs hot water to melt the sulfur from the host limestones and gypsum beds, and to transport it to the surface. Concentric pipes are emplaced in a 25-centimeter hole drilled into the sulfur-bearing rock. Hot water is passed down the outer pipe at 140°C to melt the sulfur; once molten, the sulfur is forced up the intermediate pipe by hot air forced down the inner pipe. The native sulfur recovered by the Frasch process occurs with anhydrite ($CaSO_4$) and gypsum on top of salt domes and in certain evaporite beds. Salt domes, the origin of which is described later, occur in many parts of the world, but they are especially abundant along the Gulf Coast of the United States from Alabama to Mexico (Figure 9.13). When the gypsum and anhydrite, which cap the salt domes, are brought into a near-surface environment, they are attacked by certain anaerobic bacteria.

The bacteria derive their oxygen from gypsum and their food from organic matter (commonly petroleum), and convert the gypsum into calcite ($CaCO_3$) and free sulfur. Only a small proportion of salt domes contain commercial quantities of native sulfur, but these few have served as valuable sources of sulfur. Today, Frasch sulfur production has just about ceased as it is no longer economic relative to sulfur recovery from oil and natural gas.

Since the 1940s, by-product sulfur recovered from petroleum refining, and from natural gas during treatment, has become increasingly important. Sulfur was originally removed from oil and gas merely to produce cleaner petroleum products and odorless gas, but the sulfur has proven to be useful and valuable itself. In the United States alone, about 8 million metric tons of sulfur were derived in this manner each year by the 1990s; this accounted for more than 90 percent of domestic production.

Most of the coal mined throughout the world contains some sulfur both as organically bound sulfur and as inclusions of sulfides (mostly pyrite and marcasite, both FeS_2). Burning of the coal results in the release of SO_2, which contributes to smog and acid rain (see Box 4.1.). Great effort is taken to produce "clean coal"

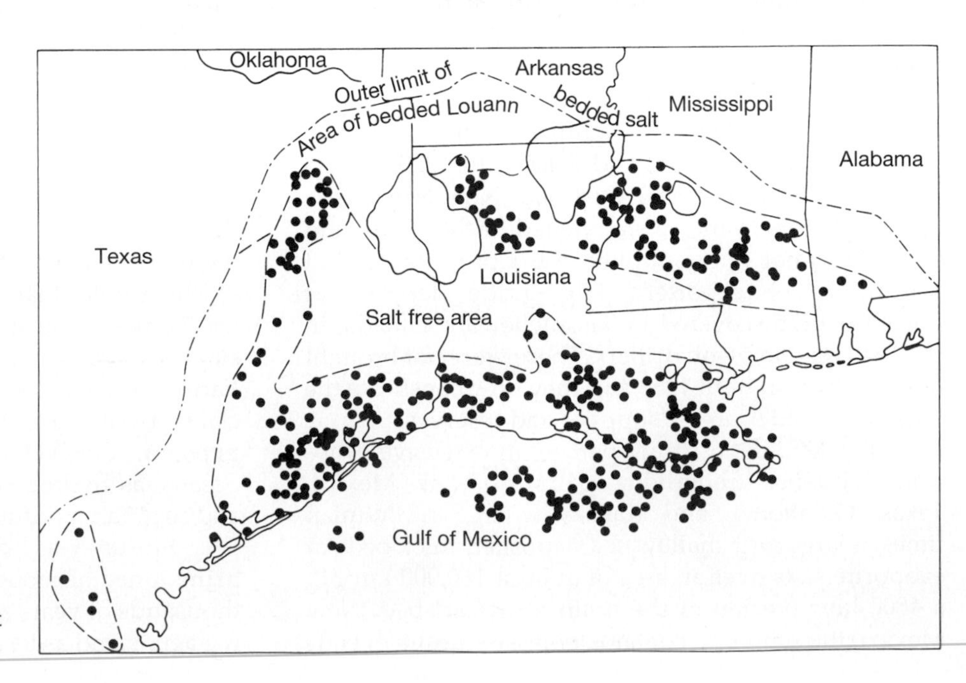

FIGURE 9.13 Salt domes, indicated by the black dots, underlie many parts of the Gulf Coast region of the United States. Individual domes may be 100 meters to more than 2 kilometers across and may have risen through as much as 12 kilometers of overlying marine sediments. The domes lie in distinct zones where the surface is underlain by the Louann Salt Bed that was deposited in an evaporite basin in the Permian age. The upturned rocks along the flanks of the salt domes have been rich sources of petroleum (see Figure 5.34) and cavities cut into the salt domes are now serving as the Strategic Petroleum Reserve. (Map courtesy of the Gulf Coast Geological Society.)

by removing the sulfur to reduce the environmental impact. Even then, the waste gases from the coal burning (mostly CO_2, SO_x, and NO_x) are passed though scrubbers to further reduce the noxious emissions. When possible, the sulfur is captured for industrial use.

Today, sulfur is used in a wide range of industrial applications and in a variety of chemical compounds. More than 90 percent of American domestic sulfur used is as sulfuric acid. The principal use of the acid is to convert insoluble natural phosphates into superphosphates that are more soluble and hence more useful for agricultural applications. Sulfuric acid or sulfur is also used in products such as soaps, rubber, plastics, acetate, cellophane, rayon, explosives, bleaches to make white paper and white titanium oxide paint pigment, leachates for copper and uranium ores, and the **pickling** or cleaning of the surface of steel products prior to further processing (Figure 9.14). New uses still in the developmental stage include sulfur-asphalt paving for highways and sulfur concretes for use in acid and brine-rich environments where salt attack leads to significant deterioration of conventional materials. World sulfur reserves are very large and will be adequate to meet needs for many years to come.

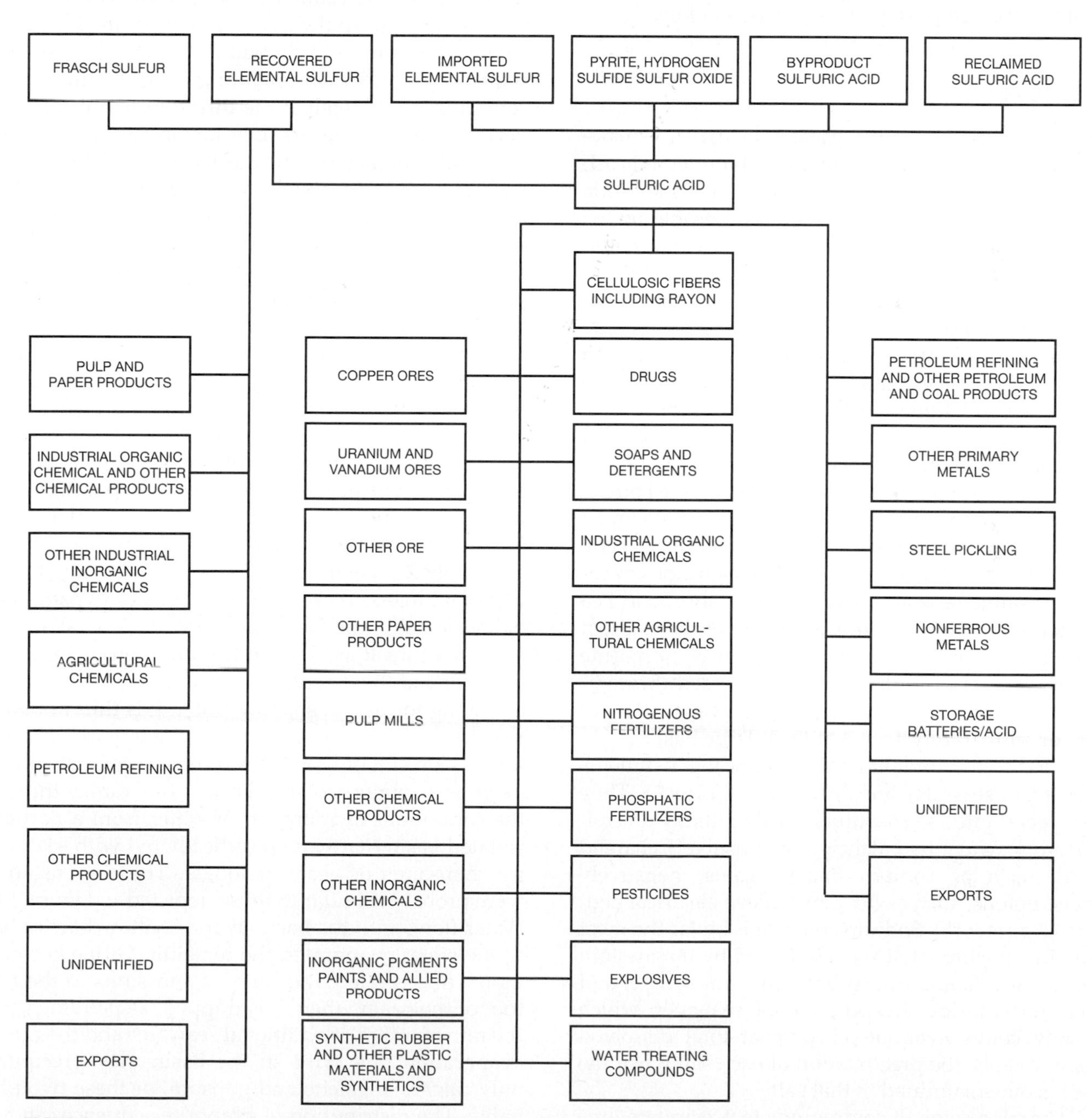

FIGURE 9.14 Sources and uses of sulfur in the United States at the beginning of the twenty-first century. Sulfur, as indicated by the complexity of this diagram, is one of the most important and widely used industrial chemicals. The largest single use, by far, is the preparation of phosphatic fertilizers. (From the U.S. Geological Survey.)

MINERALS FOR CHEMICALS

A large number of nonmetallic minerals serve as important raw material sources of elements or compounds used in the chemical industry. Many of these are little known to the public because the chemical products often bear no resemblance to the original source mineral or because the minerals are only used in processing and are not incorporated into the final product. The total list of chemical minerals is very long indeed and includes the fertilizer minerals already discussed. Table 9.2 lists several of the important chemical minerals and summarizes their uses; a few of the most important are discussed below.

Halite (NaCl)

Halite, or common salt, is a basic industrial raw material that serves as a source of sodium, chlorine, soda ash (Na_2CO_3), hydrochloric acid (HCl), caustic soda (NaOH), and other compounds indispensable in the manufacture of hundreds of other products and chemical reagents. Salt itself is important in food production and preservation, water softening, and snow and ice removal. It is essential to our diets (see Box 9.3), but only small amounts of total production are used for human consumption. The recognition that excessive salt usage is associated with hypertension has led to a reduction of salt levels in many foods. Much of the table salt used today contains about 0.01 percent potassium iodide as an additive to provide the iodine needed by the body. If the body does not have sufficient iodine, enlargement of the thyroid gland, a condition called *goiter*, can occur.

Halite (NaCl) occurs naturally in solution in seawater, in saline seas and lakes (such as the Dead Sea in Israel, the Salton Sea in California, and the Great Salt Lake in Utah), and as thick sequences of marine evaporites. Evaporite deposits have formed throughout geologic time as a result of the evaporation of ocean or, more rarely, lake waters in large basins.

The most abundant of the chemical constituents dissolved in seawater are shown in Figure 9.15. They can be recast into the constituents that actually precipitate from seawater by balancing the positively charged cations, such as sodium (Na^+), against negatively charged anions, such as (Cl^-), to achieve electrical neutrality (Figure 9.15). Sodium chloride is by far the most abundant constituent. This is followed by magnesium chloride and magnesium sulfate, calcium sulfate, and potassium chloride. Evaporation of seawater, which normally contains about 3.5 percent total dissolved salts, will cause the precipitation of each salt when the brine becomes saturated in that salt.

The succession of compounds to precipitate during progressive evaporation is shown in Figure 9.16. Calcium carbonate is the first substance to precipitate, but the quantity is very small. Once the volume of seawater has been reduced to only 19 percent of the starting amount, $CaSO_4$ or, depending on temperature, $CaSO_4{\cdot}2H_2O$, begins to precipitate. The most abundant of the dissolved salts, NaCl, the mineral halite, first begins to precipitate when the volume reaches about 9.5 percent of the original. When the volume is finally reduced to 4 percent, a complex magnesium and potassium salt called **polyhalite** ($K_2SO_4{\cdot}MgSO_4{\cdot}2CaSO_4{\cdot}2H_2O$) begins to crystallize. The amount of NaCl in solution is large to begin with, and considerably more than half of it will be precipitated during the reduction in solution volume from 9.5 to 4 percent, so the thickest layer formed during a single evaporation cycle will be the NaCl layer. The sequence of minerals separating from the final 4 percent of the brine (called the **bitterns**) is complex and variable, depending on such factors as the temperature and whether or not the final liquid remains in contact with, and hence can react with, the earlier-formed crystals. Two of the precipitates in the last stage, sylvite (KCl) and carnallite ($KCl{\cdot}MgCl_2{\cdot}6H_2O$), constitute the world's principal sources of soluble potassium used for fertilizers.

The complete evaporation of an isolated body of seawater should produce the sequence and volume of salts shown in Figure 9.16. However, examination of natural marine evaporite sequences nearly always reveals greater amounts of calcite, gypsum, and halite and only the rare presence of potassium and magnesium salts. Furthermore, complete evaporation of a body of seawater even as deep as the Mediterranean Sea, which averages about 1370 meters, would produce only 24 meters of halite and a layer of gypsum only 1.4 meters thick. However, beds of gypsum and halite several hundreds to more than 1000 meters (3200 feet) thick are known from numerous localities, and many of these contain fossil and textural evidence of having formed in shallow water. It is thus apparent that these thick marine evaporite sequences did not form as a result of a single evaporative episode in very deep and totally isolated basins, but rather through the continuous evaporation of water from a partially isolated basin that was episodically fed with seawater for thousands of years or longer. The circumstances were probably similar to those depicted in Figure 9.17. Water flows into the basin over a shallow barrier bar; as the water evaporates, the remaining brine becomes more concentrated and heavier and sinks to the bottom of the basin where it is trapped. Depending upon the rate of influx of additional seawater and the rate of evaporation, the brine in the basin may precipitate only calcite, or calcite and gypsum, or these two plus halite. The distribution of evaporite sequences shown in Figure 9.11 reveals that only in a few places has evaporative concentration been sufficient to result in

TABLE 9.2 Important mineral-derived chemicals and their uses*

Principal Element or Compound	Mineral Source	Chemical Products and Uses
Antimony	Stibnite (Sb_2S_3)	Flame retardants, batteries, glass, ceramics
	Tetrahedrite ($Cu_{12}Sb_4S_{13}$)	
Arsenic	Tennantite ($Cu_{12}As_4S_{13}$)	Wood preservatives, agricultural herbicides and desiccants, semiconductors
	Arsenopyrite (FeAsS)	
	Realgar (AsS)	
Bismuth	Bismuthite (Bi_2S_3)	Pharmaceuticals
Boron	Borax ($Na_2B_4O_7 \cdot 10H_2O$)	Glass products, detergents, fibers
	Kernite ($Na_2B_4O_7 \cdot 4H_2O$)	
	Brines	
Bromine	Brines	Gasoline additives, flame retardants
Cadmium	Minor component of sphalerite ((Zn, Fe)S)	Batteries, pigments, plastics
Chlorine	Halite (NaCl)	Plastics, water treatment, paper manufacture
	Brines	
Fluorine	Fluorite (CaF_2)	Steel and aluminum flux, welding rods, enamels, water fluoridation
Gallium	Minor component of sphalerite ((Zn,Fe)S)	Semiconductors, light-emitting diodes, lasers
Germanium	Minor component of sphalerite ((Zn,Fe)S)	Semiconductors, infrared optics, catalysts, phosphors
Indium	Residues from base metal refining	Alloys, nuclear reactor control rods, glass coating for liquid crystal displays
Iodine	Brines	Colorant in dyes, antibiotics, iodized salt
Lead	Galena (PbS)	Glass, paints, ceramics
Lime	Limestone ($CaCO_3$)	Agriculture, refractory
Lithium	Brine	Glass, ceramics, greases, batteries, aluminum production
	Lepidolite ($K(Li,Al)_3(Si,Al)_4O_{10}(F,OH)_2$)	
Mercury	Mercury (Hg)	Paints, fungicides, chlorine gas production
	Cinnabar (HgS)	
Nitrogen	Atmospheric (N_2)	Ammonia for fertilizers and chemical reagents, plastics, fibers, explosives (see text discussion)
	Nitre ($NaNO_3$)	
Phosphorus	Apatite ($Ca_5(PO_4)_3(F,OH,Cl)_2$	Phosphate fertilizers (see text discussion)
Potassium	Sylvite (KCl)	Potash fertilizers (see text discussion)
Rhenium	By-product of molybdenite (MoS_2)	Catalysts in gasoline production, thermocouples, flash bulbs
Rubidium	Minor element in lepidolite ($K(Li,Al)_3(Si,Al)_4O_{10}(F,OH)_2$)	Photochemical applications, medicines
Salt	Halite (NaCl)	Basis for many chemicals including chlorine, caustic soda, soda ash (see text discussion); food products, de-icing, water treatment, aluminum and steel manufacture, many other uses
	Brines	
Selenium	By-product of copper refining	Photocopiers, semiconductors, glasses, pigments, rubber compounds
Silver	Argentite (Ag_2S)	Black and white photographic film
	Tetrahedrite [$(Cu,Ag)_{12}Sb_4S_{13}$]	
Sodium	Halite (NaCl)	Detergents, glasses, pulp and paper manufacture
	Soda ash (Na_2CO_3)	
	Mirabilite ($Na_2SO_4 \cdot 10H_2O$)	
Sulfur	Sulfur (S)	Sulfuric acid for phosphate fertilizer production
	By-product of oil refining	
Zinc	Sphalerite (ZnS)	Alloys, paints, medical compounds

*Although most materials used primarily for alloys or building materials are omitted, there is some overlap with other chapters where, for example, metals are used both in the chemical and metallurgical industries.

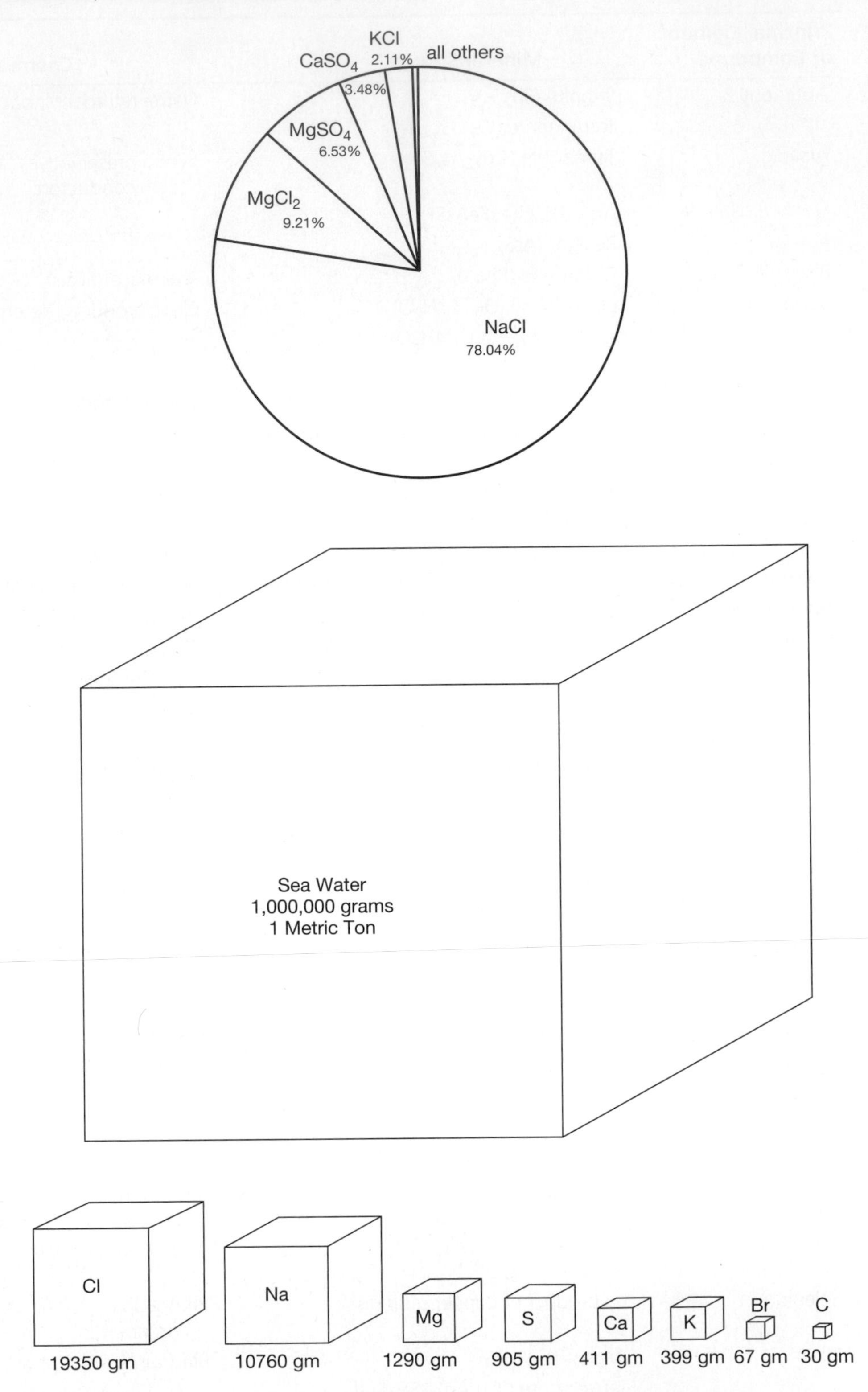

FIGURE 9.15 The constituents dissolved in seawater expressed in terms of elements and major minerals. The total amount of dissolved solids in seawater is about 3.5 percent and is remarkably uniform throughout the oceans. Sodium and chlorine are the two principal elements dissolved, and halite, NaCl, is the chief mineral deposited when seawater is evaporated to dryness.

the precipitation of potassium and magnesium salts as well as halite, calcite, and gypsum.

Evaporite deposits are widespread in the geologic record, both in time and space. High temperatures are important but not necessary for the evaporative concentration of salts. Indeed, the largest evaporites forming today are within 30° of the equator, but there are several small lakes in which salts are depositing in Antarctica. Most of these lakes, which owe their evaporative concentration to high winds and very low humidities, are rich in sodium chloride. One small Antarctic lake, named Don Juan Pond, is well known

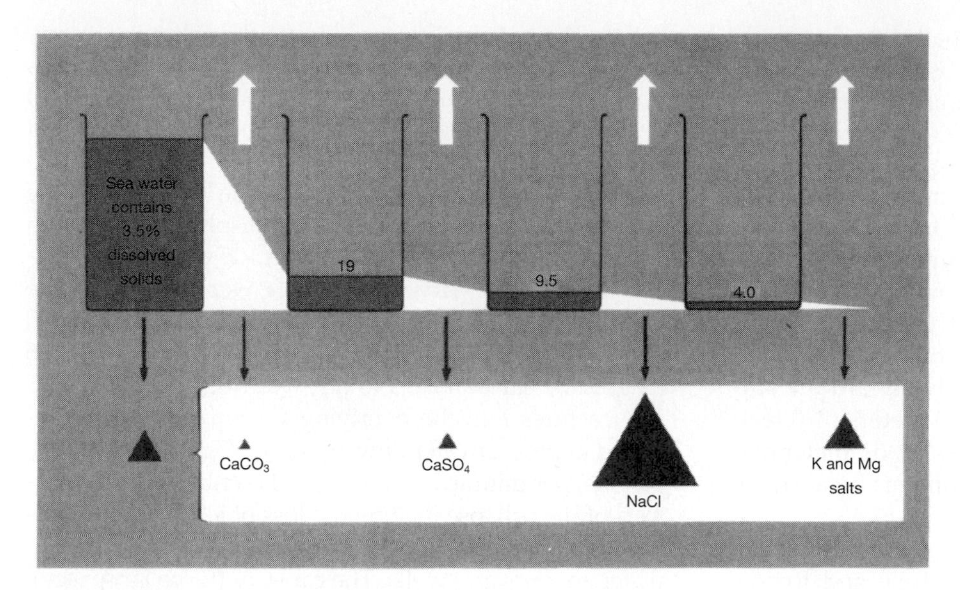

FIGURE 9.16 Schematic presentation of the sequence of minerals precipitated by the evaporation of seawater. When evaporation reduces the volume to 19 percent of the original quantity, gypsum begins to precipitate; at 9.5 percent, halite begins to precipitate; and at 4 percent, potassium and magnesium salts begin to precipitate. (After B. J. Skinner, *Earth Resources*, 3rd ed., Prentice-Hall, 1986.)

for being rich in calcium and being the only place on Earth where the evaporite mineral antarcticite ($CaCl_2{\cdot}H_2O$) exists.

There is no modern evaporite basin comparable with the major ones found preserved in the geologic record. The largest evaporite area today is the 18,000 km^2 Kara-Bogaz-Gol at the eastern edge of the Caspian Sea. At this site, the waters spilling into the Kara-Bogaz Basin are evaporatively concentrated so that sodium sulfate is precipitating on large flats. Numerous small flats, where halite is forming, exist along the shores of the Red Sea, and salt pans are well known adjacent to landlocked bodies of water such as the Great Salt Lake and the Dead Sea. Several of the large evaporite sequences mined today, no longer close to the equator, have been moved by plate tectonics from their site of origin.

The Mediterranean Sea, with its narrow inlet at Gibraltar and its location in a relatively arid region, is almost an evaporite basin. Here, evaporation increases the salt content and the density of the surface waters so that they begin to sink. But the Straits of Gibraltar are deep enough to permit these heavier waters to escape out of the Mediterranean as a westward-flowing bottom current before they are concentrated enough for salt deposition. Above westward-flowing salt-concentrated waters, less salty and less dense seawater flows into the Mediterranean as an eastward-moving surface current. If the channel at Gibraltar were shallower, the evaporatively concentrated water would not escape but would become further concentrated until salts precipitated.

Evaporite sequences containing halite occur worldwide. The problem is not availability of this resource but one of mining and shipping. Furthermore,

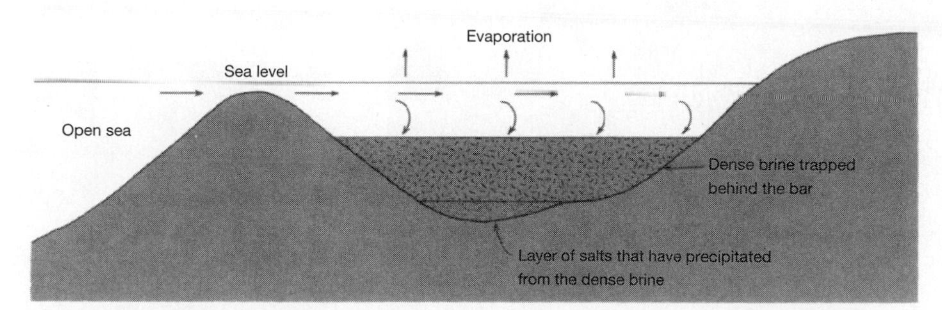

FIGURE 9.17 Cross section of a basin in which evaporite salts accumulate. Seawater containing 3.5 percent dissolved salts flows into the basin via a shallow inlet and is concentrated by evaporation. The concentrated seawater is more dense and sinks to the bottom allowing more seawater to flow into the basin. When the brines (the evaporated seawater) become sufficiently concentrated, salts begin to precipitate as described in Figure 9.16. In many of the major evaporite basins of the world, slow subsidence allowed for the constant inflow of seawater and the constant precipitation of salts for long periods of time, with the basin never being more than a few dozen meters deep. However, over long periods of time, hundreds to thousands of meters of salts could build up.

the world's oceans contain essentially inexhaustible quantities (a conservative estimate is 46×10^{15} metric tons). In the United States, salt is mined (Figure 9.12) from flat-lying evaporite sequences, such as those in Michigan, Kansas, and New Mexico, and from large remobilized salt domes in the sediments of the Gulf Coast area (Figure 9.13). Most of the bedded salt in the Gulf Coast region is too deep to mine, but, in hundreds of places, the salt has risen upward through the weak sediments as great columns. This occurs because the salt has a slightly lower density than the overlying rocks and has slowly risen buoyantly toward the surface. The salt domes range from 100 meters (330 feet) to more than 2 kilometers (1.25 miles) in diameter and have risen upward through the sediments by as much as 12 kilometers (7 miles).

Salt domes are known in many areas of the world: Europe, South America, the Middle East, and Russia. They are particularly abundant in the area on the north side of the Gulf of Mexico, as shown in Figure 9.13. Although the domes serve as an obvious site for salt mining, they also serve as major sources of sulfur extracted from the overlying cap rock by the Frasch method (described previously), and as major sources of petroleum that may occur trapped in upturned beds adjacent to the dome. The extraction of both oil and salt from the same dome has led to problems (see Box 9.2). In 1980, an oil drill operating from a barge in Lake Peigneur, Louisiana, a lake that lies directly above a buried salt dome, penetrated a salt mine some 430 meters (1410 feet) below. The entire lake drained into the salt mine, carrying with it the drill rig, several barges, many holly trees from a shoreline garden, and a tug boat. Fortunately, all drillers and miners escaped unharmed, but the well and the salt mine were lost.

In March 1994, a major collapse occurred in the largest salt mine in the Western Hemisphere, at Retsof in western New York. After more than 100 years of continuous mining in a "dry mine," there was a major failure that extended from the mine workings at a depth of 300 meters (1000 feet) to the surface; it registered as a 3.6 magnitude earthquake. No one was hurt, but the resulting fractures into the overlying water-bearing beds resulted in groundwater flow of nearly 75,000 liters (20,000 gallons) per minute into the mine. Potential long-term effects of the collapse include the loss of jobs of the miners and the widespread dispersal of salty water into freshwater aquifers and wells. The cause of the collapse is not known with certainty, but it may have been the introduction of a newer mining method, which removed 90 percent of the salt, leaving only 6-meter (20-foot) pillars supporting the central portions of mined-out rooms measuring 90 meters (300 feet) by 365 meters (1200 feet).

Salt production world-wide now exceeds 250 million metric tons per year with roughly one fifth of it coming from the United States. Production ranges from mining of salt in beds or salt domes by modern diesel equipment to the harvesting of salt from evaporating seawater or brines in large fields or terraced ponds, as shown in Figure 9.18. Approximately 45 percent of salt

(a)

(b)

FIGURE 9.18 Simple solar evaporation of brines and seawater has served as a source of salt for thousands of years. (a) Terraced evaporation ponds in the mountains of Peru. (Photograph courtesy of V. Benavides, Geological Society of America Special Paper 88.) (b) Salt being harvested in a broad evaporation pond in Colombia. (Photograph courtesy of the Colombian Government Tourist Office.)

is used in the chemical industry for the manufacture of chlorine gas and sodium hydroxide, and 20–30 percent of the salt is used for de-icing depending upon the severity of the winter. Although the most visible use of salt for most people is in their salt shakers, the consumption of salt in all food products represents less than 6 percent of total usage. Few data are available on actual reserves, but all producing countries are well-endowed. Because of the vast quantities of bedded evaporites, the enormous amounts of salt dissolved in the oceans, and the simplicity of salt extraction, the world's supply of salt is inexhaustible. A greater concern than the availability of salt is the environmental impact of salty waters released from chemical processing, or as runoff into rivers and lakes when salt is used for highway de-icing.

Soda Ash (Na_2CO_3, trona) and Sodium Sulfate (Na_2SO_4)

Sodium carbonate (trona) and sodium sulfate (thenardite) are chemical minerals widely used in the manufacture of glass, soaps, dyes, detergents, insecticides, paper, and in water treatment. Sodium bicarbonate ($NaHCO_3$) is the common household baking soda. Natural **soda ash,** as the carbonate is commonly called, was probably first derived from mineral crusts around alkaline lakes in southern Egypt in biblical times. The early Egyptians and Romans used it to make glass, as a medicine, and in bread-making. Until the eighteenth century, it was primarily obtained by leaching the ashes of burned seaweed, but in 1791 a process was developed to prepare it from halite and sulfuric acid. In the 1860s, a more efficient process was developed to prepare soda ash from salt, coke, limestone, and ammonia. This process became a major source of soda ash for many years, but the discovery of large deposits in the western United States made the recovery of natural materials cheaper than synthesis.

Sodium carbonate and sodium sulfate are evaporite minerals that form in some arid region lakes where the weathering of rocks releases abundant sodium and, sometimes, sulfur. Modern examples of such areas include the Searles, Owens, and Mono lakes of California, and Lake Magadi in Kenya. The lakes supply some of these sodium salts, but the bulk of production comes from bedded deposits formed from preexisting lakes. There are more than sixty identified sodium carbonate deposits in the world, the largest of which is in southwestern Wyoming. One unit in the Green River Formation, Wyoming, contains 42 beds of trona, 25 of which have a thickness of 1 meter or more. Eleven of these beds are more than 2 meters thick and underlie an area of more than 2850 km^2. These beds alone contain more than 52×10^9 metric tons of soda ash, of which more than 22×10^9 metric tons are reserves. These reserves are enough to meet U.S. needs for more than 700 years. The American reserves are much larger than those of any other part of the world, but several alkaline lakes in eastern Africa contain vast quantities of soda ash. These deposits and other insufficiently evaluated ones in South America and Asia will likely supply the world's needs for many years.

The reserves of sodium sulfate worldwide are much smaller than those of sodium carbonate but are also sufficient to meet global needs for at least 600 years. Commercial sources in the United States include shallow subsurface brines in west Texas, Searles Lake in California, and the Great Salt Lake in Utah.

Boron

Natural boron minerals are very restricted in their geologic occurrence. Nevertheless, we find boron compounds in glass products, insulation, laundry detergents, food preservatives, fire retardants, and ceramic glazes and enamels. The trade in boron compounds dates from the thirteenth century when Marco Polo brought borax ($Na_2B_2O_4$) crystals from Tibet to Europe. The discovery of natural boric acid (H_3BO_3, the mineral *sassolite*) in the hot springs of Tuscany, Italy, in 1771 led to the development of an industry that supplied most of the world markets from the 1820s into the 1870s. This market was superseded by Chilean production in the latter part of the nineteenth century. Borax crystals were found in springs north of San Francisco, California, in 1864, but the modern U.S. boron industry is based upon large deposits subsequently discovered in Nevada and southern California. Deposits developed in Death Valley, California, between 1881 and 1889 had their borax minerals transported by the celebrated 20-mule teams (Figure 9.19).

Although limited to a few deposits where volcanically derived fluids have concentrated and formed boron-bearing minerals, there are economic deposits in the United States, Russia, China, Turkey, Chile, and Argentina, with reserves of more than 150 million metric tons of boron oxide content. This will be sufficient to meet the world's needs far into the twenty-first century. Turkey has the largest reserves but the United States and Russia are also well endowed.

Fluorine

Fluorine compounds are of vital importance to the production of steel and aluminum, and they are used in the production of the uranium fuel for nuclear power plants. Fluorine compounds are also used in ceramics, water fluoridation agents to reduce dental cavities, Teflon to line cooking pans, and in experimental artificial blood substitutes for humans. Nearly all fluorine is derived from the mineral **fluorite** (CaF_2), which occurs in minor amounts in many hydrothermal ore deposits and in relatively rich limestone-hosted deposits formed by low-temperature, high-salinity brines forced out of

FIGURE 9.19 The celebrated 20-mule team wagons carried boron minerals from the deposits in Death Valley, California, in the 1880s. (Photograph courtesy of United States Borax & Chemical Corporation.)

sedimentary basins. The fluorite in these latter deposits is commonly associated with lead and zinc mineralization and with barite. World fluorite reserves are widespread and total about 300 million metric tons, enough to last well into the twenty-first century. The United States, though processing large fluorite resources, has very limited reserves and currently relies heavily upon Mexico, the world's largest producer.

FERTILIZER AND CHEMICAL MINERALS IN THE FUTURE

The demand for the fertilizer minerals will surely increase over the next 100 years because fertilizers will be needed to grow the food required for an expanding world population. It is also likely that the use of minerals in the chemical industry will continue to increase and to diversify into many applications in foods, supplements, toothpastes, and cosmetics (see Box 9.3 and Table 9.3). Worldwide reserves of the fertilizer and chemical mineral resources are large and certainly appear adequate to meet these needs for at least a century. It is likely, however, that production patterns will change as less developed countries increase production and become competitors for the advanced countries. Among the nonmetallic resources, the fertilizer and chemical minerals are the most vital, but the amounts used are small when compared with the building materials and industrial minerals to which we now turn.

BOX 9.2

Lake Peigneur, Where Oil and Salt Did Not Mix

It is unfortunate but true that the exploitation of a resource may have adverse effects on the environment, and that the search for one resource may impact on the availability of another. Such was the case at Lake Peigneur in southern Louisiana in November 1980, a case that provides an object lesson for care to be taken when resources are exploited.

The area in question is flat and marshy, but it has a gentle hill called Jefferson Island where a large pillar or dome of salt comes close enough to the surface to bow up the land (Figure 9.B). Adjacent to the hill lies Lake Peigneur, a shallow lake connected to the Gulf of Mexico by a canal. In 1920, a mine shaft was sunk into the salt dome. By 1980, some 50 million tons of salt had been extracted, the shaft had reached 600 meters (1800 feet) in depth, and there were many miles of mine workings. The upturned layers of sediment adjacent to salt domes commonly form oil traps (see also Figure 5.35) and many millions of barrels of oil and gas have been produced from the flanks of domes similar to the Jefferson Island dome. Oil drilling into the upturned rocks around the Jefferson Island salt dome began in November 1980 using a small platform located in Lake Peigneur. The operation was uneventful until about 5:00 a.m. on November 20 when the drill pipe

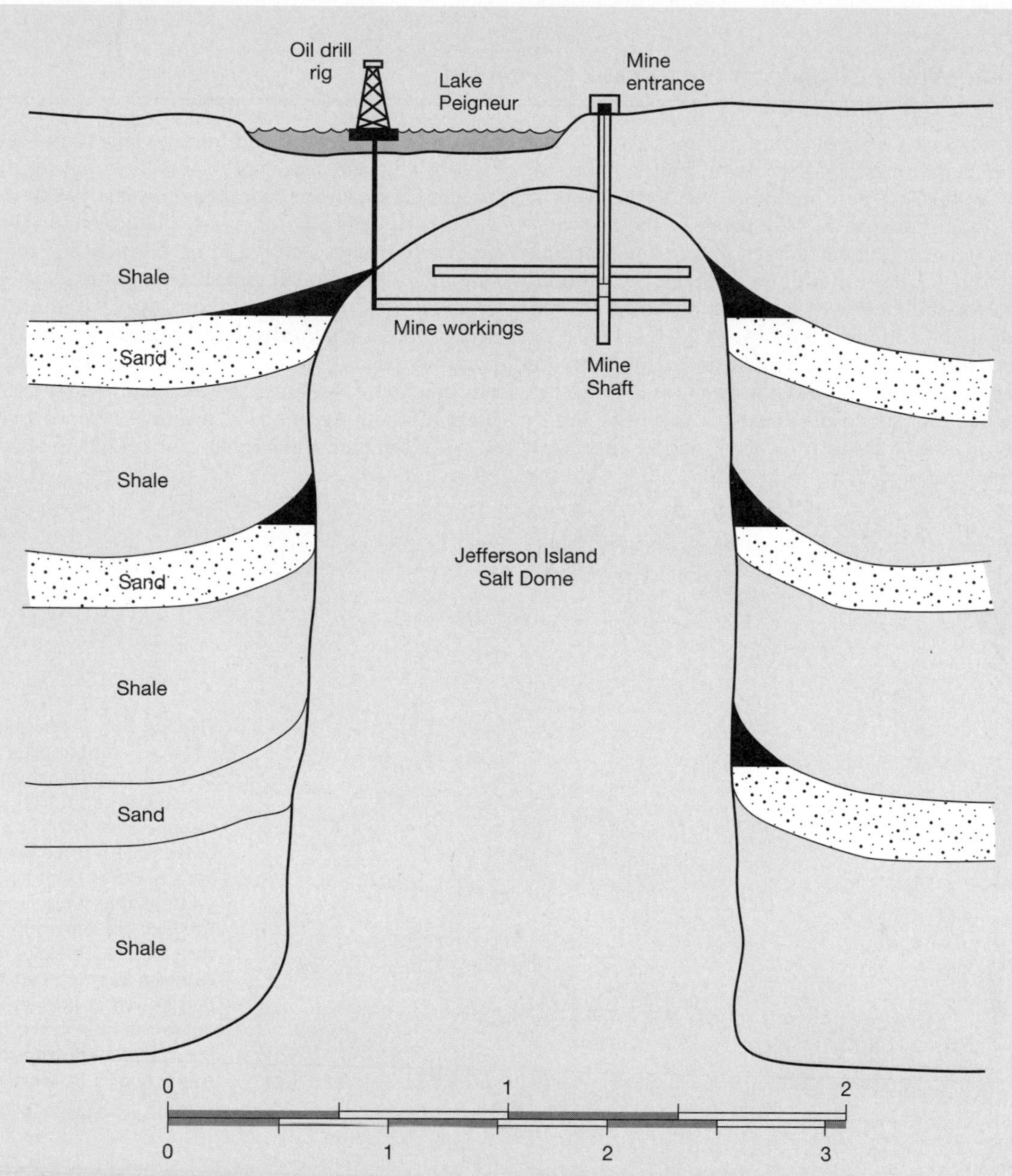

FIGURE 9.B The Jefferson Island salt dome, shown in schematic cross section, is approximately a mile in diameter and has risen more than five miles above the top of the original salt bed. The Diamond Crystal Salt Mine operated in the uppermost 1500 feet of the salt dome, and the oil drill rig in Lake Peigneur was attempting to locate oil adjacent to the dome in the upturned beds.

became stuck at a depth of about 1250 feet. As the crew attempted to pull the pipe loose, the pipe bounced, the platform shook, and then began to list. The drill crew loaded onto small boats and reached shore just in time to see the drill platform fall over and disappear from sight—in a lake that was supposed to be only 1–2 meters deep (Figure 9.C)!

At about the same time, salt miners were descending into the mine to begin the day's production. They soon found that the bottom levels of the mine were filled with muddy water that was rising rapidly. Responding to the danger, the miners exited the mine and all reached the surface safely by 9:00 a.m. Apparently, the drill either encountered a fault that extended into the mine area or penetrated the mine workings. Drilling fluids and then the lake water poured through the drill hole, enlarging it so that the flow rapidly increased. The effect was much like that of pulling a plug in a bathtub full of water.

At about 11:00 a.m., two men enjoying early morning fishing in a small boat realized that the lake level was dropping; they witnessed the development of a whirlpool into

(continued)

BOX 9.2

Lake Peigneur, Where Oil and Salt Did Not Mix (*Continued*)

which water, fish, and several moored barges were being drawn. The fisherman escaped to shore. Water normally flowed from the lake out into the canal and to the Gulf of Mexico, but the draining of the lake reversed the flow and began to draw in shrimp boats moored in the canal. A small tugboat tried to block the water flow with a barge, but both the barge and the tugboat were sucked in and disappeared down the hole in the bottom of the lake; the crew escaped by diving overboard into the mud. However, by the time the lake was drained, a dozen barges had been pulled into the swirling hole like toys down the drain in a washtub. The final loss also included some fifty acres of the Live Oak Gardens, a botanical garden on the lakeshore, along with the owner's house, which slid into the lake as the shoreline collapsed. The last gasp was a short-lived geyser that spouted air and muddy water from the mine shaft at about 2:30 p.m.

Fortunately, no one was killed or injured, but the sequence of events resulted in no oil recovery, the loss of the salt mine, one tugboat, three barges (nine of the other twelve originally lost resurfaced like corks after three days), fifty acres of botanical gardens, and one home. The drilling company was also subject to a protracted lawsuit. Although the oil company never admitted blame, it did pay $45 million to those affected. Accidents in resource exploitation are bound to occur but many, like the one at Lake Peigneur, are probably avoidable.

FIGURE 9.C Aerial photograph of Lake Peigneur after it had drained into the salt mine. The lake waters and drill rig have disappeared, leaving a mud bottom with some grounded barges. Portions of holly tree gardens that have slumped into the lake are shown in the foreground. The lake refilled within a few days, and several barges, which had disappeared down the hole at the drill site, resurfaced. (Photograph courtesy of J. D. Martinez.)

BOX 9.3

Minerals in Foods, Medicines, and Cosmetics

Many mineral commodities are very visible in their typical uses. We can all picture mineral products that are merely roughly shaped or cut, such as stone walls and carved statues, and many that are quite highly processed, such as concrete in highways, glass, and steel. In addition to these, there are many other minerals that are very widely used but are much less visible. Various minerals and their synthetic analogues are used in the production of foods, in medicines, and in cosmetics (Table 9.3).

In foods, minerals are used to enhance flavor, help preserve, impart color, or modify physical characteristics. The most common mineral substance used in foods is halite, common salt (NaCl), which has been used as a flavor enhancer for thousands of years. Salt was also very widely used as a food preservative before refrigeration was developed. Although less commonly used in that manner today, many salted products remain available. Baking soda, $NaHCO_3$ (equivalent to the mineral nahcolite) is widely used in many types of baking to help develop the desired porous texture of breads, cookies, or cakes. It is prepared from naturally occurring Na_2CO_3. More visible, but rarely recognized, is the use of synthetic anatase (TiO_2) in white powdered sugar on donuts and other foods. The anatase possesses a very fine white opacity that is preserved even when adhering to a moist donut, but it has no flavor, no calories, and is inert in the human body. Several forms of aluminum silicates and

TABLE 9.3 Minerals used in foods, medicines, and cosmetics

Foods	
NaCl—halite—common salt	Sodium and calcium phosphates
$NaHCO_3$—nahcolite—baking soda	SiO_2—silica
TiO_2–anatase	MgO—periclase
$AlxSi_yO_4$—aluminum silicates	$CaSO_4$—anhydrite
Supplements	
$CaCO_3$—calcite	$(Ca,Mg)CO_3$—dolomite
KCl–sylvite	
Toothpaste	
SiO_2—silica	$CaCO_3$—calcite
$CaHPO_4{\cdot}2H_2O$—brushite	NaF—villiaumite
$NaHCO_3$—nahcolite	mica
TiO_2—anatase	
Cosmetics	
kaolinite	zinc-oxide
montmorillonite	titanium oxide
bentonite	iron oxide
talc	mica
calcium carbonate	metal powders

silica (SiO_2) are widely used in materials such as nondairy coffee creamers to prevent them from caking into one big lump owing to absorbed moisture.

Intermediate between foods and medicines are food additives, food supplements, and toothpastes. Many stores today offer a vast array of supplements with elements such as cobalt, iron, selenium, and zinc. These elements are generally bound in some form of a soluble organic compound so that they may be readily absorbed by the body. In contrast, calcium carbonate ($CaCO_3$) and dolomite [$CaMa(CO_3)_2$] tablets, even if synthetically prepared, are virtually identical to the vast quantities of these minerals that occur worldwide in limestones and dolostones. Toothpastes, carefully designed to be gentle on mouth tissues, typically contain one or more of a variety of abrasive materials to help remove organic deposits, stains, and bacteria from dental surfaces. Among the most commonly used abrasives today are silica, calcium carbonate, several calcium salts, and sodium bicarbonate. Sparkle and brightness are sometimes added by including mica or TiO_2-coated mica. Fluoride, to help harden teeth, is added in the form of sodium fluoride or sodium fluorophosphates.

Natural and synthetic mineral compounds, especially carbonate and bicarbonates, have long been used to neutralize stomach acids and to generally control pH. Zinc oxide is widely used to promote healing either by itself or mixed with other compounds. Long before modern medicine, various peoples had discovered the medicinal properties of some petroleum products. Today, petroleum jelly is commonly used on rashes, burns, and minor skin complaints. Barite, $BaSO_4$, is not medicinal itself but is widely used in X-ray imaging to enhance the visibility of some body organs and vessels.

Thousands of years before the development of the modern cosmetic industry, our ancestors adorned their faces and bodies with a variety of colored mineral pigments, especially red and yellow oxides and hydroxides of iron. Today's cosmetics are much more complex and contain an array of special organic oils, emulsions, dyes, and scents, but they still commonly contain clays, such as kaolin, bentonite, or montmorillonite, and other minerals such as talc, calcium or sodium carbonate, calcium silicate, and silica, to provide bulk and to help to cover blemishes, absorb oils, and control pH. In addition, cosmetics often contain color additives such as titanium oxide, iron oxides, zinc oxide, ultramarines (originally prepared by powdering the gem lapis lazuli, but now usually synthetic), green chrome oxides, mica, and even powdered metals such as aluminum, copper, or bronze.

The minerals that we use in foods and medicines serve importantly to enhance the quality and flavor or promote health but are generally invisible. In contrast, the minerals used in cosmetics are intentionally visible but generally remain unrecognized in terms of their true nature.

The Sphinx and the Great Pyramids attest to the abilities of the early builders and to the durability of natural stone as a building material. (Courtesy of the Egyptian Tourist Authority.)

CHAPTER 10

Building Materials and Other Industrial Minerals

Stone has been used since the earliest days of our civilization, first as a tool or weapon, then as construction material, and later, in its crushed form as one of the basic raw materials for a wide variety of uses ranging from agriculture and chemicals to complex industrial processes.

CRUSHED STONE; STATISTICAL COMPENDIUM, *USGS, 1998*

FOCAL POINTS

- Building materials are the largest volume solid mineral commodities extracted from Earth.
- Most building materials have relatively low intrinsic value. Many different types of material are used and local materials generally provide the main supplies.
- The treatment of building materials by cutting, polishing, refining, or calcining markedly increases their value.
- Crushed stone, used primarily for roads, building foundations, and concrete, is the most widely used of building materials; limestone is the principal rock type employed, but many other types are also used.
- Cement is the most important treated rock product and is prepared by heating limestone with small amounts of clay and silica; although cement was used by the Romans, the formula was lost and not rediscovered until 1756.
- Gypsum is widely used in the manufacture of plaster of Paris and plasterboard.
- Clays of various types are used to prepare ceramic materials ranging from bricks to fine china.
- Glasses are made mainly from quartz, but various materials such as borax and alumina are added to vary the properties.
- Asbestos is one of the most widely known industrial minerals because of its association with health problems. The six asbestos minerals form strong, flexible, and nonreactive fibers that have many industrial uses. The recognition of health problems has dramatically reduced the use of asbestos.
- Gemstones are particularly unusual and rare forms of relatively common substances.
- Natural diamonds are formed in Earth's mantle at depths greater than 150 kilometers (95 miles) and have been brought to Earth's surface in kimberlite pipes that formed as explosive vents.
- Synthetic diamonds are now produced on a large scale, and they are being used in applications ranging from abrasives to coatings and electronic "chips."

INTRODUCTION

Building materials are by far the largest volume solid mineral commodity that we extract from Earth. They rank second only to the fossil fuels in value. Almost every known rock type has contributed in some way to the construction of homes, civic buildings, roads, bridges, dams, or air fields.

Most building materials, unlike metals, fuels, or fertilizers and chemical minerals, have relatively little intrinsic value. They are both abundant and widely distributed, and it is only after they are removed from Earth and processed to more useful forms that their value increases, often many times over. Thus, for example, the limestone and shale used to make cement may have values of $10 per metric ton or less in the ground, but after mining, crushing, firing, and conversion to a high-quality cement, the product is worth $100 or more per ton. As Figure 10.1 shows, the processing of building stone, or of the clays used in brickmaking and ceramic products, adds even more to their value.

Because building materials are widely distributed and of low value, the factors controlling their exploitation and marketing are very different from those for high cost commodities such as fuels or metals. Normally, mining operations for building materials are undertaken only to satisfy a local demand because the cost of transportation for any great distance is rarely justified. For many of the materials discussed in this chapter, the volume of the trade is large, but it is conducted mainly at local levels. Only in the case of special ornamental stones is the expense of distant transport justified. Therefore, discussion of the contributions made by particular nations is not relevant, as demand is commonly satisfied internally. Reserves of building materials tend to be large, and potential resources are even larger. Reserves of building stone are so vast that it is pointless to even attempt putting numbers on them. The wide distribution of building materials also means there is little point in detailed discussion of the geology of particular deposits. This is not true, however, of a few of the more specialized building materials such as vermiculite and perlite, or of some of the major industrial minerals.

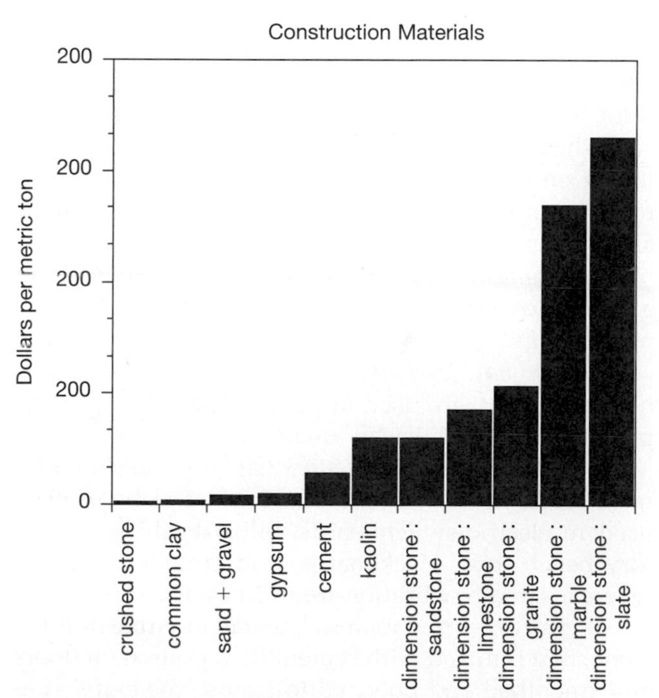

FIGURE 10.1 The natural rocks from which building materials are extracted have very little value. However, the processing of the original materials by cutting, shaping, calcining, or other means adds value. (Data from the U.S. Geological Survey.)

In this chapter, major types of building material are discussed, emphasizing the character of the raw materials, problems involved with their extraction and processing, the uses to which they are put, and their limitations. As explained in the following sections, the building materials are separated for discussion into two groups—*untreated* and *treated* rock products. Also discussed in this chapter are other major industrial minerals that do not readily fall into the categories of fertilizers and chemical minerals dealt with in the previous chapter. These include minerals valued because of particular properties they possess, such as great hardness, resistance to high temperatures, inertness, distinctive color, or high density.

UNTREATED ROCK PRODUCTS

In discussing building materials, it is convenient to distinguish between those materials that have simply been quarried or mined and either used directly or cut or crushed to a suitable size, and those materials that have been subjected to more complex physical and chemical treatments and that change the finished product into something very different from the raw material. The former we shall call *untreated* rock products and we discuss them first. *Treated* rock products are discussed later in this chapter. The two major categories of untreated rock products are the various types of **building stones,** and the **aggregates,** which include **crushed rock,** sand, and gravel, and **lightweight aggregates** (although the last of these commonly are subjected to heat treatment). Aggregates are used mainly in concrete, as highway or railroad base, as ballast materials, or as graded fill on construction sites.

Building Stone

Natural stone is one of the oldest building materials known to humans. Historically, it has had the advantages of being widely available, durable, easy to maintain, and pleasing in appearance. Consequently, natural stone has provided shelter from the elements, safety from enemies and other dangers throughout human history, and also provided infrastructure for transportation and media for artistic expression. In the simplest of shelters, humans used natural caves for protection; a natural extension was the carving out of shelters within rocks that were soft enough to cut but strong enough to withstand the weather. Numerous dwellings have been cut into the thick loess deposits in China, but perhaps the best examples are those carved into volcanic ash beds in Cappadocia in Turkey [Figure 10.2(a) and Plate 50]. Similarly, some of the world's great statuary, such as Mount Rushmore in South Dakota, has been carved into the sides of mountains. Our ancestors constructed dwellings and ceremonial sites from many materials such as reeds, wood, clay, and stone, but it was generally only the stone edifices that have survived (Figures 10.3, 10.4, and 10.5). However, throughout the twentieth century, the use of natural stone has declined in the face of competition from processed materials. In the United States, for example, at the beginning of the twentieth century, more than half of the stone produced was so-called dimension stone—stone that was quarried, cut, and finished to a predetermined size. By the 1990s, dimension stone accounted for only 0.1 percent of the stone that was quarried. Before considering the types of rock that are commonly used, the various categories of building stone, and the special names given to them, will be outlined.

FIGURE 10.2 (a) The volcanic ash in the Cappadocia region of Turkey is soft enough to carve, yet sturdy enough to last; this has been the site of hundreds of homes. (Courtesy of the Ministry of Tourism of Turkey.) (b) The great carvings of the faces of American presidents into the granites of the Black Hills in South Dakota illustrate the durability of natural building materials in their most natural state. (Courtesy of Mount Rushmore National Memorial, U.S. National Park Service.)

Rubble and *rough construction stone* refer to large blocks of rough-hewn rock used for sea walls, bridge works, and similar construction situations, which must be very resistant, and to smaller stone blocks used as wall-facing materials.

Rip-Rap is the name given to large, irregularly shaped blocks of rock (generally from about 70 to 700 kilograms (100 to 1000 lbs) in weight) used in river and harbor work, and to protect highway embankments from erosion.

Ashlar consists of rectangular pieces of building stone of nonuniform size that are set randomly in a wall. Ashlar blocks are generally 10 centimeters (4 inches) or more thick and up to a square meter (3 feet square) or more in area and may have rough-hewn or smooth faces (Figure 10.6).

Cut stone refers to all building stones that are cut to precise dimensions on all sides (hence, the term "dimension" stone), the surfaces being textured, smoothed, or polished. Most cut stone is used as a facing on exterior or interior walls of buildings and for counter tops, and is therefore applied in thin slabs as a veneer.

Monumental stone is material employed in gravestones, statues, mausoleums, or more elaborate monuments.

Flagstones, *curbing*, and *paving blocks*, as the names suggest, are used in paving and edging roads, paths, and public areas in towns and cities.

Roofing slate and *mill-stock slate* refer to slate tiles used for roofs and to smooth-finished slabs of slate used in electric switchboards, billiard tables, laboratory bench tops, blackboards, and similar situations where a smooth, vibration-free surface is required.

Terrazzo is sized material, usually marble or limestone, that is mixed with cement then poured on floors and smoothed and polished to expose the chips after the cement has hardened.

Almost every kind of rock has been used at some time or another as building stone. The important factors

FIGURE 10.3 The Stonehenge stone circle, constructed by the peoples of Wessex, Britain in about 2000 B.C., has been variously interpreted as a ceremonial, astronomical, or burial site. The great stone blocks were originally shaped and then transported as far as 150 miles (240 km). The rocks appear today just as they did 4000 years ago. (Photograph by J. R. Craig.)

governing the suitability of any particular rock for use in building are its physical properties and whether or not it is pleasing to the eye. The names used for rock types by those in the building industry, although comparable to those used by scientists, are nevertheless applied much more loosely. The most important rock types are as follows.

Commercially, the term granite is used not only to include real granites composed of feldspars and quartz but also any coarsely crystalline igneous rocks. Hence, dark colored igneous rocks such as gabbros and **norites**, rich in **ferromagnesian minerals**, are commonly marketed as "**black granites.**"

FIGURE 10.4 Natural stone was the principal construction material used by the Romans. The durability of a structure built using carefully cut and shaped blocks is evident in this amphitheater constructed in Nimes, France, by the Romans in the first century. (Courtesy of the French Government Tourist Office.)

Thus, many commercial rocks do not fit into a scientific classification of igneous rocks (Table 10.1), but they do satisfy the customers. "Granites" in this broad definition commonly serve as building and monument stones because they are attractive, break easily along joints, and resist abrasion and weathering. **Sandstone** is defined in commercial terms as consolidated sand in which the grains are chiefly of quartz and feldspar cemented by various materials that may include silica, iron oxides, **calcite,** or clay. A conspicuous feature of these clastic sedimentary rocks is their bedding or layering. *Bedding planes* are commonly planes along which the rock splits easily; when these planes are regular and evenly spaced, the rock may be a natural flagstone. **Sandstones** are also widely used for cut stone, ashlar, and rubble; a notable example is brownstone, a red sandstone of Triassic age used in the eastern United States as a building material.

Limestone is the other sedimentary rock that is most widely used in building. In commercial terms, it is a *rock of sedimentary origin composed principally of calcium carbonate or dolomite, the double carbonate of calcium and magnesium* and therefore includes true limestones and **dolostones.** Like sandstone, limestone is a well-known building material used for ashlar and cut stone and may also form a natural flagstone. It is widely used throughout Europe in the construction of large public buildings, cathedrals, and mansions. Many of the colleges of the Universities of Oxford and Cambridge were constructed from limestones of Jurassic age quarried from central England (Figure 10.7).

Marble, in scientific usage, refers to a limestone or dolostone that has been thoroughly recrystallized during metamorphism, but the term also has broader use in commercial terminology. Commercial marble includes not only true marbles but also certain crystalline limestones and even highly altered

FIGURE 10.5 The castles of Europe, such as Kilchurn Castle on Loch Awe in Scotland, have survived for hundreds of years because of the durability of the local stone that was used to construct the walls. (Photograph by J. R. Craig.)

ultramafic igneous rocks that are almost entirely made of the hydrated magnesium silicate minerals known as **serpentines** (and therefore called by scientists, *serpentinites*). All of these marbles are materials that can easily be cut (Figure 10.8), carved, or shaped, and they can take a good polish. They are the best known of all monumental stones and are greatly prized as facing materials for both exterior and interior uses. See Box 10.1 for an expanded discussion of marble and its modern uses.

Slate is a very fine-grained rock that is formed when clay-rich sediments are compressed during metamorphism. As well as hardening the rock, metamorphism imparts the cleavage whereby the rock is very readily split into thin parallel sheets. The dominant minerals are quartz, **micas,** and other platy, mica-like minerals that align parallel to the cleavage planes. Slate is best known as blackboards in schools and as a roofing material; it is still regarded by many people as the highest quality and

FIGURE 10.6 Ashlar, on the campus of Virginia Tech, is commonly used as a building facing. It consists of randomly set rectangular pieces of building stone. (Photograph by J. R. Craig.)

TABLE 10.1 Classification of some of the most important igneous rocks

	Abundance of quartz and/or light-colored (felsic) minerals			**Abundance of dark-colored (mafic) minerals, especially ferromagnesian minerals (olivine, pyroxene)**		
	FELSIC	INTERMEDIATE		MAFIC	ULTRAMAFIC	
	Mainly potash feldspar ($KAlSi_3O_8$)		Mainly plagioclase feldspar $(Na,Ca)(Al,Si)Si_2O_8$		No feldspar	
	with quartz	little or no quartz	with biotite and/or hornblende	with pyroxene and/or olivine	without olivine	with olivine
Coarse grained (plutonic)	GRANITE (granodiorite has dominant Na-plagioclase)	SYENITE	DIORITE	GABBRO (also norite)	PYROXENITE	PERIDOTITE
		MONZONITE				
Medium grained (hypabyssal)	MICRO-GRANITE	MICRO-SYENITE	MICRO-DIORITE	DIABASE (dolerite)		
Fine grained (volcanic)	RHYOLITE	TRACHYTE	ANDESITE	BASALT		

FIGURE 10.7 The Radcliff Camera with All Souls College. Many of the university buildings in Oxford are built from limestones of the Jurassic Period quarried in central England. (Photograph by D. J. Vaughan.)

most permanent form of roof covering. However, as with the other building stones, its use has largely been superseded by less expensive, processed materials. Today, the great slate quarries like those in Vermont and North Wales, the latter once supplying most of England and Wales, have been largely abandoned (Figure 10.9).

Crushed Rock

Crushed rock is the largest volume hard-rock mineral commodity used in the United States and most other countries. Annually, the United States produces nearly 2 billion metric tons of crushed rock with a value of more than $14 billion. The actual

FIGURE 10.8 High-quality marble used in construction may be mined in open quarries or in underground mines such as this mine in Vermont. The marble is cut into rough blocks in the mine and then shaped and polished for its final usage. (Courtesy of the Vermont Marble Corporation.)

value of the rock in the ground is no more than a few cents per ton; nearly all of the value comes from the costs of extraction, crushing, and transport. The principal use of crushed rock is as the base for the construction of roads, but very large quantities are also used in the foundations of buildings and as the aggregate in concrete. The quantities used are enormous; for example, the 3 million miles of paved roads in the U.S. highway system required 60 billion tons of crushed stone for construction. The total used in construction of all of the highways and cities in the United States probably exceeds 100 billion tons. Lesser amounts of crushed rock find a wide range of other uses in fertilizers, glass-making, refractories, fillers, and terrazzo surfaces. Many of these uses are discussed in detail elsewhere in this book.

BOX 10.1
Marble for the Masters

Sculptors are always very particular about the medium in which they work, desiring that the final work speak both in substance and style. For more than 2000 years, the rock of choice has been marble, and the single most highly sought after marble has been that from Carrara in the Apuan Alps of northern Italy. Quarrying apparently began in the Carrara region by 500 B.C. and gradually expanded. Major expansions came during the Renaissance in the 1400s and 1500s; ultimately, more than 650 different quarries have been cut into the sides of the mountains seeking the marble that has been transported around the world. These quarries have yielded more marble than any other place on Earth and continue to be mined at a rate of more than a million tons each year.

It is widely held that Michelangelo, like sculptors before and after him, visited Carrara to select the best stone for his works (Figure 10.A). On Monte Altissimo, he found a *statuario* marble that he believed was more uniformly white than any other marble in Italy. The Romans believed in a *living* earth and supposed that the marble regrew after excavation and would never be exhausted. There was still much from which Michaelangelo could choose in the early 1500s, but today the best of the *statuario* is nearly gone. The remaining individual blocks of *statuario* are sought by sculptors from around the world at premium prices.

FIGURE 10.A The *Pieta* by Michelangelo is a classic example of sculpture made from white marble quarried at Carrara, Italy. (Courtesy of the Italian State Tourist Office [E.N.I.T.], London.)

The marble at Carrara consists of calcite, $CaCO_3$, which initially formed as a submarine limestone. Subsequently, it was recrystallized as a result of increasing temperature and pressure as the rocks of the Alps were deeply buried and intensely deformed. The individual grains of calcite increased in size from their original hundredths or tenths of a millimeter to several millimeters. This coarseness, along with the clarity of the individual crystals, allows light to penetrate and gives a special sheen to the finished product. Colored marbles, which occur in many localities around the world, owe their shades to various amounts of impurities, such as iron oxides and clays.

Marble quarrying is tedious work because cracks and chips can weaken blocks, thus greatly reducing value. From Roman times until the nineteenth century, most of the work was done by hand. Rows of holes were chipped to define the boundaries of blocks; then wooden wedges were driven into the rock. When soaked with water, the wedges expanded and gently cracked the blocks open. Explosives were once tried for this purpose but were found to shatter and weaken the rock. Wire saws, which operate by continuously wearing their way through the soft marble, were introduced in the late 1800s. These saws reduced waste and worked well until the 1970s when large diamond-studded circular saw blades came into use.

Although little *statuario* now remains, the lesser quality marble is in high demand for ornamental construction. Consequently, Carrara is still a major supplier of the world's marble needs. Every year, Carrara hosts a marble fair. In 1991, for the first time in more than fifty years, there was a large block of marble on display from the Yule marble quarry in Colorado. The Yule quarry, the largest in the United States for many years and second only to Carrara in world production from 1909 through 1917, provided the stone for the Tomb of the Unknowns at Arlington National Cemetery in Arlington, Virginia, and for many buildings in Washington, DC. The Yule quarry was closed in 1941, but has been periodically reopened since that time to provide large blocks of high-quality marble for use around the world.

The best rocks for aggregate should be hard and inert but still easy to mine and crush. Specifications for crushed rock to be used for various purposes are usually established by national organizations and include properties such as resistance to abrasion, ability to withstand freeze–thaw conditions, size distributions of fragments, and absence of material that may react with the alkali substances in the cement matrix of a concrete. Depending on the specifications, crushed rock might be used for concrete, coarse or fine bituminous (asphaltic) concrete, **macadam** (the common "black top" of roads) or other surfacing for roads, road base, railroad ballast, fill, and a variety of other applications.

Many rock types meet the basic requirements for use as aggregate. In the United States, limestone and dolostone make up more than two-thirds of the stone quarried for this purpose because they are widely available, easy to mine and crush, and are strong in use (Figure 10.10). Granite is second in importance, making up about one-seventh of the stone quarried, along with related light-colored igneous rocks. The only other rock types that make any significant contribution to U.S. production are fine-grained, dark-colored igneous rocks, principally basalt, which is known in the trade as *traprock,* and sandstones or quartzites.

In most countries, surface exposures of rock suitable for aggregate are commonplace. Extraction in large quantities by blasting and quarrying in an open pit followed by crushing at the site of mining operations can be combined with a location close to the user to lower transport costs. For these reasons, crushed stone is one of the lowest unit-cost items of commerce, averaging less than $10 per metric ton in the United States, a country that produces nearly 2 billion tons of this commodity each year. The growth of the crushed rock industry can be gauged by the fact that it has been only about 150 years since Eli Whitney Blake invented the modern rock crusher in 1858. This was to provide crushed rock for the then ambitious project of a two-mile long macadam road from New Haven to Westville, Connecticut. Before this, all rock crushing was done by hand, even for the construction of roads, using the methods pioneered by the Scottish inventor John McAdam in 1819. Today, concrete composed primarily of crushed rock with cement as the binder is the most widely used construction material for highways and large buildings. Its strength, resistance, and moldability also allow it to function well as road surfaces and sidewalks, in bridges (where it is reinforced by internal steel rods), in walls, and in statuary (Figures 10.11 and 10.12).

Sand and Gravel

Sand and gravel mining constitute the second largest nonfuel industry in the United States and in many other countries of the world. More than 1.3 billion metric tons of sand and gravel are mined each year in the United States in operations at over 5000 locations, involving more than 3500 producers, with a total product value of $8 billion. The principal uses of sand and gravel are as aggregate for concrete and as the rock matter added to bituminous mixtures to make

FIGURE 10.9 Slate, widely used in the past as a roofing material and for a variety of other purposes, was mined in regionally metamorphosed terrains such as in North Wales in the United Kingdom. The large Bethesda quarry is typical of slate-producing operations. (Photograph by B. J. Skinner.)

FIGURE 10.11 Concrete and steel are the primary construction materials at the beginning of the twenty-first century. Steel provides the framework and concrete is poured in desired shapes to provide floors, walls, and arches; the combination can provide structures that will last hundreds of years even under the most severe weather conditions. (Photograph by J. R. Craig.)

macadam road surfaces. Sand is classified as having particle sizes less than 2 millimeters in diameter; those of gravel are larger and can range up to approximately 9 centimeters (three and one-half inches). In the United States, where the construction industry consumes more than 95 percent of the sand and gravel produced, the tonnage of gravel used is about twice that of sand.

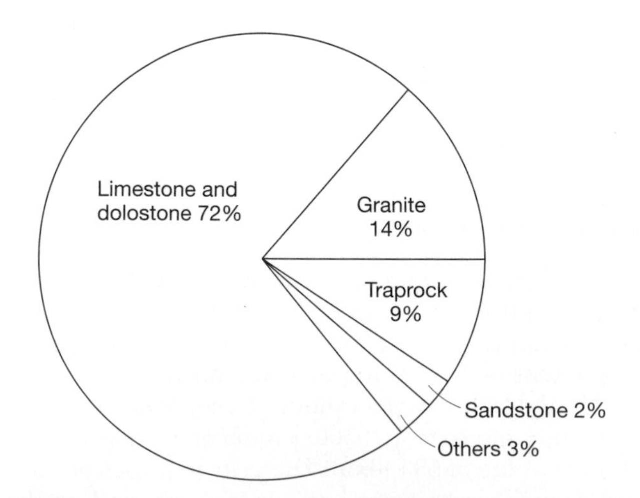

FIGURE 10.10 The rock types most widely used in the preparation of crushed stone in the United States. (From the U.S. Geological Survey.)

Commercially, sand and gravel are obtained from many locations, but the main sources are present-day or ancient river channels (Plate 46), their floodplains, or their **alluvial fans.** Here, rounded pebbles are produced by the action of transport downstream; usually only the harder and more stable rock fragments survive. There is also some separation of different sized particles, with finer grains being washed farther downstream or out to sea. The deposits that result may have properties desirable for use in concrete aggregate (rounded, hard, and stable particles), as well as being readily accessible and easily mined. The deposits of sand and gravel that were left behind following the retreat of the great ice sheets that once covered much of northern Europe, Canada, and the northern United States are also commercially important (Figure 10.13). Many beaches in the temperate latitudes are composed of sand and some contain gravel. However, the recreational value of these beaches usually precludes mining them for construction materials.

(a)

(b)

FIGURE 10.12 (a) Modern highway and bridge construction using large quantities of crushed stone, concrete, and steel have eliminated many of the problems of unpaved roads. (Photograph courtesy of the Portland Cement Association.) (b) The Cowlitz River arch span in Washington is 520 feet (170 meters) long and typifies modern road construction. (Photograph courtesy of Federal Highway Administration.)

As with crushed rock, sand and gravel are resources that are exploited in areas as close to the consumer as possible so as to reduce transport costs. However, in some regions, local resources are limited or have largely been consumed, as in the densely populated areas of Europe and northeastern North America. In these northern latitudes, extensive sand and gravel deposits occur beneath the sea on the **continental shelves,** having been deposited there by glaciers during the height of the most recent Ice Age. These submarine deposits are now being exploited by dredging off the west coast of several European countries on a substantial scale and, in a smaller way, off the shores of New Jersey, New York, and the New England states. Gravel deposits are sparse or absent off many tropical coasts; in these areas, there may also be little surface exposure of rocks suitable for crushing. In the Gulf of Mexico, for example, although sands can be found in river deltas, gravels are almost unknown. The only available large-sized aggregate materials are old shell beds and coral reefs; hence, it has become necessary to import gravel and crushed stone into southern Texas, Louisiana, and Florida. Deposits of sand and gravel are generally simple to mine using power shovels, bulldozers, and drag-lines in dry pits, or by dredging in rivers, offshore, and in

FIGURE 10.13 Sand and gravel are the products of natural geological weathering processes, and exploitable deposits occur worldwide. Glacial outwash deposits, such as those near Trondheim, Norway, can provide excellent sources of naturally washed and sorted gravels and are very easily worked. (Photograph by J. R. Craig.)

natural or artificial lakes. Processing involves little more than washing and screening to separate the different particle size fractions; there may also be crushing operations or even some form of separation process to remove unwanted impurities such as particles of shale.

The crushed stone industry and the sand and gravel industries are both increasingly subject to disputes over land use and environmental damage. The irony of this situation is that the people who benefit from the materials are the ones usually trying to close down the quarries. The pits and quarries were originally located adjacent to cities to minimize the transportation distances and keep down building costs. The materials that were extracted were used in the expansion and suburbanization of today's cities until new homes were located in close proximity to the pits and quarries. Then, the new residents objected to the noise and dust from the operations and forced their closure and relocation. Additional construction continued and repeated the series of events. Every time the pits and quarries have moved farther from the centers of the cities, the costs of transporting the materials have increased, bringing new complaints about the costs. As populations continue to expand, this situation will recur many times.

Vermiculite, Perlite, and Other Lightweight Aggregates

Lightweight aggregates include a variety of materials used chiefly in making wallboards, plaster, concrete, and insulation. Use of these materials is increasing because they are much more easily handled in construction work and because the trapped air that they contain makes them good thermal insulators and valuable in energy conservation. Examples include such natural materials as volcanic cinders and **pumice,** which are simply crushed and sized after quarrying; there are also similar synthetic materials that are by-products of industrial operations such as processed slag. Other lightweight aggregates involve treatment, usually by heating, of some product of mining operations. In some instances, clay or shale is sintered or roasted in kilns, which drives off water and causes them to expand. Two particularly interesting natural substances that expand on heating are **vermiculite** and **perlite.** For both, the expansion is so great that the end product is an *ultra*-lightweight aggregate.

Vermiculite is a mineral with a mica-like layer structure. Although the composition of vermiculite is variable, it is basically a hydrous magnesium iron aluminum silicate. When it is heated rapidly to above 230°C, the layers separate as water between the layers converts to steam. This process, called **exfoliation,** is similar to that which creates popcorn when steam causes special kinds of corn to expand. The increase in volume from the heating of vermiculite averages eight to twelve times, although individual particles may expand as much as thirty times to form worm-like pieces (the name *vermiculite* comes from the Latin *vermiculare*, "to breed worms"; see Figure 10.14). Vermiculite occurs in many places throughout the world, although the most important deposits are found in Montana and South Carolina, and Palabora, in northeastern Transvaal in South Africa. The Montana deposit is a large altered **stock** of mafic igneous rock called **pyroxenite,** while the Palabora deposit is an altered part of an igneous rock called carbonatite. Mining of the brown or greenish flakes of vermiculite is by open-pit methods, and the total annual world production is about 500,000 metric tons. Of this, an estimated 100,000 tons is produced in the United States. The exfoliation plants are usually sited close to the final markets—in 2008, there were 17 plants in the United States—because this reduces transport costs. The U.S. Geological Survey estimates the total world reserves, excluding Russia and China, at 50 million metric tons, of which 25 million metric tons are in the United States and 20 million metric tons in South Africa. Total world resources are estimated to be at least three times the reserves.

Perlite is a glassy volcanic rock containing a small amount of combined water that vaporizes on rapid heating so that the rock expands to between four and twenty times the original volume. The resulting material is a solid filled with bubbles and pores and has a characteristic white color. Perlite ore is found in belts of volcanic rock in many parts of the world, but the major producers are Greece, Hungary, Italy, Russia, and the United States. As with vermiculite, deposits are mined by open-pit methods, and the ore is crushed, dried, and screened before shipment to expanding plants near the final markets. Annual world production is about 1.8 million metric tons from estimated world reserves of at least 700 million metric tons.

FIGURE 10.14 Vermiculite exhibits the unusual characteristic of swelling when heated. The effect is shown here in the contrast of vermiculite as (a) mined and that which has been (b) heated. (Sample courtesy of the Virginia Vermiculite Corporation; photographs by S. Llyn Sharp.)

TREATED ROCK PRODUCTS

Treated rock products are materials that must be chemically processed, fired, melted, or otherwise altered after mining so that they can be set into new forms before use. Consequently, nearly all of these materials have values significantly greater than those of most *untreated* rock materials. Important examples include the raw materials for the manufacture of cement and of plaster, clay and other substances needed in the production of bricks and a wide range of ceramic products, and the raw materials for the glass industry.

Cement

Cement, a chemical binder made chiefly from limestone, has been one of the most important construction materials since the eighteenth century. It is generally mixed with sand to produce mortar, the binding agent for brick, block, or other masonry, or with sand and gravel to make concrete (a form of instant rock). The first use of cement and mortar was by the ancient Greeks and Romans; they added water to a mixture of quicklime (CaO, made by heating or calcining limestone), sand, and a finely ground glassy volcanic ash. Because the volcanic ash came from the town of Pozzuoli, near Naples, the mixture is known as **pozzolan cement.** The addition of water to this mixture causes a series of chemical reactions leading to recrystallization and hardening of the cement when it dries. The end product is then stable in air and water, as we can see from the fact that pozzolan cement was used to build the Roman Pantheon and the Colosseum, and both structures are still standing after more than 2000 years.

However, the "art" of making such cements, although further developed by these ancient civilizations, was forgotten during the Dark Ages and was only rediscovered in 1756 when a British engineer named John Smeaton was commissioned to rebuild the famous Eddystone Lighthouse off the coast of Cornwall. Smeaton searched for a hydraulic cementing material that would set and remain stable under water, and he is said to have discovered the formula when examining an ancient Latin document. Smeaton found that clay must also be present to produce hydraulic cement and that this could be introduced by calcining a limestone naturally rich in clay (hence, known as **cement rock**). Such **natural cements** were further developed during the eighteenth century, but because the compositions of cement rocks vary widely, the cements were not uniform in strength or setting times. In an effort to improve the uniformity of cement, an Englishman, Joseph Aspdin, in 1824 patented a formula for **portland cement,** so called because of a fancied resemblance to Portland Stone, a limestone widely used in British buildings. It soon supplanted all other cements, and today is the basis of modern cements and concretes—the most common construction materials in the world (Figure 10.15). Today, twice as much concrete is used as all the other structural materials combined.

The raw materials needed for portland cement manufacture are a source of lime (CaO), which is generally from limestone, a source of alumina (Al_2O_3), which is commonly a shale or a clay, and a source of silica (SiO_2), which may also come from a clay or a shale, or from sand. Because limestone is the major ingredient, outweighing the others by roughly a factor of ten to one, cement plants are usually located near limestone quarrying operations. Sometimes a natural limestone contains clay or shale impurities of the desired composition and is therefore a cement rock. Small amounts of iron-containing materials (iron ores or waste products from iron works) are also used, and either gypsum or anhydrite is added later to control setting times. The extraction of the major raw materials involves the same kinds of open-pit mining as for other forms of crushed rock.

The steps involved in the manufacture of portland cement require crushing and grinding of the raw materials and blending of the crushed products together either as dry powders or as slurries mixed with water [Figure 10.15(a)]. The blend is then fed into a long rotating kiln in which temperatures of

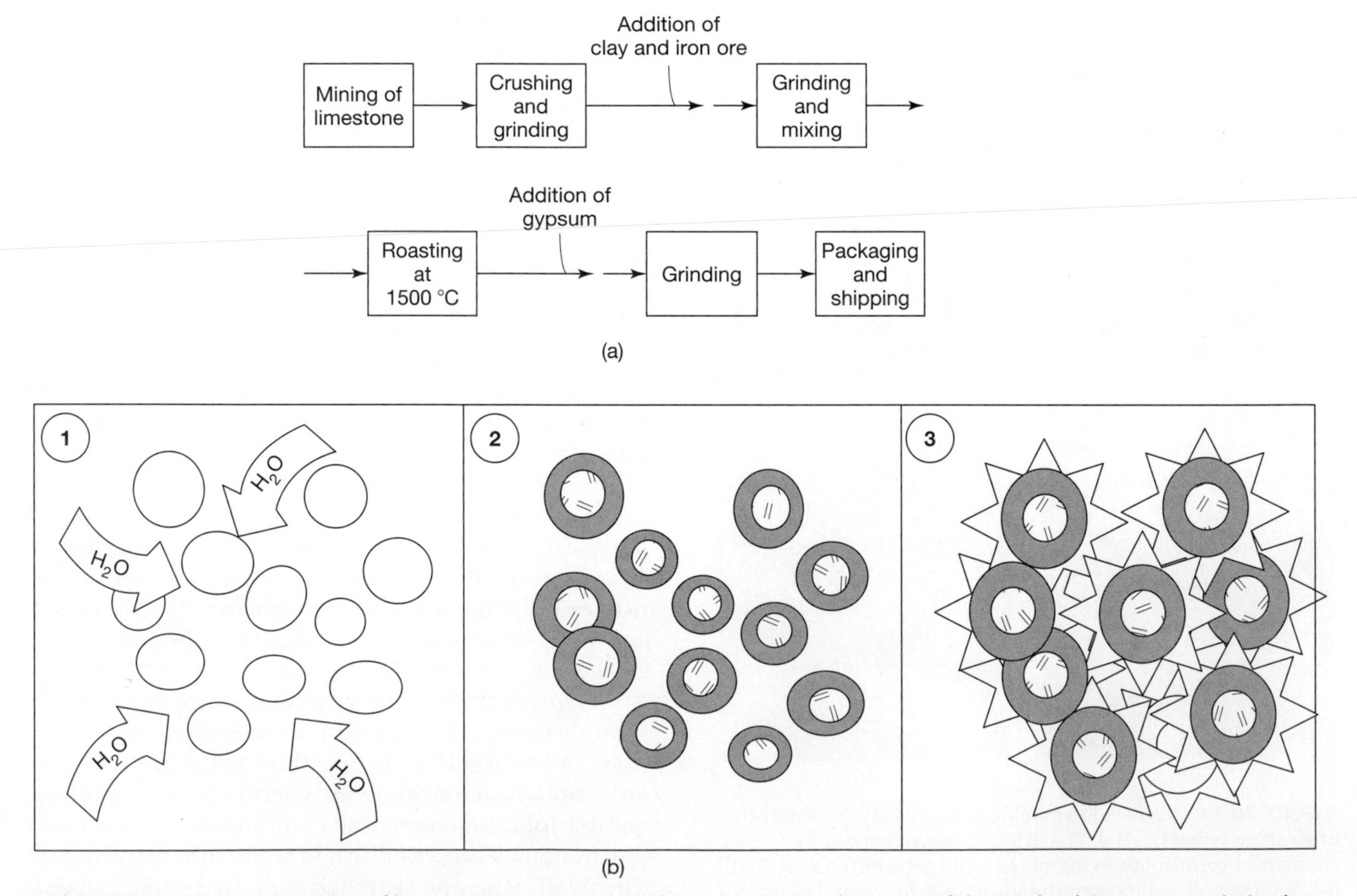

FIGURE 10.15 (a) The conversion of limestone into cement involves several steps. (Courtesy of the Portland Cement Association.) (b) The crystallization of cement begins when water is added to the clinker. The reactions result in the formation of calcium silicate crystals that interlock and give cement its durability and strength.

about 1500°C are reached. High temperature drives off the carbon dioxide and water and partly melts some of the material to a glass. The resulting clinker is ground to a powder, mixed with a small amount of gypsum (approximately 5 percent of the bulk), and is bagged for sale. The great value of cement is that it readily forms "instant rock" when water is added to the fine powder. The water reacts with the complex calcium and aluminum silicates to grow a series of artificial, but very strong and interlocking, crystals [Figure 10.15(b)]. Nearly all applications of cement are in preparing concrete, which is composed primarily of gravel or crushed stone with about 10 percent cement to serve as a "glue." The concrete is commonly further strengthened by the incorporation of iron or steel rods called *rebar*. The resulting material is hard, strong, durable, resistant to weathering, and can be formed to fit nearly any shape.

The raw materials needed for cement manufacture are widely available throughout the world, and more than 100 countries produce significant amounts of cement. China has become the world's leading manufacturer of cement, producing more than 1450 million metric tons in 2008; the world total in 2008 was 2.9 billion tons.

Not just any limestone is suitable for cement manufacturer; only limestones that are nearly pure $CaCO_3$ with very low amounts of iron, magnesium, or manganese are suitable to make cement. Hence, cement-producing plants tend to be concentrated along specific geologic formations that were created during times when very pure limestones were deposited in warm tropical seas.

The preparation of the massive quantities of cement used by modern society points out another of the complexities of the modern environment. The heating (calcining) of pure limestone reduces the calcium carbonate to calcium oxide by driving off carbon dioxide, which equals 44 percent by weight of the original rock. Hence, the calcining of one metric ton of limestone yields 560 kilograms (1230 pounds) of lime and releases 440 kilograms (970 pounds) of carbon dioxide into the atmosphere. Thus, the preparation of massive amounts of cement in countries such as the United States and China releases very large quantities of CO_2 into the atmosphere at a time when there are increasing calls to reduce CO_2 levels.

Plaster

Plaster is made by heating or calcining gypsum, the hydrated form of calcium sulfate ($CaSO_4{\cdot}2H_2O$). When gypsum is calcined at 177°C, 75 percent of the water is driven off and a new compound $CaSO_4 \cdot \frac{1}{2}H_2O$ is formed. The new compound is commonly called **plaster of Paris,** after the famous gypsum quarries in the Montmartre district of Paris from which a plaster of particularly high quality has long been produced. When plaster of Paris is mixed with water, it recrystallizes by rehydration and forms a mass of a finely interlocking gypsum crystals. The earliest known use of gypsum, and the plaster made from it, was by the Egyptian civilization some 5000 years ago; the Greek writer Theophrastus described the burning of gypsum to prepare plaster in a manner that remains little changed today. Plaster of Paris was used in England as early as the thirteenth century, but the first British manufacture of plaster dates from the late seventeenth century.

However, the use of plaster remained limited until about 1870 because of its very rapid setting time; then it was discovered that organic additives such as glue and starch could be used to retard setting. This revolutionized the industry and permitted the first large-scale use of plaster in construction. Most early building use was as hand-applied wet plaster that was spread over a wire screen or wooden laths to cover walls and ceilings. The development of prefabricated wall board in 1918 provided a way to greatly reduce labor costs while still using plaster. Today, **plasterboard** is by far the most widely used indoor wall covering in North America and Europe. It is prepared by feeding a slurry of plaster onto a rapidly moving, continuous roll of heavy paper. A second sheet of paper is fed onto the slurry to sandwich it. As the sandwich of paper and plaster travel several hundred meters on a conveyer system, the plaster sets sufficiently for the continuous slab to be cut into standard-sized sheets. These sheets are then sent slowly through a long drying kiln so that the plaster slurry hardens into solid gypsum and the boards are ready for use.

Gypsum is one of the series of minerals formed on tropical marine tidal flats as seawater evaporates (as shown in Figure 9.16). Gypsum and a related anhydrous form of calcium sulfate, called *anhydrite* ($CaSO_4$), are often found in thick bedded sequences of marine sedimentary rocks. The anhydrite can be converted to gypsum by soaking it in water after fine grinding; then, it is equally suitable for making plaster of Paris.

The United States, which possesses gypsum-producing areas that are widely distributed (Figure 10.16), is the world's largest producer and consumer of gypsum and anhydrite. In fact, as much as 10 percent of the land area in the United States is underlain by gypsiferous rock, indicating that U.S. potential resources are so vast (over 20 billion tons) as to be superabundant. The U.S. Geological Survey has, in fact, stated that reserves are sufficient for 2000 years at projected rates of production. The size of some deposits is

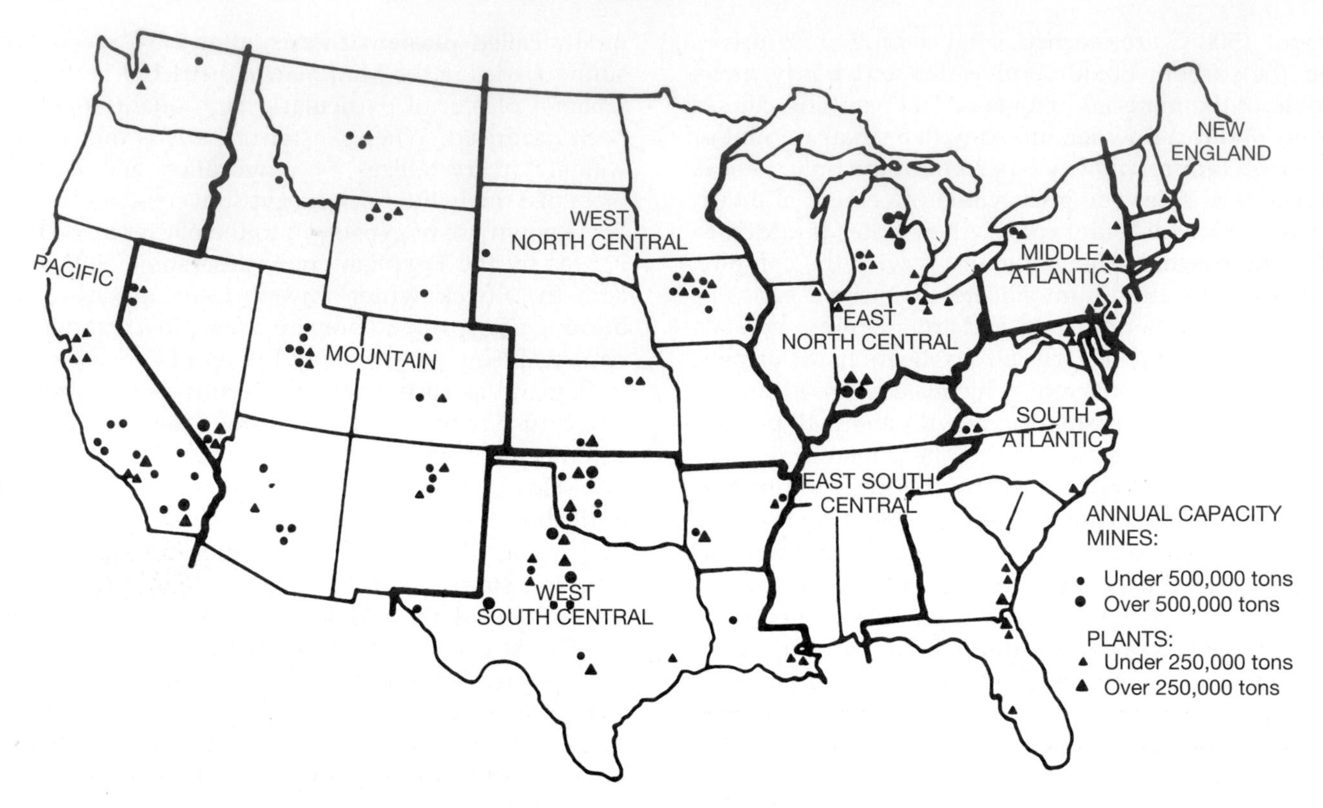

FIGURE 10.16 Gypsum is produced in a large number of areas in the United States. Gypsum deposits were formed as parts of evaporative sequences that were deposited as described in Figures 9.16 and 9.17. (Map from the U.S. Geological Survey.)

FIGURE 10.17 The great sand dunes at White Sands National Monument in New Mexico are composed of gypsum that has weathered out of the nearby mountains and accumulated by the action of the wind. (Courtesy of White Sands National Monument, National Park Service.)

impressive. For example, gypsum deposits in Texas and Oklahoma extend for more than 320 kilometers (200 miles) over a width of 32 (20 miles) to 80 kilometers (50 miles) with a thickness of up to 7 meters (25 feet). Culberson County, Texas, is underlain by a gypsum bed covering more than 1500 km^2 (600 square miles) and which is up to 20 meters (60 feet) thick. Undoubtedly, the most picturesque gypsum deposits are the 700 km^2 (275 square miles) of snow white dunes at White Sands National Monument in New Mexico (Figure 10.17). The widespread geologic occurrence and extensive use of gypsum is shown by the fact that it is produced in over sixty countries, with the largest producer being China. Other major producers include the United States, Canada, Germany, France, Russia, Great Britain, Spain, and Italy. With deposits being so widespread, and with transportation costs high relative to the value of the commodity, there is only limited world trade in gypsum. The price of gypsum today depends on the amount of treatment and energy needed to produce it; thus, raw gypsum ore has a value of less than $10 per ton, but fine powdered plaster is valued at more than $130 per ton. Despite the United States being a large producer of gypsum, it is also a large importer, most of the imports coming from Canada. Gypsum is mined using both open-pit and underground methods, and apart from crushing and grinding, there is little processing of the ore prior to calcining.

FIGURE 10.18 (a) Sun-dried adobe bricks, as used in this building in a small town in Jujuy Province in northwestern Argentina, are widely used as construction materials in many arid regions of the world. The thatch helps to carry the rain, when it does occur, away from the clay-rich bricks so that they do not break down. (b) Tiles, as those used in this Roman work in Merida, Spain, are made of the same materials as adobe but have been glazed and fired. They are then very durable and may last for thousands of years. (Photographs by J. R. Craig.)

Brick and Ceramic Products

A very large variety of raw materials and a vast range of finished products are the province of the ceramics industry. The simplest of the raw materials are clay-rich muds, which are shaped and sun dried into adobe bricks [Figure 10.18(a)]. These bricks, which often incorporate pebbles, straw, sticks, and other materials, would decompose quickly in rainy areas but are quite durable in arid regions. In most of the developed world, the principal ceramic products are made chiefly from clays that can be molded into the desired shapes before firing to hardness. Structural ceramics such as bricks, tiles [Figure 10.18(b)], and sewer pipes are a major part of that industry. In this chapter, we can do no more than mention a few examples of the raw materials used in this field, their availability, and their utilization.

Ceramic products require materials to make up the bulk of the product (skeleton formers or fillers), bonding agents that may be glass formers, fluxes that help during the firing, and various materials to give special properties such as color and durability to the product. Clays are the usual skeleton formers and so make up the major raw material used in the industry; fluxes, pigments, and glasses are discussed later in this chapter. The term **clay** refers to a group of very fine-grained hydrated minerals with layer structures at the atomic level (see also Figure 12.4). The ability of many clays to become plastic by the taking up of water enables them to be molded. Firing at high temperatures in a kiln drives off the water and melts some of the particles, which then welds the material together.

Clays can accumulate as residual deposits, as discussed in the section on aluminum (Chapter 7), or they may be transported and deposited as sediments in lakes, seas, or oceans. Clays soon transform to other solid rocks when they are dehydrated and heated by burial, so they are found only near Earth's surface, not deep in the crust. Like many of the other building materials, clays are usually recovered by large quarrying operations exploiting surficial deposits, with the quarries located as close as possible to the manufacturers and markets. In some pits, the clays are washed out by high pressure water cannons (Figure 10.19). Some of the most important clay minerals are listed in Table 10.2, along with information on their compositions, properties, and geologic occurrence.

The clays used in the manufacture of bricks and tiles are quite widespread, and reserves are so vast that few countries make attempts to estimate them. Those of the United States, for example, are more than sufficient for another century of production at anticipated rates of growth. Potential resources are even larger. Structural ceramic products such as bricks, tiles, and sewer pipes are manufactured by extruding the plastic clay mass formed by mixing the mined clay with water (10–15 percent), through a die of appropriate shape, drying the extruded mass under conditions of controlled humidity, and then firing in a kiln. The dehydration and vitrification on firing results in a resistant material made of high silica glass, **mullite** ($3Al_2O_3 \cdot 2SiO_2$), quartz, and other compounds welded together.

A relatively recent use of clays is in the manufacture of catalytic converters for automotive exhaust systems (Figure 10.20). Clays, mixed with water and other silicate starting materials, are kneaded into a soft mass and extruded through a special die system to produce a continuous, roughly cylindrical tube with window screen-like openings running the length. The tube is cut into proper lengths, fired at 1400°C (2550°F) to produce cordierite ($Mg_2Al_4Si_5O_{18}$), a mineral that has very high thermal stability, but expands very little when heated. The tube is then coated with catalytically active platinum by dipping it into solutions containing dissolved platinum salts.

A number of special types of clays, deposits of which are more restricted, are also used in ceramics. **China clay** or **kaolinite** is a particularly pure hydrated aluminum silicate much used originally for high-quality porcelain. This is now a relatively minor use compared with its importance as a filler; china clay is further discussed as regards its occurrence and utilization on page 367. **Ball clay** is also largely made up of kaolinite and is used in the manufacture of electrical porcelain, floor and wall tiles, dinnerware, and similar products. Deposits occur in the United States in Georgia, Tennessee, and Kentucky, in Devonshire in Great Britain, and in India. In the manufacture of whiteware (tableware and sanitaryware) ball clay is used to impart plasticity and dry strength, whereas china clay imparts a whiter color. Also mixed with these clays are very finely ground silica, called *potters flint,* to decrease the shrinkage that occurs during drying and firing, and feldspar ($(Na,K,Ca)Al_{1-2}Si_{3-2}O_8$) as a flux to lower the firing temperature. Clays have also been used in the preparation of cosmetics ranging from body paint to face powders (see Box 9.3).

FIGURE 10.19 Clay pits, such as those in Cornwall in southwest England, provide a raw material used chiefly in the paper industry and also in the manufacture of fine china and a variety of ceramics. (Photograph courtesy of ECC Ltd.)

Glass

The glass industry uses substantial amounts of many industrial minerals to make its products. Glass is actually made by melting together certain rocks and minerals that can be cooled in such a way that crystallization does not occur. Consequently, glass is made of atoms bonded together without the repetitive ordering characteristic of crystals, as in the liquid state [see Figure 10.21(a)].

The most important and common glass-forming material is *silica,* which is usually obtained from the quartz in sandstones. However, the high melting point of quartz (1713°C) and the high viscosity of molten silica make it difficult to melt and work, so that fused silica products are only used when their special properties, such as high softening point, low thermal expansion, and resistance to corrosion, are essential. To lower the melting temperature of silica (sometimes to as low as 500°C [930°F]), soda (Na_2O) is added. Because the resulting product (referred to as **water glass**) has little chemical durability and is soluble even in water, lime (CaO or CaO + MgO) is also added as a stabilizing agent. The first of these ingredients comes from sodium carbonate or sodium nitrate (see Chapter 9), or from the processing of

TABLE 10.2 The names, compositions, structures, and some information regarding properties, occurrence, and uses of some industrially important clay minerals. Note that most mined clays are mixtures of several clay minerals.

Name	Composition	Structure	Properties, Occurrence, and Uses
Kaolinite	$Al_4[Si_4O_{10}](OH)_8$ (The Kaolinite Group or "Kandites" includes kaolinite, dickite, nacrite, and halloysite—see below.)	One layer of linked SiO_4 tetrahedra bonded to a layer with Al in octahedral coordination (1 : 1)	Kaolinite can form the white high-purity material prized for porcelain making, as a filler in paper manufacture, and a wide range of other industries. May be of hydrothermal, residual, or sedimentary origin. Many refractory fire clays are essentially of kaolinite.
Halloysite	$Al_4[Si_4O_{10}](OH)_8 \cdot 8H_2O$	As above, but with a single layer of water molecules between the (1 : 1) sheets	Halloysite occurs in residual and hydrothermal deposits. It is used for a variety of purposes including catalysis in the oil industry.
Illite	$K_{1-1.5}Al_4[Si_{7-6.5}Al_{1-1.5}O_{20}](OH)_4$ (The Illite Group includes illite, hydro-micas, phengite, and glauconite.)	A sheet of octahedrally coordinated Al atoms sandwiched between two tetrahedral $(Si,Al)O_4$ sheets (2 : 1). K between these (2 : 1) sheets	Deposits are generally sedimentary in origin and contain other clay minerals. Used in common structural clay products such as bricks and tiles.
Montmorillonite	$Na_{0.7}(Al_{3.3}Mg_{0.7})[Si_8O_{20}](OH)4 \cdot nH_2O$ (The Montmorillonite Group, or "Smectites," includes montmorillonite, nontronite, saponite, and hectorite—see below.)	Again a (2 : 1) sheet with interlayer Na and H_2O	Montmorillonites are the essential clays in the "bentonites" produced by alteration of volcanic rocks. These are used in foundry clays and drilling muds. A moderate amount of montmorillonite is important in structural clay products to give plasticity. Also used in bleaching clays and adsorbents.
Hectorite	$(Ca,Na)_{0.66}Mg_{5.3}Li_{0.7}[Si_8O_{20}](OH)4 \cdot nH_2O$	As above, but with Mg and some Li instead of Al	Results from alteration of volcanic rocks. One use is in drilling muds.
Chlorite	$(Mg,Al,Fe)_{12}[(Si,Al)_8O_{20}](OH)_{16}$ (The Chlorite Group includes "chamosite" and clinochlore.)	A (2 : 1) sheet with a further sheet of octahedrally coordinated Al,Fe,Mg. Hence, a (2 : 1 : 1) sheet	Chlorite occurs as a component in many clays and shales mined for use in structural clay products.
Attapulgite	$Mg_5Si_8O_{20}(OH)_2 \cdot 4H_2O$ (Members of this miscellaneous group include sepiolite and palygorskite.)	A chain-type structure. Double silica chains linked together by octahedral groups of oxygens and hydroxyls containing Al and Mg atoms	Deposits of sedimentary origin occur. Attapulgite is in the clays called fuller's earth (fulling is removal of grease from wool) valued for adsorbing properties.

FIGURE 10.20 The substrate of the catalytic converter used to reduce automobile exhaust emissions is composed of synthetic cordierite ($Mg_2Al_4Si_5O_{18}$). The cordierite is made by baking a mixture of clays and additives at about 1500°C; platinum or palladium-bearing catalysts are added after synthesis. (Photograph courtesy of Corning, Inc.)

rock salt ($NaCl$) and limestone ($CaCO_3$); the second comes from crushed limestone or dolostone. A small percentage of alumina (Al_2O_3) is also often incorporated into the glass to further improve chemical resistance; feldspars from pegmatites and **aplites** are a common source of this ingredient. The resulting product is the basic soda-lime-silica glass that is used for the bulk of common glass articles such as bottles and window panes.

Other minerals are used in smaller amounts in glass manufacture to give particularly desired properties or provide glass for more specialized applications. Borosilicate glass contains approximately 10–14 percent boron oxide (B_2O_3) derived from the mining of **borax** (see Chapter 9). The borosilicate glasses are resistant to corrosion and can withstand repeated heating and cooling, so they are used in oven and tableware and for industrial glassware (such as the commercial product Pyrex). High alumina glass (15–30 percent Al_2O_3) is used in the manufacture of fibers and certain cooking utensils. As noted previously, alumina for the glass industry is obtained by

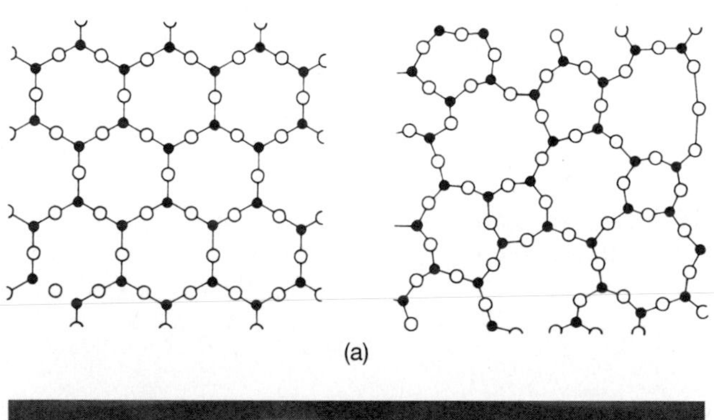

(a)

(b)

(c)

FIGURE 10.21 (a) Schematic diagrams illustrating the differences between a crystalline solid (left) and a glass (right). The crystalline material has a regular repeated atomic pattern, whereas the glass exhibits a more variable random pattern. (b) Glass is used to make thousands of items including beautiful decorative pieces. (Courtesy of Corning, Inc.) (c) Beautiful cameo glass pieces such as the vase are prepared by layering glass of various colors; the intricate designs and colors are brought out by careful etching and sand-blasting away layers to reveal the desired ones. (Photograph courtesy of Pilgrim Glass, Ceredo, West Virginia.)

the mining of feldspar-rich rocks such as pegmatites, aplites, and also nepheline syenite, an igneous rock made largely of feldspar and **feldspathoid** minerals. Lead crystal glass may contain as much as 37 percent lead oxide (PbO) for which red lead is the basic raw material, generally obtained as a by-product of base metal mining (see Chapter 8). This is the glass used for high-quality tableware and certain optical uses [Figure 10.21(b)].

Many diverse substances are used to color glass—commonly, these are the oxides of metals such as chromium, cobalt, nickel, copper, iron, vanadium, manganese, and uranium. Such coloring agents are used in successive layers in preparing special cameo glass ornaments; sand blasting, carving, or etching is then used to produce intricate designs [Figure 10.21(c) and Plate 47]. In contrast, in producing high-quality colorless glass, iron oxide present in the sand and limestone raw materials might be a problem. This is overcome by using a decolorizing agent, most commonly made up of selenium and cobalt oxide, or certain oxides of the rare earth elements. These additives have the effect of absorbing light of spectral colors not absorbed by the iron oxides, hence, cancelling their effect on the observer. The decolorizing agents are among the most expensive materials used in the glass industry. Small amounts of chemicals such as sodium sulfate (Na_2SO_4) and arsenic oxide (As_2O_3) and mineral products such as fluorite (CaF_2) and rock salt (NaCl) are also used as refining agents. The two basic raw materials used to make glass—sand and limestone—are abundant and widely distributed in Earth's crust and have low initial cost. Soda (Na_2O) is the most expensive of the major raw materials used to make the common types of glass and accounts for over 50 percent of the raw material cost per ton.

Glass production involves batch mixing of the raw materials and their melting together in pots and crucibles or large tanks, usually heated by the burning of oil or gas. After forming, glass is slowly cooled or **annealed** to reduce internal stresses, and during this stage it may be rolled out to form plate glass or subjected to a wide variety of manufacturing processes ranging from injecting the molten glass into molds to make bottles or glasses, to the traditional mouth-blowing and hand-shaping of artistic glass.

Today, in the United States alone, the production of various glass products annually consumes about 15 million metric tons of very high purity silica sand, several million tons of limestone, more than 3 million tons of various sodium compounds (such as Na_2O, $NaHCO_3$), 500,000 tons of feldspar, and more than 350,000 tons of boron oxide (B_2O_3).

Glass does occur naturally in two forms; **fulgerites,** which are rare fused, slender glass tubes that result from lightning strikes in sand, and obsidians, which are natural volcanic glasses formed when magma cools very quickly. Our ancestors found that obsidian was useful for making arrowheads, knives, jewelry, or even money. The first glass objects manufactured by humans date from about 3000 B.C., and these were actually glazes on ceramic vessels. By about 1500 B.C., vessels entirely of glass were being produced, and by 50 B.C., glass-blowing was an established art.

Little is known about glass production after the fall of the Roman Empire, but glass production continued to develop in Italy, and there was an elaborate guild system of glassworkers in Venice by the time of the Crusades in the twelfth century. Glassmaking remained primarily for artistic works, containers, and drinking vessels well into the 1700s. It was not until the early nineteenth century that an expansion in manufacturing facilities made window glass widely available. The modern industry has continued to produce new types of glass with novel properties—fiberglass insulation, photosensitive glass, special insulating and reflective glass, and fibers for fiber optics. Fortunately, the abundance of most of the materials needed to produce glass will ensure its future availability, and the likely development of glass products with new properties should result in even wider applications.

OTHER MAJOR INDUSTRIAL MINERALS

All of the substances discussed in this chapter, and in Chapter 9 on fertilizers and chemical minerals, can be classed as **industrial minerals.** This broad class of materials includes nearly all mineral derived resources except metallic ores, fuels, water, soil, and the materials just discussed. In this book, we separated fertilizers and chemical minerals to permit separate discussion of them. The remaining representatives are many and diverse, but a number of the more important examples are briefly discussed next. Each example is a mineral or rock that is valued because of a particular property it possesses. **Asbestos** minerals are fibrous materials that are strong, flexible, inert, and heat resistant; a whole range of rocks and minerals are capable of withstanding great heat and are used to make refractory products, whereas others are **fluxes** that enable the melting of materials in furnaces at lowered temperatures. Another large group of minerals and rocks are used simply as **fillers,** or bulking agents, in a wide range of products because they are inert and physically and chemically harmless. More specialized minerals are required for **pigments** or coloring agents, whereas a mineral such as **barite** finds industrial applications because of its high density and chemical inertness, diamond because of its hardness, and the **zeolites** because of their unusual crystal structures.

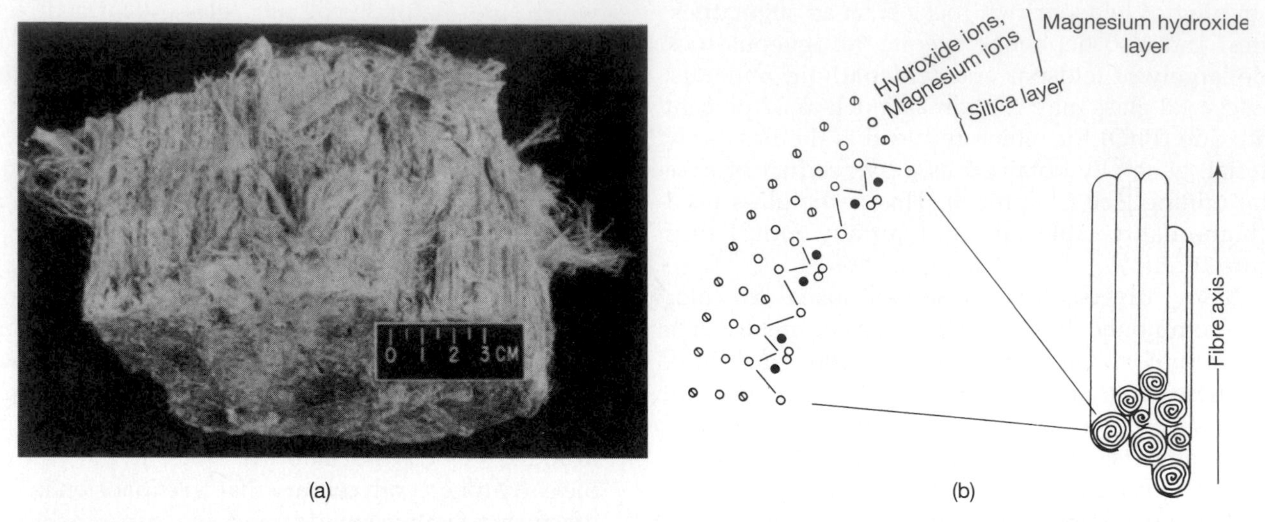

FIGURE 10.22 (a) Chrysotile asbestos has found many uses because it forms as long, flexible, and strong fibers. (Sample from Thetford, Ontario; photograph by S. Llyn Sharp.) (b) Schematic diagram of the structure of a fiber of chrysotile asbestos. Each crystal is in the form of a scroll made from a closely connected double layer with magnesium hydroxide units on its external face and silica units on its inner face. The details of a small section of the scroll show the structure of a double layer. Each fiber is made of several such scrolls. (Reproduced from A. A. Hodgson, "Chemisty and Physics of Asbestos," in L. Michaels and S. S. Chessick, *Asbestos,* vol. 1, John Wiley and Sons, Ltd., 1979. Used with permission.)

Asbestos

Asbestos is not a single mineral; the term refers to a number of silicate minerals that occur as fibrous crystals [Figure 10.22(a)]. Hence, the definition of asbestos is complicated because the term is based upon mineral habit. The former U.S. Bureau of Mines, summarizing the 1984 definition given by the American Society for Testing and Materials, said: "Asbestos is a term applied to six naturally occurring minerals exploited commercially for their desirable physical properties, which are in part derived from their asbestiform [fibrous] habit. The six minerals are the serpentine mineral chrysotile and the amphibole minerals grunerite asbestos (also referred to as amosite), riebeckite asbestos (also referred to as crocidolite), anthophyllite asbestos, tremolite asbestos, and actinolite asbestos. Individual mineral particles, however processed and regardless of their mineral name, are not demonstrated to be asbestos if the length-to-width ratio is less than 20:1." The six types of asbestos, their chemical formulae, and their properties are listed in Table 10.3.

TABLE 10.3 Types of asbestos fibers and their properties

Type	Formula	Color	Tensile Strength (p.s.i.)	Resistance to: Acids,etc.*	Resistance to: Heat**
Chrysotile	$Mg_6(Si_2O_5(OH)_4)_2$	White, gray, green, yellowish	80,000–100,000	Poor	Good
Crocidolite ("Blue asbestos")	$Na_2Fe_5(Si_4O_{11}(OH))_2$	Blue	100,000–300,000	Good	Poor
Amosite	$(Mg,Fe)_7(Si_4O_{11}(OH))_2$	Ash gray, brown	16,000–90,000	Good	Poor
Anthophyllite	$(Mg,Fe)_7(Si_4O_{11}(OH))_2$	Gray-white, brown, green	4000	Very good	Very good
Tremolite	$Ca_2(Mg,Fe)_5(Si_4O_{11}(OH))_2$	Gray-white greenish, yellow blue	1000–8000	Good	Fair to good
Actinolite	$Ca_2(Mg,Fe)_5(Si_4O_{11})(OH))_2$	Greenish	1000	Fair	—

*Refers to resistance to dissolution in 25% HCl, CH_3COOH, H_3PO_4, and H_2SO_4 (and also NaOH) at room temperature for long periods and boiling temperature for short periods.

**Refers to weight loss on heating at temperatures of ~200–1000°C for 2 hours due to loss of OH.

BOX 10.2

What Is This Page Made Of?

Most readers of this book know that paper is made from wood, but many probably do not realize that much of this page is actually mineral matter. Paper gets its name from *papyrus*, a reed-like plant that the ancient Egyptians would cut into thin slices for use as a writing material. Modern paper, however, is produced using techniques first discovered by a Chinese worker. In A.D. 105, Ts'ai Lun found that he could pound fibers of the inner bark of the mulberry tree into sheets.

Much of the paper used today, especially that used for newspapers, is still composed of the fibers from a variety of soft and hardwood trees. Logs are cut into 1–2-centimeter (half to one inch) chips that are chemically digested in sulfide or sulfate solutions to liberate the cellulose fibers. The masses of fibers, called *pulp*, are poured out on special wire-cloth belts where excess water is drained and the mass of crisscrossing fibers are then pressed between steam-heated cylinders and wound into large rolls. This paper can then be used directly on printing presses.

High-quality glossy papers, such as those used in many books and magazines, are much more complex and may contain as much as 30–50 percent mineral matter. Figure 10.B shows a cross section of a piece of paper similar to the page you are reading. The innermost zone of the paper is cellulose fiber; the second zone is calcium carbonate; and the outermost zone is the clay mineral kaolinite. These minerals provide strength, counteract acid decomposition, and provide a smooth surface that will accept the high-quality prints and pictures desired by publishers. If the surface were composed of cellulose fibers or of porous mineral matter, the printing inks would blur and would prevent high-quality image reproduction. The addition of the calcium carbonate and clay increases the quality and cost of the paper, but it also reduces the recyclability of the paper because it is so difficult to separate the mineral matter from the cellulose fibers. The presence of the mineral matter is also apparent if the paper is burned because so much ash remains.

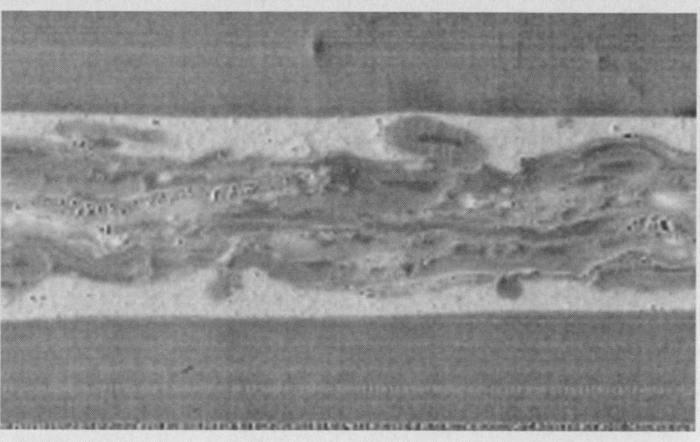

FIGURE 10.B Scanning electron microscope photograph of a cross section of coated paper such as that used in the production of most glossy pages in books and magazines. The central wood pulp portion has been coated on both sides with minerals such as calcium carbonate, clays, and titanium dioxide. The minerals are applied in a paint-like slurry and the paper then passes through a large roller press, called a *calender*, which compresses and smoothes it so it can receive high-quality images. (Courtesy of ECC Ltd.)

The application of the calcium carbonate and clay is accomplished by spreading a paint-like slurry of water, bonding agents, and mineral matter on heavy paper as it moves at a rate of several meters (feet) per second. The slurry-coated paper is then passed through a series of large metal rollers, called *calenders*, that flatten and polish the surface of the paper.

The use of mineral-coated papers has increased a great deal in recent years. Today, paper coating constitutes the second largest single use of clay minerals; the first is for the making of bricks. Before the development of papyrus, the Egyptians carved in stone or impressed cuneiform writing into clay tablets. At present, we no longer use clay tablets, but we are returning to the increased use of clay on the surfaces of our reading and writing materials.

The remarkable properties of asbestos minerals are chiefly the result of their crystal structures. For example, chrysotile [a sample of which is shown in Figure 10.22(a)] has a structure in which atomic layers of silica and magnesium hydroxide are rolled up to form scrolls, with several such scrolls making an individual fiber [Figure 10.22(b)].

Asbestos is valuable because the fibers, which are strong and flexible, can be separated, spun, and woven like organic fibers such as cotton and wool. Asbestos fibers also make materials more flexible when they are embedded in a suitable matrix. Unlike organic fibers, asbestos fibers are fire and heat resistant, stable in many corrosive environments, and are good electrical and thermal insulators. They are also resistant to wear and are strong, as the tensile strength shows. (See Table 10.3 and compare these values with that of steel piano wire, which has a tensile strength of 100,000 pounds per square inch.) Some asbestos is suitable in length and flexibility for spinning; chrysotile has the best spinnability, amosite and crocidolite are fair, and the others are poor. The length and flexibility of fibers will also vary within and between particular asbestos deposits. Asbestos is classified commercially on the basis of length of fiber and the degree of openness (or separation) of the fibers.

Asbestos has been used in a wide range of products. Some of the well-known uses are in the motor industry—for brake linings, clutch plates, and other friction materials—and in asbestos textiles used in the electrical industry. The single greatest application is in asbestos cement products where the large volume of short fibers that are unusable for spinning are bound in portland cement to make pipes, jackets, shingles, sheets, or corrugated and flat boards. The addition of asbestos increases the strength or flexibility of the product. It is also added to some papers, millboards, paints, putties, and plastics to provide desirable properties.

Chrysotile asbestos, which makes up the great majority of the world's asbestos production, is formed during the alteration of magnesium-rich (ultramafic) rocks such as peridotite and dunite. This alteration by hydrous solutions produces the serpentine minerals antigorite, lizardite, and chrysotile (which may or may not be asbestiform), all of which are hydrated magnesium silicates. Many ultramafic rocks have undergone this process of **serpentinization,** but few contain workable deposits of chrysotile asbestos. Formation of such deposits requires an unusual combination of faulting, shearing, folding, serpentinization, and metamorphism of the host rocks. Many mountain belts around the world contain bodies of ultramafic rock that have been serpentinized. Commonly, these rocks are parts of **ophiolite complexes,** a name given to masses of mafic and ultramafic rocks with associated volcanic rocks that were originally formed as part of Earth's crust beneath the oceans, and then thrust up onto the continents during mountain-building episodes.

Because only a very small fraction of serpentine bodies contain commercially exploitable asbestos, the availability of this commodity is similar to that of scarce metals. World production is dominated by Russia, China, and Kazakhstan, which together contribute about half of the total tonnage. Total world reserves are estimated to be in excess of 100 million tons and will likely be adequate for the foreseeable future. Canadian production of chrysotile comes mainly from the Eastern Townships of Quebec, where the deposits lie in a serpentine belt that extends from Newfoundland down into Vermont, where a small-scale U.S. production operates. The deposits of the Eastern Townships are a good example of the location of asbestos ore within the serpentinized rocks of an ophiolite suite. Other Canadian deposits occur in Ontario and within the Rocky Mountains in British Columbia and the Yukon Territory. Major producing areas for chrysotile in Russia occur in the Ural Mountains (Bajenova District). Anthophyllite is mined in Finland, while small deposits are worked for tremolite in Italy. Most of the mining of asbestos is by open-pit methods, with some underground mining of deeper deposits or those of unsuitable shape for surface exploitation. The ore is crushed and milled as a dry process, with screening to separate impurities. Mining of crocidolite and amosite, the most deadly forms of asbestos, has virtually ended.

Asbestos has apparently been known and used in small quantities for thousands of years, but it was not until the late nineteenth century that it became important as an industrial mineral. The modern industry grew from the processing in England and Italy of asbestos mined in Quebec, Canada. In 1900, the total production was 200,000 to 300,000 metric tons; it has been estimated that the cumulative world total production to 1930 was about 500 million metric tons. After that time, the use of asbestos accelerated so rapidly that, in 1979 alone, world production was about 5 million metric tons. However, the health hazards associated with asbestos have had a profound effect on its use in the United States, where consumption dropped from a high of 803,000 metric tons in 1973 to only 207,000 metric tons in 1985, and to about 5,000 metric tons in 2000 and less than 2000 tons in 2008. Consumption in the world reached a peak (about 5 million metric tons) in 1976 and remained above 4 million metric tons annually until 1990, but has since dropped to less than 2 million metric tons (Figure 10.23).

Although the relationship between asbestos and various diseases has received most attention over the last few decades, health problems associated with asbestos were evident before 1900 and within twenty years of the first factory production. Between 1890 and 1895, sixteen of seventeen workers in a French asbestos weaving factory had died. *Asbestosis,* a disease resulting from the inhalation of very fine particles of asbestos dust, was first recognized in 1906 and was completely described in 1927. The deposition of the asbestos fibers in the lungs causes the formation of scar tissue (fibrosis) (Figure 10.24) and results in fatigue and breathlessness after some years. Other diseases associated with asbestos exposure include lung cancer, which develops in about 50 percent of asbestosis sufferers and which is

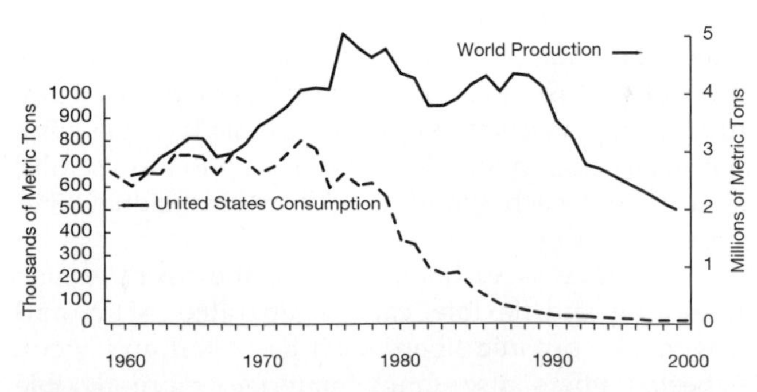

FIGURE 10.23 Worldwide production of asbestos peaked in 1976 and began to drop in the late 1980s. Consumption in the United States gradually rose until about 1973; it has since dropped dramatically because of concerns about its impact on human health. (Data from U.S. Geological Survey.)

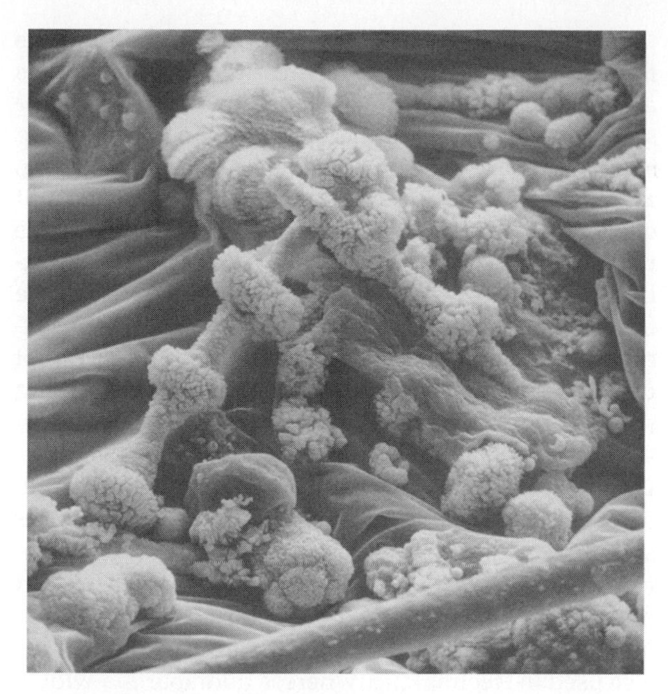

FIGURE 10.24 Scanning electron microscope photograph of asbestos fibers in an autopsied human lung. The asbestos fibers have been overgrown by iron-rich materials as the body tried to protect itself from the effects of the asbestos. (Photograph by Lesley S. Smith and Anne F. Sorling; from "Health Effects of Mineral Dusts," Mineralogical Society of America, *Review in Mineralogy*, vol. 28. Used with permission.)

also promoted by smoking, and mesothelioma, a rare cancer associated with the presence of crocidolite.

In most applications, asbestos fibers are totally encapsulated in various matrix materials; thus, they cannot be inhaled and do not pose a health risk. However, when asbestos fibers are freed into the air during mining and processing, during drilling and sawing, or during deterioration of the matrix (e.g., breakdown and peeling of old asbestos-bearing paints), they constitute a health hazard. The recognition that exposure to asbestos can have such deadly consequences has led to increasingly stringent rules concerning its use. In 1985, the U.S. Environmental Protection Agency (EPA) proposed banning the use of many asbestos-containing products (pipes, flooring, tiles) and phasing out the remaining uses of asbestos over a ten-year period.

Subsequently, there have been numerous additional EPA rulings banning the use of asbestos in various materials and some court orders overturning various regulations. One important ruling by the U.S. Occupational Safety and Health Administration (OSHA) in 1992 did remove the nonasbestiform varieties of actinolite, anthophyllite, and tremolite from the list of materials covered by legislation on asbestos under which they had been placed. It is apparent that, because of the widespread public concern over the hazards of asbestos to health and the large number of lawsuits being brought on behalf of those exposed to asbestos, its use will continue to decline worldwide. However, the health hazards of asbestos create a dilemma for some parts of the mineral industry, because no other natural or synthetic material has been found that exhibits the qualities of strength, chemical inertness, and durability of asbestos and at such low cost.

It is also true, as pointed out by Dr. Malcolm Ross of the U.S. Geological Survey, that crocidolite and amosite have been shown to be far more dangerous than the chrysotile asbestos, which accounts for about 90 percent of total usage. Careful thought must be given to the costs of removal of asbestos-containing materials and of the imposition of rigorous safety standards; for example, there is no evidence that ingestion of asbestos in drinking water causes disease. Furthermore, it is now apparent that the removal processes often liberate more asbestos dust than was present from its original applications.

Refractories, Foundry Sand, and Fluxes

Refractories are materials that can withstand high temperatures without cracking, spalling, or reacting despite contact with molten metals, slags, or other hot substances. Not only do refractories have to withstand high temperatures, but they also may be subjected to abrasion, impact, sudden temperature change, chemical attack, and high loads under these extreme conditions. They provide linings for the furnaces used in the smelting and refining of metals and the production of alloys, for the furnaces or kilns used in the ceramic industry, in glass and cement manufacture, and for coke ovens or boilers used in gas or electricity generation plants. There are a host of other uses ranging from the linings of incinerators to the manufacture of automotive spark plugs. This wide variety of applications means that many different raw materials with different refractory properties are used.

The industrial minerals now used in the greatest quantities to make refractories are clays that have traditionally been called **fire clays.** These are generally of sedimentary origin and are often found beneath coal seams in a sequence of sedimentary rocks. The dominant clay mineral present is kaolinite with various impurities that affect such properties as the plasticity of the clay. Fire clays can be molded and shaped to make bricks, tiles, or more elaborately shaped products before being fired. The heat resistance of such products ranges from 1500°C to 1650°C, and they have a wide range of applications as furnace and boiler linings. Fire clays are also used in the manufacture of refractory cements. These clays are found in England, Germany, and parts of the United States such as Pennsylvania,

Georgia, and Alabama. The U.S. production of fire clay is approximately 3 million metric tons, and there are very large reserves of this material worldwide.

Silica (SiO_2) in its various widely occurring natural forms is commonly employed to manufacture refractories. Some sandstones are directly cut into bricks; alternatively, quartzite or quartz itself may be crushed and bonded with lime to make silica bricks. Such bricks do not soften much below their melting point (approximately 1700°C) and are used in many metallurgical processes. Other refractories are high in alumina (Al_2O_3) and are made using bauxite (see Chapter 7) and other forms of natural alumina such as **diaspore** ($Al_2O_3 \cdot H_2O$). The aluminosilicate minerals **sillimanite, andalusite, kyanite,** which are all **polymorphs** of Al_2SiO_5, are also used to make high-alumina refractories. These minerals are common in metamorphic rocks; concentrations mined as commercial deposits occur in South Africa (andalusite), Sweden (kyanite), and Australia (sillimanite), as well as in the United States (in Virginia, California, and Nevada). High-alumina refractories can withstand temperatures up to 1800°C to 2000°C. Magnesia (MgO), produced from magnesium in seawater and from quarried magnesite and dolomite, is an important component in refractories used in the steel industry. Two of the best raw materials for high-temperature refractories are chromite, the nature and origin of which is discussed in Chapter 8, and **zircon** ($ZrSiO_4$). Chrome refractories are widely used in steel mills, whereas zirconia (ZrO_2), which will withstand temperatures of 2500°C, is used for refining precious metals. Zircon, providing the raw materials for these refractories, comes from beach sands in India, Brazil, Australia, and Florida.

Foundry sand is used to make the molds in which molten metals are cast. The most widely used foundry sand is known in the industry as **greensand** and is a mixture of silica (SiO_2), clay, and water. In some cases, the silica is a clay-free sand dredged from lakes or taken from dunes and then washed, graded, and dried. To this is added clay (commonly **bentonite**), water, and some cellulose to provide the necessary binding properties. There are also natural molding sands that contain sufficient clay to be used directly in the foundry. Other minerals that are sometimes used instead of silica for foundry sands include the silicate minerals olivine ($(Fe,Mg)_2SiO_4$) and zircon ($ZrSiO_4$), and chromite ($(Fe,Mg)(Cr,Al,Fe)_2O_4$). The refractory properties of chromite vary with its exact composition (Fe/Mg, Cr/Al, and Cr/Fe ratios). The particular type of sand used depends on the nature of the casting operation. In general, raw materials for producing foundry sands are widespread and not in short supply; annual production in the United States, for example, is about 30 million metric tons.

In contrast to the refractories and foundry sands that are valuable because they resist high temperatures and can be used to contain molten metals, fluxes are substances that help in the melting of material during smelting or in operations such as **soldering, brazing,** and welding. The fluxes used in smelting are aimed at efficiently separating the waste products from the metal in the form of a **slag.** The slag must have a relatively low melting and formation temperature, must be fluid at smelting temperatures, must have an appreciably lower specific gravity than the metal, and must not dissolve appreciable amounts of the metal being smelted. The most common fluxing materials are limestone, silica, and **fluorspar.** Limestone is a basic flux used in ferrous and nonferrous metallurgy; the calcium oxide formed when limestone decomposes produces slags of low specific gravity and low fusion temperatures when used in the smelting of copper and lead ores. As discussed in Chapter 7, limestone is also used in iron and steelmaking. Silica is an acid flux that is also used in steelmaking, whereas fluorspar is a widely used neutral flux employed to make slags more fusible and more fluid. Limestones, quartzites, and sandstones suitable for use as fluxing agents are relatively widespread throughout the world. Fluorspar deposits, which are formed from relatively low temperature **hydrothermal solutions**, are widespread but often small and impure.

Soldering and *brazing* are metallurgical joining operations in which a joint is formed using a filler metal of different composition and lower melting point to the joined pieces. Fluxes act to aid the spread of the solder or braze and mop up impurities (such as oxide coatings) that may hinder the formation of a successful joint. In soldering, which differs from brazing only in being a lower temperature operation (< 425°C), fluxes employed may contain various chlorides of zinc, ammonia, sodium, and tin, along with several organic compounds and acids. Brazing fluxes commonly contain borax and various borates or various fluorides and chlorides of sodium, potassium, lithium, and zinc. Fluxes are also used in welding, the joining of metals by heating to high temperatures with or without a filler metal being present. In this case, the compounds used in the fluxes include silica and oxides of manganese, titanium, aluminum, calcium, and zirconium. All industries using metals for construction and fabrication (for example, manufacturers of automobiles, ships, aircraft, household appliances) make use of these metal-joining techniques, especially welding.

Fillers and Pigments

Mineral fillers are fine particles of chemically inert, inexpensive mineral substances that are added to a great range of manufactured products to provide bulk or to modify the properties of the product by providing weight, toughness, opacity, or some other characteristic. A list of the more important mineral fillers is given in Table 10.4 with information on their compositions, useful properties, and applications. As can be seen, these materials range from the hydrated magnesium silicate **talc,** a very soft mineral best known for its use in cosmetics, through a variety of clays, to rocks such as limestone or pumice. Some major industries that use fillers are the manufacturers of paper, paint, plastics, rubber, pesticides, detergents, and fertilizers. Many of the mineral substances listed have other uses and are discussed elsewhere in the text, so their nature, occurrence, and exploitation need not be considered again here.

Several of the clay minerals (or clay rocks) provide useful fillers, being already of fine particle size. Kaolin, or china clay, although originally used in the manufacture of fine porcelain, is now used in large amounts by the paper industry (see Box 10.2 and Figure 10.B). The best quality clays, which are of high purity and whiteness, are used to coat high gloss papers. The United States and Great Britain are the world's principal producers of kaolin clays. The U.S. deposits occur as clay sediments and kaolinitic sands of late Cretaceous–early Tertiary age in Georgia and South Carolina. The kaolin was probably derived from deeply weathered bedrock and deposited by a system of rivers. The mined material is 90 percent kaolite, with quartz as the main impurity. The British deposits are in Cornwall and Devon and derive from the **hydrothermal alteration** of feldspars in granite bodies (Figure 10.19).

Clays such as **fuller's earth** and bentonite occur in sedimentary sequences, although they generally result from the breakdown and alteration of volcanic ash deposited within those sequences. Bentonite is chiefly made of the **smectite** group of clay minerals such as **montmorillonite, saponite,** and **hectorite** (see Table 10.2). Most bentonites appear to have formed from

TABLE 10.4 The more important mineral fillers

Substance	Composition	Useful Properties/Major Applications
Asbestos*	See Table 10.3	Fibrous and strong; building materials, tiles, plastics, etc.
Barite*	$BaSO_4$	Dense and inert; rubber, paint
Bentonite	Clayrock, a mixture of clay minerals (montmorillonite dominant)	Pesticides and detergents
Diatomite SiO_2	Sedimentary rock, dominantly from shells of small organisms (diatoms)	Porous and light, absorptive; paints, paper, plastics, pesticides
Fuller's earth	Clayrock, a mixture of clay minerals (dominant montmorillonite and attapulgite)	Absorptive; pesticides, greases, paper
Gypsum*	$CaSO_4 \cdot 2H_2O$	Low cost; paints, paper, cotton goods, pesticides
Kaolin* (china clay)	A pure clay mineral ($Al_4Si_4O_{10}(OH)_8$)	White color, low cost; paper, paint, adhesives, plastics, rubber, ink, pesticides
Limestone*	Rock, dominantly $CaCO_3$	Soft particles, soluble in acids, abundant and cheap; asphalt, fertilizers, insecticides, paints, rubber, plastic
Mica (muscovite)	Layer silicate ($KaI_2(AlSi_3)O_{10}(OH)_2$)	Layer structure and electrical insulation proper ties; roofing material, paint, rubber, wallpaper
Perlite*	Variable	Lightweight; fines used in paint, drilling muds, plastic
Portland cement*	Mostly calcium carbonate	
Pumice*	Vesicular volcanic rock	Stuccos, plasters, paint
Pyrophyllite	$Al_2Si_4O_{10}(OH)_2$	Soft platey structure; asphalt roofing, paint, rubber battery boxes
Rock dusts	Variable; commonly carbonates	Low cost, strong; asphalts and cheap fillers
Quartz (+ other silica)	SiO_2	Low cost, hard, inert; quartz paints, bitumens
Slate*	Rock comprised of silica and micaceous minerals	Low cost, inert, compatible with bitumens; roofing, sealing compounds, paints, hard rubber
Talc	$Mg_3Si_4O_{10}(OH)_2$	Soft platey structure, good adhesion; paints, rubber, roofing, insecticides, asphalt, paper, cosmetics, textiles
Vermiculite*	Variable	Fertilizers, pesticides

*Discussed in detail elsewhere in this book. See index for relevant sections.

volcanic ash that was transported considerable distances in the atmosphere and then deposited into the sea or lakes, where it broke down to form the constituent clay minerals. Commonly, the parent volcanic material is thought to have been a **rhyolite** or **andesite** (see Table 10.1). Some bentonites appear to have formed by hydrothermal activity and occur as irregular bodies in rocks showing other evidence of hydrothermal alteration. Hectorite deposits in California are thought to have formed by alteration of volcanic ash or **tuff** as a result of hot-spring activity in alkaline lakes. Most fuller's earth deposits are, in fact, bentonites, although there are also some that consist largely of the minerals **palygorskite (attapulgite)** and **sepiolite** (see Table 10.2). These may not have formed from alteration of volcanic materials, possibly having been precipitated from seawater evaporating in a tidal-flat environment. The name *fuller's earth* comes from the process of **fulling,** which is the removal of grease from wool or other organic fibers. Such clays, as well as being used as fillers, therefore have important applications because of their absorptive properties and are used as filters and purifiers of oils, fats, and various chemicals. **Diatomite,** a sedimentary rock made up of the siliceous skeletons of microscopic organisms called *diatoms,* also has important applications in purification and filtration (Figure 10.25). The United States is the world's leading producer from beds in California, Washington, and Oregon.

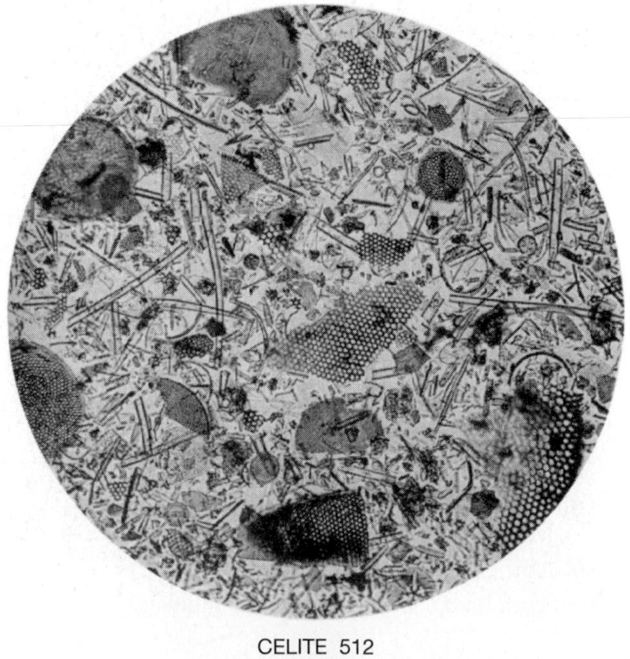

CELITE 512
500X
100 microns

FIGURE 10.25 Diatomite consists of the accumulated remains of siliceous shells of marine diatoms as shown in this photomicrograph from Lompoc, California. (Courtesy of the Mansville Sales Corporation.)

The addition of fillers will commonly modify the color of materials, whereas pigments are mineral-derived powders added solely to give color. From the cave paintings of primitive humans to the modern production of paints for industrial and domestic use, mineral pigments have been employed to provide long-lasting color. Some pigments are the untreated natural minerals, others are made by burning or subliming natural minerals, and a third group of pigments are chemically manufactured. Natural pigments include the ochers, umbers, and siennas that come from iron oxides and hydroxides (hematite, **limonite**), sometimes mixed with clays and manganese oxides, and which provide permanent reds, yellows, and oranges. Roasting of some of these mixtures gives the brown colors of burnt umber or burnt sienna. Green colors are provided by silicates rich in iron and magnesium, and white is provided by gypsum, barite, and white clays. Chemical paints were formerly made using lead compounds, but are now little used because of their toxicity. Zinc oxide, barium sulfate (barite), and titanium dioxide (rutile) are all used in white-paint manufacture, which consumes substantial amounts of these commodities. Natural pigments such as the ochers, umbers, and siennas generally form as residual surface deposits through the alteration of ores and rocks. Certain areas are noted for particular pigments: yellow ocher from France and the United States, sienna from Italy, umber from Cyprus, and red oxides from Spain are examples.

Diamond and Other Abrasives

Abrasives, of which diamond is the most important, are the materials that are used to cut, shape, grind, and polish all of the modern alloys and ceramics. Diamond—the most dense form of carbon—is the hardest natural or synthetic substance known. This hardness makes it essential for some uses and much more efficient than other abrasives for many other uses. We think of diamonds most often as beautiful cut gemstones; however, most natural diamonds are small, poorly shaped, and contain imperfections; thus, they are unsuitable for use as gems. In fact, only about 20 percent of natural diamonds are considered gemstones; the remaining 80 percent are used in industry. Such diamonds are classified using terms such as *die stones, tool stones, dresser stones, drilling stones, bort, diamond dust,* and *powder.* Their principal uses are in diamond saws, drilling bits, wire drawing bits, glass cutters, grinding wheels, and bonded and loose abrasives for lapping and polishing.

Natural diamond requires high pressures for its formation, pressures reached only at depths of 150 kilometers (95 miles) or more within Earth. Diamond-bearing rocks called **kimberlites** originate in Earth's mantle and reach the surface in narrow pipe-like vents, often no more than 50 meters (160 feet) in diameter (see Figure 10.26). The reasons for the formation and location of the pipes remain a geologic puzzle.

Kimberlite pipes are relatively rare, with only about 5000 known throughout the world (Figure 10.27). Of these, only about 500 are confirmed to contain diamonds and less than 100 contain sufficient diamonds to justify mining. South Africa remains the best-known location for diamond pipes because of its long and interesting history (see Figure 3.15), but it has now been surpassed in diamond production by several countries (see Table 10.5). The economic viability of a kimberlite pipe depends both upon the number of carats of diamond per ton and the percentage of the diamonds that are of gem quality. The kimberlites of South Africa average only about 0.3 carats per ton, but about 20 percent of the stones are of gem quality. In contrast, the Argyle mine in Australia, among the world's largest diamond producers at greater than 20 million carats annually and with a grade of 6 carats per ton, contains only about 5 percent gem-quality stones.

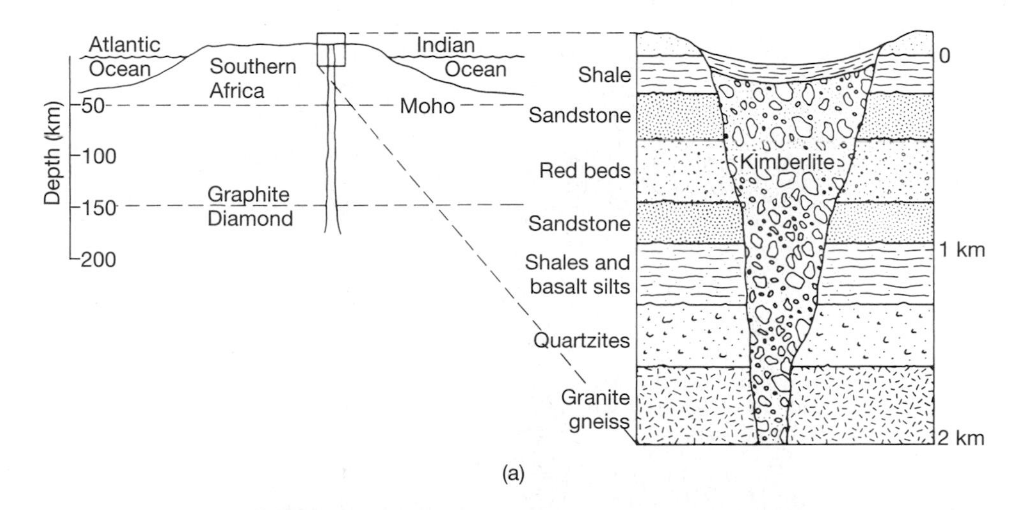

FIGURE 10.26 (a) Kimberlite pipes are the only primary sources of diamonds. The pipes formed as violent gas-rich volcanic explosions carried mixtures of volcanic rocks and fragments of the surrounding rocks upward. Diamond-bearing pipes must have originated at depths of 150 kilometers (95 miles) or more because diamonds do not form at pressure conditions encountered at shallower depths (see also Figure 3.15).
(b) Kimberlites consist of breccias with a wide variety of fragments representative of all rock types through which the pipe has intruded. Diamonds are recovered by crushing and carefully processing the Kimberlite. (Photograph by B. J. Skinner.)

FIGURE 10.27 Diamond-bearing kimberlites occur in all continents, but the greatest number of pipes are in central and southern Africa. The size of the circles shows, in a relative sense, the size of the deposits. In the later part of the twentieth century, the Argyle mine in Australia had become the world's largest producer of industrial diamonds. The discoveries of hundreds of kimberlite pipes in northwestern Canada in the 1990s have led to the development of several mines and will likely see Canada become one of the world's major diamond producers.

Because diamonds are very hard and resistant to weathering, they also accumulate in placer deposits. In fact, about 40 percent of diamonds mined today come from placers. Although placers can occur anywhere that rivers drain from kimberlite-bearing rocks, the most important today lie along several hundred miles of the coast of Namibia in southwestern Africa. The overall grades are low, but the percentages of gem-quality stones are high (up to 95 percent of the stones recovered); recovery is very easy because the stones are just mixed in with normal quartz beach sand. Diamond recovery from any source relies upon its high specific gravity of 3.52, its ability to stick to grease, and its characteristic fluorescence in an X-ray beam (it glows visibly when exposed to X-rays).

TABLE 10.5 The principal producers of natural and synthetic industrial diamond in 2008 (natural data from USGS; synthetic data estimated)

	Diamond production in millions of carats	
Country	**Natural**	**Synthetic**
United States	—	150
Australia	18	—
Botswana	8	—
China	1	30+
Ireland	—	150+
Japan	—	50+
Russia	15	100+
South Africa	9	100+
Congo	23	—
World Total	77	600+

Diamond production figures are divided into two categories: (1) industrial and (2) gems. The industrial stones include virtually all small stones (called *micros* if they are less than 0.45 millimeters) and many larger ones (if impure, cracked, or discolored). Natural industrial diamond production is dominated by Australia, Republic of the Congo, and Russia (see Table 10.5). Today, however, synthetic diamonds provide nearly 90 percent of the world's requirements for industry; consequently, all future needs can easily be supplied. See Box 10.3 for a discussion of synthetic diamonds.

Diamond synthesis was, for a long time, a goal of scientists. Once it was realized that diamonds only formed within Earth's mantle, there were many attempts to make diamonds using very high pressures and high temperatures. It was not until 1955 that the General Electric Company in the United States announced the development of a process for the synthesis of industrial diamonds. This, the first commercially viable process for the manufacture of diamonds, involved subjecting graphite to pressures greater than 1 million pounds per square inch (70,300 kilograms per square centimeter) and temperatures up to 2000°C in a sealed reaction vessel. The problem of promoting the transformation of graphite to diamond was solved by using molten nickel to dissolve the graphite and recrystallize the carbon as diamond. This process spawned an industry that by 2008 was producing over 650 million carats a year, which is nearly 10 times larger than world production of natural industrial diamonds (about 75 million carats; Table 10.5). Thus,

BOX 10.3

Synthetic Diamonds

"Diamonds are a girl's best friend," sang Marilyn Monroe in the movie *Gentlemen Prefer Blondes.* A similar sentiment is expressed by scientists and engineers who have discovered that diamonds are more than just beautiful. Diamond is by far the hardest substance known, it has the highest thermal conductivity at room temperature, it is virtually inert to attack from chemicals, it is an excellent electrical insulator, it is transparent to light and to X-rays, and can be prepared as a superior semiconductor for electronic devices. The drawbacks to its widespread use have been cost and low availability in the forms desired. Natural diamonds vary widely in size and degree of perfection and are so valued as gemstones that their extensive use in electronics is impractical.

Recognizing that both high temperature and high pressure must have been required to form the natural diamonds found in kimberlite pipes arising from Earth's mantle, scientists began serious efforts to synthesize diamonds in the early 1900s. Success came in 1955 when General Electric scientists in Schenectady, New York, heated graphite (another form of pure carbon) in the presence of a nickel or iron metal catalyst at about 1500°C and under a pressure of 50,000 to 60,000 atmospheres. Unfortunately, the crystals were small and randomly shaped (Figure 10.C) and, hence, of little value for electronics. They were, however, quite inexpensive, and production costs have averaged less than $1.00 per carat since the early 1990s. This has led to such large-scale production of this material, commonly called *bort,* that the world now produces many more synthetic diamonds for industrial use than are mined (see Table 10.5).

Scientists began to search for other methods of synthesis that could produce larger, sheet-like diamonds. Guided by a few initial successes in the 1960s, Japanese scientists in the 1980s began to solve the problems associated with a synthesis method known as *chemical vapor deposition* (CVD). In this technique, hydrogen gas is mixed with a hydrocarbon gas such as methane (CH_4) and heated to more than 2000°C using microwaves. The hydrocarbon gas decomposes, and carbon precipitates out on a substrate as a single crystal film suitable for electronics. General Electric scientists announced in 1990 that they had successfully made diamond sheets up to 30 centimeters long (12 inches) and 4 centimeters (2 inches) wide using CVD. The cost of CVD synthesis was announced as more than 100 times that of high-pressure synthesis, but the cost dropped rapidly as techniques were perfected. The CVD synthesis takes advantage of the fact that minerals can sometimes form even when they should not. Diamonds are actually only stable under the high temperature and pressure conditions of Earth's mantle. However, catalysts can sometimes promote crystallization under metastable conditions; that is, conditions under which some other mineral structure should form. Hence, CVD diamonds are created under conditions where graphite should exist.

The development of new synthesis procedures, yielding larger diamonds at lower prices, has scientists and engineers anticipating the use of diamonds in integrated circuits in computers, on tweeters in stereo speakers, and even as scratch-proof coatings on eyeglasses. In the 1990s, two other diamond synthesis breakthroughs were also announced: (1) General Electric produced gem-quality synthetic diamonds larger than one-half carat and DeBeers made a synthetic 5-carat gem diamond and (2) Scientists synthesized ultrapure diamonds, which are 99.9 percent composed of the isotope carbon-12 (natural stones are 99 percent carbon-12 and 1 percent carbon-13), and which have superior hardness and conductivity properties compared with natural stones. In the late 1990s, there was even the announcement of synthesis of a 25-carat diamond crystal, albeit slightly discolored; this offers many more potential applications. Diamonds may, indeed, be far more than just a girl's best friend.

FIGURE 10.C Synthetic diamonds now constitute the world's major source of abrasive diamonds. Synthetic diamond is also finding increasing use in coating surfaces to ensure long life and resistance to corrosion. There have been reports of the synthesis of large gem-quality diamonds, but these have not yet appeared on the market. (Courtesy of General Electric Company, Specialty Materials Department.)

about 90 percent of all abrasive diamond grit used is now artificially made, with production taking place not only in the United States, but also in Sweden, South Africa, Ireland, Japan, and Russia. Although other diamond synthesis techniques have been developed, including some using explosive charges to generate very high pressures for very short times as a shockwave, most methods result in diamonds of small size (see Figure 10.C). There have been many claims of synthetic gem-quality diamonds above 5 carats; DeBeers reported synthesis of a 34-carat stone. Despite increasing success in diamond synthesis, the natural diamond is still much preferred as a gemstone.

In recent years, there have also been developed very effective techniques to synthesize diamond films from carbon-bearing gases (especially methane, CH_4) at low temperatures using radio frequency waves. This is creating a whole new series of applications because diamond coatings can give surfaces nearly unlimited lifetimes by protecting them against wear. Furthermore, diamond has very useful electrical properties, so there are many potential applications of diamond thin-films as semiconductors and even computer chips.

Natural gem diamonds, which are priced arbitrarily according to people's desires, range from hundreds to tens of thousands of dollars per carat depending on size, color, and clarity. This makes diamonds, along with some rubies, sapphires, and emeralds, the most costly of all the mineral resources. In contrast, the prices of the industrial stones had dropped to less than $0.60 per carat in the mid-1990s, and to less than $0.25 per carat for the poorest quality today. Larger stones needed for special cutting dies are, of course, more expensive and may range up to $80 per carat.

Other important natural abrasives include **corundum, emery,** and **garnet.** Corundum, which is a hexagonal form of aluminum oxide, is the second hardest natural substance. It is used almost exclusively as a finely crushed and sized material for lapping and polishing optical glass and metals. As with diamond synthesis, methods have been developed for the synthesis of corundum, most particularly by heating bauxite in an electric arc furnace with small amounts of coke that chemically reduce impurities, especially iron with which the other impurities can then combine and sink to the bottom of the furnace.

Corundum could be completely replaced by other abrasives, but it retains some usage because it has the tendency to cut with a chisel-like edge rather than to scratch. Emery is a gray to black granular mixture of variable amounts of corundum, magnetite, spinel, hematite, garnet, and other minerals. Emery has been widely replaced by synthetic abrasives because of its variability, but it is still used in coated abrasive sheets, nonskid pavements, and in stair treads. Garnet remains a popular sheet abrasive for dressing wood and soft metals. Much more garnet is used as a sandblasting medium and as a grit and powder for optical grinding and polishing. Garnet is used almost exclusively as an abrasive, with the United States contributing more than 95 percent of world production and more than 80 percent of its use. This cubic silicate mineral that has no cleavage is extracted from metamorphic rocks, especially the very coarsely crystalline gneisses at North Creek, New York, where individual garnet crystals are often 10 to 20 centimeters (6 to 12 inches) or more in diameter (Figure 10.28).

Today, the production of more than 1 million tons of synthetic abrasives, such as alpha- and gamma-alumina, and various carbides and nitrides (materials of uniform quality, superior hardness, and comparable price to many natural abrasives) threatens the economic future of the natural abrasive industry.

Barite ($BaSO_4$)

The consumption of barite reflects the state of the world economy because barite is almost entirely used by the petroleum industry. Approximately 90 percent of world barite production is finely ground for use in drilling mud, the circulation of which lubricates the drill stem, cools the drill bit, and seals off the walls of the drill hole. The remaining 10 percent is used for a variety of chemical applications, glassmaking, and medical and pharmaceutical purposes. Barium sulfate, the mineral barite, has long been used to enhance the visibility of body organs and vessels in X-ray images, but this use is slowly declining as new imaging

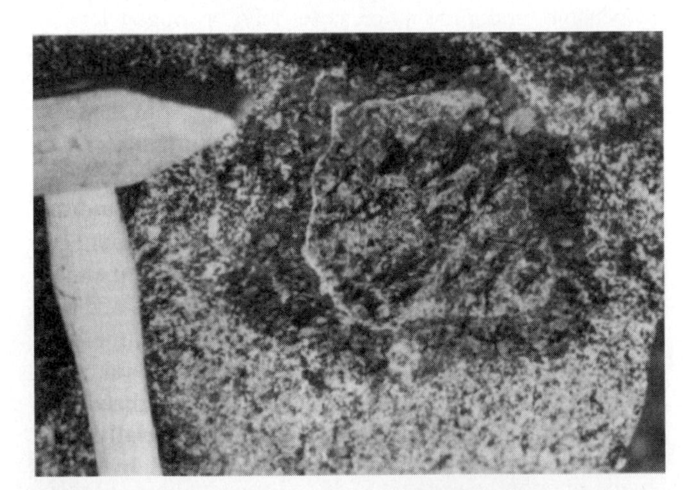

FIGURE 10.28 Garnet is a common mineral in many metamorphic rocks. It serves as an important abrasive material because it develops and retains sharp cutting edges when it is broken. Only rarely, however, are the crystals abundant or large enough to be economically exploited. This crystal is from Gore Mountain, New York, where crystals as large as 20 centimeters (8 inches) or more in diameter were long mined for use in abrasives. (Photograph by R. J. Tracy.)

procedures have been developed. Barite is very well suited for oil and gas drilling because its high density helps prevent blowouts that can occur when high pressures are met at depth. World barite production soared during the twentieth century, especially since the 1940s, as oil demand and oil drilling rapidly increased. Barite occurs worldwide in vein deposits and cavity-filling deposits, in weathered residual surface deposits, and in bedded accumulations. The origin of these deposits is not entirely clear, but most barite appears to have been deposited from hydrothermal solutions as fracture fillings or as chemical precipitates on the seafloor near volcanic vents. The world's principal barite producer is China (more than 4 million metric tons annually). Barite is mined in a great many countries, often for the individual nation's own local oil industry. The United States has long been the major consumer of barite, but the slowing of the U.S. domestic oil exploration and drilling program has seen a drop in the demand for barite from more than 2 million metric tons in 1983 to far less than 1 million metric tons by 1993. However, the rise in the demand and price of oil since that time has now raised the annual usage in the United States to more than 3 million tons and world usage to about 8 million tons.

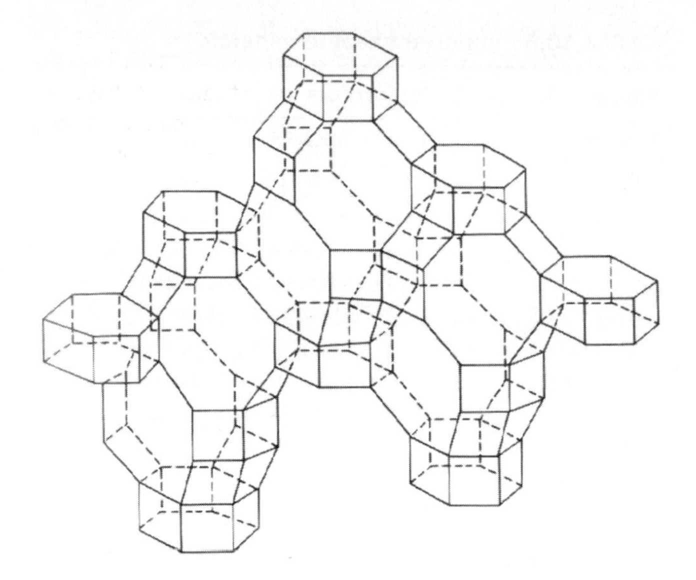

FIGURE 10.29 Schematic representation of the crystal structure of the zeolite mineral chabazite. The framework outlined consists of silicon and aluminum tetrahedra, and each framework unit contains a cavity connected to adjacent cavities by channels. This unique structure permits zeolites to serve as molecular sieves to separate organic molecules and to serve as hydrogen donors in the conversion of crude oil into gasoline. (After Breck and Smith, *Scientific American*, 200, [1959], p. 881.)

Zeolites

The zeolites are a group of hydrous aluminum silicates of sodium, calcium, potassium, and to a lesser extent, barium and magnesium. They have crystal structures in which tetrahedral clusters (SiO_4 and AlO_4) of atoms are joined together to form frameworks within which are large cavities containing water molecules (see Figure 10.29). The cavities may be interconnected in one, two, or three directions, and when zeolites are dehydrated by heating to approximately 350°C, this leaves a crystal permeated with channel systems in up to three directions. The sizes of these channels (apertures are generally approximately $2.5–7.5 \times 10^{-8}$ centimeters) are such that they will allow certain smaller molecules to pass through them but not the larger molecules. Hence, zeolites have great commercial importance as molecular sieves.

Zeolitic tuffs, formed by the alteration of volcanic ash deposits, have been used for more than 2000 years as lightweight building materials and in pozzolan cement. However, it has only been since the 1950s that the zeolite minerals themselves have been extracted to make use of their unique **ion exchange** and adsorption properties. Early applications were as water-softening agents to extract calcium and magnesium from drinking water and replace it with sodium. Subsequently, applications have greatly increased and include the selective extraction of radioactive elements such as cesium-137 from contaminated water, the extraction of poisonous ammonium ions from sewage and agricultural effluent, the extraction of sulfur and nitrogen oxides from smokestack gases, and of CO_2 and H_2S from natural gas.

Far more important today than the natural zeolites are the hundreds of thousands of kilograms (tons) of synthetic zeolites prepared from solutions of sodium hydroxide, sodium silicate, and sodium aluminates. Synthetic zeolites have larger structural cavities and may be prepared with specifically designed structural sites that allow them to be used to break down large organic molecules in the refining of oil. As a result, these synthetic zeolites now serve as the catalysts in the catalytic cracking units at every major oil refinery. The use of zeolites is more efficient than using simple thermal cracking (merely heating the oil to cause its breakup into smaller hydrocarbon units) and can be used to add hydrogen to the oil. This process, known as *hydrogenation,* results in an increased yield of gasoline from every barrel of petroleum.

Zeolites, the most important of which are listed in Table 10.6, occur in a wide variety of rock types, although prior to about 1950 most examples described came from fractures or cavities (known as **vesicles**) in altered igneous rocks. In these environments, zeolites form good crystals, readily identifiable, and of the kind sought after by museums and collectors but not occurring in the quantities necessary for economic recovery. In recent years, zeolites have been recognized as important constituents in a variety of sedimentary

TABLE 10.6 Important zeolite minerals

Name	Formula
Analcime	$NaAlSi_2O_6 \cdot H_2O$
Chabazite	$(Ca,Na)_2Al_2Si_4O_{12} \cdot 6H_2O$
Clinoptilolite	$(Na_2K_2Ca)_3Al_6Si_{30}O_{72} \cdot 24H_2O$
Erionite	$(Na_2K_2Ca)_{4.5}Al_9Si_{27}O_{72} \cdot 27H_2O$
Faujasite	$(Na_2Ca)_{1.75}Al_{3.5}Si_{8.5}O_{24} \cdot 16H_2O$
Ferrierite	$(K,Na)_2(Mg,Ca)_2Al_6Si_{30}O_{72} \cdot 18H_2O$
Heulandite	$(Ca,Na_2)_4Al_8Si_{28}O_{72} \cdot 24H_2O$
Laumontite	$Ca_4Al_8S_{16}O_{48} \cdot 16H_2O$
Mordenite	$(Na_2K_2Ca)Al_2Si_{10}O_{24} \cdot 7H_2O$
Phillipsite	$(K_2Na_2Ca)_2Al_4Si_{12}O_{32} \cdot 12H_2O$

rocks and in metamorphic rocks of low grade (i.e., formed by metamorphism under conditions of relatively low temperature and pressure). Clay minerals, feldspars, and feldspathoids can react with pore waters during metamorphism or in buried sediments to form zeolites. In sediments, the breakdown and reaction of **volcanic glass** appears to have often been the way in which zeolites formed, although they may also form by processes related to weathering or to alteration by hydrothermal solutions.

Commercial interest centers on the bedded, near-surface sedimentary zeolite deposits. These include deposits formed from volcanic material in saline lakes, which, although only a few centimeters to more than 10 feet, commonly contain nearly monomineralic chabazite and erionite. Deposits formed in marine environments or from groundwater systems can be several hundred meters thick and are characterized by clinoptilolite and mordenite. Zeolites are mined in the United States, Japan, Italy, Hungary, Yugoslavia, Bulgaria, Mexico, and Germany. The use of natural zeolites will continue to increase in the future as new applications are developed, but there will be increasing competition from synthetic zeolites that can often be tailored for special purposes.

Bituminous Materials

Bitumen is the general name for a group of materials made up of mixtures of hydrocarbons and includes petroleum, asphalts, asphaltites, pyrobitumens, and mineral waxes. Petroleum is discussed at length in Chapter 5 as a fossil fuel. It is worth emphasizing here, however, its importance as the source of a diverse range of organic products including plastics and other polymers widely used in industry. **Asphalt** is a solid (or near solid) hydrocarbon found in native form in fissures and pore spaces in rocks and even as small lakes. It probably formed by slow natural fractionation of crude petroleum at or near Earth's surface. Because of its resistant and waterproof characteristics, asphalt is widely used in road construction, flooring, roofing, and more specialized waterproofing compounds. Although most asphalt is refined from crude petroleum, rock asphalts (bituminous sandstones and limestones) are mined in parts of the United States (Kentucky, Texas, Oklahoma, Louisiana) and certain European countries (France, Germany, Italy, Switzerland) where they are used for local industries. Bitumen content is generally 3–15 percent. The best known of the much richer lake deposits is on Trinidad (West Indies), where the lake spans 114 acres and reaches a depth of 285 feet.

Asphaltenes, pyrobitumens, and mineral waxes are the other natural bitumens; the first two are dark solids mostly found in veins and fissures; mineral waxes are softer, as the name suggests. These relatively uncommon materials have a variety of uses ranging from paints, inks, and varnishes to rubber and plastic manufacturing.

GEMSTONES

The first uses of gems date from very ancient times, and they are believed by some to have been worn even before clothes. The designation gemstone is generally accepted to refer to materials appropriate for personal adornment and used for that purpose or in decorative artwork. Gems are a unique type of resource because very small amounts can have extremely high value. Hence, gems are the most valuable of Earth's resources per unit size or unit weight. A flawless diamond no more than a centimeter across can cost many tens of thousands of dollars, while a beautiful emerald can be even more expensive.

The most important properties of gem materials are color, luster, transparency, durability, and rarity. Size alone is not of great importance; thus, a perfect small stone is commonly worth far more than a large imperfect or poorly cut stone.

Of the 4000 or so known mineral species, only about 100 have attributes that allow them to be considered gems; the most important gemstones are listed in Table 10.7. Gems are commonly designated as precious (diamond, ruby, sapphire, emerald, and pearl) or semiprecious (all others) on the basis of market price, but all species exhibit wide variations in quality and value. Long used in jewelry because of their beauty (Plate 55), gems have also been viewed as endowing their wearers with mystical powers. Diamond, with a hardness greater than any other substance was considered a symbol of strength. Sapphire has been viewed as a symbol of heavenly bliss and faithfulness and as a protection for its owner against poverty and snakebites

TABLE 10.7 Principal types of precious and semiprecious gems

Name	Composition
Amber	Hydrocarbon (fossil resin)
Beryl:	$Be_3Al_2Si_6O_{18}$
Aquamarine	$Be_3Al_2Si_6O_{18}$
Emerald	$Be_3Al_2Si_6O_{18}$
Chrysoberyl:	$BeAl_2O_4$
Catseye	$BeAl_2O_4$
Corundum:	Al_2O_3
Ruby	Al_2O_3 (with trace of Cr)
Sapphire	Al_2O_3 (with trace of Ti)
Diamond	C
Feldspar:	$KalSi_3O_8$
Amazonstone	$KalSi_3O_8$
Garnet	$(Ca,Mg,Fe)_3(Al,Fe,Cr)_2(SiO_4)_3$
Jadeite	$Na(Al,Fe)Si_2O_6$
Peridot	Mg_2SiO_4
Opal	Hydrous silica
Pearl	$CaCO_3$
Quartz:	SiO_2
Agate	SiO_2
Amethyst	SiO_2
Jasper	SiO_2
Onyx	SiO_2
Spinel	$MgAl_2O_4$
Topaz	$Al_2SiO_4(F,OH)_2$
Turquoise	$CuAl_6(PO_4)_4(OH)_8 \cdot 5H_2O$

(Figure 10.30). Ruby was believed to bring peace, love, and happiness to its possessor, and emerald was thought to confer riches, fame, and wisdom to its owner. In contrast, some well-known gemstones, such as the Hope Diamond, have associated stories of curses that befall those who own or wear them.

Gems are generally measured in weight units called *carats* (equal to 0.2 grams). Small stones are also commonly measured in points, each point being $\frac{1}{100}$ of a carat. Gem sizes range from the smallest of chips that can be incorporated into jewelry up to crystals measured in hundreds of carats. The largest cut gem known is the Brazilian Princess, a 21,327-carat (about $9\frac{1}{2}$ pounds), light blue topaz found in eastern Brazil in the 1960s. Many large and beautiful diamonds have been discovered, but among the most famous are the Hope Diamond (a blue, 44-carat stone from India now in the Smithsonian Institution's Museum of Natural History) and the Cullinan, or Star of Africa (a colorless, 3106-carat stone from South Africa), that was cut to form part of the British Crown Jewels kept in the Tower of London (Table 10.8).

For many thousands of years, gems were used as they were found, without cutting or polishing; however, from about 4000 B.C. onward, gemstones have been engraved, drilled, and cut. Until the late Middle Ages, most gems were cut as flat slabs or rounded into a low dome shape called a *cabachon*. In the fifteenth century, the cutting power of diamond was discovered by gem workers in France and the Netherlands, and modern cutting and polishing techniques employing diamond abrasives were developed. Today, virtually all crystalline gems are cut and polished to provide the most effective display of light and color. Thus, diamonds are transformed from fragments and roughly equant crystals [Figure 10.31(a)]

(a)

(b)

FIGURE 10.30 (a) Naturally occurring sapphire in its typical six-sided, barrel-like crystal. (b) Sapphire, cut and polished into a form ready for mounting. (Photographs by Bart Curren; courtesy of the International Colored Gemstone Association.)

TABLE 10.8 Largest Rough and Cut Diamonds

Largest Rough Gem Diamonds

Weight (carats)	Name	When found	Where found	Cut into
3106.00	Cullinan	1905	South Africa	Cullinan I-IX; 96 others
995.20	Excelsior	1893	South Africa	21 gems (max. = 69.8 ct)
968.90	Star of Sierra Leone	1972	Sierra Leone	—
787.50	Great Mogul	1650	India	Great Mogul (280 ct)
770.00	Woylie River	1945	Sierra Leone	30 gems (max. = 31.35 ct)
726.60	President Vargas	1938	Brazil	Vargas (48.26 ct); 22 others
726.00	Jonker	1934	South Africa	Jonker (125.65 ct); 11 others
650.80	Reitz	1895	South Africa	Jubilee (245.35 ct); 1 other

Largest Cut Diamonds

Weight (carats)	Name	Shape	Location or history
530.20	Cullinan I	Pear	British Crown Jewels, London
317.40	Cullinan II	Cushion	British Crown Jewels, London
280.00	Great Mogul	Rose	Lost after sack of Delhi in 1739
277.00	Nizam	Dome	Nizam of Hyderabad, India
250.00	Indien	Pear	Reported by Duke of Brunswick 1869
245.35	Jubilee	Cushion	Paul-Louis Weiller, Paris 1971
234.50	DeBeers	Unknown	Sold to Indian Prince in 1890
228.50	Victoria 1880	Brilliant	Sold to Indian Prince in 1882
205.00	Red Cross	Square	Sold in London in 1918
202.00	Black Star of Africa	unknown	Exhibited in Tokyo in 1971
45.52	Hope Diamond	Cushion	Smithsonian Institution

into lustrous and striking cut stones [Figure 10.31(b)]. The form into which a gem is cut depends upon its original size, shape, and impurities; the most common of these cuts are shown in Figure 10.32. The most favored cut for diamond is the *brilliant cut*, which has 58 facets; it was developed in Venice about 1700 and promotes the internal reflection of light so as to enhance the appearance of the diamond.

Since about 1900, techniques for the synthesis of several types of precious and semiprecious gems have

(a)

(b)

FIGURE 10.31 (a) Uncut gem diamonds as recovered from kimberlites in South Africa. Many of the crystals exhibit a crude octahedral shape. (b) *Brilliant* and *marquise* cut diamonds ready for jewelry mounting. The diamonds are cut to take advantage of their ability to reflect light internally to increase their sparkle. (Photographs courtesy of De Beers Consolidated Mines, Ltd.)

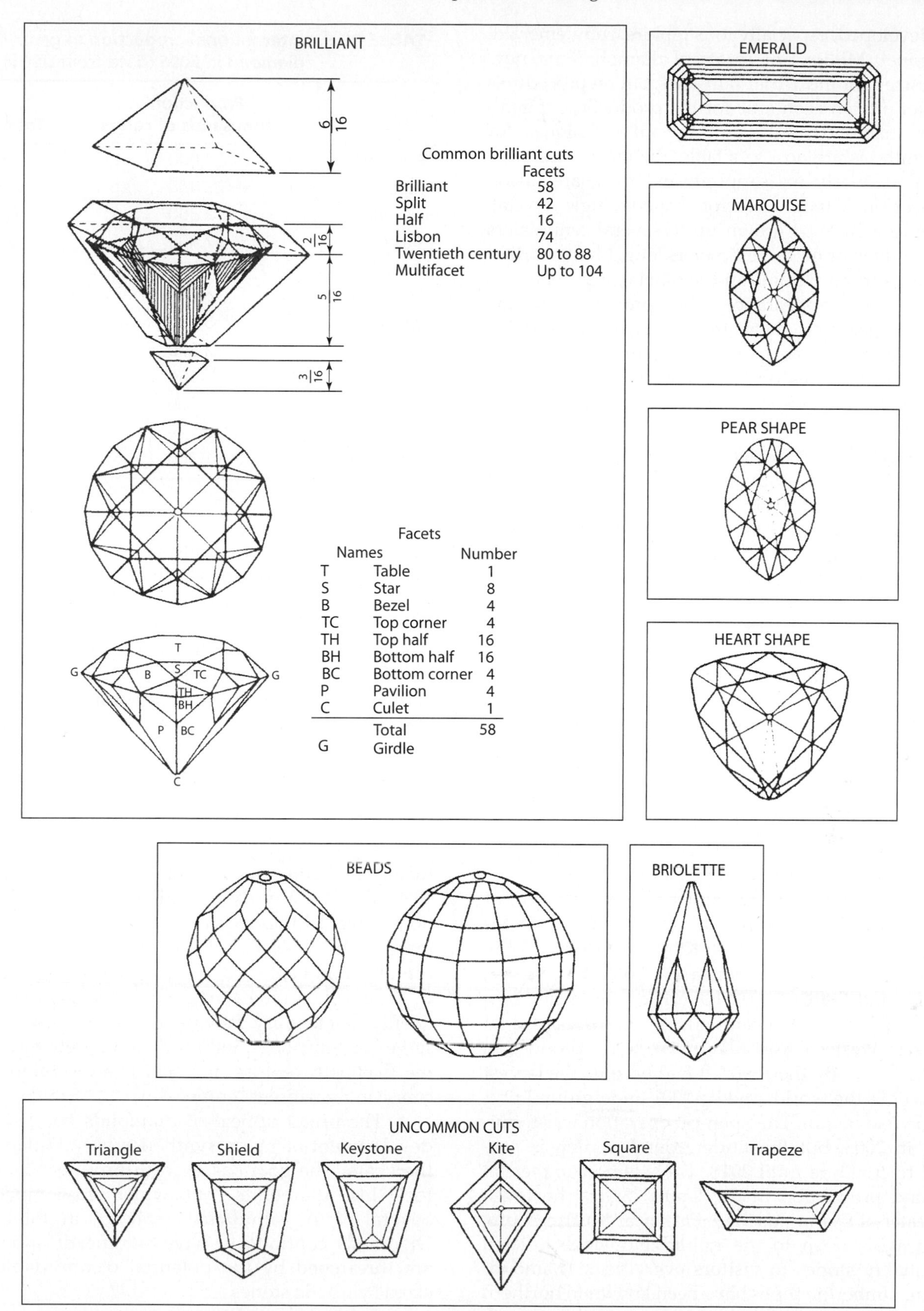

FIGURE 10.32 The principal gemstone cuts used in jewelry today. The most popular cut for diamond is the *brilliant.* (From the U.S. Bureau of Mines, *Mineral Facts and Problems,* 1985.)

been developed, especially for sapphire, ruby, emerald, and spinel. Millions of carats of diamonds are now synthesized for industrial purposes, but no procedures have yet proven economic for the production of gem-quality diamonds. Although several techniques for gem synthesis are now available, the most commonly used, particularly for sapphire and ruby, are fusion processes in which a carrot-shaped single crystal, known as a *boule,* is grown up to several centimeters across and ten or more centimeters long. Many modern gemstones, both natural and synthetic, are treated to enhance their appearance. The processes include bleaching, staining, heat treatments, and radiation treatments. Other means of enhancement include the use of foils or dyes on the backs, gluing stones together, and using gels to fill in cracks and pits.

Gemstones occur in many parts of the world and have formed in such diverse environments as Earth's mantle (diamond), high grade regional and contact metamorphic rocks (ruby and sapphire), pegmatites (emerald), oysters (pearls), and trees (amber). As far as commerce is concerned, diamonds are by far the most important of gems, and South Africa has been the most significant source of gem diamonds ever since discovery there in the 1860s. The original discoveries were of placer diamonds weathered out of diamond pipes, volcanic vents that have brought up material from Earth's mantle.

The largest diamond ever recovered was the Cullinan, which was found in 1905 in South Africa's Premier mine; it weighed 3106 carats (621.2 grams or 1.4 pounds) before cutting. One side of the Cullinan was a natural cleavage plane indicating that the stone had originally been much larger. Despite all efforts, the other portion has never been found.

Placer diamonds are still being recovered, but the major mines today are the diamond pipes themselves, such as the one at Kimberley (Figure 3.15). In recent years, however, Australia, Botswana, Russia, Canada, and South Africa have been the major producers (Table 10.9). Diamonds were discovered at Argyle in Western Australia in 1967, and production began in 1981. By the 1990s, it had become the largest producer in the world, and by 2001 underground production had begun. The open-pit operation was terminated in 2008, but the underground mining is projected to continue until 2018. There is no commercial diamond production in the United States; however, the Crater of Diamonds State Park near Murfreesboro, Arkansas, is open to the public and yields several gem-quality stones to visitors every year. Diamond-bearing kimberlite pipes have been known in northern Colorado for some years, but their grades have never been sufficient to consider mining.

Many diamonds have been found in the glaciated territories of North America, but their original sources, lying beneath the glacial tills, were not identified until 1990. Then a geologist, who had conducted geologic detective work for ten years by tracking the glacial dispersal patterns of the kinds of minerals found with diamonds, discovered a kimberlite pipe near Lac de Gras in the northern Northwest Territories of Canada. Subsequently, at least scores of additional pipes have been found in what appears to be one of the world's richest clusters of kimberlites. Some of the pipes were shown to be twice as rich as the typical South African pipes and to contain good percentages of gem-quality stones. This discovery led to a claim-staking rush more intense than at any time since the California Gold Rush. The Ekati mine, the first in Canada, opened in 1998 and has already exceeded production expectations. Two additional mines have been opened and many of the other pipes have been tested and found to contain commercial diamonds, so it is clear that Canada will become a major diamond producer in the twenty-first century. Few data exist on the world's reserves of gemstones, but production grew throughout the twentieth century and will likely continue to increase in the present century.

TABLE 10.9 International production of gem diamond in 2006 (data from USGS)

Country	Production in thousands of carats	Total %
Botswana	24,000	26.4
Russia	23,400	25.7
Canada	12,350	13.5
Australia	7,305	8.0
Angola	7,000	7.7
South Africa	6,240	6.8
Congo	5,600	6.1
Namibia	2,200	2.4
Ghana	780	0.8
Brazil	300	0.3
All others	2,125	2.3
World total	91,300	100

The broad appeal of diamonds has led to the development of many synthetics or stimulants over the years. The goal has been to produce stones that look like diamonds, but which cost much less. Several of the simulants are listed in Table 10.10. Diamonds continue to have significant appeal, but are threatened by the potential of affordable gem-sized synthetic stones.

Pearls (Figure 10.33) are unique as gems because they are harvested instead of being mined and are a truly renewable resource. They form within the shells of oysters, as the result of the deposition of concentric

TABLE 10.10 Diamond and its simulants

Material	Chemistry	Hardness
Natural diamond	C	10
Synthetic diamond	C	10
Glasses	SiO_2, Pb, Al, Tl	6
White sapphire	Al_2O_3	9
Spinel	$MgAl_2O_4$	8
Rutile	TiO_2	6
Cubic zirconia	ZrO_2	8.3
Boron nitride	BN	9
Moissanite	SiC	9

layers of calcium carbonate (in the form of the mineral aragonite) around a sand grain or other irritant. The occurrence of natural pearls is very unpredictable because only some oysters have had sand grains accidentally taken in, and the only way to discover if a pearl is in an oyster is to open it. Cultured pearls result from the intentional placing of sand grains into oysters and then allowing the oysters to live in ideal conditions for several years. Recently, the process has been accelerated by using seed spheres cut from oyster shells instead of small sand grains, as shown in Figure 10.34. In

FIGURE 10.33 Cultured pearls were traditionally produced by placing a sand grain into a young oyster and allowing the oyster to precipitate calcium carbonate layers over the grain. The external layers are identical to natural pearls, and they are produced more rapidly and reliably than the random occurrence of natural pearls.

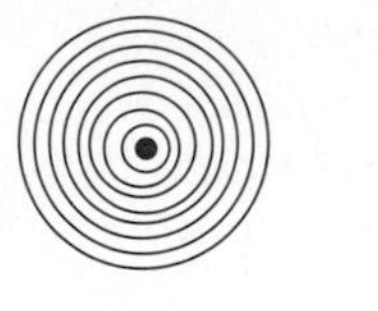

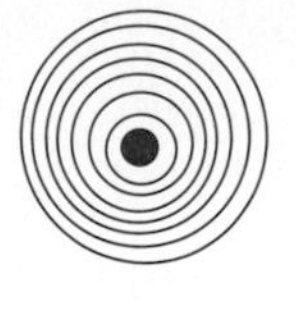

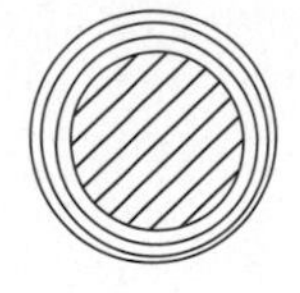

FIGURE 10.34 Cross sections of natural and cultured pearls. Natural pearls have developed when calcium carbonate layers were precipitated over sand grains that were accidentally introduced into young oysters. Cultured pearls were, for many years, very similar to natural pearls except that the sand grains were artificially introduced. In recent years, the growth time for cultured pearls has been decreased by introducing a small sphere cut from oyster shell into the young oysters instead of sand grains. The small sphere is the same composition as the pearl material and serves as a nucleation site for the overgrowth of calcium carbonate; the precipitation of only a few layers gives the pearl an appearance identical to that of natural pearls and takes much less time.

this way, the oysters require much less time to produce a pearl of larger size, and once the seed material has been coated by aragonite, the interior material is not visible. Pearls are softer than any of the other gemstones and require gentle handling so that they are not damaged. Cultured pearls, especially those that have relatively thin veneers of aragonite over a seed material, are even more vulnerable to damage.

THE FUTURE FOR BUILDING MATERIALS AND INDUSTRIAL MINERALS

The mineral commodities discussed in this chapter are essential to nearly all modern industries, directly or indirectly. The construction of roads, bridges, dams, and all kinds of buildings, the extraction of fossil and nuclear fuels and metals, and the manufacture of chemicals, plastics, ceramics, glass, paper, and processed foodstuffs all require minerals. They are essential to the complex system of dependencies on which modern industrial societies are built.

The reserves of most of the minerals and rocks discussed in this chapter are large, and the potential resources even larger. Commodities such as building stone are so abundant that it is pointless even to attempt an estimate of the reserves.

Water and Soil for Life Support

The ancients recognized four basic elements—earth, air, fire, and water—and considered them to be the controls of life. Our view today, although based on countless scientific investigations over the intervening years, is rather similar. We realize that three basic components of Earth—the soils of the lithosphere, the gases of the atmosphere, and the water of the hydrosphere—still provide the necessities to sustain life. Fire, the equivalent in our view to diverse forms of energy, is needed to make everything work.

WATER

With the exception of the oxygen in the atmosphere, no resource is more vital to human survival than water. In fact, there are no known life-forms that can survive without water. However, despite how critical water is for survival, it is commonly taken for granted, and its complex role in geological and geochemical cycles is all too often overlooked. The development of soils, with their ability to support the biosphere, is inextricably tied to the presence of water. Without water, rocks would be unchanged and their mineral components unusable by plants. A comparison of Earth with the other planets and large moons in the solar system reveals that only Earth has an oxygen-rich atmosphere plus abundant liquid water, together with soils containing organic matter and hydrated minerals capable of supporting an abundant biosphere. Other terrestrial planets may contain trace amounts of oxygen and may even contain ice in their polar regions and traces of water vapor in their atmospheres, but they are far more sterile in terms of life support than are the driest of Earth's deserts and the coldest of Earth's polar regions.

Seen from space, Earth is a blue planet because 70 percent of its surface is covered by water. Earth also has many white areas because its atmosphere contains scattered clouds, evidence that water vapor is transported from equatorial regions to polar regions, and concentrated there as ice, which constitutes the greatest amount of freshwater on Earth. Earth's hydrosphere is an astounding volume of water—some 1360 million cubic kilometers (326 million cubic miles). The movement of water through the hydrologic cycle is complex but continual. On the scale of the whole Earth, we see the flow of rivers from the continents to the sea and the torrential rains of great storms over the oceans; but on a local scale, we see the condensation of dew in the evening and the slow percolation of groundwater to the surface in a spring; all such flows are parts of the vast hydrological cycle.

The perception that Earth's crust contains huge quantities of mineral resources, but that only a very small portion of these is actually accessible, applies equally to water. More than 97 percent of Earth's water lies in the oceans and is difficult to use because of its high salinity. Another 2 percent of the world's water is fresh, but it is held in icecaps and glaciers, which means that most of it is unavailable because of its location. As a result, we humans have had to learn to survive with less than 1 percent of Earth's free water—and more than half of that is held in deep groundwater reservoirs, which are not readily accessible.

Availability of the less than one-half of 1 percent of the water that we can readily access is further complicated by large temporal and spatial variations in its distribution. Annual precipitation varies widely around the world, from more than 500 inches in parts of India and Hawaii to less than one-half inch in some of the world's great deserts such as those in North Africa, northern Chile, Namibia, and Australia. Many areas experience cyclical rainfall, with great downpours for several months, then little or nothing for the remainder of the year. The United States is fortunate in having large water supplies, but every year there are many local occurrences of droughts and floods. Sometimes human activities exacerbate the natural variations in water accessibility because we build cities in deserts where water supplies are limited (such as Phoenix, Tucson, and Las Vegas) or we build cities in areas that are periodically subject to natural flooding. We further change natural systems by draining wetlands, by increasing areas of impermeable land by building roads and parking lots which promote runoff, and by building dams, cutting channels, and allowing pollutants to contaminate waters.

Humans now make use of half of the accessible freshwater on earth. When world population approaches 12 billion, as well could happen in the twenty-first century, current water supplies will be severely taxed. The hydrologic cycle makes water a renewable resource, but human demands already exceed rates of replenishment in many places, as evidenced by falling groundwater tables and by drained rivers. A case in point is the Colorado River, which still flows through the Grand Canyon. So great is the demand for its water that for many years none of its water has reached the great delta at the head of the Gulf of California. Because rivers are sometimes international boundaries or flow through many nations, water supply problems are frequently international and the focus of political and economic issues. Every nation now recognizes that more water will be needed to supply growing populations, despite the fact that the total supplies are more or less constant.

Water is also inextricably tied to energy generation because it is one of the "green," or environmentally friendly, means of energy production. But the same dam that creates electricity without releasing any greenhouse gases can alter the natural river system and prevent fish migration. The oceans may offer nearly limitless quantities of water, but to convert seawater to usable freshwater requires the expenditure of huge amounts of energy.

SOIL

Soil, like water, is often taken for granted despite the critical role it plays in providing human food. Soil forms as the result of natural weathering processes and is highly influenced by the availability of water. The weathering process is especially effective in the decomposition of original rock materials to form the hydrated aluminosilicate clay minerals. The clays largely determine soil textures and play a major role in soil fertility by serving as the means of providing vital nutrients to plants. Moderately watered plains in temperate regions—like the American Midwest and the great farmlands of Europe and Asia—possess the world's best soils and produce much of the world's food. Desert soils, with little rainfall and high salt contents, and tropical soils, heavily leached of nutrients by high rainfall, are usually much less productive than are soils in temperate regions.

Most of the world's best agricultural lands have already been developed. Providing food for a doubling of population over the next century will require the use of marginally productive lands, more fertilizers, increased irrigation, and more genetic engineering of crops. At the same time, there are concerns about soil losses due to erosion; locally, loss of land to agriculture occurs because of other factors such as roads, buildings, and housing subdivisions. Meeting the world's needs for cropland to provide food for the growing human population may well be the greatest challenge of the twenty-first century.

CHAPTER 11

Water Resources

Three major components of Earth's hydrologic cycle are the oceans, which contain the vast majority of all water; glaciers and icecaps, which contain most of the fresh water; and the atmosphere, which serves as a conduit to transport water rapidly among the components. These are visible in this photograph of the Muir glacier on the coast of Alaska. (Photograph courtesy of Andrew Maslowski.)

Many existing sources of water are being stressed by withdrawals from aquifers and diversions from rivers and reservoirs to meet the needs of homes, cities, farms, and industries. Increasing requirements to leave water in streams and rivers to meet environmental, fish and wildlife, and recreational needs further complicate the matter.

W. B. SOLLEY, R. R. PIERCE, AND H. A. PERLMAN, (1998), ESTIMATED USE OF WATER IN THE UNITED STATES IN 1995. *U.S. GEOLOGICAL SURVEY CIRCULAR 1200.*

FOCAL POINTS

- Water is the most vital natural resource because it is essential for human survival.
- 97.2 percent of the water of Earth's hydrosphere is contained in the oceans; this water is saline, containing 3.5 percent salts, and is not potable.
- Most of the freshwater—2.15 percent of Earth's total water—is held in ice caps and glaciers and is unavailable for human use.
- The largest store of freshwater available for human use occurs as groundwater.
- The hydrologic cycle describes the constant movement of water from the oceans and the biosphere to the atmosphere by *evapotranspiration,* and back to the oceans, often via the land surface, as *precipitation.*
- The world's principal rainfall belt lies along the equator, whereas the main desert regions lie 25–30 degrees north and south of the equator and in the polar regions.
- Flooding occurs as a natural consequence of heavy rainfall, but is aggravated by human constructions on floodplains and by human activities that increase surface runoff.
- Attempts to control flooding usually involve the construction of dams to hold back water and channelization to promote rapid water flow away from an area.
- Supply systems to provide potable water for communities date from prehistory and are used on a massive scale today.
- The United States has the highest per capita water usage, averaging about 1340 gallons of freshwater per day.
- In the United States, electricity generating plants use the greatest amounts of water, most of which is recovered and recycled, but irrigation consumes the greatest amount of water. The largest irrigated crop in the United States is the turf grass of lawns, sports fields, and golf courses.
- Extraction of groundwater can lead to a lowering of the water table, ground surface subsidence, and even saltwater intrusion into aquifers in coastal areas.

- The importance of water as a resource will inevitably grow as world population increases.
- Cyclical droughts create the need for alternative water supplies for metropolitan areas and agriculture.
- Locally, the need for in-stream flow to protect endangered fish, shellfish, and other aquatic creatures limits the withdrawal of water for human consumption and agricultural needs.

INTRODUCTION

Seen from outer space, the blue oceans and white clouds make it obvious why Earth is called the water planet (Figure 11.1). Indeed, no resource is more apparently abundant or more necessary to us than the water that covers three-quarters of Earth's surface and that moves constantly about us in both seen and unseen forms. From the dawn of history, the oceans, rivers, lakes, and springs have served humans in many ways—as gathering points, as routes of transport, and as either the means of, or barriers to, our migrations. The availability of clean water is as important for the development and maintenance of modern cities as it was for the most primitive of early communities.

Yet despite its global abundance, there is an uneven distribution of water over Earth that creates local problems. These problems are often aggravated by our desire to use ever increasing amounts of water, by our modification of natural waterways, and by our contamination of surface and groundwaters. These problems highlight our need for a thorough knowledge of the distribution of water, an understanding of the effects of our activities on water availability and purity, and for long-range planning of water requirements. Indeed, water is the most critical of the resources on which we rely. The rising world population, especially in areas of limited freshwater, will strain the capabilities of supply systems; this is occurring at the same time as an increasing awareness of the need to use large quantities of water for preservation of the environment. Consequently, there is an increasing political importance given to freshwater, and that importance will grow in the years ahead.

FIGURE 11.1 Earth, the water planet, as seen from the Apollo 17 spacecraft. The abundance of water in the oceans, clouds, and ice caps gives Earth an appearance that is unique among the planets. (Photograph from NASA.)

THE GLOBAL DISTRIBUTION OF WATER

The total amount of water available in Earth's hydrosphere is approximately 1360×10^6 cubic kilometers (326 million cubic miles), or 1.36×10^{21} liters, distributed as shown in Figure 11.2. Of this total, 97.2 percent resides in the oceans and 2.15 percent is frozen in polar ice caps and in glaciers. The oceans are saline and not directly usable for most human needs; glacial and polar ice is fresh but inaccessible. Consequently, the vast majority of our requirements for freshwater must be met by the remaining 0.65 percent of the total water. Distribution of this small proportion of Earth's water at any given time is a function of the hydrologic cycle and the storage capacity of aquifers and surface landforms such as rivers and lakes. Thus, the problems of water supply are more complex than just total abundance; they also include local distribution patterns, rates of recharge and natural loss, and, increasingly, the cleanliness of the water. The availability of **potable water** (i.e., water suitable for drinking), more than any other factor will, in the future, determine the number of people who can live in any geographic province as well as many aspects of their lifestyles.

The Hydrologic Cycle

Water in the region of Earth's surface is constantly moving through the hydrologic cycle (Figure 11.3). The atmosphere is a great solar-powered "engine" that draws up water by evaporation and transpiration, transports it as vapor water, condenses it as clouds, and then discharges it as rain or snow. The precipitated water completes the hydrologic cycle by flowing via rivers, streams, and groundwater systems back to the oceans, by evaporation from the land surface back into the atmosphere, or by transpiration from

	Location	Water volume km^3	Percentage of total water
Surface water			
	Fresh-water lakes	125×10^3	.009
	Saline lakes and inland seas	104×10^3	.008
	Average in stream channels	1×10^3	.0001
Subsurface water			
	Vodose water (includes soil moisture)	67×10^3	.005
	Groundwater within depth of half a mile	4.2×10^6	.31
	Groundwater-deep lying	4.2×10^6	.31
Other water locations			
	Icecaps and glaciers	29×10^6	2.15
	Atmosphere	1.3×10^3	.001
	World ocean	1320×10^6	97.2

FIGURE 11.2 Distribution of water in various forms and locations on Earth in terms of cubic kilometers and percentages. (From the U.S. Geological Survey.)

plants. Each region of the world has a natural water budget in terms of precipitation, **evapotranspiration,** and runoff, but human activities commonly alter these budgets.

Water has the highest heat capacity (ability to absorb and hold heat with minimal temperature change) of any substance known. Consequently, the movement of massive amounts of water through the atmosphere and the oceans also involves the movement of large quantities of thermal energy; these movements play a major role in the control of the world's climates. This effect is probably best seen in the North Atlantic Ocean where the Gulf Stream, warmed by the sun in the Caribbean, flows northeastward as the North Atlantic Current, releasing heat to provide the relatively mild climates of northern Europe. Without the Gulf Stream to transport this heat, Britain and Scandinavia would be as cold as northern Canada or Siberia, regions which lie at the same latitude.

The evaporation of water from any wet surface requires the input of 540 calories for every gram of water that changes from a liquid to a vapor. Because this heat comes from the surrounding environment such as the remaining water, soil, or air, evaporation is a very effective cooling process. The condensation of water vapor to liquid water reverses the heat flow and liberates 540 calories per gram of liquid water condensed. The melting of ice to water also requires energy—80 calories per gram—and, hence, is also an effective cooling process.

Precipitation and Evaporation Patterns

Precipitation is very unevenly distributed around the world (Figure 11.4), depending upon latitude, ocean currents, and topography. The highest precipitation zone is the equatorial belt where annual precipitation generally exceeds 40 inches, and commonly exceeds 80 inches. The wettest place on Earth is close to the equator in the mountains of Colombia where approximately 524 inches of rain fall annually. Earth's wet equatorial zone is flanked by two zones, at approximately 25 to 30 degrees north and south latitudes, that contain many of the major deserts and in which precipitation is commonly less than 10 inches per year. Precipitation generally increases in the temperate regions of 35 to 60 degrees north and south latitudes, and decreases to less than 20 centimeters in the polar regions. These zones arise from Earth's receipt of solar energy and convection of air through the major atmospheric cells (Hadley cells) (Figure 11.5). High rainfall

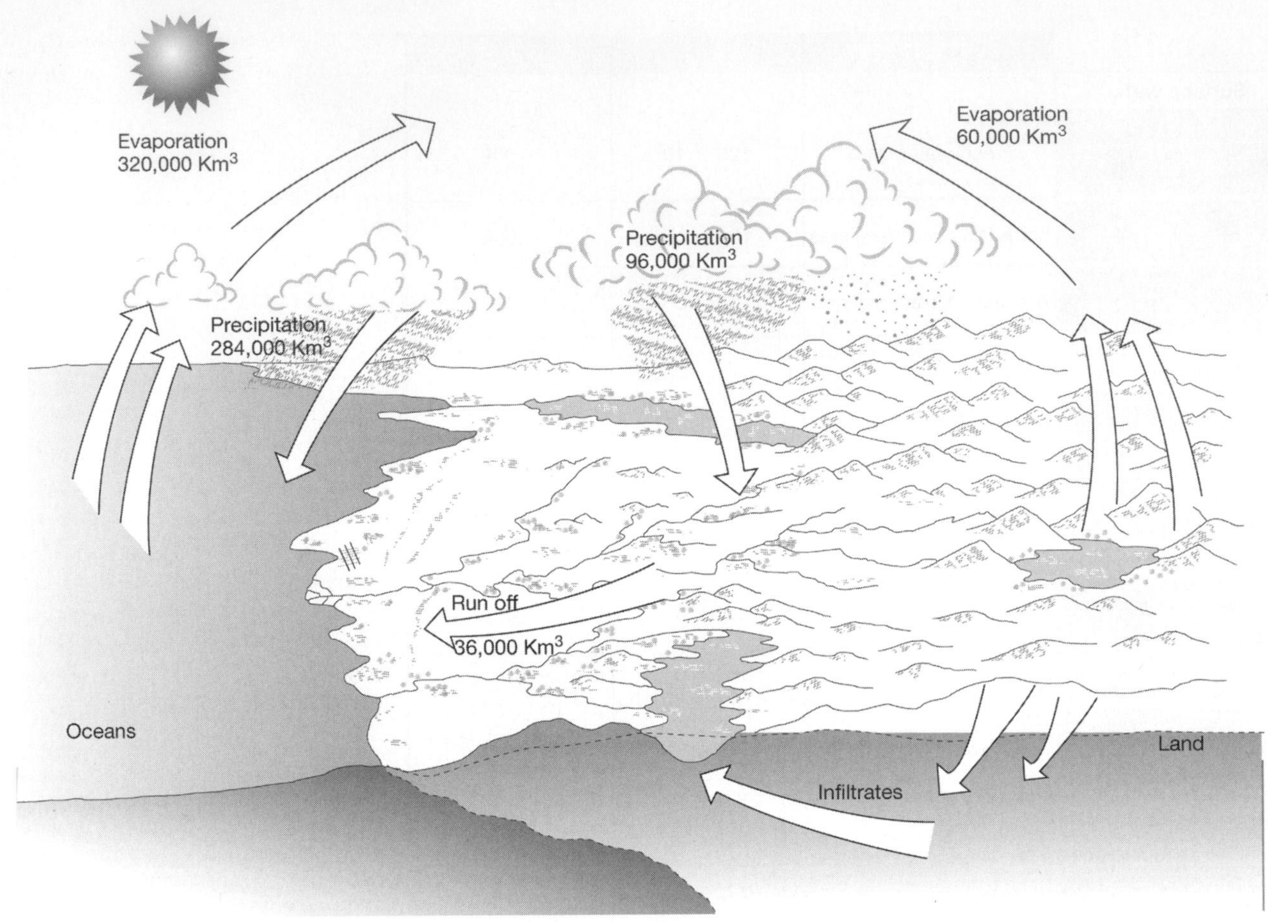

FIGURE 11.3 The general hydrologic cycle for Earth. Water is constantly moving from one reservoir (atmosphere, ocean, rivers, groundwater, etc.) to another in the processes of evaporation, precipitation, runoff, and infiltration.

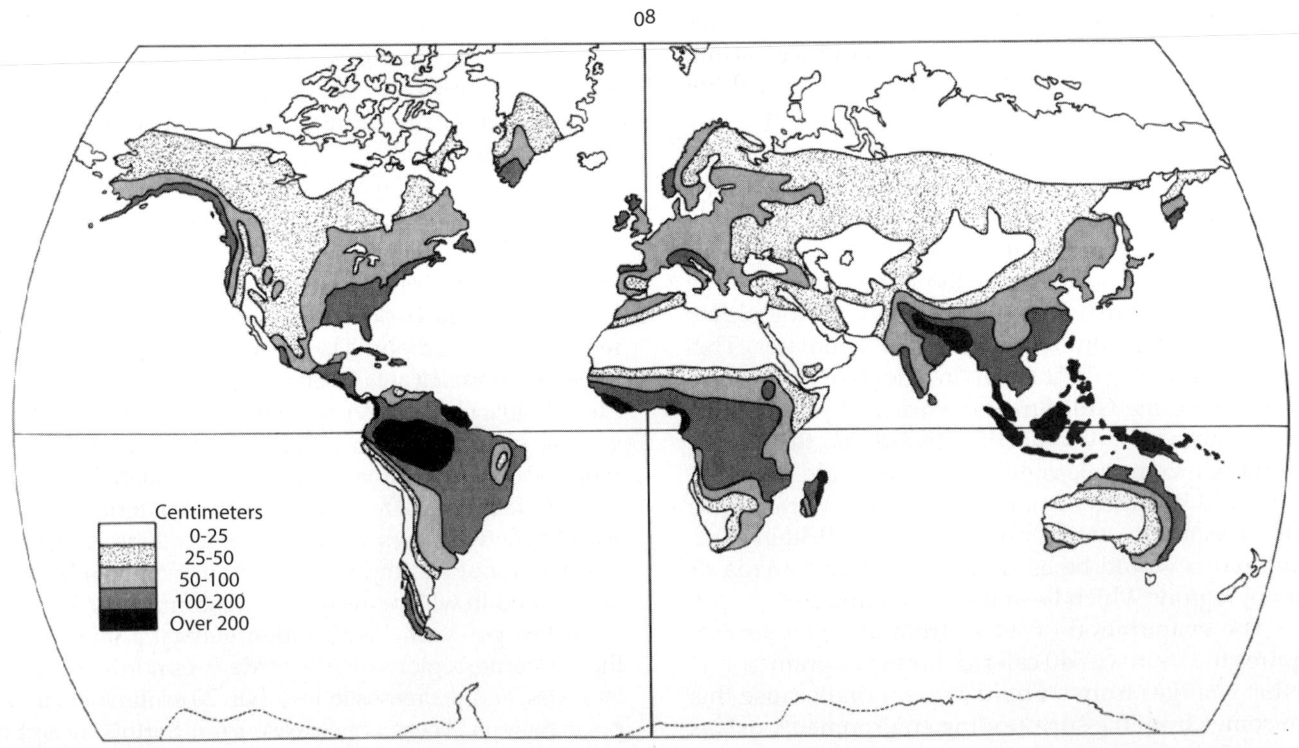

FIGURE 11.4 Worldwide precipitation patterns. Note that a zone of high rainfall lies along the equator and that more arid zones lie along belts that are 25°–30° north and south of the equator. (From B. J. Skinner, *Earth Resources,* 3rd ed., Prentice-Hall, 1986.)

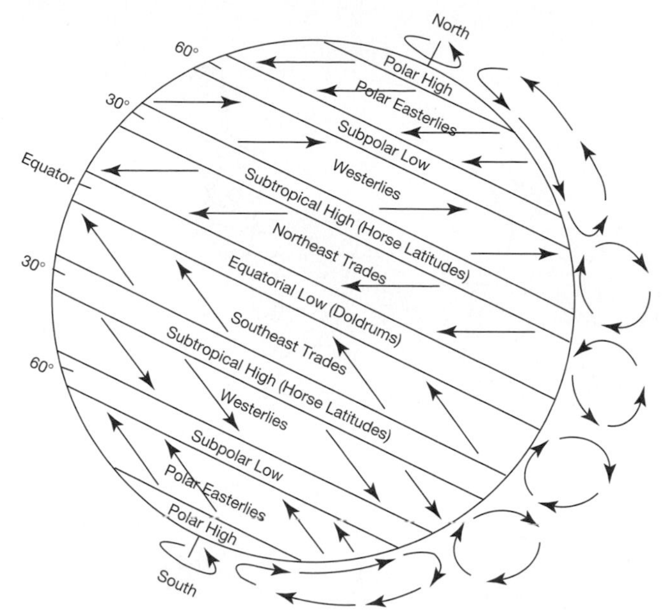

FIGURE 11.5 Idealized circulation of Earth's atmosphere showing the locations of, and air movements in, the Hadley cells. In general, the low-pressure areas are zones of high rainfall and the high-pressure areas are zones of low rainfall. The general easterly or westerly motions of the air in the broad zones (Westerlies, Northeast Trades, etc.) are controlled by the Coriolis effect, which results from Earth's rotation.

in the equatorial zone results from the rising of warm, damp air into the upper atmosphere, where cooling reduces its capacity to hold water and, thus, produces torrential rain. The arid regions that flank the equatorial zone result from the descent of the cooler and much drier air from the upper atmosphere. As the air descends, it is warmed by heat from Earth's surface; the ability of this descending air to hold water rapidly increases. Thus, instead of releasing water, the air is absorbing water, resulting in arid regions.

Global precipitation patterns are also affected by major ocean currents and by major mountain chains. For example, the Atacama Desert extends along the west coast of South America, adjacent to the Andean Mountain chain and the waters moved by Humboldt Current; similar desert regions lie on the eastern flank of the Rocky Mountains in North America. The driest place on Earth is in northern Chile, where the combined effects of atmospheric convection cells and the adjacent cold ocean current result in an average precipitation of less than 0.03 inches of rainfall per year.

The United States is, overall, an example of a country well endowed in terms of water resources; it receives an average of about 30 inches of rainfall annually. This rainfall is, however, quite irregularly distributed with annual average values ranging from more than 250 centimeters in some mountainous areas of Washington, Oregon, and North Carolina, to less than 4 inches in some desert regions of the southwest [Figure 11.6(a)]. The actual amounts of precipitation vary significantly above the average, with the greatest variations occurring in areas of lowest average precipitation. The region to the east of the Mississippi River, thanks largely to the Gulf of Mexico, enjoys an abundant supply of water and receives 65 percent of the total precipitation in the continental states, whereas the western part of the country, due largely to high mountains, has a relative deficiency of water. This geographic variation is compounded by temporal variations tied to long-term weather fluctuations, such as those created by the episodic appearance of the *El Niño* and *La Niña* phenomena in the Pacific Ocean. Thus, although water is a renewable resource, the rate of renewal is neither uniform nor totally predictable. Accordingly, the long-term availability of water to satisfy national needs requires both efficient storage systems and effective distribution systems.

Water is returned to the atmosphere from land or standing water by evaporation and by **transpiration**—the discharge of water vapor by plants directly to the atmosphere. The average annual evaporation rate for a site is calculable on the basis of weather conditions, is readily tested by simple experiments, and is well established for many areas [Figure 11.6(b)]. The rates are highest where solar insolation (radiation) and winds are greatest and, especially, where humidity is least; rates are lowest where temperatures are lowest. Transpiration is a function of the type of plants involved as well as weather conditions and can vary markedly, depending upon the vegetation cover in an area. The effects of both temperature and evaporation combine to return water into the atmosphere, to cool the surface, and to reduce the availability of free water for agricultural, domestic, or industrial use. Worldwide, the combined evapotranspiration rate is about 62 percent (see Figure 11.3) and, for the United States, it is about 70 percent. This means that on average 62 percent of the precipitation that falls worldwide is returned to the atmosphere before reaching streams, rivers, and the oceans. The percentages are much greater in areas of low rainfall and high temperature and much lower in areas of high rainfall and cooler climate. In arid countries, such as Australia, the fraction of water transferred by evapotranspiration is very large; in damp climates, such as in Great Britain, the fraction carried by this process is relatively small.

The type and density of natural vegetation reflects the availability of water in a region. In areas of low rainfall, plant cover develops to a point where all precipitation is used in evapotranspiration and none is left for stream flow; additional plant growth can only occur if there is groundwater to support it. Ephemeral (or temporary) streams may, of course, flow during periods of high rainfall. In several parts of the southwestern United States, introduced pest plants such as mesquite and cottonwood have become such major

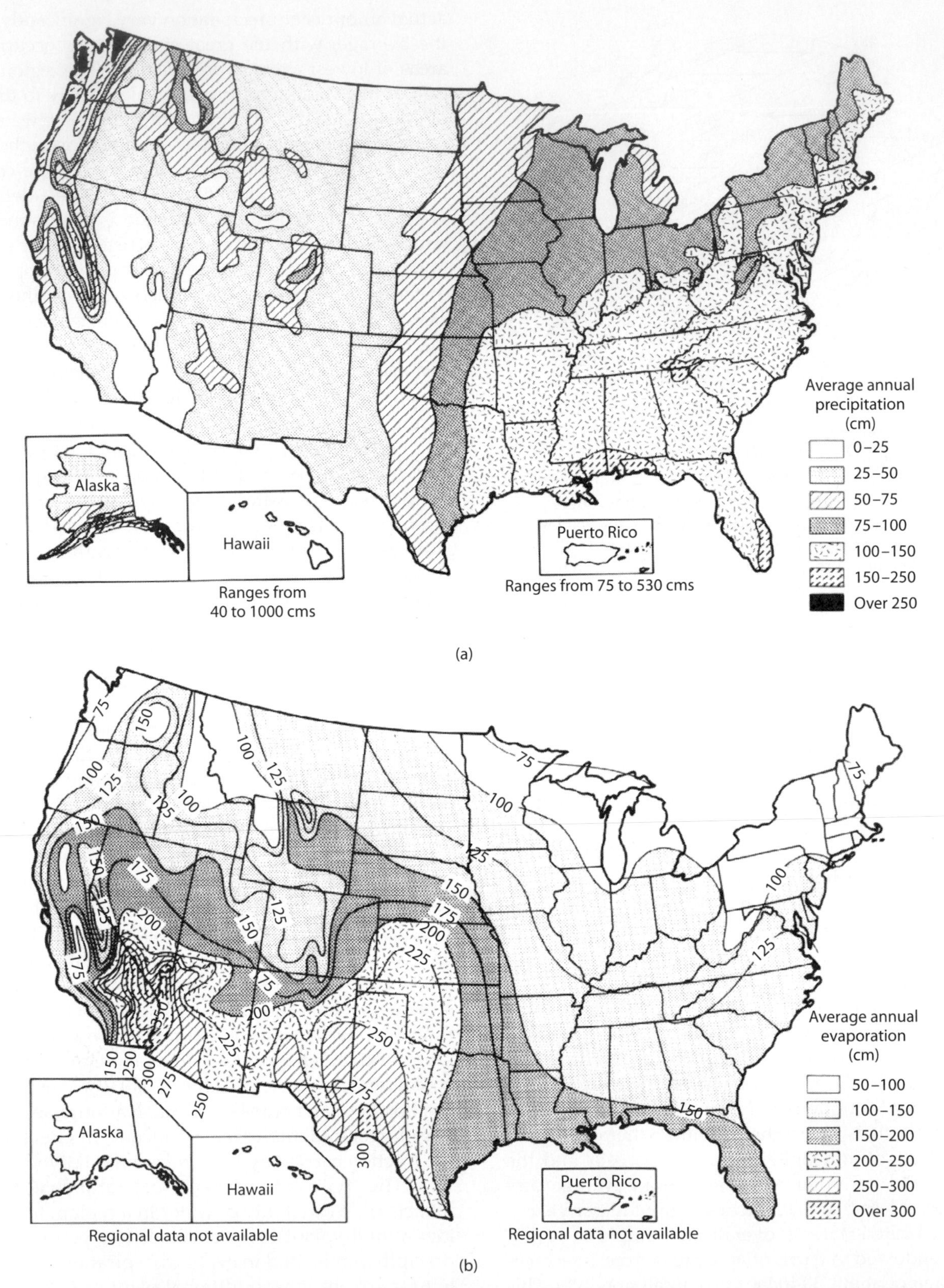

FIGURE 11.6 (a) Annual precipitation patterns for the United States. (b) Annual evaporation pattern for the United States. (c) Annual runoff patterns for the United States. (d) General water surplus–deficiency relationship in the United States. (From *The Nation's Water Resources, 1975–2000,* The Water Resources Council, 1968.)

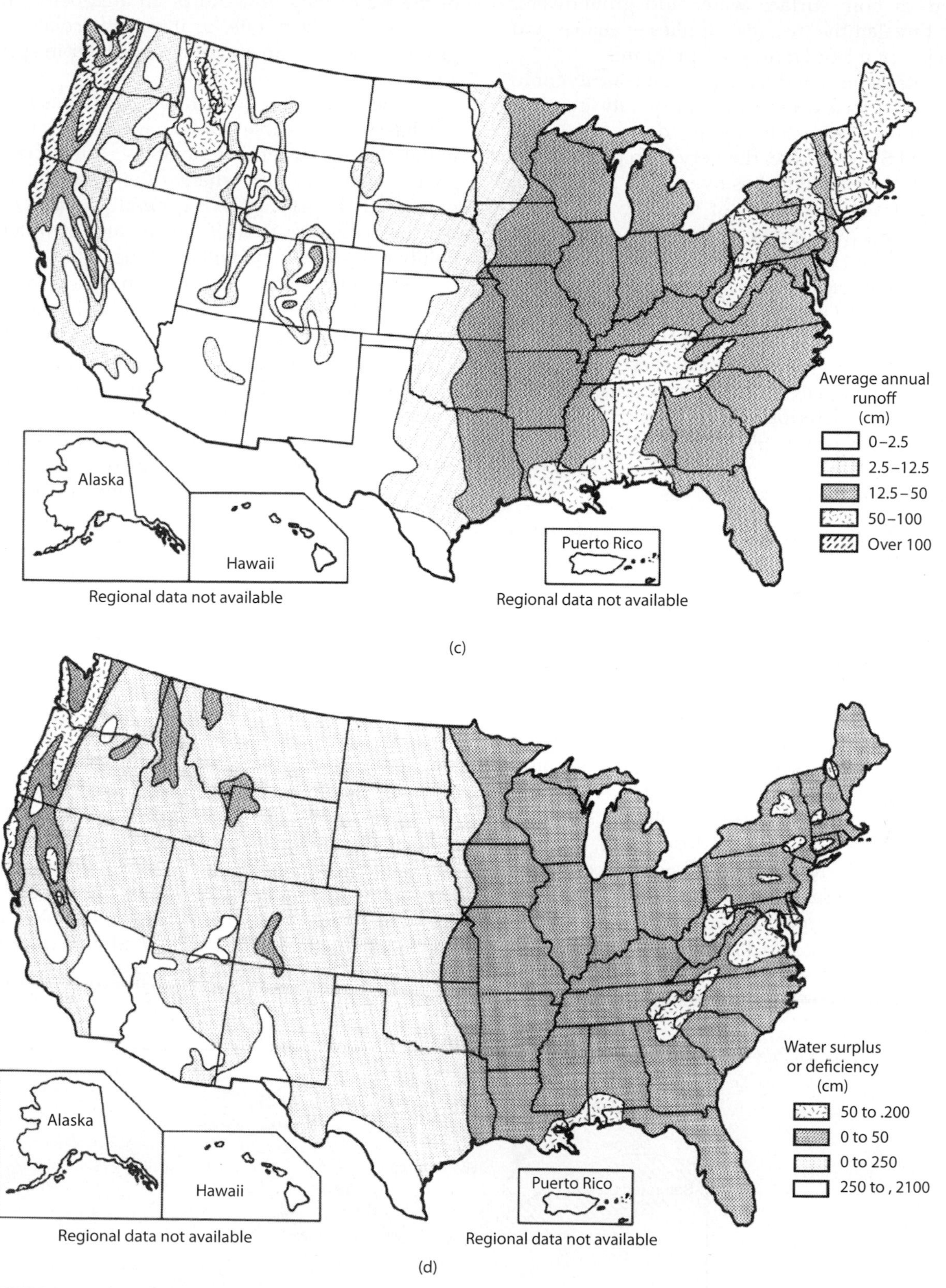

FIGURE 11.6 (cont.)

consumers of both surface water and groundwater that they threaten the meager supplies available, and are the object of major eradication programs.

Evapotranspiration is a significant contributor to problems of surface water and soil quality in many arid parts of the globe, including the desert southwest of the United States. There, the very high evaporation rates, which exceed 100 inches over large areas, result in the loss of vast quantities of stored water. These losses reduce the availability of water for domestic or industrial use, and the capacity of dammed rivers to generate hydroelectric power. In addition, evaporation from rivers and reservoirs leads to a deterioration of water quality because of the residual concentration of salts. The buildup of salts on the surface of irrigated fields in areas of high evapotranspiration has resulted worldwide in the deterioration or loss of millions of acres of previously productive cropland

Surface Water—Rivers and Lakes

The presence of rivers and lakes is an indication that the precipitation in an area exceeds the losses of water to evapotranspiration and groundwater seepage. In a very general sense, the annual runoff pattern for the United States [Figure 11.6(c)] reflects the combined effects of precipitation [Figure 11.6(a)] and evaporation [Figure 11.6(b)]. Thus, areas of high rainfall are also areas of high runoff, and large areas of low rainfall, such as parts of the western states, have essentially no runoff at all. It is, of course, important to remember that many areas in which the average annual evaporation exceeds average annual precipitation still have significant stream flow, at least for part of the year. This occurs because neither rainfall nor evaporation is constant during all seasons of the year or all times of the day. Precipitation may be seasonal, but it can occur at any hour of the day, whereas evaporation increases sharply during summer months and during afternoon hours. Furthermore, in periods of high rainfall much of the water may flow out of an area before there is time for it to evaporate, or it may percolate into the groundwater system and re-emerge later in springs or streams.

The U.S. Water Resources Council has found, on the basis of available surface water and water demand, that the eastern region of the United States is an area of water surplus, whereas the western (and geographically larger) region is generally an area of water deficiency [Figure 11.6(d)]. The pattern of water availability has played, and will continue to play, an important role in population distribution and in the pattern of land use and resource exploitation in the country. To permit accurate assessment of the regional water supply and demand, the continental United States has been subdivided into eighteen Water Resources Regions by the U.S. Water Resources Council, primarily on the basis of major surface water drainage systems. The 30 percent of precipitated water that flows via rivers and streams into the oceans in Figure 11.3 is a somewhat misleading figure because it does not show that a considerable amount of water, ultimately lost to evapotranspiration, actually first travels long distances as stream flow. Much of this water has, in fact, already been used in domestic water supplies and in industry before it returns to the atmosphere.

Groundwater

Near-surface rocks and soils serve as storage sites for quantities of water estimated to be 3000 times larger than the volume of water in all rivers at any given time, and 35 times larger than the volume of all inland lakes and seas. Although this underground water is by far the largest quantity of accessible freshwater, it is commonly a nonrenewable resource because the natural rates of movement and recharge are so slow relative to the rapid rates at which it is withdrawn (Figure 11.7). Deep groundwater commonly is water that was trapped and isolated in sediments some time in the geologic

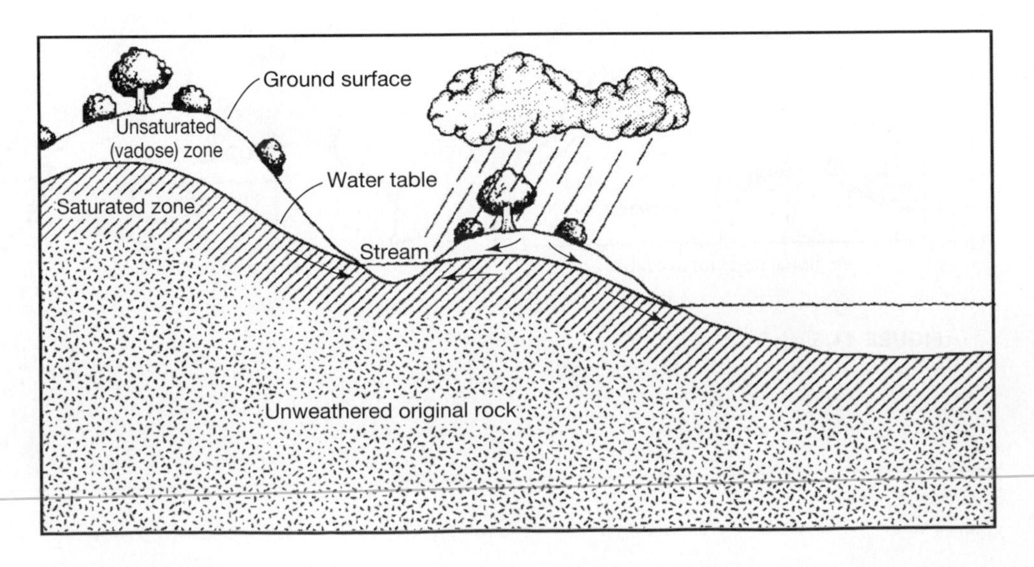

FIGURE 11.7 Cross-section of a typical soil zone showing the relationship of the water table to the ground surface, streams, and lakes.

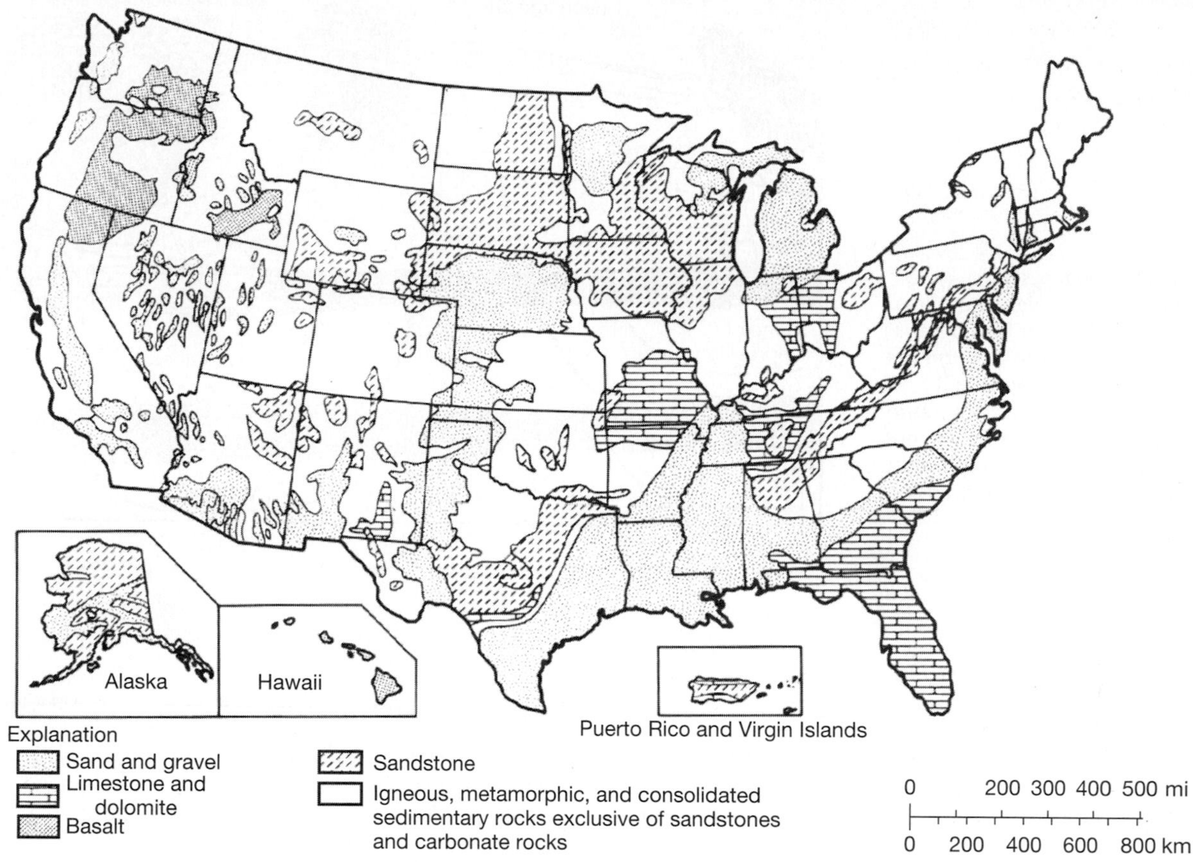

FIGURE 11.8 General occurrence of the principal types of water-bearing rocks in the United States. (From R. C. Heath, "Ground Water Regions of the United States," U.S. Geological Survey Water Supply Paper 2242, 1984.)

past. In contrast, shallow groundwater supplies are usually intimately related to surface water flow as shown in Figure 11.8. Depending upon the land surface slope, vegetation, soil depth, and rock-type, widely varying amounts of precipitated and runoff water may percolate into the intergranular pore spaces and fractures. Water percolates downward until it reaches the **water table,** the surface below which the pores and fractures are water-filled. The water table is not flat, but usually has a shape that is similar to, but smoother than, the topography of the land surface.

Above the water table is an unsaturated or **vadose** region of the soil. The upper part of this zone is filled with water when it rains, but it drains relatively quickly except for water adhering to mineral surfaces. However, even this small amount of adhering water is very important because it is the principal water supply for most plants. During periods of drought, the upper soil zone can lose much water directly to evaporation; under such conditions, moisture actually moves upward, possibly even from the water table, by means of capillary action. Somewhat deeper than the upper soil zone is a zone in which the flow of water through the unsaturated soil or rock is usually downward toward the water table. The soil and vadose water zones do not constitute direct resources of water, but they are essential for the replenishing of the groundwater zones.

It is important to realize that most streams and lakes that are in equilibrium with their surroundings represent the intersection of the groundwater table with the surface topography. It is the slow lateral seepage of groundwater that provides the water for stream flow when there has been no rain and there is no surface runoff. In humid areas, streams continue to flow, perhaps with reduced volumes, even in long periods of drought. In arid regions, where the groundwater table may lie far below the land surface, streams will often flow after rainstorms only until the water has either evaporated or percolated into the subsurface. In these areas, the high rates of evapotranspiration commonly result in the return of most water to the atmosphere. The shallow penetration of the rainfall before being evaporated often allows the water to pick up dissolved salts that are then left as a near-surface soil cement (referred to as caliche or "hard pan"); this makes the soil less permeable and reduces the value of the soil for agriculture (note Figure 12.3)

Aquifers, geologic formations that possess sufficient porosity and permeability to allow for movement of the water contained within them, underlie large areas of the United States (Figure 11.9). In fact, more than

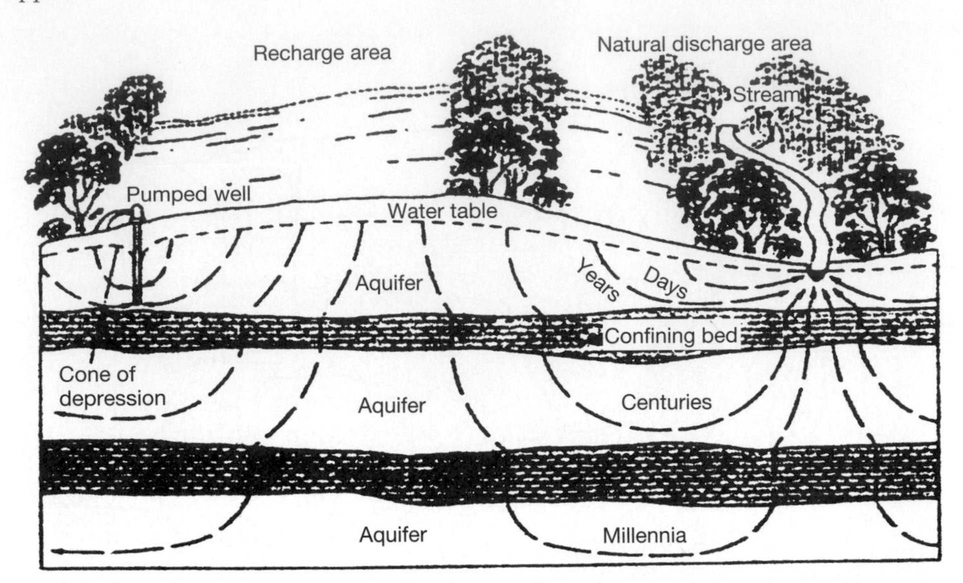

FIGURE 11.9 The rates of groundwater flow are very much less than the rates of surface flow because the water must percolate through the very small openings between soil particles or along the thin fractures in rocks. Consequently, water movement that may take only a few hours on the surface can take years to hundreds of years in the subsurface.

BOX 11.1

Restoring a River: The Kissimmee

The rapid growth of Florida's population, from fewer than 3 million in 1950 to more than 16 million in 2008, has resulted in a variety of water resource problems. The Kissimmee River flows from the Orlando area southward to Lake Okeechobee (Figures 11.A and 11.B). Prior to the 1960s, the Kissimmee flowed slowly in a meandering channel bordered by more than 50,000 acres of marshy wetlands. Episodic rains, some resulting from hurricanes, created large fluctuations in the water flow of the Kissimmee and could turn the river's floodplain into a broad sheet of shallow water. To control the flooding, between 1961 and 1971 the Army Corps of Engineers converted the original 102 mile-long meandering river into a 58 mile-long, 30 foot-deep channel with a series of dams, water control structures, drainage canals, and navigation locks. The elimination of the flooding problem and the construction of drainage devices to divert water for use by farms, orchards, and the growing cities served to convert 45,000 acres of natural marshlands into pasturelands. The value of the land rose from $400/acre to more than $4000/acre. At the time of completion of the conversion, the cost of $32 million was considered a reasonable price for the benefits returned.

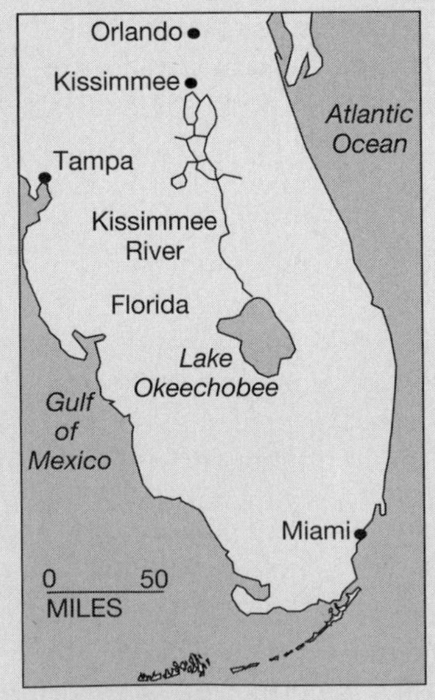

(a)

FIGURE 11.A Location map of the Kissimmee River in south-central Florida.

However, by the mid-1970s it became apparent that the channelization of the Kissimmee River had many other effects. The wetlands had served as an important water filter to remove nitrogen and phosphorous from sewage and fertilizers. In addition, the wetlands had played a key role in supplying much of the water that evapotranspired into the atmosphere causing frequent rainstorms. With the removal of the wetlands, rainfall decreased, water levels fell, water purity decreased, and the water flow in the Kissimmee and the productivity of Lake Okeechobee declined. The spawning sites for bass and other fish along the river were lost, and 90 percent of the migratory and resident bird population disappeared. Thorough studies of the hydrology of Florida also revealed that the Kissimmee River and Lake Okeechobee were integral parts of a broad flow of water that created and supported the Everglades. With the decrease in the flow of water, the Everglades rapidly began to deteriorate.

It became clear that the only way to restore the quality of the water in Lake Okeechobee, and perhaps to save the Everglades, was to return the Kissimmee River to its original condition. Although the state authorized such action as early as 1976, lack of funding and the view by the Army Corps of

FIGURE 11.B Photograph of the Kissimmee River showing the difference between the original meandering course in the foreground and straightened channel in the background. The cost of restoring the Kissimmee to its prechannelization condition is now projected to be at least $8 billion. (Photograph courtesy of South Florida Water Management District.)

Engineers that restoration was not a high priority, resulted in little action until 1985 when the state sponsored a demonstration project. The project blocked a portion of the artificial channel and forced water back into the original riverbed. The test had dramatic and rapid effects—water quality improved and natural plant and animal life returned to the floodplain.

The final 1945-page report and plan approved by the Army Corps of Engineers in December 1994 called for restoring the entire Kissimmee River system, possibly the most ambitious environmental restoration project ever attempted in Florida. The plan called for filling in 29 miles of the newer river channel, rebuilding 11.6 miles of original river channel, moving more than 50 million cubic yards of soil to fill in up to 2000 miles of canals, and returning 35,000 acres of original floodplain wetlands to their natural condition. The estimated cost was high because some 30 percent of the drained pastureland was sold into private hands and had risen a great deal in value since the 1961–1971 conversion. Unfortunately, the plan was never put into action, and the price tag proposed in 1994 to complete the restoration keeps rising with each year of delay; the original cost estimate of about $100 million had risen to $300 million by the 1980s, to $2 billion by 1995, and to $7.8 billion by 2000. In 2008, the state agreed to purchase large areas from sugar companies and convert the areas back to their natural state. However, the lack of funds due to the economic recession in 2009 threatened to undo all of the negotiations. Consequently, new uncertainties have arisen and if the project is completed by about 2015, there will have been the expenditure of perhaps $8 billion to $10 billion in order to regain what was originally free.

50 percent of the population of the United States is presently supplied by groundwater from aquifers for their domestic supplies. In many arid parts of the world, aquifers constitute the only significant source of water. Even in more humid parts of the world where surface water is present, aquifers are commonly utilized as major water sources because they provide a relatively constant flow of good quality water.

The major problems in the utilization of groundwater are the rate of water withdrawal, the rate of recharge, and the water quality. The surfaces of many parts of the continents are underlain by metamorphic or igneous rocks in which the only available quantities of groundwater are the meager quantities that lie in the fractures and joint systems or along faults. Interconnectedness of the joints allows ready movement of the water, but the quantities stored in such rocks are often very limited. Even in many areas underlain by sedimentary rocks, porosity or permeability is too low to allow for a worthwhile rate of water flow. If an aquifer is to have a sustained yield, there must be a constant replenishment from surface water through the generally slow process of percolation. It has been estimated that, in the United States, if all groundwater were removed to a depth of 2460 feet, 150 years would be required to totally recharge the system.

TABLE 11.1 National drinking-water regulations*

Constituent	Maximum Concentration, ppm
Primary and enforceable standards	
Arsenic	0.01
Barium	2
Cadmium	0.005
Chromium	0.1
Lead	0.0
Mercury	0.002
Nitrate (as N)	10
Selenium	0.05
Copper	1.3
Fluoride	4
Cyanide	0.2
Coliform bacteria	in no more than 5% of samples
Endrin	0.002
Lindane	0.0002
Methoxychlor	0.04
Toxaphene	0.003
2, 4-D	0.07
Total trihalomethanes [the sum of the concentration bromodichloromethane, dibromochloromethane, tribromomethane (bromoform), and trichloromethane (chloroform)]	0.10
Radionuclides:	
Radium 226 and 228 (combined)	3 pCi/L
Gross alpha particle activity	0.0
Gross beta particle activity	0.0

Constituent	Maximum Level, ppm
Secondary and nonenforceable standards	
Chloride	250
Color	15 color units
Silver	0.1
Dissolved solids	500
Foaming agents	0.5
Iron	0.3
Manganese	0.05
Odor	3 (threshold odor number)
pH	6.5–8.5
Sulfate	250
Zinc	5

Data from the U.S. Environmental Protection Agency, 2000 and updated regularly.

*The U.S. Environmental Protection Agency's National Interim Primary Drinking-Water Regulations and National Secondary Drinking-Water Regulations are summarized here. The primary regulations, which specify the maximum permissible level of a contaminant in water at the tap, are health related and are legally enforceable. If these concentrations are exceeded, or if required monitoring is not performed, the public must be notified. The secondary drinking-water regulations control contaminants in drinking water that affect the esthetic qualities related to public acceptance of drinking water. These secondary regulations are intended to be guidelines for the states and are not federally enforceable.

The problem of the slow recharge of aquifers has become evident in several parts of the world, including the western United States, where withdrawal rates up to 100 times the recharge rates are rapidly lowering the water tables. In areas where withdrawal exceeds recharge, water is being "mined" and is being extracted like any other nonrenewable mineral commodity. The effect of the loss of water on land value is being recognized, so that the farmers who own the land are permitted to depreciate its taxable value as the water table falls. Even in humid regions where there is abundant rainfall, the withdrawal of water from aquifers at rates exceeding those of recharge creates local problems. These may include lowering the water table such that other nearby shallow wells run dry, or even the movement of salt water into beds that previously contained only freshwater.

The third problem of aquifers is water quality. As groundwater moves through the rocks it may dissolve the more soluble constituents. The problem varies with rock type and with flow rate, and has been greatly aggravated in recent years by the introduction of contaminants from agricultural, industrial, and domestic sources. In general, water with less than 0.05 percent (500 parts per million) total dissolved solids is considered suitable for human consumption (specific requirements for potable water are listed in Table 11.1 on the previous page); however, water with up to 1 percent dissolved solids can be used for some industrial and agricultural purposes. Natural bacteria present within the soil generally cleanse slow-moving water of harmful natural biological contaminants. Unfortunately, complex synthetic chemical contaminants have seriously limited the usefulness of some aquifers, particularly where there is a rapid rate of water movement that spreads contaminants much more rapidly than they can be filtered or decomposed by bacteria. Drinking water regulations are, of course, designed to ensure safety of the general public.

However, there can also be problems in unnecessary or overly strict regulation. An example is provided from the Wisconsin lead-zinc district in the United States, a significant region for the production of these metals for the last 150 years. The EPA demanded there be no more than 0.5 parts per million (ppm) of zinc in water effluent from a mine in this district, one-tenth of the 5 ppm limit set by the U.S. Public Health Service as a maximum for this metal in drinking water. In fact, zinc is not toxic to humans even at much higher levels; this limit was based on taste tests, not possible health hazards, zinc often being considered beneficial. The operating mine's effluent actually contained 2.5 ppm—a level that was approved by Public Health Service standards, but five times above the level demanded by the EPA. Unable to comply, the mine had to close with the consequent loss of metal production and employment. The EPA's reason for setting the limit at 0.5 ppm was to permit more fish in the streams, but this has not happened since closing the mine because the natural local groundwater still contains about as much zinc as the mine effluent. An ironic footnote to this story is that the EPA office in Washington, DC, was using drinking water containing 20 ppm of zinc, forty times the level imposed as an acceptable maximum for the Wisconsin mine effluent.

Another problem resulting from the withdrawal of water from aquifers is land subsidence. This is a local, but increasingly observed phenomenon that can have serious consequences. This is discussed in greater detail later in this chapter.

Ice Caps and Glaciers

About 2.15 percent of the world's total surface water, but more than 70 percent of the freshwater, is held as ice in ice caps and glaciers. This water is mostly contained within the ice cap and glaciers of Antarctica and for all practical purposes is unavailable for human use (Figure 11.10). Proposals to tow large icebergs to water-deficient areas such as the Middle East have been proposed, but have not yet resulted in any significant financial backing or serious efforts to carry out the proposals. In the short term, the amount of water held in glaciers and ice caps may be considered constant, but during the Pleistocene Age about 25,000 years ago, at the height of the most recent Ice Age, the amount of water held as ice in these regions was as much as 50 percent greater than at present. During the major glacial advances, more of the snowfall over polar and cold temperate landmasses built up and persisted, with the result that glaciers advanced and sea level dropped as much as 330 feet below its present level. In contrast, during warmer interglacial periods, sea level rose significantly above present levels. One of the potential consequences of the rising level of atmospheric carbon dioxide and subsequent global warming is a drastic reduction of ice in the Arctic and Antarctic regions. The resulting runoff would contribute to a substantial rise in sea level. Furthermore, general

FIGURE 11.10 The world's ice caps and glaciers such as these in Victoria Land, Antarctica, contain most of the world's nonsaline water, but this water is inaccessible for human use.

worldwide warming, even by very small amounts, would also result in volume expansion of ocean waters and further contribute to sea level rise. Biologists have noted that there would be numerous secondary effects of such changes, such as shifting biological habitats and a drastic impact on animals such as polar bears. The bears rely on the presence of floating summer pack ice for their survival. All of this further emphasizes the interconnectedness and complexity of Earth's systems.

Surface Runoff, Floods, and Flood Control

Most rainfall produces some **surface runoff.** The amount of this runoff is a function of the amount of rainfall, the slope and length of the drainage basin, the rock and soil type of the drainage basin, the vegetation cover, and the extent of any impermeable areas in the basin. The runoff can range from zero to more than 90 percent of total rainfall in a given basin; the remainder evapotranspires back into the atmosphere, percolates into the groundwater system, or is held back in storage facilities. Activities such as mining, timbering, farming, and construction frequently promote an increase in the amount and rate of surface runoff. Removal of the trees and other vegetation reduces the retention of water in the biologic materials (leaves, roots, grasses, trees) and soil, and the amount of evapotranspiration (see Figure 11.36). At the same time, there is a great increase in the surface runoff and erosion.

Surface runoff can be defined by a **hydrograph** or **lag-time diagram** (Figure 11.11). This diagram depicts both the quantity and time of rainfall and the subsequent runoff from a drainage basin. Small drainage basins have lag times measurable in minutes or hours, whereas large ones may have lag times of hours to days or even weeks. Once the runoff characteristics have been determined for a basin, it is possible to predict waterflow levels and to estimate potential flood conditions.

Now, consider the effects of urbanization of a previously tree-covered or grass-covered area. Construction of a typical suburban community makes 10–30 percent of the area impermeable because of streets, driveways, houses, and sidewalks; construction of a city environment, such as a community shopping center or a large shopping mall, can make 50 to 100 percent of an area impermeable. Most of the water from an impermeable area runs off onto permeable areas, thereby subjecting the permeable sections to water conditions equivalent to added rainfall. The result of natural rainfall plus the effect of the added water is then the **equivalent rainfall.** Much of the added runoff water does not actually drain onto adjacent land but is carried by storm drains into streams, rivers, or the sea; nevertheless, that extra water will appear in some part of a drainage basin. Unfortunately, as more water runs off more rapidly, less of it is able to percolate into the soil to be added to the groundwater system, as shown in Figure 11.12. If we

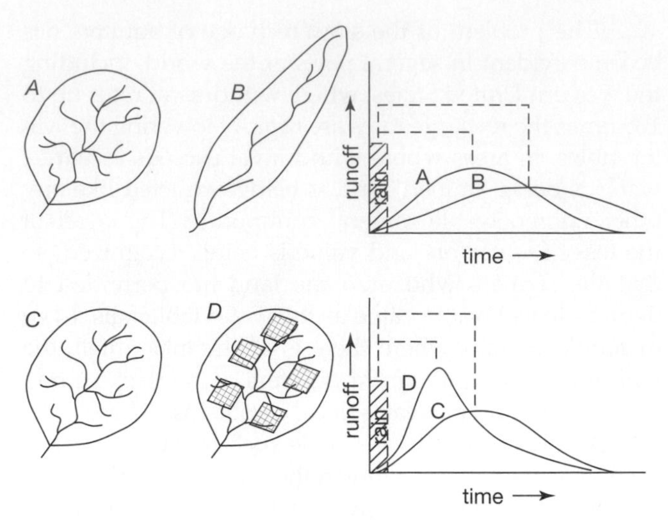

FIGURE 11.11 Hydrographs, or lag-time diagrams, show the relationship between rainfall and surface runoff in a basin. (Upper-map views of basins and hydrographs.) The two curves labeled (A) and (B) in the right-hand portion of the upper part of the diagram illustrate the effect of basin shape on the surface water runoff after a period of rainfall. Runoff from the nearly circular basin A rises during and after a period of rainfall until it peaks and then slowly returns to its normal rate of flow (curve A). Runoff from the elongate basin B rises more slowly because of the length of the basin, until it peaks and then also returns to its normal rate of flow (curve B). (Lower-map views of basins and hydrographs). Basin C is identical to basin A and has a runoff curve (C) that is the same as that for basin A in the upper diagram. Basin D is the same as basin C except that there has been significant development of impermeable areas (shaded areas) due to urbanization. Consequently, the post-development runoff curve (D) rises much more quickly and to a greater height than does the original curve C. The effect of the creation of much impermeable area is to have more water run off, more quickly; this commonly can lead to unexpected flooding and can reduce the amount of water that ultimately infiltrates into the groundwater supplies.

assume uniform rain distribution in a basin and 100 percent runoff of water from impermeable areas (something that is never quite true, but that suffices for a demonstration), converting 25 percent of a basin to an impermeable condition will result in a 33 percent increase in equivalent rainfall for the permeable areas; conversion of 33 percent to an impermeable condition will raise equivalent rainfall by 50 percent, and 50 percent impermeability will raise equivalent rainfall by 100 percent. Of course, in any specific area, soil permeability, the duration and intensity of the rain, and the type of vegetation will play important roles in the actual runoff.

The increase in runoff that inevitably results from increases in impermeable areas due to urbanization can either be permitted to contribute to normal stream flow or can be controlled. An example of such control is found on Long Island, New York, where the runoff from impermeable areas is diverted into shallow catchment basins from which the water seeps downward to increase the groundwater supply. The system provides another environmental benefit because excessive amounts of

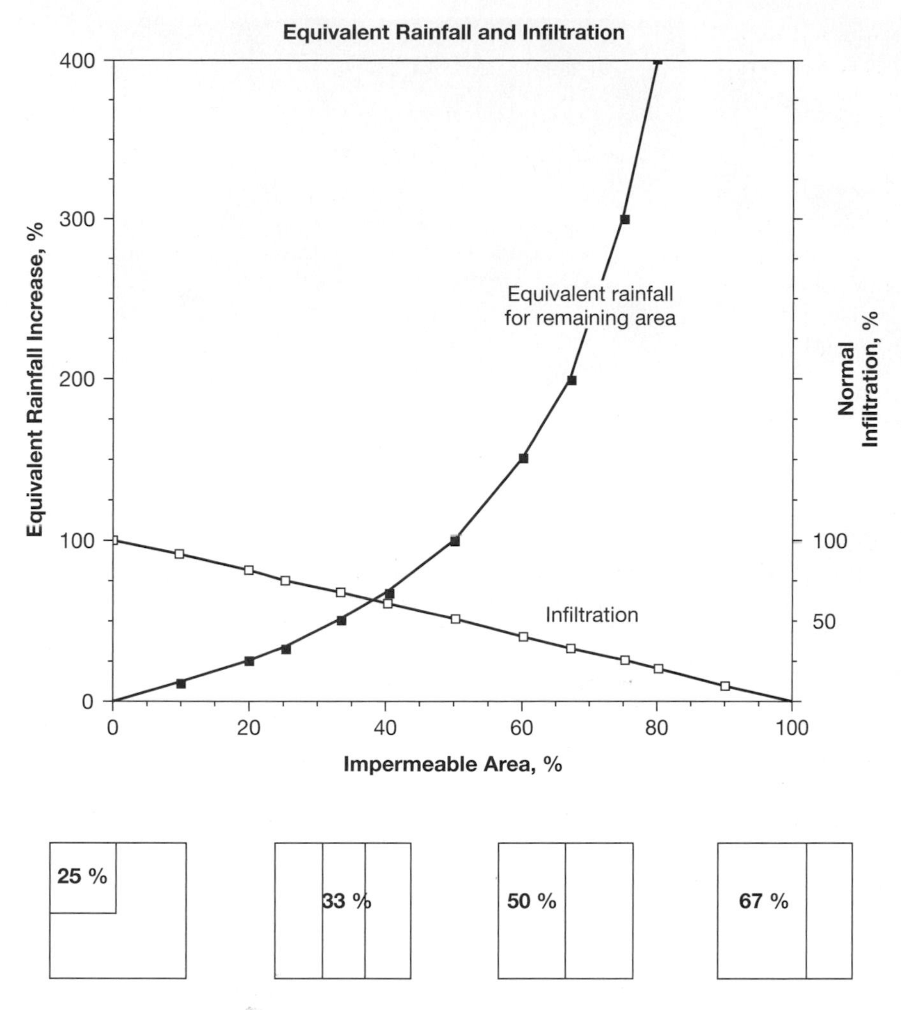

FIGURE 11.12 General relationship between equivalent rainfall and infiltration. As the amount of a natural soil-covered land surface is converted into impermeable area (usually by building houses and by paving streets and parking lots), the amount of rainwater that infiltrates the ground decreases. The increased runoff is like extra rainfall (referred to as *equivalent rainfall*) for the remaining portion of the area. The two curves illustrate, in a very general way, the effects of increasing impermeable area. The amount of infiltration decreases as the amount of impermeable area increases. The equivalent rainfall for the remaining permeable area increases as the amount of impermeable increases. Hence, if 25 percent of an area is paved, the infiltration decreases by about 25 percent, but the equivalent rainfall for the remaining permeable area increases 33 percent. If 50 percent of an area is paved, the infiltration decreases by about 50 percent and the equivalent rainfall for the remaining permeable area increases 100 percent.

freshwater do not pour into the brackish estuaries where its dilution effects would be detrimental to marine life.

Flooding occurs when surface runoff exceeds a normal stream channel's capacity and the water spreads out onto the floodplain or beyond. Flooding is a natural phenomenon triggered by intense or prolonged rainfall or the rapid melting of snow cover. It is of little or no consequence in undeveloped areas where streams and their floodplains have developed unhindered over geological ages, but our tendency to build homes, businesses, and factories on floodplains brings civilization into conflict with nature, and all too often into peril. Floodplains are widely used as farmlands because they are flat, well-watered, and fertile as a result of the soil and organic matter deposited by episodic flooding. Relatively short-duration flooding of farm fields, except at planting and harvest time, are usually not a great problem because plants are tolerant of brief submersion. But intensive flooding can result in extensive erosion, burial of crops by too much new silt, or the rotting of the crops owing to extended inundation. Although the extent to which human activities actually cause flooding is not completely understood, it is evident from the previous discussion that the removal of vegetation from large parts of the drainage basins and the subsequent expansion of impermeable surfaces must increase runoff and must contribute to the potential for flooding. Once hydrographs (Figure 11.11) have been defined, they can serve as valuable aids in predicting floods and to the timing of their rising, cresting, and falling.

The United States, like many nations, suffers some local flooding every year. Usually, the flooding is the result of intense but brief storms that drop large quantities of rain where cool and warm air masses meet or around the center of a low-pressure zone. Whether the flooding is brief and local or extended and extensive, the energy of the flowing water (with a mass more than 800 times greater than flowing air) often causes great damage to human-made structures (Figure 11.13). Widespread flooding often results from the hurricanes that strike the southeastern United States from July to November. For example, in late October 1998, Hurricane Mitch dumped up to 6 feet of rain in parts of Central America in five days. The normal flooding expected from this much rainfall was intensified because deforestation and farming practices resulted in large increases in the runoff. Parts of the American Gulf Coast were devastated by the flooding resulting from Hurricane Katrina

(a)

(b)

(c)

FIGURE 11.13 (a) A sudden flood in late 1938 swept through Louisville, Kentucky, toppling a row of houses but otherwise doing little damage to the structures. (Photograph from the American Red Cross.) (b) A brief but intense storm dropped heavy rains and resulted in a devastating flood in Big Thompson Canyon, Colorado, in 1978. Many campers were killed and numerous houses, such as the one shown, were torn apart. (Photograph from the U.S. Geological Survey.) (c) Heavy rains, such as occurred in June 1995 in central Virginia, may overwhelm river and stream channels and erode the supports for roads and bridges, making rescue and repair efforts difficult. (Photograph courtesy of Virginia Department of Transportation.)

in 2005. In Southeast Asia, the seasonal monsoon rains create massive flooding of low-lying areas annually. The worst in recent history struck Myanmar in May 2008, killing more than 150,000 people. In the spring and summer of 1993, the central United States suffered some of the most massive and extended flooding in the country's history. The "Flood of 1993," as it has become known, resulted from the persistence of a stationary front that allowed for the convergence of warm, moist air moving northward from the Gulf of Mexico and cooler air from the northwest along a band extending from Colorado to Michigan (Figure 11.14). This type of weather phenomenon is common for brief periods, but in 1993 it persisted for approximately five months. Many areas received rain virtually every day for weeks and total rainfalls through the period exceeded 200 percent of normal for the year. Because soils were saturated, runoff was nearly 100 percent, and the dams that had been built to control flooding were full. Consequently, the riverbanks overflowed onto floodplains and hundreds of communities and more than 10 million acres of farmland were under water (Plates 61 and 62). Barge traffic along the Mississippi River had to be halted, railway lines were blocked, and interstate highways were under water.

Over the years, the U.S. Army Corps of Engineers had constructed levees along many stretches of the Mississippi and other rivers to prevent flooding of towns and farmland, but the flooding in 1993 was so extensive, and the water levels were so high, that many levees were topped and some breached. This resulted in flooding of areas previously viewed as safe and that were unprepared for flooding. In some areas, homes as far as 7 miles from the major rivers were flooded. There was significant loss of crops in the flooded area, at least 48 deaths, and damages estimated to be at least $10 to $12 billion. The rains finally ended in August 1993 and floodwaters gradually subsided, but some of the effects will be permanent. Many families moved away, fearing they might face similar flooding again, and in a few instances entire towns had to be relocated to higher ground. The effectiveness of the levees and the value of the dams as flood-control devices have now been questioned. It has been recognized that the levees of many areas held the water in the main channels and, thus, allowed it to flow more rapidly downstream where, in several cases, the flooding became worse. Consequently, it was decided not to rebuild all of the levees because the absence of the levees upstream would have allowed the floodwaters to spread laterally.

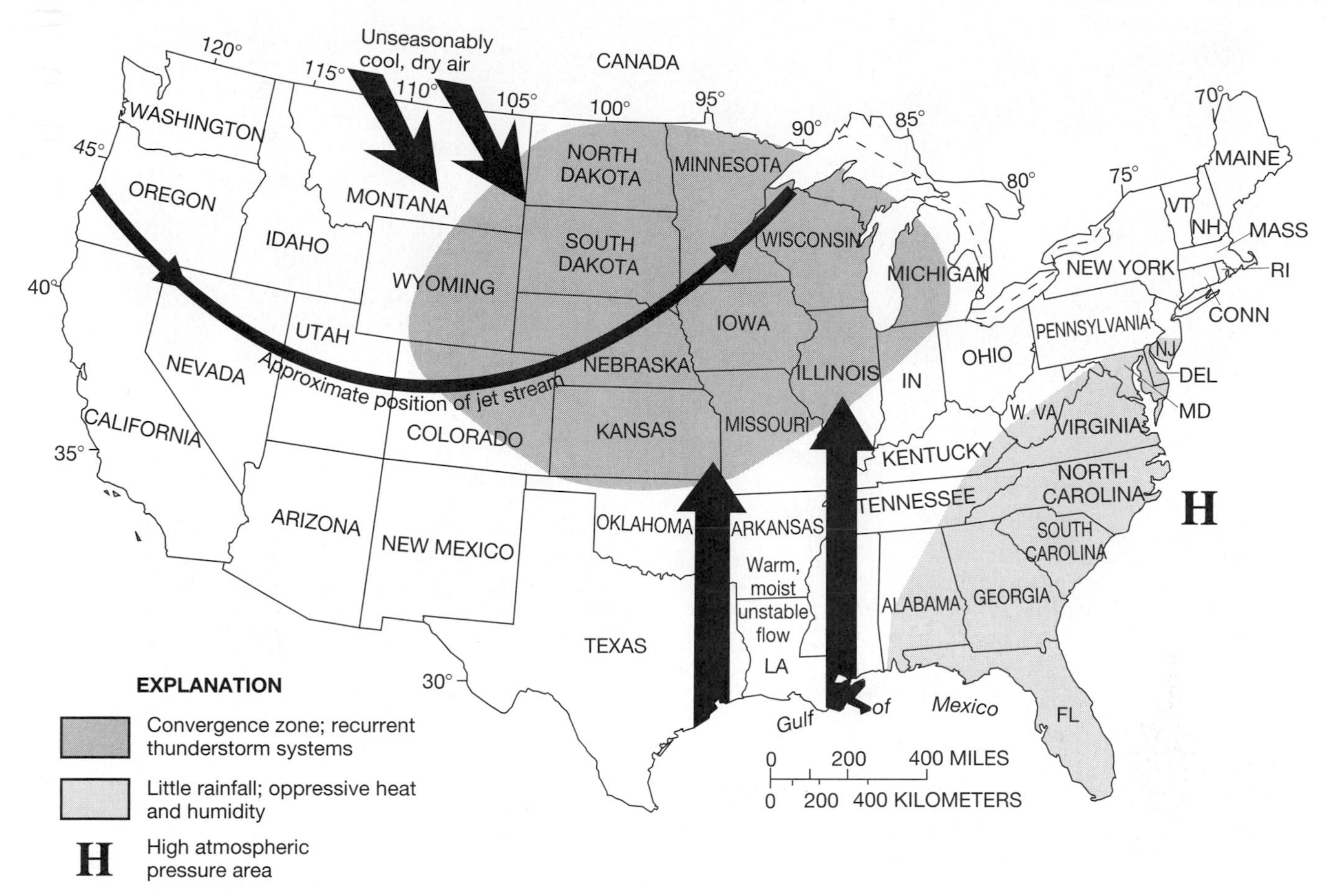

FIGURE 11.14 The dominant weather patterns in the spring and early summer of 1993 brought warm, moist air from the Gulf of Mexico into convergence with unseasonably cool, dry air along the jet stream over the American Midwest. The area of convergence (heavier shading) experienced much heavier than normal rainfall and extensive flooding (see Plate 61) beyond levels ever previously recorded. At the same time, a portion of the southeast (lighter shading) experienced severe drought. Human activities can intensify flooding or water shortage, but natural weather changes can often bring about intense variations in water availability.

This would have reduced the highest water flows in the major river channels and made the flooding downstream significantly less. Such decisions are not easy because minimizing flood damage downstream requires increasing the flood damage upstream. The entire situation highlights the problems encountered when human activities come into conflict with natural environmental processes. Unfortunately, very similar, if somewhat less widespread flooding occurred again in the spring and summer of 2008. Hurricane Katrina's assault on New Orleans in 2005 presented yet another flooding event. Torrential rains were accompanied by a tidal surge driven by hurricane winds. Not only was the land nearly flat, much of it actually was six to twelve feet below sea level. Several of the major dikes were breeched and much of New Orleans and the surrounding areas experienced flooding far worse than ever seen in the United States. Some of the neighborhoods have not yet been rebuilt and probably never will be. In a country as large as the United States, there is almost always some area suffering a deficiency of rainfall while others are suffering flooding. Furthermore, the change in conditions can occur very unexpectedly and in a short time span. For example, the Atlanta area of north Georgia suffered under record drought conditions for several years through the summer of 2009. Then, an unexpected change in rainfall patterns brought more than 22 inches of rain in a few days and created the worst flooding the area had ever seen.

In an effort to reduce the vast amounts of damage and the scores of deaths and injuries that occur annually as a result of flooding, the two procedures now most widely used are the construction of dams and the channelization of rivers. The two processes operate on different principles, but both seek to achieve the same result. **Dams** serve as temporary water barriers to hold back high flow before it reaches a vulnerable area, and hence prevent the land from flooding (Figures 11.15 and 11.16). **Channelization,** by contrast, provides an efficient means by which water can be carried out of an area so quickly that it does not rise to flood levels. The construction of levees, as noted previously, serves to dam waters from lateral movement while also serving as a formal channel for

FIGURE 11.15 The principal methods of flood control are the construction of dams and channels. Dams such as the Tennessee Valley Authority's Fontana Dam in western North Carolina have been used for flood control, recreation, and hydroelectric power generation. (Photograph courtesy of Tennessee Valley Authority.)

downstream movement. Several of the consequences of building dams and channels are summarized in Figures 11.16 and 11.17, respectively.

Dams, of course, serve many other purposes, such as water storage for irrigation, electric-power generation, recreation, and livestock watering, but in the United States a significant proportion of the more than 58,000 dams are used, at least in part, for flood control. The dams range from earthen barriers used for farm ponds to the 770-foot high Oroville Dam in California and the 14.5 mile-long Watkins Dam in Utah. Dams have been effective in the reduction of flooding and have provided the added benefit of generating large amounts of hydroelectricity. They have also provided many new lakes for recreational purposes. Unfortunately, the water requirements for recreation and power generation are often incompatible. Recreation requires access to open water, while flood control calls for the emptying of reservoirs, at least before anticipated heavy precipitation, in order for there to be ample storage capacity during the period of high runoff; power generation requires a steady water flow, or one cycled to match electricity demand; and recreation calls for lakes to remain at a constant high level. A contribution frequently overlooked is enrichment in the quantity of groundwater around dam sites; as dams fill, groundwater tables generally rise as more water percolates into the subsurface.

Against the advantages of dam construction must be weighed some disadvantages, such as sediment catchment, increased evaporation, loss of inundated land, interruption of river transport, disruption of fish migration, and other environmental damage. Construction of the Aswan High Dam in Egypt on the Nile River in the 1960s ended the annual flooding of the Nile Valley and has provided electricity generation facilities, but the reservoir that formed is now filling with the sediment that for thousands of years served

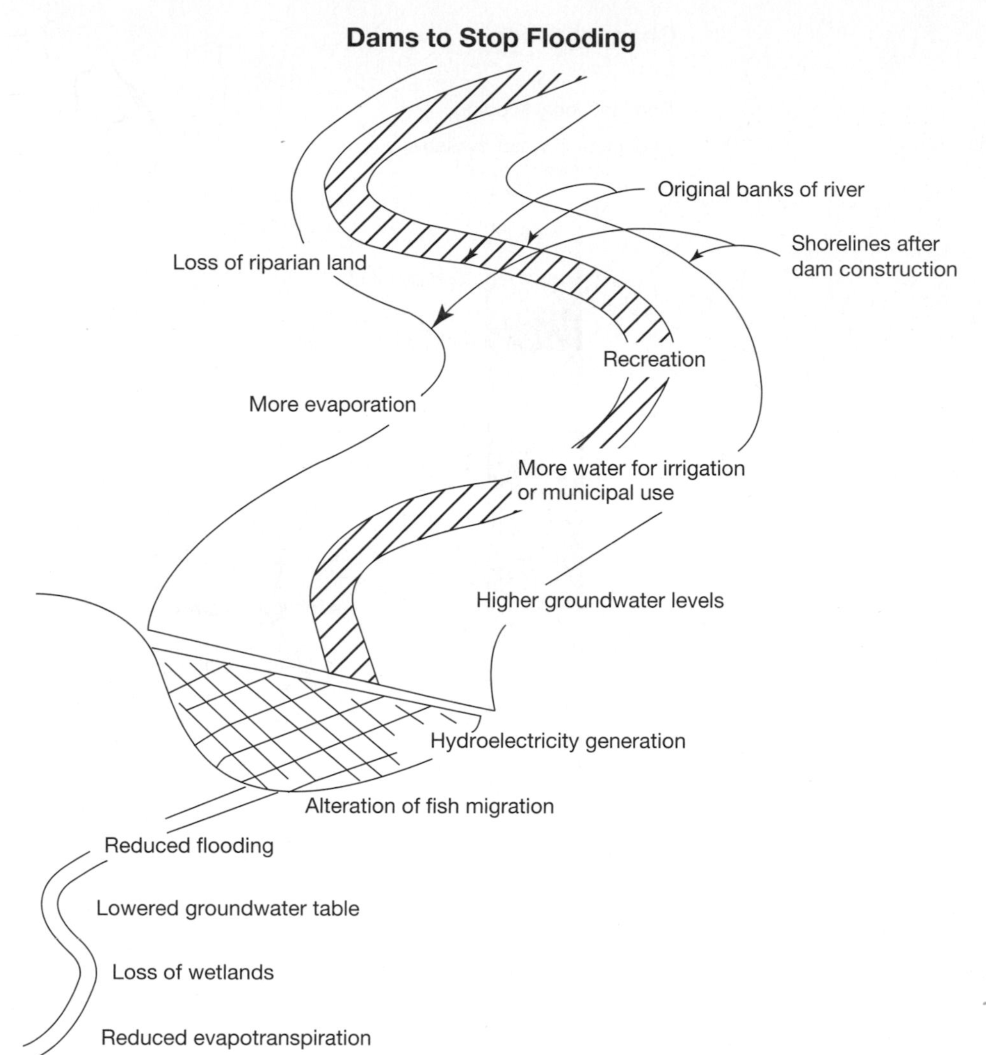

FIGURE 11.16 Construction of dams to reduce flooding has many consequences, as noted on this schematic diagram of a dam and the lake formed behind it.

as a natural fertilizer for agriculture in the Nile Valley. The Aswan High Dam has markedly reduced soil fertility along the Nile floodplain, and is rapidly leading to **eutrophication** of the reservoir behind the dam. In all arid regions, the damming of rivers provides water for many uses, but at the same time promotes evaporative water loss and the build-up of salts in the remaining waters.

For example, it is estimated that the combined natural and dam-induced evaporation of the Colorado River removes 10 percent of its total water flow. The construction of nearly every new dam meets with opposition from those whose land will be inundated and from those who do not want to see further change of the natural environment. In the 1970s, the concern for endangered species of both fish and plants in the United States nearly prevented the completion of massive dams in Tennessee and Maine. The discovery of the snail darter, a 3-inch minnow-like fish found only in the area to be flooded by the $116 million Tellico Dam in Tennessee, provided the basis for halting construction for more than a year until it was determined that these fish could, and do, live in other rivers of the area. The Furbish lousewart, a wild snapdragon-like plant that was thought to be extinct, was discovered in the valley to be flooded by the $600 million Dickey-Lincoln Dam in Maine; the discovery provided grounds to delay the construction for many months until it was determined that additional colonies of the plants existed elsewhere. Clearly, it is wise to study the potential impact on the landscape and the ecology before allowing the major changes associated with such construction projects.

The U.S. governmental agencies responsible for major dam construction and supervision have now determined that few, if any, major new power or flood-control dams will be constructed in the U.S in the next several decades. In fact, by the late 1990s, plans were underway to tear down existing dams in states from Maine to those of the West Coast, in order to restore original river environments and to promote salmon migration. Other countries, especially developing ones, continue to consider new projects to provide electrical

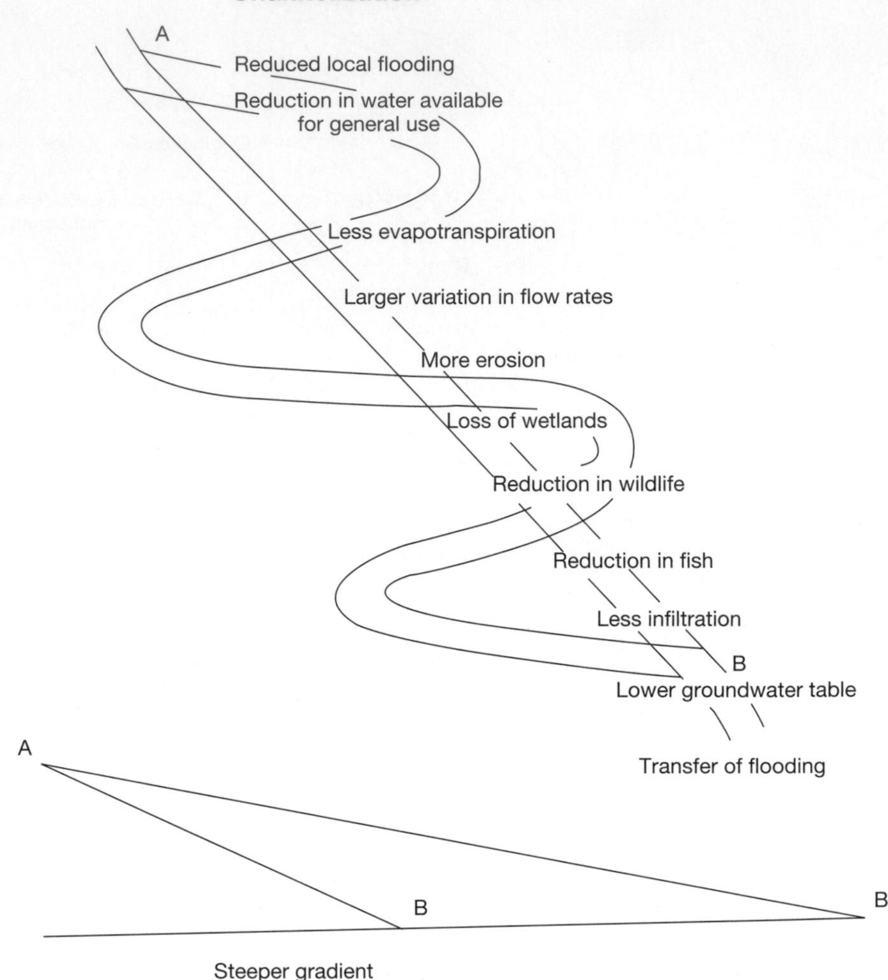

FIGURE 11.17 Channelization is the conversion of a sinuous channel, flowing from A to B, into a straight channel connecting the same points. Channelization has commonly been used to reduce flooding and is designed to allow for the very rapid movement of water away from an area so that flooding does not develop. The upper part of the diagram shows a variety of consequences that result from channelization. The lower part of the diagram is a profile showing that straightening and shortening the channel results in a steeper gradient; this results in more rapid flow of the water.

power for economic development and to control flooding problems. Thus, Brazil is considering construction of major dams on the Amazon, and China has recently built the Three Gorges Dam on the Yangtze River in central China. The Three Gorges Dam is the world's largest hydroelectric facility and will alleviate annual flooding downstream in China's largest river system. It has met with widespread opposition ever since it was first suggested in 1919 because it has altered the river ecology, flooded hundreds of square miles of farmland, and displaced more than 1 million people from their homes. In Canada, Quebec Hydro developed plans for major hydroelectric dams on the rivers that drain the sparsely inhabited area east of Hudson Bay. Even here, environmental concerns raised by potential customers in Canada and New York, as well as pressure from Native American tribes, brought about cancellation of major parts of the project. The Republic of the Congo has proposed construction of the Great Inge Dam on the Congo River. Designed to be the largest dam in the world, it would produce one-third of Africa's total current electricity needs. However, many experts feel that the combination of the environmental impact, the huge cost, the lack of necessary infrastructure, and the record of political instability of the region make its completion highly unlikely.

Channelization has provided an effective means of flood control in many areas. The principle (illustrated in Figure 11.17) is straightforward: replacement of a natural sinuous channel by a shorter and straighter one that allows for more rapid water flow out of a flood-prone area. The rate of water flow is increased because a straighter channel offers less resistance and because the gradient of the new shorter channel is steeper than a long winding one. Frequent secondary effects include a lowering of the water table and drainage of swamplands adjacent to the river; such lands then have considerable real estate value.

Although often effective, and carried out in hundreds of areas, channelization has also been found to have significant drawbacks such as increased erosion, transfer of flooding, reduced natural filtering of groundwater, a loss of wetlands habitat, and a reduction in diversity and numbers of wildlife. Unless the channelization extends to a flood-control reservoir or to the ocean, the rapid transport of water from one part of a river basin to another, only to dump the water back into its original channel farther downstream, merely transfers the problem of flooding downstream. An example of this in the United States is the Blackwater River in Johnson County, Missouri, where channelization did reduce local flooding, but it created extra flooding in adjacent counties downstream. The decrease in channel length from 33.5 miles to 18 miles nearly doubled the gradient and increased the water velocity, which, in turn, increased stream channel erosion. The original channel was 45 to 90 feet wide, but erosion broadened the channel up to 200 feet and resulted in the collapse of several bridges. The much greater rate of water flow tended to scour the channel and reduced the total amount of **biomass** production (fish, plants, algae, and insects) in the river by about 80 percent.

Channelization of the Kissimmee River in central Florida in the 1960s, and the decision in the 1980s and 1990s to restore the river to its original state, provides an informative lesson on the relationships between the channelization project, commercial interests, water needs, and environmental concerns (see the discussion in Box 11.1).

OUR USE OF WATER

Water Usage and Consumption

Water is more widely used and more essential than any other resource. The amount consumed per capita, however, varies widely as a function of the lifestyles and standards of living of societies. In discussing water usage, it is important to distinguish between **withdrawal** (sometimes called *usage*), which is the water physically extracted from its sources, and **consumption,** which is the withdrawn water that is no longer available because it has been evaporated, transpired, incorporated into products or crops, consumed by humans or livestock, or otherwise prevented from returning to its source.

Withdrawal uses of water are generally subdivided into four areas: (1) domestic–commercial; (2) industrial–mining; (3) thermoelectric power; and (4) irrigation–livestock, as illustrated in Figure 11.18. Hydroelectric power generation, in which water is actually withdrawn only to the extent that it is diverted through turbines to generate electricity, is considered a special category and is discussed separately. The amount of water withdrawn, and its division between surface sources and groundwater, varies according to the population, the type of society, the local geology, and the climatic conditions in an area. In the United States, approximately three-quarters of water usage is supplied by surface sources (Fig. 11.19), but major agricultural states such as Nebraska and Kansas, and the arid state of Arizona, draw most of their water from underground sources. Not surprisingly, by virtue of size, population, and agricultural production, California uses the most water, and Alaska, with its small population and very small agricultural production, uses the least. California, and in particular southern California, has to import water from the Colorado River, and the northern, better-watered parts of the state.

Total water usage in the United States currently amounts to about 408 billion gallons or about 1300 gallons per person per day when all of the usages are considered. If only the domestic household water supply that is delivered by public supply systems is considered, the figure is about 105 gallons per person per day. The use of water by U.S. citizens is largely taken for granted, and the large quantities required to support our modern lifestyle is overlooked. Table 11.2 presents some data on the water usage required for particular purposes by modern Western society.

Rural water withdrawn from private wells constitutes only about 1 percent of United States water usage, but in many sparsely populated areas this represents the dominant water supply. The use at any one site varies from the small amounts withdrawn for a single house to very large quantities used to supply large herds of livestock or for irrigation. In many less developed parts of the world, the rural water supply is the major water source for large numbers of the population. Because rural water is used for many agricultural purposes as well as for household needs, a somewhat larger proportion of the rural water is consumed.

Domestic and Commercial—Supplying Our Cities

Domestic and commercial water usage includes that needed by average households and for motels, hotels, restaurants, offices, stores, and businesses, as well as government and military establishments. Although it is only about 11.5 percent of total usage, domestic water is what most of us see directly each day. More than 80 percent of what we use is returned into the water systems, and most of that is through public water-treatment plants.

The growth of cities has always required the availability of continuous supplies of freshwater. Virtually all ancient and most modern cities have been established along rivers or in places where there were ample springs. As cities grew, so did their needs for water. When needs exceeded local supplies, it became necessary to find additional water and to develop ways to transport it to urban distribution centers.

The earliest water transportation systems, or **aqueducts,** were probably stream channels that had been altered or extended so that they flowed into more accessible areas. Biblical Jerusalem was served by an

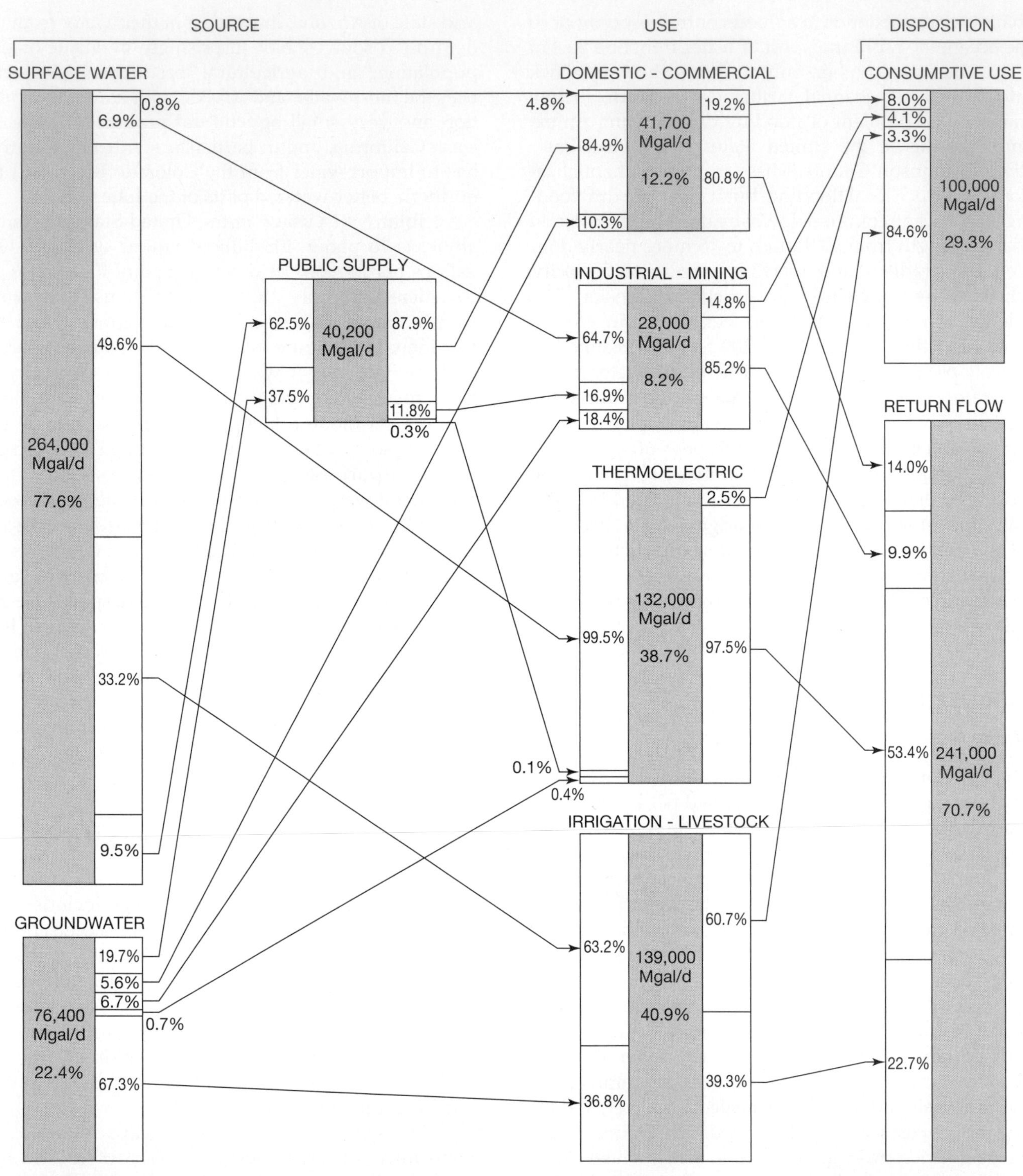

FIGURE 11.18 Schematic presentation of water in the United States in 1995 showing the source, type of use, and disposition. (From U.S. Geological Survey Circular 1200).

aqueduct consisting of limestone blocks through which a 38-centimeter (15 inch) hole had been drilled by hand. The Greeks bored tunnels up to 1280 meters (4000 feet) long at Athens, and built masonry structures to carry water. The ancient masters of the construction of aqueducts were the Romans who built nine major aqueducts that brought 85 million gallons of water a day to Rome in A.D. 97. All told, the Romans constructed aqueducts (Figure 11.20) to service nearly 200 of their cities and many of their mining efforts throughout their empire. Roman skills also built an aqueduct 373 kilometers (230 miles) long to supply the

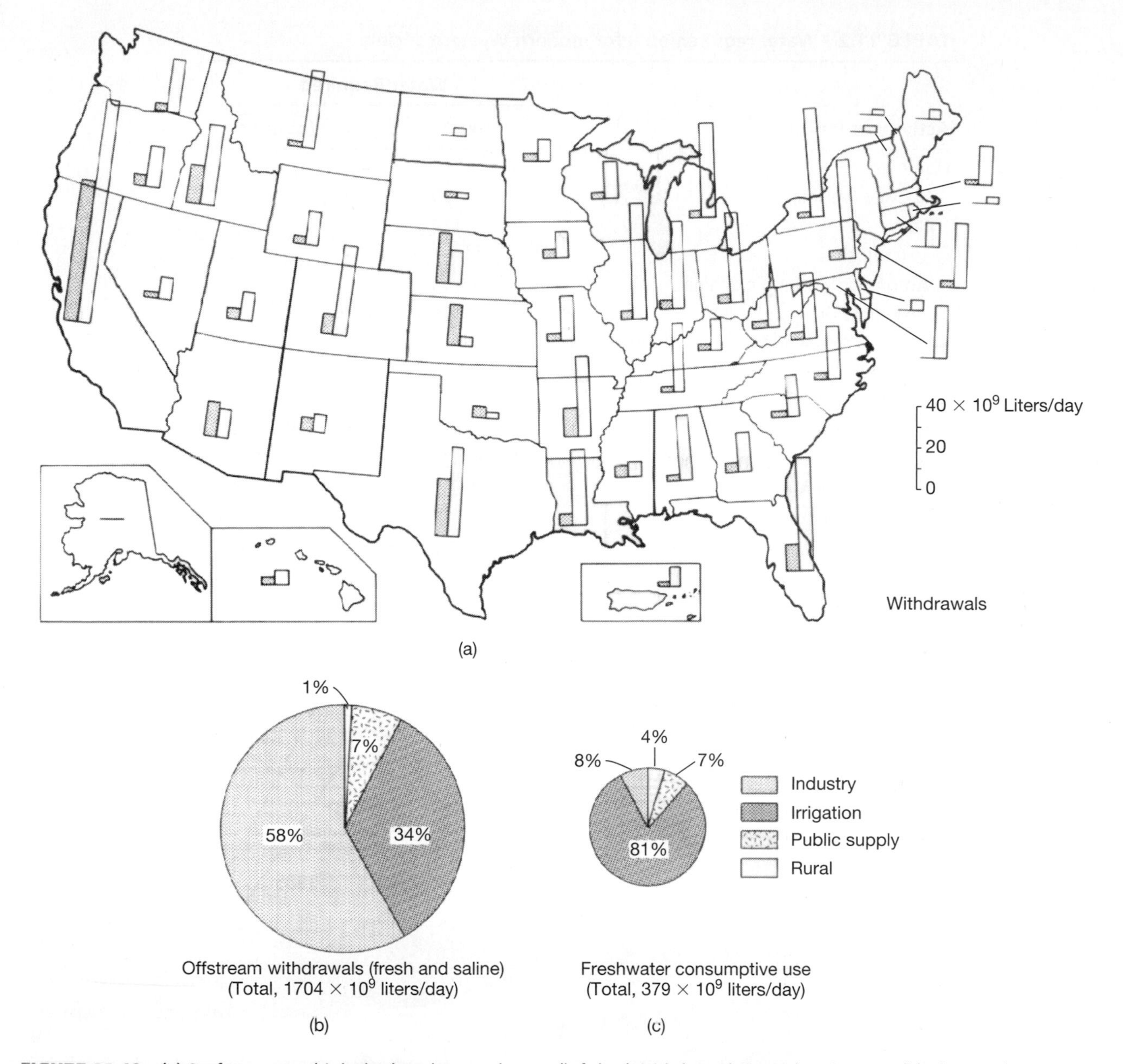

FIGURE 11.19 (a) Surface water (right bar) and groundwater (left bar) withdrawals in each U.S. state. (b) The total withdrawal of water from streams and rivers and (c) the total consumptive use of water in the United States for industry/mining, irrigation, domestic/commercial, and thermoelectric use are shown. (Updated from the U.S. Geological Survey Circular 1001, 1983.)

great city of Constantinople. After the fall of Rome and the division of the Empire into eastern and western parts, few new aqueducts were built until the late 1500s when Sir Francis Drake, then mayor of Plymouth, England, constructed one that was 24 miles long. In 1609, a 38-mile aqueduct called the New River was built to bring water to London.

In the era of modern cities, even though the demand for water has increased, the spectacular large stone aqueducts of the past have been replaced by buried steel pipes and pumping stations. A good example is in New York City, where a complex system of aqueducts links 15 major reservoirs containing more than approximately 490×10^9 gallons of water, some of the reservoirs are as much as 125 miles from the city (Figure 11.21). Despite the vastness of the system used to supply New York City, there is little problem because the abundance of rainfall in the northeastern United States provides more than adequate water for all other users as well as those in New York. However, New York, like many long-established cities, faces critical problems with its aging water transport infrastructure,

TABLE 11.2 Water requirements for modern Western society

Activity or Product	Water Required (liters)	(gallons)
Home use:		
Shower (per minute)	19	5
Bath	114	30
Toilet flush	15	4
Automatic washing machine	114	30
Hose flow per hour of lawn watering or car washing	1136	300
Food production:		
Sugar per ton	946,000	250,000
Corn per ton	946,000	250,000
Rice per ton	9,460,000	2,500,000
Milk per gallon	61,000	16,000
Beef per pound	14,000	3700
Nitrate fertilizer per ton	568,000	150,000
Industrial:		
Paper per ton	23,700	62,500
Bricks per ton	950–1900	250–500
Oil refining per 42-gal barrel	1770	468
Synthetic rubber per ton	2,500,000	660,000
Aluminum per ton	1,325,000	350,000
Iron per ton	113,600	30,000
Human survival:		
154-lb person per year	720	190

USGS pamphlet *Scientific American,* September 1963, *World Book Encyclopedia,* and Water & Industry.

FIGURE 11.20 Supplying water to cities has been a major concern since the Romans built aqueducts, such as the Pont du Gard at Nimes, France, to transport water to nearly 200 cities. This was built in the first century A.D. and carried water from two springs to Nimes. It is one of the best preserved of Roman structures. (Photograph courtesy of the French Government Tourist Office.)

as described in the discussion on page 409 (see Box 11.2, Figure 11.C, Figure 11.D, and Plate 64).

Some of the water problems of the western United States are also being addressed by means of aqueducts. Thus, the Central Arizona Project (Figure 11.22) is a major supplier of water for cities such as Phoenix and Tucson; California has constructed a complex system to supply its major cities. One of these extends for a total

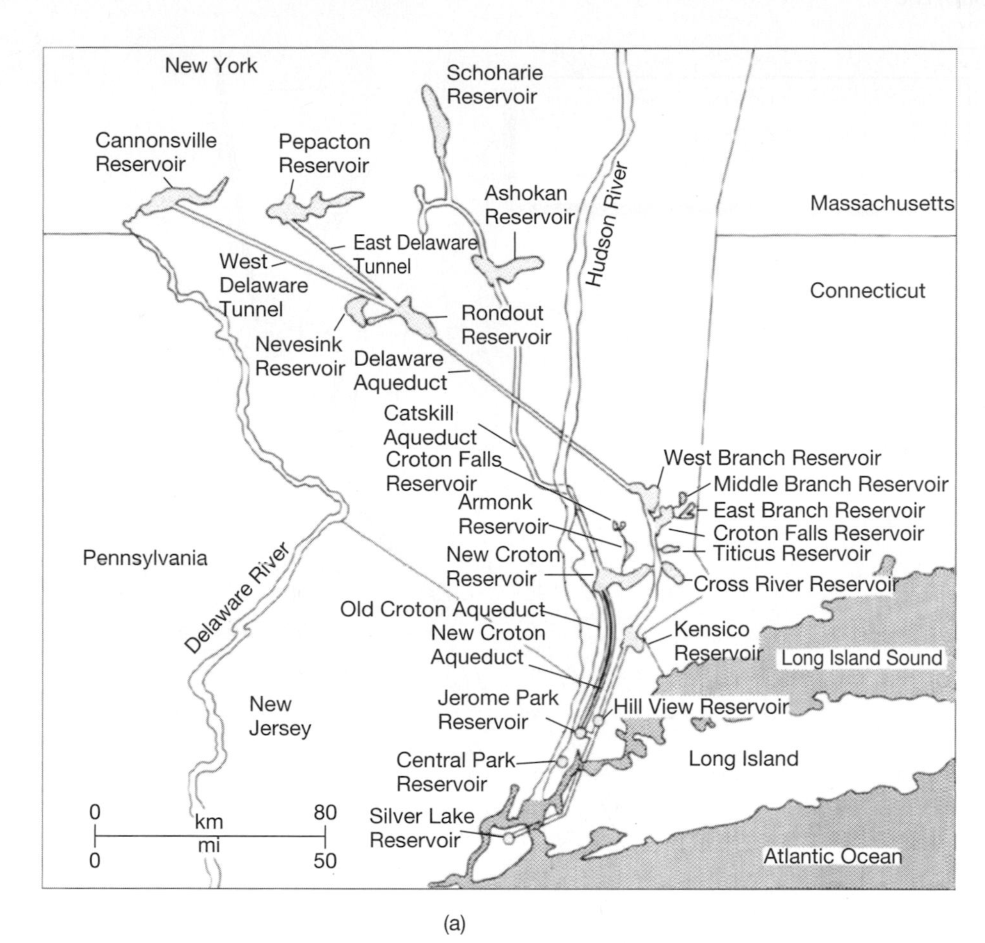

FIGURE 11.21 (a) The water supply system for the city of New York links 15 major reservoirs—some as much as 125 miles away—to meet the needs of approximately 10 million people. (After a map courtesy of the City of New York Department of Water Resources.) (b) Existing and projected New York City water supply tunnels. (Courtesy of Department of Environmental Protection of the City of New York.)

of more than 685 miles across California to bring water from many parts of the state to Los Angeles; it is discussed later in this chapter in the "Water for Drinking—The Los Angeles Aqueduct System" section (p. 434).

Irrigation

Irrigation has become an essential requirement for farming in large areas of the world where soils are sufficiently fertile, but rainfall is too low or too irregular to support the types of crops being grown. Water demand for this purpose has been rising rapidly and is now approximately 42 percent of total U.S. usage and as much as 90 percent of usage in India and Mexico. In the United States, as in many countries, the withdrawal of water for irrigation takes place on an irregular geographic distribution pattern depending upon rainfall. Thus, the eastern part of the United States uses only approximately 5 percent of its water withdrawal for irrigation, whereas the western part uses 90 percent of its water for this purpose. Irrigation systems range from simple siphons (Figure 11.23) in which gravity carries water from a main water course into the furrows, to large mechanized walking systems [Figure 11.24(a)] that can systematically distribute water in a circular pattern up to 1 mile from a central well [Figure 11.24(b)]. Depending upon the weather conditions and the crops raised, irrigation can consume vast quantities of water. For example, whereas irrigation constitutes only 42 percent of total U.S. water use, it accounts for 84 percent of water consumed. The demand for irrigation water has resulted in the building of elaborate surface water catchment and transport systems, as seen in the Lower Colorado River region (p. 420). Irrigation demand has also resulted in severe drainage of groundwater from parts of some aquifers, such as the Ogallala where 150,000 wells now draw water (p. 423) for farms along the eastern flank of the Rocky Mountains.

Discussion of water for irrigation usually brings to mind great fields of corn, cotton, wheat and the like, but the largest "crop" irrigated in the United States is turf grass. Thus, more water is used to water the grass in 40 million acres of home lawns, parks, and golf courses than any single food crop. Such usage, especially in the arid southwestern part of the United States, has led to many conflicts between environmental groups trying to conserve water, and local home owners desirous of having lush lawns and gardens. Increasingly, there is usage of recycled wastewater to water golf courses and parks in order to reduce total water consumption. On a smaller scale, more and more homes are using cisterns to capture and hold what rainwater falls on roofs or is discharged from air

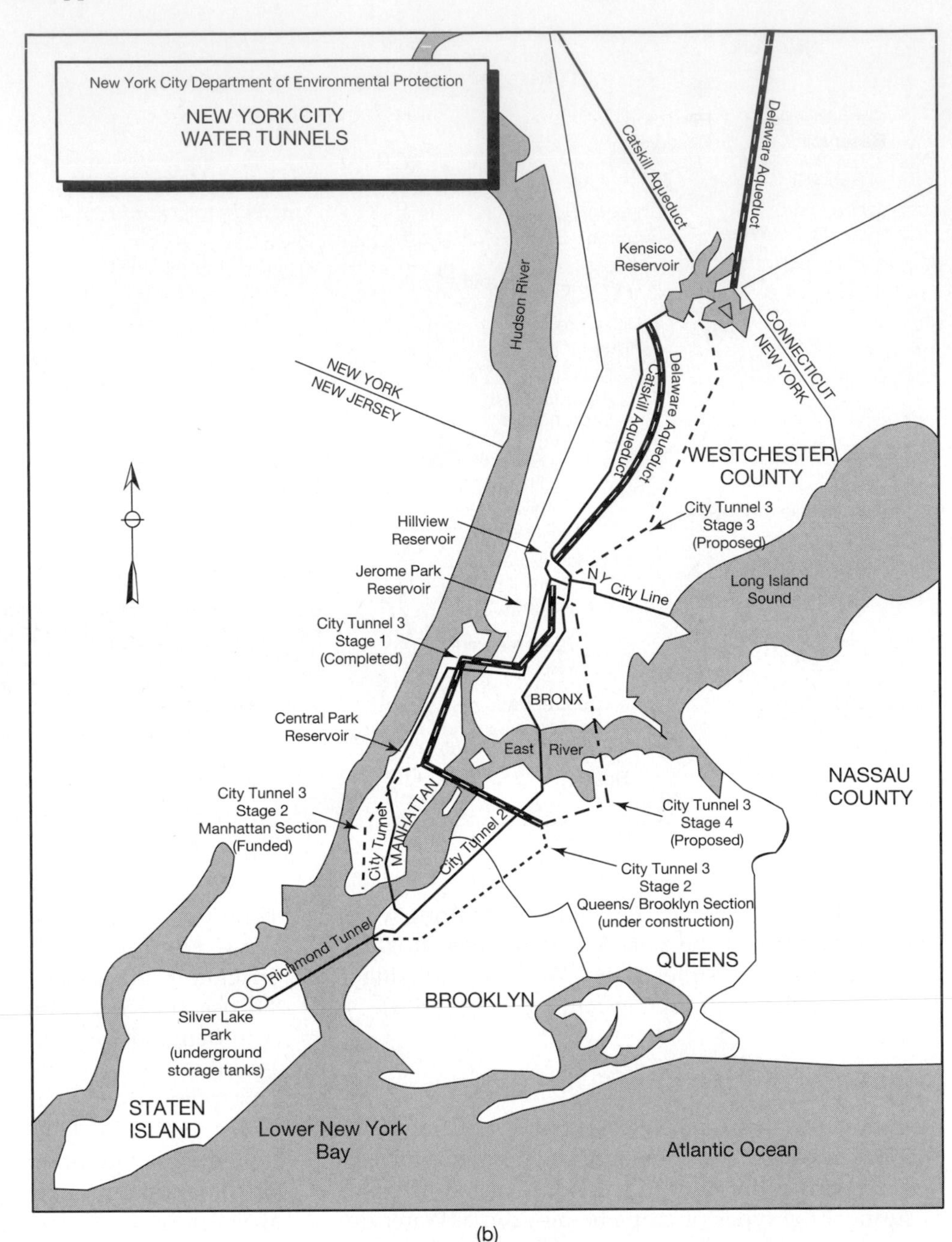

FIGURE 11.21 (cont.)

(b)

conditioning units. There is also greater emphasis being placed upon xerogardening, gardening with native plants that do not require irrigation to survive.

Water for Industry and Mining

Industrial water use includes water for processing, washing, and cooling, with some of the major users being the steel, chemical, paper, and petroleum industries. Environmental concerns about industrial water usage usually do not center upon the quantities of water used because about 85 percent of the water is returned; rather, the focus is on the cleanliness of the returned water. Prior to the 1970s, industrial pollution was widespread, but the increasingly stringent water quality regulations in most western nations, especially the United States and countries of the European Union, now require that most industries return clean water to the environment.

Mining uses water for the extraction and milling of minerals but, except for these purposes, most water at mine sites is an impediment to operations. Open-pit mines can collect large quantities of water during heavy rains, and both surface and underground mines commonly have problems of groundwater inflow. Consequently, most mines produce all the water needed on-site, and many must dispose of excess water by pumping it into rivers. Modern mining and water regulations commonly require that all water discharged from mining sites be clean enough to support

BOX 11.2

Water for New York City

How do you supply more than 1.5 billion gallons of clean water to more than 8 million people who live over an area of 300 square miles including two major islands (Manhattan and Staten Island)? The answer is with an incredible series of reservoirs, aqueducts, tunnels, and pump stations with an aggregate value of more than $8 billion. There are similar systems for every major city in the developed world and they are going to grow even larger and more complex as world population continues to increase.

The story began in a relatively simple way when the first Dutch immigrants founded New Amsterdam on Manhattan Island in the early 1600s. They met their water needs by relying on ponds, springs, and a few private wells. In 1667, shortly after the British seized the city and changed the name to New York, the first well was dug to serve as a public water supply. By the early 1700s, the combined effects of a growing population, contamination by sewage and garbage, and saltwater intrusion into the wells forced the inhabitants to begin to haul in freshwater from unspoiled springs in Brooklyn. Continued growth made these supplies inadequate and contributed to the city's inability to control major fires like the one that destroyed one-quarter of the buildings in 1776. Limited supplies of clean water also contributed to a cholera epidemic that killed 3500 people in 1832.

Citywide efforts to provide an adequate water supply really began in 1799 when the state legislature gave water delivery rights to a company that dug new wells, built new storage ponds, and installed a distribution system of wooden pipes. The company was headed by the American patriot Aaron Burr, who used its excess funds to start the Chase Manhattan Bank. More and more water was needed, so 4000 immigrants were set to work in 1837 to develop drains, reservoirs, and a 41-mile aqueduct to bring water from the Croton River north of the city. The aqueduct, moving water largely by gravity, delivered its first water to New York City during a celebration on July 4, 1842. The new water system seemed large enough for many years to come, but more people meant more demand, and the population soon outgrew the supply. By the 1880s, it was necessary to build newer, larger dams and reservoirs, and to construct the first large underground supply tunnel.

More clean water allowed the population to continue to grow. That growth, combined with the new flush toilets and household faucets, demanded yet more water. The city then looked outward again and purchased large watershed areas in the Catskill Mountains more than 100 miles to the northwest. This vastly enlarged the total capacity of reservoirs and the daily supply, but it required the construction of two large tunnel systems that were placed in service in 1917 and in 1936, 200 to 700 feet beneath the city streets.

Today, New York City is looking ahead again with the construction of a third tunnel to improve the adequacy and dependability of the entire system. The new tunnel ranks as one of the world's great engineering feats, being 60 miles long, 24 feet in diameter and, in places, lying 450 to 800 feet below the ground surface in solid rock (see Figures 11.C, 11.D, and Plate 64). When completed, anticipated to be in about 2020, this tunnel will not only help in water delivery, but will permit maintenance and repair of some of the older tunnels and pipe systems for the first time in more than a century, and will provide the first testing of some vital valve systems that have not been closed for more than fifty years. No resource is more vital than water, but most of us take the incredible infrastructure required for its delivery totally for granted.

FIGURE 11.C One of the pump stations required to supply 1.5 billion gallons of water per day to New York City.

FIGURE 11.D A part of the 24-foot diameter Tunnel No. 3 being constructed to carry water under New York City. (Photograph by Carl Ambrose; courtesy Department of Environmental Protection of the City of New York.)

FIGURE 11.22 The Central Arizona Project carries water from the Colorado River to the major population centers in Arizona. (Photograph courtesy of the U.S. Bureau of Reclamation.)

fish growth and to meet strict water standards. Because various chemicals used during mineral processing (especially cyanide used to extract gold) are very toxic, modern mining companies generally adhere to zero discharge rules and recycle all of the chemical-bearing solutions. This avoids problems of public relations and environmental impact and saves the costly chemicals. The total water usage by mining operations in the United States is only about one-eighth of the combined industry and mining figure and, hence, only about one percent of total U.S. water usage.

Thermoelectric Power

Thermoelectric power plants use water during the generation of electricity from fossil fuel, nuclear, and geothermal sources. Most of the water goes for condenser and reactor cooling, and about 98 percent of the freshwater withdrawn is returned to the rivers from which it comes. Thermoelectric power generation actually uses approximately 50 percent more water than is shown in Figure 11.18 (the total being about 195 million gallons per day) with that extra water being saltwater withdrawn from coastal estuaries. The large water requirements of thermoelectric plants necessitate that they be located where abundant water supplies exist—nearly always on large rivers or along the coast.

The water used in thermoelectric plants passes rapidly through the cooling systems; hence, there is almost never any problem of contamination. The main environmental impact results from the return of the heated discharge water into rivers and estuaries. Unless carefully monitored and remixed with sufficient quantities of cool water, the warmer water can adversely affect aquatic life.

Hydroelectric Power

The amount of water used to generate hydroelectric power in the United States dwarfs all other usage, with the total of about 3.3 trillion gallons per day being about 2.6 times as much as all of the runoff water in all the nation's rivers and streams. This apparent impossibility results from the repeated reuse of water within pumped-storage power plants (where excess electricity generation capacity is used to pump water

FIGURE 11.23 Simple gravity siphon irrigation system in which the water flows from a feed canal into furrows across the field. (Courtesy of R. B. Ross.)

back up into a reservoir so that it can be used another time), from the repeated reuse that occurs in successive hydroelectric plants along the same river, and from the use of some water before it evaporates or is consumed in irrigation. The process of hydroelectric power generation (see Figure 6.26) itself consumes very little water, but the ponding of large reservoirs behind power dams, especially in arid regions, does result in the evaporative loss of significant quantities of water.

Hydroelectric power is generated in all parts of the United States; but by far the largest producing area is the Pacific Northwest, where the tremendous flows of the Columbia and Snake Rivers pass through several dams. Hydroelectric power generation has often been promoted as a nonpolluting alternative to fossil fuel and nuclear plants. This statement is only partly true because the dams do have a large environmental impact on the areas they flood and in the modification of fish habitats and migration paths. There has been much concern expressed about major dams playing a significant role in the decline of salmon in the Columbia River and its tributaries. In attempts to increase the numbers of salmon moving upstream, special lock systems, fish ladders, and other novel approaches have been employed. However, it is clear that as long as major dams block the flow of rivers containing migratory fish, there will be conflict between the needs of power generation and various environmental concerns.

The potential for hydroelectric power generation in a country such as the United States has been largely developed. Although the potential generating capacity for the world is approximately seven times that presently generated, the development of additional hydroelectric capacity will be hampered by the remoteness of suitable areas from population centers (see also Chapter 6).

The United States is a relatively water-rich nation, but because of its size and variable climate, it has an irregularly distributed water supply. Differences in supply and consumption of the various water regions are considerable, but it is apparent that the nation is withdrawing only about one-third of the available runoff and consuming only about one-third of that withdrawn. Despite the remaining large capacity for development, local supply problems are becoming increasingly apparent; careful decisions will be needed in future years to ensure a constant high-quality supply.

The Global 2000 Report to the President, written in 1978, summarized the world's water supply situation at that time and noted that there should be adequate water available on Earth to satisfy aggregate totals of projected water withdrawals by the year 2000. However, because of the regional and temporal nature of water resources, and local demands that may not be met by supplies, shortages were predicted to become more frequent and more severe; indeed, many of the water problems forseen in 1978 are coming to pass.

Water Composition and Quality

Earth's waters range widely in composition and in suitability for human use. The most pure spring or rainwaters may have as little as 30 parts per million (ppm) (0.003 percent) of dissolved minerals, whereas the most saline waters, such as those found in the Dead Sea or Great Salt Lake, have nearly 300,000 ppm (30 percent) dissolved substances (Table 11.3). Seawater is remarkably homogeneous, with about 35,000 ppm (3.5 percent) dissolved salts. In general, waters with more than 500 ppm (0.05 percent) dissolved salts are unsuitable for human consumption, whereas waters with more than 2000 ppm (0.2 percent) dissolved solids are unsuitable for most other human uses.

The dissolved constituents in surface and groundwater are derived from the atmosphere and from the soils and rocks with which they come in contact

(a)

(b)

FIGURE 11.24 (a) Walking irrigation system used to disperse water over large fields. (b) Aerial view of a large center-well walking irrigation system, in which a central well supplies water sprinklers that continuously proceed in circular paths up to 1 mile in diameter. (Photographs courtesy of Valmont Industries, Inc.)

(see Chapter 12). Rainwater and snow generally contain a predominance of bicarbonate (from the solution of atmospheric carbon dioxide), but only a few parts per million of salts, dominantly sodium chloride from ocean spray carried aloft by winds. Other natural sources of atmospheric salts are volcanic eruptions that can release significant amounts of sulfates and chlorides into the atmosphere, wind-blown dust from

TABLE 11.3 Compositions of some typical river waters in the United States and of ocean water

Substance (ppm)	Kootenai River, Roxford, MO	Mississippi River, Cape Girardeau, MO	Arkansas River, Derby, KS	Chicorrea Creek, Hebron, NM	Colorado River, Hoover Dam, AZ	Delaware River, Philadelphia, PA	Ocean Water
Silica (SiO_2)	6.9	6.8	13	11	8.7	45	—
Iron (Fe^{2+})	0.06	0.18	—	—	0.01	—	—
Calcium (Ca^{2+})	46	47	107	225	92	18	413
Magnesium (Mg^{2+})	14	14	26	129	30	5.0	1288
Sodium (Na^{+})	3.8	11	355		106	13	10,717
Potassium (K^{+})	1.0	4.0	13	3.2	5.3	2.1	385
Bicarbonate (HCO_3^-)	160	138	249	380	159	28	—
Carbonate (CO_3^{2-})	0	0	0	0	0	0	—
Sulfate (SO_4^{2-})	45	64	217	1300	322	39	2863
Chloride (Cl^-)	2.0	12	505	48	104	19	19,275
Fluoride (F^-)	1.2	0.4	1.0	0.6	0.4	0.2	—
Nitrate (NO_3^-)	0	7.9	9.3	17	2.0	13	—
Total Dissolved Solids	215	254	1375	2220	763	128	35,000
pH	7.9	7.5	8.0	7.4	8.0	7.3	8.1

Data from "Quality of Surface Waters of the U.S.," U.S. Geological Survey Water Supply Paper 2141–2150, 1999.

continental areas, and organic aerosols released by vegetation. In recent years, there has been a growing concern about the effects of fossil fuel combustion and the release of industrial gases and particulate matter on the quality of rainwater. Numerous studies have demonstrated an increase in the acidity of rainfall in certain areas (so-called acid rain; see also Box 4.1 and Figure 4.A). This has been found to be harmful to vegetation, fish, and many terrestrial organisms and to increase the rate of weathering of building materials and natural rocks. In recent years, the pH of rainfall has dropped to 4.5–4.2 over large parts of southern Norway, southern Sweden, and the eastern United States; an example of an extreme case was a rainfall of pH 2.4, equivalent to the acidity of vinegar, in Scotland in 1974. Two primary causes of acid rain appear to be sulfur dioxide (SO_2) and nitrogen oxides (NO_x) generated by the burning of fossil fuels in power plants, industries, and motor vehicles.

Most of the dissolved substances in terrestrial waters are derived from the associated rocks, but the degree of concentration varies not only with rock type, but also with the duration of contact and the amount of evaporative concentration. Compositions of waters in several American rivers that are typical of waters worldwide are listed in Table 11.3. The differences demonstrate the effects of evaporative concentration (higher salt levels in rivers from Kansas, Arizona, and New Mexico) that occur in arid parts of the world. The partial dissolution of limestone leads to higher concentrations of calcium, magnesium, and bicarbonate; evaporation leads to higher concentrations of all substances, especially sodium chloride. Most surface waters are usable directly for most purposes, but the evaporative concentration has caused significant deterioration in some waters in arid regions. An example is the problem of the high salinity of the Colorado River as it passes from the United States into Mexico (see Figure 11.29).

The U.S. Public Health Service and the World Health Organization (WHO) have established recommended maximum limits for the concentrations of many mineral, organic, and synthetic substances in public water supplies (Table 11.1). Of particular concern is the accidental introduction of synthetic organic chemicals into water supplies because many have toxic effects even in extremely low concentrations. The maximum total dissolved solids should not exceed 500 ppm, but numerous public and private supplies, especially in arid regions and in many developing countries, yield waters that are above this limit (usually excess sodium chloride) because better quality water is not available or because the costs to produce it are prohibitive. There will be increased difficulty, both in maintaining old supplies and in developing new clean water supplies in the future, as population pressures mount and the number and complexity of possible chemical contaminants continue to grow.

A tragic case of natural water poisoning is known from Bangladesh and Bengal in Southeast Asia, and is likely to affect other regions including parts of Cambodia and Vietnam. In an attempt to help provide clean drinking water to millions of citizens who were ingesting surface water tainted by bacteria and waterborne diseases, organizations such as the WHO helped drill more than a million shallow wells. The effort seemed a success until it was revealed, in the mid-1990s, that a large percentage of the wells were contaminated by very low, but nevertheless dangerous, levels of naturally occurring arsenic derived from the underlying sediments; as many as 65 million people were at risk. Because the arsenic poisoning develops slowly and because the water from the wells tastes and appears better than contaminated surface water, many of the people have continued to use it, despite warnings. There have now been many deaths, and thousands of people have skin lesions and cancers caused by the accumulation of arsenic in their bodies, in what has been described as the "worst mass poisoning in human history."

Water Ownership

Ownership of most mineral resources is relatively straightforward because they are static materials lying on or below the land surface in some relatively easily definable form. In most areas of private land ownership, the resources are considered a part of the land that can be exploited at the discretion of the owner, subject to state and local regulations. Frequently, however, mineral rights have been separated from land ownership or have been sold or leased by the landowners to companies; the companies can exercise these rights to extract mineral resources if they comply with state and local laws regarding disturbance to overlying or adjacent properties.

The ownership of water, in its constant movement in visible surface waterways and invisible subsurface aquifers, is less well-defined than ownership of minerals. The present rules of ownership and the use of water differ from one country to another, but the complexities are perhaps best shown by considering the example of the United States, where the existence of a relatively water-rich east and a relatively water-poor west has resulted in the enactment of different kinds of laws. It is impossible to thoroughly explain the complexities of water law; hence, the following is only intended to serve as an overview, and to demonstrate the basis of modern American water laws.

RIPARIAN RIGHTS IN THE EASTERN UNITED STATES. Basic **riparian** law may be summarized as the right of every landowner to make reasonable use of a lake or stream that flows through, or borders on, his or her property as long as the use does not damage the similar rights of other landowners. Although now locally much modified by regulatory statutes to provide for cities or public utilities, the riparian principle still basically governs the use of surface water in most of the eastern states. The rights have generally functioned in a proportional manner, with the understanding that when water is plentiful, all have plenty, and when water is scarce, all share the hardship. The major exception to this is that municipal water supplies are now usually given protection of the right of eminent domain; hence, in times of shortage, cities get their quantities of water first, and riparians share what remains. The sale of riparian rights to those who do not border on streams has been allowed in some states but is not common. Because the eastern United States generally has large and continuous water supplies, the riparian system has worked well.

PRIOR APPROPRIATION IN THE WESTERN UNITED STATES. The law of **prior appropriation** grew out of the California Gold Rush when the "forty-niners" staked claims for placer gold and for the water to wash the gold from the gravel. The rights to both the gold and the water were "first come, first served." This concept grew into the formalized laws that allowed the settlers in an area to make an appropriation of a specific quantity of water for any beneficial use. The appropriations were protected on the basis that the oldest are honored first and newer appropriations are only honored as long as there is sufficient water. Thus, in times of shortage the more recent appropriations would be denied water, whereas the earliest appropriations would always have some water unless there was none at all.

In contrast to the riparian rights, which are generally held only by the landowner adjacent to a stream, appropriation rights have generally been available for sale to all those who would pay, even if they were long distances from the stream. The consequences of this are seen in California, where cities such as Los Angeles were very farsighted in the early 1900s and bought up water rights in areas hundreds of miles away, in anticipation of their needs decades later. Today, Los Angeles exercises its appropriation right to secure water that is transported by a complex series of aqueducts. Protests over the removal of water from source regions, such as the Owens Valley east of San Francisco, to Los Angeles has led to numerous lawsuits, small pitched battles, and even bombings of the aqueducts. Nevertheless, Los Angeles bought the water appropriations and will have the right to use them until or unless the courts rule otherwise.

Just as many riparian principles have been altered, appropriation rights have now been modified or encumbered by various compacts, agreements, or legislature decrees. These actions have allowed either more equitable or more economic use to be made of the water. Nevertheless, the original stamp of the appropriative right is still clearly visible in the water laws of many states of the American West.

Groundwater

However difficult or arbitrary the decisions on surface water rights have been, the decisions on groundwater rights have been even more difficult because the source of the water, its quantity, and its movement have generally been poorly known. Clarification of groundwater rights is extremely important because this source of water is so widely used, for example, in more than 50 percent of American homes. Most courts in the past, and some even today, follow what has been termed the "English rule of absolute ownership," which states that groundwater, like the rocks, belongs to the property and, thus, is the possession of the owner of the surface who can extract as much as he or she desires for any purpose. As long as wells were widely spaced and pumping was relatively limited, there were few problems; however, the advent of modern high-capacity pumps and the turning of many large cities to groundwater for part of their water supplies has resulted in the drying-up of many shallow wells. This led to widespread application of what is now called the "American rule of reasonable use," which permits unlimited extraction of groundwater for use on a plot of overlying land, but not the removal of water to distant places for sale (e.g., to cities), without compensation to farmers whose wells go dry as a result of the sale. In the western United States, many states have simply applied the law of prior appropriation to both groundwater and surface water. However, increasingly the western states have placed groundwater usage under the control of water commissions so that this valuable resource is not subject to excessive or wasteful withdrawals. Fortunately, in recent years, courts have begun to consider our growing knowledge of the limits of groundwater resources, as well the manner in which groundwater moves, and have recognized the "conjunctive" relationship between surface and groundwater as shown in Figure 11.25(a). This does not solve all problems, but it is certainly better than relying solely upon previous rulings that assumed the presence of unlimited quantities.

There are widespread misunderstandings about the amounts and the flow of groundwater. Studies usually report the saturated thickness of an aquifer, the *specific yield,* and the *safe yield.* The groundwater in aquifers actually only occupies the cracks or pores of the sediment or rock unit; only in the underground caves of some karst limestones are there actual underground rivers. Most aquifers in sedimentary rocks actually only contain 15–30 percent open pores, and the fractures in igneous or metamorphic rocks usually constitute only a few percent of the volume. Furthermore, much of the water does not drain out, but is retained as films between grains or along fractures by capillary action (this is called "specific retention"); the retained water may ultimately evaporate but it will not drain out by gravity. As a result, an aquifer that may have a reported thickness of 100 meters (300 feet) might contain only the equivalent of 25 meters (75 feet) of water, of which only to 15 meters (45 feet) can be extracted.

The withdrawal of groundwater depends not only on how much is present, but also on how fast it can move through the pores. Surface rivers and streams commonly flow at rates of 2 to 4 miles per hour, but groundwater movement is typically only centimeters or millimeters or less per hour. Thus, well water is often many years old when extracted, and the rate of groundwater recharge is very slow compared to that of a surface reservoir. The *specific yield* is the maximum rate at which one can continuously pump water from an aquifer; the *safe yield,* usually a much lower value, is the maximum rate at which water can be pumped without lowering the water table. Fractured igneous or metamorphic rocks usually contain much less water than sediments because the volume of the fractures is small. Nevertheless, the water can often flow more quickly because the fractures are larger, more continuous, and more intersecting. Furthermore, if they intersect areas of flowing streams, they may be more rapidly recharged than typical sedimentary aquifers.

Mining can lower groundwater tables locally or regionally when water is pumped out to permit mining operations beneath the level of the original water table, as shown in Figure 11.25(b). The extent of the effect depends upon the mine depth and the rate of groundwater flow. After mining has been completed, groundwater tables commonly return to their original levels, filling underground mine workings and turning open pits into lakes (see Plate 18).

Environmental Water Rights

In many countries, over the past twenty years or more there has been an increasing awareness and an emphasis on the water needs of wildlife; for example, in the United States many relatively new regulations regarding environmental water rights have been developed. In the 1950s, many ventures sought to drain wetlands to make them into farmland. Furthermore, diversionary canals were built to move water to farms and cities. Today, the situation is almost the reverse. The federal Endangered Species Act and the Wetlands Act are both specifically aimed at preserving populations of animals and the habitats necessary for their survival. The effects of withdrawing surface waters are reasonably obvious and the effects on wildlife, such as fewer fish, beavers, or water birds, can be clearly

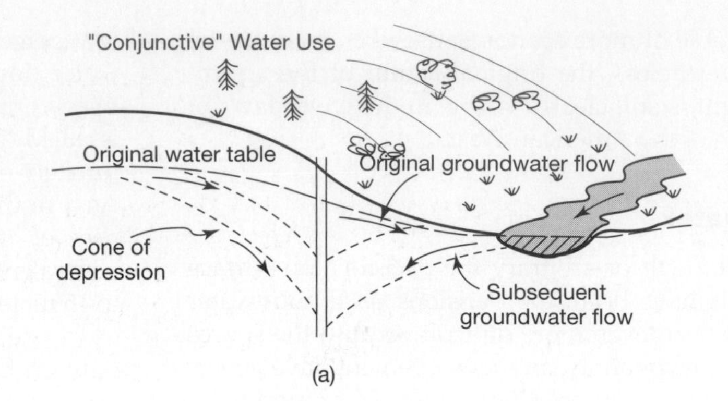

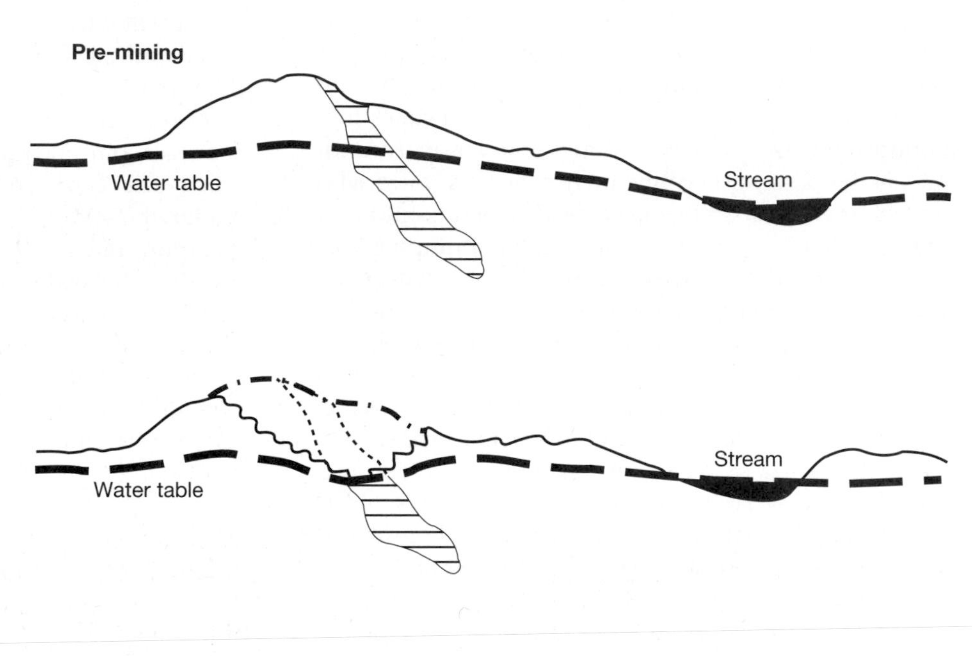

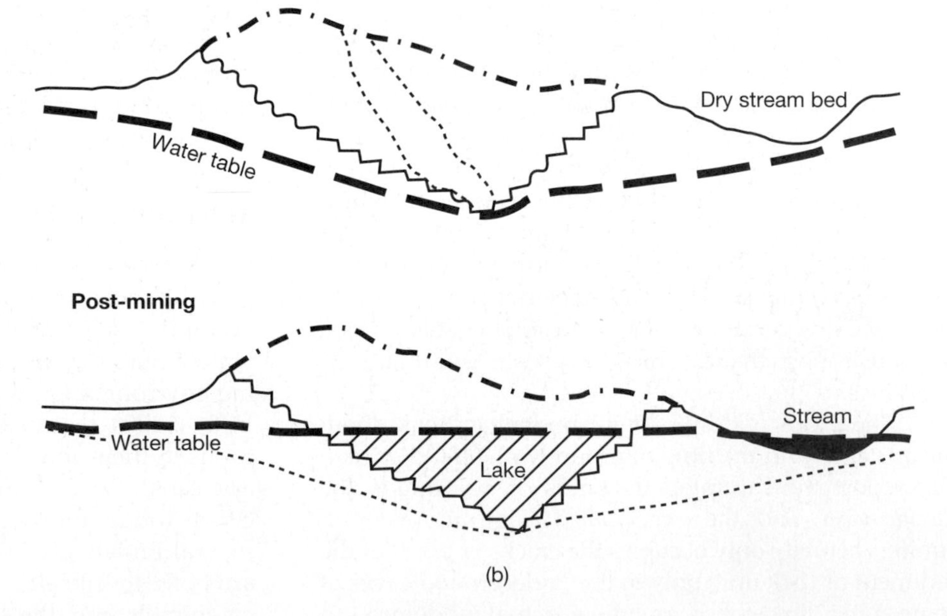

FIGURE 11.25 *Conjunctive* water relationships are those demonstrating the connection of ground and surface waters in an area. (a) Intense pumping of a well adjacent to a small stream can lower the groundwater table such that the stream water begins to infiltrate the ground, lowering the stream flow, and is actually drawn into the well. (b) Shown here are the effects of mining on the groundwater table as the mine pit is deepened and water is pumped out of the pit to permit continued mining. The first stage of mining only has a minor effect on the groundwater table in the immediate vicinity of the mine. The third stage of mining has lowered the groundwater table so much that the stream has gone dry. After mining has been completed and the groundwater table returned to it normal level, the stream again flows and the mine's open pit has become a lake.

linked with water loss. Conversely, groundwater withdrawal has less immediate and less obvious effects, but may be equally important as stream levels begin to drop, springs dry up, and marshy wetlands are gradually converted to meadows. The combined effects of surface drainage and groundwater withdrawal were especially evident along the Mississippi migratory flyway, the migratory path for water birds, where duck populations dropped dramatically from the 1950s to the 1980s. The restoration of wetlands, beginning in the 1980s, had beneficial effects as breeding and feeding areas reappeared; the population of ducks has risen steadily since the early 1990s.

Effects such as those discussed here have brought environmental water rights into the complicated factors governing water use throughout the United States and in many other countries. No longer is it sufficient merely to point to an increased need for water to grow crops or serve cities. Every new major use and every transfer of water from surface or subsurface sources must be considered in terms of environmental impact as well as need. The result is a much more careful consideration of water needs, much better conservation practices, and, frequently, higher water prices.

Desalinization of Water

Samuel Taylor Coleridge in his poem "Rime of the Ancient Mariner" identified a problem faced by large numbers of coastal cities and islands when he wrote, "Water, water everywhere nor any drop to drink." Coastal localities have access to vast quantities of water that cannot be used because it is saline. In fact, four principal methods have been developed to permit the use of the seawater, or other brines, for human consumption. The general process, called **desalinization** or **desalting,** can be accomplished by any one or a combination of the following: (1) distillation; (2) electrodialysis; (3) reverse osmosis; or (4) freezing. These processes are shown in very simple schematic form in Figure 11.26.

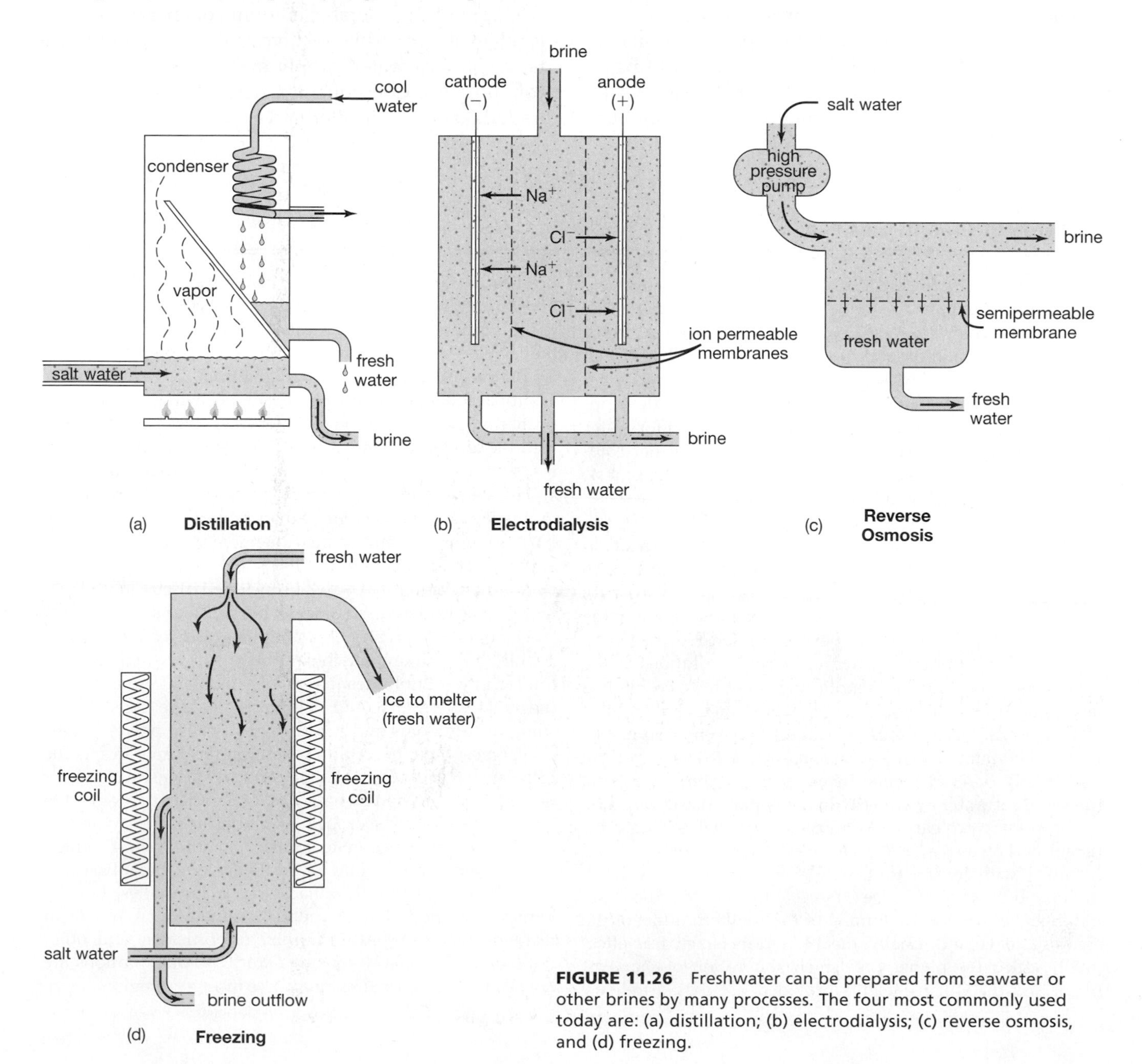

FIGURE 11.26 Freshwater can be prepared from seawater or other brines by many processes. The four most commonly used today are: (a) distillation; (b) electrodialysis; (c) reverse osmosis, and (d) freezing.

Distillation is the process long used in school chemistry laboratories to produce high-purity water. Salty water is boiled and the evolved steam is condensed into freshwater. The dissolved salt is left behind, making the remaining brine even saltier.

Electrodialysis uses two special membranes that selectively allow for the passing of sodium (Na^+) or chloride (Cl^-) ions. As salty water passes between the membranes, the sodium ions are drawn through one membrane to the cathode, and the chloride ions are drawn through the other membrane to the anode. The result is a flow of freshwater from the center of the cell and saltier brine to the lateral parts containing the electrodes.

Reverse osmosis produces freshwater from salty water by forcing water molecules through a semipermeable membrane when high pressure is placed on the salty water. The membrane has pores that allow for the relatively small free water molecules to pass through but will not allow the larger hydrated salt ions (Na^+ and Cl^-) to pass through. In a continuous process, about 30 percent of the original salty water passes through the membrane to produce freshwater, and the remainder, now saltier, is discharged.

The *freezing process* takes advantage of the fact that when salty water freezes, the ice that forms is freshwater and the salt is concentrated in the remaining brine. In a simplistic form, salty water is fed into a freezing chamber where the ice crystals that form are separated out and collected to form freshwater. The saltier brine is rinsed off the ice crystals by use of a small amount of freshwater and is allowed to drain away. Regardless of the desalinization process employed, one of the products, in each case, is a brine that can be toxic and corrosive. Safe disposal of this brine must be considered when desalinization plants are designed.

Each of the desalinization procedures requires the input of considerable amounts of energy, and each is relatively expensive relative to the usual groundwater or surface water supply systems that serve most communities. Consequently, desalinization is usually undertaken only if alternatives are not available. As

BOX 11.3

Water in the Middle East

When one thinks of critical resources and the Middle East, thoughts generally focus on petroleum. After all, this is the area of the world's greatest petroleum reserves, the site of many OPEC countries, and the place where the Gulf War of 1991 was fought. But it is another resource, one that is often taken for granted, that has emerged as critical to development and to peace in the region—water. The need for water in this arid and semiarid region has been apparent since biblical times, and its truth is clear in the Israeli comment, "Water is like blood; you can't live without it."

The dual problems of water availability and agreement on its equitable distribution have grown more acute in the past fifty years because of large population increases resulting from immigration into some areas (especially Israel), high birth rates in other areas (for example, among the Palestinian population), the need for intensive irrigation, and the desire to develop hydroelectricity. Solutions to the problems of locating and distributing water have been made even more difficult by a series of Arab–Israeli conflicts and several border and resource disputes between Arab neighbors. The water resource problems of the Middle East are but briefly described here; in many ways, they reflect the problems of water availability in many parts of the world.

Water was early recognized as a vital resource for Israel, and by the early 1950s a regional agreement was being planned with Jordan to share the waters of the Yarmouk River and the Sea of Galilee (Figure 11.E). The 1956 Suez War, followed by the 1964 damming of the southern outlet of the Sea of Galilee and, finally, the 1967 Arab–Israeli war effectively killed the plans. The Israeli annexation of the West Bank and the Golan Heights provided not only a military position, but also gave Israel control of the runoff areas that now supply two-thirds of its water. The Israelis control most of the Yarmouk River flow, while the Jordanians contend that they have never been provided with 100 million cubic meters (25 billion gallons) of water annually from the Sea of Galilee as promised. Drought combined with the extraction of water, primarily for the irrigation of water-intensive crops, has lowered the level of the Sea of Galilee to its lowest point in history and raised its salinity to concentrations that threaten its aquatic life. Jordan has been forced to rely more than ever on groundwater, but anticipates exhaustion of those supplies by 2011. Attempts to broker peace agreements between Israel and its Arab neighbors in the 1990s and the early part of the twenty-first century have always included discussions of water rights as well as discussions of borders and security.

Slightly north and east, another water resource drama is unfolding. Turkey, which controls the headwaters of both the Tigris and Euphrates Rivers, began an ambitious plan in the mid-1980s that could ultimately include constructing some twenty dams and fifteen hydroelectric power plants. The biggest dam, the Ataturk Dam on the Euphrates, is the fifth largest in the world and reduces downstream water flow to Syria and Iraq by half. The total project is designed to generate sufficient electricity to modernize Turkey and to provide irrigation water to transform more than 4 million acres of semiarid land into a Middle Eastern "breadbasket."

Turkey returns flow back to the Euphrates, but the Syrians fear pollution of the river—their main source of drinking water—by salts, fertilizers, pesticides, and other pollutants. Predictably, tensions are high, but solutions are few. Water has no rival as a resource in an arid region.

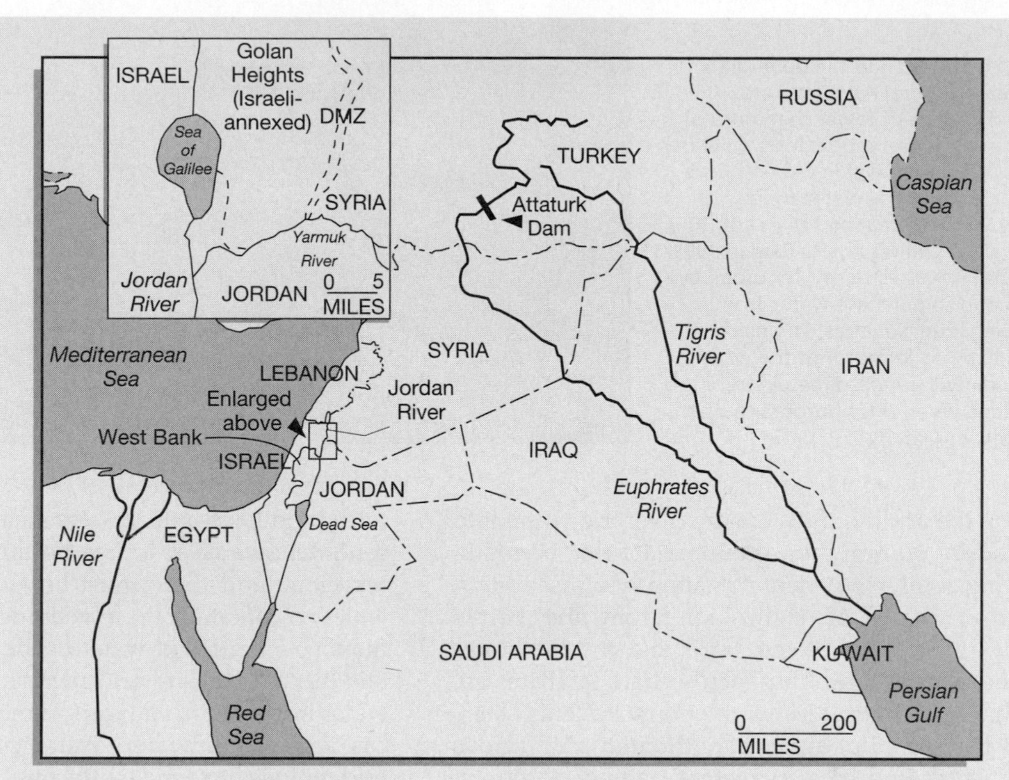

FIGURE 11.E Map of the Middle East showing the locations of the Sea of Galilee, the Dead Sea, and the Tigris and Euphrates Rivers.

freshwater supplies become more and more committed, countries and cities in arid regions are increasingly looking to desalting as a viable, although expensive, source of potable water. This is especially true in countries such as Saudi Arabia, where oil for energy is plentiful and inexpensive, but water is scarce. During the Gulf War of 1990–1991, much effort was made to protect some of the world's largest desalting plants from oil spilled in the Persian Gulf. In the United States, coastal cities in Florida and California are looking to the ocean or to subsurface brines to provide increasing proportions of their future water supplies. Catalina Island, off the coast of southern California, has turned to desalting because of an increased population, but now the community faces the problem of even more people wanting to live on the island because of the success of the desalting plant.

POTENTIAL WATER PROBLEMS

Water, like most mineral resources, is irregularly distributed over Earth's surface. Unfortunately, this distribution often does not correspond to our needs for water at a given place and time. These inconsistencies have frequently led to problems of supply and quality, and clearly suggest that such problems will increase in the years to come. In general, humid regions with more than about 30 inches of annual precipitation have sufficient surface water available in the forms of lakes, rivers, and permanent streams to meet water needs. However, in areas of intense population concentration, especially those without neighboring large rivers, the local demand can easily exceed supplies. Furthermore, normally well-watered areas can suffer periods of drought, which result in depletion of usual water supplies (Figure 11.27). Arid regions are constantly plagued by inadequate surface water supplies, with the deterioration of water quality due to the evaporative concentration of salts and, in some areas, with dwindling groundwater supplies. In some regions, water has become as important politically as it is economically because the control of water governs many other activities (see Box 11.3 for an example).

It is not possible to chronicle here examples of all current and potential water problems, but the following discussion examines some of the major problems with which we must contend in the near future.

Limited Surface Water Supplies—The Colorado River Project

Deserts, by virtue of the paucity of life-sustaining water, have always been some of the most inhospitable areas of the world for humankind (Figure 11.28). We have partly

FIGURE 11.27 Even in the generally well watered Central Atlantic states, prolonged periods of lower than normal rainfall can have dramatic effects on water supplies. The water level in Carvin's Cove reservoir, the principal water supply for Roanoke, Virginia, dropped to nearly 30 feet below the spillway of its dam in 1999, forcing the city to restrict water usage by residents and to purchase water from surrounding municipalities. The pier normally extends far out into the water but was nearly a hundred meters from the water's edge when this photograph was taken. (Photograph by J. R. Craig.)

overcome the aridity of deserts by diverting rivers into them, and by pumping up groundwater that occurs in underlying aquifers. Ancient irrigation systems brought about the spread of civilization from the Fertile Crescent—the valleys of the Tigris and the Euphrates Rivers in what is now Iraq—across Iran, Afghanistan, Pakistan, and India. More modern systems have allowed the spread of agriculture through arid regions of many lands and have converted parts of deserts in Israel and in California into some of the most productive regions in the world. The introduction of additional water supplies has allowed for the development of large population centers where naturally available surface water would not have permitted such populations.

The low-latitude desert regions of the world have offered good sites for large-scale agriculture and development because many of them permit year-round growth of crops. The extensive agricultural development of these areas does, however, call for the consumption of vast quantities of water. High evaporation rates in low latitude deserts mean that water becomes a nonrenewable resource because there can be little recycling, and there must be a constant influx of the water to maintain these activities. The sources of the massive amounts of water needed to develop and sustain our activities in arid regions have been twofold—water imported from rivers in more humid adjacent areas and groundwater. Water provision schemes for arid regions have met with considerable success as evidenced by the creation of millions of hectares of agriculturally productive land. Unfortunately, even some of the largest and most carefully planned projects have the potential for major problems.

An example of this is the well-known Colorado River Project, which supplies water to seven western states and Mexico [Figure 11.29(a) and Plate 65]. Since the late 1800s, farmers have tapped the Colorado River for its water. By the 1920s, it became apparent that the water of the Colorado was too valuable a resource to allow uncontrolled exploitation. Therefore, in 1922 the

FIGURE 11.28 The droughts that have been experienced in several parts of Africa in the past twenty years have killed large numbers of cattle and severely limited the capabilities of many people to raise crops. (Photograph from the United Nations.)

Colorado River Compact [Figure 11.29(b)], signed by the states in its drainage basin, decreed that the upper basin states of Wyoming, Colorado, New Mexico, and Utah should forever get 7.5 million acre feet (9×10^{12} liters; one **acre foot** of water is equivalent to about 1.2×10^6 liters or 3.26×10^5 gallons) of water to share annually. The lower basin states of Arizona, California, and Nevada would draw the same amount. In 1944, a treaty guaranteed Mexico 1.5 million acre feet of water annually; although the original treaty did not specify the quality of the water reaching Mexico, a subsequent agreement established that it should not contain more than about 900 ppm dissolved solids.

The problems that have arisen are threefold. The Colorado River does not generally carry as much as 15 million acre feet of water [Figure 11.29(b)]; the water reaching Mexico sometimes contains as much as 1500 ppm of salt; and the Navajo Indian reservation, never considered in the allotment schemes, has proposed a project that would claim a significant part of the Colorado River to irrigate its crops.

The original allocations of water between the upper and lower basin states of the Colorado River were based upon water flow estimates done between 1896 and 1922. Unfortunately, these estimates were made during a generally wet period, when the average annual flow was about 16.8×10^6 acre feet. Since 1931, however, the flow has only averaged about 13.1×10^6 acre feet and in 1934, the flow was only 5.6×10^6 acre feet.

To smooth annual and seasonal fluctuations and to retain waters, an elaborate scheme has been built

FIGURE 11.29 (a) The Colorado River Basin, showing the location of the major dams and lakes and the areas of the upper and lower basins. (b) The top curve represents the total flow of the Colorado River—measured at Yuma, Arizona—before the Hoover Dam was constructed in the 1930s, and calculated as the flow at Hoover Dam plus the Gila and Bill Williams Rivers after that. The actual flow reaching Yuma since 1934 is the lower curve; the rest of the water has been diverted for irrigation and municipal water supplies. The dashed lines indicate the total amounts of water promised to the Colorado River Basin states and Mexico. (Data from U.S. Geological Survey, Yuma, Arizona.)

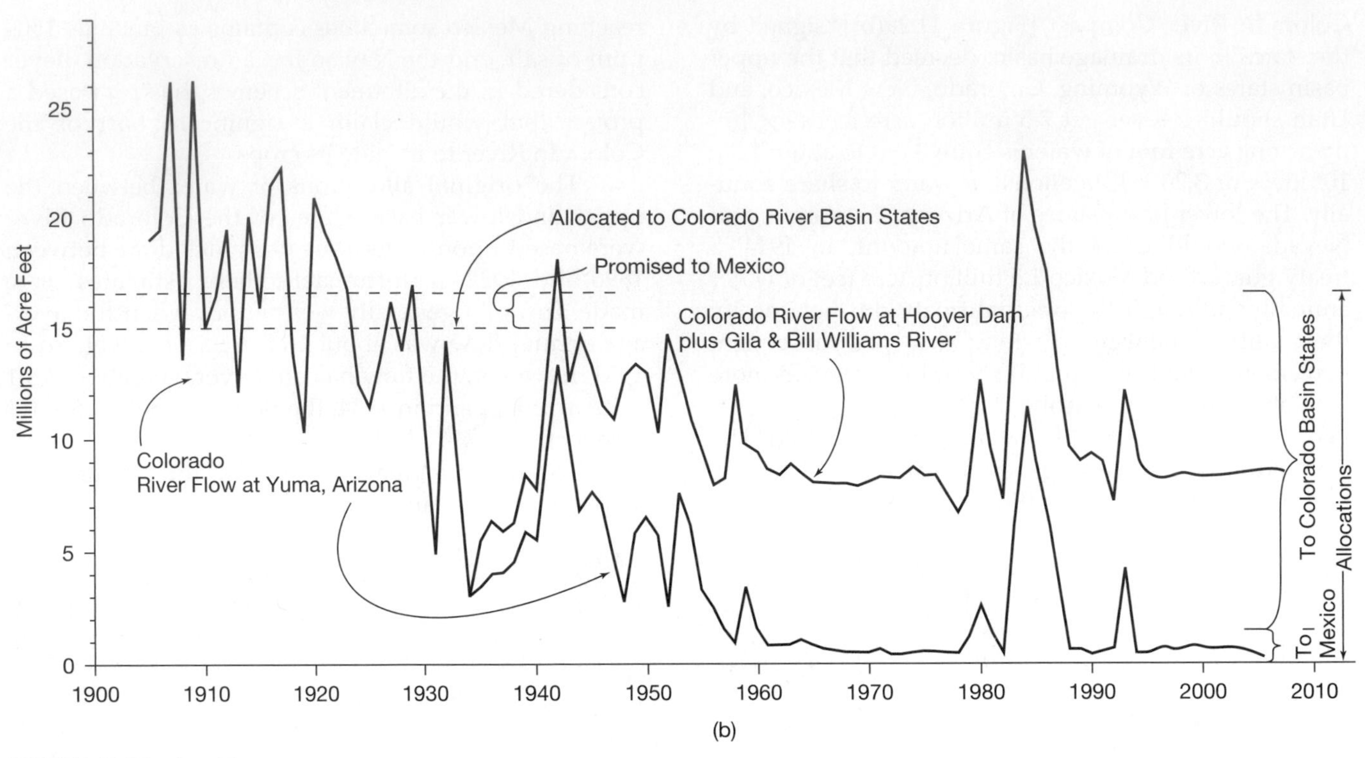

FIGURE 11.29 (cont.)

for trapping and tapping the Colorado River, as is shown in Figure 11.29(a). The dams store water for usage but also bring about an increase in the salinity by allowing extra evaporation. The problem of supply has not yet been fully felt because some states have not demanded their total allocation, and the Navajo Indians have not pressed their demands. However, the completion of the Central Arizona Project (Figure 11.22) in the 1980s resulted in Arizona using its allocation, and in California having to give up an extra million acre feet over its allocation that it had been taking to satisfy the water needs of San Diego and Los Angeles. Furthermore, projections for future water demand for agriculture and for the processing in the Colorado Basin of energy resources of coal, oil, and perhaps someday, oil shale, far exceed the river's flow.

The solution to the problem of the quality of water being passed on to Mexico has been the construction of a large desalinization plant at Yuma, Arizona. This $350 million facility, operating by **reverse osmosis,** delivers 1.5 million acre feet of water (with only about 800 ppm impurities) to Mexico for its irrigation needs. The saline by-product water from the plant, with about 8200 parts per million dissolved salts, is channeled in a diversionary canal into the Gulf of California. The cost of the desalinated water to be delivered to Mexico to meet treaty obligations has been estimated at thirty times the cost of irrigation water in California.

Additional conflicts have arisen along the Colorado River Basin over the function of the dams that hold back the large reservoirs. The dams have eliminated the problems of flooding that had previously occurred periodically along the lower Colorado; however, the dams are now often so filled with water being held for irrigation that they no longer have the excess capacity necessary to stop floodwaters. Furthermore, the floodplains below the dams are now heavily populated; population issues limit the rapid release of water, which is sometimes needed to create the excess capacity needed for flood control. Compounding these problems is the need to be able to generate hydroelectric power to meet the increasing energy demands to pump water to the various areas served by the basin.

The presence of the dams, while stopping flooding, has also altered the natural seasonal redistribution of sediment and adversely affected the fluvial habitat necessary for the native fish to survive. Consequently, the government has periodically opened the flood gates to "flush" the channels and more naturally redistribute the sediments. While environmentalists have applauded this action to preserve the fish, others have viewed this as a great waste of water that could have been held for irrigation or used to generate electricity.

Groundwater Depletion and the Problem of the High Plains Aquifer

Under normal conditions, the quantity of groundwater and the level of the water table exist in a long-term equilibrium in which the recharge is balanced by the discharge. When pumping begins, the equilibrium is

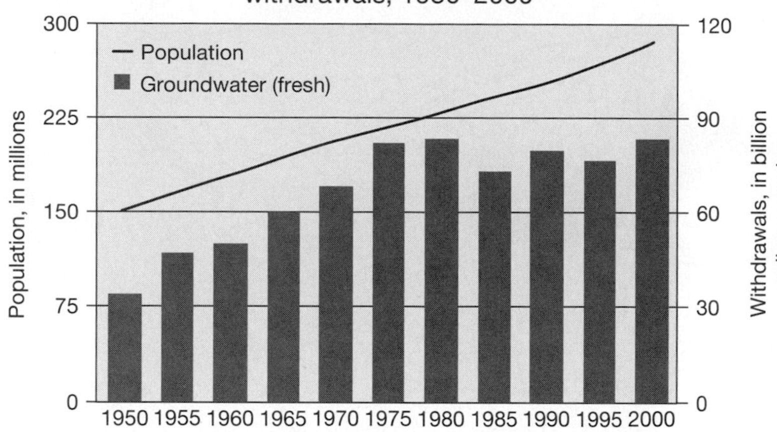

FIGURE 11.30 Trends in groundwater withdrawals in the United States, 1950 to 1995. (From U.S. Geological Survey Water Supply Papers.)

disrupted and, in general, groundwater levels fall. If pumping is only of small quantities, the decline may be local, as a **cone-of-depression** around a single well; in contrast, if pumping is of large quantities from many wells, the fall may be widespread. Pumping may also bring about decreases in the natural discharge to streams, to the sea, or in the rates of evapotranspiration.

A safe or a sustained yield is the amount of groundwater withdrawal that can be pumped for long periods of time without a continuing drop in the water table. Withdrawals in excess of a safe yield result in *water mining* and a progressive drop of the water table and, at some point, a decrease in the rate at which water can be pumped. Water mining is thus much like the mining of any other mineral resource except that there is often at least some replenishing of supplies by natural recharge.

As noted previously, commonly only 15–25 percent of the thickness of an aquifer is actually extractable water. Furthermore, as the water table drops and the saturated thickness decreases as a result of pumping, the rate of additional extraction also decreases because more of the water movement is lateral instead of flowing directly downward by the pull of gravity.

The pumping of groundwater has increased rapidly in this century in response to burgeoning populations, increased industrial demands, the expansion of irrigation into semiarid regions, and the development of high-capacity pumps. The increase in the United States since 1950 is shown in Figure 11.30; the present rate of pumping ($>80 \times 10^9$ gallons per day) approaches 10 percent of the estimated 10^{12} gallons per day of water passing through the aquifers. Unfortunately, the demand for groundwater is very unevenly distributed and often does not correspond to the rates of recharge. Hence, groundwater mining with a resultant fall in the water table has occurred in many parts of the United States, as shown in Figures 11.31 and 11.32.

The aquifers of the Atlantic and Gulf coastal plains are recharged by relatively high rainfall (>40 inches per year), but the heavy demand of dense population and industry has resulted in a falling water table in every coastal plain state. An example of the decline is the area near Houston, Texas, where the water table dropped nearly 100 meters (300 feet) between 1940 and 1970, when stabilizing measures were taken. Groundwater levels in the upper Midwest and the western parts of the United States display marked falls in many areas because the lower rates of precipitation have been unable to recharge the aquifers as rapidly as pumping for irrigation withdraws water. This problem is especially prevalent in California, the nation's principal user of groundwater. The California Department of Water Resources has determined that large declines in the groundwater are occurring in eleven basins, eight of which are in the San Joaquin Valley where agricultural irrigation is greatest. In the mid-1980s, the water table was declining as much as 6 feet annually and averaged about 2.5 feet per year. The coastal basins, serving cities as well as irrigation schemes, experienced water table drops of as much as 200 feet from 1950 to 1983. Another prime example of water table decline is in Arizona, southeast of Phoenix, where water has been withdrawn for agricultural and municipal use since 1930. The average annual drop is now about 8 feet each year, and the total decline is nearly 400 feet.

The southern High Plains of the United States, although dry, hot, and windswept on the surface, is the location of one of the country's major groundwater accumulations—the Ogallala aquifer (Figure 11.33). This Miocene deposit contains more than 24,000 cubic kilometers (6000 cubic miles) of gravel, much of which is saturated with high-quality groundwater. The southern High Plains, with an annual rainfall of 20 to 30 inches and an evaporation rate of 60 to 100 inches was the site of poor dry-land farming until the water of the

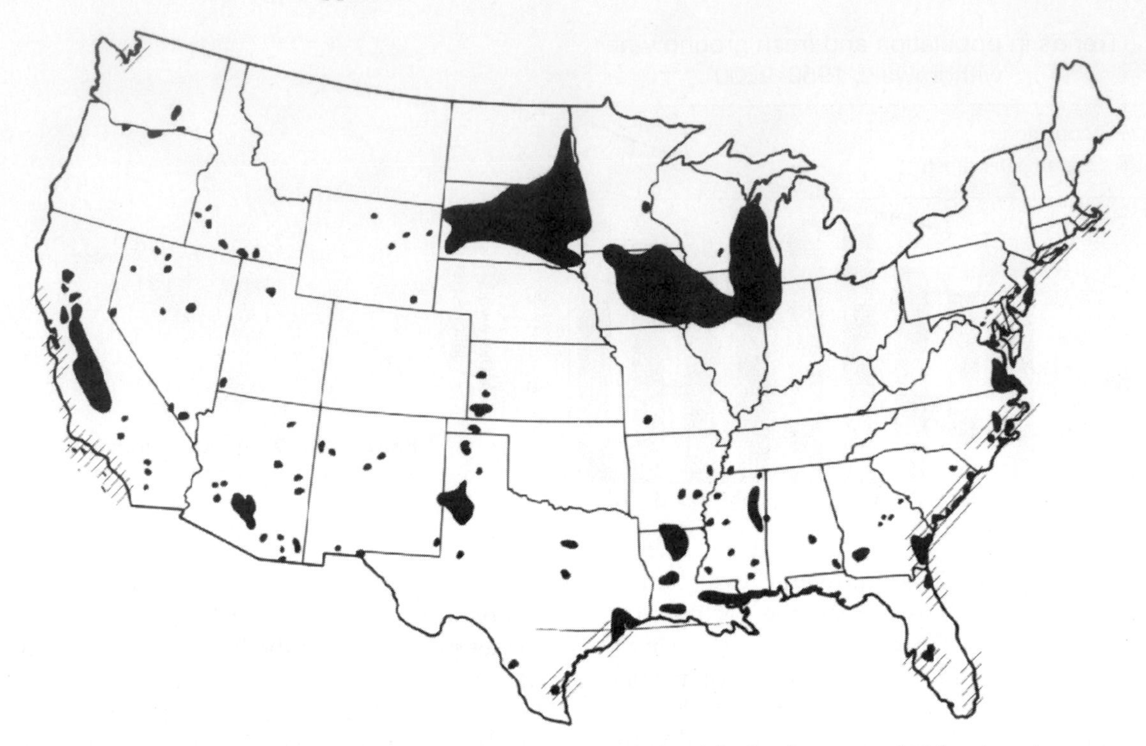

FIGURE 11.31 Areas of water table decline or artesian water level decline in excess of 40 feet in at least one aquifer since predevelopment are shown in black. Areas of saltwater intrusion into aquifers along coastal margins are shown by the striping. (From U.S. Geological Survey Water Supply Paper 2250.)

Ogallala was discovered in the 1930s. Since that time, some 150,000 wells have penetrated the aquifer to draw out millions of acre feet per year for use in irrigation. By the late 1960s, it became apparent that the water table in several parts of the aquifer was being depressed at rates as great as 1.5 to 2 meters (4 to 6 feet) each year. A few parts of the aquifer have actually registered a rise in the water table due to the addition of irrigation water, but large areas in the Texas Panhandle and in western Kansas have seen a drop in the water table of 30 meters (100 feet) or more in a half century. Accurate records only exist for about the past fifty years, so the predevelopment levels are often estimates. After 1980, the records are very detailed, and it is apparent that the areas estimated to have suffered the greatest drop in water table before 1980 have continued to suffer in this way. The mining of this water at present rates, in an area where recharge is effectively nil, will leave many parts of the Ogallala dry within a few decades. At stake are some 5 million acres in six Great Plains states, an area as large as Massachusetts; this region has been a major agricultural producer. The inevitability of the draining of the Ogallala and the nearly valueless nature of the land when there is no more water has even led the Internal Revenue Service to grant Texas High Plains' farmers depreciation on their land value as the water table drops.

The examples of groundwater depletion discussed here are representative of a problem that is growing in magnitude both in the United States and worldwide. We shall either have to find ways to live with the amounts of continuously available water in each area or be willing to pay for massive water transport systems; we shall never find a way to live without water.

Land Subsidence Due to Groundwater Withdrawal

The removal of large quantities of groundwater in some areas has resulted not only in the lowering of water tables but also in the local and significant subsidence of the land surface, as shown in Figure 11.34(a). Extraction of groundwater from most aquifers has little or no effect on the land surface because the water is only interstitial to the grains of the rock that support the entire rock column. However, in some confined or semiconfined aquifers containing fine-grained sediments, the trapped water actually partially supports the rocks. Hence, when the water is pumped out there is a slow and generally irreversible subsidence of the land surface. Occasionally, there are even sudden collapses, as depicted in Figure 11.34(b).

Although subsidence rates are rarely dramatic, the damage can be considerable and may include: (1) damage to well casings; (2) structural damage to buildings, roads, and bridges; (3) damage to buried cables, pipes, and sewers; (4) changes in the grades and efficiencies of canal and irrigation systems; and (5) increased

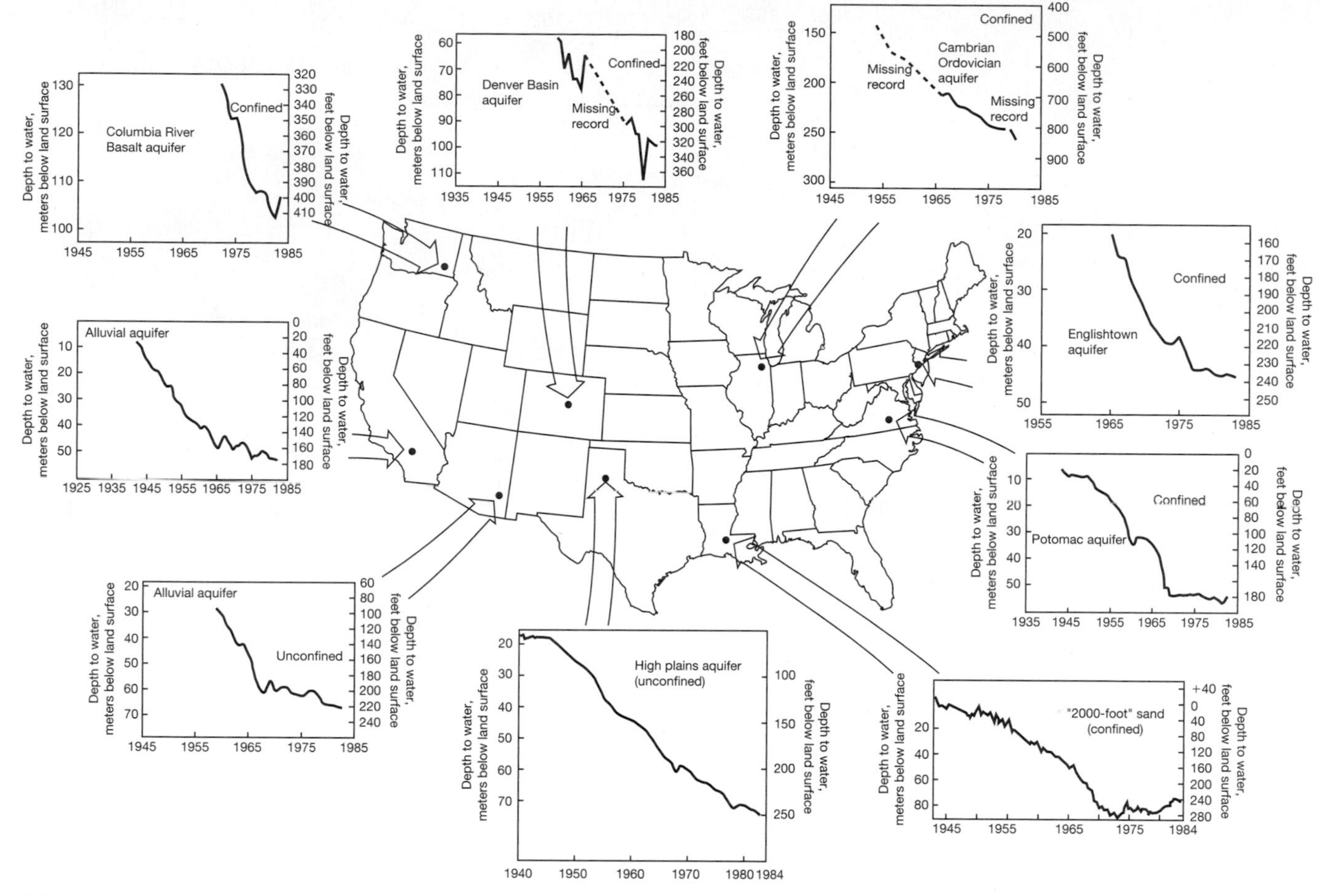

FIGURE 11.32 Examples of groundwater table depression in many parts of the United States as a result of major groundwater withdrawals. Many areas have seen continued decline and a few have stabilized at the lowest levels shown; very few, if any, have experienced a rise in groundwater levels. (From U.S. Geological Survey Water Supply Paper 2275.)

susceptibility to flooding in low-lying coastal areas. For years, there was a growing fear that New Orleans, where land subsidence left much land 2–3 meters (6 to 9 feet) below sea level, could be flooded by a hurricane. That fear was realized in 2005 when Hurricane Katrina flooded New Orleans with torrential rains and a large sea surge. The catastrophic flooding caused billions of dollars of damage, killed hundreds of people, and left some of the lowest areas uninhabitable for years after the event. There is disagreement about the effectiveness and reasonability of trying to offer below sea level neighborhoods protection by rebuilding levees and sea walls. New Orleans is not the only area with this problem. Subsidence in the Santa Clara Valley of California has lowered the land surface below sea level with resulting infrastructure costs estimated at more than $30 million. In the Central Valley of California, subsidence began in the 1920s as groundwater was utilized for irrigation. By 1964, annual groundwater withdrawals had exceeded 20 million acre feet (24×10^6 liters) and subsidence had affected about 5200 square miles. By 1970, the water table had dropped as much as 350 feet and the land surface had subsided about 26 feet (Figure 11.35); the combined effect even reversed the direction of water flow in the aquifer.

Extraction of groundwater to meet the growing needs of the Houston–Galveston area since 1915 has resulted in subsidence of 8 feet in the Brownwood subdivision of Baytown. As a consequence, most of the 450 houses of this coastal subdivision have become permanently inundated by seawater (see Plate 58).

Saltwater Intrusion into Aquifers

Under normal conditions, the slow but steady percolation of groundwater in response to the pull of gravity is sufficient along most coastal areas to keep marine saline waters from seeping inland into the aquifers. The location of the boundary, the freshwater–saltwater interface (Figure 11.36) varies from one shoreline to another depending upon the rainfall and the permeability of the rocks and sediments, and it changes slightly in any given area as a function of annual or longer-term climatic conditions.

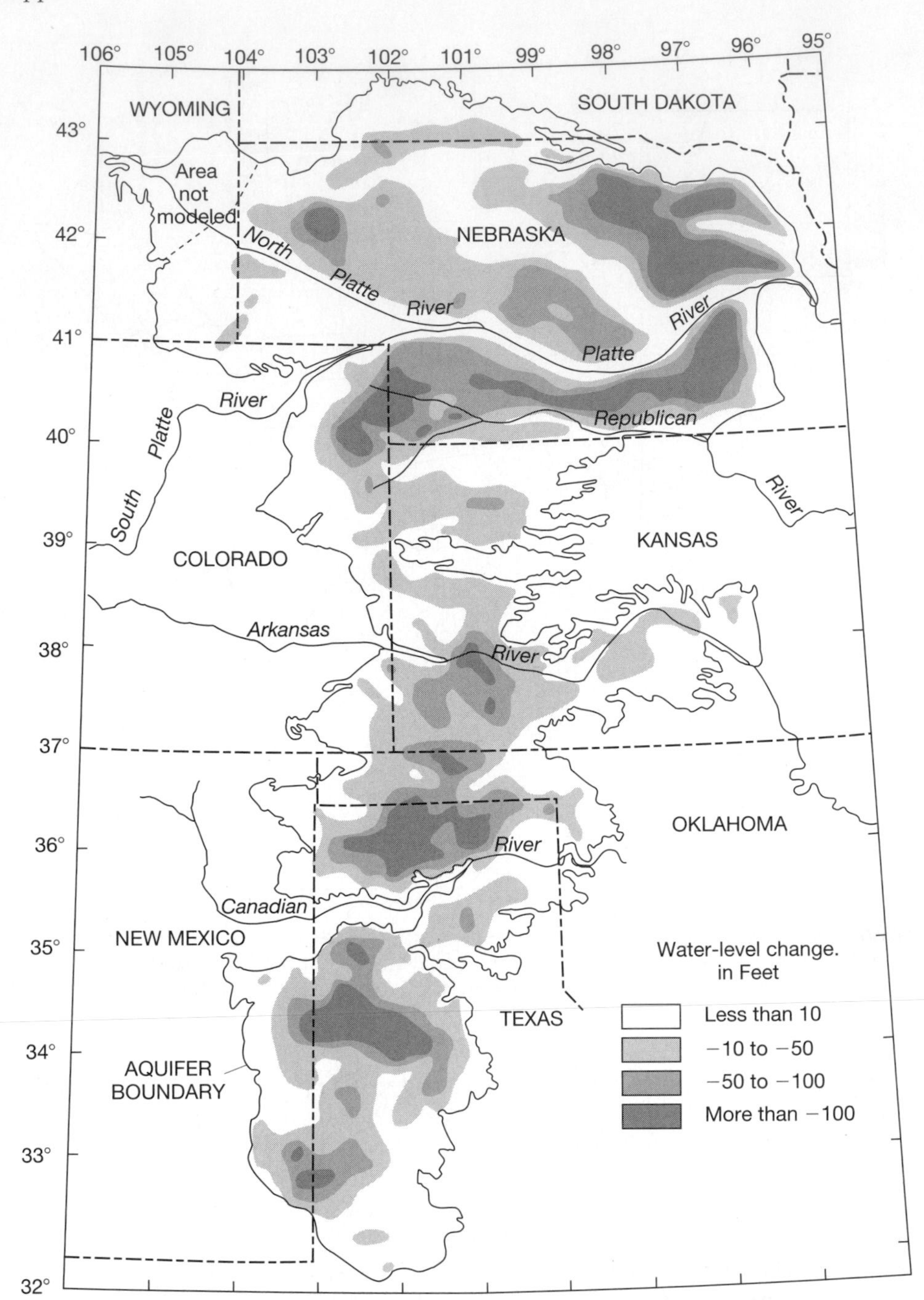

FIGURE 11.33 Groundwater level changes in the High Plains Aquifer, the Ogallala Formation, as measured and projected from pre-development to the year 2020. Note that there are large areas, especially in Texas and Nebraska, where the decline is in excess of 100 feet. In some areas, the decline is projected to reach 150 feet. On average, only about 15 percent of the total amount of the water in the aquifer is extractable. (From U.S. Geological Survey Professional Paper 1400A, and B, 1988.)

Since the 1960s, it has become apparent that the extraction of large quantities of water from many aquifers to serve growing metropolitan areas has altered the natural hydrologic balance. The consequence of removing vast quantities of freshwater that previously held back the saline waters is the landward movement of the freshwater–saltwater interface, resulting in **saltwater intrusion** into the previously freshwater aquifers. This phenomenon has been observed in many places, but is especially well-documented along the coastal areas of the United States. Thus, freshwater wells have been abandoned near Atlantic City (New Jersey), Savannah (Georgia), New York City, and Los Angeles; increases in salinity threaten usable water supplies near many rapidly growing sites. One example of the problem, as shown in Figure 11.37, results from the effects of heavy pumping of freshwater from an aquifer near New York City. The increase in salinity resulting from saltwater intrusion into the aquifer parallels the increase in pumping activity.

Although saltwater intrusion occurs primarily in coastal areas, similar problems can occur in other areas. Thus, in the Central Valley of California, an extensive body of saline water containing up to 60,000 ppm dissolved solids lies below the freshwater

FIGURE 11.34 (a) Areas of significant land surface subsidence caused by the withdrawal of groundwater. (From U.S. Geological Survey Water Supply Paper 2250, p. 56, 1984.) (b) The Giant Sinkhole, which collapsed in December 1972, left a crater 425 feet across and 150 feet deep. This and about a 1000 other sinkholes in Shelby County, Alabama, are believed to have resulted in part from natural and human-induced groundwater table lowering. (Photograph from the U.S. Geological Survey.)

aquifers. There is considerable concern about the potential upward movement of this saline water into the aquifers. In Mississippi, the re-injection of saline wastewater from oil production has led to the contamination of normally freshwater wells in areas far removed from the coasts. The saline water was re-injected so as not to pollute surface waters; unfortunately, the directions of movement of underground waters, especially those under increased pressure, are often not well known.

It is apparent that the movement of saline water, especially when promoted by human activities, poses a threat to many important water supplies. Our increasing demands on groundwater supplies are likely to intensify the problems and hence require careful consideration of the most efficient uses of this valuable resource.

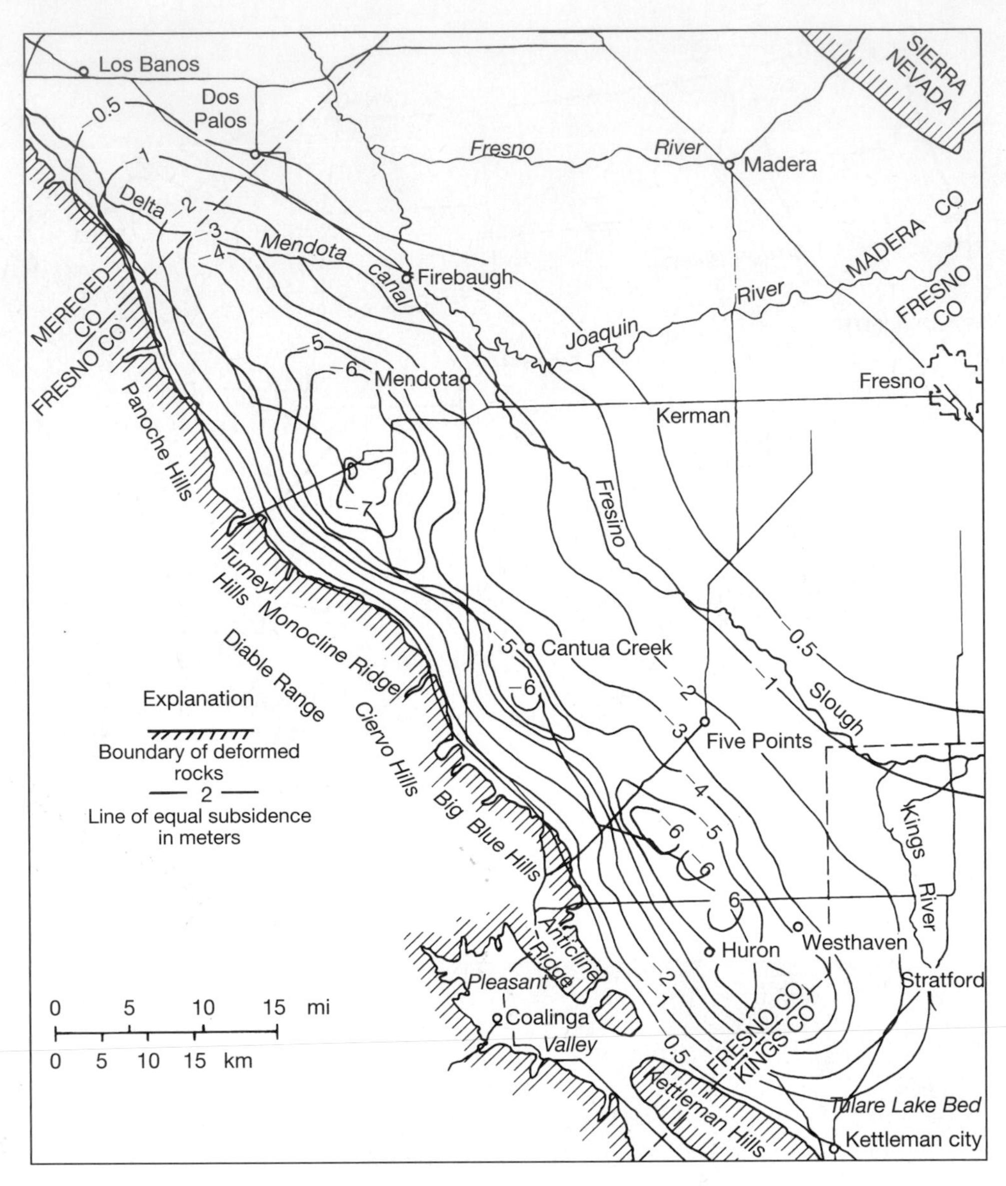

FIGURE 11.35 An example of land surface subsidence as much as 26 feet in the Los Banos–Kettleman City area of California between 1920 and 1966 as a result of groundwater withdrawal. Further subsidence has been prevented by re-injection of groundwater and the use of alternative sources. However, after subsidence, re-injection of groundwater usually does not result in any rise in elevation. (From U.S. Geological Survey Professional Paper 437-F.)

Soil Deterioration Due to Water Logging, Salinization, and Alkalinization

Irrigation in arid regions, although intended to bring unused or low productivity land into full agricultural production, has unfortunately also caused the deterioration or loss to production of an estimated 125,000 hectares of land every year. The problems arise in arid regions where irrigation systems supply water to soils faster than drainage can remove it. The excess water raises the water table to near the soil surface, causing **waterlogging,** and permits evaporation to concentrate dissolved salts. Waterlogging is a problem by itself because most crop plants are not able to survive if their roots are under water; rice is the major exception. The buildup of mineral crusts (Figure 11.38) of the halides (**salinization**) or alkali salts (**alkalinization**) impairs plant growth; furthermore, runoff of salt-laden waters into streams reduces the usefulness of that water for irrigation elsewhere.

The deterioration of soils by salt buildup as early as 2400 to 1700 B.C.E. is believed to have caused the collapse of ancient civilizations in Mesopotamia and in the Upper Nile Valley in Egypt. In 1959, it was estimated that 60 percent of Iraq's agricultural land was seriously affected by salinity. In the 1960s, the same problem arose in the Sind, one of Pakistan's major provinces, when 49 percent of all agricultural land was waterlogged. Furthermore, 50 percent of the irrigated land of the Sind was highly saline, and 25 percent was moderately saline. Argentina has 2 million hectares of irrigated land affected in this way; Peru, 300,000 hectares; and the United States potentially faces the

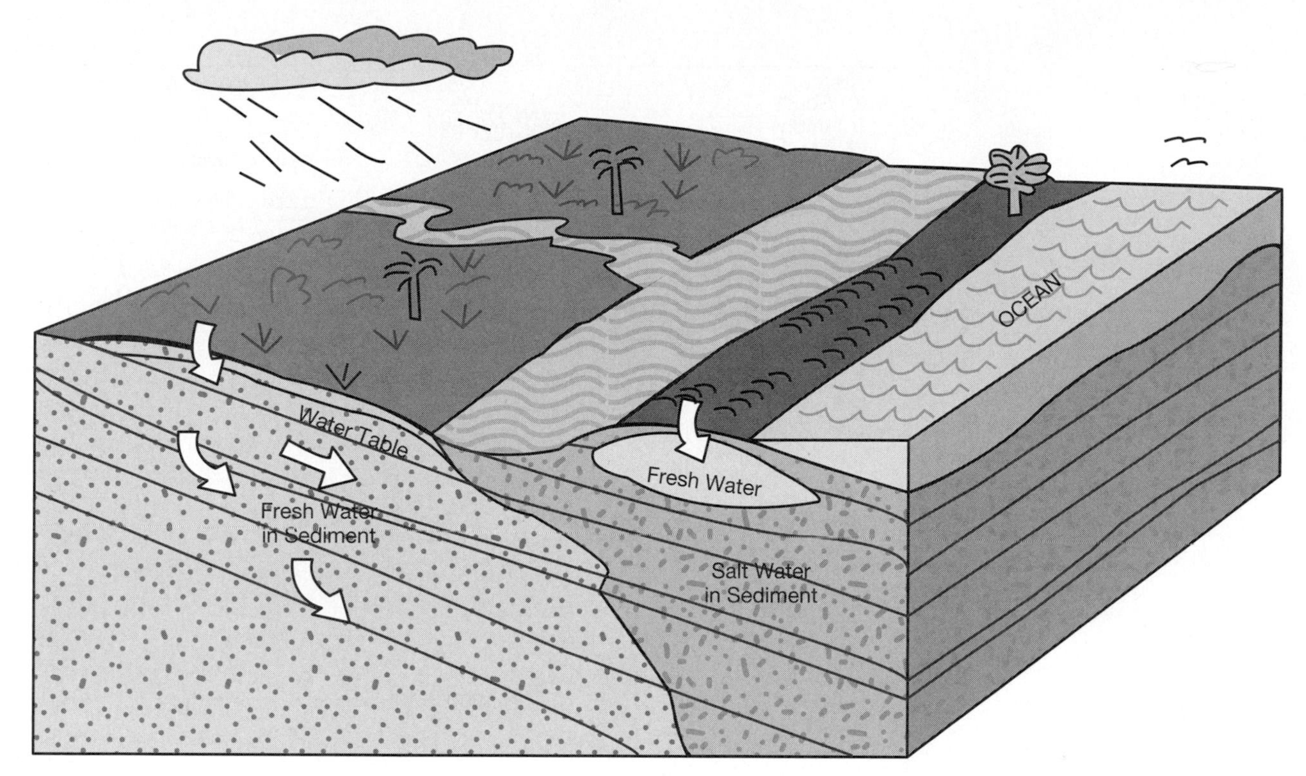

FIGURE 11.36 Groundwater conditions along a typical coastal region. The land surface, as well as a region under the barrier island, is underlain by rocks and sediments which are saturated with fresh water recharged by rainwater. The ocean and the bays behind the barrier island are underlain by sediments saturated with salt water. The boundary between the fresh and salt water may move inland (salt water encroachment) if the recharge of the fresh water is reduced or if excessive amounts of fresh water are removed by pumping.

same problem in more than 1 million hectares of the rich San Joaquin Valley in California. In the west central San Joaquin Valley, selenium weathering out of the adjacent formations has been concentrated by evaporation of water on the farmland soils. The selenium, mobilized by irrigation drainage, has accumulated in wetland ponds of the Kesterson National Wildlife Refuge, where closed basins have further concentrated the selenium to the point that it is causing very high rates of deformation of waterfowl. Box 12.3 gives the details of this problem.

Salinized land can be reclaimed by the installation of expensive subsurface drainage systems that allow the irrigation water to percolate downward through the soil. This downward movement is similar to the natural water movement in soils in humid areas and eliminates the buildup of the soluble salts at the soil surface. Such reclamation, however, has only been carried out in local areas because it is very expensive to install the drainage systems, because it requires even larger amounts of water to flush out the salt-rich soils, and because the salts that have been washed out may reach groundwater supplies or merely be deposited in downstream areas.

Desertification and Deforestation

Desertification is a term used in recent years to describe the transformation of once productive agricultural land into desert (or other forms of wasteland). Although prolonged drought may appear the obvious cause of desertification, recent studies have shown that humans are the chief culprits. Desertification can be caused by overgrazing, excessive wood-cutting, land abuse, improper soil and water management, and land disturbance. It can be anything from slight to severe in extent, and it results in reduced productivity of land and environmental degradation, which in extreme cases is catastrophic for the local population. It is not a new problem. The Greek philosopher Plato wrote 2000 years ago that Grecian Attica was "a mere relic of the original country. . . . All of the rich soft soil has molted away, leaving a country of skin and bones." The deplorable conditions in Attica were the result of tree-cutting and overgrazing, with subsequent water erosion.

The best-known modern examples of desertification center on the Sahel region of central Africa, a region that lies along the southern edge of the Sahara Desert and includes parts of Mauritania, Senegal,

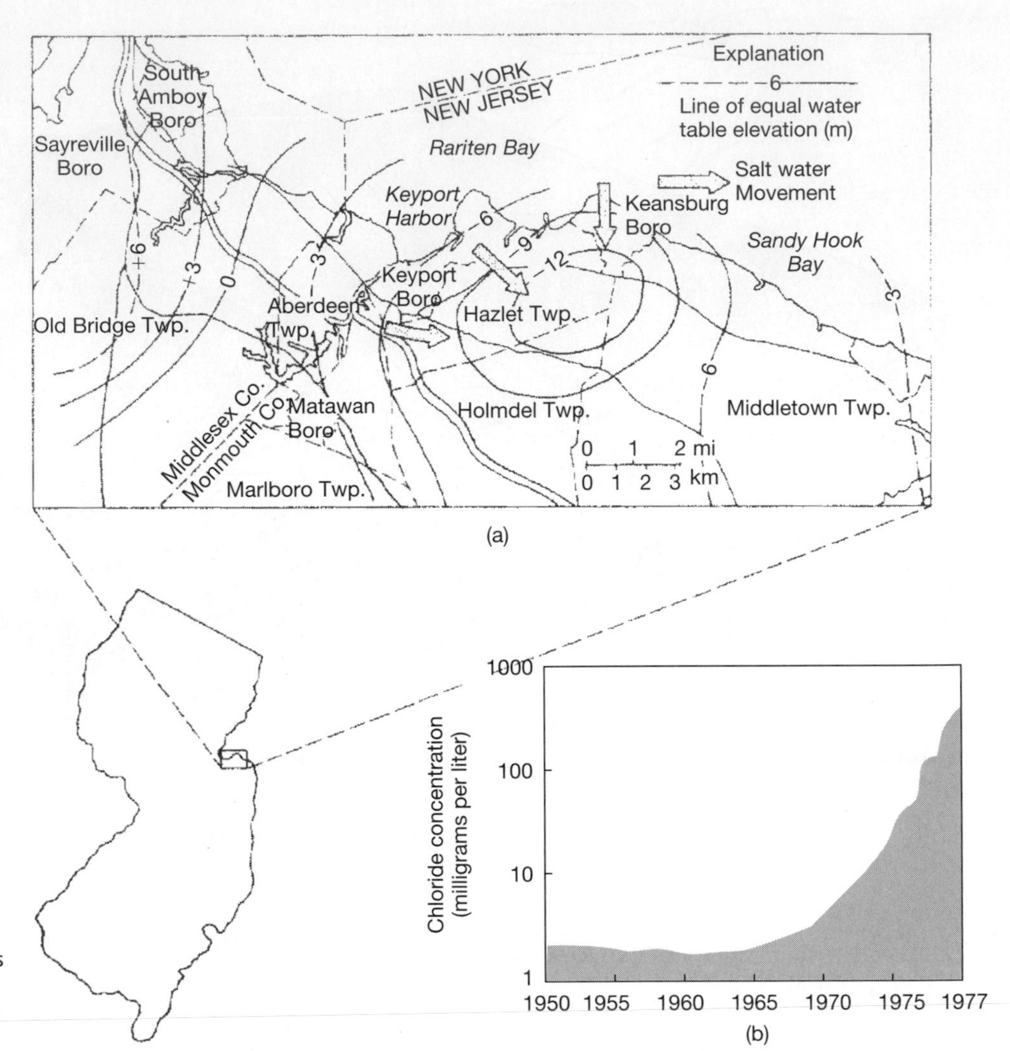

FIGURE 11.37 (a) Groundwater table depression in Monmouth County, New Jersey, as a result of increased pumping to meet the needs of a growing population. (b) The rise in chloride concentrations in water samples from the Union Beach Borough well field from 1950 to 1977 forced abandonment of the wells. Since this saltwater encroachment, the same situation has been reported from many coastal regions with rapidly growing populations. (From U.S. Geological Survey Water Supply Paper 2184.)

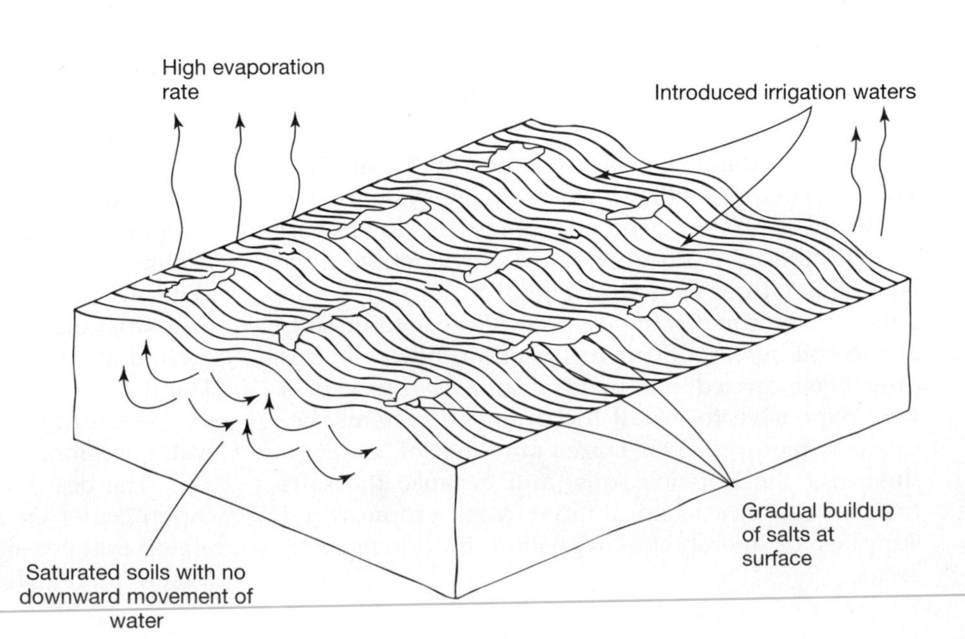

FIGURE 11.38 Salts can build up in soils as a result of water brought in for irrigation. Ultimately, the soils can become so salt-rich that they can no longer be used to grow food crops.

Gambia, Mali, Burkina Faso, Niger, Chad, Ethiopia, and the Sudan. A severe drought from 1969 to 1973, which first focused world attention on the plight of the peoples of the Sahel, put the region's agricultural resources under severe strain. Had resource management been good, then little or no permanent damage need have been done, but this was not the case. Excessive cultivation and overgrazing, in response both to drought conditions and to growing populations, led to the depletion and erosion of fertile soils. In turn, atmospheric moisture was reduced because there were fewer plants to transpire water into the air; as a result, drought conditions became more prevalent, populations were forced to rely on still smaller land areas for food, and the process of desertification accelerated. At present, desert areas in the Sahel are expanding at rates up to 14 million acres annually, and the harrowing sight of thousands of starving people has become all too familiar from television and newspaper reports across the world. Although the human suffering caused by desertification has been greatest in Africa, it is a worldwide problem, as can be seen from Figure 12.16.

Even in areas of high rainfall, the same human activities that lead to desertification can have devastating effects. Thus, in Haiti and some of the other Caribbean Islands, excessive wood-cutting, overgrazing by goats, and poor agricultural practices have denuded the hills which are now eroding and suffering a reduced agricultural productivity (Figure 12.17). Perhaps nowhere in the world are the effects of *deforestation* more pronounced than in Cherrapunji, India, where the wettest place on Earth suffers from water shortages. Cherrapunji, in the mountains of eastern India, receives more than 500 inches of rain yearly, but the cutting of all the trees now allows all the water to run off immediately, taking with it most of the soil. In the past, the dense forest held back much of the water, maintaining a high water table and flowing streams. Today, all of the vegetation is gone and streams are muddy torrents when it rains and dry when it does not. The situation is shown schematically in Figure 11.39.

Water for Sale from the Great Lakes and Alaska?

Humans have commonly constructed or modified water transportation systems to provide water to areas where demand exceeded natural supplies. Some early

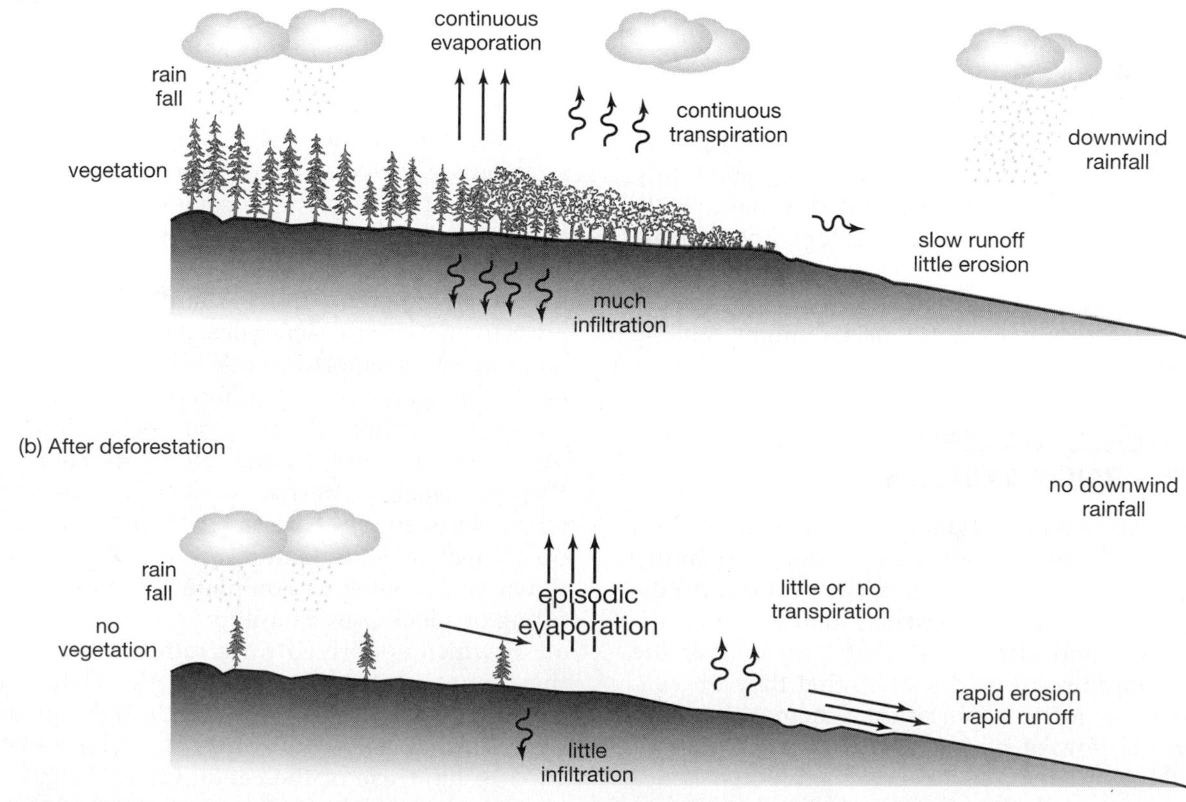

FIGURE 11.39 The effects of deforestation in reducing water retention and infiltration, increasing runoff and erosion, and altering rainfall.

efforts were aqueducts, as built by the Romans to supply cities, and others were irrigation canal systems, as constructed by the Incas and the ancient Persians. The 1990s saw many new proposals, which, although not yet in place, may offer a glimpse of the future and which attest to the value of water as a resource. One of the most serious has been the transport of water from Alaska or northern Canada to thirsty southern California. As California's water needs have risen as a result of its booming population, access to regionwide water has declined. Accordingly, several entrepreneurs have considered the possibility of large-scale transport of water from rivers in Alaska or Canada to southern California. The proposed means of transport include huge water tankers, similar to oil tankers, submarine pipelines, and large plastic bags up to 1000 feet long. The bags would be filled in Alaska and floated (because freshwater is less dense than saltwater) to California. The technical aspects of each system are quite manageable and costs would be kept reasonable by economy of scale. Spillage of freshwater would not constitute much of an environmental threat, but there are ecological concerns about the changes that might occur in the rivers and bays from where the water would be removed. None of these systems have yet been approved, but there are increasing numbers of discussions about them.

At the same time, a case has been made to transport water in large tankers from the Great Lakes to arid or drought-stricken parts of the world. The Great Lakes span the international boundary between the United States and Canada; hence, any action requires the approval of both nations. Small-scale extraction of lake water for local irrigation has always been allowed, but both countries are concerned about any precedent that might be established if large-scale water sales were permitted. Therefore, for at least the early years of the twenty-first century, water will not be taken from the Great Lakes, although special exemptions for humanitarian needs might still be considered.

LARGE-SCALE TRANSPORTATION AND DIVERSION SYSTEMS

Although we live on a planet 70 percent covered by water, we find that, in many areas, either the quantity or the quality of water is not sufficient for our needs. Consequently, we have deepened, dammed and diverted rivers and streams so that they deliver the water to more useful places, or so that they provide better avenues for transportation. Schemes that alter or divert the flow of rivers have, in this age of environmental awareness, also become emotional issues that commonly bring those who desire to maintain the *status quo* into opposition with those who see benefit from change. Because water, as a limited resource, needs to be considered in terms of municipal services, industrial production, power generation, and crop growth, its availability has broad economic implications.

It is, of course, not possible to begin to evaluate all of the types of water transportation and diversion schemes that have been constructed. The following discussion concerns three major examples.

Water for Transportation—The Panama Canal

When we think of water as a resource, we generally consider only that which is actually used or consumed in daily life, in industry, or in irrigation. Water is no less an important resource when it serves as an avenue for transportation. Thus, the world's oceans and rivers have served as trade routes since before recorded history. In Italy, coastal rivers and estuaries were modified into canals for commerce; in England, a far-reaching canal system was built to facilitate transport of the coal and iron ore needed to fuel the Industrial Revolution. In the United States, the Erie Canal and many other waterways like it were built to provide efficient and inexpensive means of large-volume transport. In a similar manner, the great St. Lawrence seaway, a channel with a series of locks, was constructed to permit the direct movement of commodities from the Great Lakes to the Atlantic Ocean along an otherwise unnavigable river. (See Box 11.4 for a discussion of the problems redirecting water into the Aral Sea in the former Soviet Union.)

Perhaps the two most famous water transport systems are the Suez Canal and the Panama Canal. The Suez, opened in 1869, provided a short sea route to the Far East from southern European ports. The Panama Canal, completed in 1914, reduced the length of the route from Atlantic ports to Pacific ports by more than 9400 miles. Although water is the medium of transport in both canals, the source and use of the water is very different. The Suez is a sea level canal without the need for locks; ships merely sail from one end to the other. In contrast, the Panama Canal is a freshwater-filled elevated canal in which ships must be raised by locks from sea level up 85 feet to Lake Gatun, and then lowered back down to sea level to complete the crossing. Every filling of a lock uses 53 million gallons of freshwater, all of which is derived from rainwater that falls on the mountains and runs into Lake Gatun. In the 1990s, an abnormal period of drought greatly reduced the normal runoff into the lake and threatened to limit use of the canal. Current plans to enlarge the locks and require more water per operation

BOX 11.4

The Death of a Lake—The Aral Sea

The Aral Sea, located astride the border between Kazakhstan and Uzbekistan in the Central Asian part of the former Soviet Union, has been referred to as "one of the planet's most serious environmental and human tragedies." In 1960, the Aral Sea was the world's fourth largest lake—a beautiful body of water with a surface area of about 26,000 square miles that supported a commercial shipping and thriving fishing industry. By 1993, the sea level had dropped more than 40 feet, the surface area had decreased by about 50 percent, the volume had fallen by 73 percent, salinity tripled, shipping traffic had ended, the fish had died, and the sea had been split in two much smaller bodies (Figure 11.F and Plate 59). What happened and why?

From the earliest times, the Aral Sea had been fed by the waters of the Syrdarya and Amudarya Rivers. However, in the 1950s and 1960s, the Soviet Union decided to dramatically increase cotton production by expanding the area of Central Asia under irrigation and cultivation. The first step was to build a 800-mile-long canal to divert large quantities of water from the Amudarya River to the southern part of the Kara Kum Desert. This was subsequently expanded into a complex network of 20 reservoirs and 60 canals that fed the water of the Amudarya and Syrdarya Rivers into 19 million acres of cropland in one of the driest regions of Asia. The rivers that sustained the Aral Sea by feeding it 12–14 cubic miles of freshwater per year had, by 1985, become totally dry. Subsequently, small amounts of inflow have occurred, but they have not been sufficient to reverse the effects of the diminished supplies.

The devastating effects of the diversion of the inflow are most visible in the change in the size and shape of the Aral Sea, as shown in Figure 11.F. Former ports, with now useless ships (Plate 59), lie as much as 20 miles from the shrinking shoreline. The rich schools of fish that once yielded 160 metric tons per day and supplied 10 to 15 percent of the Soviet Union's freshwater catch are now totally

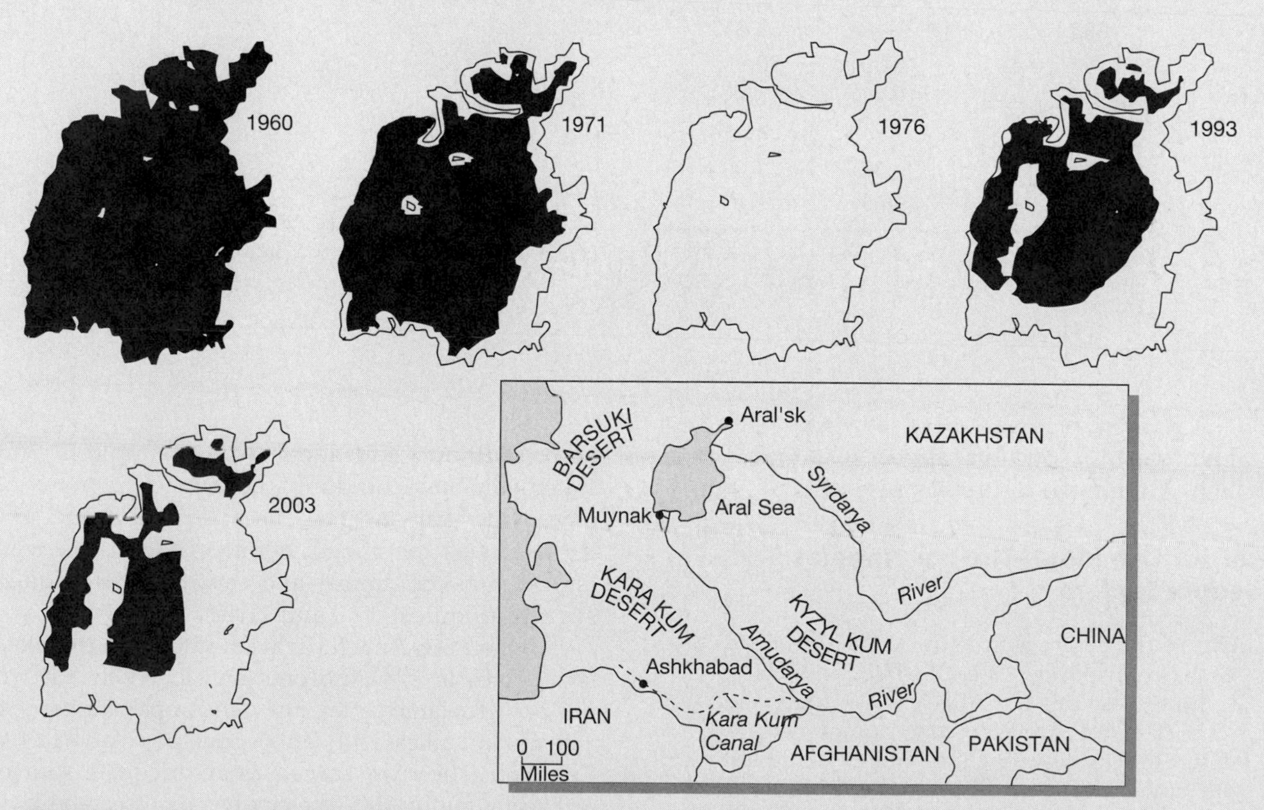

FIGURE 11.F The surface area of the Aral Sea (black) has shrunk dramatically since 1960 and is projected to continue shrinking because of the diversion of the water from the Amudarya and Syrdarya Rivers for irrigation. See also Plate 59, which shows how the drop in the level has stranded ships many kilometers from the water. (From "The Shrinking Aral Sea," P. Micklin, *Geotimes,* April 1994. Used with permission.)

(continued)

BOX 11.4

The Death of a Lake—The Aral Sea (*Continued*)

gone because the salinity is nearly equal to that of ocean water. The volume has decreased by approximately 75 percent as the water level has dropped more than 50 feet in the main body. The now-dry seafloor, subject to strong winds, has been the source of vast clouds of dust and salt that have blown onto and damaged the irrigated crops. Table 11.4 summarizes data from 1960 extrapolated through 2000.

The devastation of the Aral Sea was completely predictable and demonstrates how easily and rapidly human activities can disrupt the environment in the pursuit of a resource. The hope of many scientists is that the lessons learned from the shrinking of the Aral Sea will mean that the same mistakes will not be repeated in other areas. In the United States, diversion of water from Mono Lake in eastern California to supply growing cities resulted in many of the same effects. However, in the mid-1990s, water supplies began to be returned to Mono Lake in order to restore the lake to its previous state.

TABLE 11.4 Changes in the Aral Sea from predevelopment (1960) of irrigation projected to the year 2000

Year	Surface level (meters)	Area (sq. km)	Volume (cu. km)	Annual inflow, ave of previous 5 yrs. (cu. km)	Salinity (grams/liter)
1960	53.41	66,900	1090	56.8	10
1971	51.05	60,200	925	45.0	11
1976	48.28	55,700	763	25.3	14
1993	—	33,642	300	5.0	—
large sea	36.89	30,953	279	—	37
small sea	39.91	2689	21	—	30
2000	—	24,154	185	?	—
large sea	32.38	21,003	159	—	65–70
small sea	40.97	3152	24	—	25

Data from P. Micklin, *The Shrinking Aral Sea, Geotimes* April 1993, p. 16, except for inflow, which is from P. H. Gleick, *Water in Crisis,* Oxford University Press 1993, p. 313.

are again raising concerns about sustainability of new locks if there are future dry periods.

Water for Drinking—The Los Angeles Aqueduct System

> Out of the desert with its rocks, heaven-hued and awe inspiring, its cactus-like sentinels of solitude, rose this Los Angeles—your city and mine. The magic touch of water quickened the desert into its flowering life—our city. And lest our city shrivel and die, we must have more water, we must build a great new aqueduct to the Colorado.

This statement by William Mulholland, the long-time water czar for Los Angeles, succinctly summarized the recurring plight of that city, this time in 1925, when it was realized that further growth could not occur without additional water supplies. Similar statements have, no doubt, been made in many other cities with various responses; the response in Los Angeles was the construction of one of the world's largest, most complex—and controversial—aqueduct systems (Figures 11.40 and 11.41).

When Los Angeles was founded as a pueblo on a small river in 1781, no one envisioned that it would one day grow into a major city, encompassing more than 450 square miles with an overall population of over 7 million. The river served as an adequate source for 120 years, but by 1904 the city began to search for additional supplies to accommodate anticipated growth. Surface and groundwaters in adjacent areas were already in use, so city officials turned their attention to the Owens Valley on the eastern flank of the Sierra Nevada, 250 miles to the east. Land and water rights were acquired, sometimes by subterfuge, and construction

FIGURE 11.40 The water supply system of southern California involves the transport of water by aqueducts from various parts of the state. (From the Los Angeles Department of Water and Power.)

FIGURE 11.41 Aqueducts such as this one transport water from the mountains of California to the agricultural and urban areas of southern parts of the state where needs far exceed the local supply. (From the Los Angeles Department of Water and Power.)

began; by 1913, the $25 million aqueduct was completed and Owens Valley water flowed into Los Angeles.

By 1923, the growth of the population to more than one-half million brought the realization that still more water was needed. This time the Colorado River aqueduct system reached east and began to tap water dammed in Lake Havasu; the project was completed in 1941. Because Arizona did not use its full share of water as authorized by the Colorado River Compact (see p. 422), Los Angeles was allowed to temporarily take Arizona's unused portion. With continued growth, more water was needed, and a second Owens Valley aqueduct was added in 1970.

Two circumstances have forced Los Angeles, in the 1980s, to again seek more water, this time to the north where there are plans for a $5 billion Peripheral Canal that would take water from the Sacramento River and pass it south via the California aqueduct. First, population has continued to grow and to require increased amounts of water; second, in 1985, Arizona completed the first part of the Central Arizona Project to supply Colorado River water from Lake Havasu to Phoenix and Tucson. Consequently, Arizona is reclaiming the Colorado River water it had allowed Los Angeles to use since 1941; its right to do so was upheld by the U.S. Supreme Court.

The water supply system for Los Angeles has some similarities with that for New York City (described on p. 409), but the legal and emotional ramifications of the former are far greater. Because of the abundance of water in the northeastern United States, New York's use of water has negligible impact on the availability of water for others. In contrast, Los Angeles's needs and claims on water supersede the availability of water for many others, including those who live where the surface waters originate. This has resulted in scores of lawsuits, bombings of aqueducts and, in recent years, concern about severe environmental effects. The continued growth of major cities such as Los Angeles in relatively water-poor areas will place greater demands on scarce or distant water supplies in the years ahead.

CHAPTER 12

Soil as a Resource

The fragile nature of the soils on which we depend for food supplies is illustrated by the severe wind erosion and deposition that occurred during the Dust Bowl days in the central and western United States during the 1930s. This 1936 photograph, taken in Gregory County, South Dakota, shows a buried car and farm machinery on a previously prosperous farm. (Courtesy of Soil Conservation Service, U.S. Department of Agriculture.)

Recent increases in the human population have placed a great strain on the world's soil systems. More than 6 billion people are now using about 10 percent of the land area of the Earth to raise crops and livestock. When used for such purposes, soils can suffer various types of degradation that can ultimately reduce their ability to produce food resources.

LIVING LANDSCAPES . . . THOMPSON-OKANAGAN: PAST, PRESENT & FUTURE
<HTTP://ROYAL.OKANAGAN.BC.CA/MPIDWIRN/AGRICULTURE/EROSION.HTML>

FOCAL POINTS

- Soils form by the decomposition of all types of rocks in response to climate, vegetation, slope of the ground, and time.
- Physical weathering reduces the size of rock fragments and separates mineral grains, whereas chemical weathering changes original minerals into new soil minerals.
- Soils are generally composed of residual quartz grains, clay minerals, iron oxides, and organic matter.
- Clays are the most important soil minerals in providing nutrients to plants because the clays loosely bond the nutrient cations and readily release them to the plants.
- Soils are characterized according to color, texture (including mineral grain sizes), consistency, and structure; distinctive soil horizons form over time and may have very different characteristics.
- The potential use of a soil, including agricultural production, depends upon its properties.
- Only about 10 percent of the land surface (or about 3 percent of the total Earth surface) is suitable for crops, but these lands provide about 97 percent of the world's food supply.
- Most of the U.S. land area is held as forest, pasture, range, or cropland; the total amount used for dwellings, cities, roads, and mines is less than 4 percent.
- Earth's total amount of cropland has been slowly increasing, but newly developed land is generally not so fertile or well watered as lands lost to human development.
- Natural erosion rates vary widely depending upon topography and climate; human interventions, such as those involved in cultivation, commonly increase soil erosion.

- Human activities causing the release of nitrogen and sulfur gases into the atmosphere may result in acid rain that leaches important nutrients from soils, thus reducing the fertility.
- Desertification, the transformation of productive land into deserts, occurs in many semiarid regions as a result of overgrazing and deforestation.

INTRODUCTION

Most of Earth's land surface is covered by a continuous layer of soil that is commonly less than 2 meters thick. The term *soil,* as used in soil science, refers to a naturally formed Earth surface-layer containing living matter and capable of supporting the growth of rooted plants. Soil results from the physical and chemical weathering of underlying rocks by processes involving the **hydrosphere** and **atmosphere,** but a key role in soil formation and development is also played by living organisms of the **biosphere.** Where underlying rocks are covered by broken, unconsolidated materials devoid of living matter (as on the surface of the Moon, for example), the surface material is not a soil as the term is defined here; such a surface covering is described by the more general term **regolith.** Soil is composed of inorganic (mineral) matter together with both living and dead organic matter; it is, therefore, a complex geological *and* biological system, and because there are many components in a soil, there are many different types of soil. Although soils constitute only a minute fraction of the material of the entire Earth, they are essential for the production of food and are literally a *vital* resource.

Because soils support the growth of rooted plants, they are at the base of the life-support system on which humans depend. Soils are the resource we exploit through agriculture, and soils are renewable in the sense that they can be preserved and nurtured through careful use of fertilizers and by crop rotation. However, although soils can be made agriculturally productive through irrigation and fertilization, they can also be destroyed or irreparably damaged by natural agencies or by careless human intervention. The natural cycle of weathering, soil formation, and erosion is a delicate balance; it is much easier for us to damage or destroy soils than to create them.

In this chapter, the formation of soils, their chemistry, characteristics, classification, and distribution are reviewed; then their utilization and conservation are considered. Certain types of soils are exploited as sources of metals (e.g., aluminum is derived from bauxites), but these soils are really ores and are discussed elsewhere in this book (see Chapter 7).

SOIL FORMATION AND DISTRIBUTION

To understand the types of soil and their distribution, it is necessary to consider first how soils form and the factors that lead to the diversity of soil types.

Formation—The Major Factors

The type of soil found at any given place results from different processes acting on many different materials. However, six major factors can be identified in soil formation:

1. parent material, identified as the underlying rock or rock debris;
2. climate;
3. vegetation;
4. micro- and macro-organisms (such as bacteria, earthworms, etc.)
5. slope of the ground, which determines how quickly rainwater will drain away and how deeply it will penetrate the ground; and
6. time, in the sense of the extent to which the various processes have progressed.

Formation Processes

Most rocks exposed at Earth's surface are not chemically stable and are constantly undergoing the process of breakdown known as weathering, a process that involves both mechanical disintegration (physical weathering) and chemical decomposition (chemical weathering) (Figure 12.1). Weathering is the first stage in soil formation.

In physical weathering, rocks are broken down to smaller and smaller pieces by various natural agencies, the importance of which depends on the type of rock being weathered and the climate under which weathering occurs. Wind, rain, frost action, and the differential expansion and contraction during rapid heating and cooling all contribute to the physical breaking of rocks. Most rocks already contain planes of weakness. When material originally formed at depth is exposed at the surface, the release of the confining pressure of overlying rocks causes expansion, fracturing, and joint formation. Joint planes are also formed when an igneous rock cools. Bedding planes (original sedimentary layers) in sedimentary rocks, or fracture planes introduced into rocks when major Earth movements (**tectonism**) or minor movements occur, are other ways that planes of weakness can develop. In the broad temperate belts, frost wedging is probably the most important physical weathering agent. When water that is trapped in fractures or pore space in rocks near the surface freezes, its volume increases by about 9 percent. The maximum pressure that can be generated when confined water freezes is

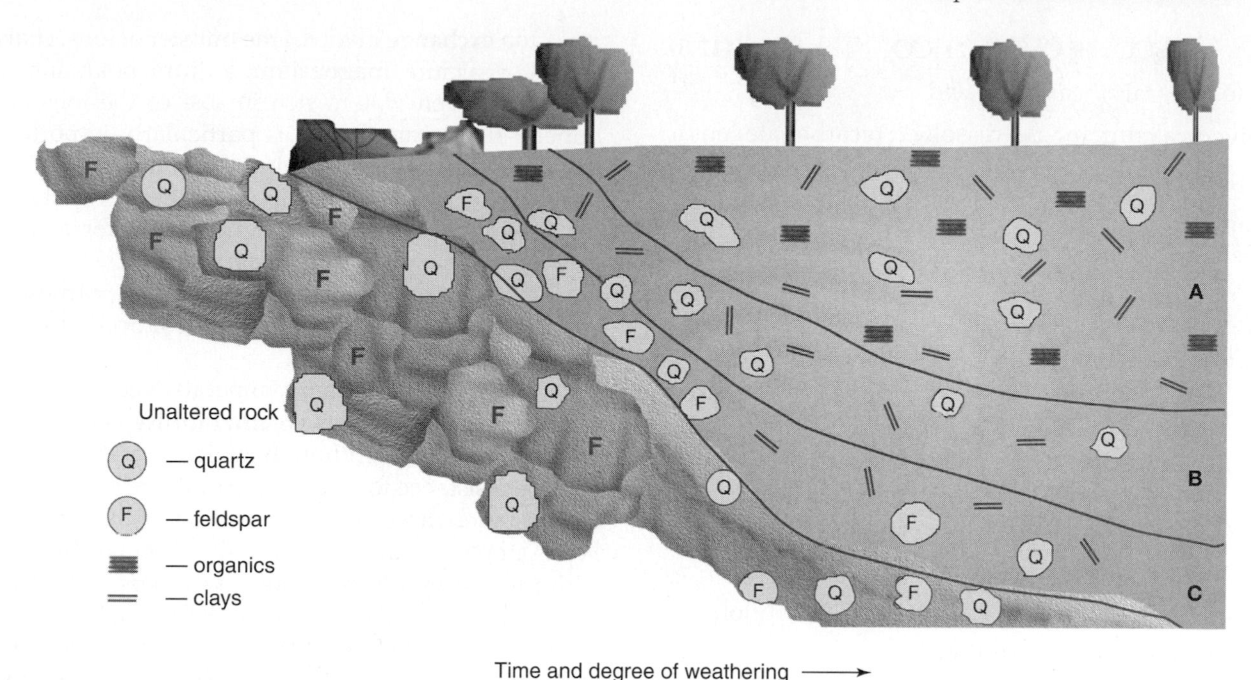

FIGURE 12.1 Soil formation is a function of time and climatic conditions. In this generalized diagram, the time that unaltered rock (granite) has been exposed to weathering increases from left to right. Hence, freshly exposed rock is unaltered and consists of quartz, feldspar, and mica. After considerable time (center of the diagram), weathering results in the partial decomposition of the original rock with partial conversion of feldspar to clays. After long periods of weathering (right side of the diagram), the feldspars and mica have been completely decomposed at the surface and have formed clay minerals; organic matter has also been introduced so that the mature soil consists of clays, residual quartz grains, and organic matter. The *A*, *B*, and *C* zones that develop display varying degrees of conversion of original minerals into soil components.

about 2100 tons per square foot (about forty times greater than the force needed to break an average granite). Although this maximum amount of pressure is rarely reached, frost wedging produces stresses capable of fracturing most rocks. This process is widely visible in the wintertime when road surfaces disintegrate with the formation of potholes. The frequent freezing and thawing of water in small cracks in the road surface breaks it down in the same way that rocks are broken. Furthermore, plants and animals can also contribute significantly to physical weathering through the wedging action of plant and tree roots and the burrowing of animals.

Chemical weathering involves the breakdown of the *primary* minerals in the rock to form new *secondary* minerals that are more stable in the surface environment or to form material that may be carried away in solution. The extent to which the breakdown has occurred can be expressed as the **index of weathering,** the ratio of a common element such as aluminum or iron present in the secondary mineral compared to the total present in the soil. Water is the essential agent in chemical weathering, either reacting with the minerals directly or carrying dissolved species such as carbon dioxide, which themselves react with the minerals. Although there are many complex reactions, they can be grouped into these major categories:

Hydrolysis, a reaction with water that decomposes minerals, is common in the major rock-forming silicate minerals; for example:

$$Mg_2SiO_4 + 4H^+ + 4OH^- \rightarrow 2Mg^{2+} + 4OH^- + H_4SiO_4 \quad \textbf{(12.1)}$$

(olivine mineral + 4 ionized water molecules → magnesium and hydroxyl in solution + silicic acid in solution).

Hydration is the addition of an entire water molecule to the mineral structure. This is especially common in clay minerals and plays an important role in rock disintegration because the addition of the water causes the minerals to swell, resulting in a progressive spalling off from the weathering rock surfaces.

Carbonation is the reaction with carbonic acid, which forms when carbon dioxide from the atmosphere dissolves in rainwater:

$$CO_2 + H_2O \rightarrow H_2CO_3 \quad \textbf{(12.2)}$$

(carbon dioxide gas + water → carbonic acid).

Carbonic acid reacts with minerals, in particular the carbonate minerals (calcite, dolomite), the principal components of limestones.

$$CaCO_3 + H_2CO_3 \rightarrow Ca^{2+} + 2HCO_3^- \quad (12.3)$$

(calcite mineral + carbonic acid →
dissolved calcium ions + dissolved bicarbonate ions).

Oxidation is the bonding of oxygen, abundantly available and dissolved in surface water, to the metallic elements (potassium, calcium, magnesium, and iron) of the primary minerals. A common example is the formation of the rusty brown and orange hydroxides of iron on the surfaces of iron-containing rocks. Thus, in the case of the olivine mineral, fayalite, ferrous iron is first released by hydrolysis:

$$Fe_2SiO_4 + 2H_2CO_3 + 2H_2O \rightarrow 2Fe^{2+} + 2OH^- + H_4SiO_4 + 2HCO_3^- \quad (12.4)$$

(olivine mineral + carbonic acid → iron and hydroxyl ions in solution + silicic acid in solution + bicarbonate ions in solution).

The ferrous iron (Fe^{2+}) in solution rapidly oxidizes because of the oxygen in the atmosphere and soil waters and usually forms iron hydroxide as follows:

$$4Fe^{2+} + O_2 + 6H_2O \rightarrow 4FeOOH + 8H^+ \quad (12.5)$$

(iron in solution + gaseous oxygen + water →
iron hydroxide mineral + hydrogen ions in solution).

The hydrogen ions generated by the oxidation of the iron to form goethite then help to promote more weathering.

Ion exchange involves the transfer of **ions,** charged atoms, of calcium, magnesium, sodium, potassium, and others, between waters rich in one of the ions and a mineral rich in another. It is particularly important in the alteration of one clay mineral to another (e.g., illite, the potassium-rich clay mineral, may lose potassium into solution and take up magnesium to form another clay mineral called *montmorillonite*).

Chelation, a reaction in which hydrocarbon molecules absorb metals, is a biological process that takes place in soil formation.

The different primary minerals weather at different rates, such that their **weatherability** is highly variable (Figure 12.2). Although variable, it is systematic and the resistance to weathering of the primary silicates can be rationalized in terms of their crystal structures. For example, olivine, a mineral that contains SiO_4 tetrahedra linked by Mg or Fe ions, weathers more rapidly than quartz, which is made up of SiO_4 tetrahedra linked by their corners to form a complete framework of these stable units. Thus, quartz (SiO_2) has a very low solubility. Because it is also hard, resists abrasion, and is commonly a primary mineral in many rocks, residual quartz grains are a common constituent of many soils and of the sands in streams and rivers.

The secondary minerals resulting from weathering processes depend directly on the nature of the primary minerals and, hence, the type of underlying bedrock. Therefore, rock type (parent material) exercises a major control over the kinds of soils that form. In the chemical weathering of granite, many reactions take place simultaneously (Table 12.1). Many of the rocks exposed at Earth's surface are sediments made

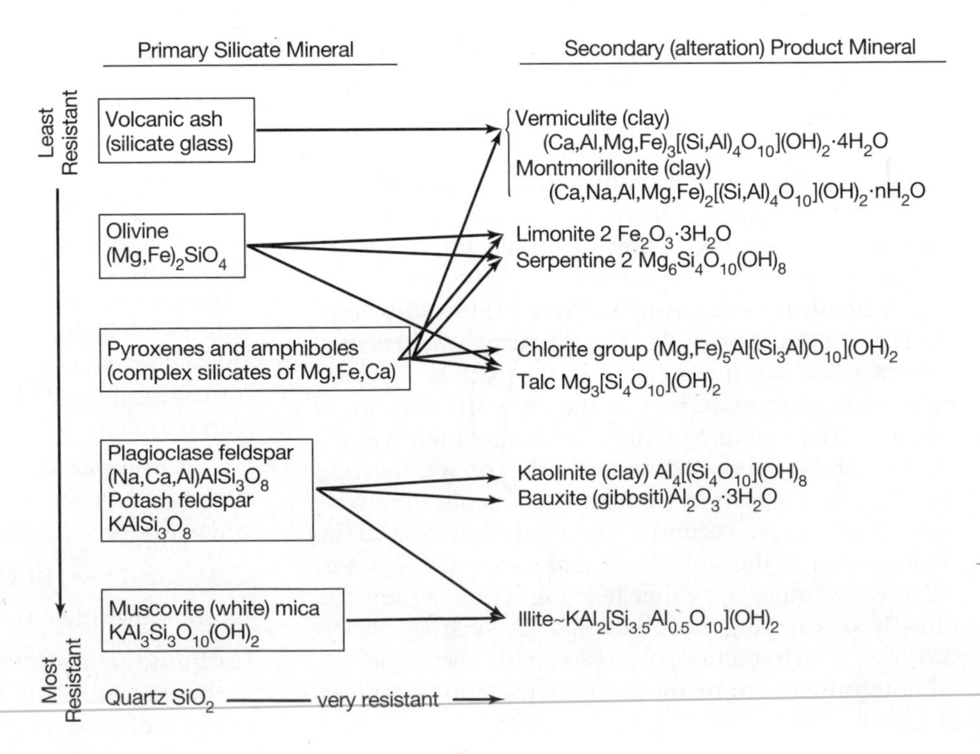

FIGURE 12.2 Weatherability of major (primary) silicate minerals and their common (secondary) alteration products.

TABLE 12.1 Products of weathering of a granite (idealized)

Mineral Component	Chemical Composition (idealized)	Products of Weathering	
		Soluble	Insoluble
Orthoclase feldspar	$KAlSi_3O_8$	K^+ (minor), soluble silica	Clay with K^+
Plagioclase feldspar	(Na, Ca) $Al_2Si_2O_8$	Na^+, Ca^{2+}, soluble silica	Clay with some Na^+, Ca^{2+}
Biotite	$K(Mg, Fe)_3(Al, Fe)Si_3O_{10}(OH, F)_2$	K^+ (minor), Mg^{2+}, soluble silica	Clay minerals, hematite (Fe_2O_3)
Muscovite	$K\,Al_2(AlSi)_3O_{10}(OH)_2$	K^+, soluble silica	Clay with K^+ and/or limonite (FeO, OH)
Quartz	SiO_2	None	Quartz grains

largely of the secondary minerals redeposited after one (or more) weathering cycles.

Climate is second only in importance to parent material as a controlling factor in soil formation. The climate controls weathering and soil formation directly through the amount of precipitation and the temperature. It also works indirectly through the kinds of vegetation that cover the land. The importance of climate can be illustrated by considering four contrasting examples.

Humid tropical climates lead to intense chemical weathering that produces soils largely made of insoluble residues—iron hydroxides (laterites) and aluminum hydroxides (bauxites). The removal of metal atoms in forming bauxite from an original igneous rock may follow a sequence of the type shown below for the breakdown of the potash feldspar present:

$$4KAlSi_3O_8 + 4H^+ + 18H_2O \rightarrow Al_4Si_4O_{10}(OH)_8 + 8H_4SiO_4 + 4K^+ \quad \textbf{(12.6)}$$

(feldspar + H ions in solution + water → kaolinite + silicic acid in solution + K ions in solution)

$$Al_4Si_4O_{10}(OH)_8 + 7H_2O \rightarrow 2Al_2O_3 \cdot 3H_2O + 4H_4SiO_4 \quad \textbf{(12.7)}$$

(kaolinite + water → gibbsite (bauxites) + dissolved silica).

Humid mid-latitude climates with seasonal freezing allow much greater accumulation of vegetational debris—a **humus** layer—and dissolved species may not be removed but may recombine to form stable clay minerals.

Hot arid climates allow for the growth of little vegetation and provide too little water to permit much chemical weathering. Consequently, such regions often do not develop true soils. Instead, salts may be left at or near the surface from the evaporation of the little available water, or a variety of rock-like crusts may form such as the calcium carbonate-rich crust known as *calcrete* (or caliche) (Figure 12.3). Weathering involves

FIGURE 12.3 A white, 1-meter (three foot) thick layer of caliche (calcium carbonate) has formed in the upper soil horizon in much of Patagonia, in Argentina. The high rate of evaporation relative to precipitation in this part of the world results in calcium carbonate accumulating near the surface when the calcium-bearing groundwater evaporates. (Photograph by J. R. Craig.)

rapid mechanical and chemical breakdown of the less resistant silicates. The clay minerals that do form are very small and tend to be blown away, leaving mostly quartz grains behind.

Cold climates, such as are found in Antarctica, may also be very dry because all the water has turned into the solid form (snow, ice, frost) and chemical weathering is exceedingly slow. The biological activity of plants and microorganisms is also much reduced, although the slow rate of decay of organic material can lead to its accumulation, forming peat bogs and the thick peat accumulations found in Canada and known as **muskeg.** Mechanical breakdown (by frost wedging) is the major weathering process.

The degree to which biological processes, chiefly involving vegetation and microorganisms, contribute to soil formation is dependent upon the temperature and available moisture. Living plants take up certain chemical elements as essential **nutrients,** but these are returned to the surface soil when the plants shed their leaves or when they die. Plants also control the moisture content of the soil by transpiring water and by serving to protect soils from erosion. Animals burrowing through the soil also play an important role. For example, earthworms rework the soil by burrowing and passing the soil through their intestinal tracts. These biological processes are, in turn, influenced by climate and by the parent rock material from which the soil forms. These processes control the development of vegetation, which, in turn, may permit animals to flourish. A soil is, therefore, a complex and constantly changing system in which many interacting physical, chemical, and biological processes proceed at the same time.

Soil Chemistry

Soils are both complex mixtures of chemical compounds and complicated systems within which chemical reactions constantly take place. Reactions occur chiefly as the result of water and air that are present. The air in pore spaces provides atmospheric gases (oxygen, nitrogen, carbon dioxide) that dissolve in the water and play important roles in reactions. Soil water contains a large number of dissolved atoms, usually in the form of positively charged **cations,** such as Al^{3+} (aluminum), Ca^{2+} (calcium), K^+ (potassium), Mg^{2+} (magnesium), Na^+ (sodium), Fe^{2+} or Fe^{3+} (iron), NH_4^+ (ammonium), H^+ (hydrogen), and negatively charged **anions,** for example, Cl^- (chlorine), SO_4^{2-} (sulfate), HCO_3^- (bicarbonate), OH^- (hydroxide), and NO_3^- (nitrate).

Soil temperature exerts an important control over both chemical and biological processes. Below 0°C (32°F) there is essentially no biological activity, and chemical processes are virtually inoperative; between 0°C and 5°C (41°F), both root growth of most plants and germination of most seeds are impossible, but water can move through the soil and chemical reactions can occur. Biological activity increases at higher temperatures; seeds of most plants do not germinate until the soil is warm (15°–24°C; 60°–75°F), and some only when it is even warmer (24°–32°C; 76°–90°F). Temperatures vary both through diurnal and annual cycles and as a function of depth beneath the surface. They lead to the recognition of various **soil temperature regimes** described by the mean annual soil temperature and by average seasonal fluctuations from that mean. Water also plays a key role in soil chemistry and biochemistry, and the amount of water generally available in the soil also leads to the recognition of **soil water regimes.** Both soil regimes are important factors when overall soil classification is considered.

Chemical processes in soils are particularly influenced by the clay minerals that are present. These minerals have crystal structures in which the atoms are arranged in layers (see Figure 12.4), and the forces bonding the layers together are much weaker than those within each individual layer. This enables water molecules and various other ions dissolved in the water to penetrate between layers and become loosely bonded into the structure. Ions can also be attached in a similar fashion to the surface of the clay particles. Ions and molecules bound to clay particles in this way can be replaced by other ions and molecules in a process of **exchange.** Most commonly, this involves **cation exchange.** Common cations selectively replace specific others in a sequence: aluminum (Al^{3+}) replaces calcium (Ca^{2+}), which replaces magnesium (Mg^{2+}), which replaces potassium (K^+), which finally replaces sodium (Na^+). The capacity of a soil to hold and exchange cations in this way is its *cation exchange capacity.* Clay minerals formed in the early stages of weathering or by less intense weathering (e.g., montmorillonite and vermiculite) generally have high cation **exchange capacities,** whereas clays formed in advanced stages and by intense weathering (e.g., kaolinite) have much lower exchange capacities. Soils with a high cation exchange capacity can usually function well in storing plant nutrients and are generally more fertile than those with low exchange capacity. See Box 12.1 for a discussion of how quickly soils form.

Another important factor in soil fertility is the acidity or alkalinity of the soil expressed in terms of pH, which is a measure of the concentration of hydrogen ions (H^+) in the soil moisture. Certain cations, in addition to H^+, especially Al^{3+} and $Al(OH)^{2+}$, promote acid

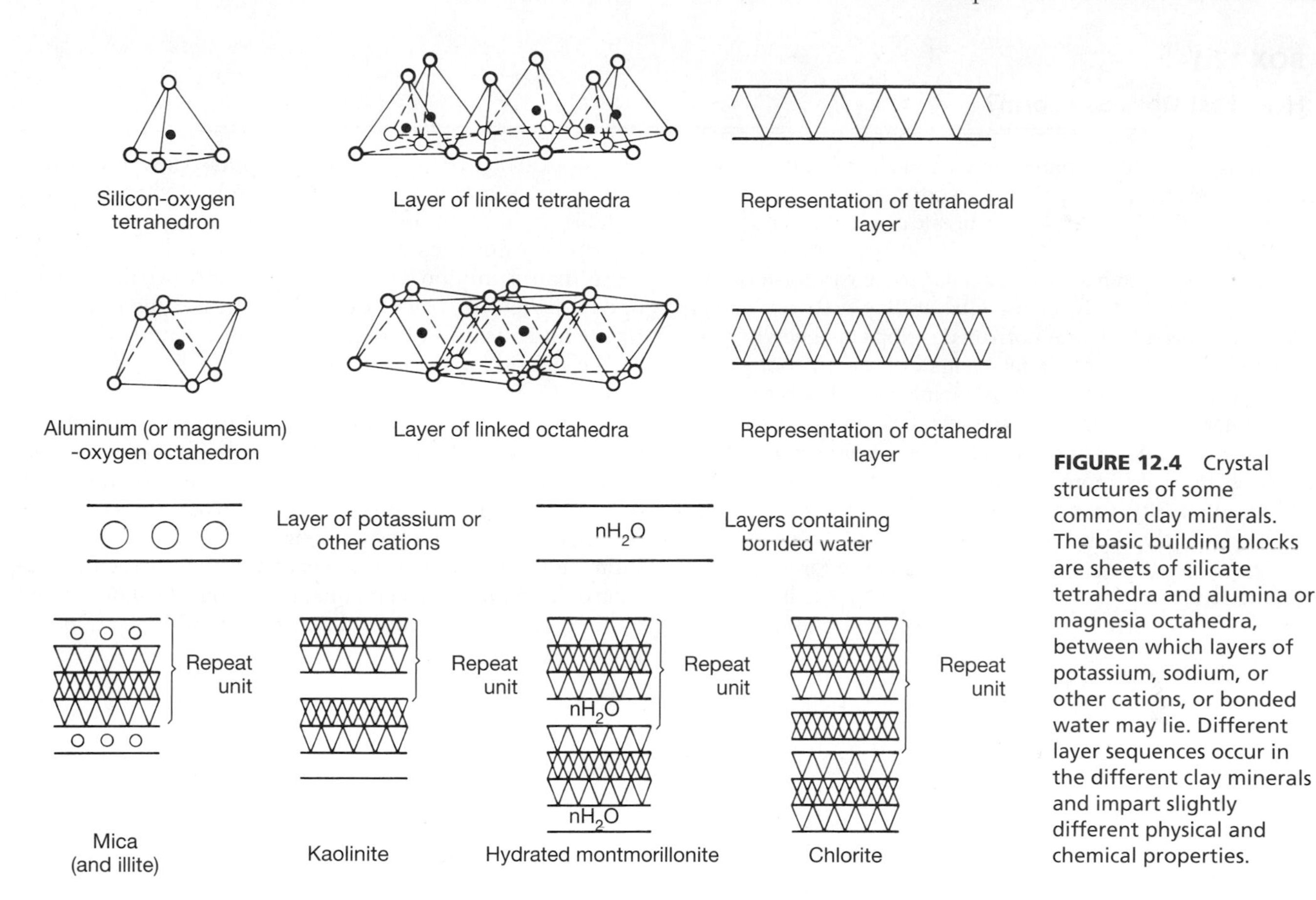

FIGURE 12.4 Crystal structures of some common clay minerals. The basic building blocks are sheets of silicate tetrahedra and alumina or magnesia octahedra, between which layers of potassium, sodium, or other cations, or bonded water may lie. Different layer sequences occur in the different clay minerals and impart slightly different physical and chemical properties.

conditions in the soil and are said to be *acid-generating;* others, such as Ca^{2+}, Mg^{2+}, K^{+}, and Na^{+}, promote alkalinity. Both types of cations must be present in the soil water, or sufficiently loosely held in the clays, to readily enter the soil water; otherwise, the cations will have no effect on soil chemistry. Certain crops require near-neutral (pH = 7) values, but other plants may show considerable preference for either acid or alkaline soils. Soils with pH less than 6 require the addition of **lime** (either calcium oxide or calcium carbonate) to bring the pH closer to neutral if most farm crops are to be successfully cultivated. Acid soils are also commonly deficient in certain nutrients and may require the addition of fertilizers, as well as limestone to raise the pH. Acid rain (discussed in Chapter 4), generated by the release of nitrogen and sulfur gases into the atmosphere, has been found to lower the pH of soils in many areas. The acidity is especially effective in reducing fertility by leaching calcium, magnesium, potassium, and many trace metals that are important for plant growth. In some cases, the precipitation is so acidic that it harms the vegetation directly and then damages the important populations of organisms that normally enrich soils.

Soil Characteristics

Before considering the differences between soil types, it is useful to emphasize some of the most important properties and characteristics of soils that might form the basis for a classification scheme.

Color is an obvious property that can be described using quantitative scales. Sometimes, color arises from the parent matter that weathers to form the soil, but commonly it arises during the soil-forming process. For example, a black color comes from organic matter; a red color comes from iron oxide.

Texture is used to classify soils following a U.S. Department of Agriculture scheme of defining the percentages of sand, silt, and clay present in the soil. The three components are defined in terms of particle size:

Sand: particle size, 2.0–0.05 mm diameter

Silt: particle size, 0.05–0.002 mm diameter

Clay: particle size, less than 0.002 mm diameter

The boundaries of these textures are drawn somewhat differently from those used by geologists in the

BOX 12.1

How Fast Does Soil Form?

Although soil development is part of the complex process of weathering, a soil profile can form in a regolith much more rapidly than it takes to break down the underlying bedrock.

In some environments, a soil profile can form quickly. For example, a study in the Glacier Bay area of southern Alaska showed that an *A* horizon develops on a newly revegetated landscape within a few years of the retreat of glaciers (see Figure 12.A). In this environment, rapid leaching of parent material occurs because of the high rainfall and moderate temperatures. Also, as the plant cover becomes more dense, the soil becomes more acid and leaching is more effective. After about fifty years, a *B* horizon appears: the combined thickness of the *A* and *B* horizons is about 10 centimeters (4 inches). Over the next 170 years, as a mature forest develops on the landscape, the *A* and *B* horizons increase in thickness to 15 centimeters (6 inches) and iron oxides accumulate in the developing *B* horizon.

Rates of soil formation are much slower in more arid climates, and it may take thousands of years for a detectable *B* horizon to appear. Thus, the ice-free polar deserts of Antarctica are so dry and cold that sediments more than a million years old have only weakly developed soils (entisols). The deep red-colored ultisols of temperate and subtropical regions probably date back to the Tertiary Period and likely took many millions of years to form.

The great length of time needed to develop a mature, productive soil highlights the potentially disastrous consequences of soil deterioration and erosion in agricultural regions. Once agricultural soils are destroyed, they can only be replaced over geologically long time intervals. Although ultimately renewable in this very long term, over the lifetime of individuals—even of nations—soils must be viewed as nonrenewable resources requiring careful utilization and preservation.

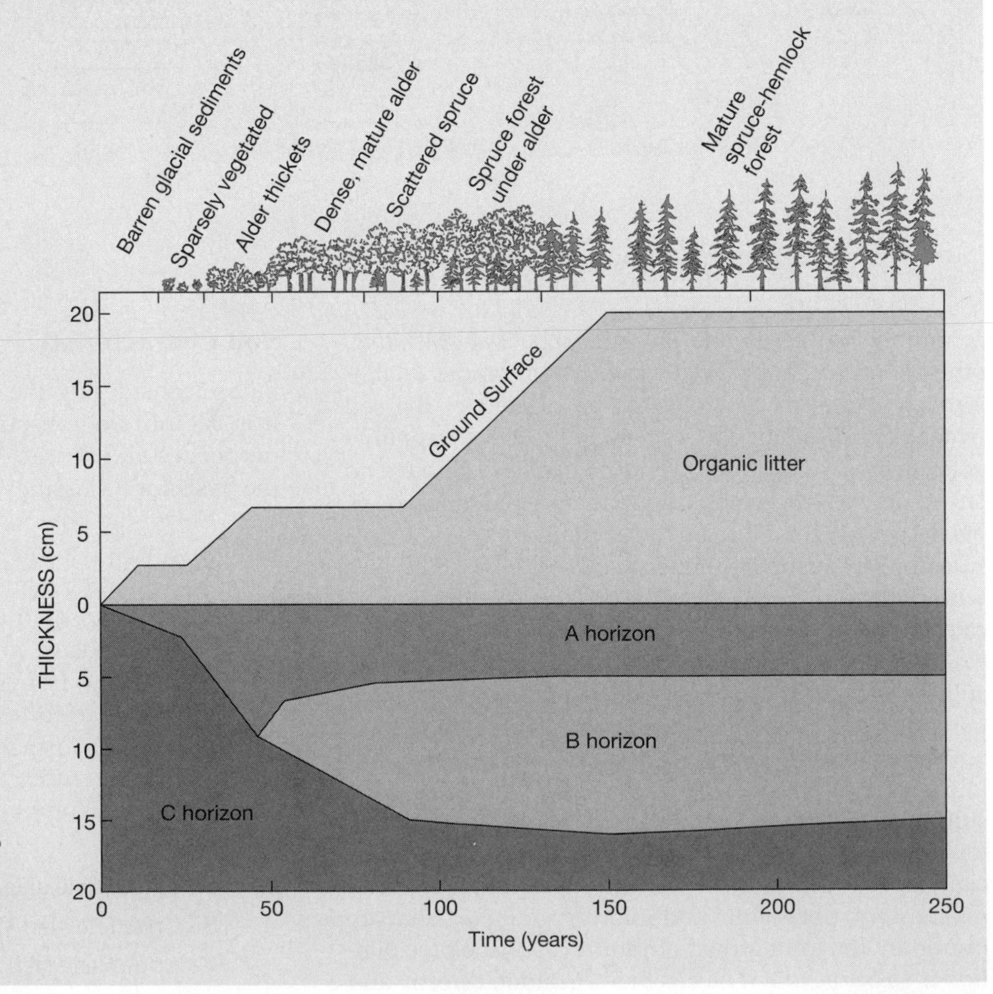

FIGURE 12.A Soil development in the area of Glacier Bay, Alaska, over the past 250 years. During the first forty years, the organic litter increased and the *A* horizon began to develop directly on the *C* horizon. The *B* horizon began to develop after about forty years and reached a stable thickness after about 100 years. (Adapted from data by F. C. Ugolini.)

classification of sedimentary rocks. It is also important to note that the term *clay* in this sense is used purely to describe particle size and does not describe the mineralogy (Figure 12.5). A mixture in which all three size-range materials occur in substantial amounts is called a **loam.** Soil texture is important because it largely determines the extent to which water is either retained or passed. Pure clay holds the most water and pure sand the least; hence, sandy soils require more frequent watering. Clay-rich soils take water very slowly, so there is a danger of losing surface runoff if irrigation is used.

Consistence is a property that relates to the stickiness of wet soil or plasticity of moist soil. Some soil horizons can become cemented through the accumulation of minerals such as silica, iron oxides, or calcium carbonate.

Soil structure refers to the presence and nature of lumps made from clusters of individual soil particles. Such a natural lump is called a **ped,** as distinct from a **clod** that is produced during plowing. Soil structure is described in terms of the shape, size, and durability of the peds and is a property of considerable importance in agriculture, affecting ease of cultivation, susceptibility to erosion, and ease of water penetration into the dry soil. The four basic types of soil structure are: *platy, prismatic, blocky,* and *spheroidal* (Figure 12.6).

Soil horizons are the distinctive horizontal layers with differing chemistry and structure that are exposed in a vertical **soil profile**. The two major classes of soil horizons are *horizons of organic matter* and *mineral horizons of differing compositions.* The former (labeled with a letter O) is made of plant and animal debris and commonly has an upper horizon (O1) of recognizable plant material underlain by decomposed material (humus) comprising the mineral horizon (O2). Mineral horizons consist of detrital particles of sand and silt-size mineral fragments and of clay minerals and other similar weathering products. The *A* horizon overlies the *B* horizon and typically contains more organic matter; the *B* horizon contains more mineral matter and is generally less friable (easily pulverized). Underlying the*B* horizon might be a *C* horizon of weathered parent material and the bedrock that is labeled the *R* horizon. In detail, the variations found in soil profiles are extensive, and the system of labeling them is suitably complex.

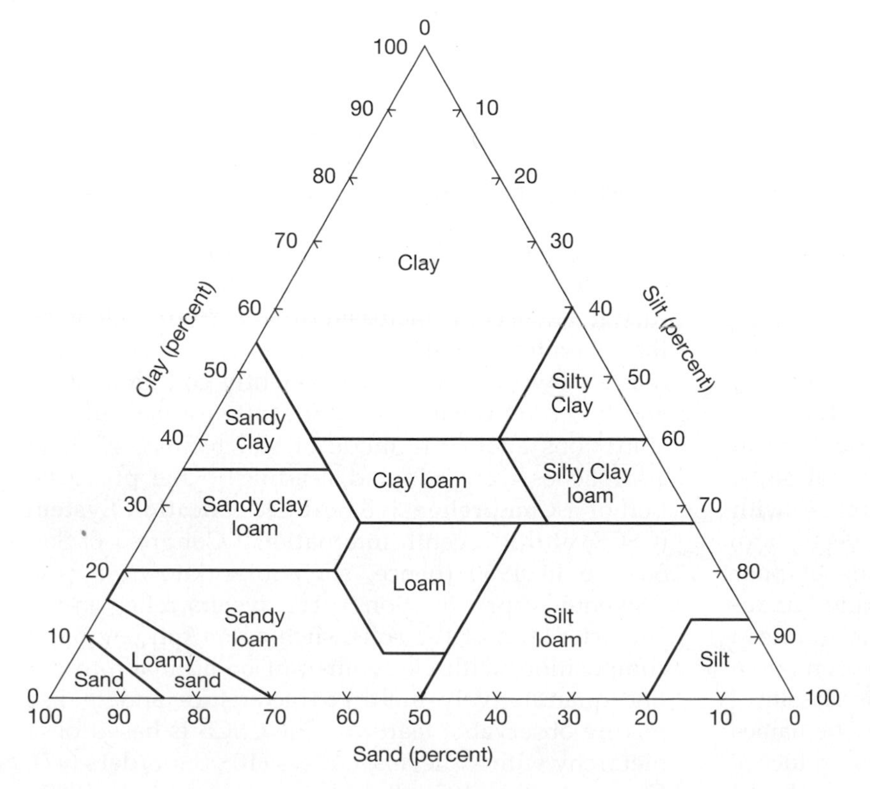

FIGURE 12.5 A triangular diagram illustrating the textural classification of soils. Note that, in this classification, the terms *clay* and *sand* refer only to the sizes of particles and not their compositions.

FIGURE 12.6 An illustration of the four basic soil structures: (a) blocky (angular) soil from a *B* horizon in western New York; (b) granular (spheroidal) soil from an *A* horizon in southwestern Kansas; (c) platy soil from an *A* horizon in central Iowa; (d) prismatic soil from a *B* horizon in central South Dakota. The scales are all in inches. (Photographs courtesy of Roy W. Simonson, Soil Conservation Service, U.S. Department of Agriculture.)

Soil Classification

Older systems of soil classification emphasized factors such as climate, relief, and parent material that control soil formation. A widely used scheme of this type was developed by the U.S. Department of Agriculture in 1938 to recognize three orders of soils: **zonal soils,** formed under conditions of good drainage with marked involvement of climate and vegetation; **intrazonal soils,** formed under conditions of poor drainage (e.g., bogs); and **azonal soils,** formed under conditions not conducive to any real soil (e.g., on steep slopes, in certain deserts, etc.). Further division into a number of suborders was followed in this classification by the recognition of *great soil groups.* The names of some of these groups reflect the important influence of Russian scientists in this field (e.g., podzol and chernozem), and other names reflect American contributions (e.g., prairie soils, chestnut soils).

However, by the latter half of the twentieth century, it had become increasingly clear to soil scientists that classification schemes of this type were inadequate. They were both insufficiently comprehensive and based on unsupported and often untestable assumptions about the mode of soil formation. New approaches were adopted, leading to the presentation of a **Comprehensive Soil Classification System** (CSCS) at the Seventh International Congress of Soil Science in 1960 (hence, sometimes known as the "Seventh Approximation"). The system relies on the characteristics of the soils, such as morphology and composition, with every attempt being made to define quantitatively these characteristics and to use readily observable features. The CSCS is based on a hierarchy with six levels: Orders (10), Suborders (47), Great groups (185), Subgroups (more than 1000), Families (more than 5000), and Series (more than 10,000). The smallest distinctive division of the soil of

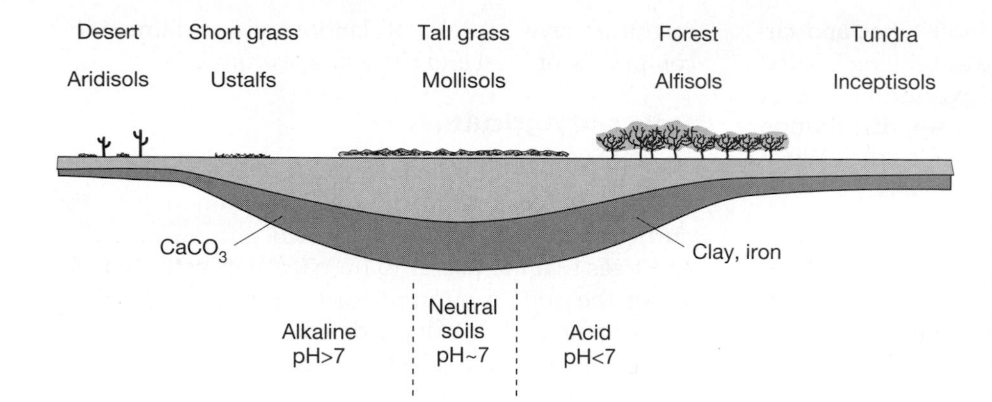

FIGURE 12.7 Idealized soil profile from the southwest to the northeast of the North American continent. Note that the thickest and richest soils occur in the Midwest. Soils in the desert Southwest, where it is arid, and in the subarctic Northeast where it is cool, are much thinner.

a given geographical area is termed the **polypedon,** and every polypedon falls within only one of the ten **soil orders.** Various criteria are used to uniquely define each of the orders; these may include gross composition (e.g., percent clay, percent organic matter); degree of development of soil horizons and presence or absence of certain diagnostic horizons; and degree of weathering of the soil minerals (possibly expressed by cation exchange capacity).

Soil Distribution

The processes of formation and the problems of classification of soils are topics that naturally lead to the distribution of soils (Figures 12.7 and 12.8). Figure 12.7 is a simplified cross section from the southwest to northeast of the North American continent and shows the differences in soil type and thickness. The map in Figure 12.8 is oversimplified, but it does show the

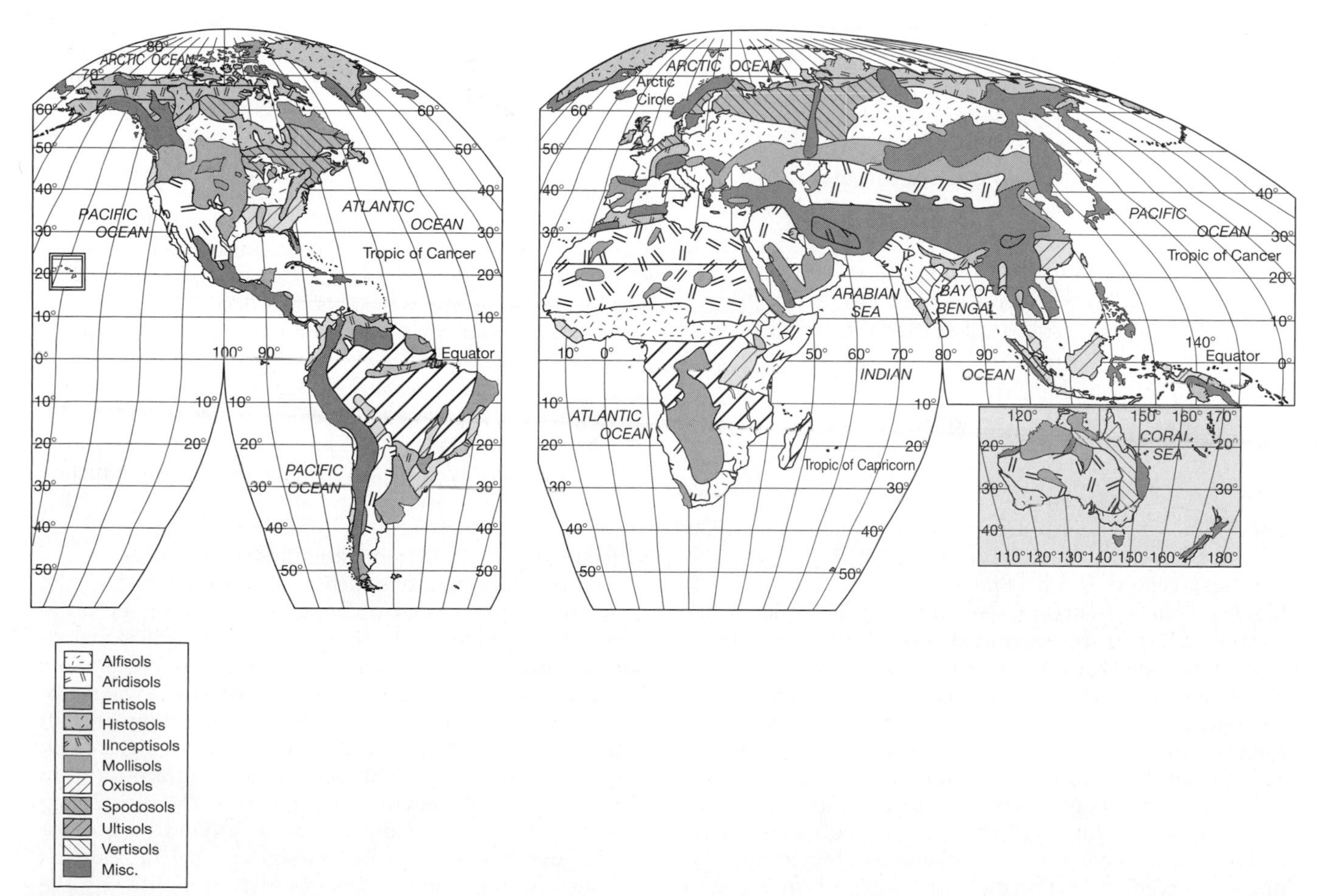

FIGURE 12.8 A generalized map showing world distribution of soils. (Modified from Soil Conservation Service, U.S. Department of Agriculture.)

correlations to be expected among soil types and climate. For example, it shows the aridisols being located in the world's great deserts and the oxisols in the great tropical zones. This overall pattern of soil distribution must be kept in mind as we now consider the utilization, management, and conservation of soils.

SOIL TYPE AND LAND USE

The soil classification systems provide the basis for discussing soil use in agriculture on a global or regional scale. The potential use of a soil depends upon its particular properties. However, suitable soils are not the only factors governing successful use of the land for agriculture—there are other factors, of which the most important is certainly the water supply. The practical exploitation of soils through agriculture can be one of several forms of utilization of the land in a particular area. Conflicts result from problems of land management and might involve individual landowners, communities, companies, or local and national governments.

Soils and Agriculture

Soils that can be used to grow crops are said to be *fertile*. This means that the soil is rich enough in the nutrients needed for the sustained growth of plants and trees that are useful to humans. The actual capacity of the soil to support such growth is called its *productivity*, and this depends on the soil being fertile and on the soil having a structure and consistency that makes it easily tilled. Tilling aerates the soil, allows passage of water, and promotes the spreading of plant roots. Despite adequate rainfall, some soil is unproductive because it drains too quickly; it is too **permeable.** Other soils may be barren because they are **impermeable** to moisture. See Box 12.2 for a discussion of "dust bowl" conditions that resulted from prolonged droughts in the United States.

BOX 12.2
The Dust Bowl

> Now the wind grew strong and hard and it worked at the rain crust in the corn fields. Little by little the sky darkened by the mixing dust, and the wind felt over the earth, loosened the dust, and carried it away. The wind grew stronger. The rain crust broke and the dust lifted up out of the fields and drove gray plumes into the air like sluggish smoke. . . . All day the dust sifted down from the sky, and . . . it settled on the corn, piled up on the tops of the fence posts, piled up on the wires; it settled on roofs, blanketed the weeds and trees.
>
> John Steinbeck, *The Grapes of Wrath*

The *Dust Bowl* is a term that was first applied in the 1930s to an area of some 20 million hectares (50 million acres) in the west central United States (Texas, Oklahoma, New Mexico, Kansas, Nebraska; see chapter opening picture and Figure 12.B). The area expanded episodically, extending as far north as the Dakotas, as far west as Arizona, and as far east as the Mississippi River. Throughout that decade, vast dust storms stripped soil from fields and devastated farms, forcing many farmers to abandon their homes. These areas of the High Plains had been natural grasslands that experienced episodic rainfall and occasional periods of prolonged drought. Although subjected to high winds, large temperature fluctuations, and violent rainstorms, the grasses served to anchor the soils and prevent excessive wind or water erosion.

The first farmers moved into the area later named the Dust Bowl after the Civil War; they were successful in rainy years, but struggled during the droughts. By the 1920s, the farms were mostly cultivating wheat, a grain that was in demand but that did not resist wind erosion nearly as well as the natural grasses. Consequently, soils began to drift. Drought conditions followed in the 1930s and erosion was rapidly accelerated. In 1934, high curtains of dust developed in the strong winds and were carried out into the Gulf of Mexico and as far east as the Atlantic Coast. The dust storms were blinding and created numerous respiratory problems. Buildings, vehicles, and even livestock were buried by migrating sand dunes constantly creeping in response to the driving winds. The devastating effects on families, many of which were forced to leave the area, were chronicled by John Steinbeck in his novel, *The Grapes of Wrath*. Rainfall returned to normal in the 1940s and ended the worst period of the Dust Bowl, but drought conditions in the 1950s and 1960s resulted in local problems similar to those of the 1930s.

Erosional conditions such as those experienced in the Dust Bowl are a combined consequence of weather variations and imprudent human activities. Drought is a periodic occurrence in many areas and weakens vegetation; hence, drought allows some increase in wind and water erosion. Arable farming, in which all vegetation is first stripped by plowing and then only selectively replaced by planting crops in rows, exposes the land surface to rapid erosion by wind or rain in times of storms. The use of deep furrows to break up wind action at the ground surface, the use of crops that better anchor the soil, and the introduction of extensive irrigation that ensures early crop growth have all reduced the effects of the wind and rain that devastated

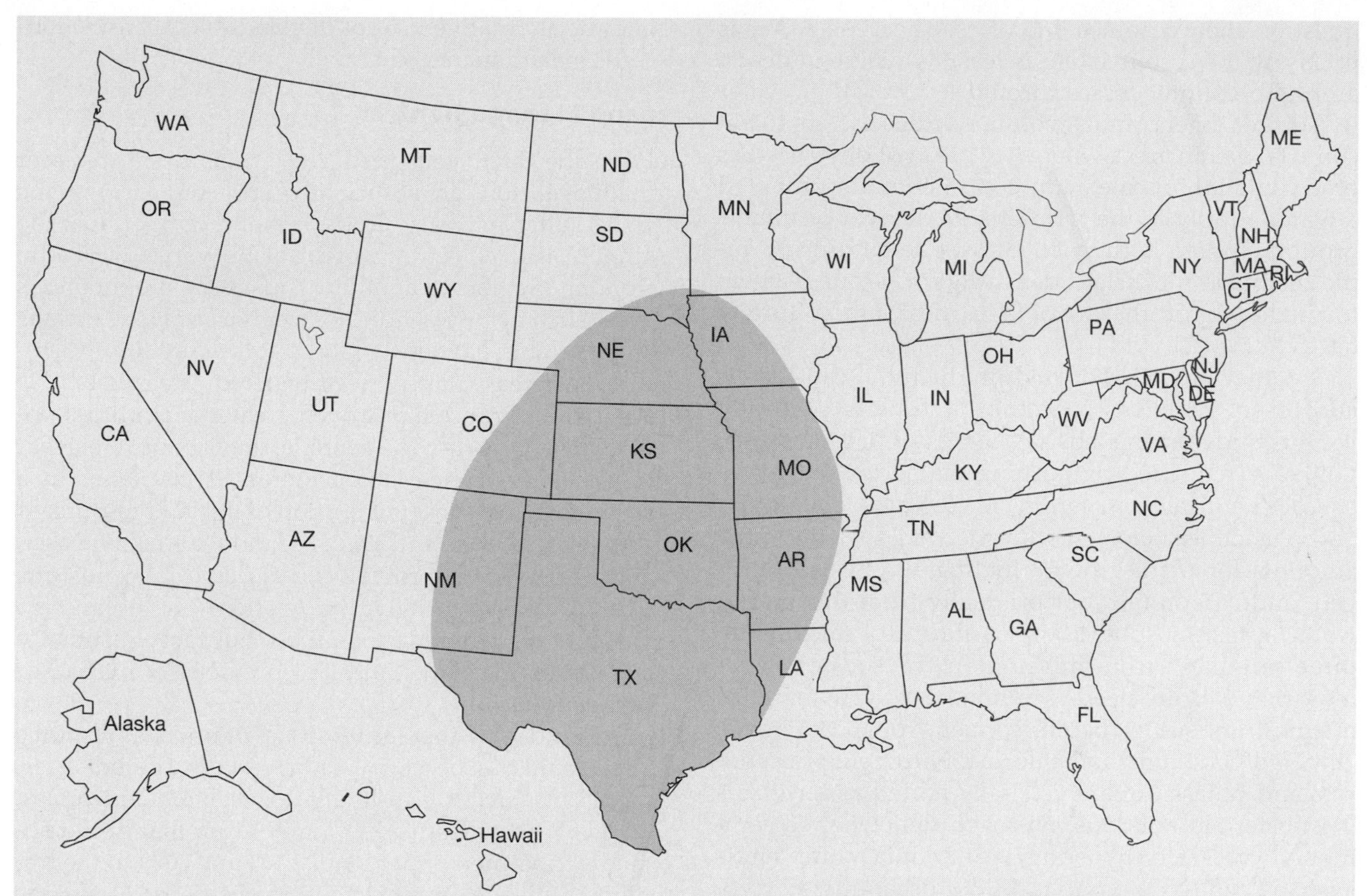

FIGURE 12.B Map showing the Dust Bowl region of the United States in the 1930s. The severe erosion and dust conditions developed from prolonged drought and were intensified by poor agriculture practices.

so many areas at the time of the Dust Bowl. In addition, farmers are increasingly turning to *no-till* agriculture, in which the seeds (especially corn) are drilled into the ground, thereby not loosening soil; erosion then does not become a severe problem.

Unfortunately, the rapidly increasing populations in many regions today are forcing people to attempt farming on marginal and semiarid lands. In several parts of sub-Saharan Africa, prolonged drought and inappropriate farming practices are causing Dust Bowl conditions to recur.

Productive soils belong only to certain categories named in the CSCS classification. In general terms, soils of mountainous areas and deserts (entisols and aridisols) are nonproductive, whereas the most productive soils include the mollisols, alfisols, and spodosols of the temperate and more humid regions of the globe. Oxisols and most ultisols, which are found in subtropical and tropical regions, are also unsuited to agriculture because the essential nutrients have been removed, leaving an infertile soil that tends to be hard and brick-like, as are many laterites and bauxites. The inceptisols of tundra areas will not support agriculture because the ground is often frozen; the organic-rich histosols of many bog-land areas are too acidic for cultivation.

Productive soils, like many other types of resources, are very unevenly distributed over Earth's surface (Figure 12.8). Indeed, on average only about 10 percent of Earth's land surface (or 3 percent of the total surface of Earth) is productive cropland. This cropland, despite the importance of the seas and oceans as sources of food, is estimated to produce 97 percent of the world's food. The percentage of cropland varies markedly from country to country, being very high in flat-lying, well-watered regions such as the Netherlands, and much less in mountainous countries such as Switzerland or in arid countries such as Saudi Arabia. Whereas desert soils are estimated to cover about 17 percent of Earth's total land area, the continents of Australia and Africa are about 44 and 37 percent desert, respectively, but Eurasia is only 15 percent desert. Do all of these desert areas need to be unproductive? In theory, the answer to this question is no. Deserts include some potentially very fertile and productive areas; the main factor preventing their utilization is the lack of water. The question then becomes one of cost because water is precisely the resource that is least available in desert areas. To counter losses through evaporation, water is required in enormous volumes if deserts are to become fertile. In areas where irrigation of deserts takes place on a large scale, nature has generally provided a water supply in the form of a great river or underground aquifer. Perhaps the most outstanding example of this is the Nile River in Egypt, around which formed one of the world's earliest agriculturally based societies.

Other problems of land productivity can also be addressed. Problems concerning the permeability of the ground, or the absence of essential nutrients, might be remedied by farming methods and by the use of chemicals and fertilizers (see also Chapter 9). Seven chemical elements are needed in substantial amounts for plant growth: hydrogen, oxygen, nitrogen, and carbon (all four originally from the air and water), plus phosphorus, potassium, and calcium (all three originally from minerals in the soil). Another nine elements are needed in minor or trace amounts: magnesium, sulfur, boron, copper, iron, manganese, zinc, molybdenum, and chlorine. Nitrogen, phosphorus, and potassium are generally added as fertilizers in substantial amounts. Nitrogen deficiency is commonly corrected by adding ammonia compounds such as NH_4NO_3 or $(NH_4)_2SO_4$, or a nitrate compound such as KNO_3 or $NaNO_3$. In the eastern and central United States, the nitrogen content of the soil to a depth of 40 inches is estimated at approximately 11,000 to 37,000 pounds per acre, making up on average about 0.2 percent of the soil by weight. Phosphorus makes up roughly 0.025 to 0.075 percent of the plowed layer of farmed land in the United States; its deficiency is remedied by the addition of phosphate fertilizers. Potassium is commonly present in the surface layers of the soil in much greater amounts than nitrogen or phosphorus and can average as much as 2 percent in productive soils. Much of the potassium may be held in mineral structures and be readily available for immediate uptake by plants. Deficiency in readily available potassium can be countered by addition of potassium chloride (KCl). The elements needed in only minor trace amounts can be easily supplied by spraying the soil with appropriate additive compounds or by addition of these elements to the major fertilizers. It has been already noted that extreme acidity or alkalinity of soils has a very adverse effect on fertility because nutrients may be destroyed or removed under such conditions. Acidic ground is, of course, treated with lime to increase the pH to nearer neutral values.

Previously unproductive land may be converted to agricultural land; the limiting factors are the cost of the process, the return on an investment of this kind, and the alternative uses of the land. These are all concerns of land management.

Land Management

From the dawn of agriculture, it has been apparent that not all soils are sufficiently fertile or lie in climatic zones suitable to raise crops. It is also obvious that the amount and percentage of cropland varies markedly from one region to another. Thus, very mountainous areas such as Switzerland or very arid places such as Saudi Arabia have little land appropriate for agriculture, whereas flatter, well-watered areas such as Belgium or France have much greater agricultural potential. As a worldwide average, only about 10 percent of the land surface is considered as cropland. In a sense, then, only a small portion of the land resource is actually a reserve, and it is this land that must be used and preserved to provide the agricultural foodstuffs for the world's growing population. The oceans are vast and do provide much food, but recent estimates suggest that it is the land that provides 97 percent of human food.

Land (and therefore soil) is an unusual resource because it can be put to a range of different uses, many of which are mutually exclusive. Land prices, tax laws, and lifestyles—ranging from that of the nomadic tribesmen of parts of Africa and Asia to the traditional village communities of old Europe, and the urban sprawl of many large cities—all influence the ways in which land is managed. If we consider the example of the United States, the total land area of the country is approximately 2300×10^6 acres, of which slightly less than half is devoted to raising crops and livestock (45 percent), while the remainder is devoted to a variety of other uses including forestry (29 percent), urban regions (3 percent), and mining (0.3 percent) (see Figure 12.9). It must also be remembered that the overall pattern of land use is constantly changing. Considering again the example of the

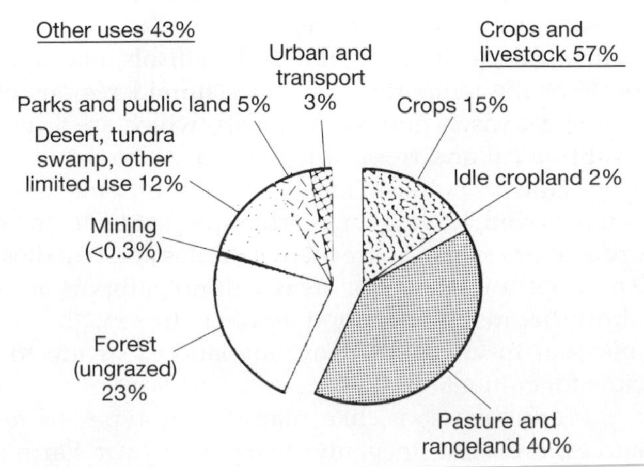

FIGURE 12.9 Land use in the United States.

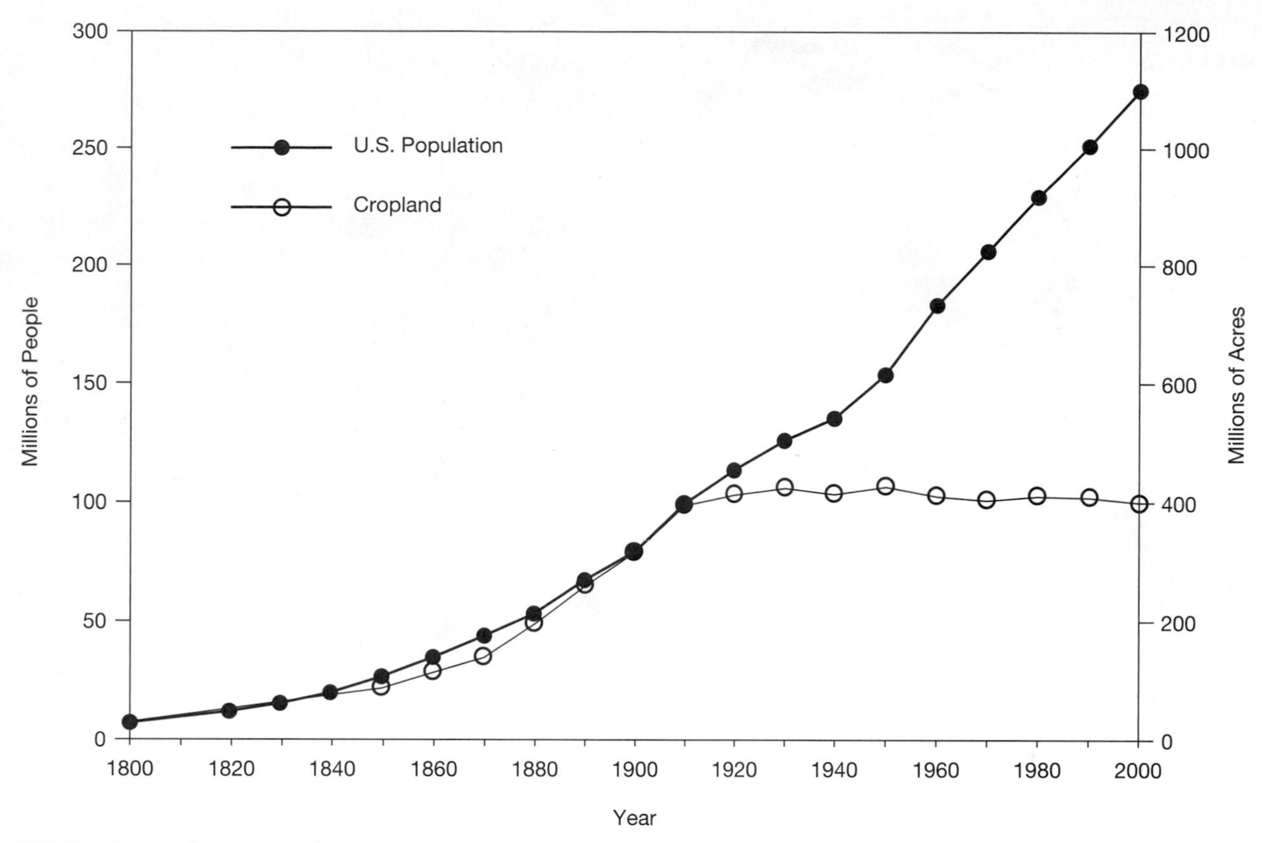

FIGURE 12.10 The U.S. population and cropland each increased approximately fourteen times from 1800 until 1900. Since the 1920s, the amount of cropland has remained nearly constant as productivity has increased through the use of improved farming practices, new hybrid crops, genetic engineering, irrigation, and more intense use of fertilizers. (Updated from D. W. MacCleery, "American Forests," FS542, U.S. Forest Service, Department of Agriculture, 1992, p. 22.)

United States, the total land area used for raising crops is approximately 470 million acres. In an average year during the last decade, roughly 350,000 acres were lost to make way for urban development and a staggering 1.9 million acres were lost to make way for roads, airports, flood-control systems, or recreational projects. Balanced against these figures, roughly 2.2 million acres of land were transformed to crop-producing areas annually through irrigation and fertilization, processes that might cost $1000 per acre for every new acre added. Much of this new cropland has also been added at the expense of former pasture- or rangeland. The pattern of U.S. population growth and cropland development since 1800 is shown in Figure 12.10. These patterns are much as expected from 1800 to 1900; the steadily rising population cleared more forest and rangeland to meet the growing agricultural needs. After about 1920, although the population continued to increase, the total amount of cropland remained more or less constant at between 450 and 400 million acres. This was because improved yields were achieved by using more intensive agricultural methods, including the use of greater amounts of fertilizer (see Figure 1.1), better hybrid seeds, and much more intensive irrigation.

There are also cases where new agricultural land has been created by reclamation from the sea. The most famous example is the extensive network of dikes in the Netherlands, a system commenced over 300 years ago, which has enabled the Dutch to reclaim 20 percent of the land area of their country from coastal submergence (Figure 12.11). The danger of storm damage to the dikes, along with the consequent flooding of the type that occurred in 1953 when 1800 people were drowned and 47,000 buildings lost, has been averted by closing large estuaries with vast dams (Figure 12.12) to put the dikes out of reach of storm tides.

Utilization of soil as an agricultural resource will inevitably conflict with the alternative uses to which a particular area of land could be put. A good illustration of this problem is provided by considering the large areas of the central United States that are underlain by coal at shallow depths, which could be removed by strip mining. Many such areas are largely or partly the croplands responsible for much of American's wheat and corn production, as clearly shown in Figure 12.13.

The encroachment of alternative forms of land utilization, whether for urban development, road

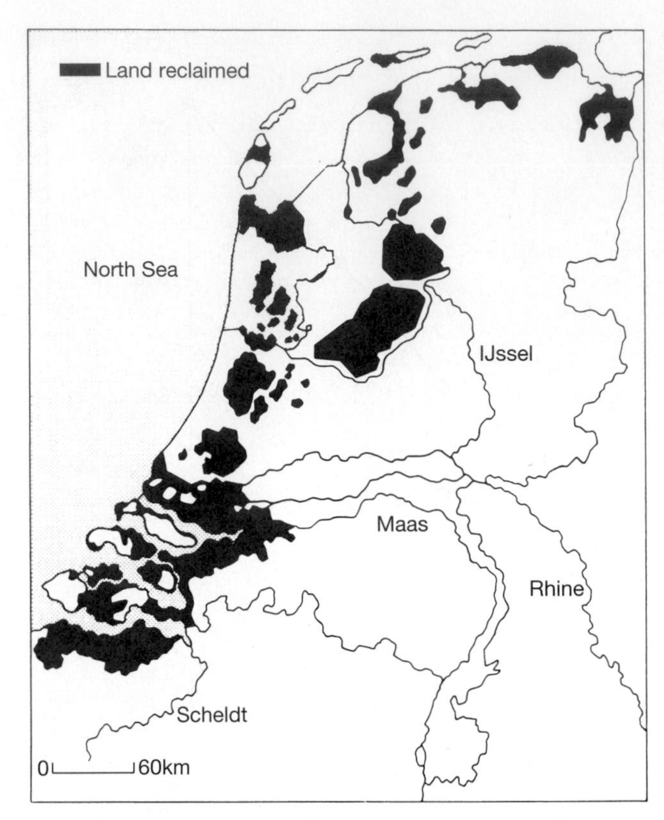

FIGURE 12.11 The darkened areas represent more than 20 percent of the land area of the Netherlands that has been reclaimed from the sea by the construction of dikes and the pumping away of water. Dikes were built as early as A.D. 1000; windmills (Figure 6.30) were used to pump water from the reclaimed areas by the early 1400s. (Map courtesy of the Royal Netherlands Embassy.)

FIGURE 12.12 Much of the area of the Netherlands lies at or below sea level. The 30-kilometer-long (19 miles) Barrier Dam, constructed in the 1930s to hold back the sea, has transformed Zuyder Zee from a seawater estuary into a freshwater lake known as IJsselmeer. It has also allowed the draining of more than 165,000 hectares (410,000 acres) of land areas for use in farming and housing. (Courtesy of the Royal Netherlands Embassy.)

construction, or mining operations, poses an obvious threat to the soil resource and food-production chain dependent upon it. However, another threat, and many would argue a much more serious one, is that posed by the degradation and deterioration of soils by natural agencies, human intervention, or both. (See Box 12.3 for a discussion of drainage and irrigation problems in the San Joaquin Valley.)

EROSION AND DETERIORATION OF SOILS

Soil Erosion

The beginning of this chapter explained that soils form through the weathering and breakdown of bedrock material. Without the binding and stabilizing effect of organic matter and vegetation, soil would soon be

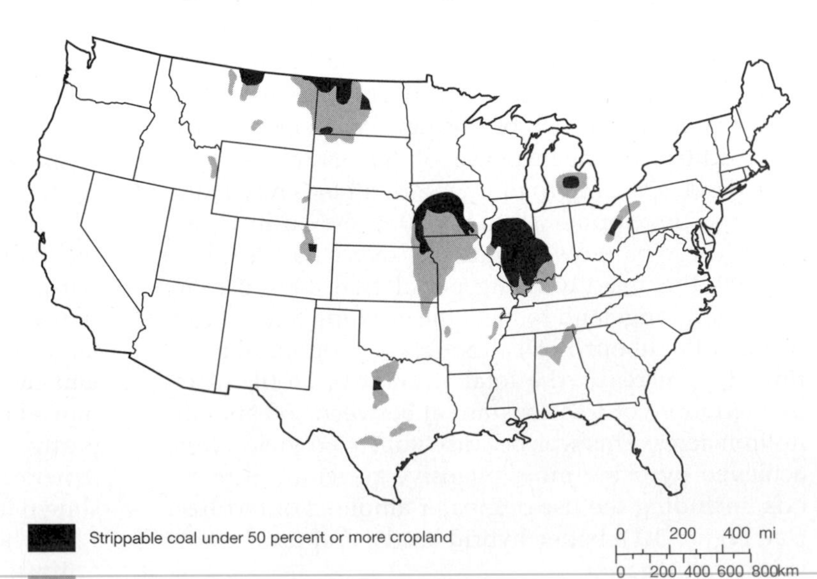

FIGURE 12.13 A map of the United States showing cropland areas overlying coal suitable for strip mining. (Based on U.S. Department of Agriculture Economic Research Service Publication no. 480, 1971.)

BOX 12.3

Selenium Poisoning in the San Joaquin Valley

Selenium is a naturally occurring nonmetallic element that is essential to human and animal nutrition in trace quantities, but is toxic in higher concentrations. The U.S. Environmental Protection Agency sets safe drinking water standards at a maximum of only 0.01 milligram of selenium per liter and classifies any solution with more than 1 milligram per liter as a hazardous waste. The U.S. Fish and Wildlife Service, alerted by deformities and a high mortality rate in newborn waterfowl, was surprised in 1982 and 1983 when it was discovered that selenium levels in farm drainwaters at the Kesterton National Wildlife Refuge in California were as high as 4.2 milligrams per liter.

This specific problem is confined to the San Joaquin Valley of California, but it has broad implications concerning irrigation practices and the problem of salt buildup in soils in arid and semiarid regions. The San Joaquin Valley (Figure 12.C) contains some of the most productive farmland in the United States, but the low rainfall in the region necessitates the irrigation of some 1.2 million acres. Agricultural activity began in the valley in the 1870s, but large-scale farming and irrigation did not occur until World War I. Increased production required the pumping of nearly 1 million acre-feet (1.2 billion cubic meters) of deep groundwater per year by about 1950. As a result, the water table of deep aquifers dropped by as much as 200 feet, and some parts of the land surface subsided by as much as 28 feet. Groundwater is contained within several aquifers that are separated from one another by thick, impermeable clay beds. Consequently, irrigation raised the water level in near-surface beds at the same time that it was being withdrawn from the deeper beds.

In 1967, surface water from the Sacramento–San Joaquin River system was imported through the California Aqueduct and began to replace groundwater. The imported waters drained the sediments on the western side of the San Joaquin Valley; these sediments contain higher than normal amounts of selenium and other salts. Because

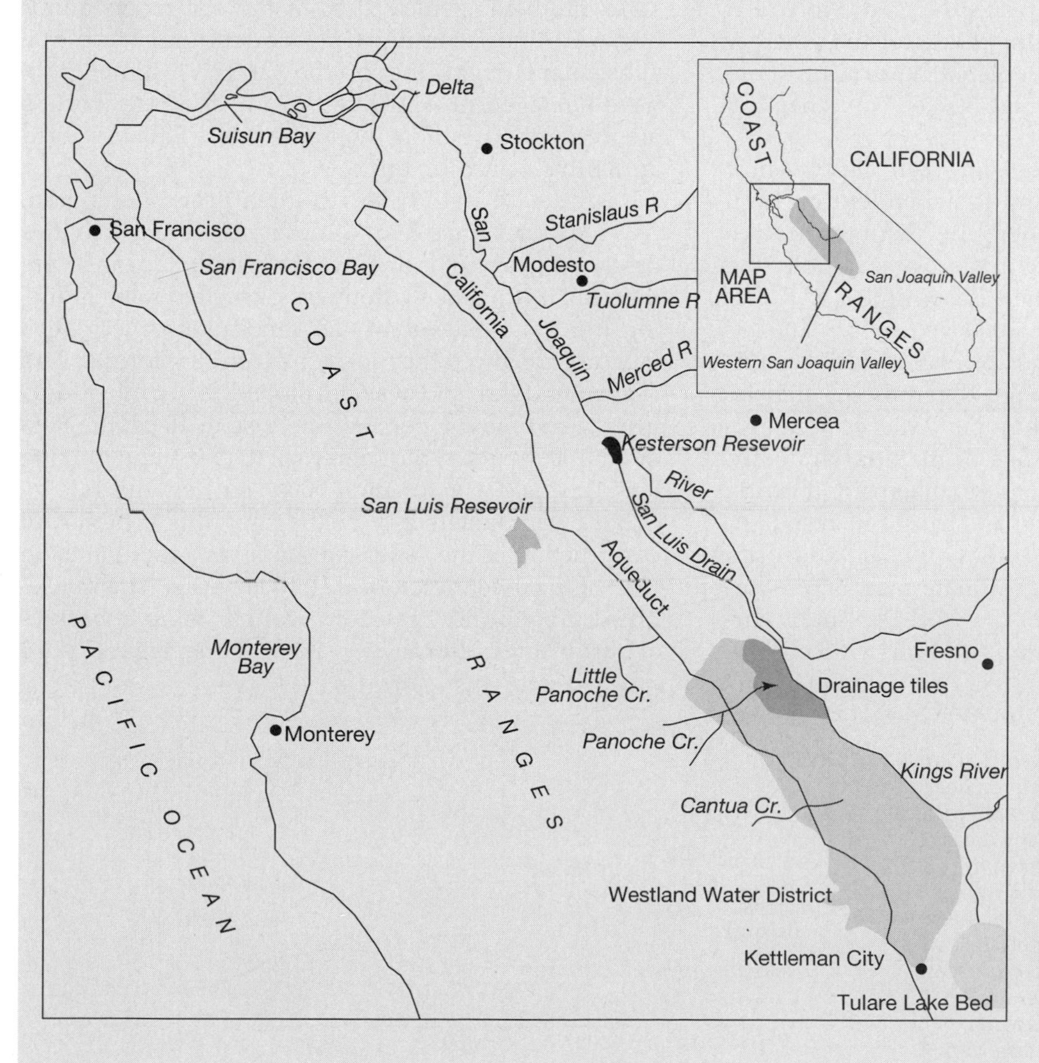

FIGURE 12.C Map of the San Joaquin Valley area of California where the concentration of selenium has created significant environmental problems for wildlife.

(continued)

BOX 12.3

Selenium Poisoning in the San Joaquin Valley *(Continued)*

of the well-known tendency for salt buildup in irrigated fields in arid regions, groundwater drainage pipes were installed in about 42,000 acres of the fields. The perforated pipes allowed excess water to drain off, thus reducing waterlogging of soils and the buildup of salts as water evaporated from the surface. The system worked well and successfully carried some 8.5 million cubic meters (6,900 acre-feet) of water annually along drainage canals into the wetlands of the Kesterton Wildlife Refuge. Unfortunately, these waters also carried dissolved salts and selenium (which is very soluble as selenate, SeO_4^{2-}) into the refuge. Constant evaporation of the water there has multiplied the concentration of selenium and resulted in its uptake by water-borne plants and by animals to levels as high as 3000 ppm. The result has been especially evident in the waterfowl, where there are high levels of mortality and birth defects.

The selenium poisoning in the San Joaquin Valley underscores how a natural pollutant can become a serious local problem due to human intervention. Similar problems have been documented in at least nine sites in eight western states and involving 1.5 million acres of farmland.

swept away by wind and rain into rivers and the sea. In fact, such erosion is the ultimate fate of all soils because it is part of the process of recycling of the material that continually renews Earth's surface. The questions that concern us are the rates at which the renewal processes are taking place, the extent to which human intervention is affecting that rate, and whether the soil resource is being depleted by erosion more rapidly than it is being regenerated by chemical weathering.

The rate of natural erosion will clearly vary from one area to another and is dependent on local geology, climate, and topography. Estimates have been made for a number of areas. For example, the Amazon River drainage basin is being lowered at a rate of 4.7 centimeters per 1000 years with the removal of 780 million metric tons of material annually; for the basin of the Congo River, the figures are approximately 2.0 centimeters per 1000 years and a loss of 133 million metric tons of material annually. Neither of these basins has, as yet, been significantly affected by human activities as far as soil resources are concerned. However, studies of this kind on a global scale have led to the estimate that, before humans appeared, the rivers of the world annually carried 9.3×10^9 metric tons of material into the oceans. After humans intervened by carrying out extensive cultivation, the amount removed is estimated to have increased by about two and one-half times to 24×10^9 metric tons per year. Studies of worldwide erosion rates reported in 1995 that 75 billion metric tons of soil were being removed from the continents annually, with most coming from agricultural land. The subsequent degrading of the land is estimated to have damaged or destroyed so much of the world's arable land that there will soon be a fall in world agricultural production.

A detailed study of one area near Washington, DC, illustrates the effects of different human activities on erosion rates. The area was originally forest (before the year 1800), and erosion was estimated to be reducing the ground level by 0.2 centimeters every 1000 years. Throughout the nineteenth century, forests were cleared and the land was developed for farming, during which time the erosion rate rose to 10 centimeters every 1000 years. A partial return to forest and grazing land in the early to mid-1900s reduced the rate to about 5 centimeters per 1000 years, until a period of construction in the 1960s caused a very rapid erosion rate of 1000 centimeters per 1000 years. The consequent urbanization has now brought the rate back to about 1 centimeter every 1000 years. Figure 12.14

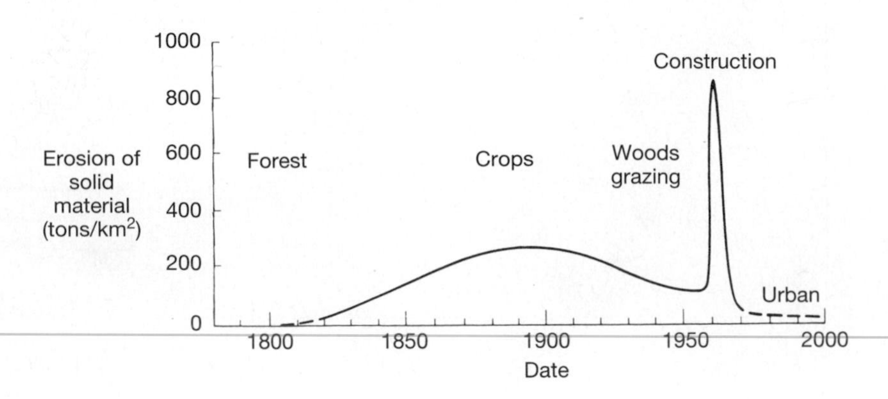

FIGURE 12.14 Variation in rate of soil erosion as a function of land usage in an area near Washington, DC, since 1800. The same general pattern has been repeated in many areas as original forest lands were converted into farmland and then into suburbs or were converted directly into suburbs. Intense erosion of soils commonly occurs during periods of development and construction but decreases to very low rates when properties and landscaping are in place. (From Judson, "The Erosion of the Land," *American Scientist*, 56 (1968), p. 356. Reprinted by permission.)

shows this sequence of erosion rates as a graph with tonnages of soil removed (per square kilometer) over the period since 1780. This study illustrates very well the pronounced effect that human activities can have on the loss of soil through erosion.

Soil Depletion, Deterioration, and Poisoning

In areas rich in natural vegetation where there is no intervention in the chemical and biological cycles involving the soil, both organic and inorganic nutrients are returned naturally to the soil when plants shed their leaves or when they die. There is little loss of nutrients unless, for climatic reasons, excess leaching occurs. However, most farmed crops are totally removed from the ground, leaving no possibility of replenishing the soil. If action is not taken to counteract this depletion, yields eventually begin to decline.

Soils may also be unable to support most or all plant growth because of the presence of certain toxic substances in relatively large amounts. An excessive amount of a substance necessary for plant growth might also prove toxic. For example, many flowering plants and trees will not grow where the soil water contains excessive amounts of sodium chloride. This is easily seen along any beach front where salt spray from the ocean severely limits the plants that can be grown. Toxic substances may be naturally present in the soils but may also be introduced as a by-product of various extraction and production industries (see Chapter 4). For example, destruction of vegetation has occurred in the vicinity of many mining areas because of the release of sulfurous fumes from smelting operations. The heavily eroded wastelands that have resulted are a stark reminder of the consequences of such abuse of the environment (see Figure 12.15).

Desertification and Deforestation

Desertification refers to the process by which fertile and watered cropland is transformed into deserts. Although prolonged droughts are locally and temporarily the cause of desertification, human activities are commonly more responsible than natural events. Overgrazing, deforestation, and poor soil management practices result in progressive and often irreversible effects that damage once productive soils, especially in areas where food supplies are already in danger. The effects occur worldwide (Figure 12.16), but are most dramatic in Africa.

Where desertification often damages soils, deforestation often totally removes them. Trees, and most other forms of vegetation, play very important roles in anchoring soils, slowing the runoff of rain, and retaining soil moisture. Stripping of the vegetation almost invariably leads to erosion, loss of soils, and reduced agricultural value of any land (see Figure 12.17). These issues have also been discussed in Chapter 11 (see p. 429 and Figure 11.39).

CONSERVATION—THE KEYWORD FOR SOIL SCIENCE

Soils are a resource upon which the ever-increasing population of the world must rely, but these soils are an unusual resource because we need to increase rather than diminish the availability of good agricultural soil along with continued use. The problem is therefore one of conservation and soil buildup, and it is a very serious problem indeed. On every continent, precious agricultural land is being lost because conservation measures are not being taken. (See Box 12.4 for a look at the problems of erosion and deforestation faced by Costa Rica.)

FIGURE 12.15 Severe erosion in the mining district of Ducktown, Tennessee, resulted initially from the cutting of trees to provide wood to roast copper ores. The open roasting of the sulfide ores released large amounts of sulfur oxides, much of which combined with rainwater to form sulfuric acid that killed the remaining vegetation. By the 1960s and 1970s, the entire Copper Basin, shown here beyond the old Burra Burra open-cut mine, was denuded, and erosion had removed nearly all of the upper soil horizons. Intense erosion of this type continues to occur in many parts of the world where natural vegetation is removed for a variety of reasons. (Photograph by J. R. Craig.)

BOX 12.4

Deforestation, Soil Erosion, and the Destruction of Environmental Assets

Future generations may regard one of the greatest disasters of the late twentieth century to have been the wholesale destruction of natural forest areas in less developed countries. Consider the example of Costa Rica in Central America. Since the end of World War II, approximately 80 percent of the forests of Costa Rica have disappeared. A particularly intense period of destruction was from the 1960s through the 1980s (see Figure 12.D); for example, between 1966 and 1989, 28 percent of the Costa Rican forests (equivalent to 2.1million acres) were destroyed despite a relatively enlightened attitude to conservation, in which a fifth of the land was set aside for national parks. Furthermore, most of the forests were simply burned to clear the land for relatively unproductive pastures and hill farms, causing the loss of tropical timber and many species of plants, animals, and insects. The largest losses were in upland and tropical wet forests and tropical moist forests, those containing the most diverse plant and animal species. Two-thirds of the deforestation affected areas in which the forest really was the most intensive sustainable use of the land. Only 14 percent of the area cleared was suitable for pasture, despite the livestock industry being a major reason for such deforestation.

Because much of Costa Rica's terrain is steeply sloping, subject to heavy rainfall, and therefore unsuitable for agriculture, the loss of forest cover led to rapid erosion. Soil erosion rates have been estimated at more than 300 tons per hectare (120 tons per acre) from land used to grow annual crops and nearly 50 tons per hectare (20 tons per acre) from pastures. In fact, between 1970 and 1989 an estimated 2.2 billion tons of soil washed away, enough to bury the Costa Rican capital city of San Jose to a depth of 12 meters (40 feet). Much of the intense flooding from the rains of Hurricane Mitch in 1998 resulted from rapid runoff from the deforested areas. The extent of deforestation per year has decreased from approximately 50,000 hectares (200,000 acres) in the 1970s to less than 5000 hectares (20,000 acres) in 2005, but enormous damage has been done over the past half-century.

This destruction of a rich and varied natural environment was undertaken in the name of economic progress. In fact, a precious natural asset was being lost, and this is rarely taken into account when politicians present their analysis of the country's economy. In Costa Rica, for example, 3.2 million cubic meters of commercial timber worth $400 million was destroyed in 1989 (equivalent to $69 per citizen of Costa Rica), and erosion of soils from farmland and pastures washed away nutrients worth 17 percent of the value of new annual crops and 14 percent of the value of livestock products. Nothing in Costa Rica's national economy reports records these losses in assets. Beyond national economic losses associated with deforestation, there are possible local and global effects on climate and on the atmosphere. Many environmental scientists believe that the cumulative effect of deforestation across the major forest areas of the world could have permanent long-term effects on climate.

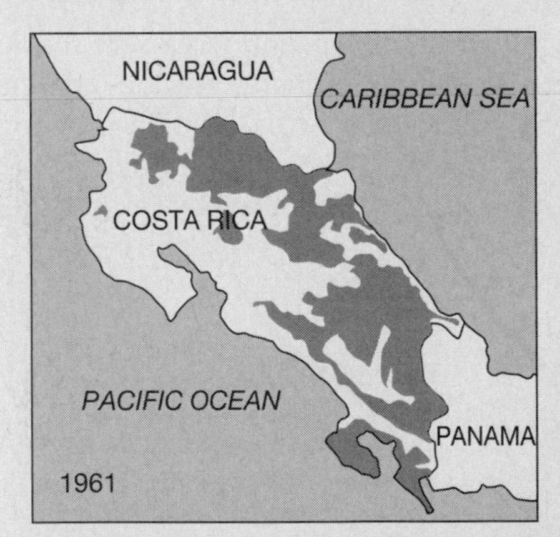

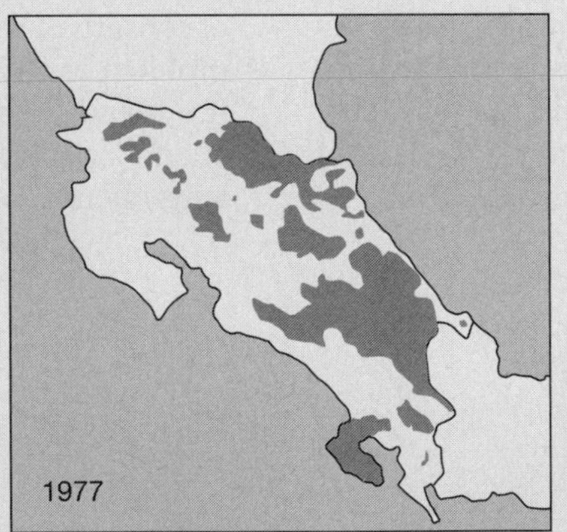

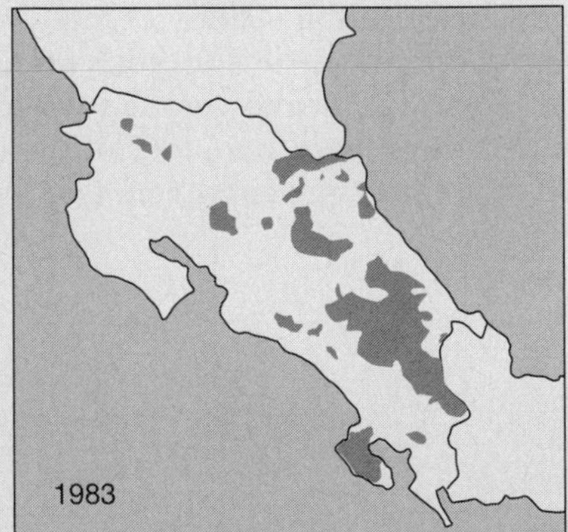

FIGURE 12.D Deforestation greatly reduced the natural forests of Costa Rica (shown as the dark areas), particularly in the 1960s to 1980s. Much of the deforestation here, as in many developing countries, was done to provide cropland for rapidly growing human populations.

Conservation is concerned with minimizing soil erosion, minimizing the loss of nutrients through leaching, preventing the buildup of excess salts or alkalis through control of drainage (see Figure 11.38), and restoring nutrients to the soil that were removed during cultivation. Certain conservation practices are well known, such as crop rotation involving the planting of a succession of different crops on the same piece of ground. The principle involved is that where a cultivated crop (e.g., potatoes, turnips) exposes the ground to maximum erosion, small-grain crops cause less exposure, and grasses protect the ground against erosion. Hence, in the northeastern United States, a common combination of rotated crops involves oats, red clover, and potatoes. Another common practice is contour plowing, where surface runoff is reduced by

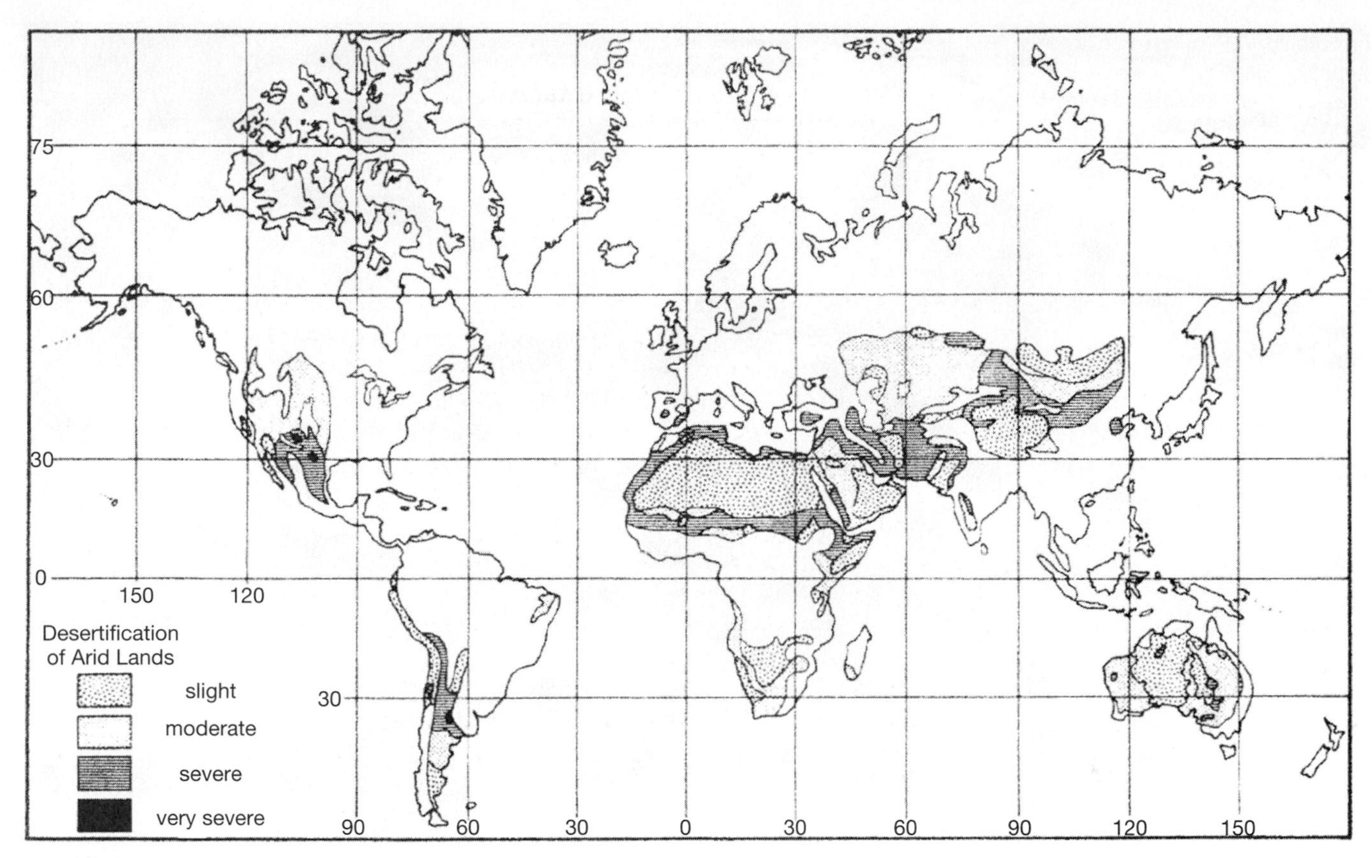

FIGURE 12.16 The desertification of arid lands around the world. The term *very severe* indicates land essentially denuded of vegetation and crop yields reduced by more than 90 percent; *severe* indicates poor range conditions and crop yields reduced by 50–90 percent; *moderate* indicates fair range conditions and crop yields reduced by 10–50 percent; *slight* indicates good range conditions and crop yields reduced by less than 10 percent. (From H. E. Dregne, *Desertification of Arid Lands,* Harwood Academic Publisher, 1983, p. 6. Used with permission.)

the furrows (see Figure 12.18). Sloping ground can also be terraced to reduce runoff and prevent gully formation, and natural channels can be controlled by damming and by ditch building. Farmland can be protected from erosion by hedges, woods, and grasses; the removal of such protection can prove disastrous because the plowed soil is very vulnerable to removal by wind and rain. In many areas, conventional plowing is giving way to seed "drilling" in which seeds are buried at the appropriate depth as a

FIGURE 12.17 Deforestation of many parts of the islands in the Caribbean has had severe effects on agricultural productivity. Goats in the Fort Charles area of Nevis have been especially effective in destroying the original vegetation, exposing topsoils to subsequent erosion. (Photograph by B. C. Richardson.)

FIGURE 12.18 Contour plowing and strip farming in Carroll County, Maryland, are effective ways to minimize the loss of soil by erosion. (Courtesy of Soil Conservation Service, U.S. Department of Agriculture.)

hole is punched or drilled. This procedure reduces the amount of energy consumed because only one pass of a tractor is needed instead of two or three, and this greatly reduces the amount of erosion. One added cost, however, that must be weighed against the benefits is the need for additional herbicides to control the weeds that are normally eliminated by being turned under by plowing.

Despite the vital importance of soil conservation, the threat to this resource is considerable. The United Nations recently reported that not only is more than one-third of Earth's land surface now desert or semidesert, but another 19 percent of the land surface spread among 150 countries is threatened. As we have already emphasized, the chief cause of this threat is people. People strip the land of trees and other cover; they overplant, overgraze, and ignore proper fertilization, irrigation, and crop rotation practices; in addition, they sell good agricultural land for urban development. A special study prepared for the United Nations recently concluded that "as a result of the unsound use of land, deserts are creeping outward in Africa, Asia, Australia, and the Americas. Worse, the productive capacity of vast dry regions in both rich and poor countries is falling."

CHAPTER 13

Future Resources

The search for resources is being carried out under increasingly difficult conditions. Drilling for oil occurs from a synthetic ice island anchored to the floor of the Arctic Ocean off the coast of Alaska. The island was constructed to prevent the drill rig from being damaged by floating ice. (Courtesy of Exxon Corporation.)

A transition is underway to a world in which human populations are more crowded, more consuming, more connected, and in many parts, more diverse, than at any time in history. Current projections envisage population reaching around 9 billion people in 2050 and leveling off at 10 to 11 billion by the end of [this century] . . . . Meeting even the most basic needs of a stabilizing population at least half again as large as today's implies greater production and consumption of goods and services, increased demand for land, energy, and materials, and intensified pressures on the environment and living resources. These challenges will be compounded to the extent that the resource-intensive, consumptive lifestyles currently enjoyed by many in the industrialized nations are retained by them and attained by the rest of humanity.

OUR COMMON JOURNEY: A TRANSITION TOWARD SUSTAINABILITY,
U.S. NATIONAL ACADEMY OF SCIENCES, 1999.

FOCAL POINTS

- Rising world population, projected to reach 9.5 billion by 2050 and 11 billion to 12 billion by 2100, technological advances, economic factors, and social pressures will determine future resource requirements.
- Gradual exhaustion of the most accessible "mineral" deposits in the most developed countries will require exploration and exploitation in more remote areas and deeper within Earth's crust.
- The exhaustion of easily accessible rich deposits, coupled with better recovery technologies, will require and permit the exploitation of lower-grade deposits.
- Fossil fuels will continue to serve as Earth's primary energy sources through the first half of the twenty-first century, but shrinking reserves, higher costs, and environmental concerns will promote the development and use of alternative energy sources such as nuclear, solar, wind, and biofuels.
- Nuclear energy will be phased out in some countries, but will be embraced as an increasingly important source of electricity in others. Because it does not contribute to atmospheric carbon dioxide, nuclear energy is viewed by many as an acceptable replacement for fossil fuel–generated electricity.
- The reserves of the abundant metals and materials derived from common rocks are so large as to be effectively inexhaustible.
- The geochemically scarce metals will continue to play vital roles in new technologies. Their reserves are adequate for at least the first half of the twenty-first century, and greater quantities will be recycled.

- Fertilizer, chemical, and building materials exist in large quantities, and reserves will be adequate for many years. Nevertheless, there are increasing concerns about secondary effects associated with their exploitation and use, such as surface runoff and water contamination. The oceans and saline lakes will serve increasingly as sources of chemical minerals.
- Population growth will require greater use of surface waters and groundwaters, and will likely cause conflicting demands for water use. There will be increasing debate about the use of dams to control sources of water and provide environmentally friendly hydroelectric power versus their potential damage to the environment caused by altering natural waterways. Desalinization of seawater and construction of water-transportation systems will increase as needs for freshwater increase.
- Technological innovation will reduce the demands for some resources much used today, while creating needs for other new or hitherto little-used resources.
- Efforts to develop "sustainability" will promote ever-greater recycling and reuse of many resources.

INTRODUCTION

The uses of renewable and nonrenewable natural resources are intimately intertwined. The rapidly growing world population, along with rising standards of living, result in demands for more and more resources. At the same time, there is a rising concern about the environmental impact of resource usage and calls for the preservation of pristine natural areas of the planet. For soil and water, the resource question is quite clear—we must learn to live with what we have. For other resources, and especially metallic ores, the case is not so clear. Our society's use of resources has developed through a combination of technological advances, economic opportunity, and social acceptance. Undoubtedly, changes will occur in the future in response to technological innovations, economic developments, or social pressures. Because we cannot predict these changes, we cannot foresee exactly how the use of resources will change. Some mineral resources may become very expensive, and others may become abundant and inexpensive. Often the best projections for the near future are based upon the consumption figures of the recent past; Figure 13.1 shows resource usage by the United States through the twentieth century and points strongly to continued increasing levels of resource demand in the twenty-first century. Recessions such as the one that began in 2007 briefly reduce the demand for some resources, but it is clear that overall demand will continue to increase through the first half of the twenty-first century.

FUTURE MINERAL RESOURCES

Important questions that arise include: "Are there enough mineral resources to meet the future needs? Will the resources be readily available and inexpensive? In exploiting and utilizing these resources, will we permanently alter the Earth's climate or pollute the Earth so much that it becomes uninhabitable?" The answer to the first question is clearly yes. The answer to the second question is more complex. Even when global supplies of materials are vast, or even inexhaustible, individual deposits are smaller and finite in size. Water and air are rapidly recycled and are constantly renewed, but most mineral resources occur in fixed amounts in any location and can become depleted. Every mine is eventually exhausted and every oil well eventually runs dry. The history of human usage of resources has been to exploit them until they are used up and then to move on to a new site or to develop some infrastructure to convey some more of the resource to where it is needed. The future will, in many ways, be the same; but what we use, how much we use, and where we use it will continually change. The answer to the third question is likely regional in scope. By 2010, the clearly measurable increase in the CO_2 content of the atmosphere appears to be causing an increase in the overall surface temperature of the Earth and altering precipitation patterns. We have also seen an increase in the number and size of areas that are so polluted that they will not sustain human inhabitants.

Mineral resources were formed through geological and geochemical cycles that commonly operated over millions of years; human exploitation of these resources follows cycles that are usually measured in tens of years (see Chapter 3, Figure 3.10). Most European countries that have been industrialized for several centuries no longer produce any of the scarce metals. Their mines are either closed or closing because the ores have been depleted. Additionally, environmental concerns and other social pressures have caused some mines to shut down prematurely or prevented exploration to discover otherwise economic deposits. The pattern observed for Europe and for the United States can be seen in other parts of the world. However, many other countries, such as Australia and Brazil, are still in the period of active exploration for mineral deposits exposed at the surface or buried by only a shallow soil cover. European nations, and increasingly the United States and Japan, have

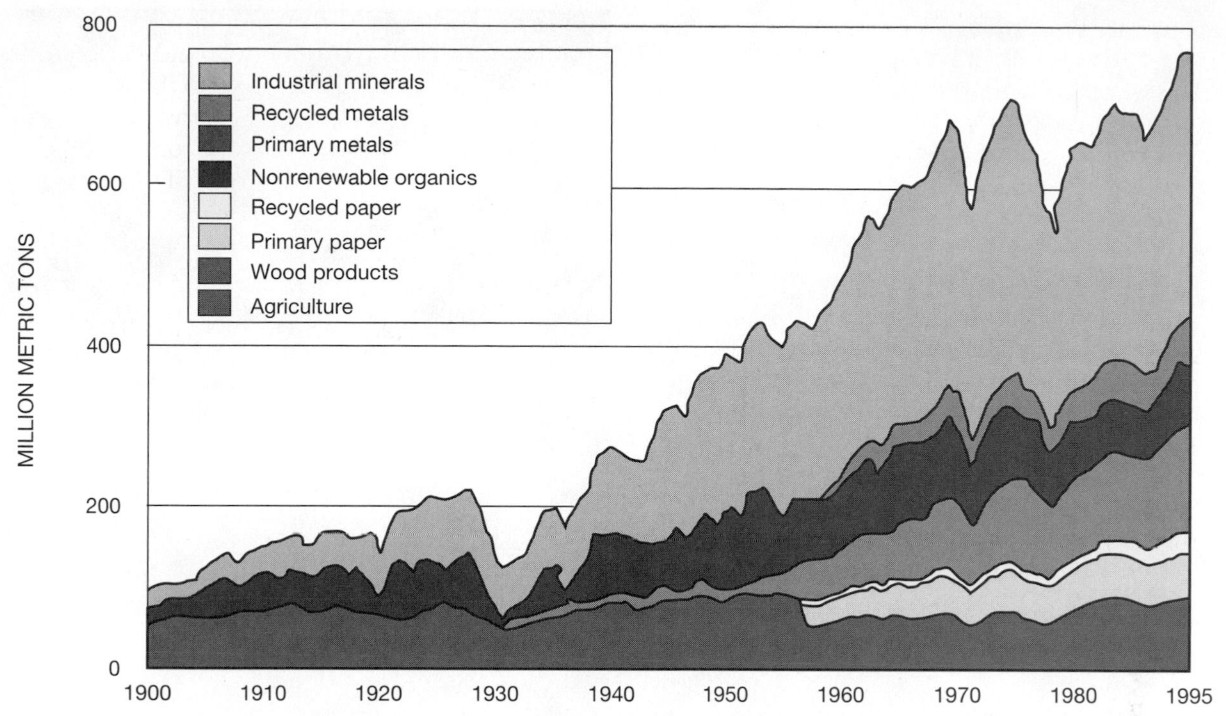

FIGURE 13.1 The quantities of raw materials consumed in the United States during the twentieth century increased more than tenfold and project to additional large increases in demand during the twenty-first century. (Most recent diagram, from the U.S. Geological Survey 2008.)

responded to a shortage of materials from their own mines by importing the balance required from countries where ore deposits are still being discovered and developed. This pattern will likely continue for the next quarter century or more because there are still large areas of the continental crust that have not been intensively prospected. The day will come, however, when all of the easily accessible deposits will have been found. Where will we turn then?

The challenge will probably be met in several ways, some of which might seem unlikely or even unreasonable by present-day standards. For example, lands previously considered off-limits, such as Antarctica, will be explored and may eventually be mined. Mining is already being carried out in Greenland and in the islands north of the Arctic Circle in Canada. This is being done under conditions of extreme difficulty (Figure 13.2). It would be a relatively small step to use similar techniques in Antarctica. Another step will surely be to explore the ocean floor. Manganese nodules are widely distributed on the deep seafloor and certain types of metal deposits form along the mid-ocean ridges at water depths in excess of 1500 meters (4500 feet). The deposits found so far are mostly small and mainly of scientific interest. Within the geological record, however, other kinds of deposits seem to have formed through the same submarine processes. This makes it likely that continuing exploration of the seafloor will reveal some of these deposits. Someday they and the manganese nodules will be mined. In a sense, we can conceive of the seafloor as another continent to explore for minerals, just like Antarctica. The difficulty of working in such inhospitable places means that we are likely to seek only the richest and largest ore bodies in such environments.

A more likely place in which to seek the ores of the future is the deeper parts of the continents. Other than the drilling of an actual test borehole, even the most sophisticated techniques used today cannot locate ore deposits beneath 500 meters (1500 feet) of barren rock. We believe that more deposits are there to be discovered because a few have been uncovered by accident or brilliant deduction; also, many deposits extend down several thousand meters in depth and are only exposed at the surface because random erosion has uncovered them. Unfortunately, that knowledge provides little help in finding deeply buried deposits (Figure 13.3). Here, obviously, is a circumstance where technological developments might be expected to play an important role.

Already, several countries—Russia, the United States, Canada, and France among them—have programs of varying intensity aimed at developing a three-dimensional picture of Earth's crust through new seismic, electrical, and magnetic exploration techniques (Figure 13.4) and drilling programs. These special programs are only the start of what someday will probably be full-scale attempts to map

(a)

(b)

FIGURE 13.2 Mining under difficult conditions. (a) The ore bodies of the Black Angel Mine in Greenland were found beneath a glacier. The mine openings were cut in a vertical cliff wall high above a fjord; all people and materials were transported by a cable system that extended across the fjord. The only water available for use in the mine is seawater, which required special corrosion-resistant equipment. (Photograph by F. M. Vokes.) (b) The Red Dog Mine, north of the Arctic Circle in Alaska, is one of the largest and richest zinc mines in the world. Mining is being carried out in open pits exposed to some of the worst weather on Earth, illustrating the need to go to regions of hostile climates to satisfy our need for metals. (Photograph courtesy of Jim Kulas, Cominco, Inc.)

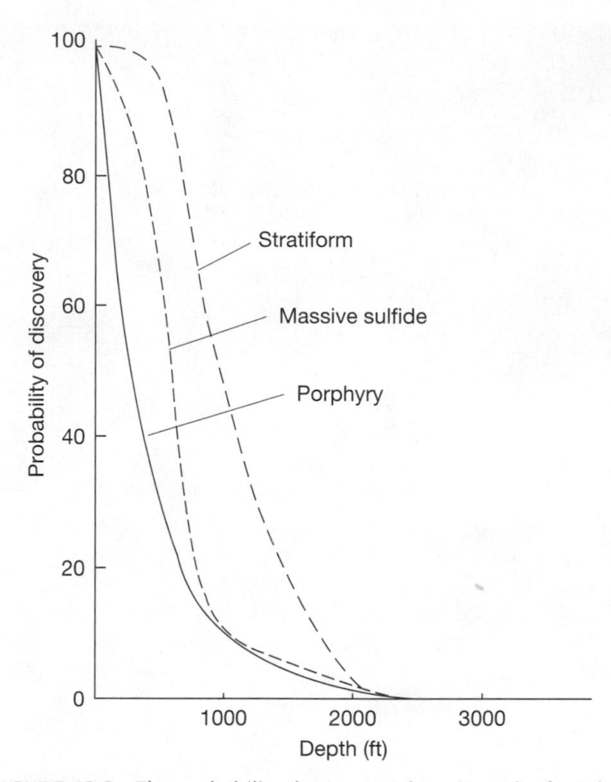

FIGURE 13.3 The probability that an ore deposit can be found decreases with depth. A probability of 100—for ore deposits outcropping at the surface—means certain discovery. A probability of 50 means that half of such deposits can be found. A stratiform deposit, because of its shape, is more likely to be found at low depth than a porphyry copper deposit of the same size.

details of Earth's crust down to 10,000 meters (30,000 feet) or more. When that has been done, a new frontier, larger than the surface of all continents, will have been opened. No one can say when, or if, we will be able to explore and mine the crust at great depths, but some believe that we might start doing so within the next twenty to thirty years. The first deep discoveries might be made in Australia and the countries of Europe and North America, where deep mapping is being carried out.

Exploitation of many deep deposits will be in truly hostile environments. The extremes of temperature and the dangers of high pressure will require some new, likely robotic, mining technologies (Figure 13.5). The great expense will limit such mining to the richest deposits and may be dependent on some significant increase in the costs of the materials recovered. Because of technological advances in mining and recovery techniques, we now mine large quantities of ores once considered as waste. This is evidenced by the decreases in the grades of copper and gold ores exploited from 1900 to 2000 (see Figures 8.2 and 8.31). Further technological advances will probably push mineable grades lower, but we must realize that we are approaching the limits where the added amount of metal recovered will no longer pay for the added cost of processing the ore. We will have reached the mineralogical barrier. To transgress that barrier and mine scarce metals

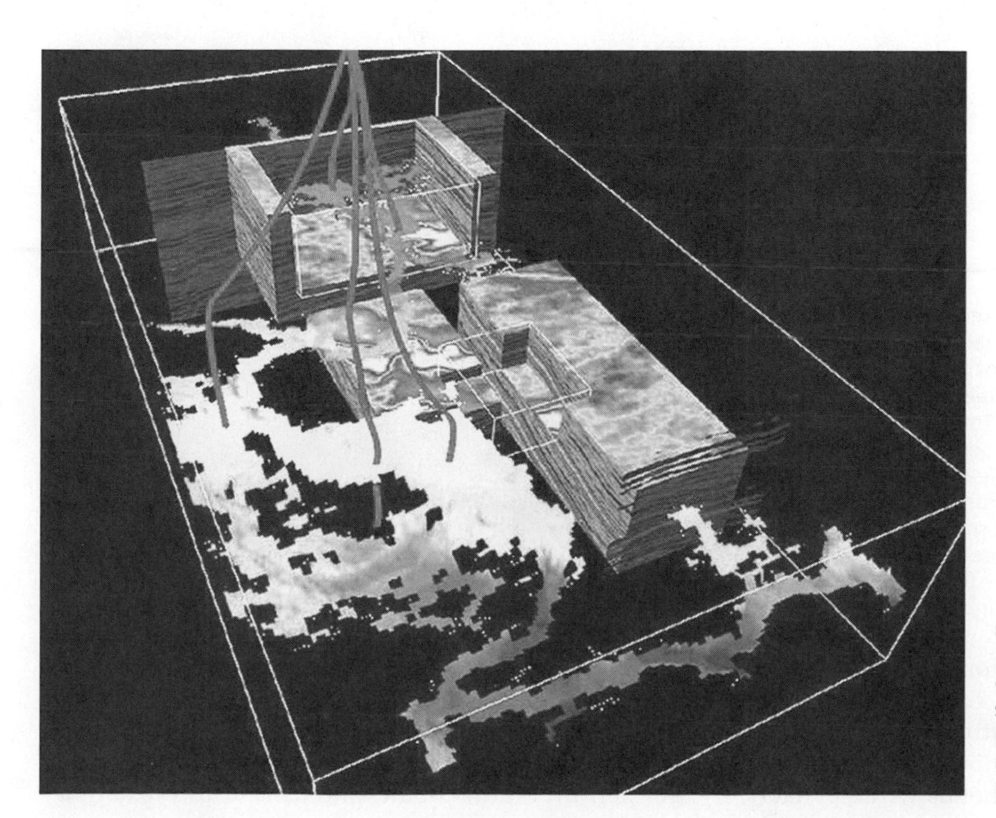

FIGURE 13.4 Modern seismic imaging techniques permit three-dimensional modeling of the subsurface, which greatly increases the probability of finding an oil-bearing reservoir or an ore body. (Courtesy of Magic Earth LLC.)

(a)

(b)

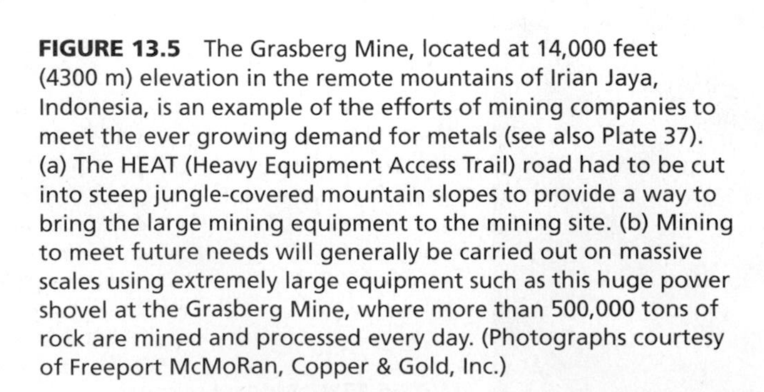

FIGURE 13.5 The Grasberg Mine, located at 14,000 feet (4300 m) elevation in the remote mountains of Irian Jaya, Indonesia, is an example of the efforts of mining companies to meet the ever growing demand for metals (see also Plate 37). (a) The HEAT (Heavy Equipment Access Trail) road had to be cut into steep jungle-covered mountain slopes to provide a way to bring the large mining equipment to the mining site. (b) Mining to meet future needs will generally be carried out on massive scales using extremely large equipment such as this huge power shovel at the Grasberg Mine, where more than 500,000 tons of rock are mined and processed every day. (Photographs courtesy of Freeport McMoRan, Copper & Gold, Inc.)

held in solid solution in common minerals will be extremely expensive (Figure 13.6). The pattern shown in Figure 13.6 is conjecture, of course, but even if it is wrong in detail, it highlights an important factor—recovery of scarce metals will inevitably be more difficult and expensive in the future. Examples of large, low-grade deposits are certain black shales that are not only rich in organic matter, but have relatively high contents of metals such as copper, uranium, cobalt, and zinc. Other examples

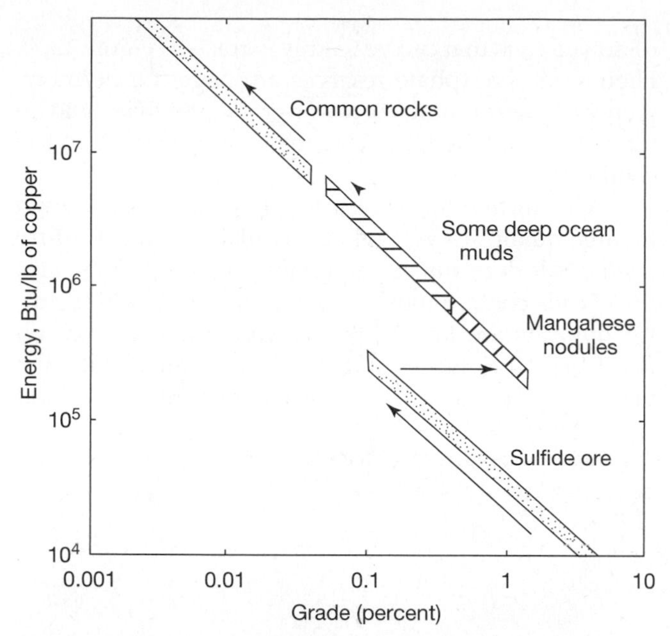

FIGURE 13.6 A hypothetical depiction of copper production in the centuries ahead. When the rich sulfide ores have been depleted, manganese nodules may be exploited. Some nodules contain more copper than certain sulfide ores, but the cost of recovery is higher for nodules. As nodules are mined, copper-rich deep-sea muds may also be worked; only as a last resort would common rocks, such as basalts, be considered as sources of metals.

include certain large igneous intrusions, such as the Duluth Gabbro in Minnesota, which contains massive tonnages of very low-grade nickel and copper resources. The large, low-grade deposits are really like most other mineral deposits, i.e., small, chemically anomalous volumes of Earth's crust.

THE ENVIRONMENT IN THE FUTURE

The large growth of human population in the twenty-first century will demand more and more living space, water, energy, and mineral resources. At the same time, the activities required to meet those needs, along with the human activities of daily life, will place greater and greater pressures on the natural environment and will likely increase pollution of the Earth's atmosphere and waters. As the human population increases, there will be a corresponding decrease in wildlife—a sort of "zero sum game" in which each new person equates to one or more fewer animals. Wildlife will increasingly be confined to game reserves and zoos.

The environmental factor of greatest general concern will likely be that of global warming. Unless there are major governmental commitments and cooperative international agreements to reduce carbon emissions, atmospheric carbon dioxide levels will continue to increase as they did during the latter parts of the twentieth century and first ten years of the twenty-first century. This will contribute significantly to the warming of Earth's atmosphere and the broad spectrum of resulting consequences from sea level rise to shifting precipitation patterns. There will be more calls for reductions of our individual "carbon footprints" and increasing conflicts between the demands for resources (especially energy) and the demands of conservationists pushing for environmental cleanup and set-aside areas for preserving wildlife.

METALS FOR THE FUTURE

Abundant Metals

Iron and aluminum have become the principal metals for the construction and transportation industries and the manufacture of many other products that are regarded as essential to our lives, and they will continue to remain so. The reserves of these metals are vast, the technologies for their extraction are well developed, their costs are low compared with other metals, and the infrastructure for their supply and recycling is in place and efficient. There are no metals that could be substituted for them in a cost-efficient manner. The other abundant metals—silicon, magnesium, manganese, titanium—also have large reserves and established use patterns that will continue. Silicon has an especially bright future because it will likely continue as the dominant material needed to manufacture the solar cells that can directly generate electricity.

Scarce Metals

The scarce metals will continue to serve in myriad uses in the twenty-first century. Some, such as copper, platinum, and beryllium, will likely expand in use associated with new electronics industry innovations, while others, especially lead and mercury, will likely see decreasing usage because of their negative environmental impact. Reserves of the scarce metals, thought to be running out twenty-five years ago, have grown to all-time high levels. This has resulted from better geologic knowledge (in part, due to theories such as plate tectonics), increased understanding of ore-forming processes, advances in the technologies of recovery, and increased recycling. The precious metals will hold their allure in jewelry and value in monetary terms, and will likely see growth in technological applications. Ferro-alloy metals will become more important as new alloys are developed. Special metals will likely see many new applications in the growing world of electronic devices.

In summary, the needs for metals will remain high, with traditional uses dominating the early parts of the twenty-first century. These uses will be accompanied by new technological applications, and perhaps partly supplanted by new technology. The strength and versatility of metals will ensure the major role of metals in society for the foreseeable future.

FERTILIZERS AND CHEMICALS FOR THE FUTURE

The growth of world population will require a large increase in the use of fertilizers to sustain both old and new types of food crops and to ensure their increased productivity. Nitrogen supplies from Earth's atmosphere are inexhaustible, but they do require significant amounts of energy (usually as natural gas) in order to be converted into usable forms. Marine potassium and phosphate deposits are extensive and sufficient to meet world needs far into the twenty-first century. Furthermore, the oceans and some saline lakes and seas contain large quantities of dissolved salts that can be readily extracted (Figure 13.7). High-grade phosphate reserves are large; furthermore, there are huge amounts of lower-grade ores that could be ultimately exploited, both on the continents and on the continental shelves.

Chemical mineral reserves, in general, are known in large quantities and will be available to meet future needs. Salt deposits are accessible in many areas, and the oceans contain inexhaustible supplies. Natural sulfur deposits are large, but our continued reliance on fossil fuels for energy and sulfide ores for metals will yield by-product sulfur in huge quantities. Beyond

(a)

(b)

FIGURE 13.7 (a) The large brine recovery plant at the Great Salt Lake. Brine evaporation ponds are visible in the background. A section of the Great Salt Lake has been isolated by dikes to permit solar evaporation to concentrate the salts for recovery. (b) Salt recovery from an evaporation site in the Great Salt Lake. Once evaporation has removed the water, the salts are scooped up and removed prior to allowing in a new flood of brine to begin the process again. (Photographs courtesy of Great Salt Lake Minerals and Chemicals Corporation.)

that, the oceans contain vast quantities of dissolved sulfur. The total magnitude of fertilizer and chemical minerals presents no problem for meeting world needs in the twenty-first century. A major concern, however, arising from the increasing use of fertilizers has been the discovery of extensive "dead zones" developing where major rivers, such as the Mississippi, carry the fertilizer nutrients into the oceans.

BUILDING AND INDUSTRIAL MATERIALS FOR THE FUTURE

Earth contains such a variety of building and industrial minerals, in such wide distribution and large quantities, that they will always be available. Local shortages can result from very high demand or from the public desire to limit the exploitation of source areas. This will happen increasingly as population centers encroach on quarries and force the quarries to relocate to more rural and distant sites. This will increase transportation costs for the products but will not seriously limit their availability. Natural materials such as dimension stone and marble will continue to compete with synthetic materials such as concrete, ceramics, glass, and plastics. Materials such as asbestos will be increasingly replaced by synthetic alternatives because of their threat to human health. Gem synthesis, already a large business, will increase, and synthetic diamonds will find new technological applications and may begin to compete with natural stones in the jewelry trade.

ENERGY FOR THE FUTURE

Fossil Fuels

There is broad agreement that the major fossil fuels—petroleum, coal, and natural gas—will continue to serve as the world's principal energy sources in the first quarter to first half of the twenty-first century. America's reliance on fossil fuels for 85 percent of its energy needs is predicted by the Energy Information Agency to drop to about 79 percent by 2030. The slowness of change results from the large infrastructure of mines, wells, pipelines, refineries, power plants, and service stations associated with the fossil fuel industries, and the billion or so gasoline or diesel vehicles in the world. At the same time, it is clear that changes, small and slow at first, are coming. Despite new exploration and recovery techniques, it is likely that world production of petroleum and natural gas will peak in the next twenty-five years. Even so, there are great concerns that the environmental impact of burning fossil fuels are irreversibly altering the carbon dioxide content of the atmosphere and accelerating global warming. Rising prices for oil, projected to reach $130 (in 2007 dollars) per barrel by 2030, combined with new technologies, will continue to change some "resources" into "reserves"; this was exemplified by the change in the status of the tar sands of western Canada in the early years of the twenty-first century. OPEC will continue to control most of the world's easily extractable petroleum, and countries such as the United States will continue to rely heavily upon imported oil.

The primary end product made from oil will likely still be gasoline for motor vehicles because it is the fuel that contains the greatest energy per unit volume and it is convenient to supply to the user. Throughout the 1990s and early years of the twenty-first century, American car manufacturers concentrated on larger, heavier, and more powerful but less fuel efficient vehicles, whereas European and Asian manufacturers built smaller, more fuel efficient vehicles. The erratic nature of the price of gasoline after 2005, with a brief rise to more than $4.00 per gallon in 2008, sent shockwaves through America and focused interest on smaller, more fuel efficient cars, including electric and hybrid models. The Energy Information Agency projects a significant increase in the numbers of flex fuel (able to use more than one kind of liquid fuel), hybrid (able to run on liquid fuel or electricity), and electric cars in the next twenty years. Biofuels production will likely increase to provide approximately 15 to 20 percent of vehicle fuel needs.

Nuclear Power for the Future

At the end of the twentieth century, the aging of America's 100 nuclear power plants and the failure of the federal government to settle on the site for a long-term, high-level waste repository, seemed to point to a gradual shutdown of nuclear power in the United States. Several countries, such as Germany and Sweden, appeared to be heading in the same direction. In contrast, other countries, such as France, Japan, and China, planned to use nuclear power as their primary electricity source.

However, the increase in environmental concerns about rising carbon dioxide levels in the atmosphere and consequent global warming, have led to a reconsideration of nuclear power as a viable major electricity source in the United States through the twenty-first century. Unfortunately, the problem of high-level nuclear waste disposal still looms, and with no resolution in sight.

Nuclear fusion, despite many clai
one day provide all of the world's en
now provides the sun's energy, is still
There is no expectation that it will bec
cially available in the first half of th
century.

RENEWABLE ENERGY FOR THE FUTURE

The "greening" of society, especially in the most developed countries, has led to increased interest in, and development of, renewable energy. This has been driven by fears over finite fossil fuel supplies, increased foreign dependency for fossil fuel supplies, erratic and sometimes high fossil fuel costs, and global warming due to rising carbon dioxide in the Earth's atmosphere. The "green" renewables—hydroelectric, solar, wind, geothermal, biofuel, ocean, and hydrogen power—are all growing on a worldwide basis, but they will not likely overtake the fossil fuels and nuclear power as the world's major energy sources until after 2025.

The "green" nature of hydroelectricity (with no gases or radiation being released) is now overshadowed in some countries (such as the United States) by environmental concerns related to the alteration of river systems and the impact on fish migration. Hence, numerous dams have now been removed from American rivers and no new ones are planned. In contrast, other countries, such as Canada, Repubic of the Congo, China, and Brazil, are in the planning or construction phases of major hydroelectric facilities that are expected to provide a major contribution to their electricity needs.

Solar energy has been increasing rapidly, but it remains a very small contributor to overall electricity supplies. Improved efficiency of solar cells and innovative construction designs that build solar cells into roofs will undoubtedly spur increased solar energy production.

Wind generation of electricity has been the most rapidly growing renewable energy source and is likely to grow even faster in the next twenty-five years. Although some large windmills have been criticized for killing migratory birds, it appears that warning systems and careful siting of the windmills (including offshore) should minimize these problems. Innovative designs continue to provide increased efficiency, but there is continuing opposition to windmill farms on the basis of aesthetics. Overall, there are optimistic projections of wind-generated electricity providing 20 percent of America's needs by 2025.

Geothermal energy will continue to grow slowly, but it will remain limited geographically. Ocean generated energy will similarly be restricted to a few coastal sites and will only grow slowly.

Biofuels, especially alcohol bio-diesel, have seen much expansion since 2000 and are expected to continue to grow. There are, however, concerns about the feedstocks cutting into world food supplies and of environmental problems. Undoubtedly, the production of biofuels will increase markedly, and is projected to reach as much as 15 percent of vehicle fuel needs by about 2030. The primary driving forces for biofuel production will be the cost of petroleum and the desire to reduce world emissions of carbon dioxide.

Hydrogen as a significant transporation fuel will remain a future prospect for many years to come. The difficulty and danger inherent in its generation, distribution, and use will restrict it to a "novelty" fuel for the near future.

Broad expansion of renewable energy sources will require the combined support of the public and of worldwide governments. The United States goverment, at one time a leader in this area, made only modest progress for the first eight years of the twenty-first century, even failing to sign on to the Kyoto Protocol (an international agreement concening the

FIGURE 13.8 The Pretonius platform, in the Gulf of Mexico, constructed between 1995 and 2000, illustrates the great efforts to which oil companies must now go to find, extract, and produce oil and gas from ever more remote sites. The platform, built by Texaco Inc. and Marathon Oil Company, is 2001 feet (610 m) tall, sits in 1724 feet (525 m) of water located 130 miles (208 km) southeast of New Orleans, and is the world's tallest free-standing structure. It has a production capacity of 60,000 barrels of oil and 100-million cubic feet of natural gas per day and is producing from a field estimated to contain 80- to 100-million barrels of oil equivalent. (Photograph ourtesy of Texaco Inc.)

limiting of emission of greenhouse gases, especially carbon dioxide). However, as we move into the second decade of the twenty-first century, there seems to be a renewed interest and support for the development of renewable energies, in participating in international energy agreements, and whatever treaties follow the Kyoto Protocol when it expires in 2012. The European Union reiterated a path to a low carbon economy in 2008 when it stated its "20-20-20" targets: 20 percent reduction of greenhouse gases and 20 percent energy produced from renewables, all by the year 2020.

WATER AND SOIL FOR THE FUTURE

Although Earth is a water planet, with 70 percent of its surface covered with oceans, the significant problem of insufficient potable water is being experienced in many places. Water is vital for life and there is no substitute; hence, we must either restrict our activities to areas with sufficient water or we must construct an infrastructure to deliver the water where we want it. Growing populations and the resulting food requirements will force us to build more transport and purification systems to population centers and to agricultural regions. Waste water recycling will become more common and irrigation systems more efficient. The increase in demand for water will stress some surface-water sources, continue to deplete groundwater supplies, and create more conflicts with environmental water rights. Water will assume a growing political importance on all scales from local to international as individuals and countries vie for surface and groundwater supplies. The oceans represent the only inexhaustible supply, and we shall inevitably expand desalinization projects, but these will require huge expenditures of energy. We shall not run out of water, if for no other reason than we cannot live without it. Water has been called "the most critical resource" of the twenty-first century.

More cropland will be required to feed the nearly doubled world population by the end of this century. As the same time, some of the best croplands have been lost to encroaching development or have suffered degradation due to erosion. New lands are constantly being developed to replace lost ones and to expand the total land available to grow crops, but these are generally of lesser quality than those lands lost and will require greater fertilization and irrigation to be as productive. Furthermore, the development of new croplands commonly leads to additional fragmentation of the world's remaining forests and grasslands and disrupts the environment for wild birds and animals.

TECHNOLOGY AND INNOVATION

Human history has witnessed increases in the number of individuals on the planet, expansion in the amounts of mineral commodities used, and social progress punctuated by the injection of new technological innovations—fire, the wheel, steam engines, electricity, nuclear power, and many others. The world's population will likely grow to at least 11 billion people during the twenty-first century and, based upon present technology, will demand nearly a doubling in the supplies of every Earth resource. That projection is very simplistic and no doubt somewhat inaccurate because we cannot easily envision the technological changes ahead. The acceleration of technological change in the past twenty-five years has surprised most of us, and imagining even more rapid change in the next twenty-five years is difficult. However, regardless of innovation, the world's demands for energy, food, water, and material goods will be staggering and will call for maximum, efficient, and innovative use of all of Earth's resources. An old but elegant example of such use was the invention of the charcoal briquette that many now use for barbecues. It resulted from the application of a simple, existing technology to a common, but incompletely used resource, by one of America's most innovative geniuses, Henry Ford.

Ford operated a saw mill in the hardwood forests around Iron Mountain, Michigan, in the years prior to 1920, to make wooden parts for his successful motor car (the Model T). Ford watched the growing piles of wood scraps with distaste and sought a way to use them. Most charcoal available up to that time was made in lumps. It was not uniform in size nor in heat output and was used primarily as an industrial fuel or for home cooking in wood-burning stoves. Ford's idea was to chip the wood into small pieces, and then, after the wood was turned into charcoal, to grind it into a powder, add a binder, and compress the mix into the now familiar pillow shape of the charcoal briquette. He called on his friend, inventor Thomas Alva Edison (Figure 13.9), to design the plant. By early 1921, Ford's plant was complete and in full operation, using every wood scrap generated and even condensing the vapors in the smoke.

Power came from a Ford-built-and-owned dam and hydroelectric facility nearby, and the wood by-products drawn off during the charring process were run through a condenser to make ketones for paints for Ford's cars and methanol for antifreeze. The briquettes were sold to industry and later to the public through his automobile agencies. E. G. Kingsford, a relative who owned one of Ford's earliest automobile sales agencies, was named manager of the briquette operation. A company town was built nearby and named Kingsford. This is the origin of the name Kingsford on one of the present-day popular brands of briquettes.

On the broader scale, the resources and the needed inventive genius are still available. Even so,

FIGURE 13.9 Henry Ford and his associates who were responsible for producing the first charcoal briquettes. From the left, Harvey Firestone, Henry Ford, Thomas Alva Edison, and E. G. Kingsford, the first manager of the briquetting plant. The development of the briquettes is an example of making maximum use of a resource. (Courtesy of the Kingsford Products Company, a subsidiary of the Clorox Company.)

the demands for, and use of, resources may still lead to dispute and conflict, as well as to disruption of the land surface and the spaces in which we live. The challenges posed by the needs and aspirations of a world population of 11 billion or more people, the uneven distribution of the natural resources required to meet their needs, and the pressures on the environment as the resources are produced and used, will make the next 100 years the most crucial and difficult ones the human race has ever faced. Every detail of our societal needs for, and uses of, resources will be examined and reexamined in the process.

Perhaps, it is appropriate to conclude this text with the reiteration of the simple statement that "the Earth is our only home." None of us writing or reading this will ever live anywhere else—we will depend upon Earth for all of our resources and need to maintain it as a suitable habitat for ourselves and our descendents.

APPENDIX: CALENDAR OF EARTH RESOURCES EVENTS

Events related to the origin, discovery, use, politics, cost, and environmental impact of Earth resources occur every day. Some events are of immediate impact (e.g., the discovery of oil or an embargo), whereas others have a delayed or more subtle impact (e.g., the birth of a great inventor or the passage of an environmental regulation). This calendar presents examples of just a few of the events in the past; new examples are occurring every day.

Jan. 1, 1901 Spindletop oil gusher ushered in the East Texas fields

Jan. 2, 1975 The first day that the American public was allowed to hold gold without a permit since 1934

Jan. 3, 1961 The only fatalities at an American nuclear plant occurred at a reactor near Idaho Falls, Idaho

Jan. 4, 1995 The Berkshire County Council, U.K., voted to permit the drilling of oil exploration well on the grounds of Windsor Castle

Jan. 5, 1993 Oil tanker *Braer* ran aground in the channel between Scotland and the Shetland Islands, spilling 25 million gallons

Jan. 6, 1974 U.S. went on daylight savings time for 2 years to conserve electricity, a response to the OPEC oil embargo

Jan. 7, 2009 The first commercial jet flight using biofuel made from algae and jatropha

Jan. 8, 1986 U.S. and Canada issued a joint report on acid rain recommending that the U.S. government help industry

Jan. 9, 1970 The market price of gold fell below the official price of $35.00 per troy ounce

Jan. 10, 1976 A gas leak caused an explosion in the basement of a hotel in Fremont, Nebraska, killing 18, injuring 50

Jan. 11, 1966 Rio de Janeiro; Brazil's heaviest rainfall resulted in flooding and landslides that killed 239

Jan. 12, 1909 A coal mine disaster at Switchback, West Virginia, killed 67

Jan. 13, 1971 U.S. Department of the Interior released a study approving construction of the Alaskan pipeline

Jan. 14, 1970 The president of General Motors predicted the production of an essentially pollution-free car by 1980

Jan. 15, 1971 The Aswan High Dam on the Nile was formally dedicated by Egyptian President Anwar Sadat

Jan. 16, 1968 ARCO announced the discovery of oil at Prudhoe Bay, Alaska

Jan. 17, 1706 Benjamin Franklin was born in Boston, Massachusetts

Jan. 18, 1980 The price of silver reached an all-time high of $50.35 per troy ounce

Jan. 19, 1813 Sir Henry Bessemer, inventor of the Bessemer steel-making process, was born in Hertfordshire, U.K.

Jan. 20, 1989 Two hundred gold miners were killed near Nazca, Peru, by explosion and fire

Jan. 21, 1980 The market price of gold reached and all-time high at $850 per troy ounce

Jan. 22, 1993 U.S. Public Health Service announced studies showing that mercury in dental amalgams is not a health hazard

Jan. 23, 1980 President Carter said that the United States was prepared to go to war to protect the oil supply routes from the Persian Gulf

Jan. 24, 1848 James Marshall found gold in the American River at Sutter's Mill; this led to the California Gold Rush

Jan. 25, 1988 Coal mine explosion and fire near Las Esperanzas, Mexico, trapped 140 miners, 34 of whom perished

Jan. 26, 1975 OPEC ministers met in Algiers to discuss how to meet and bargain with oil consumers regarding energy, raw materials, and the world economy

Jan. 27, 1880 A U.S. patent was granted to Thomas A. Edison for his "electric lamp"

Jan. 28, 1991 Iraqi troops set fire to more than 700 oil wells in Kuwait

Jan. 29, 1907 Coal mine disaster at Stuart, West Virginia, killed 84 miners

Jan. 30, 2000 Cyanide overflowed at the Baia Mare Mine in Romania into the Danube River in one of Europe's worst environmental disasters

Jan. 31, 1934 The official price of gold was raised from $20.67 to $35.00 per troy ounce

Feb. 1, 1995 Flood waters that ravaged Belgium, Germany, France, and the Netherlands began to recede and the 300,000 people displaced began to return home

Feb. 2, 1848 Mexico ceded lands of 7 western American states to the United States in the treaty of Guadalupe Hidalgo

Feb. 3, 1963 The oil tanker *Marine Sulphur* vanished off the southeast coast of the U.S. with 39 crewmen missing

Feb. 4, 1865 Nevada legislature passed an act giving 50 years of rights to build and operate tunnels to mine the Comstock silver lode

Feb. 5, 1970 Algeria announced a 4-year plan to increase oil exports by 50 percent to raise revenue to develop a more self-sufficient economy

Feb. 6, 1964 Cuba cut off water supplies to the U.S. base at Guantanamo Bay

Feb. 7, 1974 Egypt began the clearing of the Suez Canal, closed to the passage of oil tankers and freighters since the Arab–Israeli War of 1967

Feb. 8, 1973 Underground fire in West Dreifontein gold mine near Johannesburg, South Africa, killed 26

Feb. 9, 1974 As a result of gasoline shortages, several governors and U.S. congressmen call for gasoline rationing in the United States

Feb. 10, 1848 The first production of iron in the Lake Superior District by the Jackson Mining Company

Feb. 11, 1847 Thomas A. Edison born in Milan, Ohio

Feb. 12, 1983 The coal freighter *Marine Electric* sank off Chincoteague, Virginia, with 33 lost

Feb. 13, 1960 France exploded its first atomic bomb in the Sahara Desert

Feb. 14, 1971 The Persian Gulf oil-producing nations and western oil companies signed agreements of prices and marketing

Feb. 15, 1996 The oil tanker *Sea Empress* ran aground on the west coast of England, spilling 19 million gallons of crude oil

Feb. 16, 1955 General Electric announced the synthesis of diamonds

Feb. 17, 1817 A street in Baltimore became the first to be lit with natural gas supplied by America's first gas company

Feb. 18, 1965 An avalanche swept down from the Le Duc glacier near Stewart, B.C., burying a mining camp and trapping 40

Feb. 19, 2008 Petroleum closed on the world market at more than $100 per barrel for the first time

Feb. 20, 2009 Gold closed at more than $1000 per ounce for the first time

Feb. 21, 1885 The Washington Monument was dedicated

Feb. 22, 1956 President Eisenhower approved the sale of ^{235}U for peaceful atomic power production in the free world

Feb. 23, 1886 Charles M. Hall discovered the process of aluminum production

Feb. 24, 1991 United Nations forces attacked Iraqi forces in Kuwait

Feb. 25, 1977 Oil tanker *Hawaiian Patriot* burned, releasing 27 million gallons of oil

Feb. 26, 1919 U.S. Congress established the Grand Canyon National Park in Arizona

Feb. 27, 1943 U.S. Mint began production of zinc-coated steel pennies to conserve copper for the war effort

Feb. 28, 1972 British coal miners returned to work ending a seven-week strike that forced large-scale power cuts

Feb. 29, 1936 Boulder Dam on the Colorado River completed; the 221-meter (nearly 700-foot)-high dam forms 185-kilometer (115-mile)-long Lake Mead

March 1, 1977 Flooding of an anthracite mine in Pennsylvania swept away timbers, choking tunnels and killing 9 miners

March 2, 1915 A mine disaster in Layland, West Virginia, killed 112

March 3, 1847 Oil was discovered in Saudi Arabia

March 4, 1962 U.S. Atomic Energy Commission announced the operation of the first nuclear power plant in Antarctica at McMurdo Station

March 5, 2009 U.S. Department of Energy announced that Yucca Mountain, Nevada, was no longer an option for storage of U.S. high-level nuclear waste

March 6, 1626 Peter Minuit bought Manhattan Island for the Dutch from the Man-a-hat-a Indians for trinkets valued at $24

March 7, 1876 U.S. patent 174,465 granted to Alexander Graham Bell for the telephone

March 8, 1924 Mine disaster at Castle Gate, Utah, killed 171 miners

March 9, 1957 Magnitude 8.3 earthquake in the Aleutians caused more than $3 million in damage in Hawaii

March 10, 1959 President Eisenhower announced imposition of quotas on oil imports into the United States

March 11, 2000 A methane gas explosion killed 80 miners in a coal mine in the Ukraine

March 12, 2008 The U.S. EPA announced fines of $250 million to clean up asbestos at Libby, Montana

March 13, 1884 Explosion in a Pocohontas, Virginia, coal mine killed 114 and blew houses off their foundations

March 14, 1994 Fiery collision of a tanker and a freighter in the Bosphorus killed 11 and spilled 29 million gallons of oil

March 15, 1848 First printed report of the discovery of gold at Sutter's Mill appeared in the San Francisco *Californian*

March 16, 1978 Tanker *Amoco Cadiz* grounded near Portsall, France, spilling 62 million gallons of oil

March 17, 1975 U.S. Supreme Court ruled federal government, not state governments, can control continental shelf oil drilling

March 18, 1991 President Salinas ordered shutdown of the largest oil refinery in Mexico City to eliminate a source of air pollution

March 19, 1968 President Johnson signed a bill into law eliminating the necessity that 25 percent of U.S. currency be backed by gold reserves

March 20, 1973 The Shah of Iran announced the nationalization of Iran's foreign-operated oil industry

March 21, 1993 Four days of Agung volcano eruptions in the Philippines killed 1600 and destroyed 123,000 acres of cropland

March 22, 1975 A technician using a candle to check air leaks caused $100 million fire damage at Brown's Ferry Nuclear Plant, Alabama

March 23, 1989 University of Utah chemists announced discovery of "cold fusion," but experiments have never been confirmed

March 24, 1989 *Exxon Valdez* ran aground in Prince William Sound, Alaska, spilling 11 million gallons of crude oil

March 25, 1947 Coal mine disaster at Centralia, Illinois, killed 111 miners

March 26, 1999 First receipt of nuclear waste at the WIPP site in New Mexico

March 27, 1975 First section of the Trans-Alaska pipeline was laid at the Tonsina River

March 28, 1979 Meltdown at the nuclear reactor at Three Mile Island, Pennsylvania; worst U.S. nuclear accident

March 29, 1982 Dormant for centuries, El Chichon volcano in Mexico erupted, spewing 1 billion tons of ash and rock, killing 100

March 30, 1983 NYMEX introduced the concept of futures in the crude oil market for the first time

March 31, 1932 Ford Motor Company introduced the V-8 engine for automobiles

April 1, 1946 400,000 U.S. coal miners went on strike, closing most U.S. coal mines

April 2, 1792 U.S. Mint established and the minting of the gold eagle ($10) coin was authorized

April 3, 1898 Massive snowslide killed 43 near Chilkoot Pass on their way to search for Yukon and Klondike gold

April 4, 1974 Airplane crash killed 77 gold miners flying home to Malawi from the mines in South Africa

April 5, 1982 Mount Galurggung volcano in Indonesia erupted, killing 200 and destroying much cropland

April 6, 1993 Radioactive waste tank exploded and burned at the Siberian City of Tomsk-7, contaminating 2500 acres

April 7, 1869 Yellow Jacket fire at Comstock Silver Lode in Virginia City, Nevada killing 37 miners

April 8, 1952 U.S. government seized control of the nation's steel mills to avert a strike

April 9, 1968 South Africa announced that it would not sell gold in the official or free markets because prices were too low

April 10, 1973 USGS announced evidence of oil and gas 30–50 miles off the East Coast from Maine to New Jersey

April 11, 1956 President Eisenhower signed $760 million bill to build 4 dams on the Upper Colorado River

April 12, 1989 European Parliament voted in favor of U.S. 1983-type automobile emissions controls

April 13, 1979 Personnel began the final processes to permanently shut down the damaged Three Mile Island nuclear reactor

April 14, 1935 Colorado's worst "Dust Bowl" storm left drifts that stopped cars and trains

April 15, 1981 Methane gas explosion at Dutch Creek No. 1 mine in Redstone, Colorado, killed 15 miners

April 16, 1964 Formal announcement in the *Northern Miner* of the discovery hole of the Kidd Creek ore deposit at Timmins, Ontario

April 17, 2009 CO_2 declared a threat to the planet; lays the groundwork for the U.S. government to impose limits on CO_2 emissions

April 18, 1906 Magnitude 8.3 San Francisco earthquake caused more than $400 million damage and killed more than 700

April 19, 1955 Gold mine cave-in at Minas Gerais, Brazil, killed 30 miners

April 20, 1980 Iran threatened to cut off oil shipments to Japan unless Japan was willing to pay $35 per barrel; Japan refused

April 21, 1991 Coal dust explosion in a shaft killed all 147 miners in a mine in Shanxi Province, China

April 22, 1970 First Earth Day celebrated; emphasized antipollution programs and preservation of the environment

April 23, 1989 Heavy rains caused collapse of a gold mine in Burundi, killing more than 100 miners

April 24, 1970 Ships from the U.K. and Iceland rammed and fired live rounds in a dispute over fishing rights and territorial limits in the North Atlantic

April 25, 1942 World's worst mine disaster killed 1549 miners in Honkeiko coal mine in Manchuria

April 26, 1986 Explosion at the Chernobyl nuclear plant at Kiev in the Ukraine

April 27, 1978 Fall of scaffolding in a power plant cooling tower killed 51 workers at Willow Island, West Virginia

April 28, 1914 A coal mine disaster at Eccles, West Virginia, killed 181 miners

April 29, 1973 Rising floodwaters inundated more than 6 million acres of farmland south of St. Louis, Missouri

April 30, 1803 Louisiana Territory purchased by the United States from France

May 1, 1486 Christopher Columbus laid his petition before Isabella of Castile, requesting support for his voyage

May 2, 1982 Exxon announced it was terminating the Colony Shale Oil Project in Colorado after spending more than $1 billion

May 3, 1881 Fire began in two major mines of the Comstock silver lodes, Silver City, Nevada, and burned for 3 years

May 4, 1976 The Israeli government signed an agreement with a U.S. oil company to develop a new oil field in the Israeli-occupied (but Egypt-owned) Sinai Peninsula

May 5, 1964 First water flowed in the Israeli pipeline from the Sea of Galilee to irrigate the southern Negev Desert region

May 6, 1937 Airship *Hindenburg,* filled with hydrogen, exploded and burned at Lakehurst, New Jersey

May 7, 1958 The flooding of a coal mine near Nagasaki, Japan, killed 29 miners

May 8, 1979 U.S. patent issued for growth of synthetic cubic zirconia, the most widely used diamond imitation

May 9, 2008 The average price of oil on world markets reached $125 for the first time

May 10, 1869 The Central Pacific and Union Pacific railroads were joined with the golden spike at Promontory Point, Utah, providing the first cross-country railroad

May 11, 1977 A methane gas explosion in a coal mine on Hokkaido, Japan, killed 25

May 12, 1944 935 Allied planes bombed the synthetic fuel factories of Germany

May 13, 1607 Captain John Smith and 105 pilgrims in 3 ships landed in Virginia, establishing the first English settlement in the New World

May 14, 1996 The first National Windmill Day in the Netherlands

May 15, 1973 Libya, Iraq, Kuwait, Algeria halted oil flow to western countries briefly in a symbolic protest of the existence of Israel

May 16, 1971 An explosion in a coal mine near Quettz, West Pakistan, killed 32

May 17, 1985 A methane gas explosion killed 36 miners in Hokkaido, Japan

May 18, 1980 Mount St. Helens, Washington, erupted, killing 60

May 19, 2009 President Obama announced new rules to increase U.S. car mileage from about 26 mpg to 39 mpg in 2016

May 20, 1979 Swiss voters approved construction and operation of a nuclear power plant in Switzerland

May 21, 1964 The price of oil closed over $130 per barrel for the first time

May 22, 1868 "The Great Train Robbery" in Marshfield, Indiana; robbers made off with $96,000 in gold, bonds, and cash

May 23, 1966 British collier *Kaitawa* ran agound at Pandora Banks, N.Z. and sank, killing 29

May 24, 1844 Samuel Morse sent the first telegraph message, "What hath God wrought," over lines between Baltimore and Washington, DC

May 25, 1985 Tropical cyclone flooded eastern Delta region of Bangladesh, ruining crops and killing thousands

May 26, 1908 Giant oil gusher at Masjid-i-Suleima ushered in the oil prominence of Persia (present-day Iran)

May 27, 1937 Golden Gate Bridge opened at San Francisco, California

May 28, 1901 Shah Muzaffar al Din signed a 60-year lease for oil concessions in Persia with William Knox D'Arcy

May 29, 1848 San Francisco *Californian* wrote, "The rush for gold . . . the whole country resounds with the sordid cry of Gold, Gold!"

May 30, 1975 U.S. government announced that it would sell 500,000 troy ounces of gold on the open market on June 30, 1975

May 31, 1889 A 72-foot wall of water swept though Johnstown, Pennsylvania, killing 2200 and destroying the center of the city

June 1, 2008 China banned the use of plastic shopping bags because they have littered the entire country

June 2, 1973 Oil tanker *Esso Brussels* rammed by large ship while lying at anchor in New York harbor

June 3, 1979 Ixtoc 1 oil well failed and spilled more than 100 million gallons of oil into the Gulf of Mexico

June 4, 1973 The *Wall Street Journal* reported that the U.S. would build the world's largest desalinization plant to make Colorado River water usable for Mexico

June 5, 1933 U.S. went off the gold standard

June 6, 1967 Arab oil ministers call for embargo against countries friendly to Israel, one day after beginning of Arab–Israeli war

June 7, 1494 Treaty of Tordesilla signed between Portugal and Spain, dividing the lands of the New World

June 8, 2009 UN called for a worldwide ban on the manufacture and use of thin single-use plastic bags that choke marine life and pollute oceans

June 9, 1972 Fifteen Caribbean nations announced support for the setting of territorial limits at 12 miles and the establishment of 200-mile seabed natural resource limits

June 10, 1972 Flash flood damages downtown Rapid City, South Dakota

June 11, 1971 Finland became the first nation to formally sign an agreement to ensure the peaceful use of nuclear materials

June 12, 1897 Magnitude 8.7 earthquake at Cherrapunji, India, did massive damage

June 13, 1968 Hull failure of the tanker *World Glory* off South Africa spilled more than 13 million gallons of crude oil

June 14, 2000 The German government announced that all nuclear power plants will be phased out

June 15, 1752 Benjamin Franklin, with his kite experiment, proved that lightning is electricity

June 16, 2009 The U.S. Global Climate Change Research Program detailed far-reaching climate effects in the U.S., noting many are "severe, widespread, and irreversible"

June 17, 1914 Winston Churchill introduced a bill so that British government would acquire 51 percent of the Anglo-Persian Oil Company

June 18, 1971 Venezuelan Chamber of Deputies passed a bill to eventually nationalize foreign-owned oil operation

June 19, 1978 U.S. Senate voted to allow clearly justified exemptions to the Endangered Species Act

June 20, 1977 The first oil flowed into the Alaskan Pipeline at Prudhoe Bay

June 21, 1960 U.S. patent granted to General Electric Co. for synthesis of diamonds

June 22, 1969 Pollution of the Cuyahoga River in Cleveland was so bad that oil slicks and debris in the river caught fire (and this was at least the tenth time)

June 23, 2009 The U.S. Department of the Interior awarded the first five offshore wind energy leases for the construction of windmills 6 to 18 miles off the New Jersey coast

June 24, 1882 The first gold mining company, General Prospecting Co. of Burghers of the ZAR, formed in the Rand area, South Africa

June 25, 1876 George A. Custer and 225 cavalry men, protecting gold interests, killed by Indians at the Little Big Horn, Montana

June 26, 1965 Rain loosened coal cinder piles and mud buried houses and killed 24 at Kawasaki City, Japan

June 27, 1985 Explosion of gases in a sewage pipe killed 24 in Chongquing, China

June 28, 1983 A 100-foot section of Interstate 95, weakened by water flow, collapsed at Greenwich, Connecticut

June 29, 1767 British Parliament approved Townshend Revenue Act, imposing taxes on resources and materials shipped to America

June 30, 1975 U.S. government sold 500,000 troy ounces of gold at auction

July 1, 1992 European Community legislation went into effect, requiring new cars to have platinum catalytic emission converters

July 2, 1900 Count Ferdinand von Zeppelin launched his first hydrogen-filled air ship

July 3, 1990 The price of rhodium metal rose about $7000 per troy ounce

July 4, 1977 Canada approved the construction of a pipeline across its territory to transport natural gas from Alaska to the lower 48 states

July 5, 1944 A mine disaster at Belmont, Ohio, killed 66 coal miners

July 6, 1988 The Piper Alpha oil platform in the North Sea exploded, killing 167 workers

July 7, 1961 A natural gas explosion in a coal mine in Dolna Suce, Czechoslovakia, killed 108 miners

July 8, 1896 William Jennings Bryan, Democratic presidential candidate, delivered his "Cross of Gold" speech seeking restoration of bi-metal standard for U.S. currency

July 9, 1984 The largest one-day drop in the price of gold occurred—$25.50 (to that date)

July 10, 1991 President George H. W. Bush lifted U.S. economic sanctions against South Africa and allowed the importation of gold Krugerrands

July 11, 1991 ARCO announced development of a cleaner burning gasoline for use in California

July 12, 1991 A giant sinkhole, 150 feet across and 60 feet deep, swallowed a house in Florida

July 13, 1969 United Arab Republic signed an agreement to build an oil pipeline to bypass the Suez Canal, which had been closed in the 1967 war

July 14, 1967 The U.S. treasury halted sales of silver at the nominal value of $1.29 per troy ounce

July 15, 1979 President Carter outlined a 10-year national energy program to reduce U.S. dependence on foreign oil

July 16, 1955 A fire in an underground uranium mine near Aue, East Germany, killed 33 workers

July 17, 1913 Winston Churchill said, "If we cannot get oil... we cannot get a thousand and one commodities necessary for the preservation of Great Britain"

July 18, 1993 Massive flooding covered millions of acres in the Upper Mississippi Valley, displacing tens of thousands and causing millions in damages

July 19, 1985 Failure of an earthen dam at Stave, Italy, sends a wall of water and mud down an alpine valley, killing 250

July 20, 1969 The first lunar landing occurred with Neil Armstrong becoming the first human to set foot on the Moon

July 21, 1992 A Japanese research institute announced development of a pallidium-catalyst for diesel engines

July 22, 1892 The *Murex,* the first oil tanker sailed from West Hartlepool, England, to Batum on the Black Sea

July 23, 1965 U.S. Congress passed the Coinage Act, eliminating the use of silver in coins and replacing dimes and quarters with Cu-Ni clad coins

July 24, 1980 The American Petroleum Institute reported that U.S. oil imports dropped by 14 percent in the first half of 1979

July 25, 1941 U.S. government froze all Japanese financial assets preventing Japan from buying oil from the U.S.

July 26, 1956 Egypt nationalized the Suez Canal

July 27, 1866 The first undersea cable between the U.S. and Europe was completed

July 28, 1977 The first oil arrived at the *Valdez* tanker terminal, after passing the entire length of the Alaskan pipeline

July 29, 1588 The English navy defeated the Spanish Armada in the Battle of Gravelines

July 30, 1971 The Venezuelan president signed legislation to nationalize the oil industry

July 31, 1976 Twelve inches of rain fell in 6 hours, generating a 30-foot wall of water that swept down Big Thompson Canyon, Colorado, killing 130 campers

Aug. 1, 1993 The Mississippi River crested at 49.4 feet in St. Louis, Missouri, the all-time-high water level

Aug. 2, 1990 Iraq attacked Kuwait, taking control of all oil fields

Aug. 3, 1492 Christopher Columbus set sail from Palos, Spain, on his first voyage to the New World

Aug. 4, 1977 President Carter signed a measure establishing the Department of Energy

Aug. 5, 1954 International oil companies signed a 25-year production and marketing agreement with Iran

Aug. 6, 1945 80,000 to 200,000 killed when the first atomic bomb dropped on Hiroshima, Japan

Aug. 7, 1979 Highly enriched uranium was accidentally released from a top secret nuclear fuel plant at Erwin, Tennessee, contaminating up to 1000 people

Aug. 8, 1956 Fire in Casier du Bois mine at Marcinelle, Belgium, trapped 276 miners, 262 of whom died

Aug. 9, 1960 Standard Oil of New Jersey announced oil price cuts of up to $0.14/barrel, which precipitated the formation of OPEC

Aug. 10, 1993 50 tons of limestone dropped by helicopters into Friday Run in National Forest in Virginia to counteract acid rain

Aug. 11, 1807 Robert Fulton's first steamboat, *Clermont,* made a successful trial run

Aug. 12, 1976 Explosion in oil refinery tower at Chalmette, Louisiana, killed 13

Aug. 13, 1521 Spanish conquistadores conquered present-day Mexico City area from the Aztecs

Aug. 14, 2126 Swift-Tuttle comet could collide with Earth, causing much disruption of human activities

Aug. 15, 1932 Greensboro, N.C. *Daily News* reported the discovery of a 12-pound gold nugget near Charlotte, N.C.

Aug. 16, 1896 "Skookum Jim" discovered "Bonanza Creek," a rich gold-bearing tributary of the Klondike River in Canada

Aug. 17, 1959 Hebgen Lake, Montana, earthquake and landslide killed 27

Aug. 18, 1965 President Johnson declared Delaware River watershed, including New York City, a drought disaster area

Aug. 19, 1848 The discovery of gold in California reported in the *New York Herald;* the report was a leading contributor to the Gold Rush

Aug. 20, 1983 President Reagan lifted controls on the export of gas pipeline equipment to the Soviet Union

Aug. 21, 1986 1500 people and 7000 cattle killed by the sudden release of gases from Lake Nyos in a volcanic crater in Cameroon

Aug. 22, 1962 U.S. nuclear ship, *Savannah,* world's first nuclear powered cargo ship, completed its maiden voyage

Aug. 23, 1892 The *Murex,* the first oil tanker, sailed through the Suez Canal for the first time

Aug. 24, 79 A.D. The eruption of Mount Vesuvius buries Pompeii and Herculaneum in Italy

Aug. 25, 1958 Twenty-five died in New Delhi, India, from drinking contaminated water

Aug. 26, 1963 U.S. Department of the Interior proposed a 30-year $4 billion program to develop the lower Colorado River water resources

Aug. 27, 1859 The world's first oil well, drilled by Edwin L. Drake in Titusville, Pennsylvania, struck oil

Aug, 28, 1963 Explosion in the main shaft of a large potash mine in Moab, Utah, at the 2700-foot level killed 18

Aug. 29, 1974 Norway announced the discovery of a large oil and gas field in the North Sea with reserves of 2 billion barrels of oil and 50 billion cubic meters of gas

Aug. 30, 1994 Methane gas explosion in a coal mine in the southern Philippines killed miners

Aug. 31, 1987 Elevator carrying gold miners dropped to the bottom of a 4600-foot deep gold mine shaft in South Africa, killing 62

Sept. 1, 1973 Libya announced the nationalization of 51 percent of all foreign oil company operations

Sept. 2, 1977 The U.S. Energy Administration and the Canadian National Energy Board agreed to the exchange of 10,000 barrels of oil daily to help keep U.S. refineries supplied

Sept. 3, 1977 Earth tremor caused a cave-in at two South African gold mines, killing more than 20 miners

Sept. 4, 1888 George Eastman received a patent for his roll-film camera and registered his trademark, Kodak

Sept. 5, 1976 Floodwaters erode and break a 442-foot-high earth dam in Pakistan, flooding more than 5000 square miles

Sept 6, 1492 Christopher Columbus sailed west from the Canary Islands in his search for a route to the Far East

Sept. 7, 1970 Methane gas explosion blocked a coal mine entrance and trapped 34 miners at Sorrange, Pakistan

Sept. 8, 1900 More than 6000 killed when a large hurricane struck Galveston Texas in the worst loss of life in a U.S. disaster

Sept. 9, 1970 The U.S. Treasury Secretary defended the Nixon administration proposal to tax lead additives in gasolines as a way to reduce air pollution and raise revenue

Sept. 10, 1969 Oil leases to sites on Alaska's North Slope were sold in Anchorage for more than $900 million

Sept. 11, 1936 President Roosevelt dedicated Hoover Dam in Nevada by pressing a key to start hydroelectric generation

Sept. 12, 1848 The arrival of a ship carrying $2500 of California gold started a near riot in Valparaiso, Chile

Sept. 13, 1922 The highest shade temperature on Earth's surface, 136.4° F, was recorded at El Azizia, Libya

Sept. 14, 1979 The price of gold reached an unprecedented level of $345.80 per troy ounce

Sept. 15, 1977 President Carter announced a plan to set aside about one-quarter of Alaska as national parks, wilderness areas, and wildlife refuges, limiting oil exploration

Sept. 16, 1947 Reynolds introduced aluminum foil as a way to use excess aluminum after World War II

Sept. 17, 1954 Australia's first uranium plant at Rum Jungle, Northern Territory, was opened

Sept. 18, 1884 Frederick Struben exposed the gold rich "Confidence Reef" in South Africa

Sept. 19, 1844 A survey party discovered the great iron ores of the Marquette Range on the Upper Peninsula of Michigan

Sept. 20, 1519 Ferdinand Magellan set out from Spain in search of a western passage to the Spice Islands of Indonesia

Sept. 21, 1893 The Duryea brothers test drive the first gasoline-powered automobile in Springfield, Massachusetts

Sept. 22, 1898 The discovery of gold at Anvil Creek began the rush to the Nome Mining District of Alaska

Sept. 23, 1970 "Brownouts" occur along the East Coast of the U.S. as a prolonged heat wave taxed electrical power reserves

Sept. 24, 1994 A jury awarded $9.7 million to Alaskan Native Corporations as a result of damages caused by the Exxon *Valdez* oil spill

Sept. 25, 1513 Vasco Nunez de Balboa discovered the Pacific Ocean

Sept. 26, 1940 The U.S. government banned the export of all iron steel scrap to Japan to try to limit potential war buildup

Sept. 27, 1915 Explosion of a gasoline-filled railroad tank car killed 47 in Ardmore, Oklahoma

Sept. 28, 1999 Iran announced the discovery of the Azadeo oil field containing 26 billion barrels of crude reserves

Sept. 29, 1969 Heavy rains caused extensive flooding across the deserts of Algeria and Tunisia, killing hundreds and flooding phosphate mines

Sept. 30, 1882 Falling water generated the first hydroelectricity at Appleton, Wisconsin

Oct. 1, 1908 Henry Ford introduced the Model T Ford, the first mass-produced car

Oct. 2, 1979 The Argyle Diamond Pipe was discovered in western Australia

Oct. 3, 1994 U.S. Environmental Protection Agency banned the dumping of chemical wastes in the Gulf of Mexico about 230 miles south of the Florida panhandle

Oct. 4, 1955 The first solar-powered telephone call was made at Americus, Georgia

Oct. 5, 1966 A sodium cooling system leak at the Enrico Fermi demonstration nuclear reactor near Detroit caused shutdown

Oct. 6, 1848 The steam ship *California* left New York on its maiden voyage to San Francisco carrying gold seekers

Oct. 7, 1957 Fire at the Windscale plutonium production facility in England spread radiation in the surrounding countryside (a 1983 report said that 39 died)

Oct. 8, 1942 U.S. War Production Board Order L-208 went into effect, closing U.S. gold mines in an effort to boost copper production

Oct. 9, 1964 Landslide at Wan-li, Taiwan, engulfed a sulfur mine, entombing 25 miners

Oct. 10, 1913 The waters of the Pacific and Atlantic met for the first time in the Panama Canal when the Gamboa dam was blown up

Oct. 11, 1811 The first steam-powered ferryboat, the *Juliana,* was put into service between New York City and Hoboken, New Jersey

Oct. 12, 1999 The world's population was estimated to reach 6 billion individuals

Oct. 13, 1992 The U.K. government announced closure of 30 coal mines with layoffs of 30,000 miners

Oct. 14, 1913 Fire in the Mid-Glamorgan Coal Mine in Wales killed 439 miners

Oct. 15, 1927 The first commercial oil well in Iraq, Baba Gurgur No. 1, came in as a gusher flowing 95,000 barrels per day

Oct. 16, 1973 OPEC announced an immediate 70 percent increase in the price of oil

Oct. 17, 1956 The world's first full-scale nuclear power plant began generating electricity at Calder Hall on the west coast of England

Oct. 18, 1867 The United States took formal possession of Alaska after paying Russia $7.2 million

Oct. 19, 1973 Libya, an OPEC member, announced a boycott of oil sales to the United States

Oct. 20, 1973 Saudi Arabia, the largest OPEC country, announced a boycott of oil sales to the United States

Oct. 21, 1966 A huge coal waste pile failed and buried a school in Aberfan, Wales, killing 116 children and 26 adults

Oct. 22, 1913 Mine disaster at Dawson in the Northwest Territories killed 263 miners

Oct. 23, 1991 The 5th U.S. Circuit Court of Appeals overturned EPA regulations that banned most asbestos use in the U.S.

Oct. 24, 1973 Dense fog and smoke from a burning garbage dump caused a 65-vehicle pileup on a New Jersey Turnpike, killing 9

Oct. 25, 1973 The Organization for Economic Cooperation and Development met in Paris to discuss the potential of oil sharing if needed as a result of the OPEC embargo

Oct. 26, 1825 The Erie Canal was opened with a barge leaving Buffalo, New York

Oct. 27, 1979 Underground fire at the Unsong Coal Mine, Korea, killed 42 miners

Oct. 28, 1990 Heavy rains flooded the Brewer Gold Mine in South Carolina causing the release of cyanide solutions that killed more than 10,000 fish

Oct. 29, 1929 "Black Tuesday," the collapse of the New York Stock Exchange that led to the Great Depression

Oct. 30, 1973 The Netherlands imposed a ban on Sunday driving to conserve oil supplies

Oct. 31, 1916 British War Cabinet said to spare no efforts to destroy the German oil supplies in Romania

Nov. 1, 1755 Lisbon, Portugal, devastated by an earthquake and the following fire and tsunami, which killed 50,000 people

Nov. 2, 1986 A large chemical spill from a chemical plant at Schweizerhalle, Switzerland, contaminated the Rhine River

Nov. 3, 1975 A gas explosion at a coal mine in Figols, Spain, killed 27

Nov. 4, 1922 Howard Carter discovered the doorway to the burial vault of King Tutankhamun, which contained great golden archeological treasures

Nov. 5, 1969 The oil tanker, *Keo,* suffered a hull fracture and spilled 9 million gallons of oil off the coast of Massachusetts

Nov. 6, 1869 Cornelius Hendrik took a bottle of stones found by his children in the Kimberly area of South Africa to a company store and learned one was a diamond

Nov. 7, 1805 The Lewis and Clark expedition reached the Pacific Ocean at the mouth of the Columbia River

Nov. 8, 1958 The Hope diamond (45.52 ct), valued at $1.5 million, was donated to the Smithsonian Institution by Harry Winston, a New York jeweler

Nov. 9, 1963 A coal dust explosion in a mine killed 450 Japanese miners

Nov. 10, 1975 The iron ore carrier, *Edmund Fitzgerald,* with 26,000 tons of iron ore pellets, sank in 20-foot waves and 65 mph winds in Lake Superior, killing 29

Nov. 11, 2007 An oil tanker broke apart in a Black Sea storm, releasing 560,000 gallons of fuel oil and killing 30,000 birds

Nov. 12, 1970 A typhoon generated 30-foot tidal surge that killed more than 200,000 in Bangladesh, the worst human disaster in the twentieth century

Nov. 13, 1973 British Prime Minister Heath declared a national emergency and reduced England's energy supplies by 10 percent because of the OPEC embargo

Nov. 14, 1969 A mine elevator dropped 3500 feet in Salisbury, Rhodesia, killing 120

Nov. 15, 1533 Francisco Pizarro and his conquistadores rode into the Inca capital of Cuzco in what is now Peru

Nov. 16, 1973 President Nixon signed a bill authorizing construction of the Trans-Alaska oil pipeline

Nov. 17, 1869 The Suez Canal was opened, allowing ships for the first time to sail from the Red Sea into the Mediterranean without rounding the cape of Africa

Nov. 18, 1755 Salem, Massachusetts, earthquake, one of the largest ever to occur in New England

Nov. 19, 1984 An explosion at a gas storage facility at Tlalnepantla near Mexico City killed 452

Nov. 20, 1980 Lake Peigneur drained into a salt mine in the Jefferson Island Salt Dome, Louisiana, when an oil well was accidentally drilled into the mine

Nov. 21, 1978 A coal mine train derailed, going into a mine a half-mile deep, killing 7 at Doncaster, U.K.

Nov. 22, 1922 A mine disaster at Dolomite, Alabama, killed 90 coal miners

Nov. 23, 1980 4800 were killed in a series of earthquakes that devastated southern Italy

Nov. 24, 1848 A report published in the *New York Herald* stated that "California gold fever broke out in New York"

Nov. 25, 1980 The U.S. Environmental Protection Agency issued new air quality standards for National Park and Scenic areas in 36 states

Nov. 26, 1916 The Romanian government blew up their oil fields to prevent them falling into the hands of the German army

Nov. 27, 1992 Ecuador dropped out of OPEC, the first nation to do so since the formation of OPEC in 1960

Nov. 28, 1973 The U.S. Secretary of the Interior approved a commercial leasing plan for the development of oil shales on federal land in Colorado and Wyoming

Nov. 29, 1966 The iron ore carrier *Daniel J. Morrell* sank in 60 mph winds on Lake Huron with the loss of 28

Nov. 30, 1972 The Soviet Union began operation of the world's first commercial fast breeder reactor at Sherchenko, northeast of the Caspian Sea

Dec. 1, 1959 Antarctica became an international reserve, free from resource exploitation

Dec. 2, 1942 The first human-directed nuclear chain reaction was carried out at the University of Chicago

Dec. 3, 1992 The Greek tanker *Aegean Sea* ran aground at La Coruna, Spain, spilling 21.5 million gallons of oil

Dec. 4, 1970 President Nixon ordered the Interior Department to take over the states' responsibilities for oil and gas production on offshore federal lands

Dec. 5, 1848 President Polk triggered the Gold Rush of 1849 by confirming the discovery of gold in California

Dec. 6, 1884 U.S. Army Engineers completed construction of the Washington Monument

Dec. 7, 1941 The Japanese struck Pearl Harbor in Hawaii and precipitated the U.S. entry into World War II

Dec. 8, 1992 An avalanche of rain-soaked mud buried a gold mining camp at Llipi, Bolivia, killing 75 miners

Dec. 9, 1911 A coal mine disaster at Briceville, Tennessee killed 84

Dec. 10, 1993 The Princeton University fusion reactor produced 5.6 million watts of power, the greatest yield to date

Dec. 11, 1992 90-mile-per-hour winds from a severe winter storm flooded New York

Dec. 12, 1992 A 6.8 earthquake centered near Maumere, Indonesia, destroyed 80 percent of the town and generated a 25-meter high tsunami

Dec. 13, 1978 The U.S. Mint began stamping Susan B. Anthony $1 coins

Dec. 14, 1970 U.S. Secretary of State Kissinger met with King Faisal of Saudi Arabia to discuss how to get the OPEC oil embargo lifted

Dec. 15, 1917 French Prime Minister Clemenceau said gasoline was "as vital as blood in the coming battles" of World War I

Dec. 16, 1954 H. T. Hall first synthesized diamonds at General Electric using high-pressure techniques

Dec. 17, 1903 Orville and Wilbur Wright completed the first powered flight on level ground at Kitty Hawk, North Carolina

Dec. 18, 1942 The U.S. Government approved an act to produce steel cents in order to conserve copper for the war effort

Dec. 19, 1972 The tanker *Sea Star* released 32 million gallons of oil into the Gulf of Oman

Dec. 20, 1987 More than 3000 died when a passenger ship and an oil tanker collided in the Philippines

Dec. 21, 1990 The European Community Environmental ministers agreed to exhaust emissions limits for all new cars produced in their countries

Dec. 22, 2008 A dam holding 5.4 million cubic yards of coal waste and 1.1 billion gallons of tainted water broke in eastern Tennessee, blanketing 300 acres of land and spilling into the Emory River; clean up will take years and many millions of dollars

Dec. 23, 1958 A gasoline tank exploded in Brownsville, Texas, killing 4 and injuring 200

Dec. 24, 1980 The U.S. Environmental Protection Agency proposed curbs on the air emissions from diesel trucks and buses

Dec. 25, 1973 Arab petroleum producers announced that they would ease the oil embargo of western countries except for the United States and the Netherlands

Dec. 26, 1967 Oil was struck at Prudhoe Bay on the North Slope of Alaska

Dec. 27, 1975 372 miners died in a coal mine at Dhahbad, India, when explosions drained millions of gallons of water into the mine

Dec. 28, 1899 The Royal Dutch Shell Oil company struck oil in Indonesia

Dec. 29, 1973 Egypt and four oil-producing Middle Eastern nations signed a contract to construct a pipeline from the Red Sea to the Mediterranean

Dec. 30, 1853 The United States bought the 45,000 square miles of the Gadsten Purchase from Mexico

Dec. 31, 1879 Thomas A. Edison first publically demonstrated his incandescent electric light bulb

GLOSSARY

abundant metals: Metals with a geochemical abundance of at least 0.1 percent in Earth's crust.

acid mine drainage: Waters issuing from an active or abandoned mine that are made strongly acid by the decomposition of sulfide minerals, usually pyrite, FeS_2.

acid rain: Rainfall that is abnormally acid; generally attributed to the presence of nitrous and sulfur oxide pollutants in the atmosphere.

acre foot: The volume of water required to cover one acre to a depth of one foot; 325,900 gallons; 1,233,500 liters.

activated charcoal: Highly absorbent charcoal produced by heating granulated charcoal and used to absorb gases or dissolved substances, especially gold.

adit: A horizontal tunnel serving as an entrance into a mine.

age-sex pyramids: Diagrams that display the distribution of population in terms of age and sex.

aggregate: Any hard, inert construction material (e.g., sand, gravel, crushed stone) used in the preparation of concrete or as a road bed.

alchemy: The medieval science of chemistry, one objective of which was to transform base metals into gold; another was to discover a universal cure for disease and a means of indefinitely prolonging life.

Algoma-type: A type of banded iron formation whose formation can be attributed to submarine volcanic exhalation.

alkali feldspars: A series of silicate minerals involving solid solution from $KAlSi_3O_8$ (potash feldspar, orthoclase) to $NaAlSi_3O_8$ (albite).

alkalinization: The buildup of salts of calcium, sodium, and potassium in soils due to evaporation.

alloy: A substance composed of two or more metals or a metal and a nonmetal.

alluvial fan: A low gently sloping cone-like accumulation of sediment that has been deposited where a stream issues from a mountain valley onto a plain.

alluvium: Unconsolidated sediments deposited by running water.

alpha particle: A subatomic particle, having an atomic weight of 4 and a +2 charge (equivalent to a helium nucleus), released during radioactive disintegration.

alumina: An oxide of aluminum, Al_2O_3, which has numerous uses in the chemical industry and as an abrasive.

amalgamation: The formation of alloys of precious metals, generally gold, with mercury; usually done as a means of capturing small grains of the precious metal.

amorphous: A material without a regular crystal structure.

anaerobic digestion: The breakdown of organic material by microorganisms in the absence of oxygen.

andalusite: A mineral, composition Al_2SiO_5; widely used in the manufacture of ceramics and glasses.

andesite: An intermediate fine-grained igneous rock rich in plagioclase feldspar.

anhydrite: A mineral, composition $CaSO_4$; an evaporite mineral.

anion: A negatively charged atom.

annealing: The process of holding materials at high temperatures, but below their melting points, in order to change their physical properties such as brittleness and machinability by causing changes in the sizes and shapes of individual grains.

anorthosite: An igneous rock composed almost entirely of the mineral plagioclase feldspar.

anthracite: Coal of the highest rank, usually with a carbon content of 92–98 percent.

apatite: A mineral with the general formation of $Ca_5(PO_4)_3$ (F, OH, Cl), that constitutes a major source of phosphorous.

aplite: A light-colored, fine-grained igneous rock consisting largely of quartz and potassium feldspar.

aqueduct: A channel built to convey water from one place to another.

aqueous: Of, or pertaining to, water.

aquifer: A rock formation that is water-bearing.

artesian well: A well in which the water level rises above the level of the water table because it is under pressure in a confined aquifer.

asbestos: A general term applied to any of a group of fibrous silicate minerals that are widely used for industrial purposes because they are incombustible, nonconducting, and chemically resistant.

asbestosis: Chronic lung inflammation caused by prolonged inhalation of asbestos particles.

asphalt: Naturally occurring thick hydrocarbon or a similar material prepared by refining petroleum and used to bond mineral fragments in blacktop road surfaces.

asphaltene: A solid, noncrystalline black hydrocarbon residual of crude oils or other bitumen.

assay: The test of the composition of an ore or numeral, usually for gold or silver.

atmosphere: The mixture of gases that surrounds Earth; composed approximately of 79 percent nitrogen, 20 percent oxygen, 1 percent argon, and 0.03 percent carbon dioxide and variable amounts of water vapor.

atmospheric inversion: The abnormal condition in which a layer of warmer air overlies a layer of cooler air.

atomic substitution: The substitution of one element for another on the lattice sites in a crystalline solid.

attapulgite: (= palygorskite); a clay mineral of composition $(Mg,Al)_2Si_4O_{10}OH{\cdot}4H_2O$.

azonal soil: In U.S. classification systems, one of the three soil orders that lack well-developed horizons and that resemble the parent materials.

backfill: Rock debris, usually derived from mining or mineral processing, that is placed in mined-out areas of a mine.

ball clay: A light-colored, highly plastic, organic-containing refractory clay used in making ceramics; so named because of the early English practice of rolling the clay into balls approximately 35 centimeters (14 inches) in diameter for storage and shipping.

banded iron formations: The largest iron deposits. Sedimentary rocks consisting of alternating bands of iron oxide minerals, iron silicates, and silica; also called banded jaspilite and itabirite.

banded jaspilite: A synonym for banded iron formations.

barite: A mineral, composition $BaSO_4$; a heavy, soft mineral widely used in oil drilling muds and as a filler in paints, papers, and textiles.

basalt: A dark, fine-grained igneous rock composed chiefly of plagioclase feldspar, pyroxene, and olivine.

base cation: Cations such as Ca^{2+}, Mg^{2+}, K^+, Na^+.

base metal: Generally any nonprecious metal, but used today to refer to the metals such as copper, lead, zinc, mercury, and tin that are neither precious nor used as ferro-alloy metals.

batholith: A large intruded mass of igneous rock generally with a surface exposure of greater than 100 square kilometers (25 square miles) and usually composed of medium- to coarse-grained rocks.

bauxite: The principal ore of aluminum; a mixture of amorphous and crystalline hydrous aluminum oxides and hydroxides.

benches: Level, shelf-like areas in open-pit mines, where ore and waste rocks are extracted.

beneficiation: The process of producing a concentrate of valuable ore minerals through the removal of valueless gangue minerals.

bentonite: A soft, plastic, porous, light-colored rock consisting of colloidal silica and clay and that has the possibility of absorbing large quantities of water; forms as a result of the weathering of volcanic ash.

beta particle: A high-energy electron released during radioactive decay.

biochemical oxygen demand (BOD): The amount of oxygen required by microorganisms in natural waters of a river, stream, or lake.

biodegradation: The consumption and breakdown of lighter fractions of petroleum residue by bacteria.

biofuels: Fuels that generate energy either directly from organic materials or from secondary products made from biological materials.

biogenic gas: Gas formed as a result of bacterial action on organic matter.

biomass: The total amount of living organisms in a particular area, expressed in terms of weight or volume.

biosphere: The living sphere, encompassing all living species from the highest points on mountains to the deepest parts of the ocean.

biotite: Dark mica of composition $K(Mg,Fe)_3[(Al,Fe)SiO_3O_{10}](OH),F_2$.

bittern: A solution, such as seawater, that has been concentrated by evaporation until salt, sodium chloride, has begun to crystallize; bitterns typically contain high magnesium contents.

bitumen: A general term applied to dark-colored liquid to plastic hydrocarbons such as petroleum and asphalts.

bituminous coal: High-rank black coal containing 75–92 percent carbon; commonly contains several percent volatile gases.

black granite: A commercial term used to describe a variety of dark colored rocks used in construction or decoration; many are not true granites.

black smoker: A sea floor vent issuing hot fluids, which on mixing with seawater precipitate very fine-grained sulfide minerals that look like black smoke.

blast furnace: A furnace in which the combustion of the fuel is intensified by a blast of air; usually used to smelt iron.

block caving: A mining method in which a large mass (block) of ore is undermined and then fractured by blasting and allowed to collapse under its own weight. The ore is removed in a series of tunnels cut beneath the ore zone.

blowout: An oil or gas well in which very high pressures encountered during drilling were sufficient to force the drill out the top of the drill hole; this usually results in the fountaining of oil, gas, and water as a gusher.

bog iron deposits: Accumulations of soft, spongy hydrous iron oxides that form in bogs, swamps, shallow lakes, and soil zones.

boghead coal: A coal composed primarily of algal debris.

borax: A mineral, composition $Na_2B_4O_7 \cdot 10H_2O$; a light-colored compound formed during the evaporation of alkaline lakes; widely used in the preparation of soaps, glasses, ceramics, and other materials.

brass: An alloy of copper with zinc.

brazing: The process of soldering with copper and zinc alloys.

breccia: A coarse-grained clastic sedimentary rock composed of angular rock fragments set in a finer-grained matrix; also said of any type of rock that has been highly fractured by igneous or tectonic processes.

breeder reactor: A nuclear reactor that produces more fissionable material than it consumes.

brimstone: A common and commercial name for sulfur.

brine: Seawater that, due to evaporation or freezing, contains more than the usual 3.5 percent dissolved salts.

British thermal unit: *See* BTU.

bronze: An alloy of copper with tin.

Bronze Age: The period in the development of a people or region when bronze replaced stone as the material used to make tools and weapons.

brown ores: Brown-colored iron ores consisting of a mixture of amorphous and crystalline iron hydroxides.

BTU: British thermal unit, the energy required to raise 1 pound of water 1°F.

building stone: A general term applied to any massive dense rock suitable for use in construction.

by-product: Something produced in the making of something else.

cable-tool drill: A method of drilling in which the cutting bit is attached to a long steel cable. Cutting is accomplished by the bit being raised and dropped again and again.

calcining: The roasting of limestone to drive off CO_2 to make lime, CaO.

calcite: $CaCO_3$; a common mineral and the principal constituent of limestone.

caliche: A layer of calcite that forms in soils in arid and semi-arid regions as a result of the evaporation of calcium-bearing groundwaters.

calorie: A unit of energy defined as the energy required to raise the temperature of 1 gram of water 1°C.

cannel coal: A compact sapropelic coal consisting primarily of spores accumulated in stagnant water.

caprock: An impervious body of anhydrite and gypsum with minor sulfur and calcite which overlies a salt dome.

capture: The process by which an atom's nucleus absorbs a high-energy particle.

carat: A common term with two meanings. First, a standard unit, 200 milligrams, for weighing precious stones; second, a term used to define the purity or fineness of gold and meaning one twenty-fourth. Pure gold is 24 carat.

carbon cycle: The cyclical movement of carbon compounds between the biosphere, lithosphere, atmosphere, and hydrosphere.

carbon footprint: The carbon dioxide generated, and usually released into the atmosphere, by the energy consumed by a person or process on a daily or yearly basis.

carbonation: A chemical weathering process in which carbon dioxide dissolved in water converts oxides of calcium, magnesium, and iron into carbonates.

carbonatite: A carbonate rock of apparent magmatic origin; commonly a host for rare-earth elements.

cartel: A combination of independent business organizations formed to regulate production, pricing, and marketing of goods by the members.

cassiterite: A mineral, composition SnO_2; the most important ore of tin.

catagenesis: Physical and chemical changes intermediate between near-surface diagenesis and deep burial metagenesis; used especially in reference to organic matter.

catalysis: Acceleration of a chemical reaction by an element or compound that is not incorporated in the reaction products.

catalyst: A substance that accelerates a chemical reaction without remaining in the reaction products.

catalytic cracking: The use of catalysts to break heavier hydrocarbons into lighter ones.

cation: A positively charged ion.

cation exchange: The exchange of one cation for another, especially by clay minerals.

cement: A binding material. *See* portland cement.

cement rock: A limestone with a sufficient clay content that it becomes cement upon calcining.

chain reaction: Where one reaction leads on to further reactions; a controlled chain reaction (see Figure 6.3) occurs in a nuclear reactor; an uncontrolled chain reaction occurs in a nuclear weapon.

chalcopyrite: Mineral, $CuFeS_2$, which is a major source of copper.

channelization: The straightening, and sometimes deepening and lining, of a stream or river channel so that the water flows more rapidly; commonly used to alleviate flooding.

chelation: The retention of a metallic ion by two or more atoms of a single organic molecule.

chemical flooding: The injection of chemicals into an oil well to promote the release of oil trapped in the rocks.

chemical weathering: The process of weathering by which chemical reactions convert the original minerals into new mineral phases.

china clay: A commercial term for kaolin used in the manufacture of chinaware.

chlorofluoromethane: Compounds such as $CFCl_3$, used as propellants in aerosol cans, and which can damage the ozone layer in Earth's upper atmosphere.

chromite: A mineral, composition $FeCr_2O_4$; the principal ore mineral of chromium.

clastic: An adjective describing rocks or sediments composed of fragments derived from preexisting rocks.

clay: A term with two common meanings. First, a natural rock with fragments smaller in diameter than 1/256 millimeters; second, any of a group of hydrous sheet structure silicate minerals.

cleavage: The general tendency of a mineral or rock to split along natural directions of weakness.

Clinton type: A fossiliferous sedimentary iron ore rich in hematite and goethite of the Clinton or correlative formations in the Silurian sandstones of the eastern United States.

clod: A lump of soil produced by artificial breakage such as plowing. *See* ped.

coal: A combustible rock containing more than 50 percent by weight and more than 70 percent by volume of carbonaceous matter derived from accumulated plant remains.

coalification: The process by which plant material is converted into coal.

coke: A combustible material consisting of the fused ash and carbon of bituminous coal, produced by driving off the volatile matter by heating in the absence of oxygen.

col: An old British term for coal.

cold-working: Shaping of metals at room temperature by hammering or rolling; a process that hardens and strengthens the metal.

comminution: The process of crushing and grinding ores in order to break ore minerals loose from the valueless gangue minerals.

concentrate: Ground and beneficiated product that consists of one or more ore minerals that have been selectively removed from the original mixture of ore and gangue minerals.

concrete: A construction material consisting of pebbles, sand, or other fragments in a cement matrix.

conduction (of heat): The process by which heat is transferred by molecular impact without transfer of matter itself. The principal manner by which heat is transmitted through solids.

cone-of-depression: The depression in the water table that develops around a well from which water is being pumped.

confined aquifer: An aquifer bounded above and below by impermeable beds.

conglomerate: A coarse-grained clastic rock composed of coarse rounded fragments set in a finer matrix.

consumption (of water): The use of water such that it is not returned to the groundwater or surface water source from which it was drawn.

contact metamorphism: The thermal, and sometimes introduced, chemical effects occurring in a rock resulting from the intrusion of an adjacent igneous body.

continental shelf: That portion of the continental margin that lies between the shoreline and the continental slope.

convection: The movement of material, gaseous or liquid, wherein the hotter portion rises and cooler portions descend as a consequence of differences in density.

convert (metallurgical): The process of passing oxygen through a molten mass of sulfides in order to convert iron sulfides into iron oxides so that the iron may be more readily separated from the slag.

corundum: A mineral, composition Al_2O_3; the second hardest mineral after diamond. In clear, colored crystals, it is known as sapphire or ruby if blue or red, respectively.

cracking: The process by which heavier hydrocarbons are broken into lighter ones.

critical mass (in a nuclear reaction): The amount of uranium required to maintain a chain reaction in a nuclear reactor.

critical(ity) (in a nuclear reactor): The condition at which a nuclear reactor is maintained in order to sustain a chain reaction.

crosscut: Passageways of a mine that are cut perpendicular to the long dimension of the deposit.

crushed rock: Any rock material that has been crushed for use as fill, for road beds, or for construction aggregate.

crushing (of rock): The process of breaking rock into smaller fragments to facilitate the separation of the ore minerals from the gangue.

cupellation: The process of freeing silver or gold from base metals by using a small bone ash cup, called a cupel, and lead.

dam: A barrier built across a river or stream to hold back water.

decay (radioactive): Spontaneous, radioactive transformation of one nuclide to another.

deoxyribonucleic acid: DNA, the substance within the chromosomes of living cells that carries hereditary instructions and directs the production of proteins.

depletion allowance: A tax deduction on mineral resources initiated as an incentive for the producers to explore for new deposits to replace the present materials being depleted.

desalinization: *See* desalting.

desalting: Process by which dissolved salts are removed from water.

desertification: The expansion of desert-like conditions as a result of natural climatic changes or human-induced activities such as overgrazing or farming.

deuterium: An isotope of hydrogen containing one proton and one neutron in its nucleus.

diagenesis: Physical and chemical changes that occur in a sediment after deposition, and during and after lithification, but not including weathering or metamorphism.

diamond: A cubic form of carbon; the hardest mineral, widely used in jewelry and as an industrial abrasive.

diaspore: A mineral, composition AlO(OH); a light-colored compound that occurs in bauxite.

diatomite: A rock or unconsolidated earthy material composed of accumulated siliceous tests of diatoms; single-celled marine or freshwater plants.

dimension stone: Building stone that is quarried and shaped into blocks according to specifications.

diorite: Igneous rocks generally composed of amphibole, plagioclase, pyroxene, and sometimes minor amounts of quartz.

direct shipping ore: Ore of sufficiently high grade that it can be profitably shipped to a smelter without first requiring beneficiation.

dispersed sources: Multiple widely separated sources of pollutants (e.g., automobiles).

distillation (or fractionation): The process of separating crude oil into various liquids and gases of different chemical and physical properties.

dolomite: A mineral, composition $CaMg(CO_3)_2$; a common sedimentary carbonate mineral; also commonly, but incorrectly, used to refer to a rock composed of dolomite.

dolostone: A rock composed of dolomite.

doping: The process of introducing trace amounts of an element into another element or compound to produce desirable electrical or other properties.

doré bar: The mass of gold and silver bullion recovered from the refining of ores.

dredging: The excavation of ore-bearing or waste materials by a floating barge or raft equipped to bring up and process, or transport, the materials.

drifts: The passageways of a mine that are cut parallel to the long dimension of a deposit.

dry steam: Natural geothermal systems dominated by water vapor (steam) with little or no liquid in the system.

dunite: A rock composed nearly entirely of olivine.

electroplating: The process using electrical current to deposit a coating of a metal on another substance.

emery: A granular mixture of corundum and varying amounts of iron oxides (magnetite or hematite); used as an abrasive.

energy: The capacity to do work.

equivalent rainfall: The amount of water, including rainfall and runoff, to which an area of land is subjected.

erosion: Removal of material by chemical or physical processes.

eutrophication: The process by which waters become deficient in oxygen, due to an increased abundance of dissolved nutrients and decaying plant matter.

evaporation: The process by which water is converted from a liquid to a vapor.

evaporite: Sedimentary rocks that form as a result of the evaporation of saline solutions.

evapotranspiration: The transfer of water in the ground to water vapor in the atmosphere through the combined processes of evaporation and transpiration.

exchange: The process by which a mineral, especially a clay, gives up one cation bound to its lattice for another cation in solution.

exchange capacity: The quantitative ability of a mineral to exchange ions with a solution.

exfoliation: The process by which thin concentric shells or flat layers of a rock or mineral are successively broken from the outer surface of a larger rock.

face: The wall in a mine where ore is being extracted. To remove the ore, holes are drilled into the face; those are filled with explosives and detonated to break the rock so that it can be removed.

fast breeder reactor: A nuclear reactor in which fuel is made or "bred" in a blanket of ^{238}U wrapped around the core.

fast reactor: Nuclear reactor that utilizes high-velocity neutrons to maintain the chain reaction in enriched fuel rods.

fault: A surface or zone of rock fracture along which there has been displacement.

feldspar: A group of abundant rock-forming minerals of the general formula $MAl(Al,Si)_3O_8$, where M is K, Na, Ca, Ba, Rb, Sr.

feldspathoid: A group of aluminosilicate minerals of sodium, potassium, or calcium and having too little silica to form feldspar.

ferric: Referring to the oxidized form of iron, Fe^{3+}.

ferro-alloy metal: Any metal that can be alloyed with iron to produce a metal with special properties.

ferromagnesian mineral: Iron- and magnesium-containing minerals.

ferromanganese: An alloy of iron and manganese used in iron smelting.

ferromanganese nodules: Rounded, concentrically laminated masses of iron and manganese oxides and hydroxides that form on the floors of oceans and some lakes.

ferrosilicon: A synthetic phase FeSi used in the steel industry as a means of removing oxygen from iron and steel during smelting.

ferrous: Referring to the reduced form of iron, Fe^{2+}.

fertilizer: Natural or synthetic substances used to promote plant growth.

filler: A mineral substance added to a product to increase the bulk or weight, to dilute expensive materials, or to improve the product.

fire clay: A siliceous clay, rich in hydrous aluminum silicates, capable of withstanding high temperatures without deforming; hence, used in the manufacture of refractory cements.

fission (nuclear): The process by which a heavy nuclide is split into two or more lighter nuclides by the addition of a neutron to the nucleus.

fissure: A surface or fracture in rock along which there has been distinct separation.

flint: A dense, fine-grained form of silica, SiO_2, that was commonly used in the making of stone tools and weapons.

fluid inclusion: Small droplet of fluid trapped within a crystal during initial growth or during recrystallization.

fluorite: A mineral, composition CaF_2; a common and variably colored substance widely used in the preparation of glasses, the manufacture of hydrofluoric acid, and the smelting of aluminum.

fluorspar: An alternate name for fluorite.

flux: Any substance that serves to promote a chemical reaction; also, the number of radioactive particles in a given volume of space multiplied by their mean velocity.

fly ash: Fine particulates that are formed during the burning of fossil fuels, especially coal.

forsterite: A mineral, composition Mg_2SiO_4; a member of the olivine series of minerals.

fossil fuel: A general term for any hydrocarbon deposit that may be used for fuel—petroleum, natural gas, coal, tar, or oil shale.

fractional crystallization: Crystallization process in igneous rocks that results in progressive changes in the composition of the remaining magma.

fractionation: *See* distillation.

Frasch process: A method of sulfur mining in which superheated water is forced down a well to melt sulfur that is then pumped to the surface for recovery.

fuel element (or fuel rod): The long rod-like assemblies that contain the U_3O_8 pellets used as fuel in a nuclear fission reactor.

fulgerite: Sand fused by a lightning strike.

fuller's earth: A fine-grained earthy substance (usually a clay) possessing a high-absorptive capacity; originally used in fulling woolen fabrics, the shrinking and thickening by application of moisture.

fulling: The process of removing grease from organic fibers.

fusion (nuclear): The combination of two light nuclei to form a heavier nucleus; a reaction accompanied by the release of large amounts of energy.

gabbro: A dark-colored, coarse-grained igneous rock composed primarily of plagioclase feldspar, pyroxene, and olivine.

galena: A mineral, PbS, that serves as the major source of lead.

galvanizing: The coating of zinc placed on iron or steel to prevent rusting.

gangue: A general term for the nonuseful minerals and rocks intermixed with valuable ore minerals.

garnet: A group of minerals of general formula $A_3B_2(SiO_4)_3$, where A = Ca, Mg, Fe^{2+}, and Mn^{2+} and B = Al, Fe^{3+}, Mn^{3+}, and Cr.

garnierite: Low grade nickel ore formed by the weathering of nickel-bearing igneous rocks.

gasohol: A mixture of gasoline and alcohol used as a fuel for automobiles.

geochemical balance: The distribution of chemical elements and chemical compounds among various types of rocks, waters, and the atmosphere.

geochemical cycling: The cyclical movement of chemical elements through Earth's lithosphere, hydrosphere, and atmosphere.

geopressured zone: A rock unit in which the fluid pressure is greater than that of normal hydrostatic pressure.

geothermal energy: Useful heat energy that can be extracted from naturally occurring steam, or hot rocks or waters.

geothermal field: An area where there is the development, or potential development, of geothermal energy.

geothermal gradient: The rate of increase in temperature in the earth as a function of depth; the average is 25°C per kilometer.

geyser: A natural hot spring that intermittently ejects water or steam.

glass: Solid supercooled liquid that does not contain a regular structure.

Global 2000 Report to the President: A large report on the status of the world's resources, population, and environment from 1975 to 2000 prepared for President Carter in 1980.

gneiss: A coarse-grained, layered metamorphic rock.

goethite: The hydrated ferric oxide mineral, FeO·OH.

gossan: The iron oxides and hydroxides that form when iron sulfides are exposed to weathering at or near Earth's surface.

grade: The content of a metal or a mineral in a rock; usually expressed as a percentage by weight for most ores.

granite: A coarse-grained igneous rock consisting mainly of quartz and potassium feldspar, usually accompanied by mica, either muscovite or biotite.

granodiorite: A coarse-grained igneous rock consisting mainly of quartz, potassium feldspar, plagioclase, and biotite.

graphite: A mineral, composition C; a soft, black compound with a pronounced cleavage, widely used as a lubricant.

gravel: A general term for both naturally occurring and artificial ground rock particles in the size range 2–20 millimeters in diameter.

greenhouse effect: The warming of Earth's atmosphere brought about by an increase in the CO_2 content.

greensand: A foundry sand consisting of a mixture of silica, clay, and water.

guano: Accumulated bird or bat excrement; mined locally as a source of fertilizer.

gusher: An oil or gas well in which the high pressures encountered during drilling are sufficient to cause fountaining of the oil, gas, and accompanying water at the surface.

gypsum: A mineral, composition $CaSO_4{\cdot}2H_2O$; formed by evaporation of seawater and used to make plaster of Paris.

Haber-Bosch process: A process perfected in Germany in the early 1900s in which nitrogen from Earth's atmosphere is fixed into ammonia so that it can be used in fertilizers and chemicals.

half-life (of an isotope): The time required for half of the quantity of a naturally radioactive isotope to decay to a daughter product.

halite: A mineral, NaCl, and the most abundant material dissolved in seawater.

heap leaching: A process by which a solvent, such as a cyanide solution or an acid, is allowed to percolate through a pile (or heap) of crushed rock to dissolve out a valuable mineral resource (such as gold or copper).

hectorite: A calcium, sodium, magnesium, lithium clay mineral formed as a result of the weathering of volcanic rock.

heliostat: An assemblage of mirrors programmed to automatically track the Sun in order to constantly focus the sunlight on a central receiver.

hematite: A mineral, composition Fe_2O_3; an important ore mineral used as a source of iron, as a polishing powder, and a cosmetic (rouge).

high-level waste: Radioactive waste that is more than one million times more radioactive than the level considered environmentally acceptable.

high-quality energy: Energy derived from solar radiation where the temperature is more than 100°C.

homogeneous reactor: A nuclear reactor in which the fuel and moderator are intimately mixed.

horsepower: A unit for measuring power, originally derived from the pulling power of a horse. The rate at which energy must be expended in order to raise 55 pounds at a rate of one foot per second.

hot dry rock: Potential geothermal area in which rocks near Earth's surface are hot enough to yield useful heat energy, but where there are no natural fluids to bring the heat to the surface.

humic coals: Coal derived from peat by the breakdown of plant matter by organic acids.

humus: The generally dark, more or less stable part of the organic matter of the soil so well decomposed that the original sources cannot be identified.

hydration: The process by which water is chemically bound in a chemical compound.

hydraulic mining: The use of high-pressure jets of water to dislodge unconsolidated rock or sediment so that it can be processed.

hydroelectricity: Electricity generated by water-driven turbines.

hydrogenation: A chemical process in which hydrogen is added to complex hydrocarbons to yield less complex molecules that have a higher H to C ratio.

hydrograph: A diagram recording the relationship between time and the quantity of water leaving a drainage basin.

hydrolysis: A chemical process by which a compound incorporates water into its structure.

hydrometallurgy: Process that makes use of aqueous solutions to extract metals.

hydro-mulching: The application of a soil covering to prevent evaporation and erosion by means of a high-pressure hose.

hydro-seeding: The application of seed to barren soil surfaces by means of a high-pressure hose.

hydrosphere: The waters of Earth.

hydrothermal alteration: Mineralogic changes in rocks resulting from interactions with hydrothermal solutions.

hydrothermal (fluids) solutions: Hot, aqueous solutions, some of which transport and deposit ore minerals.

hydrothermal vein/deposit: Mineralized zone or ore that was precipitated in a fracture or fault by hot water solutions.

igneous: A term applied to a rock or mineral that has solidified from magma.

ilmenite: A mineral, composition $FeTiO_3$; a principal ore mineral of titanium.

impermeable: Referring to a rock, sediment, or soil that does not permit the passage of fluids.

inclines (in mines): Drifts or shafts in mines that are at an angle to the horizontal.

index of weathering: Ratio of an element in a secondary mineral to the total amount of that element in a soil.

industrial mineral: Any rock, mineral, or other naturally occurring substance of economic value, exclusive of metallic ores, mineral fuels, and gem stones.

inertial confinement: A means of confining the plasma in a fusion nuclear reactor.

ingot: A mass of cast metal as it comes from a mold or a crucible.

in situ leaching: The extraction of metals or salts by passing solutions through rocks that have been fractured but not excavated.

intermediate level waste: Radioactive waste that is between 1000 and 1 million times more radioactive than is considered environmentally acceptable.

intrazonal soil: One of the soil orders. All soils with more or less well-developed soil characteristics reflecting the dominant influence of relief, parent rock, or age over that of climate.

ion: Any charged atom.

ion exchange: The reversible replacement of certain ions by others, without change in the crystal structure.

Iron Age: The period that began about 1100 B.C. with the widespread use of iron for tools and weapons. It followed the Bronze Age and, in a sense, continues today.

ironstones: Sedimentary rocks of large lateral extent that contain significant amounts of iron oxides, hydroxides, and silicates as coatings on, and replacements of, sedimentary mineral fragments and fossils.

irrigation: Process of supplying water to the land to promote the growth of crops.

isotopes: Species of the same chemical element having the same number of protons but differing numbers of neutrons in the nucleus.

itabirite: A metamorphosed banded iron formation consisting of thin bands of hematite and silica.

joule: A unit of energy equal to 0.24 calorie, or the flow of one ampere of electrical energy for one second at a potential of one volt.

kaolinite: A mineral, composition $Al_2Si_2O_5(OH)_4$; a common, light-colored clay mineral.

kerogen: Fossilized, insoluble organic material found in sedimentary rocks. Can be converted by distillation to petroleum products.

kiln: An oven used to harden, burn, or dry substances; especially to convert clay products into ceramics.

kilowatt hour: A unit of electrical power consumption indicating the total energy developed by a power of one kilowatt acting for one hour.

kimberlite: The rock type in which diamonds occur; a porphyritic alkalic peridotite containing olivine, mica, and chromium-garnet.

kyanite: A mineral, composition $Al_2Si_2O_5$; a compound found in certain metamorphic rocks and used in the manufacture of ceramics and glass.

lag-time diagram: A diagram that illustrates the relationship between rainfall and surface runoff of a drainage basin in terms of time.

Lake Superior type: Banded iron ores of the type found and long mined in the Lake Superior district of North America.

laterite: A highly leached soil zone in tropical regions that is rich in iron oxides.

leaching: Removal of chemical elements or phases by fluids.

leach pad: The impermeable layer of material placed beneath a heap leach pile to allow for the collection of metal-bearing fluids.

leachate: A watery solution that has drained out of a landfill or a heap leach pile.

leucoxene: A general term for fine-grained alteration products of ilmenite, $FeTiO_3$.

level (in a mine): A main underground passageway leading out from a shaft that provides access to mine workings.

liberation (of minerals): Valuable mineral particles freed from valueless gangue.

lightweight aggregate: Aggregate of appreciably lower specific gravity than normal rock or aggregate; prepared by using very lightweight clays or porous materials.

lime: The compound CaO; usually prepared by calcining limestone.

limestone: A bedded sedimentary rock composed largely of the mineral calcite, $CaCO_3$.

limonite: A general term for amorphous brown, naturally occurring hydrous ferric oxides with a general composition of approximately $2Fe_2O_3 \cdot 3H_2O$.

lipids: Fats or fatty oils.

liquid immiscibility: The inability of two liquids to mix and form a single, homogeneous liquid. Oil and water are immiscible.

lithosphere: The rocks forming the surface of Earth to a depth of about 60 kilometers and that behave as rigid plates.

loam: A rich, permeable soil composed of a friable mixture of roughly equal proportions of clay, silt, and sand and naturally containing organic matter.

lode: Metal-bearing deposits that are mined from their original encompassing rocks—as opposed to placer deposits in which the metals have weathered out of the original rock.

low-level waste: Radioactive waste that is up to 1000 times more radioactive than is considered environmentally acceptable.

low-quality energy: Energy derived from solar radiation where the temperature is less than 100°C.

macadam road: A road made by the addition of successive layers of finer and finer pulverized rock. Named for the developer of the process, John McAdam of Scotland.

macerals: The organic components of coal. Macerals are to coal what minerals are to a rock.

mafic: A term applied to igneous rocks composed primarily of one or more ferromagnesian minerals (most mafic rocks are also basic, i.e., having SiO_2 contents less than 54 percent).

magma: Molten igneous rock that lies beneath Earth's surface.

magmatic differentiation: The changes that occur in the composition of a magma during processes of crystallization.

magnesia: The compound MgO; the rare mineral periclase has this composition. MgO is widely used in refractories.

magnesite: A mineral, composition $MgCO_3$; an ore mineral of magnesium and of the raw material used to produce MgO.

magnetite: A mineral, composition Fe_3O_4; an important ore of iron.

malleability: The property of a metal that allows it to be plastically deformed under compressive stress, such as hammering.

manganese nodules: *See* ferro-manganese nodules.

mantle: That zone of Earth that lies below the crust and above the core [from approximately 10–30 kilometers (6–20 miles) to 3480 kilometers (2200 miles) deep].

marble: A coarse-grained rock composed of calcite; usually formed by metamorphism of a limestone.

marginal reserve: That part of the reserve base of a mineral resource that borders on being economically producible.

marsh gas: *See* swamp gas.

matte: A mixture of metal sulfides and oxides produced by melting ore mineral concentrates.

metagenesis: Physical and chemical changes that occur in response to the high temperatures and pressures of deep burial; used especially in reference to organic material.

metal: An element or alloy possessing high electrical and/or thermal conductivity that is malleable and ductile.

metamorphic: Pertaining to rocks in which the minerals have undergone chemical and structural changes due to changes in temperature and pressure.

metamorphism: The mineralogical and structural changes of solid rocks in response to the changes in temperature and pressure resulting from burial or adjacent igneous intrusion.

metasomatism: Change in the character of a rock as in metamorphism, but when chemical constituents are added or removed in the process.

meteoric water: Rainwater.

methane gas: A colorless, odorless, flammable gas, CH_4; the principal constituent of natural gas.

mica: A group of sheet silicate minerals with a general formula of $(K,Na,Ca)\ (Mg,Fe,Li,Al)_{2\text{-}3}(Al,Si)_4O_{10}(OH,F)_2$.

milling: The crushing and grinding of ores so that the useful materials can be separated from gangue materials.

mineral resource: The sum of a group of valuable minerals in a given volume of crust.

Minette type: A variety of sedimentary iron ore. The European equivalent of the North American Clinton-type ores.

mining: The process of extracting mineral substances from Earth, usually by digging holes or shafts.

Mississippi Valley type: A term applied to a class of mineral deposits widespread in the drainage basin of the Mississippi River. Zinc and/or lead sulfide ores that occur in carbonate rocks.

moderator (in a nuclear reactor): The medium, such as graphite, that moderates the flux of neutrons produced during radioactive decay.

Mohs scale: A standard of 10 minerals by which the relative hardness of a mineral can be rated. From softest to hardest, they are: talc, gypsum, calcite, fluorite, apatite, orthoclase, quartz, topaz, corundum, and diamond.

molybdenite: MoS_2, the principal ore mineral of molybdenum.

monazite: A rare-earth phosphate mineral, $(Ce,La,Nd,Th)(PO_4,SiO_4)$.

montmorillonite: A common clay mineral with the general formula, $R_{0.33}Al_2Si_4O_{10}(OH)_2 \cdot nH_2O$ where R is Na^+, K^+, Mg^{2+}, Ca^{2+}.

mountaintop mining: Coal mining process in which rocks are removed from mountaintops and dumped into ravines to expose coal beds.

mullite: $Al_6Si_2O_{13}$; a rare mineral, but common synthetic material in ceramic products.

muscovite mica: White mica of composition $KAl_2[AlSi_3O_{10}](OH)_2$.

muskeg: A bog with deep accumulations of organic material forming in poorly drained areas in northern temperate or arctic regions.

natural cement: Limestones that contain impurities of alumina and silica such that when they are calcined, they form cements when water is added.

natural gas: A mixture of hydrocarbon gases, principally methane.

nepheline syenite: An igneous rock composed essentially of plagioclase feldspar and the feldspathoid mineral nepheline, $(Na,K)AlSiO_4$.

niter: Naturally occurring potassium nitrate; saltpeter.

nitrogen: Chemical element number 7; 78 percent of Earth's atmosphere and one of the most important fertilizer elements.

noble metal: A metal with marked resistance to chemical reaction; a term often applied to gold, silver, mercury, and the platinum metal group; synonymous with precious metal.

nonferrous metal: A general term referring to metals that are not normally alloyed with iron.

nonmetallic minerals: A broadly used term for minerals that are extracted other than for use of the metals they contain or for use as fuel.

nonpoint sources: Sources of pollution that are dispersed such as farm fields, road surfaces, etc.

nonrenewable resources: Resources that are fixed in total quantity in Earth's crust.

norite: A coarse-grained igneous rock composed of plagioclase and an orthopyroxene.

nuclear fission: The breakdown of a large nucleus (e.g., of uranium) to smaller nuclei with the emission of large amounts of energy.

nuclear fuel cycle: The steps involved in the natural concentration of uranium into mineable deposits, the mining and refining of the ore, the manufacture of fuel rods, use of the fuel rods in nuclear reactors, and the ultimate disposal of the fuel rods.

nuclear fusion: The joining together of the nuclei of very light elements (hydrogen, lithium) to form heavier elements with the release of large amounts of energy.

nuclear reactor: The vessel in which nuclear fuels are reacted to generate heat, in turn used to raise steam and drive turbines.

nugget: A small solid lump, especially of gold.

obsidian: Volcanic glass; usually black, but also red, green, or brown.

ocean thermal energy conversion: A system that utilizes the differences in the temperature of surface ocean water and the temperature of deep ocean water.

oil: *See* petroleum.

oil mining: The process of mining oil-bearing rock so that it can be processed to extract the oil.

oil shale: A fine-grained sedimentary rock containing much bituminous organic matter incorporated when the sediment was deposited.

oil window: The set of temperature and pressure conditions, developed during burial of a sediment, which lead to the conversion of organic matter into petroleum.

olivine: A mineral, composition $(Fe,Mg)SiO_4$; an igneous mineral used in making refractories.

open-pit mining: Mining from open excavations.

ophiolite complex: A sequence of mafic and ultramafic igneous rocks including metamorphic rocks, whose origin is associated with the early phases of ocean floor rifting.

ore: Resources of metals that can now be economically and legally extracted.

ore deposit: Equals "reserve" when referring to metal-bearing concentrations.

ore mineral: Broadly used to include any mineral from which metals can be extracted.

osmotic pressure: The pressure resulting from the movement of molecules or ions in a fluid through a semipermeable membrane as they seek to establish the same concentrations on both sides of the membrane.

overburden: The valueless rock that must be removed above a near-surface ore deposit to permit open-pit mining.

oxidant: A compound or element that brings about oxidation.

oxidation: Combination with oxygen; more generally, any reaction in which there is an increase in valence resulting from a loss of electrons.

ozone: O_3, a form of oxygen produced by lightning and by solar radiation interacting with the upper atmosphere; important in reducing the penetration of UV radiation through Earth's atmosphere.

palygorskite: A mineral, composition $(Mg,Al)_2Si_4O_{10}OH \cdot 4H_2O$; a variety of clay that is sometimes fibrous and used as asbestos.

parabolic reflector: A concave reflector so shaped that the impinging Sun's rays are focused by reflection onto a central tube that becomes heated; as a result, the tube contains a fluid that transports the heat for use elsewhere.

peat: An unconsolidated deposit of semicarbonized plant remains accumulated in a water-saturated environment such as a swamp or bog. The early stage of coal formation.

ped: A naturally formed granule, block, crumb, or aggregate of a soil. *See* clod.

pegmatite: An exceptionally coarse-grained igneous rock; sometimes contains rich accumulations of rare elements such as lithium, boron, fluorine, niobium, tantalum, uranium, and rare earths.

pellets: Small, solid particles of material, especially those formed of iron oxide grains and used as a feedstock to smelt iron.

per capita: Per unit of population; for each person.

per capita use: The amount of something used by each person during a standard time period, generally per day or per year.

peridotite: A coarse-grained igneous rock composed chiefly of olivine and pyroxene.

perlite: A volcanic glass with a rhyolitic composition, a high water content, and a characteristic cracked pattern.

permeable: A rock, sediment, or soil with the capacity of transmitting a fluid.

petroleum: A naturally occurring complex liquid hydrocarbon that after distillation yields a range of combustible fuels, petrochemicals, and lubricants.

pH: A measure of acidity expressed numerically from 0 to 14; neutral is 7, with lower values representing more acid conditions. Specifically, the negative logarithm of the H^+ concentration.

Phanerozoic type: A type of iron ore deposit formed during the Phanerozoic Eon.

phosphate(s): Compounds, including some minerals, containing phosphorus in the form of the phosphate (PO_4) anion.

phosphorus: Chemical element number 15; one of the most important fertilizer elements.

photocell: A layered chemical cell that produces electricity directly from light energy.

photochemical conversion: A chemical conversion that proceeds by the addition of energy in the form of electromagnetic radiation.

photochemical reaction: A chemical reaction promoted by the presence of electromagnetic radiation.

photochemical smog: Intense smoke or fog intensified by sunlight.

photoelectrochemical conversion: The chemical process active in a photogalvanic cell.

photogalvanic: A term used for a chemical reaction in which solar energy is converted directly into electrical energy.

photosynthesis: The process by which green plants use the radiant energy from the Sun to create hydrocarbons and release oxygen.

photovoltaic cell: *See* photocell.

pickling (of metals): The use of an acid bath to cleanse the surface of metal castings, sheet metal, etc.

pig iron: The raw iron produced during the smelting of iron ore.

pigment: A coloring agent.

pitchblende: A massive, brown to black, fine-grained variety of uraninite, UO_2; a term commonly applied to any black uranium ore.

placer: A surficial mineral deposit formed by mechanical concentration of mineral particles from weathered debris.

plagioclase feldspars: A series of silicate minerals involving a solid solution from $CaAl_2Si_2O_8$ (anorthite) to $NaAlSi_3O_8$ (albite).

plasma: A fourth state of matter (solid, liquid, gas, plasma) capable of conducting magnetic force, usually generated by application of extremely high temperatures.

plasterboard: Large panels made from gypsum and commonly used in construction.

plaster of Paris: Partially dehydrated gypsum, $CaSO_4 \cdot \frac{1}{2} H_2O$.

podiform: Referring to ore bodies with an elongate lenticular shape; especially some chromite ores.

point source: A single point, such as a smokestack or pipe, from which pollution emanates.

pollution: The presence of abnormal substances or abnormally high concentrations of normal substances in the natural environment.

polyhalite: Complex hydrated sulfate salt containing calcium, magnesium, and potassium.

polymorph: One form of a mineral that is known to exist in more than one crystallographic form; e.g., graphite and diamond are polymorphs of carbon.

polypedon: A three-dimensional body of soil consisting of more than one recognizable soil type.

porosity: The property of containing many holes.

porphyrin: A class of organic compounds that are capable of complexing with metals such as vanadium.

porphyry: An igneous rock that contains large crystals embedded in a fine-grained groundmass.

porphyry copper deposit: A large low grade copper-bearing deposit usually associated with the intrusion of a porphyry.

portland cement: A calcium alumino silicate produced by calcining limestone and clay; this finely ground product will recrystallize and set when water is added.

potable water: Water that is safe for human use.

potash: A term locally used for potassium oxide or potassium hydroxide or to define the potassium oxide content of minerals.

potassium: Chemical element number 19; one of the most important fertilizer elements.

pot line: Large vats used in the production of aluminum.

power: The measure of energy produced or used as a function of time. *See* horsepower.

pozzolan cement: A cement formed by grinding together hydrated lime and pozzolana, a natural volcanic glass capable of reacting with the lime at ordinary temperatures to form cement compounds.

precious metal: The scarce metals that have high value—traditionally, gold, silver, and the platinum group metals.

primary mineral: A mineral formed at the same time as the rock enclosing it, usually by igneous or hydrothermal processes.

primary recovery: Petroleum production that occurs as a result of natural flow or pumping.

prior appropriation (of water): The law that permits the buying and selling of specified amounts of water from a stream for beneficial use. The appropriations are honored in order of the oldest first.

pumice: A light-colored, vesicular, glassy rock formed by the eruption of gas-rich lava from a volcano.

pumped-water storage system: Hydroelectric power systems in which excess electricity at low demand times is used to pump water into a storage dam so that it can subsequently be used to generate electricity.

pyrite: A mineral, FeS_2; "fools gold."

pyrolysis: Chemical decomposition by the action of heat.

pyrometallurgy: The metallurgical process involved in separating and refining metals where heat is used, as in roasting and smelting.

pyroxene: A group of silicate minerals of general formula $WSiO_3$ (or $XYSi_2O_6$), where W = Mg, Fe; XY = Mg, Ca, Fe, Na, Li, etc.

pyroxenite: A rock composed primarily of pyroxene.

pyrrhotite: A mineral, composition $Fe_{1-x}S$; a common iron sulfide compound.

quarry: An open surface working usually dug for the extraction of building stone.

quartz: A mineral, composition SiO_2; a very common compound that is hard, lacks cleavage, and does not weather rapidly.

quartzite: A metamorphic rock derived from sandstone and composed primarily of quartz.

quick silver: A term for mercury, Hg.

radiation: Electromagnetic energy transmitted in the form of waves or photons.

radioactivity: *See* decay (radioactive).

raise (in a mine): A vertical opening connecting two levels of a mine.

rank: A coal classification based upon physical, chemical, and thermal properties.

rare-earth elements: The 15 elements from atomic numbers 57 to 71, including, for example, lanthanum (La), cerium (Ce), neodymium (Nd), and europium (Eu).

recycling: The reuse of metals or other materials.

refining: Metallurgically: the process of extracting pure metals from their mineralogical forms; petroleum: the process of distilling and cracking crude oil in order to produce a wide variety of separate hydrocarbon liquids and gases.

refractory: A term used for unreactive materials with high melting points used to line the furnaces in which metals are smelted.

regolith: A general term for the surface layer of loose material that forms as a result of the weathering of rock.

renewable resources: Resources that are naturally replenished by processes active in or on Earth's crust.

reserve (of minerals): Mineral resources that can now be economically and legally extracted.

reserve base: That part of an identified resource that meets certain minimum physical and chemical criteria to present economic potential and that has a reasonable potential for becoming economic within planning horizons.

reservoir rock: Any rock with adequate porosity and that contains liquid or gaseous hydrocarbons.

resource: Naturally occurring concentrations of liquids, gases, or solids in or on Earth's crust in such form and amount that economic extraction of a commodity is currently or potentially feasible.

retort: A furnace-like chamber used to distill volatile materials or to carry out the destructive distillation of coal or oil shale. Heat is usually applied externally, and the decomposition products are collected by cooling the gases so that different compounds condense at different temperatures.

return flow: Water that reaches a ground—or surface—water source after release from the point of use and thus becomes available for further use.

reverse osmosis: Process utilizing pressure to separate freshwater from saltwater by forcing it through a semipermeable membrane.

rhyolite: A fine-grained extrusive igneous rock consisting largely of quartz and potassium feldspar.

ribonucleic acid: RNA, a substance similar to DNA. It carries out DNA's instructions for making proteins.

Richter Scale: A scale used in the quantitative evaluation of the energy released by earthquakes.

riparian: Pertaining to the shoreline areas of a body of water. Riparian law allows landowners to draw from a lake or stream adjacent to their property if their use does not harm other users.

roasting: Heating of an ore to bring about some change, usually oxidation, of the sulfide or other minerals.

rock cycle: The cyclical progression of rocks from sedimentary to metamorphic as a result of increased temperature and pressure, to igneous as a result of melting, and back to sedimentary as a result of weathering and erosion.

rock salt: Coarsely crystalline halite, naturally occurring or synthetically prepared.

room and pillar mining: A mining method in which rock, coal, or ore is removed from a series of openings with a series of intervening columns (pillars) left to support the overlying rocks.

rotary drills: The most common method of drilling; a hydraulic process in which a hard-toothed drill bit is attached to a rotating drill pipe. As the pipe turns, the bit grinds into the rock; the loose pieces are carried to the surface by fluid circulated down the center of the pipe.

rutile: A mineral, composition TiO_2. The principal ore of titanium; used as a white paint pigment.

saline water: Water that contains 1000 or more milligrams of dissolved solids per liter, especially NaCl.

salinization: The buildup of salts, usually NaCl, in soils as a result of evaporation.

salt dome: Dome or pinnacle-like structure of rock salt, halite, which has risen through sediments above a bed of salt due to differences in densities.

saltpeter: Naturally occurring potassium nitrate; niter.

saltwater intrusion: The movement of saltwater into an aquifer, usually as a result of excessive extraction of freshwater near coastal areas.

sand: Detrital rock fragments 1/16 to 2 millimeters in diameter; natural sands are composed almost entirely of quartz.

sandstone: A medium-grained, clastic sedimentary rock composed of sand-sized particles (commonly quartz).

saponite: A mineral, composition $(Ca/2,Na)_{0.33}(Mg,Fe)_3(Si_{3.67},Al_{0.33})O_{10}(OH)_2 \cdot 4H_2O$; a soft, soapy light-colored clay.

sapropelic coal: Coal derived from organic residues (finely divided plant debris, spores, and algae) in stagnant or standing bodies of water.

scarce metals: Metals whose average crustal abundance is less than 0.1 percent.

secondary enrichment: A zone of minerals formed later than the enclosing rock and usually at the expense of earlier formed primary minerals.

secondary recovery: Oil production resulting from procedures, such as the injection of water, steam, or chemical compounds into a reservoir in order to increase oil production beyond primary production.

sedimentary: Pertaining to rocks formed from the accumulation of fragments weathered from preexisting rocks, or by precipitation of materials in solution in lakes or seawater.

seismograph: An instrument that records Earth's vibrations, especially those from earthquakes.

sepiolite: A mineral, composition $Mg_4(Si_2O_5)_3(OH)_2 \cdot 6H_2O$; a common clay that is widely used for ornamental carvings; also known as meerschaum.

sequestration: The trapping of carbon dioxide in underground formations to prevent its release into the atmosphere.

serpentine: A group of minerals, general composition $(Mg,Fe)_3Si_2O_5(OH)_4$; widely formed in metamorphism with varieties including gems (jade) and asbestos (chrysotile).

serpentinite: A rock composed primarily of serpentine group minerals and formed through the alteration of preexisting ferromagnesian minerals such as olivine and pyroxene.

shaft (of a mine): A vertical entrance into a mine.

shale: A fine-grained, indurated, detrital sedimentary rock formed by the compaction of clay, silt, or mud and with a partially developed rock cleavage.

silane: Silicon-hydrogen compounds.

silicon chip: A chip of silicon metal to which trace amounts of other elements have been added in order to affect the electronic properties.

sillimanite: A mineral, composition $Al_2Si_2O_5$; a compound found in metamorphic rocks and used in the manufacture of ceramics and glass.

skarn: An assemblage of lime-bearing silicates derived from limestones and dolomites by the introduction of silicon, iron, and magnesium, usually adjacent to an igneous intrusion.

slag: The nonmetallic top layer that separates during the smelting of ores; it is usually rich in silica, alumina, lime, and any other materials used to flux the smelting.

slate: A compact, fine-grained metamorphic rock formed from shale; it possesses the property of rock cleavage whereby it can be readily parted along parallel planes.

smectite: A term applied to the montmorillonite group of clay minerals.

smelting: The process of melting ore minerals to separate the metals from the nonvaluable phases.

smog: Fog that has become mixed and polluted with smoke.

soda ash: Sodium carbonate.

soil: The unconsolidated earthy material that overlies bedrock and that is a complex mixture of inorganic and organic compounds; the natural medium to support the growth of plants.

soil order: One of ten major subdivisions in the "Comprehensive Soil Classification System."

soil profile: A vertical section through a soil that reveals the different physical and chemical zones that are present.

soil temperature regime: The changes in temperature experienced by a soil in a normal annual cycle.

soil water regime: The changes in the amounts of water present in a soil during a normal annual cycle.

solar cell: Electronic device which yields an electric current when exposed to light.

solar energy: The total energy in the Sun's radiation.

solder: Metal alloy that melts at low temperature and that is used to bond metals together.

solid solution: A solid crystalline phase in which the composition may vary by one or more elements replacing others, e.g., Fe replacing Mg in the olivine minerals Mg_2SiO_4-Fe_2SiO_4.

solution mining: The extraction of resources by solutions instead of conventional mining procedures.

soot: A black substance consisting mainly of carbon from the smoke of wood or coal.

sour gas: Hydrogen sulfide, H_2S, that commonly occurs in minor quantities with petroleum or natural gas.

source rocks: Sedimentary rocks containing the organic matter that under heat, pressure, and time is transformed into liquid or gaseous hydrocarbons.

special metal: Metals such as tantalum and beryllium that are used increasingly because of unusual properties important to industry.

sphalerite: A mineral, ZnS, that serves as the principal ore of zinc.

spinel: A mineral, composition $MgAl_2O_4$; a common accessory mineral in many rocks; also refers to a crystal structure, common to some ore minerals, in which the external shape of an octahedron is often seen.

spot market: The buying and selling of commodities for immediate delivery at a price agreed upon at the time of sale.

stainless steel: An iron-based alloy containing enough chromium to confer a superior corrosion resistance.

stannite: A mineral, composition Cu_2FeSnS_4; an ore of tin.

steam flooding: The injection of superheated steam down an oil well in order to promote the release of oil trapped in the rocks.

steel: An iron-based alloy; other metals or substances are alloyed with the iron to impart specific properties such as hardness or strength.

stock: Exposure of plutonic igneous rock over an area of less than 40 square miles.

stockworks: Closely spaced intertwined veins that serve as feeder zones of ore deposits.

Stone Age: The period in human culture when stone was used for the making of tools. It began with the first humans and ended at various times in different places (e.g., about 3000 B.C. in Egypt and Mesopotamia) when bronze was used to make tools.

stope: The room-like area in a mine where ores are extracted.

stratification: The layer-like nature of a sedimentary rock.

stratiform: Referring to an ore deposit that is layered and parallel to the enclosing strata.

stratigraphic traps: Sedimentary units such as sandstone lenses into which oil migrates and is prevented from further movement by surrounding impermeable layers.

strip mining: The removal of coal or other commodities by surface mining methods in which extraction is carried out in successive strips of land.

structural traps: Folded or faulted rocks into which oil migrates and from which further migration is prevented.

subduction: The process in which one crustal plate descends beneath another.

subeconomic resource: That part of identified resources not meeting the economic criteria of reserves or marginal reserves.

subsidence: The naturally or artificially induced dropping of the land surface resulting from the removal of the underlying rocks either by mining or groundwater solution.

superalloy: Mixture of metals that exhibits one or more properties that is greater than either of the metals being used.

superphosphate: A soluble mixture of calcium phosphates produced by reaction of phosphate rock with sulfuric acid; used as fertilizer.

surface runoff: The water that flows directly off the land surface and the water that, after infiltrating into the ground, is discharged into the surface.

swamp gas: Methane, CH_4, produced during the decay of organic matter in stagnant water.

syenite: An igneous rock composed largely of potassium feldspar, any of the feldspathoid minerals, and an amphibole. Quartz is rare or absent.

SYNROC: Synthetic rock-like material used to incorporate radioactive waste.

taconite: A term used in the Lake Superior district for laminated iron ores consisting of iron oxides and silica or iron silicates.

tactite: An alternate name for skarn.

tailings: The valueless materials discarded from mining operations.

talc: A mineral, composition $Mg_3Si_4O_{10}(OH)_2$; it is extremely soft, making it a valuable lubricant and cosmetic ingredient.

tar: A dark, oily, viscous mass of hydrocarbons.

tar sand: Sand deposits in which the interstices of the grains are filled with viscous hydrocarbons.

tectonism: Movement of a portion of Earth's crust, usually related to mountain building.

thermal cracking: The application of heat to break heavier hydrocarbons into lighter ones.

thermal maturation: The progressive change in organic matter in sedimentary rocks resulting from increasing temperature as the depth of burial increases.

thermal pollution: Abnormal heating of the environment, usually rivers, caused by the combustion of fossil fuels, by nuclear power generation, or industrial processing.

thermogenic gas: Gas formed as a result of the thermal breakdown of organic matter.

thermosetting plastics and resins: Plastics and resins that require heating in order to become solidified.

tidal energy: Energy derived from the movement of water during the rise and fall of the tides.

transpiration: The release of water vapor by plants during their normal respiration.

traps: Rock structures or beds in which oil accumulates and is prevented from further migration.

tritium: A synthetic isotope of hydrogen containing one proton and two neutrons in its nucleus; its half-life is 12.5 years.

tuff: A compacted deposit of volcanic ash.

ultramafic: Said of igneous rocks composed chiefly of one ferromagnesian mineral (most ultramafic rocks are also ultrabasic, i.e., having SiO_2 contents less than about 44 percent).

Uravan District: Mining district spanning the Colorado–Utah boundary and that was mined primarily for ores of uranium and vanadium.

vadose: Referring to water in the uppermost soil, the zone in which most intergranular interstices are air filled.

vein: A sheet-like infilling of a fracture by hydrothermally deposited minerals often containing ore minerals.

vermiculite: A group of clay minerals characterized by the tendency to undergo extreme expansion when heated above 150°C; widely used as an insulator and as a component in lightweight construction materials.

vesicles: Cavities in a lava formed by the evolution of gas during the rapid cooling of a molten lava.

vitrification: The formation of a glassy or noncrystalline substance.

volcanic glass: Natural glass formed in volcanic eruptions when lava cools quickly; commonly called obsidian.

volcanogenic massive sulfide: Ore deposit containing more than 50 percent ore minerals and formed as a result of precipitation of the ore minerals from hydrothermal fluids on the sea floor.

Wabana type: A regional name applied to some iron ore deposits in North America.

water flooding: The injection of water into an oil well to promote the release of oil trapped in the rock.

water glass: A water soluble form of sodium silicate.

waterlogging: The situation where irrigation raises the water table such that it covers the roots of crops and either reduces productivity or kills the plants.

water rights: The right to draw and use (and sometimes sell) water drawn from a lake, river, or underground source.

water table: The surface in a soil or rock below which all the voids are water filled.

watt: A unit of power equal to work done at a rate of one joule per second; approximately equal to 1/746 horsepower.

weatherability: The rate at which weathering affects the physical and chemical nature of a rock or mineral.

weathering: The progressive breakdown of rocks, physically and chemically, in response to exposure at or near Earth's surface.

wet steam: Natural geothermal systems dominated by hot waters with associated steam.

wind farm: An array of windmills used to generate electricity.

winze: A vertical opening connecting two levels of a mine. The same as a raise.

withdrawal: The extraction of water from a surface or groundwater source.

work: Usually the result of applied force, defined as the product of the force and the displacement. Usually expressed in foot-pounds, joules, or kilowatt-hours.

yellowcake: A general term for yellow oxidized uranium oxide arising from concentration in the mining and processing of uranium ore.

zeolite: A group of hydrous aluminosilicate minerals characterized by their easy exchange of water and cations; used as catalysts in oil refining.

zircon: A mineral, composition $ZrSiO_4$; a common accessory mineral in many rocks; the main source of zirconium.

zonal soil: One of the soil orders that have well-developed characteristics that reflect the agents of soil genesis, especially climate and the action of organisms.

INDEX

C

H

N

Q

T

X

Y

Z

NOTES

NOTES

NOTES

NOTES

Earth Statistics

Earth Dimensions and Mass

Earth radius—equatorial	6378 km	3963 mi
—polar	6357 km	3950 mi
—average sphere	6371 km	3959 mi
Crustal thickness—continents	25–30 km	16–9 mi
—oceans	10–15 km	6–9 mi
Mantle thickness	2900 km	1802 mi
Outer core thickness	2420 km	1504 mi
Inner core thickness	1050 km	652 mi
Mass of Earth	5.98×10^{21} metric tons	
Mass of ice	$25\text{–}30 \times 10^{15}$ metric tons	
Mass of oceans	1.4×10^{18} metric tons	
Mass of crust	2.5×10^{19} metric tons	
Mass of mantle	4.05×10^{21} metric tons	
Mass of core	1.9×10^{21} metric tons	

Areas of the Major Continents and Oceans

Continents		
Asia	44,120,650 km^2	17,035,000 mi^2
Africa	30,134,650 km^2	11,635,000 mi^2
North America	24,436,650 km^2	9,435,000 mi^2
South America	17,767,400 km^2	6,860,000 mi^2
Antarctica	13,209,000 km^2	5,100,000 mi^2
Europe	9,971,500 km^2	3,850,000 mi^2
Australia	7,705,300 km^2	2,975,000 mi^2
Greenland	2,175,600 km^2	840,000 mi^2
Total major continents	149,520,700 km^2	57,730,000 mi^2

Some Selected Countries		
former Soviet Union	22,403,500 km^2	8,650,000 mi^2
Canada	9,976,700 km^2	3,852,000 mi^2
United States	9,520,800 km^2	3,676,000 mi^2
South Africa	1,224,000 km^2	472,600 mi^2
United Kingdom	244,000 km^2	94,200 mi^2

Major Oceans and Seas		
Pacific Ocean	165,721,000 km^2	63,985,000 mi^2
Atlantic Ocean	81,660,000 km^2	31,529,000 mi^2
Indian Ocean	73,445,000 km^2	28,357,000 mi^2
Arctic Ocean	14,351,000 km^2	5,541,000 mi^2
Mediterranean Sea	2,966,000 km^2	1,145,000 mi^2
South China Sea	2,318,000 km^2	895,000 mi^2
Bering Sea	2,274,000 km^2	878,000 mi^2
Caribbean Sea	1,943,000 km^2	750,000 mi^2
Gulf of Mexico	1,813,000 km^2	700,000 mi^2